AF386603

Probability Theory and Stochastic Modelling

Volume 108

Probability Theory and Stochastic Modelling publishes cutting-edge research monographs in probability and its applications, as well as postgraduate-level textbooks that either introduce the reader to new developments in the field, or present a fresh perspective on fundamental topics.

Books in this series are expected to follow rigorous mathematical standards, and all titles will be thoroughly peer-reviewed before being considered for publication.

Probability Theory and Stochastic Modelling covers all aspects of modern probability theory including:

- Gaussian processes
- Markov processes
- Random fields, point processes, and random sets
- Random matrices
- Statistical mechanics, and random media
- Stochastic analysis
- High-dimensional probability

as well as applications that include (but are not restricted to) :

- Branching processes, and other models of population growth
- Communications, and processing networks
- Computational methods in probability theory and stochastic processes, including simulation
- Genetics and other stochastic models in biology and the life sciences
- Information theory, signal processing, and image synthesis
- Mathematical economics and finance
- Statistical methods (e.g. empirical processes, MCMC)
- Statistics for stochastic processes
- Stochastic control, and stochastic differential games
- Stochastic models in operations research and stochastic optimization
- Stochastic models in the physical sciences

Probability Theory and Stochastic Modelling is a merger and continuation of Springer's Stochastic Modelling and Applied Probability and Probability and Its Applications series.

Paweł Lorek · Tomasz Rolski

Lectures on Monte Carlo Theory

 Springer

Paweł Lorek
Mathematical Institute
University of Wrocław
Wrocław, Poland

Tomasz Rolski
Mathematical Institute
University of Wrocław
Wrocław, Poland

ISSN 2199-3130   ISSN 2199-3149   (electronic)
Probability Theory and Stochastic Modelling
ISBN 978-3-032-01189-3   ISBN 978-3-032-01190-9   (eBook)
https://doi.org/10.1007/978-3-032-01190-9

Preface

Monte Carlo methods are a wide-ranging set of computational techniques based on repeated random sampling to compute numerical outcomes.

In this book, we present pseudocode for many Monte Carlo algorithms; however, our emphasis lies in the underlying principles, which are rooted in probability theory, mathematical statistics, and elements of statistical physics and computer science. Our goal is to present these foundations in an accessible manner. Although the ultimate aim is to compute numerical results for problems arising in diverse scientific fields, the presentation of algorithms and simulation results is also intended to develop probabilistic intuition. This approach allows us to illustrate what might otherwise seem a hermetic branch of mathematics: probability theory.

The random samples used in Monte Carlo methods are generated by computer algorithms. These are not truly "random" in a mathematical sense, and verifying their quality involves a broad class of statistical methods. At the same time, one must prove that the algorithms converge to the desired result—this is where probability theory plays a critical role. Together, these components form the theoretical foundation of what is broadly called *Monte Carlo simulation*. The term *stochastic simulation* is also sometimes used in this context.

The term *stochastic simulation* generally refers to the broader practice of simulating random phenomena—ranging from generating pseudorandom numbers to modeling complex stochastic processes—while *Monte Carlo theory* focuses on the probabilistic foundations and algorithmic principles behind such simulations. Together, these components draw on ideas from probability theory, statistics, and computer science.

The theory originated in nuclear physics computations but has since been applied to problems in many fields, including physics, engineering, finance, and operations research. Three names are particularly important: Stanisław Ulam, who introduced the idea of using randomness in computation; John von Neumann, who collaborated with Ulam in its early development; and Nicholas Metropolis, the first author of the seminal paper that gave the method its name—Monte Carlo.

This book has grown out of courses taught to master's students in mathematical sciences at the University of Wrocław. The core course, taught by PL and TR, was titled *Stochastic Simulations and Monte Carlo Theory*, and it covered topics including pseudorandom number generators, generation of random variables from various distributions, development of simulation results, variance reduction techniques, and selected applications in fields such as financial analysis and operations research. A complementary course taught by PL focused on more advanced topics, such as Markov chain Monte Carlo methods and their algorithmic applications in physics, optimization, approximate counting, and related areas.

Some theorems in the book are rigorously proven, others are sketched, and many are referenced from the literature. In every case, our focus is on the core ideas. Theoretical results are followed by algorithms to reinforce understanding, especially for students with a background in computer science. Numerical examples are illustrated using tables and figures.

Most of the algorithms in this book have been implemented in Python. The accompanying scripts allow readers to reproduce all simulation results—both numerical and graphical—exactly as they appear in the book. Running the scripts with default settings reproduces these results; alternative outputs can be generated by adjusting parameters or random seeds. All code, together with supplementary materials, is available in the GitHub repository:

https://github.com/lorek/Lectures_on_Monte_Carlo_Theory

Any errata found in the text will also be recorded in this repository.

In designing the lecture content, our goal was to integrate rigorous probability theory with statistical and computational methods. We deliberately avoided certain formal frameworks—such as measure theory—that, while important in general probability, are not essential here. Moreover, a significant portion of the material, particularly on Markov chain Monte Carlo methods, is developed in the setting of finite state spaces. This allows us to include formal proofs while avoiding technical details that might obscure the main ideas.

Of course, this book is one among many in the existing literature on Monte Carlo methods. During a visit to the Eurandom center at TU Eindhoven, TR encountered Jacques Resing's lecture entitled *Simulation*, which influenced parts of our presentation, though our approach ultimately diverges. Another important source was the elegant little book by Madras [138]. We were also strongly influenced by the monograph by Asmussen and Glynn [8], although it is more advanced, especially in its treatment of stochastic processes. The chapter on Markov chain Monte Carlo methods owes much to our reading of Häggström's book [73]. In addition to these inspirations, the book also reflects insights developed through our own research in probability, stochastic modeling, and simulation. And naturally, our work draws on a vast body of scientific literature published since the inception of Monte Carlo methods in the 1950s.

Wrocław, Poland   Paweł Lorek
                                                                    Tomasz Rolski

Declarations

Competing Interests The authors have no competing interests to declare that are relevant to the content of this manuscript.

Contents

Notation

Mathematical Notation

$\sim$	(a) $f(x) \sim g(x)$ (or shortly $f \sim g$) means that f equals g asymptotically, i.e., $\lim_{x\to\infty} f(x)/g(x) = 1$; (b) $X \sim \mathcal{U}[0,1)$ means that a random variable X has $\mathcal{U}[0,1)$ distribution
O	*Big 'O' notation:* $f(x) = O(g(x))$ (as $x \to \infty$) if there exists a positive real number N and a real number x_0 such that $\lvert f(x) \rvert \leq N g(x)$ for all $x \geq x_0$
o	*Little 'o' notation:* $f(x) = o(g(x))$ (as $x \to \infty$) if for every positive constant ε there exists a positive number x_0 such that $\lvert f(x) \rvert \leq \varepsilon g(x)$ for all $x \geq x_0$. Equivalently, $f(x) = o(g(x)) \iff \lim_{x\to\infty} f(x)/g(x) = 0$.
$\mathbb{1}$ or $\mathbf{1}$	The indicator function. Used as $\mathbb{1}_A$, $\mathbb{1}(x \in A)$ or $\mathbf{1}_A$, $\mathbf{1}(x \in A)$ respectively
$\approx$	Approximation
$\xrightarrow{P}$	Convergence in probability
$\propto$	Proportional to
$\xrightarrow{\mathcal{D}}$	Convergence in distribution
$\stackrel{\mathcal{D}}{=}$	Equality in distribution
$\mathbb{P}(A)$	Probability of A
$\mathbb{E}X$	Expectation of a random variable X
$\mathbb{E}_{\boldsymbol{\theta}} X,\ \mathbb{P}_{\boldsymbol{\theta}}(\cdot)$	Expectation/probability computed for $X \sim f_{\boldsymbol{\theta}}$ (p.d.f. parametrized via $\boldsymbol{\theta}$)
$\mathbb{E}_f X,\ \mathbb{P}_f(\cdot)$	Expectation/probability computed for $X \sim f$
$\lvert \cdot \rvert$	(a) absolute value; (b) the number of elements (cardinality) of a set
$[n]$	The set $\{1, \ldots, n\}$
$\overline{M}$	The set $\{0, \ldots, M-1\}$
$\mathbb{N}$	Natural numbers $\{1, 2, \ldots\}$

$$\sum_{i=0}^{-1} = 0 \qquad \text{Standard notation for empty sums}$$

$$\prod_{i=0}^{-1} = 1 \qquad \text{Standard notation for empty products}$$

Distributions

$\mathcal{N}(\mu, \sigma^2)$	The univariate normal distribution with mean μ and variance σ^2
$\mathcal{N}(\boldsymbol{\mu}, \boldsymbol{\Sigma})$	The multivariate normal distribution with mean $\boldsymbol{\mu}$ (column vector) and covariance matrix $\boldsymbol{\Sigma}$
$\mathcal{U}[a, b)$	The uniform distribution on an interval $[a, b)$

Abbreviations

B&D	Birth-and-death (process)
i.i.d.	Independent, identically distributed
c.d.f.	Cumulative distribution function
CLT	Central limit theorem
MCMC	Markov chain Monte Carlo
NIST	National Institute of Standards and Technology
p.d.f.	Probability density function
PRNG	Pseudorandom number generator
r.v.	Random variable
r.h.s.	Right-hand side
SACV	Squared asymptotic coefficient of variation

Chapter 1
Introduction

1.1 About the Book

By Monte Carlo theory, we refer to the theoretical foundations underlying computational algorithms that employ random experiments and the analysis of simulation results. This theory provides tools to plan simulations and design methods that enable the solution of specific tasks. Its applications span a wide range of areas, including optimization, integral computation, statistics, and generating samples from probability distributions. More recently, Monte Carlo methods have become increasingly popular for risk assessment and financial analysis.

In contrast to numerical methods, where algorithms provide deterministic control over errors, calculations using stochastic methods yield random results. It is crucial to understand how to interpret these results, define errors meaningfully, and apply appropriate measures to evaluate them. For this purpose, concepts and methods from mathematical statistics play a vital role.

Monte Carlo theory, named after the Monte Carlo Casino, was developed by John von Neumann and Stanisław Ulam during their work on the atomic program in the 1940s. Its development was heavily influenced by the emergence of fast computers and the need for practical solutions to complex problems. Stanisław Ulam, reflecting on this period, recounted the origins of the method:

'The first thoughts and attempts I made to practice [the Monte Carlo Method] were suggested by a question that occurred to me in 1946 as I was convalescing from an illness and playing solitaire. The question was: what are the chances that a Canfield solitaire laid out with 52 cards will come out successfully? After spending a lot of time trying to estimate them by pure combinatorial calculations, I wondered whether a more practical method than 'abstract thinking' might not be to lay it out one hundred times and simply observe and count the number of successful plays. This was already possible to envisage with the beginning of the new era of fast computers, and I immediately thought of problems of neutron diffusion and other questions of mathematical physics, and more generally, how to change processes described by certain differential equations into an equivalent form interpretable as a succession of random operations. Later [in 1946], I described the idea to John von Neumann, and we began to plan actual calculations.' (Quoted after [46])

P. Lorek, T. Rolski, *Lectures on Monte Carlo Theory*, Probability Theory and Stochastic Modelling 108, https://doi.org/10.1007/978-3-032-01190-9_1

This recollection not only highlights the genesis of the Monte Carlo method but also underscores its groundbreaking role in translating abstract mathematical problems into practical computational techniques. The collaboration between Ulam and von Neumann was instrumental in establishing Monte Carlo as a foundational approach in computational science. It also demonstrated the transformative potential of combining theoretical insights with computational power, a principle that remains highly relevant in modern computational science.

The book primarily focuses on simulating various tasks in probability theory. It introduces readers to topics that are challenging to understand without specialized knowledge of probability and stochastic processes. While detailed expertise in these areas is not required, a basic understanding from introductory university courses in probability and mathematical statistics is expected. By the end of this course, the authors aim for readers to not only comprehend stochastic simulation and Monte Carlo theory but also become acquainted with many classical problems in probability theory, modern financial mathematics, and stochastic models in operations research, telecommunications, and related fields.

In the contemporary literature, a distinction is often made between random number generators and pseudorandom number generators (PRNGs). This text focuses on pseudorandom numbers, which are generated through mathematical recursion. In contrast, random number sequences can be produced through physical means, such as coin tossing, dice rolling, observing specific atmospheric phenomena, or recording hardware register values. Although the distinction between hardware-based random number generators and software-based generators is recognized in some contexts, it will not be explored here. In addition to pseudorandom numbers, we introduce quasirandom numbers—numerical sequences with the property of "low discrepancy."

The book consists of chapters covering:

- an overview of pseudorandom number generator theory,
- methods for generating random variables with specified distributions,
- fundamental concepts of errors, significance levels, replication counts, and their interrelationships,
- techniques for reducing the number of replications required for a given error level,
- example tasks solved using stochastic simulation methods,
- methods based on Markov chain theory (Markov chain Monte Carlo), primarily for studying probabilistic algorithms and sampling from probability distributions over large state spaces,
- optimization methods based on Markov chain Monte Carlo theory,
- methods useful in operations research, such as discrete event simulation and telecommunications modeling.

An appendix at the end of the book surveys distributions and their properties.

1.2 Monte Carlo: Examples of Simple Probabilistic Tasks

As an introduction to stochastic simulations, we begin by discussing how pseudorandom numbers are generated on a computer and exploring basic models. For example, in Python, the NumPy library provides a simple way to generate pseudorandom numbers, a feature we will use extensively throughout this text. The following examples demonstrate key functions:

```python
import numpy as np
np.random.seed(31415)              # initialize PRNG with seed
np.random.rand(k)                  # k random floats from (0,1)
np.random.randint(low=0, high=M, size=k) # k random ...
                              # integers from {0,...,M-1}
```

Successive numbers are generated by repeatedly executing these commands. The sequence of pseudorandom numbers mimics the theoretical sequence of independent random variables with the same uniform distribution, either $\mathcal{U}[0, 1)$ or $\mathcal{U}\{0, \ldots, M - 1\}$.

However, modern NumPy versions recommend using default_rng() for better flexibility and reproducibility. Below is an example using the default_rng() interface:

```python
from numpy.random import default_rng, PCG64
prng = default_rng(PCG64(seed=31415)) # initialize PRNG with
                                  # PCG64 and seed
prng.random(size=k)                # k random floats from (0,1)
prng.integers(low=0, high=M, size=k)# k random integers from
                              # {0,...,M-1}
```

This method ensures reproducibility and allows for explicit control over the underlying pseudorandom number generator (e.g., PCG64). We will use this method for most examples throughout the book. More details on using default_rng() are described in Sect. 2.3.1.

1.2.1 Properties of the Simple Symmetric Random Walk

Random walks are fundamental models in probability theory, essential for understanding diverse phenomena such as stock market fluctuations or particle diffusion. Beyond theoretical interest, they serve as tools to visualize probability theory and stochastic processes, enabling researchers to develop intuition and test hypotheses.

To illustrate, consider the classic coin toss problem from Chapter 3 of Feller's book [52]. Two players, A and B, bet on heads and tails, respectively. Player A wins $+1$ from B if heads appear, and vice versa for tails. Starting with zero fortune, i.e., $S_0 = 0$, player A's fortune after n tosses is represented as S_n. For convenience, consider $2N$ tosses, where ties can only occur at even steps $2, 4, \ldots, 2N$.

A straight line connects each $(i-1, S_{i-1})$ and (i, S_i). A segment lies above the x-axis if $S_{i-1} \geq 0$ or $S_i \geq 0$.

We can pose questions about the realization $(S_n)_{n=1,\ldots,2N}$, such as:

Q1: Let $L_{2N} = \max\{i : S_i = 0\}$ denote the last moment of a tie. What is its distribution?
Q2: What is the probability $P(\alpha, \beta)$ (where $\alpha, \beta \in [0, 1)$ and $\alpha < \beta$) that player A is winning between fractions α and β of the time?
Q3: How do the oscillations of $(S_n)_{0 \leq n \leq 2N}$ behave?
Q4: What is the probability that one player is always winning?

These tasks are formalized as follows, enabling simulation. Let $X_1, X_2, \ldots$ be a sequence of i.i.d. random variables where $\mathbb{P}(X_j = 1) = \mathbb{P}(X_j = -1) = 1/2$. Define

$$S_0 = 0, \quad S_n = \sum_{j=1}^{n} X_j, \quad n = 1, \ldots$$

The sequence $S_0 = 0, S_1, \ldots$ is known as the *simple symmetric random walk*. Formally, define:

$$L_{2n} = \sup\{m \leq 2n : S_m = 0\}$$

and

$$L_{2n}^+ = \#\{j = 1, \ldots, 2n : S_{j-1} \geq 0 \text{ or } S_j \geq 0\}.$$

Player A wins strictly at step m if $S_m > 0$, and the total time player A is strictly winning is

$$\#\{1 \leq k \leq 2N : S_k > 0\}.$$

An example simulation of the simple symmetric random walk $S_0 = 0, S_1, \ldots, S_N$ is provided in Listing 1.1.

Python Listing 1.1 Simple symmetric random walk

```python
import numpy as np
from numpy.random import default_rng, PCG64
import matplotlib.pyplot as plt
N=500                          #    Number of steps   (2N points)
prng = default_rng(PCG64(seed=31415))    # set seed
S = np.zeros(2*N + 1)
steps = 2*prng.integers(0, 2, size=2*N)-1 # random steps
S[1:] = np.cumsum(steps)                  # random walk
                                          # with S[0]=0

plt.figure(figsize=(8, 4))                # plot the figure
plt.plot(S, linewidth=1, color='blue')
plt.grid(True, linestyle='--', alpha=0.7)
plt.show()
```

A sample plot of a random walk is presented in Fig. 1.1. This example is implemented in `ch1_simple_random_walk.py`.

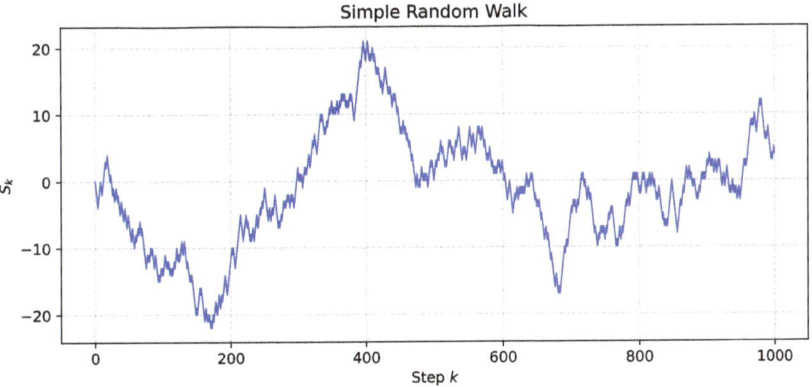

Fig. 1.1 A sample realization of a simple symmetric random walk; $2N = 1000$

Analytical answers in the form of limit theorems—when $N \to \infty$—are known for all questions Q1–Q4; see Feller [52].

We now discuss how the realizations of S_n behave. Clearly, $|S_n| \le n$. However, large values of $|S_n|$ occur with small probability, so, in practice, the values of S_n lie within a smaller range. The weak and strong laws of large numbers imply that $\frac{S_n}{n} \to 0$ (in probability and almost surely). This indicates that deviations of S_n from 0 grow slower than linearly. On the other hand, the Central Limit Theorem (CLT) states that $\frac{S_n}{\sqrt{n}} \xrightarrow{\mathcal{D}} \mathcal{N}(0, 1)$, where $\xrightarrow{\mathcal{D}}$ denotes convergence in distribution. Thus, the fluctuations of S_n eventually exceed the interval $[-\sqrt{n}, \sqrt{n}]$, as the 0-1 law of Kolmogorov implies that $\limsup_{n \to \infty} \frac{S_n}{n} = \infty$.

However, the fluctuations can be estimated more accurately. *The law of the iterated logarithm* (see [100], cf. Chapter VIII.5 in [52]) states that for a random walk S_n, we have

$$\mathbb{P}\left(\liminf_{n\to\infty}\frac{S_n}{\sqrt{2n\log\log n}}=-1\right)=\mathbb{P}\left(\limsup_{n\to\infty}\frac{S_n}{\sqrt{2n\log\log n}}=+1\right)=1.$$

Thus, dividing S_n by n is *too strong*, while dividing it by $\sqrt{n}$ is *too weak*. The fluctuations of S_n from 0 grow proportionally to $\sqrt{2n\log\log n}$, as shown in Fig. 1.2.

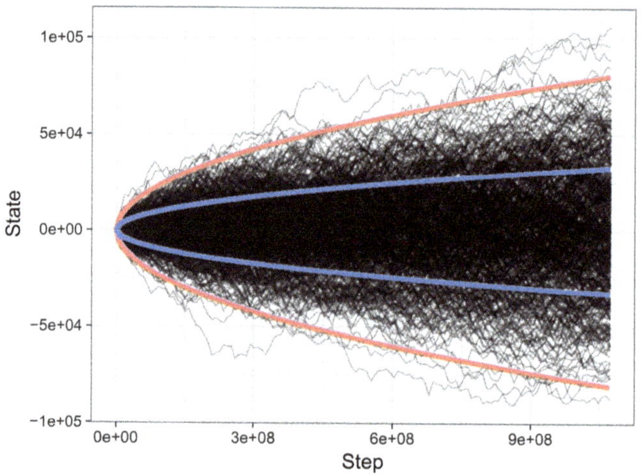

Fig. 1.2 500 trajectories of random walks of length $n = 2^{30}$. *Blue plot*: $\pm\sqrt{n}$, *red plot*: $\pm\sqrt{2n\log\log n}$

For example, it is known (see Feller [52], p. 82, or Durrett [43], p. 198) that

$$\mathbb{P}(\alpha \le L_{2n}^+/2n \le \beta) \to \int_\alpha^\beta \pi^{-1}(x(1-x))^{-1/2}\,dx, \qquad (1.2.1)$$

for $0 \le \alpha < \beta \le 1$. This is called the *arcsine law*. The name comes from the fact that

$$\int_0^t \pi^{-1}(x(1-x))^{-1/2}\,dx = \frac{2}{\pi}\left(\arcsin\left(\sqrt{t}\right)\right).$$

A similar result holds for L_{2n}; see Durrett [43], p. 197.

We will attempt to answer the questions via simulations. Using an algorithm for simulating a random walk, we propose creating histograms for the repetitions of L_{2N}^+—the total time player A is winning—and L_{2n}. It is evident that when $n \to \infty$, both L_{2n}^+ and L_{2n} diverge to infinity. However, the relevant quantities are the time fractions, i.e., $L_{2n}^+/(2n)$ and $L_{2n}/(2n)$.

The Monte Carlo method relies on repeating the experiment many times. Let R denote the number of experiments performed, called *replications* in Monte Carlo theory.

In Fig. 1.3, we show a histogram for L_{1000}^+ computed from $R = 10^5$ replications. In this book, we also use the arcsine law in Sect. 2.5 to construct a test for the randomness of pseudorandom number generators.

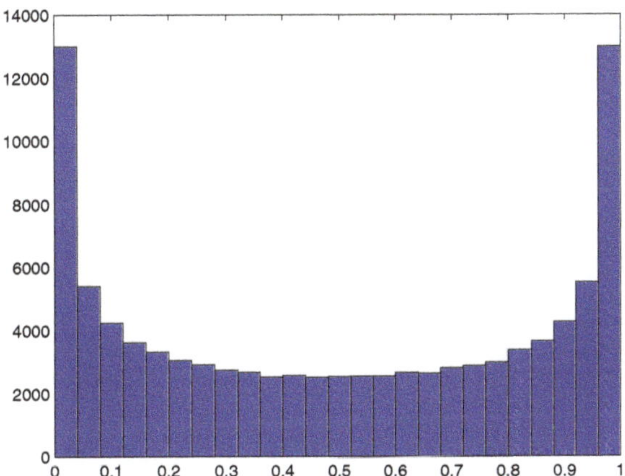

Fig. 1.3 Histogram for L_{1000}^+; $R = 10^5$ replications, 25 boxes

The ability to generate and analyze large numbers of replications is not only fundamental to verifying randomness but also essential for solving complex optimization problems. One such example is the *travelling salesman problem*, where Monte Carlo methods play a critical role in approximating solutions efficiently.

1.2.2 Travelling Salesman Problem

The salesman must find a route that visits all cities $1, 2, \ldots, n$ exactly once and returns to the starting city. The distances between city i and city j are provided in the matrix $\mathbf{M}$ of size $n \times n$. If there is no route from i to j, then $c_{ij} = \infty$. On the map in Fig. 1.4, 13 USA cities are presented—the actual distances are provided in the matrix $\mathbf{M}$ in Fig. 7.3.

Fig. 1.4 13 US cities used in the TSP example from https://developers.google.com/optimization/ routing/tsp. The figure was generated using `folium` and map tiles from OpenStreetMap

For a given permutation σ, let $l(\sigma)$ denote the total distance traveled:

$$l(\sigma) = \sum_{k=1}^{n-1} \mathbf{M}(\sigma_k, \sigma_{k+1}) + \mathbf{M}(\sigma_n, \sigma_1).$$

The task is to find the permutation σ^* that minimizes this distance:

$$\sigma^* = \operatorname*{argmin}_{\sigma} l(\sigma).$$

For large n, solving this problem deterministically becomes computationally infeasible. In fact, finding σ^* belongs to the class of NP-complete optimization problems. This implies that there is likely no efficient algorithm (with runtime polynomial in n) for finding the exact solution.

However, randomized algorithms have proven to be remarkably effective. They can identify an approximate solution $\hat{\sigma}$ such that $l(\hat{\sigma}) \approx l(\sigma^*)$. We will explore optimization algorithms based on Markov chain Monte Carlo methods in Chap. 7.

1.2.3 Computing Integrals with Monte Carlo Methods

The Monte Carlo (MC) method is widely used for computing integrals, especially for multidimensional integrals over complex spaces where simple analytical formulas are unavailable. In such cases, standard numerical methods are often infeasible.

Suppose we aim to estimate:

$$I = \int_{[0,1)^d} k(\boldsymbol{u}) \, d\boldsymbol{u},$$

where the function k is integrable.

The MC method relies on the strong law of large numbers. Specifically, if $\boldsymbol{U}_1, \boldsymbol{U}_2, \ldots, \boldsymbol{U}_n$ are i.i.d. random vectors in $[0, 1)^d$, with each coordinate of $\boldsymbol{U}_i$ following a uniform distribution $\mathcal{U}[0, 1)$, then with probability 1:

$$\hat{Y}_n = \frac{1}{n} \sum_{i=1}^{n} k(\boldsymbol{U}_i) \to \mathbb{E}k(\boldsymbol{U}) = I. \tag{1.2.2}$$

Thus, $\hat{Y}_n$ serves as an estimator for the integral. It is crucial to remember that $\hat{Y}_n$ is a *random* quantity, and the result must be interpreted accordingly. Typically, a confidence interval is provided, which requires knowledge of or an estimate for the variance of the estimator. These details are covered later in Chap. 4. In that chapter, we will also explore alternative methods to estimate such integrals by sampling from non-uniform distributions. For example, if Y_i, $i = 1, \ldots, n$, are i.i.d. random variables such that $\mathbb{E}Y_i = I$, then $\hat{Y}_n' = \frac{1}{n} \sum_{i=1}^{n} Y_i$ is another estimator for I. Later, in Chap. 5, we present a series of techniques to "improve" estimators, often by averaging samples that are not i.i.d.

Rate of Convergence of the Monte Carlo Method Most problems addressed by Monte Carlo methods can be reduced to calculating integrals with respect to arbitrary distributions. Revisiting the estimator in (1.2.2), Chebyshev's inequality shows that the convergence rate is $O(n^{-1/2})$. Notably, this rate is independent of the dimensionality of the integral, making Monte Carlo methods particularly effective for high-dimensional problems in fields such as physics, finance, and operations research.

To analyze the convergence rate further, we use the concept of discrepancy. Consider an arbitrary sequence $\mathbf{x}_1, \mathbf{x}_2, \ldots, \mathbf{x}_n \in [0, 1)^d$, and let $\mathcal{B}$ denote a family of subsets of $[0, 1)^d$. The *discrepancy* of the sequence $\mathbf{x}_1, \mathbf{x}_2, \ldots$ with respect to $\mathcal{B}$ is defined as:

$$D(\mathbf{x}_1, \ldots, \mathbf{x}_n; \mathcal{B}) = \sup_{A \in \mathcal{B}} \left| \frac{\#\{1 \leq i \leq n : \mathbf{x}_i \in A\}}{n} - \text{vol}(A) \right|, \tag{1.2.3}$$

where $\text{vol}(A)$ represents the volume of $A \subset \prod_{j=1}^{d}[0, 1)$. For instance, $\mathcal{B} = \mathcal{B}^*$ might consist of subsets of the form $\prod_{j=1}^{d}[0, u_j)$, where $0 < u_j < 1$. Other subset families are also studied in the literature.

If $\mathbf{x}_1, \mathbf{x}_2, \ldots$ is a realization of i.i.d. random vectors $U_1, U_2, \ldots$, uniformly distributed as $\mathcal{U}(\prod_{j=1}^{d}[0, 1))$, then with probability 1:

$$\limsup_{n \to \infty} \frac{(2n)^{1/2}}{\log \log n} D(\mathbf{x}_1, \ldots, \mathbf{x}_n; \mathcal{B}) = 1.$$

This result implies that the discrepancy for the sequence of i.i.d. random vectors $U_1, U_2, \ldots$ scales as $\log \log n / (2n)^{1/2}$, regardless of the dimension d. Consequently, the rate of convergence in Monte Carlo methods is of order $1/\sqrt{n}$ (more precisely, of order $o(n^{-1/2+\epsilon})$).

In specific applications, such as finance, this convergence rate may be insufficient. To address such limitations, the quasi-Monte Carlo (quasi-MC) method was developed.

1.3 Quasi-Monte Carlo Methods

An alternative to the Monte Carlo (MC) method, which uses pseudorandom numbers, is the quasi-MC method , which uses quasirandom numbers. This method is particularly useful for computing integrals of the form

$$\int_{[0,1)^d} k(\boldsymbol{u}) \, d\boldsymbol{u}.$$

Let us begin with the definition of a *low-discrepancy sequence*. A sequence $\mathbf{x}_1, \mathbf{x}_2, \ldots \in [0, 1)^d$ is called a low-discrepancy sequence if:

$$D(\mathbf{x}_1, \ldots, \mathbf{x}_n) = O\left(\frac{(\log n)^d}{n} \right).$$

Instead of generating a sequence of pseudorandom numbers, the quasi-MC method uses a deterministic sequence $\mathbf{x}_1, \mathbf{x}_2, \ldots$ with the low-discrepancy property. The integral is then computed using the fact that:

$$\frac{1}{n} \sum_{i=1}^{n} k(\mathbf{x}_i) \to \int_{[0,1)^d} k(\boldsymbol{u}) \, d\boldsymbol{u}.$$

Sequences with the low-discrepancy property are referred to as quasirandom sequences. For any quasirandom sequence, the rate of convergence is of order $O(n^{-(1-\epsilon)})$, where $0 < \epsilon < 1$ is arbitrary. Naturally, certain assumptions about the function k are required, but these are typically satisfied in practice.

Quasi-MC methods are widely used for valuing financial products.

In the literature, several quasirandom sequences are described, such as the Halton sequence, the Sobol sequence, and the Faure sequence. For each such sequence of dimension d, it can be shown that:

$$D(\mathbf{x}_1, \ldots, \mathbf{x}_n) \leq C_d \frac{(\log n)^d}{n} + O\left(\frac{(\log n)^{d-1}}{n}\right).$$

In `Python`, the procedures for generating Halton and Sobol sequences are available in the library `scipy.stats.qmc`.[1]

Hugo Steinhaus observed that the so-called golden sequences are useful for computing integrals. A golden sequence is defined as:

$$x_n = \{n\alpha\}, \quad \text{where } \alpha = \frac{\sqrt{5} - 1}{2},$$

the golden ratio, i.e., the solution of the equation $z^2 + z - 1 = 0$. Here, $\{x\}$ denotes the fractional part of x. The golden ratio has a continued fraction expansion of the form:

$$\alpha = \cfrac{1}{1 + \cfrac{1}{1 + \cfrac{1}{1 + \cdots}}}.$$

The sequence (x_n) exhibits the low-discrepancy property.

Similarly, the so-called Kronecker sequences, defined as:

$$x_n = \{n\beta\},$$

where β has a continued fraction expansion of the form:

$$\beta = \cfrac{1}{b_1 + \cfrac{1}{b_2 + \cdots}},$$

with $b_j \leq M$ being bounded, are also low-discrepancy sequences.

It remains unknown whether multidimensional Kronecker sequences of the form $(\{n\beta_1\}, \ldots, \{n\beta_d\})$ have this property. However, numerical results suggest they might.

---

[1] https://docs.scipy.org/doc/scipy/reference/stats.qmc.html.

1.4 Some Examples of Simulations that Reveal a Difference Between the MC and Quasi-MC Method

Although we do not study quasirandom numbers in this book, we now present a few examples comparing the use of quasirandom and pseudorandom numbers to illustrate the differences between their possible application domains.

Example 1.4.1 (Estimating π) More details on estimating π are provided in subsequent sections. Here, we briefly describe one of the simplest methods: we "throw" points chosen "randomly" on the square $[0, 1)^2$ and compute the fraction of points that fall within a quarter of a unit circle. To be more precise, let us take $2R$ i.i.d. numbers $U_{1,1}, U_{1,2}, \ldots, U_{j,1}, U_{j,2}, \ldots, U_{R,1}, U_{R,2}$ and compute the fraction:

$$\hat{Y}_R = \frac{1}{R}\sum_{j=1}^{R} Y_j, \quad \text{where } Y_j = 4\mathbb{1}(U_{j,1}^2 + U_{j,2}^2 \le 1).$$

We have that $\mathbb{E}Y_j = \pi$ for uniformly distributed numbers, and thus $\mathbb{E}\hat{Y}_R = \pi$.

We simulated four sets of numbers:

$$U_{1,1}^{\mathsf{A}_i}, U_{1,2}^{\mathsf{A}_i}, \ldots, U_{j,1}^{\mathsf{A}_i}, U_{j,2}^{\mathsf{A}_i}, \ldots, U_{R,1}^{\mathsf{A}_i}, U_{R,2}^{\mathsf{A}_i}, \quad i = 1, 2, 3, 4,$$

where

- $\mathsf{A}_1 = $ PCG64—pseudorandom numbers using built-in `Python`'s `NumPy` library,
- $\mathsf{A}_2 = $ `LatinHypercube`—quasirandom numbers using `LatinHypercube` algorithm,
- $\mathsf{A}_3 = $ `Halton`—quasirandom numbers using `Halton` algorithm,
- $\mathsf{A}_4 = $ `Sobol`—quasirandom numbers using `Sobol` algorithm,

In the last three cases, we used `Python`'s quasirandom numbers library `scipy.stats.qmc`.

The resulting points for $R = 1024$ (algorithm `Sobol` requires that the number of points to be generated is a power of 2) are depicted in Fig. 1.5. An implementation is available in `ch1_estimate_pi_quasirandom.py`.

The results of

$$\hat{Y}_R^{\mathsf{A}_i} = \frac{1}{R}\sum_{j=1}^{R} Y_j^{\mathsf{A}_i}, \quad \text{where } Y_j^{\mathsf{A}_i} = 4\mathbb{1}(U_{j,1}^{\mathsf{A}_i})^2 + (U_{j,2}^{\mathsf{A}_i})^2 \le 1)$$

as well as the error $|\hat{Y}_R^{\mathsf{A}_i} - \pi|$ (where as π we took its constant value from the `NumPy` library) for $R \in \{2^8, 2^9, 2^{10}, 2^{11}\}$ are presented in Table 1.1. As we can see, in this example, using quasirandom numbers yields a much better approximation. For example, for $R = 2^{11}$ pseudorandom numbers from PCG64 yielded the worst result.

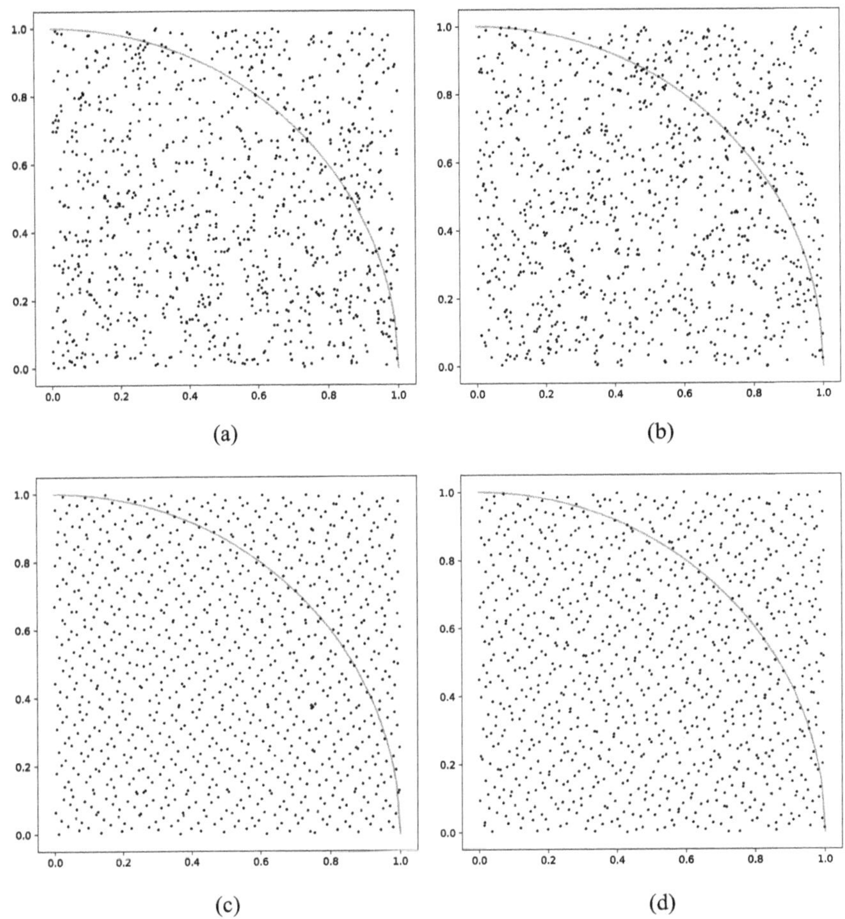

Fig. 1.5 1024 pseudorandom and quasirandom from $[0, 1)^2$. (**a**) Pseudorandom (PCG64). (**b**) LatinHypercube. (**c**) Sobol. (**d**) Halton

Remark 1.4.2 Although the discrepancy of a quasirandom sequence decreases faster than that of a realization of a uniform i.i.d. sequence, its use in solving probabilistic problems via quasi-Monte Carlo methods is mostly restricted to the computation of integrals. In Monte Carlo theory, for a sequence $U_1, U_2, \ldots$ of i.i.d. uniform random variables, we can consider the sequence of d-vectors $(U_1, \ldots, U_d), (U_{d+1}, \ldots, U_{2d}), \ldots$ as a sequence of random vectors. This is not true for quasirandom numbers. For a low-discrepancy sequence $x_1, x_2, \ldots,$ the sequence of d-vectors $(x_1, \ldots, x_d), (x_{d+1}, \ldots, x_{2d}), \ldots$ does not necessarily have low discrepancy. For each dimension d, a quasirandom sequence must be constructed separately because grouping into vectors results in the loss of the low discrepancy. ∎

Table 1.1 Estimations of π using pseudorandom numbers (PCG64) and quasirandom numbers (LatinHypercube, Sobol and Halton). The best result in each row is bolded

R		PCG64	LatinHypercube	Sobol	Halton
2^8	$\hat{Y}_R^A$	3.09375	3.07812	**3.14062**	3.18750
	$|\hat{Y}_R^A - \pi|$	0.04784	0.06347	**0.00967**	0.04591
2^9	$\hat{Y}_R^A$	3.1875	3.19531	**3.15625**	3.17188
	$|\hat{Y}_R^A - \pi|$	0.04591	0.05372	**0.01466**	0.03028
2^{10}	$\hat{Y}_R^A$	3.17188	3.11328	**3.14062**	3.16406
	$|\hat{Y}_R^A - \pi|$	0.03028	0.02831	**0.00097**	0.02247
2^{11}	$\hat{Y}_R^A$	3.10352	3.13867	**3.14062**	3.14648
	$|\hat{Y}_R^A - \pi|$	0.03808	0.00292	**0.00097**	0.00489

Remark 1.4.3 In Asmussen and Glynn's book ([8], p. 271), there is an illustrative counterexample to the use of the quasi-Monte Carlo method. Consider a random walk with increments of the form $X - Y$, where X and Y are exponentially distributed with parameters 1 and 1/2, respectively. The probability that this random walk exits the interval $(-4, 2)$ to the right is

$$I = (1 - e^2/2)(2e - e^{-2}/2) = 0.174,$$

in contrast with the result obtained by quasi-Monte Carlo using a 4-dimensional Halton sequence, which yielded $I = 0.118$. ∎

The reader must be cautioned that quasirandom numbers typically perform well for computing integrals, but they often fail in other tasks, as illustrated in the following example. In the sequel, we present a more extensive experiment to demonstrate both the strengths and weaknesses of quasirandom numbers. This example is as follows:

Example 1.4.4 (Winning Probability in a Two-dimensional Game) Consider the following game: we are playing against two other players. Our initial assets are (i_1, i_2), and the assets of the other players are $(N_1 - i_1, N_2 - i_2)$, where N_j is the total amount of assets for player j. With probability $p_j(i_j)$, we win one dollar against player j, and with probability $q_j(i_j)$, we lose one dollar. With the remaining probability $1 - (p_1(i_1) + q_1(i_1) + p_2(i_2) + q_2(i_2))$, nothing happens. Once we completely defeat player j (i.e., $i_j = N_j$), we no longer play with them (i.e., $q_j(N_j) = 0 = p_j(N_j)$).

We win the entire game if we defeat all players; we lose the game if we are defeated by at least one player. Thus, the states of the game are described as (i_1, i_2), where $i_j \in \{1, \ldots, N_j\}$, and an additional state, **LOSE**. The game is depicted in Fig. 1.6.

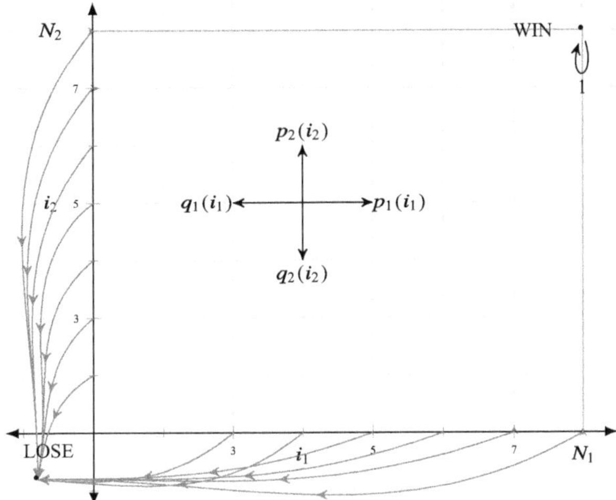

Fig. 1.6 A two dimensional game. See description in text

Let us define

$$\rho((i_1, i_2)) = \mathbb{P}(\text{we win the game} \mid \text{we started in } (i_1, i_2)).$$

For a more formal definition of the game (and, more generally, when we play against d other players), see [129], where a formula for the winning probability is presented. In our case, the winning probability is given by

$$\rho((i_1, i_2)) = \frac{\prod_{j=1}^{2}\left(\sum_{n_j=1}^{i_j}\prod_{r=1}^{n_j-1}\left(\frac{q_j(r)}{p_j(r)}\right)\right)}{\prod_{j=1}^{2}\left(\sum_{n_j=1}^{N_j}\prod_{r=1}^{n_j-1}\left(\frac{q_j(r)}{p_j(r)}\right)\right)}. \tag{1.4.1}$$

Consider a specific symmetric example, where $N_1 = N_2 = 4$ and

$$q_j(i_j) = p_j(i_j) = \frac{1}{4}, \quad j = 1, 2, \ i_j = 1, 2, 3.$$

If we start with initial assets $(2, 3)$, formula $(1.4.1)$ shows that we will win with probability

$$\rho((2, 3)) = \frac{3}{8} = 0.375.$$

We aim to estimate $\rho((2, 3))$ through simulations. We will use pseudo-random numbers from the PCG64 (denoted by $\mathbf{A}_1$) or quasirandom numbers:

`LatinHypercube` (denoted by A_2), `Sobol` (denoted by A_3), and `Halton` (denoted by A_4).

We simulate R games, and the result of the j-th game is denoted by $Y_j^{A_i} \in \{0, 1\}$, where 1 denotes winning and 0 losing. Then, the estimator is

$$\hat{Y}_R^{A_i} = \frac{1}{R} \sum_{j=1}^{R} Y_j^{A_i}.$$

Say that the current assets are (i_1, i_2). We simulate the following moves with the following probabilities:

$$(i_1, i_2) \quad \rightarrow \quad (i_1 + 1, i_2) \quad \text{with prob. } p_1(i_1),$$

$$(i_1, i_2) \quad \rightarrow \quad (i_1 - 1, i_2) \quad \text{with prob. } q_1(i_1),$$

$$(i_1, i_2) \quad \rightarrow \quad (i_1, i_2 + 1) \quad \text{with prob. } p_2(i_2),$$

$$(i_1, i_2) \quad \rightarrow \quad (i_2, i_2 - 1) \quad \text{with prob. } q_2(i_1),$$

$$(i_1, i_2) \quad \rightarrow \quad (i_2, i_2) \quad \text{with prob. } 1 - (p_1(i_1) + q_1(i_1) + p_2(i_2) + q_2(i_1)),$$

where if any coordinate of the new state is 0, we identify it with **LOSE**, and if both coordinates are 1, we identify it with **WIN**. To simulate the move, we could use a random variable $U \in [0, 1)$ and split the interval $[0, 1)$ into five subintervals with corresponding lengths. However, since we aim to use two-dimensional quasirandom numbers, we employ two random variables $U_1, U_2 \in [0, 1)$.

Using U_1, we first decide whether we will change the first coordinate, the second coordinate, or neither. This is achieved as follows:

Change 1st coordinate $\quad$ If $\quad U_1 \in [0, p_1(i_1) + q_1(i_1))$,

Change 2nd coordinate $\quad$ If $\quad U_1 \in [p_1(i_1) + q_1(i_1), p_1(i_1) + q_1(i_1) + p_2(i_2) + q_1(i_2))$,

none $\quad$ if $\quad U_1 \in [1 - (p_1(i_1) + q_1(i_1) + p_2(i_2) + q_2(i_1)), 0)$.

If we are to change one of the coordinates, we proceed as follows using U_2:

If 1st coord.:

$$(i_1, i_2) \quad \rightarrow \quad (i_1 + 1, i_2) \quad \text{if } U_2 \in [0, p_1(i_1)/(p_1(i_1) + q_1(i_1))),$$

$$(i_1, i_2) \quad \rightarrow \quad (i_1 - 1, i_2) \quad \text{if } U_2 \in [p_1(i_1)/(p_1(i_1) + q_1(i_1)), 1).$$

If 2nd coord.:

$$(i_1, i_2) \quad \rightarrow \quad (i_1, i_2 + 1) \quad \text{if } U_2 \in [0, p_2(i_2)/(p_2(i_2) + q_2(i_2))),$$

$$(i_1, i_2) \quad \rightarrow \quad (i_1, i_2 - 1) \quad \text{if } U_2 \in [p_2(i_2)/(p_2(i_2) + q_2(i_2)), 1).$$

We used pairs

$$(U^{A_i}_{1,1}, U^{A_i}_{1,2}), (U^{A_i}_{2,1}, U^{A_i}_{2,2}), \ldots, \quad i = 1, 2, 3, 4.$$

For A_1 (i.e., PCG64), we took consecutive outputs from the PRNG, whereas in the remaining cases, the pairs $(U^{A_i}_{j,1}, U^{A_i}_{j,2})$ were two-dimensional quasirandom numbers. In each case, we simulated $R = 5000$ games (note that each game required a variable number of pairs since the simulation continued until a win or loss occurred). However, when using quasirandom numbers generated by the Sobol method, the game never terminated. To address this, we added an independent normally distributed random variable (scaled by 0.1) to each $U^{A_3}_{j,1}$ and $U^{A_3}_{j,2}$. For further details, see the implementation. The complete code is provided in the Python file ch1_2d_game_winning_prob.py.

As shown in Fig. 1.7, when using quasirandom numbers, the estimators converge to very different values. In particular, the Halton and Sobol sequences yield significantly worse results. Thus, while these sequences are highly effective for estimating integrals or intervals, caution must be exercised when using them for other simulation tasks, as they may not yield reliable results.

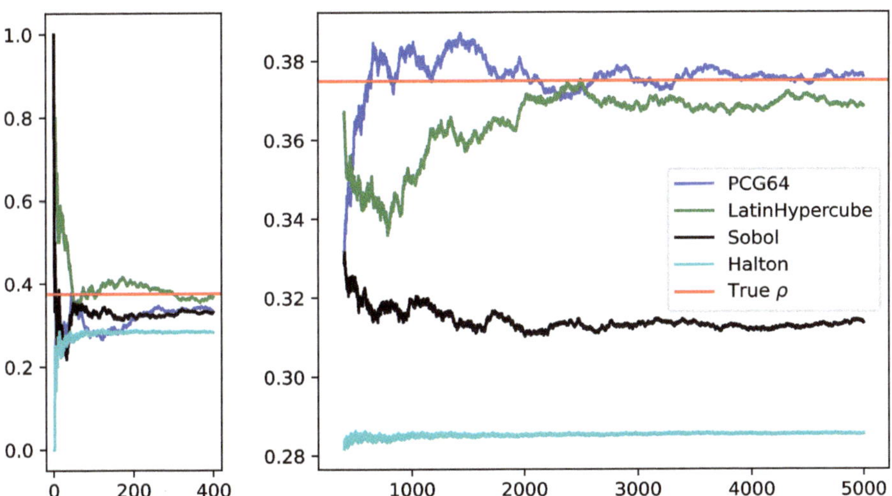

Fig. 1.7 Estimating $\rho((2, 3)) = 0.375$ (denoted above by ρ) using pseudorandom numbers (PCG64) and quasirandom numbers (LatinHypercube, Sobol, Halton). The simulation ran for 5000 steps—with steps 1–400 shown on the left and the remaining steps on the right. The *red line* indicates the true value of $\rho((2, 3))$

Table 1.2 Estimating $\rho((2, 3)) = 0.375$ using pseudorandom numbers (PCG64) and quasirandom numbers (LatinHypercube, Halton, Sobol). Values of $\hat{Y}_R^{A_i}$ estimators $i = 1, 2, 3, 4$ for $R = 5000$ and average number of steps of a single game

	PCG64	LatinHypercube	Sobol	Halton
$\hat{Y}_R^A$	0.3760	0.3686	0.3136	0.2856
Avg nr of steps	8.4088	8.4798	13.6588	11.5652

In Table 1.2, the values of estimators as well as average the number of steps of each game is presented.

In Figs. 1.8 and 1.9 we presented:

- Top rows: first 1000 points used in simulations, 15 of which are numbered.
- Bottom rows: histograms of sums $U_{k,1}^{A_i} + U_{k,2}^{A_i}$. *Blue line*—density f of a sum of two independent $\mathcal{U}[0, 1)$ random variables.

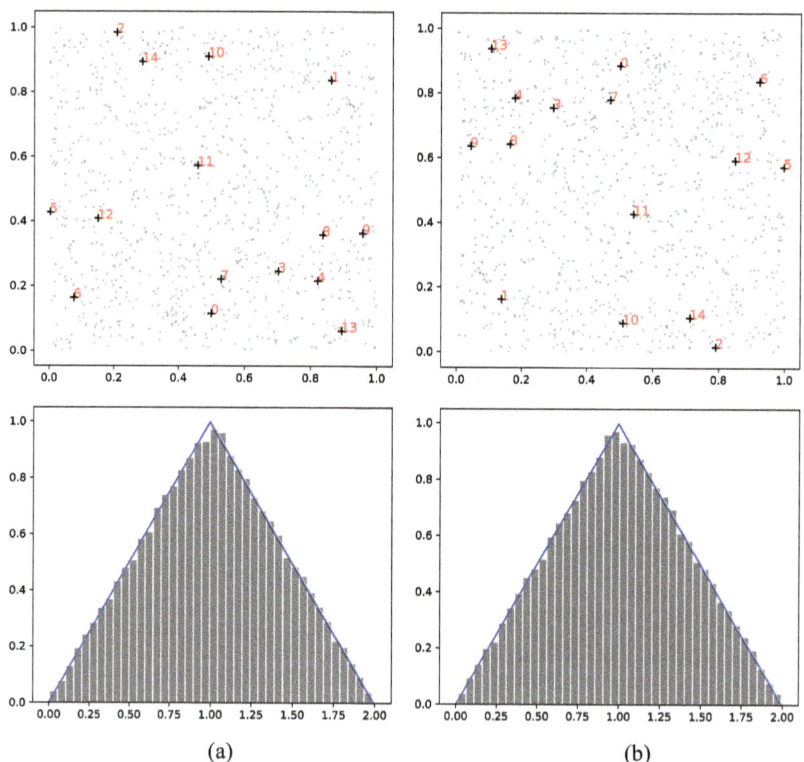

(a)   (b)

Fig. 1.8 PCG64 and LatinHypercube. Top row: 1000 points used in simulations, first 15 points numbered. Bottom row: histogram of $U_{k,1}^{A_i} + U_{k,2}^{A_i}$. *Blue line*—density f of a sum of two independent $\mathcal{U}[0, 1)$ random variables given in (1.4.2). (**a**) Pseudorandom (PCG64). (**b**) LatinHypercube

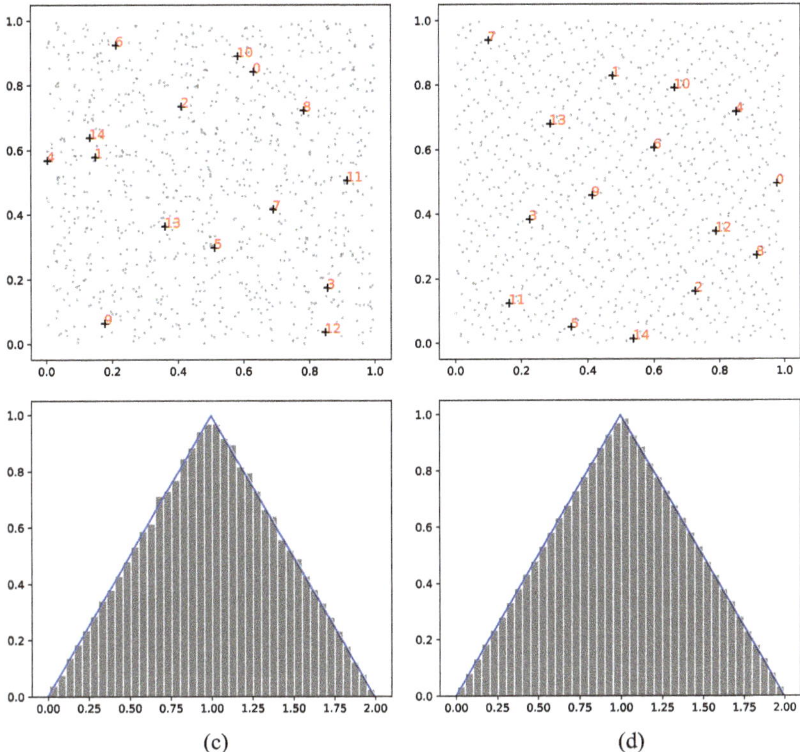

Fig. 1.9 Sobol and Halton. Top row: 1000 points used in simulations, first 15 points numbered. Bottom row: histogram of $U_{k,1}^{A_i} + U_{k,2}^{A_i}$. *Blue line*—density f of a sum of two independent $\mathcal{U}[0, 1)$ random variables given in (1.4.2). (**a**) Sobol. (**b**) Halton

As observed, there is a clear linear dependence between certain points. For instance, in the Halton sequence, points such as $(U_{3,1}^{A_4}, U_{3,2}^{A_4})$, $(U_{9,1}^{A_4}, U_{9,2}^{A_4})$, $(U_{6,1}^{A_4}, U_{6,2}^{A_4})$, and $(U_{4,1}^{A_4}, U_{4,2}^{A_4})$ lie along a single line (almost).

The distribution of the sum $U_1 + U_2$, where U_1 and U_2 are independent $\mathcal{U}[0, 1)$ random variables, follows a symmetric triangular distribution given by:

$$f(x) = \begin{cases} x & \text{if } x \in [0, 1), \\ 2 - x & \text{if } x \in [1, 2), \\ 0 & \text{otherwise.} \end{cases} \qquad (1.4.2)$$

The bottom rows of Figs. 1.8 and 1.9 display the histograms of all points used in the simulations. Notably, the histograms for quasirandom numbers appear *"too perfect"*, reflecting their structured nature.

1.5 Bibliographical Notes

The origins of Monte Carlo methods trace back to the work of Stanisław Ulam and John von Neumann during the 1940s, where they developed probabilistic techniques for nuclear physics simulations. The foundational aspects of Monte Carlo methods are well documented in the classical books by Metropolis and Ulam [150] or by Hammersley and Handscomb [77].

The reader can find aspects of simulation theory in Knuth's pioneering mono-graph *The Art of Computer Programming* (Vol. 2) [107], which discusses numerical methods, including random number generation and simulation techniques.

There are many excellent textbooks on stochastic simulation methods and Monte Carlo theory. For an elementary introduction, Ross [182] provides a broad coverage of simulation techniques, while more advanced treatments can be found in Ripley [175] and Madras [138].

Books that employ a modern probabilistic framework and leverage the theory of stochastic processes include Asmussen and Glynn [8], Fishman [55], and Rubinstein and Kroese [186], which explore Monte Carlo simulation from both theoretical and applied perspectives.

For field-specific applications of Monte Carlo methods, Glasserman [63] pro-vides an extensive treatment of simulation techniques in financial engineering, while Fishman [56] discusses applications in discrete event simulations.

Quasi-Monte Carlo (QMC) methods, which improve upon standard Monte Carlo by replacing random sampling with low-discrepancy sequences, were first explored in the works of Sobol [195] and Niederreiter [166]. Niederreiter's monograph remains a central reference on QMC theory, with additional insights provided by Glasserman [63]. A survey of QMC methods, with a primary focus on high-dimensional integration, is provided by L'Ecuyer and Lemieux [120]. Steinhaus [198] (in Polish) observed the low discrepancy of the golden sequence, a result that later influenced the development of low-discrepancy sequences in numerical integration.

For an introduction to random number generation and its role in Monte Carlo methods, the classic reference remains Knuth [107]. A survey of pseudorandom number generators and their statistical testing is given in L'Ecuyer [115], while Matsumoto and Nishimura [147] introduced the widely used Mersenne Twister generator.

Chapter 2
The Theory of Generators

Randomness is an important, sometimes crucial, element of computational science, enabling simulations, cryptographic protocols, and statistical modeling across diverse fields. While physical randomness (e.g., from quantum processes) is theoretically ideal, it is rarely practical in computational settings. Instead, pseudorandom number generators (PRNGs) provide a practical solution, producing deterministic sequences that mimic the properties of true randomness, particularly being uniformly distributed on $[0, 1)$ and, in reality, on $\bar{M} = \{0, 1, \ldots, M - 1\}$. These sequences are fundamental to ensuring reproducibility and efficiency in computational tasks. The importance of PRNGs becomes apparent in a wide range of applications. For instance:

- In simulations, PRNGs allow us to model random processes, from financial markets to physical phenomena like molecular dynamics.
- In Monte Carlo methods, PRNGs enable us to estimate quantities of interest, such as integrals or expectations, through repeated sampling. These estimators are themselves random, and their precision depends, among other factors, on the quality of the underlying PRNG.
- In cryptography, PRNGs generate keys, tokens, and other elements critical for secure communication, ensuring that outputs cannot be predicted or distinguished from truly random sequences.

One of the critical challenges in working with PRNGs is validating their quality. Poor-quality PRNGs can introduce bias, correlations, or periodicity, undermining the reliability of simulations and security systems. To address this, statistical tests have been developed to evaluate PRNGs against the properties of uniformity, independence, and randomness. These tests, explored extensively in this chapter, form the foundation for selecting or designing appropriate PRNGs for various applications.

This chapter bridges the historical and modern development of PRNGs, presenting them as an essential tool for computational science. By the end of this chapter,

P. Lorek, T. Rolski, *Lectures on Monte Carlo Theory*, Probability Theory
and Stochastic Modelling 108, https://doi.org/10.1007/978-3-032-01190-9_2

readers will understand the principles of PRNG design, the importance of statistical tests, and the role PRNGs play in simulations, cryptography, and beyond.

2.1 Introduction

The foundation of stochastic simulation lies in the concept of a random number. While we will not delve deeply into the philosophical debates surrounding randomness, it is worth reflecting on its nature as highlighted in the following statement by Donald Knuth in his seminal work [107], page 2:

> "In a sense, there is no such thing as a random number; for example, is two a random number?"

However, we can rigorously define an *independent sequence of random variables* with a specified *distribution*. This essentially means that each number in the sequence is selected entirely at random, without any dependencies on the selection of other numbers, and that each number conforms to a specified probability distribution. In this text, we will use the terms *random number* and *random variable* interchangeably.

The primary distribution in this theory is the uniform distribution $\mathcal{U}[0, 1)$.

It is important to note that a computer cannot generate all possible numbers within the interval $[0, 1)$. Instead, the numbers generated are of the form

$$0, \frac{1}{M}, \frac{2}{M}, \ldots, \frac{M-1}{M}$$

for some fixed $M = 2^k$, and we expect them to be uniformly distributed within this discrete set. In this case, throughout this text, we may omit the explicit mention of the distribution name. For instance, we might refer to a random variable U with the distribution $\mathcal{U}[0, 1)$ simply as a random number U. Symbols representing distributions, along with their definitions and key characteristics, are provided in Appendix A.2

Before the advent of the computer age, tables of random numbers were created. The most straightforward table can be made, for example, by drawing balls from a box, throwing a dice, etc., then recording these results. You can also use a 'dice' that gives results from 0 to 9; see Fig. 2.1, and produce through consecutive throws *random digits*.

Then, by grouping a fixed number of digits, preceded by zeros and periods, we obtain the desired table of random numbers. Such tables often have "good properties" (what we understand by this we will clarify later) but when lots of random numbers are needed (e.g., hundreds of thousands) the method is useless.

Today computers are used to create a sequence of random numbers (or, as it is often said, *to generate* them). Probably the first to do so was John von Neumann around 1946. He proposed to create the next number by squaring the previous number and cutting out the middle numbers. The "middle-square method" turned out to be not very good; it has a rutting tendency—becoming locked into short repeating cycles. For example, a sequence of two-digit numbers starting with 43 according to this algorithm is (we always make the resulting number 4 digits long, padding zeros if needed): 84, 05, 25, 62, 84, 05, 02, 00, 00, ... Namely, we have $\mathbf{43^2 = 1\underline{84}9}$, $\mathbf{84^2 = 7\underline{05}6}$, $\mathbf{05^2 = 0\underline{02}5}$, $\mathbf{02^2 = 0\underline{00}4}$, $\mathbf{00^2 = 0\underline{00}0}$.

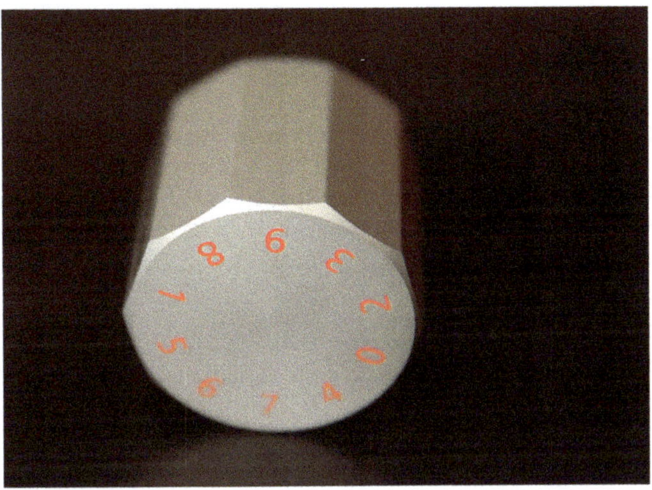

Fig. 2.1 Eurandom dice

A comprehensive theory has emerged regarding generators, i.e., algorithms producing sequences of "random numbers", and the generator's "goodness" analysis. This theory uses the achievements of modern algebra and statistics.

Unfortunately, a computer cannot generate a perfect sequence of independently uniformly distributed random variables. Therefore, such numbers are often referred to as computer-generated pseudorandom numbers. Understanding these limitations, we will mainly refer to them as sequences of random variables with the uniform distribution.

Definition 2.1.1 A pseudorandom number generator (PRNG) is a 5-tuple (S, s_0, f, V, g), where

- S is a finite state space,
- $s_0 \in S$ is the initial value of the recurrence (*seed*),
- $s_{i+1} = f(s_i)$, where $f : S \to S$,
- V is a finite set of values, and
- $g : S \to V$ (maps the generator's state into the output) and $u_i = g(s_i)$ ($i = 0, 1, \ldots$) is a generated sequence of pseudorandom numbers.

It is straightforward to observe that the sequence generated by $s_{n+1} = f(s_n)$ will eventually fall into a loop: there is a cycle that keeps repeating. The length of this cycle is called the *period*. The quality of the generator is strongly influenced by the period length, with longer periods being preferable. Note that, theoretically, the cycle length cannot exceed the cardinality of the set S.

2.1.1   Formal Definition of a Sequence of Random Numbers; Pseudorandom Numbers

By a *sequence of random numbers*, we mean a sequence of i.i.d. random variables with the same uniform distribution. We will consider (and extensively use) random variables on three types of state spaces: on $[0, 1)$, on $\{0, 1\}$, or on $\bar{M} = \{0, 1, \ldots, M - 1\}$ for some natural number $M \geq 2$. Respectively, we will talk about random variables U_j, B_j, or Y_j, for $j = 1, 2, \ldots$. We now recall formal mathematical definitions.

- By a *sequence of random numbers* from $[0, 1)$, we understand a sequence $U_1, U_2, \ldots$, where each U_j is a random variable taking values in $[0, 1)$ such that

  (i) Each U_j has the uniform distribution, i.e.,

$$\mathbb{P}(U_j \leq t) = \begin{cases} 0 & t < 0, \\ t & 0 \leq t < 1, \\ 1 & t \geq 1, \end{cases}$$

  (ii) $U_1, U_2, \ldots$ are independent and identically distributed (i.i.d.), i.e., for $n = 1, 2, \ldots$,

$$\mathbb{P}(U_1 \leq t_1, \ldots, U_n \leq t_n) = t_1 t_2 \cdots t_n,$$

  for any $0 \leq t_i \leq 1$, $i = 1, \ldots, n$.

- By a *sequence of random bits*, we understand a sequence $B_1, B_2, \ldots$, where each B_j is a random variable taking values in $\{0, 1\}$ such that

  (i) $\mathbb{P}(B_j = 0) = \mathbb{P}(B_j = 1) = \frac{1}{2}$ for each j,
  (ii) $B_1, B_2, \ldots$ are i.i.d., i.e., for $n = 1, 2, \ldots$,

$$\mathbb{P}(B_1 = b_1, B_2 = b_2, \ldots, B_n = b_n) = \frac{1}{2^n}.$$

- By a *sequence of random numbers* from $\{0, 1, \ldots, M - 1\}$, we understand a sequence $Y_1, Y_2, \ldots$, where each Y_j is a random variable taking values in $\{0, 1, \ldots, M - 1\}$ such that

  (i) $\mathbb{P}(Y_j = i) = \frac{1}{M}$ for $i = 0, \ldots, M - 1$, for each j,
  (ii) $Y_1, Y_2, \ldots$ are i.i.d., i.e., for $n = 1, 2, \ldots$,

$$\mathbb{P}(Y_1 = y_1, Y_2 = y_2, \ldots, Y_n = y_n) = \frac{1}{M^n}.$$

Note an important and non-trivial property of such sequences of random numbers. If $n_1 < n_2 < \cdots$ is an increasing sequence of natural numbers, then $U_{n_1}, U_{n_2}, \ldots$ is a sequence of random numbers. The same is true for a random sequence of bits.

A subject of further consideration in the book will be sequences of pseudorandom numbers (bits), i.e., those that imitate sequences of random numbers (bits). Such sequences can be obtained in various ways, but we will deal with those generated by appropriate algorithms on computers. As mentioned, these are called pseudorandom number generators (PRNGs). We will often omit the word *pseudo*. In the literature, it is often written simply *random number generator* (RNG), which, of course, must be a *pseudorandom* number generator. Humans are not able to generate a perfect sequence of numbers which is a realization of i.i.d. random variables.

Continuing our terminological considerations, note that when referring to R *replications* of a random number U, then if nothing else is said, we have a sequence $U_1, \ldots, U_R$ of i.i.d. random variables with the distribution $\mathcal{U}[0, 1)$.

2.2   Random Permutations

Let $\mathcal{S}_n$ be the set of all permutations of $[n] = \{1, \ldots, n\}$. By a random permutation, we understand a random element taking all values in $\mathcal{S}_n$ with equal probability $1/n!$. In other words, it is a random element on $\mathcal{S}_n$ with distribution $\mathcal{U}(\mathcal{S}_n)$.

2.2.1   An Algorithm for a Random Permutation

We now describe an algorithm for generating a random permutation of $\{1, \ldots, n\}$, such that each result is obtained with probability $1/n!$. We assume access to a random number generator, i.e., $U_1, U_2, \ldots$, a sequence of i.i.d. random variables with the uniform distribution $\mathcal{U}[0, 1)$. The procedure Fisher-Yates$(n, (U_1, \ldots, U_n))$ generates a (uniformly) random permutation of the set $\{1, \ldots, n\}$.
The algorithm has a complexity of $O(n)$ (why?). This algorithm is a direct consequence of Exercise 2.T.8.

Algorithm 1 Fisher-Yates (Random permutation)

Require: Integer n; sequence of pseudorandom numbers $\boldsymbol{u} = (u_1, \ldots, u_n)$, $u_i \in [0, 1)$
Ensure: Permutation of the set $\{1, \ldots, n\}$
  1: $S[i] = i$, $i = 1, \ldots, n$
  2: **for** $i = n$ downto 1 **do**
  3:     $k = 1 + \lfloor i u_i \rfloor$
  4:     **Swap**($S[k]$, $S[i]$)
  5: **end for**
  6: **Return** S

Algorithm 2 is another simple procedure for generating a random permutation, which involves sorting (denoted **Perm-sort**(n, $(U_1, \ldots, U_n)$)).

Algorithm 2 Perm-sort (Generating random permutation by sorting)

Require: Integer n; sequence of pseudorandom numbers $\boldsymbol{u} = (u_1, \ldots, u_n)$, $u_i \in [0, 1)$
Ensure: Permutation of the set $\{1, \ldots, n\}$
  1: $S[i] = i$, $i = 1, \ldots, n$
  2: Sort the indices in S such that S satisfies $u_{S[1]} \leq u_{S[2]} \leq \cdots \leq u_{S[n]}$
  3: **Return** S

2.2.2 *Shuffling Cards*

Shuffling cards is a procedure that leads to a random permutation. Assume we have a deck of 52 cards. We want to arrange them *randomly*, that is, so that each configuration has probability $1/52!$. We now formally define shuffling.

A deck of n cards can be identified with numbers $[n] = \{1, \ldots, n\}$, which can be arranged in $n!$ ways, these arrangements being called *permutations*. A random permutation is a random element X taking values in $\mathcal{S}_n$ such that $\mathbb{P}(X = \sigma) = 1/n!$ for any permutation $\sigma \in \mathcal{S}_n$. A card shuffling scheme is defined by its *one step*. For example, the Top-To-Random card shuffling scheme is defined as follows: take the top card and insert it randomly into the deck (details below). Let X_0 be an initial permutation (say, the identity). Then X_k is the permutation obtained after k steps of a card shuffling scheme. For reasonably defined shuffling, the distribution of X_k converges, as $k \to \infty$, to the uniform distribution on $\mathcal{S}_n$ (using fundamental Markov chain theory; details are in Sect. 6.2 and Exercise 6.T.3).

A key question is how far the distribution of X_k is from the uniform distribution. This can be quantified using the total variation distance

$$d_k = \frac{1}{2} \sum_{\sigma \in \mathcal{S}_n} \left| \mathbb{P}(X_k = \sigma) - \frac{1}{n!} \right|.$$

Another measure is the *mixing time*, which determines the number of steps k needed for the total variation distance to become small:

$$\tau^{\text{mix}}(\varepsilon) = \inf_{k \geq 0}\{d_k \leq \varepsilon\}.$$

Top-To-Random In each step, the top card is inserted randomly into one of n possible positions (including the top and bottom of the deck). Each position is chosen with equal probability $1/n$.

Random-To-Random Transposition Given a permutation $\sigma = (\sigma_1, \ldots, \sigma_n)$, choose two cards (or positions) independently and uniformly at random (each with probability $1/n$). Swap the two cards unless they are the same, in which case no change occurs.

Riffle Shuffle and its Time Reversal Riffle Shuffle is one of the most widely studied shuffling methods (believed to model real shuffling, e.g., in casinos). Suppose we have n cards. Split them into two piles by selecting $Y \sim \text{Bin}(n, 1/2)$. If $Y = k$, the first k cards form the left pile, and the remaining $n - k$ cards form the right pile (preserving order within each pile). Next, interleave the piles by dropping cards one by one, with the probability of dropping a card proportional to the number of cards remaining in each pile. Continue until all cards are dropped. Repeating this process multiple times ensures that the deck approaches a uniform distribution.

For $n = 52$ cards, Table 2.1 (taken from Diaconis [34]) shows the total variation distance d_k after k shuffles:

As seen, $k = 7$ steps suffice to make the distribution close to uniform, a fact popularized by the New York Times article [109] titled *"In shuffling cards, 7 is winning number"*.

Time Reversed Riffle Shuffle The time reversed Riffle Shuffle provides an alternative perspective on the riffle shuffle process. One step of the procedure is as follows: Given a deck of n cards, assign a random bit (0 or 1) independently to each card. Cards assigned bit 0 are moved to the top of the deck, while those assigned bit 1 remain below, with the relative order of cards within each group preserved.

This procedure is called the *time reversal of Riffle Shuffle* because of the following property (see Exercise 2.T.19): the probability that a Riffle Shuffle transforms a permutation σ into σ' is exactly equal to the probability that a time reversed Riffle Shuffle transforms σ' into σ.

Notably, the total variation distance between the distribution of the deck after k time reversed Riffle Shuffles and the uniform distribution on $\mathcal{S}_n$ is identical to

Table 2.1 Total variation distance between the distribution of the deck after k shuffles and the uniform distribution on $\mathcal{S}_{52}$

k	1	2	3	4	5	6	7	8	9	10
d_k	1.000	1.000	1.000	1.000	0.924	0.614	0.334	0.167	0.085	0.043

that for k steps of the standard Riffle Shuffle. Thus, for a standard deck of $n = 52$ cards, the values of d_k in Table 2.1 remain valid.

The simplicity of the time reversed Riffle Shuffle makes it particularly appealing for simulations and theoretical analysis. Using a *strong stationary time* argument, it can be shown that $2 \log_2 n$ steps are sufficient for the time-reversed shuffle to mix. However, Bayer and Diaconis [13] refined this result, showing that the "actual" number of steps needed for effective mixing is approximately $\frac{3}{2} \log_2 n$. This observation highlights the *cutoff phenomenon*, where performing significantly fewer steps results in a distance d_k close to 1, while slightly exceeding the threshold yields a d_k close to 0.

A common feature of both the Riffle Shuffle and its time-reversed counterpart is their natural alignment with Markov chain theory. While the convergence of these shuffles to a uniform distribution follows directly from fundamental theorems in this theory, determining the precise rate of convergence remains an active area of research, particularly with regard to the implications of the cutoff phenomenon.

2.3　Pseudorandom Number Generators

Pseudorandom number generators (PRNGs) are essential tools in computational science, with applications spanning simulations, cryptography, statistics, stochastic modeling, and randomized algorithms. They provide deterministic algorithms to produce sequences of numbers that mimic true randomness, making them integral to both theoretical analyses and practical implementations. PRNGs are particularly valued for their ability to ensure reproducibility in simulations while maintaining computational efficiency.

In this chapter, we explore the importance of PRNGs as a foundation for many tasks addressed in this book. In the next chapter, we will use PRNGs to simulate random variables with given distributions, a critical step in many simulation-based methods. Later, we will build on this by estimating quantities of interest, often expressed as expectations of random variables. These estimators rely on outputs generated by PRNGs, and their quality—such as unbiasedness and precision—depends heavily on the underlying randomness. A good PRNG is thus indispensable for accurate and reliable results.

This section is structured to reflect both the historical evolution and the modern applications of PRNGs. We begin with a discussion of modern generators, focusing on their implementation in Python (Sect. 2.3.1) and their role in secure crypto-graphic applications (Sect. 2.3.2). The latter includes an illustrative example of the RC4 stream cipher, once widely used but now deprecated. Finally, we discuss simple and historical methods such as linear congruential generators and bit generators in Sect. 2.4, which, while no longer prevalent in modern applications, remain valuable from an educational and historical perspective.

2.3.1 Pseudorandom Generators in Python

We extensively use `Python` and the `NumPy` library (which stands for Numerical Python), particularly for generating random numbers. Throughout this book, all examples are based on `Python` version 3.12.2 and `NumPy` version 1.26.4. To generate pseudorandom numbers, we first import the `NumPy` library:

```python
import numpy as np
```

The simplest way to generate random numbers is to use `np.random.rand(k)`, which generates k random numbers uniformly distributed in $[0, 1)$. The command `np.random.rand()` generates a single random number. This function is part of the older `RandomState` API in `NumPy`, which internally uses the Mersenne Twister (MT19937) generator. However, this API is now considered legacy and maintained primarily for backward compatibility.

Since `NumPy` version 1.17.0 (released in July 2019), a new approach for random number generation has been introduced. The modern `default_rng()` function is recommended as it allows for flexibility in choosing PRNG algorithms, ensures better reproducibility, and offers improved performance in multithreaded environments. A comparative overview of several PRNGs, including performance and statistical properties, is available on the *PCG Random website* [167] in the "At-a-Glance Summary" section.
`NumPy` supports four main PRNGs:

- PCG64: The default generator, based on the Permuted Congruential Generator, with a period of 2^{128}. Designed for general-purpose applications, it is highly efficient, offers excellent statistical quality, and takes advantage of 64-bit arithmetic [167, 168].
- `Philox`: A counter-based generator with a period of $2^{256}-1$. The name originates from the Greek word $\Phi\acute{\iota}\lambda o\varsigma$ ("Philos"), meaning "friend," with "x" representing multiplication. It is parallel-friendly, provides high-quality randomness, and also utilizes 64-bit arithmetic.
- SFC64: A Small Fast Chaotic generator with an average period of 2^{255}. It is lightweight, very fast, and optimized for tasks that do not require stringent randomness properties, leveraging 64-bit operations.
- MT19937: The Mersenne Twister generator with a period of $2^{19937} - 1$. It is a 32-bit generator that can run on 64-bit systems but does not fully utilize 64-bit processing. It is primarily used for reproducing results from older versions of `NumPy` (pre-1.17).

The recommended approach is to use the `default_rng()` function to create a new generator object, specifying a PRNG and a seed if needed:

```python
from numpy.random import default_rng, MT19937, PCG64
prng_pcg64   = default_rng(PCG64(seed=31415))
prng_mt19937 = default_rng(MT19937(seed=31415))
```

```python
# Generate  4 random numbers (floats) from [0, 1):
print("PCG64:    ", prng_pcg64.random(size=4))
print("MT19937: ", prng_mt19937.random(size=4))
```

This code produces the following output:

```
PCG64:     [0.49968933 0.11520254 0.86217723 0.83740069]
MT19937:  [0.49643292 0.34995806 0.36475593 0.39614581]
```

The PCG64 generator is the default and can be invoked directly using `rng = default_rng(seed=10)`. If no seed is specified, the generator uses a system-dependent random source, such as the current time or entropy sources, to ensure non-deterministic randomness by default.

Beyond generating basic uniform random numbers, NumPy provides various functions for sampling, shuffling, and random selection:

```python
rng.integers(low=0, high=10, size=5)   # 5 random integers
                                       # from 0 to 9
rng.permutation([1, 2, 3, 4, 5]) # random permutation of an array
rng.choice([10, 20, 30], size=3) # random sample with replacement
rng.shuffle(arr)                 # shuffle array in-place
```

This library also has some basic distributions built-in, for example:

```python
rng.normal(loc=0, scale=1, size=5) # samples from normal distr.
rng.binomial(n=10, p=0.5, size=5)  # samples from binomial distr.
```

For more advanced statistical distributions, using the `scipy.stats` library is recommended.

Each generator supports different seeding options:

- PCG64: Accepts an integer seed (0 to $2^{128} - 1$), an array of such integers, or None. If None, the seed is drawn from `/dev/urandom` or derived from the current time.
- Philox: Accepts an integer seed (0 to $2^{64} - 1$), an array of such integers, or None. Similar behavior as PCG64.
- SFC64: Requires four unsigned 64-bit numbers for the seed.
- MT19937: Accepts an integer seed (0 to $2^{32} - 1$), an array of such integers, or None. It uses legacy seeding behavior compatible with older versions of NumPy.

MT19937 is primarily retained for legacy support, whereas PCG64 is the recommended default for modern applications due to its performance and statistical robustness. Additionally, although not included in NumPy, the Xoshiro256 generator [15] is available in the RandomGen library and is gaining popularity for its speed and high-quality randomness.

2.3.2 Cryptographic Pseudorandom Number Generators

Cryptographic pseudorandom number generators (PRNGs) are vital in secure communications, cryptographic protocols, and data encryption. Unlike traditional PRNGs, cryptographic PRNGs must satisfy stricter security requirements to ensure their outputs cannot be distinguished from truly random sequences by feasible statistical or computational tests. In this section, we provide an overview of cryptographic PRNGs, including their definition and key features, and discuss the RC4 scheme as an illustrative example of both their utility and vulnerabilities.

The main differences between traditional and **cryptographic PRNGs** are:

- The functions f and g (as in Definition 2.1.1) must be *efficiently computable*, meaning they run in polynomial time with respect to the input size.
- A PRNG is considered cryptographic only if no weaknesses have been found using statistical tests or computational methods.

Statistical tests for cryptographic PRNGs often take the form of *polynomial-time distinguishers*, which are algorithms attempting to differentiate the PRNG's output from a truly random sequence. Distinguishers can be probabilistic, meaning they may use randomness internally. The definition of a cryptographic PRNG also involves the notion of a *negligible function*, ensuring that the difference between the probabilities of identifying true randomness and PRNG-generated sequences decreases rapidly with the input size.

We say that a function $\mathtt{negl} : \mathbb{N} \to \mathbb{R}$ is *negligible* if, for every positive polynomial $p(n)$, there exists an integer N such that:

$$\forall n \geq N \quad |\mathtt{negl}(n)| < \frac{1}{p(n)}.$$

Definition 2.3.1 Let $\ell(\cdot)$ be a polynomial, and let G be a deterministic polynomial-time algorithm such that for any input *seed* from $\{0, 1\}^n$, the algorithm G outputs a string of length $\ell(n)$. We say that G is a (cryptographic) PRNG if:

- For every n, we have $\ell(n) > n$.
- For all probabilistic polynomial-time distinguishers D, there exists a negligible function $\mathtt{negl}$ such that:

$$|\mathbb{P}(D(x) = 1) - \mathbb{P}(D(G(y)) = 1)| \leq \mathtt{negl}(n),$$

where x is chosen uniformly from $\{0, 1\}^{\ell(n)}$, y is a seed chosen from $\{0, 1\}^n$, and probabilities are over the randomness of D, x, and y.

If a distinguisher is found for a PRNG, it is no longer considered cryptographic. For example, RC4, once widely used (in 2014, around 80% of `https` traffic was encrypted using this cipher), is now regarded as insecure due to several discovered vulnerabilities. These weaknesses have led to its removal from widely used security protocols, including TLS (Transport Layer Security), which ensures secure communication over the internet.

TLS is a cryptographic protocol widely used to secure communication over the internet. It ensures confidentiality, integrity, and authentication for applications such as web browsing, email, and instant messaging. PRNGs are critical in TLS to generate encryption keys, session tokens, and other random data necessary for secure operations.

NIST Standards and Statistical Tests National Institute of Standards and Technology (NIST) has developed a comprehensive suite of tests for assessing randomness in cryptographic contexts [187]. These tests are widely used to evaluate the quality of cryptographic PRNGs, ensuring their resistance to statistical attacks and adherence to rigorous security standards.

AES The Advanced Encryption Standard (AES) is one of the most widely used cryptographic algorithms and is also considered an excellent cryptographic pseudorandom number generator (PRNG). AES has been rigorously analyzed, and no practical attacks against it have been discovered, making it a robust choice for secure applications. The algorithm operates on fixed-size blocks of 128 bits and supports key lengths of 128, 192, or 256 bits. AES is supported by NIST, which has developed extensive guidelines and standards for its usage. It is employed in numerous cryptographic protocols, including secure internet communications, such as Transport Layer Security (TLS). For more details on the algorithm, readers can consult the cryptographic literature or reliable online resources.

Other Cryptographic PRNGs Several cryptographic PRNGs have been proposed over time. They fall into two categories:

- *Special designs*:

  - ChaCha20: Replaced RC4 in OpenBSD, NetBSD, FreeBSD, and Linux, offering stronger security and better performance.
  - ISAAC: A variant of the RC4 cipher designed for cryptographic applications.
  - ANSI X9.17 standard: Used in financial cryptographic systems.

- *Number-theoretic designs*:

  - Blum-Blum-Shub: Based on the quadratic residuosity problem; secure but inefficient.
  - Blum-Micali: Relies on the discrete logarithm problem; also inefficient.
  - Dual_EC_DRBG: Based on the Diffie-Hellman assumption. It is believed to have a *kleptographic* National Security Agency (NSA) backdoor (as reported by *The Guardian* and *The New York Times* in 2013).

As a curiosity we mention that although the possibility of a backdoor in Dual_EC_DRBG was known, some companies still used it. For example, RSA Security, Inc. received 10 mln USD from the NSA to keep using it. Despite concerns about certain cryptographic PRNGs, algorithms such as AES are considered robust and are widely supported by organizations like NIST.

2.3.2.1 The RC4 Stream Cipher

RC4 is a symmetric stream cipher-key, which can be used as a pseudorandom number generator (PRNG). Roughly speaking, if Bob and Alice share a secret key K, then they may use RC4 to generate the same sequence of bits in a deterministic way—K plays the role of a seed. Then, if Alice wants to encrypt the message, she xors the bit representation of a message with the output of RC4 (for details on the xor operation see Sect. 2.4.2). Since for any bits $m, r \in \{0, 1\}$ we have $(m\,\mathbf{xor}\,r)\,\mathbf{xor}\,r = m$, Bob can decrypt it simply xoring the ciphertext with the output of RC4.

RC4 consists of two algorithms:

(i) **KSA (Key Scheduling Algorithm)** uses a secret key to transform the identity permutation of n cards into some other permutation (one can model KSA as a card shuffle). It starts with the identity permutation (of "a deck of cards"), then uses a card shuffle with random choices replaced by a secret key (assumed to be random) to obtain a "random" permutation.

(ii) **PRGA (Pseudo Random Generation Algorithm)** starts with a permutation generated by KSA and outputs random bits from it, updating the permutation at the same time.

We will consider the RC4 algorithm, which generates numbers $x_r \in \bar{n} = \{0, \dots, n - 1\}$, where $n = 2^k$. In internal memory it stores (S, i, j), where S is a permutation of $\bar{n}$ and $i, j \in \bar{n}$. In the standard version $k = 8$, it is recommended that the length of the key is between 40 and 128 bits. The KSA and PRGA algorithms are given in Fig. 2.2.

Remark 2.3.2 The idea of KSA was to produce a "random" permutation. Consider a case where a secret key K has length $L \geq n$, and each $K[i], i = 0, \dots, L - 1$ is chosen uniformly from the set $\{0, \dots, n - 1\}$. Then, the line

$\qquad j := (j + S[i] + K[i \bmod L]) \bmod n$

of Algorithm 3 can be replaced (for analysis of the randomness of the resulting permutation) with

$\qquad j := \mathbf{random}(\{0, \dots, n - 1\})$.

In such a case, the Key Scheduling Algorithm (KSA) corresponds to the *Cyclic-To-Random Transposition* card shuffling scheme (at step i, exchange the i-th card with a uniformly chosen one). It is known that it takes $\Theta(n \log n)$ steps to mix (Mironov, while studying RC4, proved an upper bound $O(n \log n)$; a matched lower bound was proven in [158], and in [134], it is conjectured (Conjecture 1) that the mixing time, measured in the separation distance, is asymptotically $n \log n$ as $n \to \infty$).

However, in RC4, only n steps are performed, which means that KSA does not produce a random (uniform) permutation. This is one of the main weaknesses of RC4. Attacks exploiting weaknesses in both the PRGA and KSA, or in the way RC4 was used in specific systems, are presented in [5, 57, 58, 68, 139].

As a result, in 2015, RC4 was prohibited in Transport Layer Security (TLS) Microsoft, Mozilla and Internet Engineering Task Force (IETF)—an organization

Algorithm 3 KSA	**Algorithm 4** PRGA
Require: n, key $K = K[0], \ldots, K[L-1]$	**Require:** $S = S[0], \ldots, S[n-1]$
Ensure: Permutation $S = S[0], \ldots, S[n-1]$	**Ensure:** Bytes $x_1, x_2, \ldots$

Algorithm 3 KSA:

> **for** $i := 0$ **to** $n-1$ **do**
> $S[i] := i$
> **end for**
>
> $j := 0$
> **for** $i := 0$ **to** $n-1$ **do**
> $j := (j + S[i] + K[i \bmod L]) \bmod n$
> $\mathbf{swap}(S[i], S[j])$
> **end for**
> $i, j := 0$

Algorithm 4 PRGA:

> $i := 0$
> $j := 0$
> $r := 0$
>
> **while** GeneratingOutput **do**
> $i = (i+1) \bmod n$
> $i = (j + S[i]) \bmod n$
> $\mathbf{swap}(S[i], S[j])$
> $x_r = S[S[i] + S[j] \bmod n]$
> $r = r + 1$
> **end while**
> $i, j := 0$

Fig. 2.2 KSA and PRGA: main ingredients of the RC4 algorithm

that develops and promotes standards for the internet, including cryptographic protocols like TLS. The IETF played a significant role in deprecating insecure RC4 ensuring the adoption of more secure alternatives. ∎

2.4 Simple Generators: Historical Perspective

This section covers some simple pseudorandom number generators, such as linear congruential generators (LCGs) and their generalizations, which are primarily of historical interest. These generators are easy to implement and may still be useful for educational purposes or lightweight tasks, but they are typically 32-bit generators with limited periods, poor randomness quality, and vulnerability to predictability. While no longer suitable for most practical applications, they provide valuable insights into the principles and challenges of PRNG design, offering a foundation for understanding the advancements in modern generators.

2.4.1 Congruent Generators

The simplest example of a pseudorandom number generator is a sequence u_n given by x_n/M, where

$$x_{n+1} = (ax_n + c) \bmod M. \tag{2.4.1}$$

We need to choose

$$
\begin{array}{lll}
M, \text{ a modulus;} & 0 < M,\\
a, \text{ a multiplier;} & 0 \le a < M,\\
c, \text{ an increment;} & 0 \le c < M,\\
x_0, \text{ an initial value (seed);} & 0 \le x_0 < M.
\end{array}
$$

To obtain a sequence from the interval $[0, 1)$, we take $u_n = x_n/M$. A sequence (x_n) has a period not longer than M. This method of obtaining pseudorandom numbers is called a *linear congruence method*.

The following theorem (see Theorem A on page 17 in Knuth [107]) gives the conditions for a linear congruence to have a period of length M.

Theorem 2.4.1 *Let* $b = a - 1$. *For a sequence* (x_n) *given by* (2.4.1) *to have a period of length M it is necessary and sufficient that:*

- *c is coprime with M,*
- *b is a multiple of p for every prime number p dividing M,*
- *if 4 is a divisor of M then a $= 1$ mod 4.*

We now give examples of pseudorandom number generators obtained by the linear congruence method.

Example 2.4.2 One may take $M = 2^{31} - 1 = 2147483647, a = 7^5 = 16867, c = 0$. This choice was popular on 32-bit computers. Another example of parameters having good properties is: $M = 2147483563$ and $a = 40014$. For 36-bit computers choosing $M = 2^{35} - 31, a = 5^5$ turns out to be good. $\diamondsuit$

Example 2.4.3 (Resing) If $(a, c, M) = (1, 5, 13)$ and $X_0 = 1$, then the sequence (x_n) is as follows: 1,6,11,3,8,0,5,10,2,7,12,4,9,1,..., which has a period 13. If $(a, c, M) = (2, 5, 13)$ and $x_0 = 1$, then we will obtain a sequence $1, 7, 6, 4, 0, 5, 2, 9, 10, 12, 3, 11, 1, \ldots$ with period 12. If, however $x_0 = 8$, then all the sequence elements are equal to 8 (thus, the period is 1). $\diamondsuit$

If $c = 0$ in (2.4.1), then the generator is called a *linear multiplicative congruential generator*. In Lewis et al. [126] the following parameters were suggested: $M = 2^{31} - 1, a = 1680$.

Generalized method of linear congruence generates sequences using the recurrence:

$$
x_n = (a_1 x_{n-1} + a_2 x_{n-2} + \cdots + a_k x_{n-k} + c) \bmod M.
$$

If M is a prime number, an appropriate choice of constants a_j allows for achieving a period of M^k. In practice, $M = 2^{31} - 1$ is commonly chosen on 32-bit computers.

Terminology and Abbreviations Generators based on the linear congruence method are termed linear congruential generators, denoted by LCG (M, a, c). In the case of generators employing the generalized linear congruential method, they will be denoted by GLCG $(M, (a_1, \ldots, a_k), c)$.

Following L'Ecuyer [118], we now list generators that were popular in commercial packages, including those used in Java, Visual Basic, and Excel. These generators are simple and easy to implement, but they do not exhibit good statistical properties. They are primarily of historical value and may serve as useful examples for experiments.

Java

$$x_{i+1} = (25214903917 x_i + 11) \bmod 2^{48},$$

$$u_i = \left(2^{27} \lfloor x_{2i} / 2^{22} \rfloor + \lfloor x_{2i+1} / 2^{21} \rfloor\right) / 2^{53}.$$

There was a UNIX PRNG called rand48 with the same recurrence for x_i, setting $u_i = x_i / 2^{48}$.

Visual Basic

$$x_i = (1140671485 x_{i-1} + 12820163) \bmod 2^{24},$$

$$u_i = \frac{x_i}{2^{24}}.$$

Excel

$$u_i = (0.9821 u_{i-1} + 0.211327) \bmod 1.$$

There are also generators based on higher-order congruences, such as square or cubic forms. Examples of such generators can be found in Rukhin et al. [187].

Another class of congruential generators is the quadratic generators, which fulfill a recurrence of the form:

$$x_{i+1} = a x_i^2 + b x_i + c \bmod M,$$

where a, b, c are non-negative integers from $\{0, 1, \ldots, M - 1\}$.

Quadratic Congruential Generator I (QCG I) The sequence is defined by the recurrence relation:

$$x_{i+1} = x_i^2 \bmod M,$$

where M is a 512-bit prime number, specified in hexadecimal as: $M =$

987b6a6bf2c56a97291c445409920032499f9ee7ad128301b5d0254aa1a9633

fdbd378d40149f1e23a13849f3d45992f5c4c6b7104099bc301f6005f9d8115e1

Quadratic Congruential Generator II (QCG II) The sequence (x_k) is defined by
the recurrence:

$$x_{i+1} = 2x_i^2 + 3x_i + 1 \bmod 2^{512}.$$

Cubic Congruential Generator (CCG) The sequence (x_k) is defined by the
recurrence:

$$x_{i+1} = 2x_i^3 \bmod 2^{512}.$$

Fibonacci Generator The sequence (x_n) is defined by the recurrence:

$$x_n = x_{n-1} + x_{n-2} \bmod M \quad (n \geq 2).$$

Subtractive Lagged Fibonacci Generator A modification that improves the
"independence" property is the so-called lagged Fibonacci generator, given by the
recurrence:

$$x_n = x_{n-k} + x_{n-l} \bmod M \quad (n \geq \max(k, l)).$$

Knuth proposed this generator with $k = 100$, $l = 37$, and $M = 2^{30}$.

Many problems require PRNGs with very long periods. Such periods are
achieved, for example, by the so-called mixed generators.

L'Ecuyer Mixed Generator This generator combines two congruential generators
to produce sequences with very long periods. It is defined as follows:

1. $x_n = (A_1 x_{n-2} - A_2 x_{n-3}) \bmod M_1$,
2. $y_n = (B_1 y_{n-1} - B_2 y_{n-3}) \bmod M_2$,
3. $z_n = (x_n - y_n) \bmod M_1$,
4. If $z_n > 0$, then $u_n = z_n/(M_1 + 1)$; otherwise, $u_n = M_1/(M_1 + 1)$.

Parameters for this generator are:

$$M_1 = 4\,294\,967\,087, \quad M_2 = 4\,294\,944\,443, \quad A_1 = 1\,403\,580, \quad A_2 = 810\,728,$$

$$B_1 = 527\,612, \quad B_2 = 1\,370\,589.$$

Some of the initial values for x and y should be specified. L'Ecuyer designed this
algorithm [117] with parameters optimized using computer simulations to ensure
the generator's statistical quality.

2.4.2 Bit Generators

Sequences of $\{0, 1\}$-valued pseudorandom numbers are referred to as pseudorandom bits. Many pseudorandom number generators (PRNGs) produce sequences of pseudorandom bits, which can then be converted into other formats, such as numbers in the range $[0, 1)$, integers, or other data types. Conversely, a random sequence of bits can be derived from sequences of independent random variables distributed as $\mathcal{U}(\bar{M})$, provided $M = 2^k$ (see Exercise 2.T.3). Additionally, pseudorandom bits are foundational in cryptographic applications.

It is worth noting that almost all pseudorandom number generators can be adapted to produce sequences of pseudorandom bits. For example:

- Generators producing numbers in $[0, 1)$ can be converted to bit sequences by representing each number in binary form.
- Integer generators that output values in $\{0, 1, \ldots, M - 1\}$ can be converted into bit sequences if $M = 2^k$.
- Specialized generators, such as those designed for cryptography, inherently operate on bits and output bit-level sequences directly.

Register Operations Bit sequences $\{0, 1\}^n$ are closely tied to computer design, where they are stored in registers. Commonly, these registers are 32-bit or 64-bit. Before we present specific examples of bit generators, let us describe two basic register operations.

Consider two sequences of bits $\mathbf{p} = (p_1, \ldots, p_n)$ and $\mathbf{q} = (q_1, \ldots, q_n)$, where $p_i, q_i \in \{0, 1\}$ for $i = 1, \ldots, n$. The following operations are fundamental:

- **XOR (exclusive OR, $\oplus$).** The XOR operation returns 1 if and only if an odd number of inputs is 1. Its behavior for two bits is summarized in the following truth table:

$$
\begin{array}{cc|c}
p & q & p \oplus q \\
\hline
0 & 0 & 0 \\
0 & 1 & 1 \\
1 & 0 & 1 \\
1 & 1 & 0 \\
\end{array}
$$

For sequences of bits, XOR is applied elementwise:

$$\mathbf{p} \oplus \mathbf{q} = (p_1 \oplus q_1, \ldots, p_n \oplus q_n).$$

- **Bit shifts.** The left shift operation is denoted by $\ll$, and the right shift by $\gg$. The second argument specifies the shift magnitude. For instance:

$$x = y \ll 2$$

indicates that x is obtained by shifting y two bits to the left. In left shifts, the vacated positions are filled with zeros.

Below, we present specific examples of bit generators that make use of the register operations.

Exclusive OR Generator (XORG) The Exclusive OR Generator is one of the simplest bitwise generators, leveraging the XOR operation to produce sequences of pseudorandom bits. The generator follows the recurrence:

$$x_i = x_{i-1} \oplus x_{i-127}, \quad i \geq 128$$

initialized with a 127-bit seed:

$$x_1, \ldots, x_{127} =$$

0001011011011001000101111001001010011011011010001000000100101111

110101001000010101101100000000000100110000101110011111111100111.

Linear Feedback Shift Register (LFSR) LFSR is a widely used bit generator in hardware implementations, such as in embedded systems and digital circuits. The output bit is derived from a linear combination of bits in the register, determined by feedback taps:

$$x_i = (a_1 x_{i-1} \oplus a_2 x_{i-2} \oplus \ldots \oplus a_k x_{i-k}),$$

where $a_j \in \{0, 1\}$ are the feedback coefficients. The sequence eventually repeats, but by choosing appropriate feedback taps, the generator can achieve a maximal period of $2^k - 1$.

Blum-Blum-Shub (BBS) Generator Cryptographic PRNGs, such as the Blum-Blum-Shub (BBS) generator, are specifically designed to output secure pseudorandom bits. The BBS generator follows the recurrence:

$$x_{i+1} = x_i^2 \bmod M,$$

where $M = pq$ is the product of two large primes, and the output bit sequence is given by the least significant bit of each x_i. BBS is provably secure under standard cryptographic assumptions but is slower compared to non-cryptographic PRNGs.

Shift-XOR Generator (SXR) This generator combines bit shifts ($\ll, \gg$) and XOR operations ($\oplus$) to produce pseudorandom numbers efficiently. It is particularly suitable for generating 32-bit or 64-bit pseudorandom numbers. The generator operates as follows:

1. Start with a seed value $x_0 \in \{0, 1, \ldots, 2^n - 1\}$, where n is the bit length (e.g., $n = 32$ or $n = 64$).

2. Define the recurrence relation:

$$x_{i+1} = (x_i \oplus (x_i \ll a)) \oplus (x_i \gg b),$$

where a, b are shift parameters.
3. Output x_{i+1} as the next pseudorandom number.

A specific example of a 32-bit SXR generator is following. Use parameters $a = 13$ and $b = 17$. Starting with a seed $x_0 = 0xABCDE123$, the sequence evolves as:

$$x_1 = (x_0 \oplus (x_0 \ll 13)) \oplus (x_0 \gg 17),$$
$$x_2 = (x_1 \oplus (x_1 \ll 13)) \oplus (x_1 \gg 17),$$
$$\vdots$$

The resulting numbers are interpreted as pseudorandom integers. It has the following properties: (i) Simple implementation with efficient bitwise operations; (ii) A short period, depending on the chosen parameters a and b; (iii) Good randomness properties for simple use cases but may fail advanced statistical tests.

These examples demonstrate the versatility of pseudorandom bit generation. While some generators are specialized for bit-level output (e.g., LFSR and BBS), others can be adapted through conversion to produce bit sequences efficiently.

2.5   Testing Good Properties of PRNGs

The quality of pseudorandom number generators (PRNGs) directly impacts the reliability of simulations, estimations, and secure communications. To ensure PRNGs meet the necessary standards, rigorous statistical testing is essential for identifying biases, correlations, or other deficiencies that could compromise their performance.

Two types of tests assess the quality of a PRNG: theoretical and statistical. Theoretical tests examine the internal structure of the generator under consideration. Classic examples are *lattice tests* [141] and *spectral tests* [107] (Section 3.3.4). See also [116] for some standard tests from this family. We encourage the reader to perform the following experiment. Consider a sequence generated by a linear congruence

$$x_j = (137x_{j-1} + 187) \bmod 256$$

with period 256. Generate 256 pairs $(x_0, x_1), (x_1, x_2), \ldots, (x_{255}, x_0)$ and triples $(x_0, x_1, x_2), (x_1, x_2, x_3), \ldots, (x_{255}, x_0, x_1)$, then plot the points. What can you see in the plots?

The second class of tests are statistical tests that use probabilistic methods. The PRNG structure is not important here, but the sequence $\mathbf{u} = (u_1, u_2, \ldots, u_n)$ is considered a simple sample, and we test if it comes from a random sequence. Recall that by a simple sample, we mean a sequence of i.i.d. random variables.

The sequence of pseudorandom numbers obtained from a PRNG is supposed to "look" like a sequence of independent random variables with a uniform distribution. After generating a sample of several thousand, it is not easy to check what it looks like, and therefore, statistical methods must be used. Preparing tests to check the "randomness" of a sequence of numbers has been in focus since the beginning of the theory of generators. However, the approach has gained importance since its utility in cryptography was noticed. The ideal situation is when the encrypted string appears completely random, as deviations from randomness can increase the chances of breaking a cryptographic protocol or discovering a private key.

Due to the above-mentioned demand, batteries of tests have been prepared. The most famous are:

- **Diehard tests:** A test suite developed by George Marsaglia in 1995.
- **TestU01:** A suite developed by Pierre L'Ecuyer and Richard Simard in 2007.
- **NIST Test Suite:** Developed by the National Institute of Standards and Technology, primarily for cryptographic applications.

Among others, the following types of tests are commonly used:

- **Uniformity tests:** These tests check whether the generated numbers are evenly distributed between zero and one (e.g., Kolmogorov-Smirnov test and chi-square goodness-of-fit test). For pseudorandom bit generators, any tested subsequence should—on average—contain half zeros and half ones.
- **Balls and boxes scheme:** Includes the serial test, which is similar to uniformity tests but applied to consecutive m-tuples.
- **Spacing test:** Measures distances between numbers that do not belong to the interval $[a, b]$ (for some fixed $0 \leq a < b \leq 1$) and compares them to the theoretical geometric distribution.
- **Tests based on classic combinatorial tasks.**
- **Tests based on properties of random walks.**

Furthermore, we require the following properties:

- **Scalability:** A test applied to the entire sequence should produce the same results for any selected subsequence.
- **Consistency:** Generator quality should be tested for arbitrary seeds. This is particularly important in cryptographic applications. There are known generators (e.g., the Mersenne Twister) whose "good" properties depend on a randomly selected seed.

2.5.1 Preliminary Remarks

We will denote random variables with capital letters. In particular:

- U: A random variable uniformly distributed as $\mathcal{U}[0, 1)$.
- B: A random variable uniformly distributed as $\mathcal{U}[\{0, 1\}]$ (a 'random bit').
- Y: A random variable uniformly distributed as $\mathcal{U}(\bar{M})$, i.e., over $\{0, 1, \ldots, M-1\}$.

Realizations are written as small letters, e.g., u, b, y, respectively.

2.5.1.1 Converting Numbers

Different tests require different types of random variables: distributed on $[0, 1)$, distributed on $\bar{M}$, or a sequence of random bits. Similarly, PRNGs can return numbers from $[0, 1)$, $\{0, \ldots, M - 1\}$, or a sequence of bits. Therefore, it is essential to be able to convert numbers between these formats.

Assume we have a string of numbers $u_1, u_2, \ldots$ from the interval $[0, 1)$, then we may convert them to numbers in the set $\bar{M}$ by

$$y_i = \lfloor M u_i \rfloor. \tag{2.5.1}$$

Conversely, numbers $y_1, y_2, \ldots \in \bar{M}$ can be converted to numbers from $[0, 1)$ by

$$u_i = \frac{y_i}{M}.$$

On a computer, there are no true "continuous" numbers. When a PRNG returns a sequence $u_1, u_2, \ldots$ from $[0, 1)$, these numbers usually have a fixed *granularity*. This is often a binary expansion with a fixed length, which is why many suggested parameters used in tests are powers of 2. For example, consider the PRNG from Visual Basic (Sect. 2.4.1):

$$x_i = (1140671485 x_{i-1} + 12829163) \bmod 2^{24},$$
$$u_i = x_i / 2^{24}.$$

Each u_i can be expressed as $(0.b_1 b_2 \ldots b_{24})_2$, where there are 2^{24} possible values of u_i. Thus, the best option is to convert it to numbers from $\{0, \ldots, 2^{24} - 1\}$ as:

$$y_i = \lfloor 2^{24} u_i \rfloor.$$

Caution is needed when performing such conversions. For instance, consider a test that requires a sequence of bits as input. If the PRNG outputs $u_i \in [0, 1)$, a straightforward conversion might be:

$$
b_i = \begin{cases} 0 & \text{if } u_i < 0.5, \\ 1 & \text{if } u_i \geq 0.5. \end{cases}
$$

However, in this approach, if $u_i = (0, b_1 b_2 \ldots b_{24})_2$, we only take the most significant bit b_1 and discard all others. This method could be exploited to "cheat" such a test by making only b_1 random and setting e.g., $b_2 = \cdots = b_{24} = 0$.

A more robust method is to first convert $u_i \in [0, 1)$ into $y_i \in \{0, \ldots, M - 1\}$, and then use the binary expansion of y_i to generate the input bits. Note that this approach requires M to be a power of 2.

In subsequent sections, we will work with numbers from $[0, 1)$, $\{0, \ldots, M - 1\}$, or sequences of bits. Always ensure proper conversion is applied when required.

2.5.1.2 Grouping Numbers into Multidimensional Vectors

To verify the hypothesis that the sequence (u_j) is a realization of i.i.d. uniform $\mathcal{U}[0, 1)$ random variables, we must examine both one-dimensional and multidimensional uniformity. For this purpose, we represent the sequence under study as a sequence of vectors. This representation will be used in the following sections.

Assume we have a sequence $u_1, \ldots, u_n$ of $n = mr$ numbers from $[0, 1)$. We group these numbers into r m-tuples as follows:

$$
\boldsymbol{u}_1 = (u_1, \ldots, u_m), \quad \boldsymbol{u}_2 = (u_{m+1}, \ldots, u_{2m}), \tag{2.5.2}
$$

$$
\ldots, \quad \boldsymbol{u}_r = (u_{(r-1)m+1}, \ldots, u_{rm}).
$$

In this way, we obtain r vectors $\boldsymbol{u}_i$, $i = 1, \ldots, r$, each belonging to the hypercube $[0, 1)^m$. Similarly, by first applying the conversion in (2.5.1), i.e.,

$$
y_i = \lfloor M u_i \rfloor,
$$

we can group these integers into r vectors from the discrete hypercube $\{0, 1, \ldots, M - 1\}^m$:

$$
\boldsymbol{y}_1 = (y_1, \ldots, y_m), \quad \boldsymbol{y}_2 = (y_{m+1}, \ldots, y_{2m}), \tag{2.5.3}
$$

$$
\ldots, \quad \boldsymbol{y}_r = (y_{(r-1)m+1}, \ldots, y_{rm}).
$$

For completeness, if we have a sequence of $n = mr$ bits $b_1, \ldots, b_n$, we can group them into r vectors from $\{0, 1\}^m$ in a similar manner:

$$
\boldsymbol{b}_1 = (b_1, \ldots, b_m), \quad \boldsymbol{b}_2 = (b_{m+1}, \ldots, b_{2m}), \tag{2.5.4}
$$

$$
\ldots, \quad \boldsymbol{b}_r = (b_{(r-1)m+1}, \ldots, b_{rm}).
$$

2.5.1.3 The Concept of p-Value

Randomness is tested using a statistic T, a function of a sequence generated by a PRNG. By *randomness*, we mean that the considered sequence is a realization of a uniformly distributed simple sample, which forms our null hypothesis $\mathcal{H}_0$. This hypothesis, referred to as the *randomness hypothesis*, applies regardless of whether the sequence consists of $U_1, U_2, \ldots$ (numbers from $[0, 1)$), $Y_1, Y_2, \ldots$ (numbers from $\bar{M}$), or $B_1, B_2, \ldots$ (bits).

Let G be the distribution function of $T = T(U_1, \ldots, U_n)$ under the null hypothesis $\mathcal{H}_0$. Suppose that an observed sequence (u_j) is tested to determine whether it is a realization of an i.i.d. sequence of uniformly distributed random variables (U_j). If $T(\mathrm{obs}) = T(u_1, \ldots, u_n)$ is the observed value of the test statistic, the p-value is computed as follows:

- For a one-sided right-tail test: $p = \mathbb{P}(T \geq T(\mathrm{obs})) = 1 - G(T(\mathrm{obs}))$.
- For a one-sided left-tail test: $p = \mathbb{P}(T \leq T(\mathrm{obs})) = G(T(\mathrm{obs}))$.
- For a two-sided test (symmetric distribution of T): $p = \mathbb{P}(|T| \geq |T(\mathrm{obs})|) = G(-T(\mathrm{obs})) + 1 - G(T(\mathrm{obs}))$.

For the more general case of a two-sided test, see Exercise 2.T.27. The p-value concept, with additional intuitions, is explored in Sect. 2.5.6.

Frequency (Monobit) Test Consider a sequence of bits $B_1, \ldots, B_n$. Transform the sequence into $\{-1, +1\}$ format using $X_i = 2B_i - 1$. Define the sum $S_n = X_1 + \cdots + X_n$. In this case, our null hypothesis $\mathcal{H}_0$ is that the (B_i) are i.i.d. random variables with distribution $\mathbb{P}(B_i = 0) = \mathbb{P}(B_i = 1) = 1/2$. Equivalently, the (X_i) are i.i.d. with distribution $\mathbb{P}(X_i = -1) = \mathbb{P}(X_i = 1) = 1/2$. From the central limit theorem (CLT), we know that $S_n/\sqrt{n}$ is approximately $\mathcal{N}(0, 1)$. Thus, we have a two-sided test with test statistic $T = |S_n/\sqrt{n}|$, which follows the so-called *half-normal distribution*. Let $Z \sim \mathcal{N}(0, 1)$. For observed bits $b_1, \ldots, b_n$, compute:

$$x_i = 2b_i - 1, \quad s_n = \sum_{i=1}^{n} x_i, \quad T(\mathrm{obs}) = \left| \frac{s_n}{\sqrt{n}} \right|.$$

Large values of $T(\mathrm{obs})$ are unlikely under $\mathcal{H}_0$. The p-value is calculated as:

$$p = \mathbb{P}\left(\left| \frac{S_n}{\sqrt{n}} \right| \geq T(\mathrm{obs}) \right) \overset{(*)}{\approx} \mathbb{P}(|Z| \geq T(\mathrm{obs}))$$

$$= 2\mathbb{P}(Z \geq T(\mathrm{obs})) = 2\left(1 - \Phi(T(\mathrm{obs}))\right),$$

where $\Phi(\cdot)$ is the standard normal cumulative distribution function. The p-value is illustrated in Fig. 2.3:

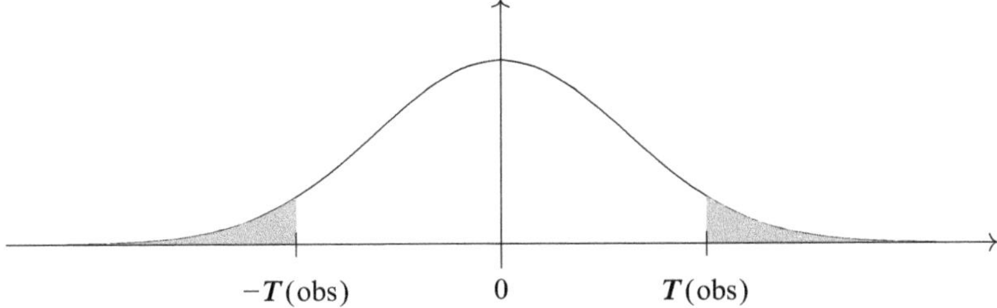

Fig. 2.3 Density of standard normal $Z \sim \mathcal{N}(0, 1)$ random variable. Shaded area is equal to p-value $\mathbb{P}(|Z| \geq T(\text{obs}))$

This test is part of a broader class of tests based on random walks. A detailed discussion, including p-value computation using the error function erfc, is provided in Sect. 2.5.4. As a rule of thumb, the approximation in $(*)$ is valid for $n \geq 30$. In practice, n is often significantly larger, minimizing approximation errors. However, for statistics with non-normal distributions, even n on the order of 1000 may not suffice, requiring corrections, as discussed in the context of the Kolmogorov-Smirnov goodness-of-fit test.

2.5.2   *Two Fundamental Goodness-of-Fit Tests: Kolmogorov-Smirnov and Chi-Square*

In this section, we describe two essential general-purpose tests for assessing the randomness of a sequence: the Kolmogorov-Smirnov (KS) test and the chi-square test.

To illustrate these tests, consider a small example. Figure 2.4 presents three sets of $n = 50$ numbers: set A (blue), set B (red), and set C (green). Additionally, we define a partition $\mathcal{P}$ of $[0, 1)$ as follows:

$$P_1 = (0, 0.15), \quad P_2 = [0.15, 0.35),$$
$$P_3 = [0.35, 0.6), \quad P_4 = [0.6, 0.8), \quad P_5 = [0.8, 1). \tag{2.5.5}$$

Notably, the sequence C originates from a quasirandom number generator (see Sect. 1.3 for details).

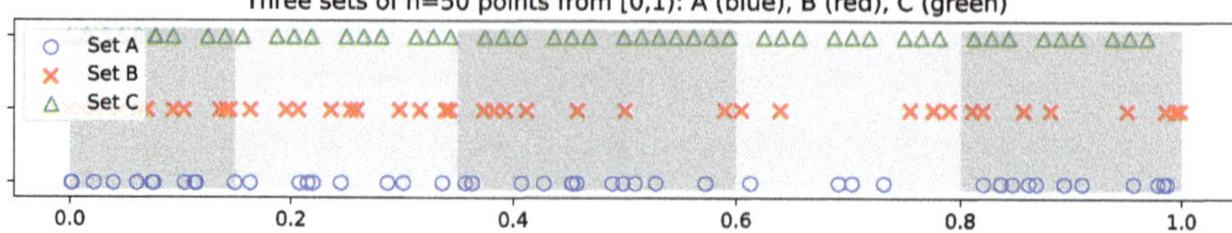

Fig. 2.4 Three sets of $n = 50$ points from $[0, 1)$: set A (blue), set B (red), and set C (green). The partition $\mathcal{P}$ defined in (2.5.5) is represented by gray-shaded regions

In the following subsections, we introduce the chi-square and Kolmogorov-Smirnov tests. Using the example above, we aim to address the question: *Are the numbers in set A (or B or C) "random"?*

This example is described in `ch2_prng_sets_A_B_C.py`. It generates the sets A, B and C, performs the Kolmogorov-Smirnov and the chi-square tests, and also applies second-level testing (which will be described later).

2.5.2.1 Kolmogorov-Smirnov Goodness-of-Fit Test (KS Test)

Let $U_1, \ldots, U_n$ be random variables. We now test the hypothesis that the marginal distribution of U_i is $\mathcal{U}[0, 1)$. For the sequence under consideration, we compute the corresponding empirical distribution function:

$$\hat{F}_n(t) = \frac{1}{n} \sum_{i=1}^{n} \mathbb{1}(U_i \leq t).$$

Under the uniform distribution hypothesis $\mathcal{U}[0, 1)$, the cumulative distribution function is $F(t) = t$ for $t \in [0, 1)$. The test statistic (often called the KS statistic) is:

$$D_n = \sup_{0 \leq t \leq 1} |\hat{F}_n(t) - t| .$$

According to the Glivenko-Cantelli theorem, if U_i are uniformly distributed, then with probability one, the sequence D_n converges to zero. Additionally, the normalized variables $\sqrt{n} D_n$ converge in distribution to the λ-Kolmogorov distribution, whose distribution function is:

$$K(t) = \lim_{n \to \infty} \mathbb{P}(\sqrt{n} D_n \leq t) = \sum_{j=-\infty}^{\infty} (-1)^j \exp(-2j^2 t^2)$$

$$= 1 - 2J(\exp(-2t^2)), \qquad t > 0, \tag{2.5.6}$$

where:

$$J(t) = \sum_{j=1}^{\infty} (-1)^{j-1} \exp(-2j^2 t^2) \,.$$

Tables for the λ-Kolmogorov distribution are widely available in programming libraries, such as `Python`'s `scipy.stats.kstest`.[1] For example, the critical values for $\alpha = 0.1, 0.05, 0.01$ are:

$$\lambda_{0.1} = 1.224, \quad \lambda_{0.05} = 1.358, \quad \lambda_{0.01} = 1.628,$$

where:

$$1 - K(\lambda_\alpha) = \alpha \,.$$

Note that K is the limiting distribution, and we use it to approximate $\sqrt{n}D_n$. For small n, the approximation may not be accurate; for example, with $n = 1000$, the error can reach 0.9%. Using the correction:

$$t + \frac{1}{6\sqrt{n}} + \frac{t-1}{4n} \tag{2.5.7}$$

in place of t improves accuracy for $n = 1000$ (100, 10) to 0.003% (0.027%, 0.27%) (see [209]).

To compute D_n, let $U_{(1)}, \ldots, U_{(n)}$ be the order statistics of the sample $U_1, \ldots, U_n$ (i.e., the sorted sequence). Define:

$$D_n^+ = \max_{1 \le i \le n} \left(\frac{i}{n} - U_{(i)} \right), \qquad D_n^- = \max_{1 \le i \le n} \left(U_{(i)} - \frac{i-1}{n} \right).$$

Then:

$$D_n = \max(D_n^+, D_n^-).$$

For a given sequence $u_1, \ldots, u_n$, we compute the statistic $D_n(\text{obs})$ and then calculate the p-value using the formula:

$$p = 1 - K\left(\sqrt{n}D_n(\text{obs}) + \frac{1}{6\sqrt{n}} + \frac{\sqrt{n}D_n(\text{obs}) - 1}{4n} \right), \tag{2.5.8}$$

where the correction (2.5.7) is included to improve the accuracy.

---

[1] https://docs.scipy.org/doc/scipy/reference/generated/scipy.stats.kstest.html.

Example 2.5.1 (Kolmogorov-Smirnov Test for Sets A, B, and C from Fig. 2.4)
The empirical distribution functions for points from sets A, B, and C are:

$$\hat{F}_n^A(t) = \frac{1}{n} \sum_{i=1}^n \mathbb{1}(u_i^A \leq t), \quad \hat{F}_n^B(t) = \frac{1}{n} \sum_{i=1}^n \mathbb{1}(u_i^B \leq t), \quad \hat{F}_n^C(t) = \frac{1}{n} \sum_{i=1}^n \mathbb{1}(u_i^C \leq t).$$

These are shown in Fig. 2.5.

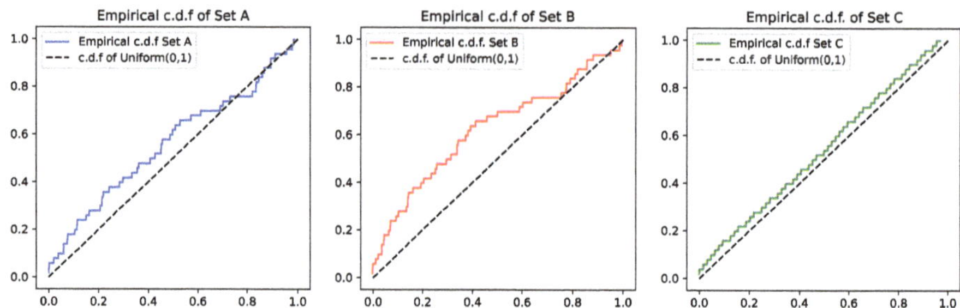

Fig. 2.5 Empirical distribution functions $\hat{F}_n^A(t)$ (blue, left), $\hat{F}_n^B(t)$ (red, middle), and $\hat{F}_n^C(t)$ (green, right) for sets A, B, and C. Each plot includes the c.d.f. $F(t) = t$ for $\mathcal{U}[0, 1)$ (black)

As shown in Fig. 2.5, all empirical distribution functions lie mostly above $F(t)$, the cumulative distribution function of the uniform distribution $\mathcal{U}[0, 1)$. By visual inspection, we observe that $\hat{F}_n^B(t)$ appears to deviate further from $F(t)$ compared to $\hat{F}_n^A(t)$, while $\hat{F}_n^C(t)$ is very close to $F(t)$. These observations are corroborated by the computed KS statistics:

$$D_n^A(\text{obs}) = \sup_{0 \leq t \leq 1} \left| \hat{F}_n^A(t) - F(t) \right| = 0.1192,$$

$$D_n^B(\text{obs}) = \sup_{0 \leq t \leq 1} \left| \hat{F}_n^B(t) - F(t) \right| = 0.2266,$$

$$D_n^C(\text{obs}) = \sup_{0 \leq t \leq 1} \left| \hat{F}_n^C(t) - F(t) \right| = 0.0462.$$

Next, we compute the corresponding p-values using the formula given in (2.5.8):

$$p^A = 1 - K\left(\sqrt{n} D_n^A(\text{obs}) + \frac{1}{6\sqrt{n}} + \frac{\sqrt{n} D_n^A(\text{obs}) - 1}{4n} \right) = 44.17\%,$$

$$p^B = 1 - K\left(\sqrt{n} D_n^B(\text{obs}) + \frac{1}{6\sqrt{n}} + \frac{\sqrt{n} D_n^B(\text{obs}) - 1}{4n} \right) = 0.99\%,$$

$$p^C = 1 - K\left(\sqrt{n} D_n^C(\text{obs}) + \frac{1}{6\sqrt{n}} + \frac{\sqrt{n} D_n^C(\text{obs}) - 1}{4n} \right) = 99.97\%.$$

Based on these results, the KS test leads us to the following conclusions:

- For set A, we fail to reject the hypothesis that the points were sampled from $\mathcal{U}[0, 1)$. The computed p-value of 44.17% indicates that observing a KS statistic of at least $D_n^A(\mathrm{obs})$ is fairly likely under the uniform distribution.
- For set B, we reject the randomness hypothesis. A p-value of 0.99% implies that it is highly unlikely for points sampled from $\mathcal{U}[0, 1)$ to produce a KS statistic as large as $D_n^B(\mathrm{obs})$.
- For set C, the hypothesis cannot be rejected. The extraordinarily high p-value of 99.97% suggests that the observed statistic is exceptionally close to what would be expected under $\mathcal{U}[0, 1)$.

It is noteworthy that points from set C are not pseudorandom but quasirandom numbers. The suspiciously large p-value highlights that while the KS test does not reject the randomness hypothesis, further investigation is warranted to understand the properties of the underlying sequence. $\Diamond$

2.5.2.2 Chi-Square Goodness-of-Fit Test

We begin by recalling the basic ideas of the chi-square Pearson test for the multinomial distribution $\mathrm{M}(r, p_1, \ldots, p_k)$. There are k mutually exclusive outcomes, say $1, \ldots, k$, with probabilities $p_1, \ldots, p_k$ respectively, and r independent trials. Let O_i be the number of outcomes of type i in the sequence of r trials, and let $E_i = \mathbb{E}O_i = rp_i$. The chi-square statistic is defined as:

$$X_k^2 = \sum_{i=1}^{k} \frac{(O_i - rp_i)^2}{rp_i}.$$

Using the properties $\sum_i O_i = r$ and $\sum_i p_i = 1$, we can rewrite this as:

$$X_k^2 = \sum_{i=1}^{k} \frac{O_i^2}{rp_i} - r.$$

As r approaches infinity, the statistic $X_k^2 \xrightarrow{\mathcal{D}} \chi_{k-1}^2$, where χ_{k-1}^2 denotes the chi-square distribution with $k-1$ degrees of freedom. Note that this is an approximation of the X_k^2 statistic.

For a given dataset, the observed occurrences O_i for each outcome type $i = 1, \ldots, k$ are counted, and the corresponding χ^2 statistic and p-value are calculated.

It is important to note that, unlike the Kolmogorov-Smirnov test, the chi-square goodness-of-fit test is not a specific test but rather a general approach. The choice of k, r, and their interpretation depends on the specific application (see Sect. 2.5.3).

Rule of Thumb Each outcome type should have at least five expected occurrences, on average. To meet this requirement, the number of trials r should satisfy:

$$r \geq \left\lceil \frac{5}{\min\limits_{1 \leq i \leq k} p_i} \right\rceil. \tag{2.5.9}$$

When $k \gg r$, alternative tests based on arrangements of balls and boxes may be more appropriate (see Sect. 2.5.3).

Example 2.5.2 (Chi-Square Test for Points in Sets A, B, and C from Fig. 2.4) Consider the partition given in (2.5.5). Here, the outcome types correspond to the intervals P_i, $i = 1, \ldots, 5$, where $k = 5$, and $p_i = |P_i|$. Define:

$$O_i^A = \#\{u_j^A : u_j^A \in P_i\}, \quad O_i^B = \#\{u_j^B : u_j^B \in P_i\}, \quad O_i^C = \#\{u_j^C : u_j^C \in P_i\}.$$

The least likely interval is P_1, where $p_1 = |P_1| = 0.15$. According to the rule of thumb (2.5.9), we require:

$$\left\lceil \frac{5}{0.15} \right\rceil = 34,$$

which is satisfied ($r = 50$). The chi-square statistics for sets A, B, and C are calculated as:

$$X_A^2(\text{obs}) = \sum_{i=1}^{k} \frac{(O_i^A - rp_i)^2}{rp_i}, \quad X_B^2(\text{obs}) = \sum_{i=1}^{k} \frac{(O_i^B - rp_i)^2}{rp_i},$$

$$X_C^2(\text{obs}) = \sum_{i=1}^{k} \frac{(O_i^C - rp_i)^2}{rp_i}.$$

The results are summarized in Table 2.2.

Table 2.2 Chi-square test for points from sets A, B, and C from Fig. 2.4 using the partition $\mathcal{P}$ given in (2.5.5)

P_i	np_i	Set A		Set B		Set C	
		O_i^A	$\frac{(O_i^A - rp_i)^2}{rp_i}$	O_i^B	$\frac{(O_i^B - rp_i)^2}{rp_i}$	O_i^C	$\frac{(O_i^C - rp_i)^2}{rp_i}$
$[0, 0.15)$	7.5	12	2.7	17	12.03	8	0.06
$[0.15, 0.35)$	10	9	0.1	11	0.1	10	0.00
$[0.35, 0.6)$	12.5	12	0.02	7	2.42	13	0.05
$[0.6, 0.8)$	10	4	3.6	6	1.6	9	0.07
$[0.8, 1]$	10	13	0.9	9	0.1	9	0.07
Chi-square			7.32		16.25		0.24

All numerical values are rounded to two decimal places

Assuming the numbers are uniformly distributed on $[0, 1)$, the statistics X_A^2, X_B^2, and X_C^2 should follow (approximately) a $\chi^2(4)$ distribution. The corresponding p-values are:

$$p^A = 1 - F_{\chi_4^2}(X_A^2(\mathrm{obs})) = 11.99\%,$$

$$p^B = 1 - F_{\chi_4^2}(X_B^2(\mathrm{obs})) = 0.27\%,$$

$$p^C = 1 - F_{\chi_4^2}(X_C^2(\mathrm{obs})) = 99.35\%.$$

Thus, the test does not reject the hypothesis that points from set A are uniformly distributed, with a p-value of 11.99%. Similarly, the hypothesis for set C cannot be rejected ($p = 99.35\%$). However, the hypothesis for set B is rejected, as the p-value is only 0.27%, indicating strong evidence against uniformity at typical significance levels (5% or 1%). $\Diamond$

Note that the distribution of χ_k^2 is $\mathrm{Gamma}(k/2, 1/2)$. For $k \geq 50$ types of observations, it is recommended to use the central limit approximation to the χ_k^2 distribution. Recall that

$$\chi_k^2 \overset{\mathcal{D}}{=} Z_1^2 + \cdots + Z_k^2.$$

Since $\mathbb{E}\chi_k^2 = k$ and $\mathbb{V}\mathrm{ar}\chi_k^2 = 2k$, from the central limit theorem we have

$$\frac{\chi_k^2 - k}{\sqrt{2k}} \overset{\mathcal{D}}{\to} \mathcal{N}(0, 1).$$

Remark 2.5.3 (On Examples 2.5.1 and 2.5.2) Concerning sets A, B, and C from Fig. 2.4, the points in set A were sampled randomly from $[0, 1)$ using the PCG64 generator built into NumPy in Python. Neither set B nor C is random. Set B is the following transformation of points from A:

$$u_i^B = \frac{e^{u_i^A - 1} - m}{M - m}, \quad \text{where } m = \min_i e^{u_i^A - 1} - \varepsilon, \ M = \max_i e^{u_i^A - 1} + \varepsilon,$$

with $\varepsilon = 10^{-7}$.

Set C is a sequence of quasirandom numbers (see Sect. 1.3). Both the chi-square and KS tests detected the "nonrandomness" of set B, whereas they failed to detect the "nonrandomness" of set C. Moreover, both tests concluded that it is very likely that set C is random (the p-values were $> 99\%$). Intuitively, the set of points in C is "too perfect," as can be seen in the plot of the empirical distribution function in Fig. 2.5 (right). This highlights the *importance* of applying **second-level testing**, described later in Sect. 2.5.6. We continue the example therein, where the "nonrandomness" of set C is easily detected.

The "nonrandomness" of set C is also identified by the *frequency of pairs test* (see Example 2.5.5). However, if we shuffle the points in C ("randomly"), the frequency of pairs test no longer detects it.   ∎

Spacing Test For a sequence $U_1, U_2, \ldots$, we consider the *spacings* between occurrences of $U_j \in (\alpha, \beta]$, where $0 \le \alpha < \beta \le 1$. Formally, we define:

$$\{i = 1, 2, \ldots : U_i \in (\alpha, \beta]\} = \{S_1, S_2, \ldots\},$$

where $S_1 < S_2 < \ldots$ The consecutive spacings are $C_1 = S_1, C_2 = S_2 - S_1, \ldots$

Under the null hypothesis $\mathcal{H}_0$, which assumes (U_i) is an i.i.d. sequence uniformly distributed $\mathcal{U}[0, 1]$, the sequence $C_1, C_2, \ldots$ is i.i.d. with probabilities:

$$\mathbb{P}(C_1 = l) = \mathbb{P}(U_1 \notin (\alpha, \beta], \ldots, U_{l-1} \notin (\alpha, \beta], U_l \in (\alpha, \beta]) = (1 - \delta)^{l-1}\delta^l,$$

for $l = 1, 2, \ldots$, where $\delta = \beta - \alpha$.

In practice, we deal with finite sequences $U_1, \ldots, U_n$. If no $U_j \in (\alpha, \beta]$ for $j = 1, \ldots, n$, we set $C_1 = n$ and $K = 1$. Otherwise:

$$\{i = 1, 2, \ldots, n : U_i \in (\alpha, \beta]\} = \{S_1, S_2, \ldots, S_K\},$$

where $S_1 < S_2 < \cdots < S_K$, and the spacings are:

$$C_1 = S_1, \quad C_2 = S_2 - S_1, \ldots, \quad C_K = S_K - S_{K-1}.$$

We propose the following test. For given numbers $u_1, \ldots, u_n$ and fixed α, β (hence δ), compute the spacings $C_1, \ldots, C_K$ (note K depends on the sequence). Define the boxes:

$$A_0 = \{0\}, \ A_1 = \{1\}, \ \ldots, \ A_{s-1} = \{s - 1\}, \ A_s = \{s, s + 1, \ldots\}.$$

Choose:

$$s \ge \max\left\{5, \left\lceil \frac{5(1 - \delta)}{\delta} \right\rceil\right\}.$$

For $\delta \le 1/2$, the Chebyshev inequality provides:

$$P(C \ge s) \le P\left(C \ge \frac{5(1 - \delta)}{\delta}\right) = P\left(C \ge \frac{(1 - \delta)}{\delta} + 4\frac{(1 - \delta)}{\delta}\right)$$

$$\le P\left(C \ge \frac{(1 - \delta)}{\delta} + \frac{1}{2} \cdot 4\frac{\sqrt{1 - \delta}}{\delta}\right) = P(C \ge \mathbb{E}C + 2\mathbb{V}\mathrm{ar}C) \le \frac{1}{4}.$$

Thus, at least $3/4$ of the probability mass lies in $A_0, \ldots, A_{s-1}$, with at most $1/4$ in A_s. For $\delta > 1/2$, set $s = 5$.

Under $\mathcal{H}_0$, we expect Kp_i spacings in each box A_i, $i = 0, \ldots, s$, where:

$$p_i = \delta(1 - \delta)^i, \ i = 0, \ldots, s - 1, \quad p_s = 1 - \sum_{i=0}^{s-1} p_i.$$

This can be tested using the chi-square goodness-of-fit test. Compute the observed counts:

$$O_i = \#\{j : C_j \in A_i\}, \quad i = 0, \ldots, s,$$

and the test statistic:

$$X^2(\text{obs}) = \sum_{i=0}^{s} \frac{(O_i - Kp_i)^2}{Kp_i}.$$

Under $\mathcal{H}_0$, this statistic approximately follows a χ^2 distribution with s degrees of freedom, allowing the computation of a p-value.

As the following example demonstrates, the choice of boxes in the chi-square goodness-of-fit test is a delicate matter.

Example 2.5.4 (Spacing Test for Old MATLAB Generators) In earlier versions of MATLAB, the default generator exhibited unusual behavior. Using the generator called state, one observes peculiar patterns in the spacings of values within interval of length $\delta = \beta - \alpha = 0.01$. This experiment was conducted with $R = 5 \cdot 10^7$ replications. The histogram for spacings of small values ($\alpha = 0$, $\beta = 0.01$) is shown in Fig. 2.6a, while the histogram for large values ($\alpha = 0.99$, $\beta = 1$) is presented in Fig. 2.6b.

The observed spacing lengths should approximate the function $\delta(1 - \delta)^k$ (for large R) in both cases. However, for $(\alpha, \beta] = (0, 0.01]$, a mysterious deviation is visible at value $k = 26$. Notably, no such deviation appears for $(\alpha, \beta] = (0.99, 1]$.

For comparison, the same experiment was performed using the twister generator, and no such deviation was observed; see Fig. 2.7.

In Table 2.3, we summarize the results of the chi-square test for this spacing experiment with boxes $A_0 = \{0\}$, $A_1 = \{1\}, \ldots, A_{103} = \{103\}$, $A_{104} = \{104, 105, \ldots\}$. After generating $R = 5 \cdot 10^7$ points, the spacings between numbers in $(\alpha, \beta]$ were recorded. The total number of spacings, K, is random and varied in the experiment between 499,539 and 500,522. The experiments used the generators rand('state',0) and rand('twister',0) for $(\alpha, \beta]$ set to $(0, 0.01]$ and $(0.99, 1]$, respectively.

The chi-square statistic $X^2(\text{obs})$ revealed a significant anomaly in the state generator for $(\alpha, \beta] = (0, 0.01]$, with a value of 758.9, as compared to much smaller values (e.g., 83.160 and 94.856) in cases without anomalies. Specifically, box A_{26} had 2237 observations, where $Kp_{26} = 3847.512$ was expected. This discrepancy

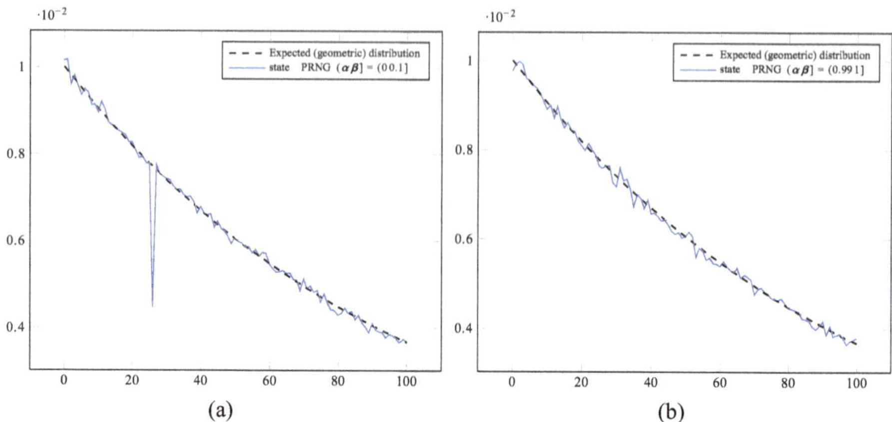

Fig. 2.6 MATLAB: $R = 5 \cdot 10^7$ numbers invoked by `rand('state',0)`; (**a**) Histogram of small values $(\alpha, \beta] = (0, 0.01]$; (**b**) histogram of large values $(\alpha, \beta] = (0.99, 1]$

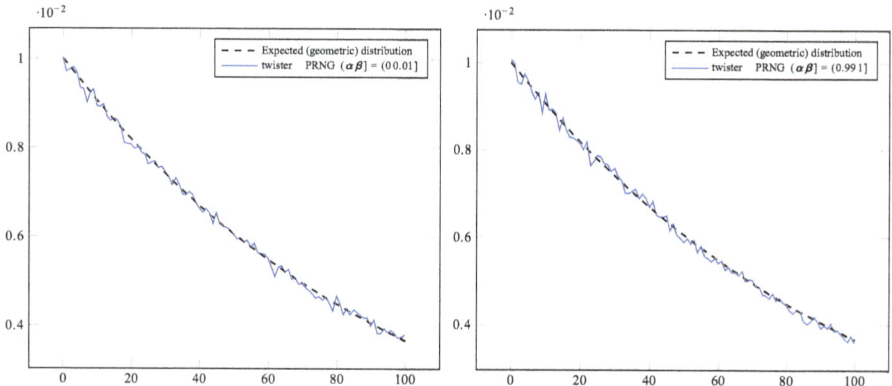

Fig. 2.7 MATLAB: $R = 5 \cdot 10^7$ numbers invoked by `rand('twister',0)`; (**a**) histogram of small values $(\alpha, \beta] = (0, 0.01]$; (**b**) histogram of large values $(\alpha, \beta] = (0.99, 1]$

resulted in an extremely large contribution to the chi-square statistic:

$$\frac{(O_{26} - Kp_{26})^2}{Kp_{26}} = 674.137,$$

leading to a minuscule p-value of $4.12 \cdot 10^{-100}$ (highlighted in bold in Table 2.3). A simpler grouping was also considered, with 27 larger boxes:

$$A'_0 = \{0, 1, 2, 3\}, \ A'_1 = \{4, 5, 6, 7\}, \ldots, \ A'_{25} = \{100, 101, 102, 103\},$$
$$A'_{26} = \{104, \ldots\}.$$

Table 2.3 Chi-square results for spacing test with boxes $A_i = \{i\}, i = 0, \ldots, 103, A_{104} = \{104, \ldots\}$

	rand('state',0)		rand('twister',0)	
K	499649	499558	500522	499539
$(\alpha, \beta]$	(0, 0.01]	(0.99, 1]	(0, 0.01]	(0.99, 1]
$\frac{(O_{26}-kp_{26})^2}{kp_{26}}$	674.137	0.012	0.554	1.554
X^2(obs)	758.90	83.160	94.856	97.540
p-value	**$4.12 \cdot 10^{-100}$**	0.934	0.728	0.659

Table 2.4 Chi-square results for spacing test with larger boxes

	rand('state',0)		rand('twister',0)	
$(\alpha, \beta]$	(0, 0.01]	(0.99, 1]	(0, 0.01]	(0.99, 1]
X^2(obs)	178.25	24.125	17.624	26.503
p-value	**$1.18 \cdot 10^{-24}$**	0.568	0.889	0.435

The results are shown in Table 2.4. For the `state` generator with $(\alpha, \beta] = (0, 0.01]$, the anomaly persisted, with a p-value of $1.18 \cdot 10^{-24}$ (highlighted in bold in Table 2.4).

The histograms for the larger bins (for both `state` and `twister` generators) in the case $(\alpha, \beta] = (0, 0.01]$ are presented in Fig. 2.8. The final box A'_{26} is not shown.

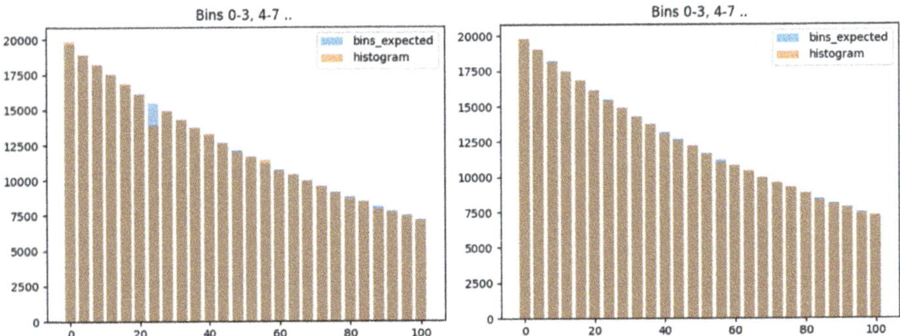

Fig. 2.8 Histograms for bins of size 4 for `state` and `twister` generators (for $(\alpha, \beta] = (0, 0.01]$) with expected observations per bin. The last bin A'_{26} is not shown

The PRNG 'state' is (was) a simple linear congruential generator (LCG). The commands `rand('state', 0)` and `rand('twister', 0)` were the primary methods for controlling the random number generator in MATLAB versions prior to R2008b. These commands are now considered *obsolete*, and their usage displays warnings in newer versions.   ◇

Serial Test (m, L) [Knuth [107], p. 60.] The goal of the serial test is to verify whether a sequence of numbers $u_1, \ldots, u_n$ from $[0, 1)$ is a realization of i.i.d. random variables with the uniform distribution $\mathcal{U}[0, 1)$.

Consider a hypercube in $[0, 1)^m$, and divide each of its m sides into $L \in \mathbb{N}$ equal segments, $[i/L, (i+1)/L)$ for $i = 0, \ldots, L-1$. This partitioning results in $k = L^m$ sub-hypercubes, referred to as boxes.

Assume $n = mr$. The sequence $u_1, \ldots, u_n$ is then grouped into r vectors from $[0, 1)^m$ as follows:

$$\boldsymbol{u}_1 = (u_1, \ldots, u_m), \quad \boldsymbol{u}_2 = (u_{m+1}, \ldots, u_{2m}), \tag{2.5.10}$$

$$\ldots, \quad \boldsymbol{u}_r = (u_{(r-1)m+1}, \ldots, u_{rm}).$$

Each vector $\boldsymbol{u}_i$ is mapped to one of the $k = L^m$ sub-hypercubes based on its components. This mapping identifies the box (or sub-hypercube) containing $\boldsymbol{u}_i$.

The serial test assesses whether the distribution of vectors $\boldsymbol{u}_1, \ldots, \boldsymbol{u}_r$ across the boxes is consistent with the expected uniform distribution under the null hypothesis of i.i.d. uniform random variables. A practical example with $m = 3$ and $L = 3$ is shown in Fig. 2.9.

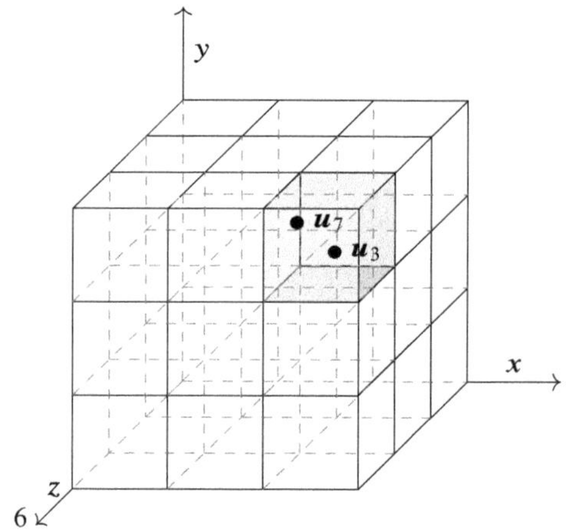

Fig. 2.9 Sample cube $[0, 1)^m$, $m = 3$, each side split into $L = 3$ segments obtaining $k = L^3 = 27$ smaller sub-hypercubes. Sample points $\boldsymbol{u}_3 = (0.8, 0.73, 0.7)$ and $\boldsymbol{u}_7 = (2.1, 2.6, 2.35)$ are presented together with the corresponding box $A_{2,2,2} = \left\{ (v_1, v_2, v_3) : v_i \in \left(\frac{2}{3}, 1 \right), i = 1, 2, 3 \right\}$, or correspondingly $\boldsymbol{y}_3 = \boldsymbol{y}_7 = (2, 2, 2)$ and $A'_{2,2,2} = (2, 2, 2)$ (see description in the text)

The sub-hypercubes are defined as:

$$A_{q_1,\ldots,q_m} = \left\{ (v_1, \ldots, v_m) : v_i \in \left[\frac{q_i}{L}, \frac{q_i + 1}{L} \right), i = 1, \ldots, m \right\}, \tag{2.5.11}$$

where $q_i \in \{0, \ldots, L - 1\}$.

For example, in Fig. 2.9, $A_{2,2,2} = \left\{ (v_1, v_2, v_3) : v_i \in \left(\frac{2}{3}, 1 \right), i = 1, 2, 3 \right\}$. It is sometimes convenient to enumerate the sub-hypercubes as $1, \ldots, k$, allowing us to say that the j-th point $\boldsymbol{u}_j$ is in the l-th sub-hypercube ($1 \leq j \leq r$, $1 \leq l \leq k$).

To facilitate computations, each u_i can first be converted into $y_i = \lfloor L u_i \rfloor$, following (2.5.3), yielding r vectors:

$$\boldsymbol{y}_1 = (y_1, \ldots, y_m), \quad \boldsymbol{y}_2 = (y_{m+1}, \ldots, y_{2m}),$$

$$\ldots, \quad \boldsymbol{y}_r = (y_{(r-1)m+1}, \ldots, y_{rm}) .$$

In this representation, sub-hypercubes become straightforward:

$$A'_{q_1,\ldots,q_m} = (q_1, \ldots, q_m),$$

as shown in Fig. 2.9.

The Testing Procedure The PRNG output sequence of length $n = mr$ is grouped into r vectors $\boldsymbol{u}_1, \ldots, \boldsymbol{u}_r$ from $[0, 1)^m$. For each sub-hypercube $A_{q_1,\ldots,q_m}$, count the vectors:

$$O_{q_1,\ldots,q_m} = \#\{i : \boldsymbol{u}_i \in A_{q_1,\ldots,q_m}\}, \quad (q_1, \ldots, q_m) \in \{0, \ldots, L-1\}^m.$$

With $k = L^m$, compute the chi-square statistic:

$$X^2(\text{obs}) = \sum_{(q_1,\ldots,q_m)\in\{0,\ldots,L-1\}^m} \frac{(O_{q_1,\ldots,q_m} - r/k)^2}{r/k}.$$

Alternatively, if sub-hypercubes are enumerated $1, \ldots, k$, this becomes:

$$X^2(\text{obs}) = \sum_{i=1}^{k} \frac{(O_i - r/k)^2}{r/k}.$$

Under the null hypothesis $\mathcal{H}_0$ (that the sequence is sampled from $\mathcal{U}[0, 1)$), for large r, X^2 is approximately chi-square distributed with $k - 1$ degrees of freedom.

Known Results (See Holst [83], L'Ecuyer et al. [122])

- If $r \to \infty$ and k is fixed, $X^2 \to \chi^2_{k-1}$.
- If $r, k \to \infty$ with $r/k \to \gamma$ $(0 < \gamma < \infty)$, then

$$T = \frac{X^2 - \mu_{r,k}}{\sigma_{r,k}}, \quad \mu_{r,k} = \mathbb{E}X^2, \quad \sigma^2_{r,k} = \mathbb{V}\text{ar}X^2,$$

converges to $\mathcal{N}(0, 1)$. It is recommended that $0 < \gamma < 5$. For $\gamma \geq 5$, the chi-square distribution better approximates X^2.

In Exercise 2.T.17, we recommend demonstrating that $\mathbb{E}X^2 = k - 1$; however, proving that $\mathbb{V}\text{ar}X^2 \sim 2(k - 1)$ is more challenging.

To decide whether to accept or reject $\mathcal{H}_0$, compute the p-value:

$$p\text{-value} = 1 - F_{\chi^2_{k-1}}(X^2(\text{obs})).$$

Second-level testing is recommended (see Sect. 2.5.6).

The reader is encouraged to adapt this procedure for sequences $y_1, y_2, \ldots$, where $y_i = \lfloor Lu_i \rfloor$.

Due to the rapid growth of $k = L^m$ for large L or m, the serial test is typically applied for $m = 1$ (*frequency test*) or $m = 2$ (*frequency of pairs test*). For larger m, other tests like the *collision test* or *birthday spacing test* are recommended.

The serial test is part of a broader family of tests based on the "balls and boxes" scheme (see Sect. 2.5.3).

Frequency Test (Serial Test with $m = 1$) This is a special case of the serial test with $m = 1$. In this test, we study whether the marginal distribution is the uniform distribution $\mathcal{U}[0, 1)$. Let us split the interval $[0, 1)$ into L intervals of equal length. We identify the resulting intervals $A_j = \{u : u \in [j/L, (j + 1)/L)\}$, $j = 0, \ldots, L - 1$, with *boxes* and the numbers $u_1, u_2, \ldots$ with *balls*. In our balls and boxes scheme, we thus have $m = 1$ and $k = L$ boxes, each having, under the randomness hypothesis $\mathcal{H}_0$, equal probability $\mathbb{P}(U_j \in A_j) = 1/k$. Using the chi-square test, we may test whether the number of points (balls) u_j within each box has a multinomial distribution $M(n, (1/k, \ldots, 1/k))$.

For convenience, instead of real numbers $u_1, \ldots, u_n \in [0, 1)$, we may consider a sequence $y_1, \ldots, y_n \in \{0, 1, \ldots, k - 1\}$ using the transformation $y_j = \lfloor ku_j \rfloor$. Then, simply, the boxes are $A'_j = j$, $j = 0, \ldots, k - 1$, and y_j denotes the number of the box with the j-th "ball." Depending on computer architecture, we can choose k to be some power of 2; for example, $L = 64 = 2^6$ (recall, $m = 1$, thus $k = L$), then y_j represents the 6 most significant bits in a binary representation of u_j. Thus, for each $i \in \{0, \ldots, k - 1\}$, we compute

$$O_i = \#\{j : y_j = i\}$$

and apply the chi-square test

$$X^2(\text{obs}) = \sum_{i=0}^{k-1} \frac{\left(O_i - \frac{n}{k}\right)^2}{\frac{n}{k}}$$

with $k - 1$ degrees of freedom.

Frequency of Pairs Test (Serial test with $m = 2$). [[107], p. 61.] This is a special case of the serial test with $m = 2$. Assume that we have $n = 2r$ numbers $u_1, \ldots, u_m$. This time, we will look at pairs (u_{2i-1}, u_{2i}), $i = 1, \ldots, r$. For an i.i.d. sequence $U_1, U_2, \ldots, U_n$ from the distribution $\mathcal{U}[0, 1)$, the sequence (U_{2j-1}, U_{2j}), $j = 1, 2, \ldots, r$, is a sequence of i.i.d. pairs, and (U_1, U_2) are independent uniformly

$\mathcal{U}([0, 1)^2)$ distributed random variables. In other words, we will test the distribution of r points in a unit square $[0, 1)^2$. We split each side of a square into L intervals of equal lengths and construct $k = L^2$ boxes:

$$A_{s,t} = \{(v, w) : v \in [s/L, (s + 1)/L), w \in [t/L, (t + 1)/L)\}.$$

Afterward, we count the number of pairs (u_{2i-1}, u_{2i}) falling into each box $A_{s,t}$. Similarly as in the frequency test, it is convenient to use the transformation

$$y_j = \lfloor Lu_j \rfloor,$$

then the boxes are $A'_{s,t} = \{(s, t)\}$. We compute the number of pairs equal to (s, t):

$$O_{s,t} = \#\{j : y_{2j-1} = s, y_{2j} = t\}$$

and apply the chi-square test

$$X^2(\text{obs}) = \sum_{(s,t)\in\{0,\dots,L-1\}^2} \frac{\left(O_{st} - \frac{r}{k}\right)^2}{\frac{r}{k}} = \sum_{(s,t)\in\{0,\dots,L-1\}^2} \frac{\left(O_{s,t} - \frac{r}{L^2}\right)^2}{\frac{r}{L^2}}$$

with $k - 1 = L^2 - 1$ degrees of freedom.

Example 2.5.5 We continue the example of three sets A, B, and C of $n = 50$ points given in Fig. 2.4. Let us take $L = 3$, so we group the sequence into $r = 25$ pairs $(u_1, u_2), \dots, (u_{49}, u_{50})$. This results in $k = L^2 = 9$ sub-boxes in the unit square $[0, 1)^2$, with each side divided into three intervals of equal length. Under the null hypothesis of uniformity, we expect, on average, $r/k = 25/9 \approx 2.77$ pairs to fall into each sub-box.

The expected number of observations, observed counts for each set, and the corresponding chi-square statistics are presented in Fig. 2.10. Each observed count is compared with the expected value of 2.77, and the deviations are used to compute the chi-square statistic:

$$X^2(\text{obs}) = \sum_{(s,t)\in\{0,\dots,L-1\}^2} \frac{\left(O_{s,t} - \frac{r}{k}\right)^2}{\frac{r}{k}}.$$

The chi-square statistics for each set are calculated, and their corresponding p-values are determined.

Recall that set C consists of quasi-random numbers. As evident in Fig. 2.10, half of the boxes are empty, suggesting the frequency of pairs test should reject the null hypothesis of uniformity for C. The corresponding p-values are:

$$p_A = 1 - F_{\chi_8^2}(X_A^2(\text{obs})) = 84.24\%,$$

$$p_B = 1 - F_{\chi_8^2}(X_B^2(\text{obs})) = 7.57\%,$$

$$p_C = 1 - F_{\chi_8^2}(X_C^2(\text{obs})) = 0.00\%.$$

2.77	2.77	2.77
2.77	2.77	2.77
2.77	2.77	2.77

Expected number of observations

4	2	2
4	1	4
4	2	2

6	0	3
5	3	1
5	0	2

0	10	8
0	0	7
0	0	0

4	3	2
2	2	2
3	6	1

$X_A^2(\text{obs}) = 4.16$ $X_B^2(\text{obs}) = 14.24$ $X_C^2(\text{obs}) = 51.68$ $X_{C'}^2(\text{obs}) = 6.32$

set A set B set C set C'

Fig. 2.10 Expected and observed values for the frequency of pairs test for sets A, B, C, and shuffled C'. The chi-square statistics and corresponding p-values are calculated for each set

Thus, we would reject the hypothesis that the points in C are realizations of uniformly distributed random variables. Interestingly, when the points in C are randomly shuffled (denoted as C'), the p-value changes significantly:

$$p_{C'} = 1 - F_{\chi_8^2}(X_{C'}^2(\text{obs})) = 61.14\%.$$

This highlights the importance of considering ordering in such tests. The shuffled points from C' no longer show evidence of non-randomness, aligning with the intuition that the quasi-random sequence in C is "too uniform." Further investigation through second-level testing (see Sect. 2.5.6) is recommended to confirm non-randomness. $\diamond$

2.5.3 Tests Based on "Balls and Boxes" Schemes

This is a group of tests based on the arrangement of r balls in k boxes. Balls and boxes are numbered from 1 to r and from 1 to k, respectively. Our null hypothesis

$\mathcal{H}_0$ is that the sequence of numbers to be tested (an output of a PRNG) is a realization of i.i.d. random numbers with the uniform distribution.

In this scheme, we assume that under $\mathcal{H}_0$, we know $p_i = 1/k$, the probability of placing each ball (independently of all the other balls) in box i, where $i = 1, \ldots, k$. The arrangement is visualized in Fig. 2.11.

We want to emphasize that the "balls and boxes" scheme is quite general. The choice of "balls" and "boxes" defines a specific test, and how to convert a sequence (u_j) into this scheme will be described in Sect. 2.5.3.1.

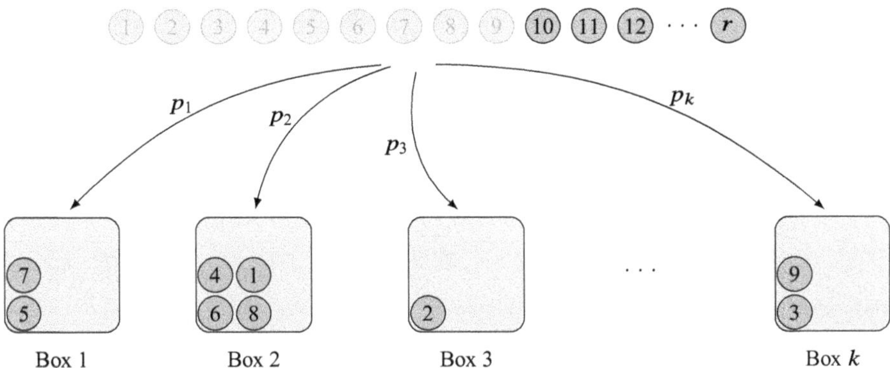

Fig. 2.11 A general procedure of placing r balls into k boxes. Each ball is placed (independently of others) in box i with probability p_i, $i = 1, \ldots, k$. Nine of the balls were already placed in boxes

We begin with a classic anecdote:

Birthday Problem In Section 2 of Feller [52], the birthday problem is presented. Consider a group of r individuals (e.g., students), and record their birthdays. Each birthday corresponds to a box ($k = 365$ total), and each individual represents a ball. Assuming all 365 birthdays are equally likely, we get an arrangement of r balls into k boxes. The classic question is: What is the probability of at least one **multiple-birth**, i.e., more than one ball in at least one box? The probability that all birthdays are unique is

$$\mathbb{P}(\text{all birthdays unique}) = \frac{(365)_r}{365^r},$$

where $(365)_r = 365 \cdot 364 \cdots (365 - r + 1)$. Thus, the probability of at least one multiple-birth is

$$p = 1 - \frac{(365)_r}{365^r} = 1 - \prod_{j=1}^{r-1}\left(1 - \frac{j}{365}\right).$$

It turns out that a group of at least 23 people makes this probability greater than 1/2.

This result can be generalized for arbitrary k and r. The probability of a multiple-birth is

$$p_{k,r} = 1 - \prod_{j=1}^{r-1}\left(1 - \frac{j}{k}\right).$$

An approximation for large k and r is given by:

$$p_{k,r} \approx 1 - \exp\left(-\frac{r^2}{2k}\right). \tag{2.5.12}$$

This approximation is derived as follows:

$$\log(1 - p_{k,r}) = \sum_{j=1}^{r-1} \log\left(1 - \frac{j}{k}\right),$$

and using $\log(1 - x) \approx -x$ for small x,

$$\log(1 - p_{k,r}) \approx -\sum_{j=1}^{r-1} \frac{j}{k} = -\frac{(r-1)r}{2k}.$$

Exponentiating both sides gives (2.5.12).

The concept of a multiple-birth translates to a **collision**, where a collision occurs if a ball lands in an already-occupied box. Denote by $C_{k,r}$ the number of collisions after r balls are placed into k boxes. Note that $p_{k,r} = \mathbb{P}(C_{k,r} > 0)$. This result can form the basis for a formal test, though extensions such as the *collision test* and *birthday spacing test* are more commonly used and are described later.

2.5.3.1 Converting a Sequence (u_j) into the Scheme of Balls and Boxes

Suppose we want to test whether a sequence (u_j), where $u_j \in [0, 1)$, satisfies randomness properties. While one-dimensional uniformity can be verified with tests such as the Kolmogorov-Smirnov or chi-square tests (see Sect. 2.5.2), independence properties are often checked using serial (m, L) tests. However, for large L, the number of boxes can become prohibitively large, and the "balls and boxes" scheme provides a practical alternative.

As described in Sect. 2.5.1.2, the sequence (u_j) is first converted into r vectors $u_1, \ldots, u_r$, which represent the balls, and the boxes are sub-hypercubes as defined in (2.5.11).

Collision Test   When the number of boxes is significantly larger than the number of balls ($k \gg r$), as in the birthday problem, the *collision test* is recommended. Assume we have Rn numbers from a PRNG, split into R groups of $n = mr$ numbers. These are then partitioned into r m-dimensional vectors, as in Sect. 2.5.1.2. The collision test is based on the statistic $C_{k,r}$, which counts the number of collisions after placing r balls in k boxes. A collision is recorded whenever a ball is placed into a box that is already occupied. The probability of exactly c collisions is given by:

$$\mathbb{P}(C_{k,r} = c) = \frac{k(k-1)\cdots(k-r+c+1)S_2(r, r-c)}{k^r}, \tag{2.5.13}$$

where $S_2(r, s)$ is the Stirling number of the second kind, representing the number of ways to partition a set of r labeled objects into s non-empty unlabeled subsets. Let us explain the reasoning behind formula (2.5.13):

- *No collisions ($c = 0$):* All r balls are placed in separate boxes, resulting in the pattern:

$$\{*\}, \{*\}, \ldots, \{*\} \quad (r \text{ subsets}).$$

- *One collision ($c = 1$):* One box contains two balls, and the remaining $r - 1$ balls occupy separate boxes:

$$\{*, *\}, \{*\}, \ldots, \{*\} \quad (r - 1 \text{ subsets}).$$

- *Two collisions ($c = 2$):* This can occur in two distinct patterns:

$$\{*, *, *\}, \{*\}, \ldots, \{*\} \quad \text{or} \quad \{*, *\}, \{*, *\}, \ldots, \{*\} \quad (r - 2 \text{ subsets}).$$

In general, there are c collisions if and only if the set of r balls is partitioned into $r - c$ subsets, which corresponds to $S_2(r, r - c)$ such partitions. These partitions are then distributed across k boxes. This can be done in $k(k-1)\cdots(k-(r-c)+1)$ ways. The total number of possible placements of r balls into k boxes is k^r. Combining these elements gives formula (2.5.13). We recommend the reader complete the above arguments in Exercise 2.T.24.

Poisson Approximation   Computing $\mathbb{P}(C_{k,r} = c)$ or $\mathbb{P}(C_{k,r} \le c)$ directly using (2.5.13) can be computationally challenging. Under the conditions $r, k \to \infty$ and $r^2/(2k) \to \lambda$, with $0 < \lambda < \infty$, it can be shown that:

$$\mathbb{P}(C_{k,r} = c) \to \frac{\lambda^c}{c!} e^{-\lambda}, \qquad c = 0, 1, \ldots \tag{2.5.14}$$

While $r^2/(2k)$ is commonly used as an approximation for λ, it may differ from the true value of $\lambda = \mathbb{E}C_{k,r}$, which can be computed as:

$$\mathbb{E}C_{k,r} = r - k\left(1 - \left(1 - \frac{1}{k}\right)^r\right). \qquad (2.5.15)$$

For example, with $k = 2^{20}$ and $r = 2^{14}$, we find $r^2/(2k) = 128$, while $\lambda = \mathbb{E}C_{k,r} = 127.3282$. The probabilities $\mathbb{P}(C_{k,r} \le c)$ and $\mathbb{P}(C_{k,r}^* \le c)$ (for a Poisson random variable $C_{k,r}^*$ with mean λ) can be compared for various c, as shown in Table 2.5.

For completeness, the values of $\mathbb{P}(\widetilde{C}_{k,r} \le c)$ are also included in the table, where $\widetilde{C}_{k,r}$ is a random variable with a Poisson distribution with mean $r^2/(2k) = 128$. It is important to note that $C_{k,r}^*$ provides a much better approximation of $C_{k,r}$ compared to $\widetilde{C}_{k,r}$.

Example 2.5.6 We proceed to use the collision test for a sequence of bits as follows. Note that we compute only one number of collisions from r balls and k boxes. To proceed, we need to repeat the procedure several times and verify whether the obtained collision counts are consistent with the actual distribution of $C_{k,r}$ (or its approximation $C_{k,r}^*$). This will be accomplished using the chi-square goodness-of-fit test. The recommendation is for $r = 2^{14}$, $k = 2^{20}$, and $L = 2$.

We could construct bins (and corresponding probabilities) using Table 2.5. However, we recommend following the suggestions provided in Koçak [108]. The bins (and their boundaries) and exact probabilities are presented in Table 2.6. For significantly larger values of r and k, we suggest using the approximation $C_{k,r}^*$.

The rule of thumb (2.5.9) for the chi-square goodness-of-fit test is to have (on average) at least 5 observations in the least probable box, which is 119–121 in this case. Thus, the number of experiments should be at least $\lceil 5/0.088373 \rceil = 56$. To perform the test, we need $56 \cdot 2^{14} \cdot 20 = 18\,350\,080 = 2.29376$ MB of bits from a PRNG. In other words, we assume that we have a sequence of bits $b_1, \ldots, b_{Rn}$, where $n = mr$, with $R = 56$, $m = 20$, and $r = 2^{14}$. First, split the bits into R subsequences, each of length mr. Denote the j-th sequence by $(b_1^{(j)}, \ldots, b_{20r}^{(j)})$. Further, split the sequence into r 20-tuples as in (2.5.4):

$$\boldsymbol{b}_1^{(j)} = (b_1^{(j)}, \ldots, b_{20}^{(j)}), \quad \boldsymbol{b}_2^{(j)} = (b_{21}^{(j)}, \ldots, b_{40}^{(j)}),$$

$$\ldots, \quad \boldsymbol{b}_r^{(j)} = (b_{20(r-1)+1}^{(j)}, \ldots, b_{20r}^{(j)}).$$

Table 2.5 Exact values of $\mathbb{P}(C_{k,r} \le c)$ and its Poisson approximations $\mathbb{P}(C_{k,r}^* \le c)$ (with mean 127.3282) and $\mathbb{P}(\widetilde{C}_{k,r} \le c)$ (with mean 128) for $k = 2^{20}$ and $r = 2^{14}$

c	101	108	119	126	134	145	153
$\mathbb{P}(C_{k,r} \le c)$	0.009	0.043	0.244	0.476	0.742	0.946	0.989
$\mathbb{P}(C_{k,r}^* \le c)$	0.0092	0.0448	0.2461	0.4766	0.7401	0.9439	0.9881
$\mathbb{P}(\widetilde{C}_{k,r} \le c)$	0.0078	0.0396	0.2281	0.4530	0.7205	0.9367	0.9860

Table 2.6 10 bins and their exact probabilities for the collision test for $r = 2^{14}$ and $k = 2^{20}$

Box nr s	Range of collisions	Probability p_s of box s
1	0–113	0.106253
2	114–118	0.109894
3	119–121	0.088373
4	122–124	0.100719
5	125–127	0.106608
6	128–130	0.104997
7	131–133	0.096632
8	134–137	0.106367
9	138–142	0.091574
10	≥ 143	0.088913

Each $\boldsymbol{b}_i^{(j)} \in \{0, 1\}^{20}$ represents a ball. Boxes are defined as $A_{q_1,\ldots,q_{20}} = (q_1, \ldots, q_{20})$, where $q_i \in \{0, 1\}$, $i = 1, \ldots, 20$, and there are $k = 2^{20}$ of them. For each $j = 1, \ldots, R = 56$, compute the number of collisions $C_{k,r}^{(j)}$ and then determine the counts of collisions in each box:

$$O_s = \#\{j : C_{k,r}^{(j)} \text{ is in the } s\text{-th box}\}.$$

Finally, compute the chi-square statistic:

$$X^2(obs) = \sum_{s=1}^{10} \frac{(O_s - Rp_s)^2}{Rp_s},$$

where the boxes and probabilities p_s are given in Table 2.6. The corresponding p-value can then be computed.

Alternatively, we can represent the boxes differently. Again, we are given a sequence of bits $b_1, \ldots, b_{Rn}$, where $n = mr$, $R = 56$, $m = 20$, and $r = 2^{14}$. First, split the bits into R subsequences, each of length mr. Denote the j-th sequence by $(b_1^{(j)}, \ldots, b_{mr}^{(j)})$. Then, group consecutive pairs of bits $(b_{2i-1}^{(j)}, b_{2i}^{(j)})$ as vectors from $\{0, 1, 2, 3\}$, and form sequences of length 10:

$$y_1^{(j)} = (b_1^{(j)}, \ldots, b_{10}^{(j)}), \quad y_2^{(j)} = (b_{11}^{(j)}, \ldots, b_{20}^{(j)}),$$

$$\ldots, \quad y_r^{(j)} = (b_{10(r-1)+1}^{(j)}, \ldots, b_{10r}^{(j)}).$$

In this case, $m = 10$, but still $k = 4^{10} = 2^{20}$, so the same procedure applies. The values in Table 2.6 remain valid, but the resulting p-value may differ. $\Diamond$

Noticing that in Example 2.5.6 we have a Poissonian approximation with a large mean, one might consider a CLT approximation of $C_{r,k}$. For the scheme of balls and boxes with the number of empty boxes $E_{k,r}$, there is a known version of the CLT; see, e.g., [210]. We will provide more details at the end of the following section.

Theoretical Results on the Collision Test Let us enumerate boxes with consecutive integers $1, \ldots, k$. Suppose X_i is the number of balls in the i-th box. We have $\mathbb{P}(X_i = 0) = (1 - \frac{1}{k})^r$, and

$$
\begin{aligned}
C_{k,r} &= \sum_{i=1}^{k} (X_i - 1)\mathbb{1}(X_i \geq 1) \\
&= \sum_{i=1}^{k} X_i \mathbb{1}(X_i \geq 1) - \sum_{i=1}^{k} \mathbb{1}(X_i \geq 1) \\
&= r - k + \mathcal{E}_{k,r},
\end{aligned}
\tag{2.5.16}
$$

where $\mathcal{E}_{k,r} =$ #empty boxes. Hence,

$$
\mathbb{E}C_{k,r} = r - \mathbb{E}\#\{\text{non-empty boxes}\} = r - k\left(1 - (1 - \tfrac{1}{k})^r\right).
\tag{2.5.17}
$$

Now expand

$$
\left(1 - \frac{1}{k}\right)^r = 1 - r\frac{1}{k} + \binom{r}{2}\frac{1}{k^2} + o\left(\frac{r^2}{k^2}\right),
$$

hence

$$
\mathbb{E}C_{k,r} = r - k\left(1 - \left(1 - \frac{r}{k} + \binom{r}{2}\frac{1}{k^2} + o\left(\frac{r^2}{k^2}\right)\right)\right) \sim \frac{r^2}{2k} + o\left(\frac{r^2}{k}\right).
$$

For $r = 2^{14}$ and $k = 2^{20}$, we obtain $\mathbb{E}C_{k,r} \approx \frac{r^2}{2k} = 128$, while the true value is $\lambda = 127.3282$.

We now sketch a proof of the Poissonian asymptotics (2.5.14). We will need the following lemma, which can be found in Durrett [43].

Lemma 2.5.7 *Suppose that for an array $(c_{r,j})$ we have $\max_{1 \leq j \leq r} |c_{r,j}| \to 0$, $\sum_{j=1}^{r} c_{r,j} \to \lambda$, and $\sup_r \sum_{j=1}^{r} |c_{r,j}| < \infty$. Then*

$$
\prod_{j=1}^{r} (1 + c_{r,j}) \to e^{\lambda}.
$$

Lemma 2.5.8 *Suppose that $\frac{r^2}{2k} \to \lambda$. Then*

$$
\frac{k(k-1)\cdots(k-r+c+1)}{k^{r-c}} \to e^{-\lambda}.
$$

Proof We have

$$\frac{k(k-1)\cdots(k-r+c+1)}{k^{r-c}} = \prod_{i=1}^{r-c-1}\left(1-\frac{i}{k}\right).$$

Using Lemma 2.5.7 with $c_{r-c-1,j} = -\frac{i}{k}$, we have

$$\sum_{i=1}^{r-c-1} c_{r-c-1,j} = -\frac{1}{k}\sum_{i=1}^{r-c-1} i = -\frac{1}{k}\frac{(r-c-1)(r-c)}{2} \sim -\frac{r^2}{2k}.$$

$\square$

To complete the proof of (2.5.14), we use the following asymptotics for $S_2(r, r-c)$:

$$S_2(r, r-c) \sim \frac{(r-c)^{2c}}{2^c c!}.$$

Hence, assuming $\frac{r^2}{2k} \to \lambda$,

$$\frac{S_2(r, r-c)}{k^c} \sim \left(\frac{(r-c)^2}{(2k)^c}\right)\frac{1}{c!} \to \frac{\lambda^c}{c!}.$$

$\square$

Recall that $\mathcal{E}_{k,r}$ denotes the number of empty boxes after a scheme of balls and boxes is applied. We compute the mean and variance of $\mathcal{E}_{k,r}$. Consider an i-th box and a single throw of a ball. The probability that it does not fall into the i-th box is $1 - 1/k$, and after the series of r balls, the probability that the i-th box is empty is $(1 - 1/k)^r$. Define

$$a_1 = (1 - 1/k)^r, \qquad a_2 = (1 - 2/k)^r.$$

We introduce

$$Y_i = \begin{cases} 1 & \text{the } i\text{-th box is empty,} \\ 0 & \text{otherwise.} \end{cases}$$

Then,

$$\mathbb{E}\mathcal{E}_{k,r} = \mathbb{E}\sum_{i=1}^{k} Y_i = k(1 - 1/k)^r = ka_1.$$

For the variance, we know $\mathbb{V}\mathrm{ar} Y_i = a_1 - a_1^2$ and

$$\mathrm{Cov}\,(Y_i, Y_j) = \mathbb{E}[Y_i Y_j] - (\mathbb{E} Y_i)^2 = (1 - 2/k)^r - (1 - 1/k)^{2r}.$$

Hence,

$$\mathbb{V}\mathrm{ar} \sum_{i=1}^{k} Y_i = \sum_{i=1}^{k} \mathbb{V}\mathrm{ar} Y_i + \sum_{i \neq j} \mathrm{Cov}\,(Y_i, Y_j)$$

$$= k(a_1 - a_1^2) + (k^2 - k)(a_2 - a_1^2).$$

We now consider a limiting approximation for the distribution of $\mathcal{E}_{k,r}$ in the following regime:

(A*) $r, k \to \infty$ and $k = \alpha r$ for some $\alpha > 0$.

Under condition (A*), we have

$$e_{k,r} = \mathbb{E}\mathcal{E}_{k,r} = k\left(1 - \frac{\alpha}{k}\right) \sim k e^{-\alpha}, \qquad (2.5.18)$$

$$\sigma_{k,r}^2 = \mathbb{V}\mathrm{ar}\mathcal{E}_{k,r} = k(a_1 - a_1^2) + (k^2 - k)(a_2 - a_1^2)$$

$$\sim k e^{-\alpha}(1 - (1 + \alpha)e^{-\alpha}). \qquad (2.5.19)$$

Proposition 2.5.9 (Weiss [210]) *Under condition (A*), we have*

$$\frac{\mathcal{E}_{k,r} - e_{k,r}}{\sigma_{k,r}} \xrightarrow{\mathcal{D}} \mathcal{N}(0, e^{-\alpha}(1 - (1 + \alpha)e^{-\alpha})).$$

Using the asymptotics (2.5.18) and (2.5.19) and the relation (2.5.16), noticing that variances of $C_{k,r}$ and $\mathcal{E}_{k,r}$ are the same, we can write:

$$\frac{\mathcal{E}_{k,r} - k e^{-\alpha}}{\sqrt{k}} \xrightarrow{\mathcal{D}} \mathcal{N}(0, e^{-\alpha}(1 - (1 + \alpha)e^{-\alpha}))$$

and

$$\frac{\mathcal{E}_{k,r} - r + k e^{-\alpha}}{\sqrt{k}} \xrightarrow{\mathcal{D}} \mathcal{N}(0, e^{-\alpha}(1 - (1 + \alpha)e^{-\alpha})).$$

The above normal approximation provides the possibility of testing generators in regimes other than $r^2/2k \to \lambda$; namely, $k/r \to \alpha$. Examples of the use of such a regime can be found in the work of Tsang et al. [205].

Birthday Spacing Test Another well-known method for testing the quality of PRNGs is the so-called *birthday spacing test*. As its author Marsaglia [145] writes, the test is difficult for many generators to pass. Assume we have Rn numbers from a PRNG. We split them into groups of n numbers. Further, having $n = mr$ numbers, we split them into r vectors (m-dimensional), as in Sect. 2.5.1.2. The test is based on the K statistic defined as follows (thus, later, we will have R values of the statistics, $K^{(1)}, \ldots, K^{(R)}$). Below, we describe how to compute one statistic K for given $n = mr$ numbers, from which we construct $Y_1, \ldots, Y_r$ vectors (m-dimensional), each $Y_i \in [k] = \{1, \ldots, k\}$. We arrange them in a non-decreasing order $Y_{(1)} \le \cdots \le Y_{(r)}$. Then, we define the so-called *spacings* by

$$S_1 = Y_{(2)} - Y_{(1)}, \ldots, S_{r-1} = Y_{(r)} - Y_{(r-1)}, \quad S_r = k - Y_{(r)} + Y_{(1)},$$

which we sort again in non-decreasing order $S_{(1)}, \ldots, S_{(r)}$. Finally, K is the number of equal spacings, i.e.,

$$K = \#\{j \in \{2, \ldots, r\} : S_{(j-1)} = S_{(j)}\}.$$

For illustration, in Table 2.7, we generated 14 birthdays in a one-hundred-day year.

In this example, we have two groups with equal spacings: $2, 2, 2$ and $4, 4$, thus $K = 2 + 1 = 3$. In Fig. 2.12, we present another example illustrating the spacings on a circle.

Table 2.7 Birthday spacings example: $r = 14, k = 100$

	Birthday	Sorted birthday	Spacings	Sorted spacings
i	Y_i	$Y_{(i)}$	S_i	$S_{(i)}$
1	92	4	14	0
2	80	18	22	1
3	96	40	26	2
4	66	66	0	2
5	4	66	2	2
6	85	68	3	3
7	94	71	4	4
8	68	75	1	4
9	76	76	4	5
10	75	80	5	7
11	40	85	7	8
12	66	92	2	14
13	18	94	2	22
14	71	96	8	26

Table 2.8 Exact values of $\mathbb{P}(K = s)$ and its Poisson approximations $\mathbb{P}(K^* = s)$ (with mean $\lambda = r^3/(4k) = 1$) for $k = 2^{25}$ and $r = 2^9$

$s =$	0	1	2	≥ 3
$\mathbb{P}(K = s)$	0.3688016	0.3690335	0.1834712	0.786920
$\mathbb{P}(K^* = s)$	0.3679	0.3679	0.1839	0.8003

For the birthday spacing test, the exact distribution is difficult to obtain. It turns out that for large r, k, if the value of $\lambda = r^3/(4k)$ is small, then under the randomness hypothesis $\mathcal{H}_0$, K approximately follows a Poisson distribution with parameter λ.[2] We denote this Poisson random variable by K^*.

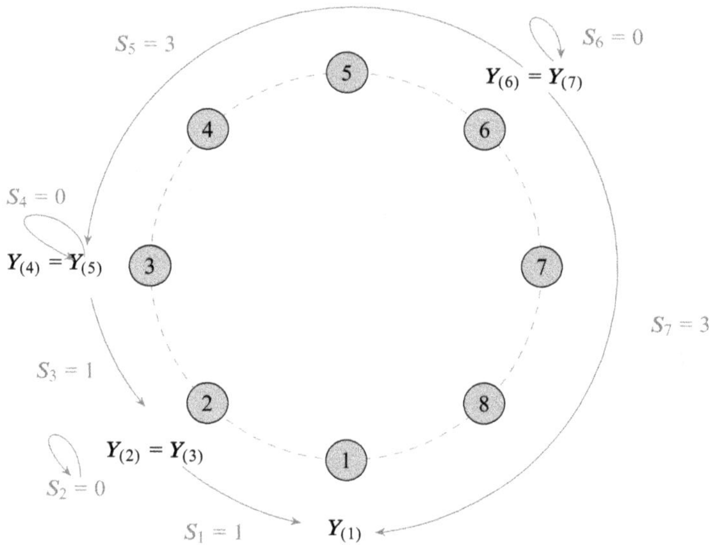

Fig. 2.12 Case: $k = 8, r = 7, Y_1 = 2, Y_2 = 3, Y_3 = 3, Y_4 = 6, Y_5 = 2, Y_6 = 1, Y_7 = 6$

Knuth [107] (p. 72) provided exact values of $\mathbb{P}(K = s)$ for $s \leq 2$ when $r = 2^9 = 512$ and $k = 2^{25}$, which gives $\lambda = 1$. The values (together with the Poisson approximation K^* of K) are provided in Table 2.8.

This Poisson approximation is discussed in Exercises 3.3.2–28–30 in Knuth [107].

Testing Procedure for $r = 2^9, k = 2^{25}$ Assume a sequence of bits $b_1, \ldots, b_{Rn}$ is given, with $n = mr$, where $m = 25$ and $r = 2^9$. Following the steps of the collision test:

---

[2] The error in this approximation is of order $O(\lambda^2/r)$. In the paper by L'Ecuyer and Simard [121], it is suggested that $r^3 \leq k^{5/4}$.

- *Sequence splitting:* Split the bits into R subsequences, each of length mr. Denote the j-th sequence by:

$$\left(b_1^{(j)}, b_2^{(j)}, \ldots, b_{mr}^{(j)} \right).$$

- *Grouping into tuples:* Further split each subsequence into r 25-tuples:

$$\mathbf{y}_1^{(j)} = \left(b_1^{(j)}, \ldots, b_{25}^{(j)} \right), \quad \mathbf{y}_2^{(j)} = \left(b_{26}^{(j)}, \ldots, b_{50}^{(j)} \right), \ldots,$$

$$\mathbf{y}_r^{(j)} = \left(b_{25(r-1)+1}^{(j)}, \ldots, b_{25r}^{(j)} \right).$$

- *Assigning birthdays and days:* Treat the $r = 2^9$ 25-tuples as birthdays (balls) and the $k = 2^{25}$ possibilities as days (boxes). For $\mathbf{y}_1^{(j)}, \ldots, \mathbf{y}_r^{(j)}$, compute the value of the statistic $K^{(j)}$.
- *Box assignments:* Perform the test over R subsequences and count the values of these statistics within the following boxes:

$$A_0 = \{0\}, \quad A_1 = \{1\}, \quad A_2 = \{2\}, \quad A_3 = \{3, 4, \ldots\}.$$

Compute the counts:

$$O_s = \#\{j : K^{(j)} \text{ is in box } A_s\}, \quad s = 0, 1, 2, 3.$$

- *Chi-square test:* Compute the chi-square statistic:

$$X^2(\text{obs}) = \sum_{s=0}^{3} \frac{(O_s - Rp_s)^2}{Rp_s},$$

where the probabilities $p_s = \mathbb{P}(K = s)$ are provided in Table 2.8.
Under $\mathcal{H}_0$, the above statistic follows a chi-square distribution with 3 degrees of freedom. The corresponding p-value can then be computed.

Marsaglia and Tsang [145] recommend other parameter values, such as: $r = 2^{12}, k = 2^{32}, \lambda = 4$, or $r = 2^{10}, k = 2^{24}, \lambda = 16$.

Now we proceed to tests based on the classical combinatorial problems of probability.

Poker Test Similarly to the frequency of pairs test, we consider a sequence of numbers $y_1, y_2, \ldots, y_n \in \bar{M}$, where $n = 5r$, which—assuming $\mathcal{H}_0$—is a realization of a sequence of i.i.d. random variables $Y_1, Y_2, \ldots$ with the uniform distribution $\mathcal{U}[\bar{M}]$.

We group the numbers into r 5-tuples (similarly as in (2.5.2)):

$$\boldsymbol{y}_1 = (y_1, \ldots, y_5), \quad \boldsymbol{y}_2 = (y_6, \ldots, y_{10}), \quad \ldots, \quad \boldsymbol{y}_r = (y_{(r-1)5+1}, \ldots, y_{5r}).$$

Following [107], each 5-tuple $\boldsymbol{y}_i$ can be classified into one of the following simplified poker ranks:

5 different: all different,
4 different: one pair,
3 different: two pairs or three of a kind,
2 different: full house or four of a kind,
1 different: five of a kind.

In our balls-and-boxes notation, the $\boldsymbol{y}_i$'s are balls, and we have $k = 5$ boxes (each box corresponds to "s different"). Note that each poker rank is possible. For actual poker ranks, we recommend Exercise 2.T.18, although using them for testing may be cumbersome.

We now compute the theoretical probability p_s that a 5-tuple $(Y_1, \ldots, Y_5)$ consists of s different numbers (as defined above) under $\mathcal{H}_0$, i.e., assuming that Y_i is chosen uniformly from $\bar{M}$. Let $S_2(5, s)$ denote the number of possible ways to partition a 5-element set into s non-empty subsets (these are the Stirling numbers of the second kind). We leave it to the reader to verify:

$$S_2(5, 1) = 1, \quad S_2(5, 2) = 15, \quad S_2(5, 3) = 25, \quad S_2(5, 4) = 10, \quad S_2(5, 5) = 1.$$

The number of different s-element sequences is $M(M - 1) \cdots (M - s + 1)$, so the desired probability is:

$$p_s = \frac{M(M - 1) \cdots (M - s + 1) S_2(5, s)}{M^5}.$$

Summarizing, we compute the number of 5-tuples within each box:

$$O_s = \#\{j : \boldsymbol{y}_j \in \{s \text{ different}\}\},$$

and compute the statistic:

$$X^2(\text{obs}) = \sum_{s=1}^{5} \frac{(O_s - rp_s)^2}{rp_s}.$$

Under $\mathcal{H}_0$, this statistic has a χ^2 distribution with 4 degrees of freedom.

2.5.4 Tests Based on Random Walks

Let $b_1, b_2, \ldots$ be a sequence of bits (i.e., $b_i \in \{0, 1\}$), which under $\mathcal{H}_0$ is a realization of a sequence of random bits $B_1, B_2, \ldots$. Define:

$$X_i = 2B_i - 1, \quad S_0 = 0, \quad S_k = \sum_{i=1}^{k} X_i, \quad k = 1, \ldots \qquad (2.5.20)$$

The sequence X_i takes values in $\{-1, +1\}$ ($\mathbb{E}X_i = 0$, $\mathrm{Var}X_i = 1$), and the process (S_k) is called a *symmetric random walk*. Some properties of (S_k) have already been discussed in Sect. 1.2.1.

Frequency (Monobits) Test (Using the erfc Function) From n bits $B_1, \ldots, B_n$, compute X_i and S_i, $i = 1, \ldots, n$, as in (2.5.20). Under the randomness hypothesis $\mathcal{H}_0$, the central limit theorem implies that $S_n/\sqrt{n}$ is approximately $\mathcal{N}(0, 1)$-distributed, and $T = |S_n/\sqrt{n}|$ follows the half-normal distribution G:

$$G(z) = \lim_{n \to \infty} \mathbb{P}\left(\left|\frac{S_n}{\sqrt{n}}\right| \le z\right) = \Phi(z) - \Phi(-z),$$

where

$$G(z) = \frac{1}{\sqrt{2\pi}} \int_{-z}^{z} e^{-\frac{x^2}{2}} \, dx = \frac{2}{\sqrt{\pi}} \int_{0}^{\frac{z}{\sqrt{2}}} e^{-y^2} \, dy = \mathrm{erfc}\left(\frac{z}{\sqrt{2}}\right).$$

For a given sequence $b_1, b_2, \ldots, b_n$, compute $s_n = \sum_{j=1}^{n} x_j$, where $x_j = 2b_j - 1$, and the test statistic

$$T(\text{obs}) = \left|\frac{s_n}{\sqrt{n}}\right|.$$

Then the p-value is given by:

$$p = \mathrm{erfc}\left(\frac{T(\text{obs})}{\sqrt{2}}\right).$$

Frequency (Monobit) Test in Blocks Consider a sequence of random bits $B_1, B_2, \ldots, B_n$ of length $n = mr$, split into r blocks, each of length m. This is an extension of the frequency (monobit) test, leveraging the CLT applied to a random walk within each block. Let $S_m^{(i)}$ denote the value of the random walk at step m within block i, i.e.,

$$S_m^{(i)} = \sum_{j=1}^{m} X_{(i-1)m+j} = \sum_{j=1}^{m} (2B_{(i-1)m+j} - 1), \quad i = 1, \ldots, r.$$

Clearly, $\mathbb{E}S_m^{(i)} = 0$ and $\mathbb{V}\mathrm{ar}\,S_m^{(i)} = m$. Define:

$$Z_i = \frac{S_m^{(i)}}{\sqrt{m}} = \frac{1}{\sqrt{m}} \sum_{j=1}^{m} (2B_{(i-1)m+j} - 1), \quad i = 1, \ldots, r.$$

For large m, the variables $Z_1, \ldots, Z_r$ are approximately i.i.d. $\mathcal{N}(0, 1)$-distributed. Hence,

$$\frac{1}{m} \sum_{i=1}^{r} \left(S_m^{(i)}\right)^2 \stackrel{\mathcal{D}}{=} \sum_{i=1}^{r} Z_i^2$$

under $\mathcal{H}_0$ follows a chi-square distribution with r degrees of freedom. To summarize, for a given sequence of bits $b_1, b_2, \ldots, b_n$ (with $n = mr$), compute:

$$T(\mathrm{obs}) = \frac{1}{m} \sum_{i=1}^{r} \left(S_m^{(i)}\right)^2 = \frac{1}{m} \sum_{i=1}^{r} \left(\sum_{j=1}^{m} (2b_{(i-1)m+j} - 1)\right)^2,$$

and the corresponding p-value.

Arcsine Law Based Test   For a random walk (S_k), define:

$$D_k = \mathbb{1}(S_k > 0 \vee S_{k-1} > 0), \quad k = 1, 2, \ldots,$$

where D_k equals 1 if the number of ones exceeds the number of zeros at step k or $k - 1$, and 0 otherwise. Consider a sequence $B_1, \ldots, B_n$ of even length n. The number of segments that lie above the x-axis is denoted by:

$$L_{2r}^{+} = \sum_{j=1}^{2r} D_j.$$

The arcsine law states that for any $w \in [0, 1)$:

$$\lim_{r \to \infty} \mathbb{P}\left(\frac{L_{2r}^{+}}{2r} \le w\right) = \frac{1}{\pi} \int_0^w \frac{dt}{\sqrt{t(1 - t)}} = \frac{2}{\pi} \arcsin \sqrt{w}.$$

The probability $\mathbb{P}(L_{2r}^{+} \le w \cdot 2r)$ represents the chance that the random walk was above the x-axis for at most a fraction w of the time. The limiting distribution (see Fig. 2.13 for this distribution and the corresponding c.d.f.) shows that the fractions of time spent above and below the x-axis are more likely to be unequal than close to each other.

This may seem counterintuitive since $\mathbb{E}S_k = 0$. However, this expectation is over all paths, and a single path will most likely spend most of the time either above or below the x-axis.

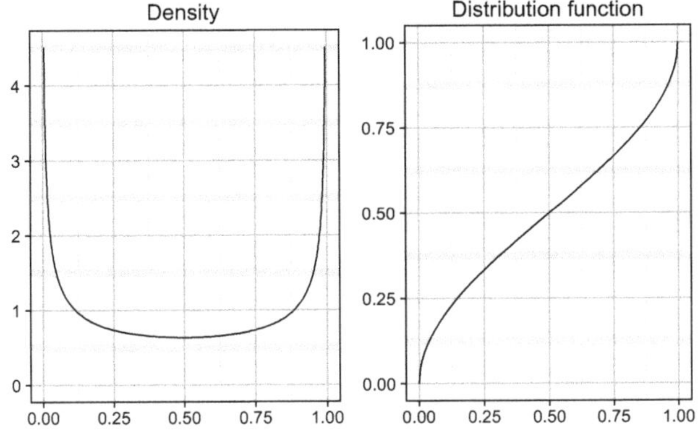

Fig. 2.13 Density function $\frac{2}{\pi}\arcsin\sqrt{w}$ and the corresponding c.d.f

Given a sequence of bits $b_1, b_2, \ldots, b_n$ (for even n) from a PRNG, compute:

$$s_k = \sum_{i=1}^{k} x_i, \quad x_i = 2b_i - 1, \quad d_k = \mathbb{1}(s_k > 0 \vee s_{k-1} > 0), \quad k = 1, \ldots, n,$$

and the fraction of time instants at which the number of ones exceeds the number of zeros:

$$T(\mathrm{obs}) = \frac{1}{n}\sum_{k=1}^{n} d_k.$$

Under the hypothesis $\mathcal{H}_0$ (that $b_1, b_2, \ldots, b_n$ are i.i.d. Bernoulli random variables), the limiting distribution of $T_n = \frac{1}{n}\sum_{k=1}^{n} d_k$ is provided by the arcsine law. Thus, compute the corresponding p-value as:

$$p = \mathbb{P}(T_n > T(\mathrm{obs})) \approx 1 - \frac{2}{\pi}\arcsin\sqrt{T(\mathrm{obs})}.$$

A detailed analysis, as well as the second-level test (see Sect. 2.5.6) based on the arcsine law, can be found in [135]. The analysis of errors in approximations is also provided therein.

Random Excursion Test For a sequence of $\{0, 1\}$-valued random variables $B_1, \ldots, B_n$, let S_k denote the corresponding random walk (2.5.20), and define J_n as the number of zeros among S_k, $k = 1, \ldots, n$, i.e., the number of recursions of the random walk to zero. Assuming the randomness hypothesis $\mathcal{H}_0$, which states that $B_1, \ldots, B_n$ are i.i.d. Bernoulli random variables, it is known that:

$$\lim_{n \to \infty} \mathbb{P}\left(\frac{J_n}{\sqrt{n}} \leq t\right) = \sqrt{\frac{2}{\pi}} \int_0^t e^{-x^2/2}\, dx. \tag{2.5.21}$$

According to the NIST report [187], the test rejects the randomness if J_n is too small. Let t_0 be such that:

$$0.01 \approx \sqrt{\frac{2}{\pi}} \int_0^{t_0} e^{-x^2/2}\, dx.$$

We have approximately $t_0 = 0.01253346951$, which defines a one-sided left-tail test. For a given sequence of bits $b_1, \ldots, b_n$, compute the corresponding $s_0 = 0$, $s_k = \sum_{i=1}^k (2b_i - 1)$, $k = 1, \ldots, n$, and determine the number of zeros in these sums, denoted by $J_n(\text{obs})$. If $J_n(\text{obs}) \leq 0.008\sqrt{n}$, the randomness hypothesis is rejected. If $J_n(\text{obs})$ is the observed number of recursions, the p-value is computed as:

$$p = \mathbb{P}(J_n \leq J_n(\text{obs})) = P\left(\frac{1}{2}, \frac{J_n^2(\text{obs})}{2n}\right),$$

where $P(a, x) = \frac{\gamma(a,x)}{\Gamma(a)}$ and $\gamma(a, x) = \int_0^x t^{a-1} e^{-t}\, dt$.

A piece of the random walk between a visit to zero and the next return to zero is called an excursion; that is, if $S_k = 0$ and $S_{k+1} \neq 0$, $S_{k+2} \neq 0, \ldots, S_{l-1} \neq 0$ and $S_l = 0$, then the excursion is $S_{k+1}, \ldots, S_{l-1}$. The 0-th excursion occurs when $k = 0$. Let $L(x)$ be the number of visits to x during the 0-th excursion. Then

$$\mathbb{P}(L(x) = k) = \begin{cases} 1 - \frac{1}{2|x|}, & \text{for } k = 0, \\ \frac{1}{4x^2}\left(1 - \frac{1}{2|x|}\right)^{k-1}, & \text{for } k \geq 1. \end{cases} \tag{2.5.22}$$

As an exercise, verify:

$$\mathbb{E}L(x) = 1, \quad \mathbb{V}\text{ar}L(x) = 4|x| - 2,$$

and

$$\mathbb{P}(L(x) \geq a + 1) = 2|x|\mathbb{P}(L(x) = a + 1) = \frac{1}{2|x|}\left(1 - \frac{1}{2|x|}\right)^a, \quad a = 0, 1, \ldots.$$

Define:

$$\pi_0(x) = \mathbb{P}(L(x) = 0) = 1 - \frac{1}{2|x|},$$

$$\pi_k(x) = \mathbb{P}(L(x) = k) = \frac{1}{4x^2}\left(1 - \frac{1}{2|x|}\right)^{k-1},$$

$$\pi_5(x) = \mathbb{P}(L(x) \geq 5) = \frac{1}{2|x|}\left(1 - \frac{1}{2|x|}\right)^4.$$

For a representative collection of x-values (e.g., NIST recommends $1 \leq x \leq 7$), the values of $\pi_i(x)$, $i = 0, \ldots, 7$ are presented in Table 2.9.

For a given input, record $J_n(\text{obs})$ excursions, and count $E_k(x)$, the number of excursions with k visits to x. For each x, compute:

$$X^2(\text{obs}) = \sum_{k=1}^{5} \frac{(E_k(x) - J_n(\text{obs})\pi_k(x))^2}{J_n(\text{obs})\pi_k(x)},$$

which, under $\mathcal{H}_0$, has a chi-square distribution with 5 degrees of freedom. Note that n must be large enough to ensure $J_n(\text{obs})\pi_k(x) \geq 5$ for all k and x, that is, $J_n(\text{obs}) \geq 1220$. We leave it to the reader to estimate how long the input must be.

2.5.5   Permutation Tests

Derangement Permutation Test   Consider a random permutation σ of $[m] = \{1, 2, \ldots, m\}$ and define $X = \#\{i : \sigma(i) = i\}$, i.e., the number of fixed points. A *derangement* of a permutation is a permutation of the elements of a set such that no element appears in its original position. In other words, a derangement is a permutation with no fixed points. Let $\mathcal{D}_m$ denote the set of all derangements of $\{1, 2, \ldots, m\}$, and let $d_m = |\mathcal{D}_m|$.

Lemma 2.5.10   *There are* $d_m = \sum_{k=0}^{m}(-1)^k \binom{m}{k}(m-k)!$ *derangements of a set* $\{1, \ldots, m\}$.

Proof   Recall the inclusion-exclusion principle. For any sets $A_1, \ldots, A_m$, we have:

$$\left| \bigcup_{i=1}^{m} A_i \right| = \sum_{k=1}^{m}(-1)^{k+1} \left(\sum_{1 \leq i_1 < \cdots < i_k \leq m} |A_{i_1} \cap \cdots \cap A_{i_k}| \right).$$

To calculate the number of permutations of $\{1, 2, \ldots, m\}$ with at least one fixed point, denote it by d'_m. Let A_i be the set of permutations σ such that $\sigma(i) = i$. For

Table 2.9 Frequencies of number of visits to x

x	$\pi_0(x)$	$\pi_1(x)$	$\pi_2(x)$	$\pi_3(x)$	$\pi_4(x)$	$\pi_5(x)$
1	0.5000	0.2500	0.1250	0.0625	0.0312	0.0312
2	0.7500	0.0625	0.0469	0.0352	0.0264	0.0791
3	0.8333	0.0278	0.0231	0.0193	0.0161	0.0804
4	0.8750	0.0156	0.0137	0.0120	0.0105	0.0733
5	0.9000	0.0100	0.0090	0.0081	0.0073	0.0656
6	0.9167	0.0069	0.0064	0.0058	0.0053	0.0588
7	0.9286	0.0051	0.0047	0.0044	0.0041	0.0531

fixed $1 \le i_1 < \cdots < i_k \le m$, we have $|A_{i_1} \cap \cdots \cap A_{i_k}| = (m - k)!$. Thus:

$$
d'_m = \left| \bigcup_{i=1}^{m} A_i \right| = \sum_{k=1}^{m} (-1)^{k+1} \left(\sum_{1 \le i_1 < \cdots < i_k \le m} (m - k)! \right)
$$

$$
= \sum_{k=1}^{m} (-1)^{k+1} \binom{m}{k} (m - k)!.
$$

Then, the number of derangements is:

$$
d_m = m! - d'_m = \sum_{k=0}^{m} (-1)^k \binom{m}{k} (m - k)!.
$$

$\square$

The value of d_m can be rewritten as:

$$
d_m = m! \sum_{k=2}^{m} (-1)^k \frac{1}{k!}.
$$

Thus, the probability that a random permutation has no fixed points (i.e., the probability of a derangement) is:

$$
\frac{d_m}{m!} = \sum_{k=2}^{m} (-1)^k \frac{1}{k!},
$$

which approaches e^{-1} as $m \to \infty$. Using Taylor's formula, the error is bounded by:

$$
\left| e^{-1} - \sum_{j=2}^{m} (-1)^j \frac{1}{j!} \right| \le \frac{1}{(m + 1)!}.
$$

This rough error decreases rapidly and is of order less than 2^{-m}.

Based on this result, we propose the following test. Consider a sequence $u_1, u_2, \ldots, u_n$ of length $n = mr$. Split the sequence into r vectors of length m (as earlier in (2.5.2)):

$$
\boldsymbol{u}_1 = (u_1, \ldots, u_m), \quad \boldsymbol{u}_2 = (u_{m+1}, \ldots, u_{2m}),
$$

$$
\ldots, \boldsymbol{u}_r = (u_{(r-1)m+1}, \ldots, u_{rm}).
$$

Using Algorithm 1 (Fisher-Yates algorithm) from Sect. 2.2.1, for each r-tuple $\boldsymbol{u}_i = (u_{(i-1)m+1}, \ldots, u_{im})$, compute the pseudorandom permutation σ^i of $\{1, 2, \ldots, m\}$.

For each permutation σ^i, define:

$$
\xi_i = \begin{cases} 1 & \text{if } \sigma^i \text{ is a derangement,} \\ 0 & \text{otherwise.} \end{cases}
$$

Under the randomness hypothesis $\mathcal{H}_0$, which states that $u_1, \ldots, u_n$ is a realization of an i.i.d. sequence of $\mathcal{U}[0, 1)$ random variables, the ξ_i are i.i.d. with $\mathbb{P}(\xi_i = 1) = e^{-1}$. Perform a variant of the monobit test by computing:

$$
T(\text{obs}) = \left| \frac{\left(\sum_{j=1}^{r} \xi_j \right) - re^{-1}}{\sqrt{re^{-1}(1 - e^{-1})}} \right| ,
$$

and calculate the corresponding p-value using the complementary error function:

$$
p = \text{erfc}\left(\frac{T(\text{obs})}{\sqrt{2}} \right) ,
$$

where $\text{erfc}(z)$ is the complementary error function (see Exercise 2.T.9).

The derangement permutation test is summarized in Algorithm 3. Recall that the Fisher-Yates algorithm was given in Algorithm 1 in Sect. 2.2.1.

Algorithm 3 Derangement permutation test

Require: $u = (u_1, u_2, \ldots u_n), n = mr$
1: $\xi = 0$
2: **for** $i = 1$ to r **do**
3:      $\sigma = \text{Fisher-Yates}(m, (u_{(i-1)m+1}, \ldots, u_{im}))$
4:      **if** σ is a derangement **then**
5:          $\xi = \xi + 1$
6:      **end if**
7: **end for**
8: $T(\text{obs}) = \dfrac{|\xi - re^{-1}|}{\sqrt{re^{-1}(1-e^{-1})}}$
9: **Return** p-value $= \text{erfc}\left(\frac{T(\text{obs})}{\sqrt{2}} \right)$

Fixed Point Permutation Tests For a random permutation σ of $\{1, \ldots, m\}$, let X denote the number of fixed points in the permutation (as defined in the derangement permutation test). The probability distribution of X is given by:

$$
P(X = k) = \frac{1}{m!} \#\{\sigma \in \mathcal{S}_m : \sigma \text{ has exactly } k \text{ fixed points}\}.
$$

Using combinatorial arguments, this can be expressed as:

$$P(X = k) = \frac{1}{m!} \binom{m}{k} d_{m-k},$$

where d_{m-k} is the number of derangements of a set of size $m - k$. For large m, we have $d_{m-k} \sim (m - k)! e^{-1}$, and thus:

$$P(X = k) \to \frac{1}{m!} \frac{m!}{k!(m - k)!}(m - k)! e^{-1} = \frac{1}{k!} e^{-1}, \quad \text{as } m \to \infty.$$

This implies that $X \xrightarrow{\mathcal{D}} N$, where N is a Poisson random variable with parameter 1.

Similarly to the derangement test, we consider a sequence $u_1, u_2, \ldots, u_n$ of length $n = mr$ and split it into r vectors $\boldsymbol{u}_1, \ldots, \boldsymbol{u}_r$, each being an m-tuple. For each $\boldsymbol{u}_i$, we construct a permutation σ^i, $i = 1, \ldots, r$, using the **Fisher-Yates** algorithm (Algorithm 1) from Sect. 2.2.1. Define:

$$\xi_i = \#\{j : \sigma^i(j) = j\},$$

i.e., the number of fixed points of σ^i. Next, consider the following boxes:

$$A_0 = \{0\}, \quad A_1 = \{1\}, \quad \ldots, \quad A_9 = \{9\}, \quad A_{10} = \{10, 11, \ldots\}.$$

For $j = 0, \ldots, 10$, compute O_j, the number of ξ_i values (or equivalently, the number of corresponding vectors $\boldsymbol{u}_i$) that fall into box A_j:

$$O_j = \#\{i \in \{1, \ldots, r\} : \xi_i \in A_j\}.$$

Under the randomness hypothesis $\mathcal{H}_0$, we expect rp_j permutations (and thus vectors) to have j fixed points, where:

$$p_j = \frac{e^{-1}}{j!}, \quad j = 0, \ldots, 9, \quad \text{and } p_{10} = 1 - \sum_{j=0}^{9} p_j.$$

To test the hypothesis, use the chi-square test with the statistic:

$$X^2(\text{obs}) = \sum_{j=0}^{10} \frac{(O_j - rp_j)^2}{rp_j}.$$

Under $\mathcal{H}_0$, this statistic follows a chi-square distribution with 10 degrees of freedom, and the corresponding p-value can be computed.

2.5.6 *p-Values and Second-Level Testing*

More on p-Values This section provides further insight and details regarding the use of p-values, as previously mentioned and applied. Let us start with an informal example. Assume that a PRNG produces n bits, and our task is to assess whether they are "random." Under the assumption of randomness, these bits can be thought of as the outcomes of n independent flips of a fair coin. In this case, each sequence of bits is equally likely, with a probability of $1/2^n$. For instance, if $n = 100$, the probability of observing a sequence of all zeros is $7.88 \cdot 10^{-31}$, which is exceedingly small. While such a sequence is not impossible, it is highly unlikely. Observing such a sequence would lead us to conclude that the PRNG is not good, and we would be correct with very high confidence.

The above reasoning motivates the following approach to testing: *Beforehand,* we identify an event whose probability under randomness is small, typically defined by a significance level α (e.g., 0.05 or 0.01). If the event occurs, we reject the hypothesis that the PRNG produces random numbers. For example, consider the event:

- $E = \{$there will be at least 63 zeros among 100 bits$\}$.

The probability of E under the assumption of randomness can be computed as:

$$\mathbb{P}(E) = \sum_{k=63}^{100} \binom{100}{k} \frac{1}{2^{100}} = 0.00601.$$

If we set $\alpha = 0.01$ as the threshold, observing at least 63 zeros would lead us to reject the randomness of the PRNG, as the event is "too unlikely" under $\mathcal{H}_0$.

Now, let us formalize these ideas using statistical methods. Instead of events, we use a test statistic T. The null hypothesis $\mathcal{H}_0$ is that the sequence under analysis is a realization of a random sequence, i.e., independent random numbers uniformly distributed. The alternative hypothesis $\mathcal{H}_1$ assumes that the randomness property does not hold.

Types of Errors In hypothesis testing, two types of errors are possible:

- **Type I error:** Rejecting $\mathcal{H}_0$ when it is true.
- **Type II error:** Failing to reject $\mathcal{H}_0$ when it is false.

The probability of a Type I error is denoted by α, the significance level (e.g., $\alpha = 0.05$ or 0.01). While testing, we define:

- $\mathcal{G}$: The acceptance region for the test statistic.
- $\mathcal{G}^c$: The critical region, where $T \notin \mathcal{G}$ leads to rejection of $\mathcal{H}_0$.

For a continuous random variable T, under $\mathcal{H}_0$, we have $\mathbb{P}(T \notin \mathcal{G}) = \alpha$. For discrete distributions, the inequality $\mathbb{P}(T \notin \mathcal{G}) \geq \alpha$ holds.

When evaluating the randomness of sequences generated by a PRNG, it is common to use p-values rather than a fixed significance level. If T has a continuous distribution function $F(t)$ and the test is one-sided (e.g., a right-tail test), the p-value is defined as:

$$p = 1 - F(T(\text{obs})), \tag{2.5.23}$$

where $T(\text{obs})$ is the observed value of the test statistic. Larger p-values generally indicate better PRNG performance, while p-values below the significance level α suggest that the PRNG produces non-random sequences.

Important Property of p-Values An essential property of p-values is that, under $\mathcal{H}_0$, if the probability density function (p.d.f.) of T is continuous, then $F(T)$ follows a $\mathcal{U}[0, 1]$ distribution. Consequently, $1 - F(T)$ is also uniformly distributed. This result will be formally proved in Sect. 3.1.

The First-Level Testing is the procedure we applied in previous examples: for one (usually very long) sequence of numbers (the output of a PRNG) we compute one p-value and then try to infer whether the sequence was randomly generated or not. However, it is crucial to be aware that p-values for statistics with continuous distribution functions are uniformly distributed $\mathcal{U}[0, 1)$ (see Corollary 3.1.3 Thus, if we set the significance level to say $\alpha = 0.05$ and we obtain, e.g., p-value 0.0093, usually (in first-level testing), we would reject $\mathcal{H}_0$. But note that on average, in 1 out of 20 p-values, we expect it to be smaller than 0.05. The same applies for large p-values. Recall Examples 2.5.1 and 2.5.2, where for points from set C we obtained p-values $> 99\%$. Again, in first-level testing, it means that we have no reason to reject $\mathcal{H}_0$, but the intuition is that they are *suspiciously* good. Indeed, if we compute p-values for such numbers several times (i.e., for different sets $C_1, C_2, \ldots,$ produced in a similar way to C), then the p-values would always be "*very good*" (>99%), which is in obvious contradiction to the uniformity of these p-values. This is exactly what the second-level testing procedure does (computes several p-values and checks their uniformity). It easily spots non-randomness in the above mentioned Examples 2.5.1 and 2.5.2.

Remark 2.5.11 Note that all p-values in previous sections were computed for distributions with a **continuous** distribution function—typically the chi-square distribution or, in some cases, the normal distribution. For such cases, p-values are uniformly distributed under the null hypothesis $\mathcal{H}_0$. However, p-values computed for **discrete** distributions are **not** uniformly distributed. In these cases, instead of testing the uniformity of p-values (which is not valid for discrete distributions), we test the distribution of the obtained statistics, typically using goodness-of-fit tests. This approach was used in the collision and birthday spacing tests. Specifically, we computed R statistics—e.g., $C_{k,r}^{(1)}, \ldots, C_{k,r}^{(R)}$ in the collision test or $K^{(1)}, \ldots, K^{(R)}$ in the birthday spacing test—and subsequently tested whether they followed the expected distribution (Poisson in both cases) using the chi-square test.

It is left for the reader in Exercise 2.T.23 to prove that p-values for discrete distributions are not uniformly distributed. ∎

The Second-Level Testing The statistic T is computed for a random sequence, and therefore, the p-value (see Eq. (2.5.23)) is itself a random variable. Suppose T is a continuous random variable with a strictly increasing cumulative distribution function (c.d.f.) F. For $t \in [0, 1]$, we have:

$$\mathbb{P}(p \leq t) = \mathbb{P}(1 - F(T) \leq t) = 1 - \mathbb{P}(F(T) \leq 1 - t)$$

$$= 1 - \mathbb{P}\left(F^{-1}(F(T)) \leq F^{-1}(1 - t)\right)$$

$$= 1 - \mathbb{P}\left(T \leq F^{-1}(1 - t)\right) = 1 - F\left(F^{-1}(1 - t)\right) = t.$$

This implies that under the null hypothesis $\mathcal{H}_0$, the distribution of the p-value is uniform on $[0, 1)$. Notably, this result holds as long as T is a continuous random variable, meaning the c.d.f. F is continuous. A formal justification for this fact is given in Proposition 3.1.2 (b).

This observation forms the basis of the *second-level testing* methodology. Instead of calculating a single p-value for a very long sequence from a PRNG, the procedure involves:

 (i) Splitting the sequence into shorter subsequences;
 (ii) Calculating p-values for each subsequence (using the statistic T);
 (iii) Testing the uniformity of the resulting p-values.

To describe this process in detail, assume we are testing the output of a PRNG of length Rn, split into R subsequences of length n each. For the i-th subsequence, compute the statistic $T^{(i)}$ ($i = 1, \ldots, R$) and the corresponding p-value, $p^{(i)} = 1 - F(T^{(i)})$. Under the null hypothesis $\mathcal{H}_0$, the sequence $p^{(1)}, \ldots, p^{(R)}$ consists of i.i.d. random variables uniformly distributed on $[0, 1)$. The uniformity of these p-values is then tested.

One common approach is the chi-square uniformity test, as recommended by NIST. Define a partition $\mathcal{P}^s = \{P_i\}_{i \in \{1,\ldots,s\}}$ of the interval $[0, 1)$ into s subintervals of equal length:

$$P_i = \left[\frac{i-1}{s}, \frac{i}{s}\right), \quad 1 \leq i \leq s. \tag{2.5.24}$$

For $1 \leq i \leq s$, let O_i denote the observed number of p-values in P_i, and E_i the expected number under uniformity:

$$O_i = |\{j : p^{(j)} \in P_i, \ 1 \leq j \leq R\}|, \quad E_i = \frac{R}{s}.$$

The chi-square statistic is then computed as:

$$X_{\text{final}}^2(\text{obs}) = \sum_{i=1}^{s} \frac{(O_i - E_i)^2}{E_i}.$$

Under $\mathcal{H}_0$, this statistic approximately follows a chi-square distribution with $s - 1$ degrees of freedom. The *final* p-value is computed as:

$$p_{\chi^2,\text{final}} = 1 - F_{\chi^2,s-1}(X_{\text{final}}^2(\text{obs})), \qquad (2.5.25)$$

where $F_{\chi^2,s-1}$ is the c.d.f. of the chi-square distribution with $s - 1$ degrees of freedom. This procedure is illustrated in Fig. 2.14.

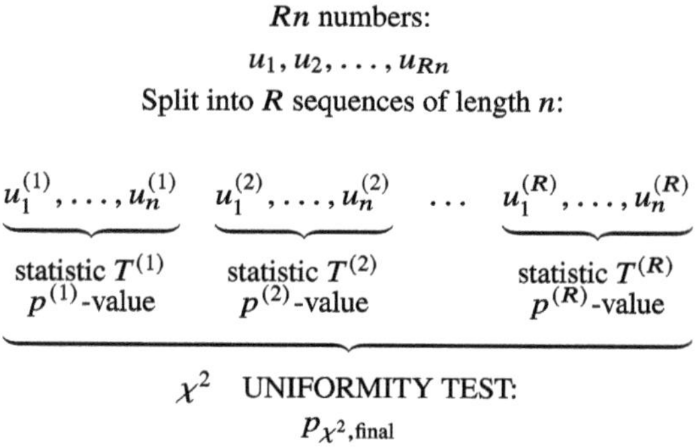

Fig. 2.14 Sketch of second-level testing for tests with continuous statistic T

$$Rn \text{ numbers:}$$
$$u_1, u_2, \ldots, u_{Rn}$$

Split into R sequences of length n:

$$u_1^{(1)}, \ldots, u_n^{(1)} \qquad u_1^{(2)}, \ldots, u_n^{(2)} \qquad \cdots \qquad u_1^{(R)}, \ldots, u_n^{(R)}$$

statistic $T^{(1)}$ statistic $T^{(2)}$ statistic $T^{(R)}$
$p^{(1)}$-value $p^{(2)}$-value $p^{(R)}$-value

$$\chi^2 \quad \text{UNIFORMITY TEST:}$$
$$p_{\chi^2,\text{final}}$$

In practice, it is recommended to use $1000 \leq R \leq 10000$ sequences, each of length $2^{20} \leq n \leq 2^{32}$.

This experiment can be repeated for increasing $n = 2^l$, observing whether there is any "breakdown" in the final p-value. NIST [187] suggests using a significance level $\alpha \in [1/1000, 1/100]$ for the final p-value.

Second-level testing is generally *more reliable* than first-level testing, though the choice of n and R requires careful consideration. Detailed discussions of reliability can be found in [170] and [169]. For specific tests, such as those based on the arcsine law, reliability analysis is provided in [135].

One scenario where second-level testing is particularly useful is when a p-value from a single sequence appears suspiciously large. For example, with quasirandom numbers, single-sequence tests using the Kolmogorov-Smirnov or chi-square tests often yield p-values exceeding 99% (see Examples 2.5.1 and 2.5.2). The second-level test reveals this anomaly by aggregating results across multiple sequences, as shown in Example 2.5.12.

Example 2.5.12 We continue the example of PRNGs that generated the sets of points A, B, and C presented in Fig. 2.4. Recall that in Example 2.5.1, the Kolmogorov-Smirnov test indicated no reasons to reject points A and C as random, while points B were rejected as random. This result was confirmed in Example 2.5.2 using the chi-square test with the partition defined in (2.5.5).

In both cases, the p-values for set C were "very good," exceeding 99%. However, this was "too good," which can be easily detected by the second-level testing approach. The procedure is as follows: we produce $R = 200$ sequences, each of length $n = 50$ (note that in Fig. 2.4, only one such sequence per PRNG is presented). Specifically:

- We generate $A_1, \ldots, A_R$ sets using Python's `NumPy` built-in PCG64 generator. Each set A_i is produced using a unique `seed` i. Specifically, in a loop over i, we invoke `rng_i = default_rng(PCG64(seed=i))` followed by `A_i = rng_i.random(size=n)`.
- We derive $B_1, \ldots, B_R$ by transforming $A_1, \ldots, A_R$ as follows: for a point $u_j^{A_i}$ in set A_i, compute

$$u_j^{B_i} = \frac{e^{u_j^{A_i}} - m_i}{M_i - m_i},$$

  where $m_i = \min_k\left(e^{u_k^{A_i}} - 1 - \varepsilon\right)$, $M_i = \max_k\left(e^{u_k^{A_i}} - 1 - \varepsilon\right)$, with $\varepsilon = 10^{-17}$.
- We produce quasirandom sets $C_1, \ldots, C_R$ using the `Halton` object from Python's `scipy.stats.qmc` library, with C_i generated using `seed` i.

Next, we perform the Kolmogorov-Smirnov test for each set, obtaining p-values $p_{\mathrm{KS}}^{A_i}$, $p_{\mathrm{KS}}^{B_i}$, and $p_{\mathrm{KS}}^{C_i}$ for $i = 1, \ldots, R$. Similarly, we apply the chi-square test with the partition defined in (2.5.5), obtaining p-values $p_{\chi^2}^{A_i}$, $p_{\chi^2}^{B_i}$, and $p_{\chi^2}^{C_i}$ for $i = 1, \ldots, R$. The resulting p-values are presented in Figs. 2.15 and 2.16.

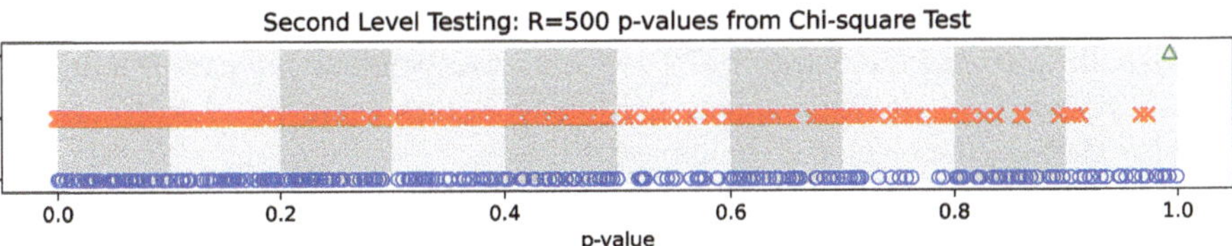

Fig. 2.15 Three sets of R p-values $p_{\chi^2}^{A_i}$ (blue), $p_{\chi^2}^{B_i}$ (red), $p_{\chi^2}^{C_i}$ (green), $i = 1, \ldots, R$: Partition $\mathcal{P}^{10}$ given in (2.5.5) grayed-out

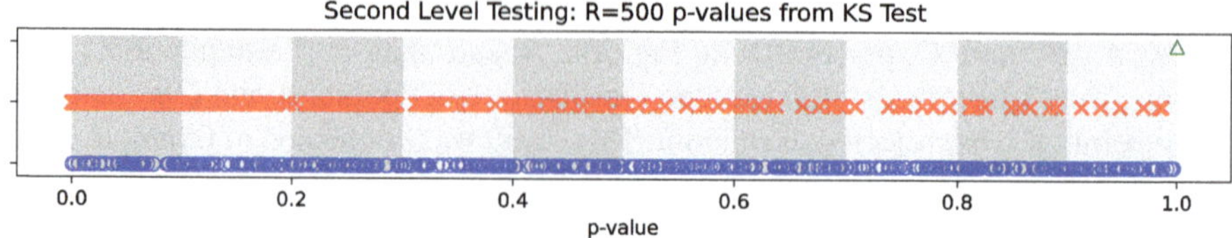

Fig. 2.16 Three sets of R p-values $p_{\text{KS}}^{A_i}$ (blue), $p_{\text{KS}}^{B_i}$ (red), $p_{\text{KS}}^{C_i}$ (green), $i = 1, \ldots, R$: Partition $\mathcal{P}^{10}$ given in (2.5.24) grayed-out

Table 2.10 Number of p-values in intervals P_i, $i = 1, \ldots, 10$ for the partition $\mathcal{P}^{10}$ given in (2.5.24)

		P_1	P_2	P_3	P_4	P_5	P_6	P_7	P_8	P_9	P_{10}
Chi-square test	A	57	51	52	52	52	38	57	38	55	48
	B	189	79	57	45	41	18	29	27	9	6
	C	0	0	0	0	0	0	0	0	0	500
KS-test	A	44	48	50	51	54	53	39	51	58	52
	B	261	78	53	28	28	13	13	9	10	7
	C	0	0	0	0	0	0	0	0	0	500

Table 2.11 Comparison of χ^2 and KS-test final statistics and p-values for sets A, B, and C

		A	B	C
Chi-square test	$\chi^2_{\text{final}}(\text{obs})$	8.56	518.56	4500.0
	p_{final}	0.479	$6.09 \cdot 10^{-106}$	0.0
KS-test	$\chi^2_{\text{final}}(\text{obs})$	5.12	1083	4500.0
	p_{final}	0.824	$2.16 \cdot 10^{-227}$	0.0

As suspected, all the p-values $p_{\text{KS}}^{C_i}$ and $p_{\chi^2}^{C_i}$ were "too perfect"—clustered around 99%. The number of p-values within each interval (box) is provided in Table 2.10.

It is "clear" that p-values $p_{\text{KS}}^{C_i}$ and $p_{\chi^2}^{C_i}$ are not realizations of an i.i.d. sequence with the distribution $\mathcal{U}[0, 1]$. Concerning the p-values $p_{\text{KS}}^{B_i}$ and $p_{\chi^2}^{B_i}$, note that there are very few in P_{10} (7 and 6, respectively), whereas many are in P_1 (261 and 189, respectively). This indicates that they are also not uniformly distributed. We test this hypothesis using the partition $\mathcal{P}^{10}$, which corresponds to the partition defined in (2.5.24) for $s = 10$. We compute the final p-values using Eq. (2.5.25) and the $\chi^2(\text{obs})$ statistic, the results are in Table 2.11.

In conclusion, using the second-level approach, we can only accept the method for producing points from the sets A_i as a good PRNG. The other methods are identified as poor PRNGs, which aligns with our expectations given the manner in which the numbers were generated. $\Diamond$

An Alternative (to the NIST's Recommendation of Using the Chi-Square Test) Method of Testing Uniformity of p-Values in the Second-Level Tests There is another approach to analyzing the sequence $p_1, \ldots, p_R$. We will focus on $\alpha = 1/100$. For the i-th subsequence, define *fail* if $p_i \leq \alpha$ and then set $B_i = 0$;

otherwise, define *pass* and set $B_i = 1$. We have $\mathbb{P}(B_i = 1) = 1 - \alpha$, $\mathbb{E}B_i = 1 - \alpha$, and $\mathbb{V}\text{ar}\,B_i = \alpha(1 - \alpha)$. The fraction of passes is then

$$\hat{B} = \frac{\sum_{i=1}^{R} B_i}{R}.$$

We have $\mathbb{E}\hat{B} = 1 - \alpha$ and $\mathbb{V}\text{ar}\,\hat{B} = \frac{\alpha(1-\alpha)}{R}$. Using the CLT and the three-sigma rule (i.e., for a standard normal random variable Z, we have $\mathbb{P}(|Z| > 3) \approx 0.0028$), we may write

$$\mathbb{P}\left(\hat{B} \notin [\mathbb{E}\hat{B} \pm 3\sqrt{\mathbb{V}\text{ar}\hat{B}}]\right) = \mathbb{P}\left(\hat{B} \notin \left[1 - \alpha \pm 3\frac{\sqrt{\alpha(1-\alpha)}}{\sqrt{R}}\right]\right) \approx 0.0028.$$

Thus, if $\hat{B} \notin \left[1 - \alpha \pm 3\frac{\sqrt{\alpha(1-\alpha)}}{\sqrt{R}}\right]$, then we do not accept $\mathcal{H}_0$.

Remarks on Errors in Approximations Note that so far, we assumed we know the distribution of a statistic T_n (i.e., the distribution of a statistic under consideration, assuming the randomness hypothesis). We add the subscript n to emphasize its dependence on sequence length. However, in practice, we often know only an *approximation* of the distribution. For example, if $U_i \in \{0, 1\}$, then $X_i = 2U_i - 1 \in \{-1, +1\}$, and the statistic

$$T_n = \frac{1}{\sqrt{n}} \sum_{i=1}^{n} X_i$$

is *approximately* distributed as the standard normal $\mathcal{N}(0, 1)$. Designing a *reliable* test requires bounds on the error of such approximations. For details on designing reliable tests, see [169, 170] (where the definition of a *reliable* test is also given).

Let $X_1, X_2, \ldots$ be mean-zero i.i.d. random variables with $E|X_1|^3 < \infty$ and $E[X_1^2] = \sigma^2$. The central limit theorem states that the limiting c.d.f. F_n^Y of

$$Y_n = \sum_{i=1}^{n} \frac{X_i}{\sigma \sqrt{n}}$$

is the c.d.f. Φ of the standard normal distribution. This means that for large n, we can approximate Y_n by $\mathcal{N}(0, 1)$, and the approximation error is bounded by the Berry-Esseen inequality:

$$\sup_{x} |F_n^Y(x) - \Phi(x)| \leq \frac{C_0 E|X_1|^3}{\sigma^3 \sqrt{n}},$$

where C_0 is a positive constant (in the original paper [49], $C_0 \leq 7.59$; in [206], it was shown that $C_0 \leq 0.4785$). See [169, 170] for detailed examples exploiting the

Berry-Esseen inequality in the binary matrix rank test. The reliability analysis for the test based on the arcsine law is also provided in [135]. Apart from errors caused by using asymptotic distributions, additional errors can arise from poor numerical procedures used to compute values of limiting distributions. For instance, accurately calculating the incomplete gamma function for large values of a can pose significant challenges.

2.6 Bibliographical Notes

The fundamental reference on random number generators remains Knuth's monograph *The Art of Computer Programming* (Vol. 2) [107], which provides a historical perspective on PRNGs, their algorithms, and statistical properties. A comprehensive survey on random numbers and their testing is presented in Ripley [174].

For specific issues related to PRNG implementation, the non-random behavior of MATLAB's generator state is discussed in Savicky [189]. The example of a poor PRNG from Example 2.5.4 is taken from MathWorks' online documentation [146].

Testing the quality of PRNGs is a vast field today. The primary source remains Knuth [107], with additional statistical testing methods covered in Marsaglia's battery of tests DIEHARD [142] (for DIEHARDER see [23]) and NIST's cryptographic test suite [187]. The history of uniform random generators can be found in L'Ecuyer [119]. The concept of second-level testing was introduced by Pareschi et al. [169, 170], and in [170] the authors apply it to NIST SP800-22. The Mersenne Twister algorithm, developed by Matsumoto and Nishimura [147], was long the default PRNG in many libraries due to its long period and equidistribution properties. However, newer generators like PCG64 [167, 168] and Xoshiro256 [15] are now preferred in many applications for their better statistical properties and performance.

Some specific tests are considered in L'Ecuyer [116, 117], L'Ecuyer and Simard [121], L'Ecuyer et al. [122], Kim [101], Kim et al. [102], Koçak [108], Lorek et al. [133, 135], Matsumoto and Nishimura [148], Takishima [199], Takishama and Keizo [200]. The small-sample correction to the Kolmogorov-Smirnov test was presented in Vrbik [209]. The use of normal approximations for the number of collisions in PRNG testing is taken from Tsang et al. [205]. Many tests together with the formula for the corresponding p-value (but without its justification) are provided in NIST's document [187].

Section 2.2.2 on card shuffling is based on the mathematical analysis by Diaconis [34]. For a broader discussion on mixing times and the convergence of Markov chains, the book by Levin et al. [124] is recommended.

2.7 Exercises

Theoretical Exercises

2.T.1 Prove the following inequalities:

Markov inequality: For a non-negative random variable X we have

$$\mathbb{P}(X \geq a) \leq \frac{\mathbb{E}X}{a}, \quad \text{for any } a > 0. \tag{2.7.1}$$

Chebyshev inequality: For a real-valued random variable X we have

$$\mathbb{P}(|X - \mathbb{E}X| \geq b) \leq \frac{\mathbb{V}\mathrm{ar}X}{b^2}, \quad \text{for any } b > 0. \tag{2.7.2}$$

2.T.2 Suppose that $B_1, B_2, \ldots$, is a sequence of random bits, and for a given k, define a sequence of strings $Y_1, Y_2, \ldots$ from $\bar{M}$, where $M = 2^k$, by

$$Y_1 = \sum_{j=1}^{k} B_j 2^{j-1}, \ Y_2 = \sum_{j=1}^{k} B_{k+j} 2^{j-1}, \ldots .$$

Show that $Y_1, Y_2, \ldots$ is a sequence of random numbers from $\bar{M}$.

2.T.3 Let $Y_1, Y_2, \ldots$ be independent random variables with the same distribution $\mathcal{U}(\bar{M})$, where M is a natural number and $\bar{M} = \{0, 1, \ldots, M - 1\}$. Assume that $M = 2^k$ and let $b(x)$ be a binary expansion of $x \in \bar{M}$ in k-digit representation (with padding zeros when needed). Show that the sequence of zero-one random variables defined by

$$B_1, B_2, B_2, \ldots = b(Y_1), b(Y_2), \ldots$$

is a random sequence of bits. Show that this property does not hold if M is not a power of 2.

2.T.4 Let $U_1, \ldots$ be a sequence of random independent variables with the same distribution $\mathcal{U}[0, 1)$ and $M \in \{2, \ldots\}$. Show that

$$X_i = \lfloor M U_i \rfloor, \quad i = 1, \ldots$$

is a random sequence with values in $\bar{M} = \{0, 1, \ldots, M - 1\}$.

2.T.5 Suppose that $B_1, B_2, \ldots$ is a sequence of random bits and for a given n, let $\mathbf{\Pi} = (\Pi_1, \ldots, \Pi_n)$ be a random permutation. We assume that $\mathbf{\Pi}$ and $B_1, B_2, \ldots$ are independent. Show that the sequence

$$B_{\Pi_1}, \ldots, B_{\Pi_n}$$

is also a sequence of random bits.

2.T.6 Suppose that $B_1, B_2, \ldots$ is a sequence of random bits and $x_1, x_2, \ldots$ is a deterministic sequence of bits ($x_i = 0, 1$). Show that the sequence $x_1 \oplus B_1, x_2 \oplus B_2, \ldots$ is also a sequence of random bits. Prove that for bits x and b we have $(x \oplus b) \oplus b = x$.

2.T.7 Verify that the following PRNG's fulfill Definition 2.1.1 (show S, f, V and g): $\text{LCG}(M, a, c)$, $\text{LCG}(M, a_1, \ldots, a_k, c)$, Java, quadratic congruential generator (QCG I), lagged Fibonacci generator, L'Ecuyer mixed generator.

2.T.8 Verify that the Fisher-Yates procedure generates a random permutation by proving the following fact. We inductively define a sequence of n permutations of the set $[n]$. We set σ_0 to be the identity permutation. For $k = 1, 2, ..., n - 1$ we define σ_k inductively, where σ_{k-1} is as follows: swap the element in position k with the element in position J_k, where J_k is uniformly distributed on $\{k, \ldots, n\}$ and independent of $J_1, \ldots, J_{k-1}$. Show that σ_{n-1} is a random permutation.

2.T.9 The *error function* $\text{erf}(x)$ is defined as

$$\text{erf}(x) = \frac{2}{\sqrt{\pi}} \int_0^x e^{-t^2}\, dt, \qquad x \in \mathbb{R}$$

and the *complementary error function* $\text{erfc}(x)$

$$\text{erfc}(x) = 1 - \text{erf}(x)\frac{2}{\sqrt{\pi}} \int_x^\infty e^{-t^2}\, dt.$$

The inverse of the error function is denoted by $\text{erfinv}(x)$, where $x \in (-1, 1)$, that is $\text{erf}(\text{erfinv}(x)) = x$. Similarly, define the inverse of the complementary error function $\text{erfcinv}(x)$. Show that $\text{erfcinv}(x) = \text{erfinv}(1 - x)$. Prove that if $\Phi(x)$ is the c.d.f. of the standard normal distribution, then

$$\Phi(x) = \frac{1}{2} + \frac{1}{2}\text{erf}(x/\sqrt{2}) = \frac{1}{2}\text{erfc}(-x/\sqrt{2}),$$

and

$$\Phi^{-1}(x) = \sqrt{2}\,\text{erfinv}(2x - 1).$$

2.T.10 Let

$$\gamma(a, x) = \int_0^x t^{a-1} e^{-t}\, dt$$

and $\Gamma(a) = \gamma(a, \infty)$. There are two *incomplete gamma functions* (igamf)

$$P(a, x) = \frac{\gamma(a, x)}{\Gamma(a)}, \qquad Q(a, x) = 1 - P(a, x).$$

Let $F(x)$ be the c.d.f. of the gamma distribution Gamma(a, λ) and $G(x) = 1 - F(x)$.

(a) Express F and G in terms of $P(a, x)$ and $Q(a, x)$.
(b) Let W have the chi-square distribution with k degrees of freedom, which is the gamma distribution Gamma$(k/2, 1/2)$ with density function

$$\frac{x^{\frac{k}{2}-1} e^{-\frac{x}{2}}}{2^{\frac{k}{2}} \Gamma\left(\frac{k}{2}\right)}.$$

Suppose statistic W is used in a one-sided right-tail test. After conducting the experiment, we have obtained $W = W(\mathrm{obs})$. Show that the p-value is expressed by $p = Q(k/2, 2W(\mathrm{obs}))$.

2.T.11 Show that

$$\frac{2}{\sqrt{\pi}} \int_0^y e^{-x^2}\, dx = P(1/2, y^2).$$

2.T.12 Let N have a Poisson distribution with parameter $\lambda > 0$. Show that the tail distribution function $\bar{F}_\lambda(k) = \sum_{j=k+1}^{\infty} \lambda^j \exp(-\lambda)/j!$ can be expressed as

$$P(N > k) = \bar{F}_\lambda(k) = P(k + 1, \lambda) = 1 - Q(k + 1, \lambda).$$

Assume $N(\mathrm{obs})$ events are observed. Express the p-value in terms of the incomplete gamma function.

2.T.13 Let $\bar{f} = \int_{[0,1]^d} f(\boldsymbol{u})\, d\boldsymbol{u}$ and $\sigma_f^2 = \mathbb{V}\mathrm{ar}\, f(\boldsymbol{U})$, where $\boldsymbol{U}$ is uniformly distributed on $[0, 1] \times [0, 1]$. Show that for a sequence $\boldsymbol{U}_1, \boldsymbol{U}_2, \dots$ of i.i.d. random vectors distributed as $\boldsymbol{U}$, we have

$$\mathbb{P}\left(\left| \frac{1}{n} \sum_{j=1}^n f(\boldsymbol{U}_j) - \bar{f} \right| < \frac{\sigma_f}{\sqrt{\delta n}} \right) \leq 1 - \delta.$$

2.T.14 [Feller [52], p. 101] In the classical random arrangement problem of r balls in k boxes, each arrangement has probability k^{-r}. Let $E_{k,r}$ be the number of empty boxes. Show that the probability of exactly c empty boxes $p_c(r, k) = \mathbb{P}(E_{k,r} = c)$ is

$$\binom{k}{c} \sum_{j=0}^{k-c} (-1)^j \binom{k - c}{j} \left(1 - \frac{r + j}{k}\right)^r.$$

Prove that if k, r increase to ∞ in a way that $\lambda = ke^{-\frac{r}{k}}$ remains bounded, then for $c = 0, 1, \ldots$

$$p_c(r, k) \to \frac{\lambda^c}{c!} e^{-\lambda}.$$

2.T.15 Continuing Exercise 2.T.14. Recall that $C_{k,r}$ is the number of collisions in the classical random arrangement problem of r balls in k boxes, each arrangement has probability k^{-r}. Consider regimes:

(A') $k \exp(-\frac{r}{k}) = \lambda$ for $k, r \to \infty$ for some $\lambda > 0$.
(A'') for $r, k \to \infty$ and $r^2/(2k) = \lambda$, for some $0 < \lambda < \infty$.

Show that under regime (A') we have $\mathbb{E}C_{k,r} \to \lambda$ for some $0 < \lambda$ if and only if under regime (A'') we have $\mathbb{E}E_{k,r} \to \lambda$.

2.T.16 Consider the variance $\sigma^2_{k,r} = \mathbb{V}\mathrm{ar}E_{k,r}$. Show that for $r = \alpha k$ for some $\alpha > 0$ and $k \to \infty$ we have

$$\sigma^2_{k,r} \to e^{-\alpha}(1 - (1 + \alpha)e^{-\alpha}).$$

2.T.17 Consider the serial test, and let O_i $(i = 1, \ldots, k)$ be the number of balls falling into the i-th box. Suppose that hypothesis $\mathcal{H}_0$ holds. Argue that the random vector $(O_1, \ldots, O_r)$ has multinomial distribution $M(r, 1/k, \ldots, 1/k)$. Define $X^2_k = \sum_{i=1}^k \frac{(O_i - r/k)^2}{r/k}$. Show that $\mathbb{E}(X^2_k) = k - 1$.

2.T.18 In a poker card game one draws 5 cards from a deck of 52. The number of all hands is $W = \binom{52}{5} = 2\,598\,960$. Justify the following results:

- one pair: $\frac{1\,098\,240}{W} = 0.422569 = 1/24$
- two pairs: $\frac{123\,552}{W} = 0.047539 = 1/21$
- three of a kind: $\frac{54\,912}{W} = 0.0211 = 1/44$
- full house: $\frac{3\,744}{W} = 0.00144 = 1/694$
- four of a kind: $\frac{624}{W} = 0.00024 = 1/4165$
- flush: $\frac{5\,148}{W} = 0.00198 = 1/50$
- straight: $\frac{9 \cdot 4^5}{W} = \frac{9216}{W} = 0.003546 = 1/282$
- straight flush: $\frac{9 \cdot 4}{W} = \frac{36/W}{W} = 0.0000138 = 1/72193$.

2.T.19 (Riffle Shuffle and its time reversal) Let $p_{RS}(\sigma, \sigma')$ be the probability that a single step of a **Riffle Shuffle** procedure (see page 27) transforms permutation σ into σ'. Similarly, $p_{TR-RS}(\sigma, \sigma')$ denote the probability that a single step of a Time Reversed Riffle Shuffle (see page 27) transforms σ into σ'. Show that for any $\sigma, \sigma' \in S_n$ we have $p_{RS}(\sigma, \sigma') = p_{TR-RS}(\sigma', \sigma)$.

2.T.20 (Geometric interpretation of **Riffle Shuffle**) Let $x_1, \ldots, x_n$ be realizations of i.i.d. $\mathcal{U}[0, 1)$ random variables and denote by $x_{(1)} \leq \cdots \leq x_{(n)}$ their sorted values. For each $x_{(i)}$ consider the transformation

$$T(x) = 2x \bmod 1$$

(the fractional part of $2x$) and let $y_i = T(x_{(i)})$. The y_i may not be in order, rearrange them (permute via σ) as

$$y_{\sigma(1)} \leq \cdots \leq y_{\sigma(1)}.$$

Show that the probability of obtaining a permutation σ is equal to $p_{RS}(id, \sigma)$, where id is an identity permutation and p_{RS} is given in Exercise 2.T.19.

In other words, the induced permutation has the same distribution as the one in **Riffle Shuffle** scheme.

2.T.21 Let σ be a random permutation of $\{1, \ldots, M\}$ and define $H = \sum_{i=1}^{n} H_i$, where $H_i = \mathbb{1}(\sigma(i) = i)$. Calculate $\mathbb{E}H$, $\mathrm{Cov}\,(H_i, H_j)$ and $\mathbb{V}\mathrm{ar}H$.

2.T.22 Let σ be a random permutation of $\{1, \ldots, M\}$ and let C_m be the number of cycles of length $m \leq M$. The total number of cycles is given by $C = \sum_{i=1}^{M} C_m$.

- On average, how many cycles of length j are in a random permutation? I.e., compute $\mathbb{E}C_j$.
- On average, how many cycles (of any length) are in a random permutation? I.e., compute $\mathbb{E}C$.

2.T.23 Suppose that statistic T takes values $\{a_1, a_2, \ldots, a_n\}$ (n can be ∞) with the corresponding probabilities $w_1, w_2, \ldots, w_n$. Let F be the corresponding distribution function of T. We define the one-sided right-tail p-value similarly as in the continuous case:

$$p = 1 - \mathbb{P}(T \leq T(\mathrm{obs})) = 1 - F(T(\mathrm{obs})).$$

The possible values of p are

$$p \in \{r_1, r_2, \ldots, r_n\}, \quad \text{where } r_k = \sum_{i=k}^{n} w_i.$$

- Show that $\mathbb{P}(p = r_k) = w_k$.
- Compute the distribution of one-sided left-tail p-values.
- Compute the distribution of two-sided p-values.

2.T.24 Let $C_{k,r}$ be the number of collisions in the collision test, with k boxes and r balls. Show that

$$\mathbb{P}(C_{k,r} = c) = \frac{k(k-1)\cdots(k-r+c+1)S_2(r, r-c)}{k^r},$$

where $S_2(r, d)$ is the Stirling number of the second kind, which is the number of ways to partition a set of r labeled objects into d non-empty non-labeled subsets.

2.T.25 Under $(A*)$ (page 68), demonstrate asymptotics (2.5.18) and (2.5.19). Hint: Use use $(1-x)^a = \sum_{j=1}^{\infty} \binom{a}{j}(-x)^j$, where $\binom{a}{j} = a(a-1)\cdots(a-j+1)/j!$.

2.T.26 Let X be a discrete random variable taking values $\{a_1, a_2, \ldots, a_n\}$ (n can be ∞) with the corresponding probabilities $w_1, w_2, \ldots, w_n$, and denote the distribution function of X by F_X. Let Y be a random variable with a continuous distribution function F_Y. Suppose that statistic T is a mixture of X and Y, i.e., for some $q \in (0, 1)$ we have

$$F_T(t) = q F_X(t) + (1-q)F_Y(t).$$

We define the one-sided right-tail p-value as usual: $p = 1 - \mathbb{P}(T \leq T(\text{obs})) = 1 - F_T(T(\text{obs}))$. Indicate all possible values of p-values.

2.T.27 For a statistic T assuming values in $\mathbb{R}$ in the case of a two-sided test with non-symmetric distribution of T, one defines the p-value by

$$p = 2 \min \{\mathbb{P}(T \geq T(\text{obs})), \mathbb{P}(T \leq T(\text{obs}))\}. \tag{2.7.3}$$

Assume that the c.d.f. G of T is continuous.

(a) Show that p is uniformly distributed $\mathcal{U}[0, 1)$.
(b) Show that if T is symmetric about zero, then

$$p = \mathbb{P}(|T| \geq |T(\text{obs})|).$$

2.T.28 Consider the example of flipping coins 100 times provided in the beginning of Sect. 2.5.6. Statistic T has a binomial distribution. Consider a one-sided right-tail test and also a two-sided test.
Compute $2 \min \{\mathbb{P}(T \leq 64), \mathbb{P}(T \leq 36)\}$. Note that under $\mathcal{H}_0$, the statistic T is symmetric about its mean.

- Compute numerically its exact value.
- Use the CLT approximation. Is it good?

Lab Exercises

2.L.1 Implement LCG with parameters $M = 2^{11} - 1, a = 5, c = 1$. Generate 10^4 numbers $u_1, \ldots, u_{10^4}$ starting with a seed of your choice, and plot points $(i, u_i), i = 1, \ldots, 10^4$.

- Perform the Kolmogorov-Smirnov test (use a built-in one) to check the uniformity of the numbers.
- What is a period of this PRNG? (Implement a function for computing it.) Does it depend on a seed? Are the assumptions of Theorem 2.4.1 fulfilled?
- Answer the question above for $a = 2048$.

2.L.2 Consider the Fisher-Yates Algorithm 1 with $n = 52$ and a modification (call it Algorithm B) where line 3 is replaced with:

$$k = \mathbf{random}(\{1, 2, \ldots, n\}).$$

In other words, in Algorithm B at step i the element at position i is swapped with an element in a random position. Generate $R = 10^4$ permutations with Fisher-Yates and Algorithm B.

- Perform the fixed point permutation test (report p-values) for both algorithms with boxes $A_0 = \{0\}, A_1 = \{1\}, \ldots, A_9 = \{9\}, A_{10} = \{10, 11, \ldots\}$. Plot histograms of the fixed points distribution of both algorithms and overlay the expected distribution of fixed points for visual comparison.
- Perform the derangement test for both algorithms, report the p-values.

2.L.3 Consider the Mersenne Twister PRNG (use the built-in generator) and your implementation of LCG with $M = 2^{31} - 1, a = 16807, c = 1$. Generate $n = 10^5$ numbers $y_1, \ldots, y_n$ using both generators. Compute the correlation between consecutive numbers—compute the Pearson correlation coefficient between sequences $(y_1, \ldots, y_{n-1})$ and $(y_2, \ldots, y_n)$, i.e.,

$$\rho = \frac{\sum_{i=1}^{n-1}(y_i - \bar{y})(y_{i+1} - \bar{y})}{\sqrt{\sum_{i=1}^{n-1}(y_i - \bar{y})^2}\sqrt{\sum_{i=1}^{n-1}(y_{i+1} - \bar{y})^2}}, \quad \text{where } \bar{y} = \frac{1}{n}\sum_{i=1}^{n}y_i.$$

Which generator produces "better" numbers (i.e., produces consecutive numbers with lower correlation)?

2.L.4 Simulate a random walk by generating randomly (use some built-in procedures) bits $b_1, b_2, \ldots$.

- Plot the position of a random walk for the first 1000 steps, i.e., plot (k, D_k), where $D_k = \sum_{i=1}^{k} b_i$.
- Repeat the simulations 1000 times and record the final positions $D_{1000}^1, D_{1000}^2, \ldots, D_{1000}^{1000}$. Plot scaled values, dividing each D_{1000}^i by $\sqrt{1000}$.
- Perform the frequency monobit test, report the final p-value. According to this test, are the generated bits random?
- What would change if you sort each group of 1000 bits first?

2.L.5 Consider the following numbers $u_i = \frac{i}{M} \bmod 1$ with $M = 100$, $i = 1, \ldots, 10^6$. Perform Kolmogorov-Smirnov and chi-square goodness-of-fit tests (use the setups from Sect. 2.5.2). Perform both first and second-level testing, and present your conclusions.

2.L.6 Consider the setup from Exercise 2.T.17. Justify by simulations that $\mathbb{V}\mathrm{ar}(X_k^2) \sim 2(k-1)$.

2.L.7 Using the library `mpmath` in Python, we can calculate π, $\sqrt{2}$, and $\exp(1)$ with arbitrary precision. The code provided in Listing 2.1 generates `pi_nr`, `sqrt2_nr`, and `exp1_nr`, each containing 10^5 digits. It also converts these numbers into NumPy arrays of digits and computes their binary expansions (fractional parts) as strings of '0's and '1's:

- `pi_nr_np`, `sqrt2_nr_np`, and `exp1_nr_np` store the corresponding decimal digits in NumPy arrays.
- `pi_nr_bits`, `sqrt2_nr_bits`, and `exp1_nr_bits` store the binary expansions of the fractional parts as strings.

Python Listing 2.1 pi, sqrt2, exp1(1) – arbitrary precision

```python
from mpmath import mp

def bin_exp(nr, nr_of_bits):
    """Compute the binary expansion of a number up to ...
        the given number of bits."""
    result = []
    x = nr
    for _ in range(nr_of_bits):
        x *= 2
        bit = int(x)
        result.append(str(bit))
        x -= bit
    return ''.join(result)

prec = 100000
mp.dps = prec   # Set precision

sqrt2_nr = mp.sqrt(2)
exp1_nr = mp.e
pi_nr = mp.pi

pi_nr_np = [int(d) for d in str(pi_nr) if d.isdigit()]
exp1_nr_np = [int(d) for d in str(exp1_nr) if d.isdigit()]
sqrt2_nr_np = [int(d) for d in str(sqrt2_nr) if d.isdigit()]
```

```
pi_nr_bits   = bin_exp(pi_nr - mp.floor(pi_nr), prec)
exp1_nr_bits = bin_exp(exp1_nr - mp.floor(exp1_nr), prec)
sqrt2_nr_bits = bin_exp(sqrt2_nr - mp.floor(sqrt2_nr), prec)
```

Analyze the "randomness" of these numbers using at least four different tests. Consider, for example, the Kolmogorov-Smirnov test, frequency test, collision test, and birthday spacing test.

Chapter 3
Generating Random Variables

This chapter explores methods for generating random numbers—commonly referred to as random variables—with specified distributions. While the discussion is firmly rooted in probability theory, focusing on random variables defined on a probabilistic space, the practical importance of this topic cannot be overstated.

In a preceding chapter, we examined techniques for generating numbers that are approximately uniformly distributed on $[0, 1]$ using pseudorandom number generators (PRNGs). This provided the foundational ability to simulate $U[0, 1)$ random variables efficiently on a computer, along with statistical tests to verify their uniformity and independence. Building on this, the methods in the current chapter show how to transform such uniform random variables into ones following arbitrary distributions.

Why is this critical? In subsequent chapters, we will consider techniques for estimating quantities of the form $I = \mathbb{E}(Y)$, where Y is a random variable. For example, Y might represent the payoff of a financial option under a geometric Brownian motion model, or it could correspond to a multidimensional random variable modeling Brownian motion paths. Monte Carlo methods, which are central to such estimations, rely heavily on our ability to generate random variables with given distributions. Generating Y with the correct distribution is essential for the validity of these estimates, making this chapter foundational for understanding and implementing advanced simulation-based techniques.

In this chapter, we start with methods for generating one-dimensional random variables, such as the inverse transform method and rejection sampling. These provide practical algorithms for working with distributions like the exponential, Gamma, Pareto, and normal distribution. We then extend the discussion to sampling from joint distributions, the multivariate normal distribution, and cases where variables are dependent, such as those modeled using copulas. By the end, readers will gain both theoretical insights and practical tools for simulating a wide range of distributions—a skill crucial for subsequent topics in simulation, optimization, and computational finance.

P. Lorek, T. Rolski, *Lectures on Monte Carlo Theory*, Probability Theory
and Stochastic Modelling 108, https://doi.org/10.1007/978-3-032-01190-9_3

3.1 Inverse Transform Method (ITM)

We aim to generate a random variable X with cumulative distribution function (c.d.f.) $F(x) = \mathbb{P}(X \leq x)$ using the Inverse Transform Method (ITM). For this purpose, we employ U, a random variable with a uniform distribution $\mathcal{U}[0, 1)$. In the sequel, we will continue without explanation to write $U_1, U_2, \ldots$ for the sequence of i.i.d. random variables with the common distribution $\mathcal{U}[0, 1)$. Let $F^{\leftarrow}(u)$ be the *generalized inverse function*, defined as

$$F^{\leftarrow}(u) = \inf\{x :\ u \leq F(x)\} . \tag{3.1.1}$$

In the case where the c.d.f. $F(x)$ is a strictly increasing and continuous function, $F^{\leftarrow}(u)$ is the inverse function of $F(x)$. In Fig. 3.1, situations for constructing the generalized inverse function are sketched (in Fig. 3.2, the corresponding generalized inverse function is plotted).

Let us point out some of the properties of the generalized inverse function $F^{\leftarrow}(u)$:

(P1) $F^{\leftarrow}(u)$ is non-decreasing, left-continuous at u, and admits a limit from the right at u.

(P2) If F is strictly increasing and $u \in F(\mathbb{R})$, then the following equivalence holds: $u = F(x)$ if and only if $F^{\leftarrow}(u) = x$.

(P3) If the condition that $u \in F(\mathbb{R})$ is not satisfied, then an example can be given such that $F^{\leftarrow}(u) = x$, but $u < F(x)$ (see $x_1 = F^{\leftarrow}(u_1)$ in Fig. 3.1).

(P4) $F^{\leftarrow}(u) = \inf\{x : u \leq F(x)\} = \sup\{x : u > F(x)\}$.

(P5) $F^{\leftarrow}(F(x)) \leq x$. If F is strictly increasing, then $F^{\leftarrow}(F(x)) = x$.

(P6) $F(F^{\leftarrow}(u)) \geq u$. We leave it to the reader to find an example in a figure where $F(F^{\leftarrow}(u)) \neq u$.

We will denote the *tail* of a c.d.f. $F(x)$ by

$$\overline{F}(x) = 1 - F(x).$$

For a tail distribution c.d.f. $\overline{F}$, define

$$\overline{F}^{\leftarrow}(u) = \inf\{x : \overline{F}(x) \leq u\}$$

$$= \inf\{x : 1 - u \leq F(x)\} = F^{\leftarrow}(1 - u).$$

The following lemma outlines some useful properties of the generalized inverse function $F^{\leftarrow}(u)$.

Lemma 3.1.1 *The generalized inverse function $F^{\leftarrow}(u)$ has the following properties:*

(a) $u \leq F(x)$ if and only if $F^{\leftarrow}(u) \leq x$.
(b) If $F(x)$ is continuous, then $F(F^{\leftarrow}(u)) = u$.

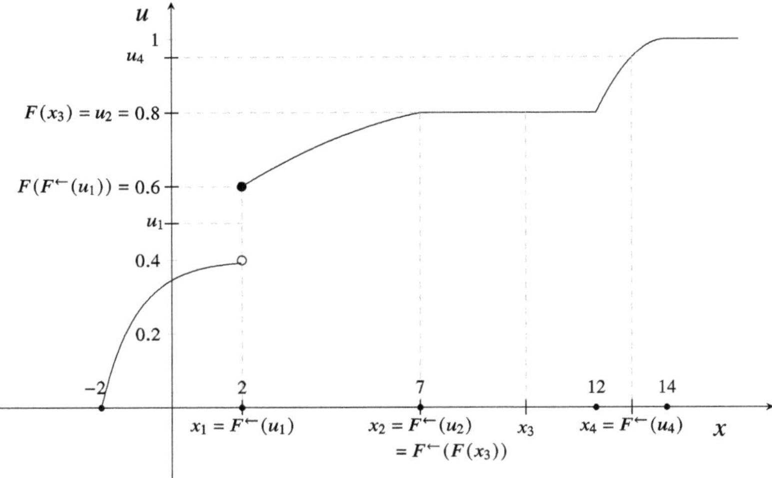

Fig. 3.1 An example of a distribution function F and some properties of the generalized inverse function $F^{\leftarrow}$

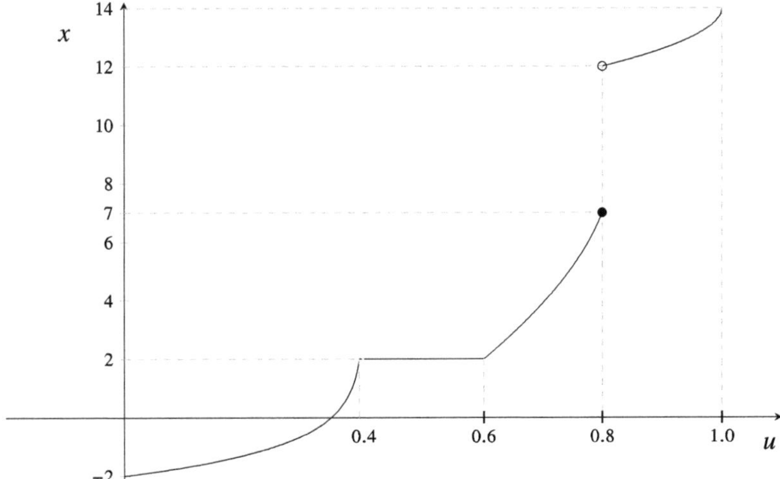

Fig. 3.2 The generalized inverse function $F^{\leftarrow}$ of F from Fig. 3.1

Proof

(a) See Fig. 3.1

(b) We always have $F(F^{\leftarrow}(u)) \geq u$, see property (P6). On the other hand, take any $u \in (0, 1)$ and let x be such that $u = F(x)$ (we can always find such x because

F is continuous). Notice the relationship $F^{\leftarrow}(u) \leq x$. Hence

$$F(F^{\leftarrow}(u)) \leq F(x) = u,$$

which completes the proof of part (b).

$\square$

Proposition 3.1.2 *The following statements hold:*

(a) Suppose U is a uniformly distributed random variable. Then $F^{\leftarrow}(U)$ is a random variable with the distribution F.
(b) If F is continuous and $X \sim F$, then $F(X)$ has the uniform distribution $\mathcal{U}[0, 1)$.

Proof

(a) From Lemma 3.1.1 (a) we have

$$\mathbb{P}(F^{\leftarrow}(U) \leq x) = \mathbb{P}(U \leq F(x)) = F(x) \,.$$

(b) From Lemma 3.1.1 (a) we have

$$\mathbb{P}(F(X) \geq u) = \mathbb{P}(X \geq F^{\leftarrow}(u)) \,.$$

Since a random variable X has a continuous c.d.f., so it is atomless, we can thus write

$$\mathbb{P}(X \geq F^{\leftarrow}(u)) = \mathbb{P}(X > F^{\leftarrow}(u)) = 1 - F(F^{\leftarrow}(u)) \,,$$

which by virtue of Lemma 3.1.1 (b) is equal to $1 - u$.

$\square$

As a consequence, we have a direct corollary:

Corollary 3.1.3 *Let the c.d.f. F of T be continuous. Then, the p-value has the uniform $\mathcal{U}[0, 1)$ distribution.*

Proof As usual $U \sim \mathcal{U}[0, 1)$. There are three scenarios for the p-value as defined in Sect. 2.5.1.3.
One-sided, Right-tail   In this case p-value

$$p = 1 - F(T) \overset{\text{Proposition 3.1.2 (b)}}{=} 1 - U \overset{\mathcal{D}}{=} U.$$

One-sided, Left-tail

$$p = F(T) \overset{\text{Proposition 3.1.2 (b)}}{=} U.$$

Two-sided, Symmetric   In this case, the distribution of T is

$$F_{|T|}(t) = \mathbb{P}(-t \leq T \leq t) \overset{\text{continuity of } F}{=} \mathbb{P}(-t < T \leq t) = F(t) - F(-t),$$

which is continuous. Hence

$$p = F_{|T|}(|T|) \overset{\text{Proposition 3.1.2 (b)}}{=} U.$$

$\square$

We are now in a position to write the following algorithm for generating a random variable with a given continuous c.d.f. F. The algorithm (and whole method) is called ITM (Inverse Transform Method). In many software packages, $\mathtt{rand}$ is the call for a random number uniformly distributed $\mathcal{U}[0, 1)$.

Algorithm 6 ITM; generating $X \sim F$

Require: c.d.f. F
Ensure: $X \sim F$
 1: Generate $U \sim \mathcal{U}[0, 1)$
 2: Set $X = F^{\leftarrow}(U)$
 3: **return** X

Another variant of the ITM algorithm is to set $Y = \overline{F}^{\leftarrow}(U)$.

It is also helpful to note an additional property that will be useful later. If $U_1, U_2, \ldots$ is a sequence of i.i.d. uniformly distributed random variables, then $U_1', U_2', \ldots$ is also a sequence of i.i.d. uniformly distributed random variables, where $U_j' = 1 - U_j$. Explicit formulas for $F^{\leftarrow}(u)$ or $\overline{F}^{\leftarrow}(u)$ are available only in a few cases; some of these will be discussed below.

We start with the exponential random variable, a fundamental distribution widely used in modeling waiting times and as a building block for the Poisson process, which underpins many applications in queuing theory and reliability analysis.

Exponential Distribution Exp(λ)   In this case

$$F(x) = \begin{cases} 0 & \text{for } x < 0, \\ 1 - e^{-\lambda x} & \text{for } x \geq 0. \end{cases} \tag{3.1.2}$$

If $X \sim \text{Exp}(\lambda)$, then $\mathbb{E}X = 1/\lambda$ and $\mathbb{V}\text{ar}X = 1/\lambda^2$. It is a light-tail distribution because

$$\mathbb{E}e^{sX} = \frac{\lambda}{\lambda - s}$$

is well defined for $0 \le s < \lambda$. We have

$$\overline{F}^{\leftarrow}(u) = -\frac{1}{\lambda}\log(u) \, .$$

Hence, we can generate an exponentially distributed random variable $\text{Exp}(\lambda)$ using the following simple algorithm.

Algorithm 7 ITM-Exp; generating $X \sim \text{Exp}(\lambda)$

Require: λ
Ensure: $X \sim \text{Exp}(\lambda)$
 1: Generate $U \sim \mathcal{U}[0, 1)$
 2: Set $X = -\log(U)/\lambda$
 3: **return** X

We now consider the Pareto distribution, an example of a heavy-tailed random variable often used to model phenomena such as wealth distribution, insurance claims, and internet traffic, where extreme values occur with non-negligible probability.

Pareto Distribution Par(α) The c.d.f. of the Pareto distribution, denoted $\text{Par}(\alpha)$, is given by

$$F(x) = \begin{cases} 0 & \text{for } x < 0, \\ 1 - \dfrac{1}{(1+x)^\alpha} & \text{for } x \ge 0 \, . \end{cases} \tag{3.1.3}$$

If $X \sim \text{Par}(\alpha)$, then $\mathbb{E}X = 1/(\alpha - 1)$ provided that $\alpha > 1$, and $\mathbb{E}X^2 = 2/((\alpha - 1)(\alpha - 2))$ provided that $\alpha > 2$. This distribution is considered heavy-tailed because

$$\mathbb{E}e^{sX} = \infty$$

for all $s > 0$. The generalized inverse function of $\overline{F}$ is

$$\bar{g}(u) = u^{-1/\alpha} - 1.$$

Note that this definition of the Pareto distribution includes only one parameter. To create a more flexible family accommodating both a specified mean m and shape parameter α, we use the fact that the random variable $m(\alpha - 1)X$ has mean m (assuming $\alpha > 1$). Then, $m(\alpha - 1)X$ has the tail distribution

$$\mathbb{P}(m(\alpha - 1)X > x) = \left(\frac{1}{1 + x/(m(\alpha - 1))}\right)^\alpha = \frac{(m(\alpha - 1))^\alpha}{(m(\alpha - 1) + x)^\alpha} \, .$$

Setting $c = (\alpha - 1)m$, we introduce a two-parameter family of the Pareto distribution, denoted $\mathrm{Par}(\alpha, c)$. We say that a random variable Y has a Pareto $\mathrm{Par}(\alpha, c)$ distribution if its c.d.f. is

$$F(x) = \begin{cases} 0 & \text{for } x < 0 \\[2ex] 1 - \dfrac{c^{\alpha}}{(c+x)^{\alpha}} & \text{for } x \geq 0 \,. \end{cases} \qquad (3.1.4)$$

It is important to note that there is no single, universally accepted definition of the Pareto distribution in the literature. Therefore, various parametrizations, including those presented here, may be encountered.

Algorithm 8 ITM-Par; generating $X \sim \mathrm{Par}(\alpha)$

Require: α
Ensure: $X \sim \mathrm{Par}(\alpha)$
 1: Generate $U \sim \mathcal{U}[0, 1)$
 2: $X = U^{(-1/\alpha)} - 1$
 3: **return** X

3.2   ITM Variants for Lattice Distributions; Geometric Distribution

Lattice distributions, characterized by their discrete nature, are essential for modeling phenomena where outcomes occur at specific intervals, such as inventory levels, queuing systems, or network traffic. Adapting the inverse transform method (ITM) for such distributions enables efficient and exact sampling, crucial for accurate simulations in these contexts.

Suppose a distribution is concentrated on $\{0, 1, \ldots\}$. Then it suffices to know the probability function $(p_k, \ k = 0, 1, \ldots)$.

Proposition 3.2.1 *Let U be uniformly distributed random variable $\mathcal{U}[0, 1)$. The random variable*

$$Y = \min\left\{ k : U \leq \sum_{i=0}^{k} p_i \right\}$$

has the probability function (p_k).

Proof Notice that

$$\mathbb{P}(Y = l) = \mathbb{P}\left(\sum_{i=0}^{l-1} p_i < U \leq \sum_{i=0}^{l} p_i \right),$$

and that interval $\left(\sum_{i=0}^{l-1} p_i, \sum_{i=0}^{l} p_i\right]$ has length p_l. Recall the convention that $\sum_{i=0}^{-1} = 0$. □

Hence we have the following algorithm for generating lattice random variables Y.

Algorithm 9 ITM-d; generating $Y \sim (p_k)$

Require: (p_k)
Ensure: $Y \sim (p_k)$
 1: $Y = 0$
 2: Generate $U \sim \mathcal{U}[0, 1)$
 3: $S = p_0$
 4: **while** $S < U$ **do**
 5: 　　$Y = Y + 1,\ S = S + p_Y$
 6: **end while**
 7: **return** Y

The number of calls in the loop is equal to 1 when $U \leq p_0$, equal to 2 when $p_0 < U \leq p_0 + p_1$, etc. Hence we have the following proposition.

Proposition 3.2.2 *Let N be the number of calls in the loop while in Algorithm 9. Then N has the probability function*

$$\mathbb{P}(N = k) = p_{k-1}, \qquad k = 1, 2, \ldots,$$

and hence $\mathbb{E}N = \mathbb{E}Y + 1$.

We thus see that the algorithm ITM-d for lattice distributions with infinite (or very large) mean is not good.

Geometric Distribution Geo(p) The geometric distribution has the probability function

$$p_k = (1 - p)p^k, \qquad k = 0, 1, \ldots$$

A random variable $X \sim \text{Geo}(p)$ has mean $\mathbb{E}X = p/q$ and

$$X = \lfloor \log U / \log p \rfloor$$

has distribution Geo(p), since

$$\mathbb{P}(X \geq k) = \mathbb{P}(\lfloor \log U / \log p \rfloor \geq k) = \mathbb{P}(\log U / \log p \geq k)$$
$$= \mathbb{P}(\log U \leq k \log p) = \mathbb{P}(U \leq p^k) = p^k .$$

$(a, b, 0)$ **Class of Distributions**   Suppose that $(p_0, p_1, \ldots)$ is a probability function. In some cases, the following quotient

$$c_{k+1} = \frac{p_{k+1}}{p_k}$$

has a simple form. In actuarial mathematics, a popular class of probability functions (p_k) is the $(a, b, 0)$ class of distributions satisfying the recursion $p_k = p_{k-1}(a + \frac{b}{k})$ $(k = 1, 2, \ldots)$ for some a, b, and $p_0 > 0$ (remember that $p_0 + \cdots = 1$). See also Exercise 3.T.21, where the $(a, b, 1)$ class of distributions on $1, 2, \ldots$ is defined. We will now give examples.

- $X \sim B(n, p)$ with probability function $p_k = \binom{n}{k} p^k (1 - p)^{n-k}$. Then

$$p_0 = (1 - p)^n, \quad c_{k+1} = \frac{p(n - k)}{(1 - p)(k + 1)}, \qquad k = 0, \ldots, n - 1.$$

- $X \sim \text{Poi}(\lambda)$ with probability function $p_k = e^{-\lambda} \lambda^k / k!$. Then

$$p_0 = e^{-\lambda}, \quad c_{k+1} = \frac{\lambda}{k + 1}, \qquad k = 0, 1, \ldots$$

- $X \sim \text{NB}(r, p)$, for $r > 0$ and $0 < p < 1$, has probability function

$$p_k = \frac{\Gamma(r + k)}{\Gamma(r)k!} (1 - p)^r p^k, \qquad k = 0, 1, \ldots .$$

Using the generalized binomial coefficients, we can write

$$p_k = \frac{\Gamma(r + k)}{\Gamma(r)\Gamma(k + 1)} (1 - p)^r p^k$$

$$= \binom{k + r - 1}{k} (1 - p)^r p^k, \quad k = 0, 1, \ldots .$$

Then

$$p_0 = (1 - p)^r, \quad c_{k+1} = \frac{(r + k)p}{k + 1}, \qquad k = 0, 1, \ldots$$

Above, we used the following property of the gamma function $\Gamma(r + k + 1) = (r + k)\Gamma(r + k)$.

In cases where the function c_{k+1} is known and has a simple form, we can use an alternative to the ITM-d algorithm called ITR. Suppose X has probability function (p_n). Then $X = l$ if and only if

$$p_0 + \cdots + p_{l-1} \le U < p_0 + \cdots + p_{l-1} + p_l$$

or equivalently,

$$p_0 + \cdots + p_{l-1} \le U < p_0 + \cdots + p_{l-1} + p_{l-1}\left(\frac{p_l}{p_{l-1}}\right).$$

Algorithm 10 ITR; generating $X \sim (p_k)$

Require: p_0 and $c_{k+1}, k = 0, 1, \ldots$
1: $S = p_0$
2: $P = p_0$
3: $X = 0$
4: Generate $U \sim \mathcal{U}[0, 1)$
5: **while** $U > S$ **do**
6:      $X = X + 1$
7:      $P = P \cdot c_X$
8:      $S = S + P$
9: **end while**
10: **return** X

3.3 Ad Hoc Methods

We now present several methods that utilize special properties of specific distributions.

3.3.1 Binomial B(n, p)

Let $X \sim B(n, p)$. Then

$$X \stackrel{\mathcal{D}}{=} \sum_{j=1}^{n} \xi_j,$$

where $\xi_1, \ldots, \xi_n$ is a sequence of Bernoulli trials, i.e., i.i.d. random variables such that $\mathbb{P}(\xi_j = 1) = p$ and $\mathbb{P}(\xi_j = 0) = 1 - p$.

Algorithm 11 DB; generating $X \sim \mathrm{B}(n, p)$

Require: p and n – success probability and number of trials
Ensure: $X \sim \mathrm{B}(n, p)$
1: $X = 0$
2: **for** $i = 1 : n$ **do**
3:      Generate $U \sim \mathcal{U}[0, 1)$
4:        **if** $U \le p$ **then**
5:            $X = X + 1$
6:        **end if**
7: **end for**
8: **return** X

We recommend that the reader compare the efficiency of the specialized Algorithm DB with the general Algorithm ITR.

3.3.2   Erlang Erl(n, λ)

Let $Y_1, \ldots, Y_n$ be i.i.d. random variables with a common exponential distribution $\mathrm{Exp}(\lambda)$. Then

$$X = Y_1 + \cdots + Y_n$$

has the Erlang distribution $\mathrm{Erl}(n, \lambda)$. The Erlang distribution is a special case of the Gamma(n, λ) distribution with p.d.f.

$$f(x) = \frac{1}{\Gamma(n)} \lambda^n x^{n-1} e^{-\lambda x}, \qquad x \ge 0 \, .$$

Since a random variable with the exponential distribution $\mathrm{Exp}(\lambda)$ is generated by $-\log(U)/\lambda$, a random variable $X \sim \mathrm{Erl}(n, \lambda)$ can be generated by

$$X = -\frac{1}{\lambda}(\log U_1 + \cdots + \log U_n) = -\frac{\log(U_1 \cdot \cdots \cdot U_n)}{\lambda} \, .$$

where $U_1, \ldots, U_n$ are i.i.d. uniformly distributed random variables. Note that there is a risk of floating-point error when n is large.

Moreover, the tail of the Erlang distribution $\mathrm{Erl}(n, \lambda)$ can be expressed as follows (see Exercise 3.T.16):

$$\bar{F}(x) = 1 - F(x) = \mathbb{P}(X > x)$$

$$= e^{-\lambda x} \left(1 + \lambda x + \frac{(\lambda x)^2}{2!} + \cdots + \frac{(\lambda x)^{n-1}}{(n-1)!} \right) . \tag{3.3.1}$$

This identity will be useful shortly.

3.3.3   Poisson Poi(λ)

Let $\tau_1, \tau_2, \ldots$ be i.i.d. random variables with the common distribution Exp(1). Consider the following process $\sigma_1 = \tau_1, \sigma_2 = \tau_1 + \tau_2, \ldots$ and count the number N of σ_j points falling into the interval $(0, \lambda]$; see Fig. 3.3.

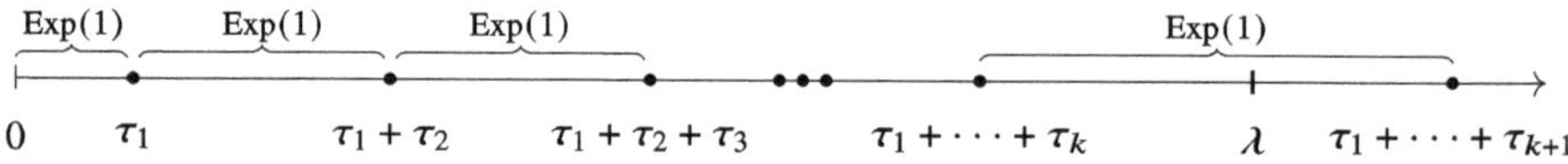

Fig. 3.3 Renewal process and the Poisson distribution, see description in the text

By the way, if $\tau_1, \tau_2, \ldots$ are i.i.d. non-negative random variables with a general distribution G, the process $\sigma_1 = \tau_1, \sigma_2 = \tau_1 + \tau_2, \ldots$ is said to be a *renewal process*. Formally, we define

$$
N = \begin{cases} \max\{i \geq 1 : \tau_1 + \cdots + \tau_i \leq \lambda\} & \text{if } \tau_1 \leq \lambda, \\ 0 & \text{if } \tau_1 > \lambda. \end{cases}
\tag{3.3.2}
$$

Proposition 3.3.1 *We have*

$$
N \overset{\mathcal{D}}{=} \mathrm{Poi}(\lambda).
$$

Proof We have $\mathbb{P}(N = 0) = \mathbb{P}(\tau_1 > \lambda) = e^{-\lambda}$. Moreover, note the following equivalence

$$
\{N \leq k\} = \left\{ \sum_{j=1}^{k+1} \tau_j > \lambda \right\}.
$$

Since (see (3.3.1))

$$
\mathbb{P}\left(\sum_{j=1}^{k+1} \tau_j > x \right) = e^{-x}\left(1 + x + \frac{x^2}{2!} + \cdots + \frac{x^k}{k!} \right)
$$

we have

$$
\mathbb{P}(N = k) = \mathbb{P}(\{N \leq k\}) - \mathbb{P}(\{N \leq k - 1\})
$$

$$= \mathbb{P}\left(\sum_{j=1}^{k+1} \tau_j > \lambda\right) - \mathbb{P}\left(\sum_{j=1}^{k} \tau_j > \lambda\right)$$

$$= \frac{\lambda^k}{k!} e^{-\lambda}.$$

$\square$

Based on the above, a procedure for sampling a random variable with Poisson distribution is given in Algorithm 12.

Algorithm 12 DP; generating $X \sim \text{Poi}(\lambda)$

Require: $\lambda > 0$
Ensure: $X \sim \text{Poi}(\lambda)$
 1: $X = 0$
 2: Generate $U \sim \mathcal{U}[0, 1)$
 3: $S = -\log(U)$
 4: **while** $S < \lambda$ **do**
 5:     Generate $U \sim \mathcal{U}[0, 1)$
 6:     $Y = -\log(U)$
 7:     $S = S + Y$
 8:     $X = X + 1$
 9: **end while**
10: **return** X

Algorithm 12 can be further simplified using the following calculations. If $\tau_1 > \lambda$, then $-\log U_1 < \lambda$, which is equivalent to $U_1 < \exp(-\lambda)$. Furthermore formula (3.3.2) can be rewritten as follows:

$$N = \max\{n \geq 0 : \tau_1 + \cdots + \tau_n \leq \lambda\}$$

$$= \max\{n \geq 0 : (-\log U_1) + \cdots + (-\log U_n) \leq \lambda\}$$

$$= \max\{n \geq 0 : \log \prod_{i=1}^{n} U_i \geq -\lambda\}$$

$$= \max\{n \geq 0 : \prod_{i=1}^{n} U_i \geq e^{-\lambda}\}.$$

In the above, the convention is that if $n = 0$, then $\tau_1 + \cdots + \tau_n = 0$ (and $\prod_{i=1}^{0} u_i = 1$).

3.4   Uniform Distribution

Consider a random vector $\mathbf{X}$ taking values in $A \subset \mathbb{R}^d$. Let Vol D denote the volume of a subset D. For example, if $d = 1$, it represents the length; if $d = 2$, it represents the area, and so on. We assume that Vol $A < \infty$. A random vector $\mathbf{X}$ is said to have a uniform distribution $\mathcal{U}(A)$ in $A \subset \mathbb{R}^d$, if

$$\mathbb{P}(\mathbf{X} \in D) = \frac{\text{Vol } D}{\text{Vol } A}, \qquad D \subset A .$$

Sometimes, this can be expressed in terms of the p.d.f. as

$$f(\mathbf{x}) = \frac{1}{\text{Vol } A} \mathbb{1}(\mathbf{x} \in A).$$

In this case, we write $\mathbf{X} \sim \mathcal{U}(A)$ and say that $\mathbf{X}$ is uniformly distributed in A. Recall that we have access to a random number U uniformly distributed as $\mathcal{U}[0, 1)$ in the interval $(0, 1)$. If a random number $V \sim \mathcal{U}[a, b)$, uniformly distributed in the interval (a, b), is needed, we can linearly transform U as follows:

$$V = (b - a)U + a .$$

Furthermore, if a random vector $V = (V_1, \ldots, V_d)$ uniformly distributed in $\prod_{i=1}^{d}(a_i, b_i)$ is needed, we can use independent random variables $V_i \overset{\mathcal{D}}{=} (b_i - a_i)U_i + a_i$ $(i = 1, \ldots, d)$, where $U_1, \ldots, U_d$ are i.i.d. $\mathcal{U}[0, 1)$ distributed random variables. We leave the reader with a brief justification for this fact.

To generate a random vector uniformly distributed in a general area A, e.g., within a ball or an ellipsoid, we make the following observation. Suppose that A and B are subsets of $\mathbb{R}^d$ for which Vol $B = \int_B d\mathbf{x} < \infty$ and Vol $A = \int_A d\mathbf{x}$, with $A \subset B$. The key to constructing an algorithm for generating $\mathbf{Y} \sim \mathcal{U}(A)$ is the following fact.

Proposition 3.4.1 *Let $A \subset B$ and* Vol $B < \infty$. *If $\mathbf{X}$ is uniformly distributed as* $\mathcal{U}(B)$, *then $\mathbf{Y} \overset{\mathcal{D}}{=} (\mathbf{X} \mid \mathbf{X} \in A)$ is uniformly distributed as $\mathcal{U}(A)$.*

Proof We use that for $D \subset A$

$$\mathbb{P}(\mathbf{Y} \in D) = \mathbb{P}(\mathbf{X} \in D | \mathbf{X} \in A) = \frac{\mathbb{P}(\mathbf{X} \in D, \mathbf{X} \in A)}{\mathbb{P}(\mathbf{X} \in A)} = \frac{\frac{\text{Vol } D}{\text{Vol } B}}{\frac{\text{Vol } A}{\text{Vol } B}} = \frac{\text{Vol } D}{\text{Vol } A} .$$

$\square$

Hence we have the following algorithm for generating a random vector $\mathbf{X} = (X_1, \ldots, X_d)^T$ uniformly distributed $\mathcal{U}(A)$, where A is a bounded subset of $\mathbb{R}^d$. The boundedness of the set A means that it can be contained in a rectangle $B = [a_1, b_1] \times \cdots \times [a_d, b_d]$. We can easily generate a random vector $(V_1, \ldots, V_d)$

uniformly distributed $\mathcal{U}(B)$, because components V_i are independent and uniformly distributed $U[a_i, b_i]$ $(i = 1, \ldots, d)$ respectively.

Algorithm 13 $\mathcal{U}(A)$; generating $X \sim \mathcal{U}(A)$

Require: A rectangle $B = [a_1, b_1] \times \cdots \times [a_d, b_d]$ containing A
Ensure: $X \sim \mathcal{U}(A)$
 1: Generate i.i.d. $U_1, \ldots, U_d, U_i \sim \mathcal{U}[0, 1)$
 2: **for** $i = 1$ **to** d **do**
 3:     $V_i = (b_i - a_i) \cdot U_i + a_i$
 4: **end for**
 5: **if** $V = (V_1, \ldots, V_d) \in A$ **then**
 6:     $X = V$
 7: **else**
 8:     **goto** 1
 9: **end if**
10: **return** X

Uniform Distribution in the Ball B_d Algorithm 13 is inefficient for sampling a point from a unit ball when the dimension d is large. Define the d-ball as

$$B_d = \{(x_1, \ldots, x_d) : x_1^2 + \cdots + x_d^2 \le 1\},$$

and the d-cube as $A_d = [-1, 1]^d$. The volume of the d-ball is

$$\text{Vol } B_d = \frac{\pi^{d/2}}{\Gamma\left(\frac{d}{2} + 1\right)}.$$

For even $d = 2k$, we have

$$\text{Vol } B_d = \frac{\pi^k}{k!}, \quad \text{Vol } A_d = 2^{2k},$$

and the probability that $V \in A$ (line 5 in Algorithm 13) is $\frac{\pi^k}{2^{2k}k!}$. Note that using Stirling's formula, $k! \sim \sqrt{2\pi k}\left(\frac{k}{e}\right)^k$, this probability is asymptotically $\frac{1}{\sqrt{2\pi k}}\left(\frac{\pi e}{4k}\right)^k$, meaning it converges to 0 exponentially fast. For example, it is approximately 2.0524×10^{-14} for $d = 30$.

3.5 Acceptance-rejection Method

We now show a general method for generating a random number, introduced by John von Neumann. This is the most adaptable method for sampling from complicated distributions under some mild assumptions. The method, called the *acceptance-rejection method*, is often highly efficient, especially when the proposed distribution closely approximates the target distribution. It serves as a foundational concept for more advanced algorithms, such as the Metropolis algorithm (studied later in Sect. 6.3.1) and its variants in Markov chain Monte Carlo (MCMC), which extend the acceptance-rejection principle to generate samples iteratively.

3.5.1 AR-d Method

We begin by presenting the method for discrete distributions, which we will abbreviate by AR-d (Acceptance-Rejection-discrete). Suppose we want to generate a random number X with probability function $(p_j, j \in E)$, where E is a finite or denumerable space of values of X (for example $\mathbb{Z}$, $\mathbb{Z}_+$, $\mathbb{Z}^2$, etc.), while we know how to generate another random number Y with a probability function $(q_j, j \in E)$ such that there exists $c > 0$ for which

$$p_j \leq cq_j, \qquad \forall(j \in E).$$

If E is finite, then such c always exists. We leave it to the reader to think about infinite cases when such c does not exist. We will also see them in due course. Clearly, $c > 1$ unless both the probability functions are the same (why?).

In the algorithm below, we will generate independently $U \sim \mathcal{U}[0, 1)$ and Y with a probability function $(q_j, j \in E)$ and define the so-called *acceptance region* by

$$A = \{cU q_Y \leq p_Y\} = \left\{ U q_Y \leq \frac{1}{c} p_Y \right\}.$$

The basis of our algorithm will be the following fact.

Proposition 3.5.1 *We have*

$$\mathbb{P}(A) = 1/c \text{ and } \mathbb{P}(Y = j|A) = p_j \quad \text{for} \quad j \in E.$$

Proof Without loss of generality we may assume that $q_j > 0$ for all $j \in E$, since if for some j we had $q_j = 0$, then we would also have $p_j = 0$, and this state can be eliminated. Furthermore, we have

$$\mathbb{P}(Y = j|A) = \frac{\mathbb{P}(\{Y = j\} \cap \{cU q_Y \leq p_Y\})}{\mathbb{P}(A)} = \frac{\mathbb{P}(Y = j, U \leq p_j/(cq_j))}{\mathbb{P}(A)}$$

$$= \frac{\mathbb{P}(Y = j)\mathbb{P}(U \leq p_j/(cq_j))}{\mathbb{P}(A)} = q_j \frac{p_j/(cq_j)}{\mathbb{P}(A)}$$

$$= \frac{p_j}{c\mathbb{P}(A)} .$$

Now we have to take advantage of the fact that

$$1 = \sum_{j \in E} \mathbb{P}(Y = j | A) = \frac{1}{c\mathbb{P}(A)},$$

which means that $\mathbb{P}(A) = 1/c$ and thus $\mathbb{P}(Y = j | A) = p_j$. $\qquad\square$

The following algorithm gives a recipe for generating a random number X with probability function $(p_j, j \in E)$.

Algorithm 14 AR-d; generating $X \sim (p_j)$

Require: Constant $c > 0$ such that $p_j \le cq_j$ and an algorithm for generating r.v. Y with probability function (q_j)
Ensure: $X \sim (p_j)$
1: **repeat**
2:     Generate $Y \sim (q_j)$
3:     Generate $U \sim \mathcal{U}[0, 1)$ (independent from Y)
4: **until** $cUq_Y \le p_Y$
5: **return** $X = Y$

We can see that the number L of repetitions in the loop has a probability function $\mathbb{P}(L = k) = \frac{1}{c}(1 - \frac{1}{c})^{k-1}, k = 1, 2, \ldots$ (this is the geometric distribution shifted to start at 1). The expected number of repetitions in the loop is c, so we have to look for the least possible c.

Example 3.5.2 We want to simulate X with distribution $p_j = \frac{6}{\pi^2}\frac{1}{j^2}, j = 1, \ldots$. As a proposal distribution, let us take Y with distribution $q_j = \frac{1}{j(j+1)}, j = 1, \ldots$, which can be simply generated as $Y = \lfloor U^{-1} \rfloor$ for $U \sim \mathcal{U}[0, 1)$ (see Exercise 3.T.18). We have

$$\frac{p_j}{q_j} = \frac{6}{\pi^2}\frac{1}{j^2} \cdot j(j + 1) = \frac{6}{\pi^2}\frac{j + 1}{j}.$$

We take $c = \max(\frac{p_j}{q_j}) = 12/\pi^2$. Let us expand step 4 of the AR-d Algorithm 14:

$$cUq_Y \le p_Y \quad \Longleftrightarrow \quad \frac{12}{\pi^2}U\frac{1}{Y(Y + 1)} \le \frac{6}{\pi^2}\frac{1}{Y^2} \quad \Longleftrightarrow \quad 2UY \le Y + 1$$

Summarizing, the procedure is given in Algorithm 15. $\qquad\diamond$

Algorithm 15 AR-d for Example 3.5.2

Ensure: $X \sim (p_j)$, $p_j = 6/(\pi^2 j^2)$
 1: **repeat**
 2:      Generate independent $U, U_0 \sim \mathcal{U}[0, 1)$ and set $Y = \lfloor U_0^{-1} \rfloor$
 3: **until** $2UY \leq Y + 1$
 4: **return** $X = Y$

3.5.2 AR-c Method

We now present a continuous version of the acceptance-rejection method (AR-c). Suppose we want to generate a random number X with a p.d.f. $f(x)$ on E (for example, $E = \mathbb{R}_+, \mathbb{R}, \mathbb{R}^d$, etc.), while we know how to generate a random number Y assuming values also in E with a p.d.f. $g(x)$. Let $c < \infty$ be such that

$$f(x) \leq cg(x), \qquad x \in E.$$

Clearly, $c > 1$ unless $f = g$. Similarly, as in the discrete case, without loss of generality, we may assume that if for some $x \in E$, $g(x) = 0$, then also $f(x) = 0$.

In the algorithm below, we generate $U \sim \mathcal{U}[0, 1)$ and independently Y with the p.d.f. $g(x)$. Similarly (as in the discrete case), we define the *acceptance region* as

$$A = \{cUg(Y) \leq f(Y)\} = \left\{ Ug(Y) \leq \frac{1}{c} f(Y) \right\}.$$

The basis of our algorithm will be the following fact, which we state for $E \subset \mathbb{R}^n$. Proposition 3.5.1 for the continuous case has the following form:

Proposition 3.5.3

$$\mathbb{P}(A) = 1/c \quad \text{and} \quad \mathbb{P}(Y \in B|A) = \int_B f(y)\, dy \qquad \text{for} \quad B \subset E.$$

Proof The joint distribution of (U, Y) has a p.d.f. $du \times g(y)\, dy$ on $[0, 1) \times E$. We have

$$\mathbb{P}(Y \in B|A) = \frac{\mathbb{P}(Y \in B, cUg(Y) \leq f(Y))}{\mathbb{P}(A)}$$

$$= \frac{\int_E \int_{[0,1)} \mathbb{1}(y \in B, cug(y) \leq f(y))\, du\, g(y)\, dy}{\mathbb{P}(A)}$$

$$= \frac{\int_B (f(y)/(cg(y)))\, g(y)\, dy}{\mathbb{P}(A)}$$

$$= \int_B f(y)\, dy\, \frac{1}{c\mathbb{P}(A)}\,.$$

Substituting $B = E$, we can see that $\mathbb{P}(A) = 1/c$. Since the equation holds for all $B \subset E$, we conclude that f is the desired p.d.f. $\square$

The following algorithm gives a recipe for generating a random number X with p.d.f. $(f(x), x \in E)$.

Algorithm 16 AR-c; generating $X \sim f(x)$

Require: Constant $c > 0$ such that $f(x) \le cg(x)$ and an algorithm for generating r.v. Y with
   p.d.f. $g(x)$
Ensure: $X \sim f(x)$
 1: **repeat**
 2:     Generate $Y \sim g$ and $U \sim \mathcal{U}(0, 1]$
 3: **until** $cUg(Y) \le f(Y)$
 4: **return** $X = Y$

Example 3.5.4 (Generating a Normal Random Number) A popular method for generating a standard normal random number $\mathcal{N}(0, 1)$ uses the AR-c algorithm. We first generate X with the half-normal distribution

$$f(x) = \frac{2}{\sqrt{2\pi}} e^{-\frac{x^2}{2}}, \qquad 0 < x < \infty \,. \tag{3.5.1}$$

It is the p.d.f. of $|Z|$, where Z has a standard normal distribution. We can take Y to be exponentially distributed $\mathrm{Exp}(1)$, which has the p.d.f. $g(x) = e^{-x}, x > 0$. We need to find the maximum of a function

$$\frac{f(x)}{g(x)} = \frac{2}{\sqrt{2\pi}} e^{x - \frac{x^2}{2}}, \qquad 0 < x < \infty \,,$$

which is equivalent to finding the maximum of $x - x^2/2$ in the region $x > 0$. The maximum is achieved at $x = 1$, which gives us

$$c = \max \frac{f(x)}{g(x)} = \sqrt{\frac{2e}{\pi}} \approx 1.32 \,.$$

The acceptance region is defined by $U \leq f(Y)/(cg(Y))$. Notice that in the simulation, we can use

$$\frac{f(x)}{cg(x)} = e^{x - x^2/2 - 1/2} = e^{-(x-1)^2/2} .$$

To generate a standard normal random number $Z \sim \mathcal{N}(0, 1)$, we need first to generate X with p.d.f. $f(x)$ and then draw $+1$ with probability $1/2$ and -1 with probability $1/2$ (see Exercise 3.T.30). Based on this, we can write the following algorithm to generate a random number $\mathcal{N}(0, 1)$. The functions $f(x)$, $g(x)$ and $cg(x)$ are depicted in Fig. 3.4. $\qquad\qquad\diamond$

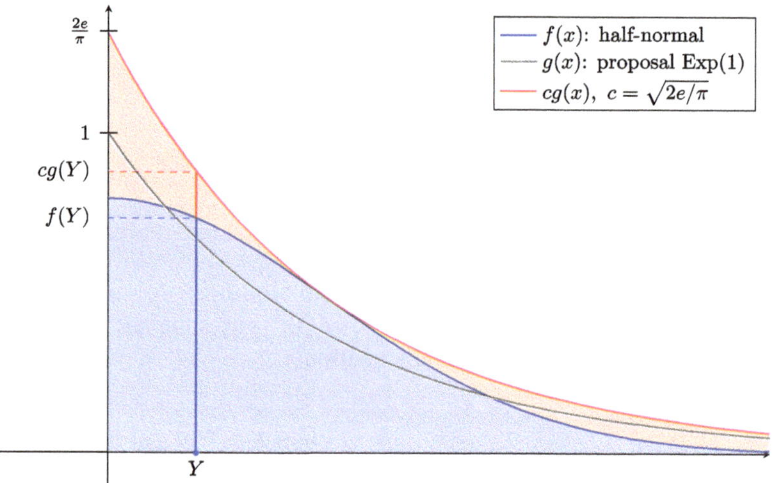

Fig. 3.4 Sampling X from half-normal distribution (with density $f(x)$): sample Y from $g \sim \text{Exp}(1)$ and accept it with probability $cg(Y)/f(Y)$

Algorithm 17 AR-N ; generating $X \sim \mathcal{N}(0, 1)$

Ensure: $X \sim \mathcal{N}(0, 1)$
 1: **repeat**
 2:      Generate $U \sim \mathcal{U}[0, 1)$
 3:      $Y = -\log(U)$
 4:      Generate $U \sim \mathcal{U}[0, 1)$
 5: **until** $U \leq \exp(-(Y - 1)^2/2)$
 6: Generate $U \sim \mathcal{U}[0, 1)$
 7: Generate $I = 2 \cdot \text{floor}(2 \cdot U) - 1$
 8: **return** $Z = IY$

Example 3.5.5 (Simulating Distribution Gamma(α, λ); Madras [138]) Consider the following cases:

Case $\alpha = 1$ This is the $\mathrm{Exp}(\lambda)$ distribution; we know how to simulate it.

Case $\alpha < 1$ Without loss of generality, we may assume that $\lambda = 1$, since $X = X'/\lambda \sim \mathrm{Gamma}(a, \lambda)$ for $X' \sim \mathrm{Gamma}(a, 1)$ (see Exercise 3.T.12). Let us consider the following p.d.f.:

$$g(x) = \begin{cases} Kx^{\alpha-1} & \text{for } 0 < x < 1, \\ Ke^{-x} & \text{for } x \geq 1. \end{cases}$$

The normalizing condition $\int_0^\infty g(x)\,dx = 1$ yields

$$K = \frac{\alpha e}{\alpha + e} \, .$$

It is convenient to write

$$g(x) = K(x^{\alpha-1}\mathbb{1}(0 < x < 1) + e^{-x}\mathbb{1}(x \geq 1)).$$

Set $g_1(x) = \alpha x^{\alpha-1}\mathbb{1}(0 < x < 1)$, $g_2(x) = ee^{-x}\mathbb{1}(x \geq 1)$, since

$$\int_0^1 x^{\alpha-1}\,dx = \frac{1}{\alpha}, \qquad \int_1^\infty e^{-x}\,dx = \frac{1}{e}.$$

We have

$$g(x) = p_1 g_1(x) + p_2 g_2(x) \, ,$$

where $p_1 = e/(\alpha + e)$ and $p_2 = \alpha/(\alpha + e)$. We can thus simulate Y with p.d.f. g in the following way: sample $U \sim \mathcal{U}([0, 1))$, if $U \leq p_1$ then sample from $g_1(x)$, otherwise sample from $g_2(x)$ (this is a simple case of a composition method presented later in Sect. 3.6). The reader is invited to write a procedure for generating a random number with p.d.f.s $g_1(x)$ and $g_2(x)$ respectively. Notice that there exists c such that

$$f(x) \leq cg(x), \qquad x \in \mathbb{R}_+.$$

To compute c we will look for

$$\min_{x>0} \frac{g(x)}{f(x)} = \min_{x>0} \frac{K(x^{\alpha-1}\mathbb{1}(0 < x < 1) + e^{-x}\mathbb{1}(x \geq 1))}{x^{\alpha-1}e^{-x}/\Gamma(\alpha)}$$

$$= \min_{x>0} K\Gamma(\alpha)(e^x\mathbb{1}(0 < x < 1) + x^{1-\alpha}\mathbb{1}(x \geq 1)) = K\Gamma(\alpha).$$

Hence $c = \frac{1}{K\Gamma(\alpha)}$ and so we can apply the AR-c algorithm to generate X. The reader is invited to complete the details and write the procedure.

Case $\alpha > 1$ Consider $g(x) = be^{-bx}\mathbb{1}(0 \le x < \infty)$ (i.e., $\mathrm{Exp}(b)$) with $b < 1$. In this case

$$c = c(b) = \sup_{x \ge 0}\frac{f(x)}{g(x)} = \sup_{x \ge 0}\frac{1}{b\Gamma(\alpha)}x^{\alpha-1}e^{-(1-b)x} < \infty\,.$$

We have first to find $x(b)$ maximizing $x^{\alpha-1}\exp(-(1-b)x)$, which is $x(b) = (\alpha-1)/(1-b)$. Hence

$$c = c(b) = \frac{1}{b\Gamma(\alpha)}\frac{(\alpha-1)^{\alpha-1}}{(1-b)^{\alpha-1}}e^{-(\alpha-1)} = \frac{1}{\Gamma(\alpha)}(\alpha-1)^{\alpha-1}e^{-(\alpha-1)}\frac{1}{b(1-b)^{\alpha-1}}\,.$$

Next, the constant b should be chosen such that $c(b)$ is minimal and so maximizing the numerator $b(1-b)^{\alpha-1}$ we obtain $b = \alpha^{-1}$. Hence the optimal c is $\alpha^{\alpha}e^{-(\alpha-1)}/\Gamma(\alpha)$.
$\lozenge$

Example 3.5.6 (Central Chi-square Distribution) For i.i.d. random variables $Z_1, \ldots, Z_d$ with a common $\mathcal{N}(0, 1)$ distribution, we have

$$X^2 \overset{\mathcal{D}}{=} Z_1^2 + \cdots + Z_d^2. \tag{3.5.2}$$

Since we showed how to sample standard normal random variables and gamma distributed random variables, we have two methods for sampling the central chi-square distribution:

(a) Method 1: we simply sample $\mathcal{N}(0, 1)$ distributed i.i.d. $Z_1, \ldots, Z_d$ (as in Example 3.5.4) and compute X^2 using (3.5.2).
(b) Method 2: A simple argument shows that Z_i^2 has distribution $\mathrm{Gamma}(1/2, 1/2)$, and thus

$$X^2 \sim \mathrm{Gamma}(d/2, 1/2). \tag{3.5.3}$$

In Example 3.5.5 we have shown how to sample a random variable with distribution $\mathrm{Gamma}(a, \lambda)$ for any $a > 0, \lambda > 0$. Note that in the case of even $d = 2k$, we are to sample from the distribution $\mathrm{Gamma}(k, 1/2)$ which is the $\mathrm{Erl}(k, 1/2)$ distribution, which we showed how to sample in Sect. 3.3.2 (roughly speaking, it is a sum of independent exponential random variables).

Note the following consequence of Method 2. To sample X with distribution $\mathrm{Gamma}\left(j + \frac{1}{2}, \frac{1}{2}\right)$ it is enough to sample $d = 2j+1$ random i.i.d. standard normal $\mathcal{N}(0, 1)$ random variables $Z_1, \ldots, Z_{2j+1}$ (this is because $\frac{d}{2} = j + \frac{1}{2}$) and take the sum of their squares. In other words, we have

$$Z_1^2 + \cdots + Z_{2j+1}^2 \sim \text{Gamma}\left(j + \frac{1}{2}, \frac{1}{2}\right). \tag{3.5.4}$$

$$\diamond$$

Unknown Constant(s) In many (practical) scenarios, we only know f, or f and g, up to normalizing constants, i.e.,

$$f(x) = \frac{1}{c_f} f^*(x), \quad g(x) = \frac{1}{c_g} g^*(x),$$

where $f^*(x)$ and $g^*(x)$ are known, but c_f and c_g are unknown (or infeasible to compute). In the method we need to have $c \in (0, \infty)$ such that $f(x) \le cg(x)$ for all x, which is equivalent to $f^*(x) \le c^* g^*(x)$ with $c^* = \frac{c \cdot c_f}{c_g}$. In other words, the acceptance region is

$$A = \left\{ U \le \frac{f(Y)}{cg(Y)} \right\} = \left\{ U \le \frac{f^*(Y)}{c^* g^*(Y)} \right\}.$$

Practically, it means that we can ignore the normalizing constants from the beginning: If we can find c^* such that

$$f^*(x) \le c^* g^*(x),$$

then it is correct to accept with probability $\frac{f^*(x)}{c^* g^*(x)}$. Note that the mean number of accept/reject trials is equal to c again (which equals $c_g c^* / c_f$). For discrete distributions the method works in a similar way, which we present in Algorithm 18.

Algorithm 18 AR-d; generating $X \sim (p_j)$, $p_j = \frac{1}{c_p} p_j^*$

Require: Constant $c^* > 0$ such that $p_j^* \le c^* q_j^*$ (where $q_j = \frac{1}{c_q} q_j^*$) and an algorithm for generating r.v. Y with probability function (q_j)
1: **repeat**
2:     Generate $Y \sim (q_j)$
3:     Generate $U \sim \mathcal{U}[0, 1)$ (independent from Y)
4: **until** $c^* U q_Y^* \le p_Y^*$
5: **return** $X = Y$

Example 3.5.7 Let

$$f(x) = \frac{1}{c_f} e^{-x^2/2} \left(1 - e^{-\sqrt{x^2+1}} \right), \quad x \in \mathbb{R}$$

(c_f - a normalizing constant, we do not know it). Define

$$f(x) = \frac{1}{c_f} f^*(x)$$

and take $g(x) = (2\pi)^{-1/2} e^{-x^2/2} := \frac{1}{c_g} e^{-x^2/2}$ (the p.d.f. of $\mathcal{N}(0, 1)$, which we know how to simulate). We have

$$\frac{f^*(x)}{g^*(x)} = 1 - e^{-\sqrt{x^2+1}} \le 1 = c^*.$$

As noted, the condition $U \le \frac{f(Y)}{cg(Y)}$ is equivalent to

$$U \le \frac{f^*(Y)}{c^* g^*(Y)} = \frac{e^{-Y^2/2}\left(1 - e^{-\sqrt{Y^2+1}}\right)}{e^{-Y^2/2}} = 1 - e^{-\sqrt{Y^2+1}}.$$

Algorithm 19 is a summary of the method. $\Diamond$

Algorithm 19 AR for Example 3.5.7

1: **repeat**
2:     Generate $Y \sim \mathcal{N}(0, 1)$
3:     Generate $U \sim \mathcal{U}[0, 1)$
4: **until** $U \le 1 - e^{-\sqrt{Y^2+1}}$
5: **return** Y

At this point, we encourage the reader to devise a procedure for simulating an r.v. with a Beta distribution using AR-c method.

In context of the Acceptance-Rejection method distribution p (the p.d.f. f) is often called a *target distribution*, whereas distribution q (the p.d.f. $g(x)$) is often called a *proposal distribution*.

3.6   Composition Method

The composition method is a versatile approach used when the distribution of a random variable X can be expressed as a mixture of simpler distributions. When X has a probability density function (p.d.f.) $f(x)$, it can be written as

$$f(x) = \sum_{j=1}^{n} p_j g_j(x),$$

where $g_j(x)$ are p.d.f.s (with n possibly infinite), and (p_j) is a probability function. To generate a random variable X with p.d.f. $f(x)$, first sample J according to the probability function (p_j). If $J = j$, then generate a random variable from $g_j(x)$; this result becomes the sampled value for X.

The composition method is widely applicable in simulation, especially in cases where

$$f(x) = \sum_{j=0}^{\infty} p_j(\lambda) g_j(x), \qquad (3.6.1)$$

where $p_j(\lambda) = \frac{\lambda^j}{j!} e^{-\lambda}$ represents the Poisson probability function and $g_j(x)$ is the p.d.f. of a gamma distribution, Gamma(α_j, β).

Example 3.6.1 (Non-central Chi-squared Distribution) The chi-squared distribution is widely used in fields like statistics and financial mathematics. While the standard chi-squared distribution has one parameter (degrees of freedom d), the non-central chi-squared distribution has two parameters: the non-centrality parameter $\lambda > 0$ and degrees of freedom $d > 0$.

For integer d, the non-central chi-squared distribution describes the distribution of $(Z_1 + a_1)^2 + \cdots + (Z_d + a_d)^2$, where $Z_1, \ldots, Z_d$ are i.i.d. standard normal random variables, and the non-centrality parameter is $\lambda = a_1^2 + \cdots + a_d^2$. The p.d.f. of a non-central chi-square distribution with non-centrality parameter $\lambda > 0$ is given by

$$f(x) = \frac{1}{2} e^{-(x+\lambda)/2} \left(\frac{x}{\lambda}\right)^{\frac{d}{2}-1} I_{\frac{d}{2}-1}(\sqrt{\lambda x}),\ x \geq 0, \qquad (3.6.2)$$

where $I_v(\cdot)$ is the modified Bessel function of the first kind with order v given by

$$I_v(y) = \left(\frac{y}{2}\right)^v \sum_{j=0}^{\infty} \frac{\left(\frac{y^2}{4}\right)^j}{j!\,\Gamma(v + j + 1)}.$$

It turns out (see, e.g., [138]) that the p.d.f. $f(x)$ can be represented in the form given by (3.6.1) with

$$p_j(\lambda) = \frac{\left(\frac{\lambda}{2}\right)^j}{j!} e^{-\lambda/2}, \qquad g_j(x) \sim \text{Gamma}\left(j + \frac{d}{2}, \frac{1}{2}\right). \qquad (3.6.3)$$

The non-central chi-square distribution can thus be represented as

$$\chi'^2 = X_N,$$

where $N \sim \text{Poi}(\frac{\lambda}{2})$ and $X_j \sim \text{Gamma}\left(j + \frac{d}{2}, \frac{1}{2}\right)$, with N independent of X_j. Thus, we can simulate it in two ways: either by directly computing $(Z_1 + a_1)^2 + \cdots + (Z_d + a_d)^2$ for i.i.d. standard normal random variables, or by sampling $2j + d$ i.i.d. standard normal random variables $Z_1, \ldots, Z_{2j+d}$ and computing $Z_1^2 + \cdots, Z_{2j+d}^2$ (since it has $\text{Gamma}\left(a, \frac{1}{2}\right)$ distribution with $a = \frac{2j+d}{2} = j + \frac{d}{2}$, see (3.5.3)).

It turns out that $f(x)$ given in (3.6.2) is a density even if d is not an integer—we then speak of a non-central chi-square distribution with non-centrality parameter $\lambda > 0$ and positive real-valued degrees of freedom d. It can still be represented in the form (3.6.1) with $p_j(\lambda)$ and $g_j(x)$ given in (3.6.3), thus we may simulate it using the above composition method—sampling $\text{Gamma}(a, b)$ with any positive parameters was provided in Example 3.5.6. $\diamond$

Example 3.6.2 (Dagpunar [28]) In atomic physics, the following p.d.f. is often used:

$$f(x) = C \sinh(\sqrt{xc})e^{-bx}, \qquad x \geq 0,$$

where $b, c > 0$ are fixed parameters, and

$$C = \frac{2b^{3/2}e^{-c/(4b)}}{\sqrt{\pi c}}.$$

Expanding $\sinh((xc)^{1/2})$ into a series gives

$$f(x) = C \sum_{j=0}^{\infty} \frac{(xc)^{(2j+1)/2}e^{-bx}}{(2j+1)!}$$

$$= C \sum_{j=0}^{\infty} \frac{\Gamma\left(j + \frac{3}{2}\right)c^{j+\frac{1}{2}}}{b^{j+\frac{3}{2}}(2j+1)!}g_j(x),$$

where $g_j(x)$ is the p.d.f. of $\text{Gamma}\left(j + \frac{3}{2}, b\right)$. Using the fact that $\Gamma(j + \frac{3}{2}) = (j - 1 + \frac{3}{2}) \ldots \left(\frac{3}{2}\right)\Gamma(\frac{3}{2})$ (see Abramowitz and Stegun [1]), and noting $\Gamma(\frac{3}{2}) = \frac{\pi^{1/2}}{2}$, we have

$$f(x) = C \sum_{j=0}^{\infty} \frac{\pi^{1/2}c^{j+\frac{1}{2}}}{2^{2j+1}j!b^{j+\frac{3}{2}}}g_j(x)$$

$$= \sum_{j=0}^{\infty} \frac{\left(\frac{c}{4b}\right)^j}{j!}e^{-c/(4b)}g_j(x),$$

resulting in a mixture of Gamma($j + \frac{3}{2}, b$) densities $g_j(t)$ with Poisson $c/(4b)$ mixing weights

$$p_j = \frac{\left(\frac{c}{4b}\right)^j}{j!} e^{-c/(4b)}.$$

To simulate the p.d.f. $g_j(t)$ consider $Y' = \text{Gamma}\left(\frac{2j+3}{2}, \frac{1}{2}\right)$. Note that $Y' \stackrel{\mathcal{D}}{=} Z_1^2 + \cdots Z_{2j+3}^2$ for i.i.d. standard normal random variables $Z_1, \ldots, Z_{2j+3}$—see (3.5.3)— thus we already know how to sample Y'. Finally, taking $Y = Y'/(2b)$ yields the Gamma($j + \frac{3}{2}, b$) distribution (cf. Exercise 3.T.12 (c)). $\Diamond$

Another variation occurs when the probability function $(p_n(\lambda))$ includes a parameter λ, as in the following example.

Example 3.6.3 Let $p_n(x) = \frac{x^n}{n!} e^{-x}$ for $n = 0, 1, \ldots$ represent the Poisson distribution with parameter x. Suppose this parameter is derived from a random variable $X > 0$ with a Gamma(α, λ) distribution. Then, the unconditional probability function is given by

$$p_n = \int_0^\infty \frac{x^n}{n!} e^{-x} \frac{\lambda^\alpha x^{\alpha-1}}{\Gamma(\alpha)} e^{-\lambda x} \, dx, \qquad n = 0, 1, \ldots.$$

From the definition of the gamma distribution, we obtain

$$p_n = \frac{\Gamma(\alpha + n)}{\Gamma(\alpha)\Gamma(n+1)} \left(\frac{1}{1+\lambda}\right)^\alpha \left(\frac{1}{1+\lambda}\right)^n$$

$$= \binom{n+\alpha-1}{n} \left(\frac{1}{1+\lambda}\right)^\alpha \left(\frac{1}{1+\lambda}\right)^n, \qquad n = 0, 1, \ldots,$$

demonstrating that the mixture follows an NB($\alpha, 1/(1 + \lambda)$) negative binomial distribution. $\Diamond$

3.7 Generating Gaussian Random Numbers

3.7.1 Box-Muller and Marsaglia-Bray Methods

We present two variants of the Box-Muller method for generating pairs of independent random variables Z_1, Z_2 with the common normal distribution $\mathcal{N}(0, 1)$. The approach is based on polar coordinates, as shown in the lemma below. Recall that if $\mathbf{X}$ is a d-dimensional random vector with density $f_\mathbf{X}(\mathbf{x})$, and $\mathbf{Y} = T(\mathbf{X})$ is a transformation of $\mathbf{X}$ through a smooth, invertible function T, then the density of $\mathbf{Y}$ is given by:

$$f_{\mathbf{Y}}(y) = f_{\mathbf{X}}(T^{-1}(y)) \cdot \left| \det \mathbf{J}_{T^{-1}}(y) \right|,$$

where $\mathbf{J}_{T^{-1}}(y)$ is the Jacobian matrix of T^{-1} at y. For details, refer to Proposition A.2.1.

Lemma 3.7.1 *Let Z_1 and Z_2 be independent random variables with a common standard normal distribution $\mathcal{N}(0, 1)$, and let (D, Θ) be their polar coordinate representation. Then D and Θ are independent, where D has the probability density function*

$$f(r) = re^{-r^2/2} \quad for\ r > 0, \tag{3.7.1}$$

and Θ is uniformly distributed over $\mathcal{U}[0, 2\pi)$.

Proof We have

$$Z_1 = D \cos \Theta,$$

$$Z_2 = D \sin \Theta.$$

The joint p.d.f. of the vector (Z_1, Z_2) is $\frac{1}{2\pi} \exp\left(-\frac{x_1^2 + x_2^2}{2}\right)$. Consider the transformation $x_1 = r \cos \theta$ and $x_2 = r \sin \theta$. The Jacobian of this transformation is

$$\det \begin{pmatrix} \frac{\partial x_1}{\partial r} & \frac{\partial x_1}{\partial \theta} \\ \frac{\partial x_2}{\partial r} & \frac{\partial x_2}{\partial \theta} \end{pmatrix} = \begin{vmatrix} \cos \theta & -r \sin \theta \\ \sin \theta & r \cos \theta \end{vmatrix} = r.$$

By formula (A.1), the random vector (D, Θ) has the joint p.d.f.

$$f_{(D,\Theta)}(r, \theta) = \frac{1}{2\pi} e^{-r^2/2} r \quad \text{for } 0 < \theta \le 2\pi,\ r > 0.$$

$\square$

The distribution with the p.d.f. in (3.7.1) is called the *Rayleigh distribution*. In the next lemma, we show how to generate random numbers with the Rayleigh distribution.

Lemma 3.7.2 *If D has the Rayleigh distribution, then $Y = D^2$ is exponentially distributed with $Exp(1/2)$. Consequently, if U is uniformly distributed as $\mathcal{U}[0, 1)$, then the random variable $(-2 \log U)^{1/2}$ has the Rayleigh distribution.*

Proof Notice that $Y = D^2$ is exponentially distributed as $\mathrm{Exp}(1/2)$ because

$$\mathbb{P}(D^2 > x) = \mathbb{P}(D > \sqrt{x}) = \int_{\sqrt{x}}^{\infty} r e^{-r^2/2} \, dr = e^{-x/2}, \qquad x > 0.$$

Since $(-2 \log U)$ is exponentially distributed as $\mathrm{Exp}(1/2)$, the square root of this random variable has the Rayleigh distribution. $\qquad\square$

As a result, we have the following fact, which forms the basis of the Box-Muller algorithm.

Proposition 3.7.3 *Let U_1 and U_2 be two independent random variables with the common uniform distribution $\mathcal{U}[0, 1)$, and define*

$$Z_1 = (-2 \log U_1)^{1/2} \cos(2\pi U_2),$$
$$Z_2 = (-2 \log U_1)^{1/2} \sin(2\pi U_2).$$

Then Z_1 and Z_2 are independent random variables with a common normal distribution $\mathcal{N}(0, 1)$.

Proof We have $Z_1 = D \cos \Theta$ and $Z_2 = D \sin \Theta$, where D and Θ are independent. Here, D has the Rayleigh distribution, and Θ is uniformly distributed over $(0, 2\pi]$. $\qquad\square$

To implement the Box-Muller algorithm, generate a random number Θ uniformly distributed on $(0, 2\pi]$ (straightforward) and D with the p.d.f. $r e^{-r^2/2}$ for $r > 0$, which can be generated using Lemma 3.7.2.

Algorithm 20 BM; Generate Z_1, Z_2 with a standard normal distribution $\mathcal{N}(0,1)$

1: Generate U_1, U_2 i.i.d. $\mathcal{U}[0, 1)$ random variables
2: Compute $D = -2 \log(U_1)$, $V = 2\pi U_2$
3: Set $Z_1 = \sqrt{D} \cos V$, $Z_2 = \sqrt{D} \sin V$
4: **return** Z_1, Z_2

The second method is a Box-Muller modification attributed to Marsaglia and Brey. It is based on the following fact.

Proposition 3.7.4 *Let*

$$Y_1 = \{-2 \log(V_1^2 + V_2^2)\}^{1/2} \frac{V_1}{(V_1^2 + V_2^2)^{1/2}},$$

$$Y_2 = \{-2 \log(V_1^2 + V_2^2)\}^{1/2} \frac{V_2}{(V_1^2 + V_2^2)^{1/2}},$$

where V_1 and V_2 are independent random variables with a common uniform distribution $\mathcal{U}[-1, 1)$. Then

$$(Z_1, Z_2) \overset{\mathcal{D}}{=} ((Y_1, Y_2) \mid V_1^2 + V_2^2 \leq 1)$$

is a pair of independent random variables with a common normal distribution $\mathcal{N}(0, 1)$.

The proof of the above is based on the following lemma. Let B_2 be a ball with radius 1 centered at the origin of $\mathbb{R}^2$.

Lemma 3.7.5 *If $(W_1, W_2) \sim \mathcal{U}(B_2)$, and (D', Θ') is the polar coordinate representation of (W_1, W_2), then D' and Θ' are independent, D'^2 has a uniform distribution $\mathcal{U}[0, 1)$, and Θ' is uniformly distributed $\mathcal{U}[0, 2\pi)$.*

Proof The joint distribution of (W_1, W_2) has p.d.f.

$$\frac{1}{\pi} \mathbb{1}((w_1, w_2) \in B_2),$$

and in polar coordinates

$$W_1 = D' \cos \Theta', \qquad W_2 = D' \sin \Theta'.$$

Using Proposition A.2.1, we establish a one-to-one mapping $\boldsymbol{g} : B_2 \setminus \{(0, 0)\} \rightarrow (0, 1) \times (0, 2\pi)$ with inverse $\boldsymbol{h} = \boldsymbol{g}^{-1}$. In this case, the Jacobian $J_{\boldsymbol{h}}(r, \theta) = r$ for $0 < r < 1$ and $0 \leq \theta \leq 2\pi$.

Consider the set $A = \{t_1 < \theta \leq t_2, \, a < r \leq b\}$; see Fig. 3.5 (left). Then

$$
\begin{aligned}
\mathbb{P}(t_1 < \Theta' \leq t_2, \, a < D' \leq b) &= \mathbb{P}((W_1, W_2) \in \boldsymbol{h}(A)) \\
&= \frac{1}{\pi} \int_{B_2} \mathbb{1}((x, y) \in \boldsymbol{h}(A)) \, dx \, dy \\
&= \frac{1}{\pi} \int_{t_1}^{t_2} \int_{a}^{b} r \, dr \, d\theta \\
&= \frac{1}{2\pi} (t_2 - t_1) \times (b^2 - a^2) \\
&= \mathbb{P}(t_1 < \Theta' \leq t_2) \mathbb{P}(a < D' \leq b).
\end{aligned}
$$

This shows that D' and Θ' are independent, and that Θ' is uniformly distributed as $\mathcal{U}[0, 2\pi)$. Furthermore, $\mathbb{P}(D' \leq x) = x^2 \mathbb{1}(0 \leq x \leq 1)$, meaning $(D')^2$ is uniformly distributed as $\mathcal{U}[0, 1)$. $\qquad\square$

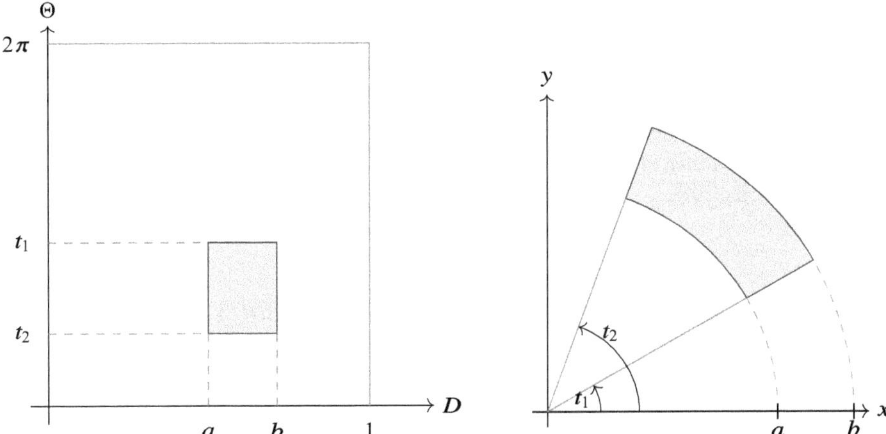

Fig. 3.5 Illustration for Lemma 3.7.5: space of parameters (D, Θ) (left) and their transformations (right)

The proof of Proposition 3.7.4 is based on the observation that, with the notation from Lemma 3.7.5, under the condition $V_1^2 + V_2^2 \leq 1$, we have $(D')^2 = V_1^2 + V_2^2 \sim \mathcal{U}[0, 1)$ and $\cos \Theta' = V_1/(V_1^2 + V_2^2)^{1/2}$. This leads to

$$\left((V_1^2 + V_2^2)^{1/2}, \frac{V_1}{(V_1^2 + V_2^2)^{1/2}} \right) \overset{\mathcal{D}}{=} (D', \cos \Theta'),$$

where $(D')^2$ has distribution $\mathcal{U}[0, 1)$ and $\Theta' \sim \mathcal{U}[0, 2\pi)$, thus Y_1 and Y_2 are essentially the same as Z_1 and Z_2 from Lemma 3.7.3. The Marsaglia-Bray method is summarized in Algorithm 21

Algorithm 21 MB (Marsaglia-Bray); generate Z_1, Z_2 with a standard normal distribution $\mathcal{N}(0,1)$

1: **repeat**
2:    Generate U_1, U_2 i.i.d. $\mathcal{U}[0, 1)$ random variables
3:    Compute $V_1 = 2U_1 - 1$, $V_2 = 2U_2 - 1$
4:    Set $X = V_1^2 + V_2^2$
5: **until** $X \leq 1$
6: $Y = \sqrt{(-2 \log X)/X}$
7: Set $Z_1 = V_1 \cdot Y$, $Z_2 = V_2 \cdot Y$
8: **return** Z_1, Z_2

Both the original Box-Muller and Marsaglia-Bray methods use logarithmic functions, but only the first one uses trigonometric functions. In general, the Marsaglia-Bray method is more efficient than the original Box-Muller method for most applications due to the computational savings from avoiding those trigonometric functions. The rejection step has a high acceptance rate (approximately 78.5% acceptance probability), so the method does not suffer significantly from efficiency loss due to rejections.

3.7.2   ITM Method for Normal Random Variables

To generate standard normal random numbers $\mathcal{N}(0, 1)$ using the ITM method, we need to invert the c.d.f. $\Phi(x)$. Unfortunately, there is no explicit formula for Φ^{-1}, so numerical methods are required. Various algorithms exist for computing $\Phi^{-1}(u)$. As is typical with numerical algorithms, it is important to consider their potential errors. The task is to solve the equation $\Phi(x) = u$ for a given $u \in (0, 1)$. A suitable starting point for this is Newton's root-finding algorithm for the function $x \to \Phi(x) - u$. We can make some simplifications here, specifically noting that since

$$\Phi^{-1}(1 - u) = -\Phi^{-1}(u),$$

it suffices to search for u in either $[1/2, 1)$ or $(0, 1/2)$. Newton's method is an iterative approach given by the recursion

$$x_{n+1} = x_n - \frac{\Phi(x_n) - u}{\phi(x_n)} = x_n + (u - \Phi(x_n))e^{0.5x_n^2 + c}, \tag{3.7.2}$$

where $c = \log \sqrt{2\pi}$. A recommended initial point is

$$x_0 = \pm\sqrt{\left|-1.6\log(1.0004 - (1 - 2u)^2)\right|}.$$

The sign $\pm$ depends on whether $u \geq 1/2$ (positive) or $u < 1/2$ (negative). This method requires the evaluation of both the function Φ and the exponential function.

In Algorithm 22, another approach, known as the Beasley-Springer-Moro algorithm for approximating the inverse of the normal distribution, is presented (following Glasserman [63]).

Algorithm 22 Beasley-Springer-Moro algorithm for approximating the inverse normal $x = \Psi^{-1}(u)$

Require: $u \in (0, 1)$
1: $a_0 = 2.50662823884;\ a_1 = -18.61500062529;\ a_2 = 41.39119773534;$
2: $a_3 = -25.44106049637;\ b_0 = -8.47351093090;\ b_1 = 23.08336743743;$
3: $b_2 = -21.06224101826;\ b_3 = 3.13082909833;\ c_0 = 0.3374754822726147$
4: $c_0 = 0.3374754822726147;\ c_1 = 0.9761690190917186;\ c_2 = 0.1607979714918209;$
5: $c_3 = 0.0276438810333863;\ c_4 = 0.0038405729373609;\ c_5 = 0.0003951896511919;$
6: $c_6 = 0.0000321767881768;\ c_7 = 0.0000002888167364;\ c_8 = 0.0000003960315187;$
7: $y = u - 0.5$
8: **if** $|y| < 0.42$ **then**
9: $\quad r = y^2$
10: $\quad x = \dfrac{((a_3 r + a_2)r + a_1)r + a_0}{(((b_3 r + b_2)r + b_1)r + b_0) + 1}$
11: **else**
12: $\quad r = u$
13: $\quad$ **if** $(y > 0)$ **then**
14: $\quad\quad r = 1 - u$
15: $\quad$ **end if**
16: $\quad r = \log(-\log(r))$
17: $\quad x = c_0 + r(c_1 + r(c_2 + r(c_3 + r(c_4 + r(c_5 + r(c_6 + r(c_7 + rc_8)))))))$
18: $\quad$ **if** $(y < 0)$ **then**
19: $\quad\quad x = -x$
20: $\quad$ **end if**
21: **end if**
22: **return** x

To further refine the result from the recursion in (3.7.2), an additional iteration can be applied, starting from the obtained value $x = \Phi^{-1}(u)$. Glasserman [63] notes that this step reduces the error to within 10^{-15}.

Inverse of the c.d.f. of a Standard Normal Random Variable in Python To evaluate $\Phi^{-1}(u)$ numerically for a given value of u, we can use the `ppf` function (which stands for "percent-point function") from the `scipy.stats.norm` library. An example of its usage is shown in Listing 3.1.

The specific algorithm used by `scipy.stats.norm.ppf` is based on Algorithm AS241 by Michael J. Wichura [214], published in 1988, which is widely used in statistical software for computing the inverse of the c.d.f. of the standard normal random variable.

Python Listing 3.1 Computing the inverse of the c.d.f. of the standard normal distribution

```python
from scipy.stats import norm
p = 0.95
quantile = norm.ppf(p)
print(quantile)
```

In brief, for probabilities p that are not too close to 0 or 1, rational functions (ratios of polynomials) are used to approximate the inverse of the c.d.f.; these are fast and accurate for most of the probability range. For values of p very close to 0 or 1 (the tails of the distribution), the algorithm switches to polynomial approximations specifically tailored for high precision in the tails. All methods are well-optimized, making the algorithm fast, reliable, and capable of achieving high precision across the entire probability range $[0, 1]$.

3.7.3   Generating Random Vectors $\mathcal{N}(\boldsymbol{\mu}, \boldsymbol{\Sigma})$

The following describes a method for generating random vectors with a multivariate normal distribution, denoted $\mathcal{N}(\boldsymbol{\mu}, \boldsymbol{\Sigma})$, where $\boldsymbol{\mu}$ is the mean vector and $\boldsymbol{\Sigma}$ is the covariance matrix. A random vector $\mathbf{X} \in \mathbb{R}^d$ is said to follow a multivariate normal distribution $\mathcal{N}(\boldsymbol{\mu}, \boldsymbol{\Sigma})$ if any linear combination $\sum_{j=1}^{d} a_j X_j$ has a univariate normal distribution $\mathcal{N}(\boldsymbol{a}^T \boldsymbol{\mu}, \boldsymbol{a}^T \boldsymbol{\Sigma} \boldsymbol{a})$, where $\boldsymbol{a}^T = (a_1, \ldots, a_d)$. Consequently, if $\mathbf{B}$ is a $d \times d$ matrix and $\mathbf{Y} = \mathbf{BX}$, then $\mathbf{Y}$ is also multivariate normal because any linear combination $\boldsymbol{a}^T \mathbf{Y} = \boldsymbol{a}^T \mathbf{BX} = (\boldsymbol{a}^T \mathbf{B})\mathbf{X}$ will be normally distributed. For a d-dimensional vector $\mathbf{X}$, $\boldsymbol{\mu}$ is a d-dimensional vector and $\boldsymbol{\Sigma}$ is a $d \times d$ covariance matrix, which is symmetric and non-negative definite.

Example 3.7.6 For $d = 2$, the covariance matrix $\boldsymbol{\Sigma}$ can be represented as:

$$\boldsymbol{\Sigma} = \begin{pmatrix} \sigma_1^2 & \rho\sigma_1\sigma_2 \\ \rho\sigma_1\sigma_2 & \sigma_2^2 \end{pmatrix},$$

where $\rho = \mathrm{Corr}(X_1, X_2)$ and $\sigma_i^2 = \mathrm{Var}(X_i)$ for $i = 1, 2$. If $\boldsymbol{\Sigma}$ is non-singular, then

$$\boldsymbol{\Sigma}^{-1} = \mathbf{C} = \frac{1}{1 - \rho^2} \begin{pmatrix} \frac{1}{\sigma_1^2} & -\frac{\rho}{\sigma_1\sigma_2} \\ -\frac{\rho}{\sigma_1\sigma_2} & \frac{1}{\sigma_2^2} \end{pmatrix}.$$

To generate such a two-dimensional normal vector $\mathbf{X} \sim \mathcal{N}(\mathbf{0}, \boldsymbol{\Sigma})$, let

$$X_1 = \sigma_1(\sqrt{1 - |\rho|}Z_1 + \sqrt{|\rho|}Z_3), \quad X_2 = \sigma_2(\sqrt{1 - |\rho|}Z_2 + \mathrm{sign}(\rho)\sqrt{|\rho|}Z_3),$$

where Z_1, Z_2, Z_3 are independent standard normal random variables, $\mathcal{N}(0, 1)$. The fact that $(X_1, X_2)^T$ has a two-dimensional normal distribution is straightforward. Verification of the covariance matrix of this vector is left to the reader. $\qquad \diamond$

For general d, unless the covariance matrix has a special structure, we can use the following procedure to generate a random vector $\mathbf{X} = (X_1, \ldots, X_d)^T \sim \mathcal{N}(\boldsymbol{\mu}, \boldsymbol{\Sigma})$.

Every symmetric, non-negative definite covariance matrix $\mathbf{\Sigma}$ can be factorized as

$$\mathbf{\Sigma} = \mathbf{A}\mathbf{A}^T \tag{3.7.3}$$

for some matrix $\mathbf{A}$.

Remark 3.7.7 Factorization (3.7.3) is not unique. If we require that $\mathbf{A}$ is positive-definite, then we denote it by $\mathbf{\Sigma}^{1/2}$, which can be computed by the diagonalization method, as discussed later in Sect. 4.1.2.1. ∎

Proposition 3.7.8 *Let* $\mathbf{Z} = (Z_1, \ldots, Z_d)^T$ *be a vector of independent standard normal variables,* $N(0, 1)$. *For* $\mathbf{A}$ *from (3.7.3), the random vector*

$$\mathbf{X} = \mathbf{A}\mathbf{Z} + \boldsymbol{\mu} \tag{3.7.4}$$

follows the distribution $N(\boldsymbol{\mu}, \mathbf{\Sigma})$.

Proof The random vector $\mathbf{A}\mathbf{Z} + \boldsymbol{\mu}$ has a joint multivariate normal distribution since every linear combination of its elements is normally distributed. The mean vector is given by

$$\mathbb{E}(\mathbf{A}\mathbf{Z} + \boldsymbol{\mu}) = \mathbf{A}\mathbb{E}(\mathbf{Z}) + \boldsymbol{\mu} = \boldsymbol{\mu},$$

and the covariance matrix, omitting $\boldsymbol{\mu}$, is

$$\mathbb{E}(\mathbf{A}\mathbf{Z})(\mathbf{A}\mathbf{Z})^T = \mathbb{E}(\mathbf{A}\mathbf{Z}\mathbf{Z}^T\mathbf{A}^T) = \mathbf{A}\,\mathrm{Id}\,\mathbf{A}^T = \mathbf{\Sigma}.$$

$\square$

To compute a 'root' matrix $\mathbf{A}$ for the covariance matrix $\mathbf{\Sigma}$, one can use procedures available in many programming packages, often based on Cholesky decomposition, resulting in a lower triangular matrix. For example, one can verify that $\mathbf{\Sigma} = \mathbf{A}\mathbf{A}^T$ for

$$\mathbf{\Sigma} = \begin{pmatrix} 1 & 1 & 1 \\ 1 & 4 & 1 \\ 1 & 1 & 8 \end{pmatrix}, \quad \mathbf{A} = \begin{pmatrix} 1 & 0 & 0 \\ 1 & \sqrt{3} & 0 \\ 1 & 0 & \sqrt{7} \end{pmatrix}.$$

In `Python`, the Cholesky decomposition is available in the `NumPy` library and can be computed using the command `A = np.linalg.cholesky(Sigma)`. For the covariance matrix given above, this numerically yields:

$$\mathbf{A} = \begin{pmatrix} 1 & 0 & 0 \\ 1 & 1.732 & 0 \\ 1 & 0 & 2.645 \end{pmatrix}.$$

Note that Cholesky decomposition is not the only way to represent $\boldsymbol{\Sigma} = \mathbf{A}\mathbf{A}^T$. If $\boldsymbol{\Sigma}$ is positive-definite and hence non-singular, the density function of $\mathbf{X} \sim \mathcal{N}(\mathbf{0}, \boldsymbol{\Sigma})$ is

$$f(\mathbf{x}) = f(x_1, \ldots, x_d) = \frac{1}{\sqrt{(2\pi)^d \det(\boldsymbol{\Sigma})}} e^{-\frac{1}{2}\sum_{j,k=1}^{d} c_{jk}x_j x_k}, \tag{3.7.5}$$

where $\mathbf{C} = (c_{jk})_{j,k=1}^{d} = \boldsymbol{\Sigma}^{-1}$.

For an i.i.d. standard normal vector $\mathbf{Z} = (Z_1, \ldots, Z_d)$, its density function is

$$\frac{1}{(2\pi)^{d/2}} \exp\left(-\frac{z_1^2 + \cdots + z_d^2}{2}\right) = \frac{1}{(2\pi)^{d/2}} \exp\left(-\frac{\mathbf{z}^T \mathbf{z}}{2}\right).$$

If $\boldsymbol{\Sigma} = \mathbf{A}\mathbf{A}^T$, then $\mathbf{X} = \mathbf{A}\mathbf{Z}$ has a multivariate normal distribution $\mathcal{N}(\mathbf{0}, \boldsymbol{\Sigma})$. The Jacobian of the mapping $\mathbf{z} = \mathbf{A}^{-1}\mathbf{x}$ is $\det \mathbf{A}$, and we can use Proposition A.2.1 to derive (3.7.5).

Example 3.7.9 (Bivariate Normal) Consider a bivariate normal distribution (X_1, X_2) with mean vector $(0, 0)$ and covariance matrix

$$\boldsymbol{\Sigma} = \begin{pmatrix} \sigma_1^2 & \rho \\ \rho & \sigma_2^2 \end{pmatrix},$$

where $\sigma_1^2 = \mathbb{V}\mathrm{ar}(X_1)$, $\sigma_2^2 = \mathbb{V}\mathrm{ar}(X_2) = 1$, and ρ is the correlation between X_1 and X_2. Figure 3.6 shows 500 samples from this distribution for three cases: (a) the standard bivariate normal; (b) with $\sigma_1^2 = 3$, $\sigma_2^2 = 1$, and $\rho = 0.6$; and (c) with $\sigma_1^2 = 3$, $\sigma_2^2 = 1$, and $\rho = 0.9$. The example is implemented in `ch3_bivariate_normal.py`.   ◊

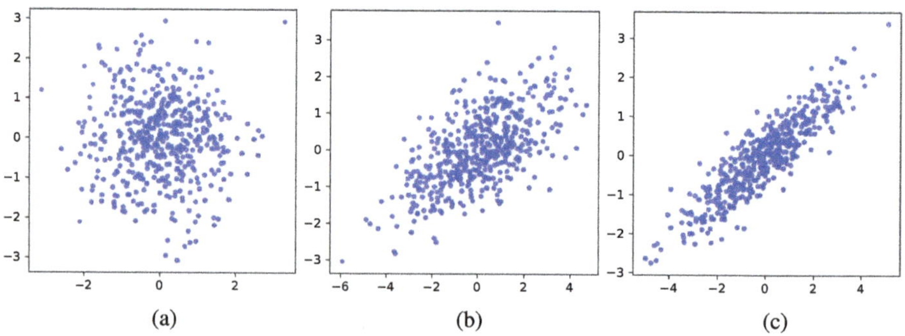

Fig. 3.6 Scatter plots of 500 samples from bivariate normal distributions under three different settings. **(a)** Standard normal ($\sigma_1^2 = \sigma_2^2 = 1$, $\rho = 0$). **(b)** $\sigma_1^2 = 3$, $\sigma_2^2 = 1$, $\rho = 0.6$. **(c)** $\sigma_1^2 = 3$, $\sigma_2^2 = 1$, $\rho = 0.9$

3.8 More on Uniform Distributions On: Balls, Spheres, and Ellipsoids

In this section, we explore methods for generating random points within various geometric objects, such as the d-ball, $(d-1)$-sphere, and d-ellipsoid, avoiding the use of acceptance-rejection methods.

We consider geometric objects in $\mathbb{R}^d$. For d-dimensional objects, their measure is the volume, denoted by $\mathrm{Vol}(\cdot)$. Let $B \subset \mathbb{R}^d$ be a subset with $0 < \mathrm{Vol}(B) < \infty$. Following the concepts introduced in Sect. 3.4, a random point $X \in B$ is said to be uniformly distributed on B, denoted $\mathcal{U}(B)$, if its p.d.f. is

$$f(x) = \frac{1}{\mathrm{Vol}(B)}\mathbb{1}(x \in B),$$

implying $\mathbb{P}(X \in D) = \frac{\mathrm{Vol}(D)}{\mathrm{Vol}(B)}$ for all $D \subset B$. However, challenges arise when dealing with objects in $\mathbb{R}^d$ of lower dimensions, such as those of dimension $d - 1$.

A typical problem is generating a uniformly distributed point in the d-ball defined as:

$$B_d = \left\{ \mathbf{x} = (x_1, \ldots, x_d) : \sum_{j=1}^{d} x_j^2 \le 1 \right\},$$

which has volume

$$\mathrm{Vol}(B_d) = \frac{\pi^{d/2}}{\Gamma(d/2 + 1)},$$

or on the $(d-1)$-sphere defined as

$$S_{d-1} = \left\{ \mathbf{x} = (x_1, \ldots, x_d) : \sum_{j=1}^{d} x_j^2 = 1 \right\},$$

which has surface area

$$\mathrm{Area}(S_{d-1}) = \frac{d\pi^{d/2}}{\Gamma(d/2 + 1)} = \frac{2\pi^{d/2}}{\Gamma(d/2)}.$$

A random point on S_{d-1} is said to have the uniform distribution $\sigma_{d-1}(d\mathbf{x})$ if its p.d.f. is $f(\mathbf{x}) = \frac{\Gamma(d/2)}{2\pi^{d/2}}$ for $\mathbf{x} \in S_{d-1}$.

Change of Variables Formula: General Theory We present an extension of the well-known change of variables formula (A.1) previously applied in this chapter. We consider subsets $C_k \subset \mathbb{R}^d$ with dimension $k = d - 1$ or $k = d$. For $k = d - 1$, these subsets are surfaces, such as $(d-1)$-spheres (hyper-spheres), and for $d=2$, they represent curves. For $k=d$, they correspond to d-balls. Parametric representations are used:

$$\mathbf{x}(\mathbf{t}) = (x_1(t_1, \ldots, t_k), \ldots, x_d(t_1, \ldots, t_k)), \qquad \mathbf{t} = (t_1, \ldots, t_k) \in \Delta_k,$$

where $\Delta_k \subset \tilde{\Delta}_k$ and $\mathbf{x}(\mathbf{t})$ is smooth on $\tilde{\Delta}_k$. For $k = d-1$, the measure is the surface area, denoted Area$(\cdot)$. Integrals over these objects are expressed as $\int_C f\, dV$ for $k = d$ and $\int_C f\, dA$ for $k = d-1$. Note that dV is an element of volume and dA is an element of surface (or line).

Let

$$\boldsymbol{J}(\mathbf{t}) = \left(\frac{\partial x_j(\mathbf{t})}{\partial t_i}\right)_{i=1,\ldots,k;\; j=1,\ldots,d} \tag{3.8.1}$$

be the *matrix of partial derivatives* (of size $k \times d$), and set

$$\sqrt{g(\mathbf{t})} = \sqrt{|\det(\boldsymbol{J}(\mathbf{t})\boldsymbol{J}^T(\mathbf{t}))|}.$$

The key formula here for a change of variable for a continuous function f on C_k when $k = d-1$ is:

$$\int_{C_{d-1}} f\, dA = \int_{\Delta_{d-1}} f(\mathbf{x}(\mathbf{t}))\sqrt{g(\mathbf{t})}\, d\mathbf{t}. \tag{3.8.2}$$

For $k = d$, $\sqrt{g(\mathbf{t})}$ is the absolute value of a determinant of a classical Jacobian, thus

$$\int_{C_d} f\, dV = \int_{\Delta_d} f(\mathbf{x}(\mathbf{t}))\sqrt{g(\mathbf{t})}\, d\mathbf{t} = \int_{\Delta_k} f(\mathbf{x}(\mathbf{t}))|\det(\boldsymbol{J}(\mathbf{t}))|\, d\mathbf{t}. \tag{3.8.3}$$

For $d = 2$, we consider the length of curves ($k = d-1$). The integral in this context is referred to as a line integral. Suppose a curve or arc C_k is given in parametric form $(x(t), y(t)) \in \mathbb{R}^2$ for $t \in [t_1, t_2]$. We have

$$\boldsymbol{J}(t) = \left(\frac{dx(t)}{dt}, \frac{dy(t)}{t}\right),$$

and thus the length of such a curve or arc is given by

$$\text{Area}(C_k) = \int_{C_k} 1\, dA = \int_{t_1}^{t_2} \sqrt{\left(\frac{dx(t)}{dt}\right)^2 + \left(\frac{dy(t)}{dt}\right)^2}\, dt.$$

Example 3.8.1 Consider the unit circle

$$S_1 = \{(x_1, x_2) : x_1^2 + x_2^2 = 1\},$$

parametrized by $x_1 = \cos\theta$, $x_2 = \sin\theta$, where $\theta \in [0, 2\pi)$. The matrix of derivatives is $\boldsymbol{J} = (-\sin\theta, \cos\theta)$, and

$$\boldsymbol{J}^T(\theta) = \begin{pmatrix} -\sin\theta \\ \cos\theta \end{pmatrix}.$$

Then $dA = \sqrt{g(\theta)}\, d\theta$, so

$$dA = \sqrt{(-\sin\theta)^2 + (\cos\theta)^2}\, d\theta = d\theta.$$

The full arc length of the circle is 2π. Denote by $\sigma_1(\cdot)$ the uniform distribution on S_1, with p.d.f.

$$f(x_1, x_2) = \begin{cases} \frac{1}{2\pi} & \text{if } (x_1, x_2) \in S_1, \\ 0 & \text{otherwise.} \end{cases}$$

Thus, $d\sigma_1(s) = \frac{1}{2\pi}\, d\theta$. This identification of a segment C on S_1 by a central angle θ complements the material introduced in Sect. 3.7. $\Diamond$

Example 3.8.2 Consider a line segment $[A_0, A_1]$ in $\mathbb{R}^2$, where $A_0 = (x_0, y_0)$ and $A_1 = (x_1, y_1)$. Without loss of generality, set $x_0 = 0$, $y_0 = 0$. Parameterize this line segment by $y = \frac{y_1}{x_1}t$, where $t \in [0, x_1]$, i.e., $\mathbf{x}(t) = (t, \frac{y_1}{x_1}t)$. Then the matrix of derivatives is $\boldsymbol{J} = (1, \frac{y_1}{x_1})$ and hence

$$\sqrt{g(t)} = \sqrt{\boldsymbol{J}(t)\boldsymbol{J}^T(t)} = \sqrt{1 + \left(\frac{y_1}{x_1}\right)^2}.$$

The length between points A_0 and $\frac{y_1}{x_1}x$ is

$$L(x) = \int_0^x \sqrt{1 + \left(\frac{y_1}{x_1}\right)^2}\, dt = \frac{x}{x_1}l_1, \quad x \le x_1,$$

where $l_1 = \sqrt{(x_1)^2 + (y_1)^2}$, which is the length of the segment. The random point $\mathbf{X}$ on the line segment $[A_0, A_1]$ is (Ux_1, Uy_1), where $U \sim \mathcal{U}[0, 1)$. Introduce a natural parametrization $t = (x/x_1)l_1$ on the line segment $[A_0, A_1]$, which is the distance of t from A_0. In this parametrization, the random point is Ul_1, which has p.d.f. $f(t) = 1/l_1$ for $t \in [A_0, A_1]$. Consider the c.d.f. $F(t) = \frac{L(t)}{l_1}$ for $t \in [0, l_1]$

and a random number U uniformly distributed as $\mathcal{U}[0, 1)$. Then $F^{-1}(U)$ is the desired random point, i.e., the solution of $U = x/l_1$ or $x = Ul_1$. $\Diamond$

Example 3.8.3 Continuing with Example 3.8.2, let C be a piecewise broken line $[A_0, A_1], [A_1, A_2], \ldots, [A_{n-1}, A_n]$, where $A_i = (x_i, y_i)$. The length of the i-th segment is l_i. Let t be the distance of a point along the line starting from the point A_0. For $t \in [l_1 + \cdots + l_{i-1}, l_1 + \cdots + l_{i-1} + l_i]$, it lies on the line segment $[A_{i-1}, A_i]$. Then a composition method can express the random point $\mathbf{X}$ on C. Choose the i-th segment with probability $l_i/(l_1 + \cdots + l_n)$ and place $\mathbf{X}$ uniformly on this segment. Thus, it has p.d.f.

$$f(t) = \sum_{i=1}^{n} \frac{l_i}{l_1 + \cdots + l_n} \frac{1}{l_i} = \frac{1}{l_1 + \cdots + l_n}, \quad t \in C.$$

$\Diamond$

3.8.1 Generating a Random Point in B_3 and on S_2

Consider the case $d = 3$. For a ball B_3, its volume is $4\pi/3$. Here, *spherical coordinates* are given by

$$x = r \cos \theta \sin \phi, \tag{3.8.4}$$
$$y = r \sin \theta \sin \phi,$$
$$z = r \cos \phi,$$

where $r > 0$, $\theta \in [0, 2\pi)$, and $\phi \in [0, \pi)$. The Jacobian is $r^2 \sin \phi$, so the volume element is $dV = r^2 \, dr \sin \phi \, d\phi \, d\theta$.

On the sphere S_2, we have spherical coordinates

$$x = \cos \theta \sin \phi, \tag{3.8.5}$$
$$y = \sin \theta \sin \phi,$$
$$z = \cos \phi.$$

Here $\theta \in [0, 2\pi)$ and $\phi \in [0, \pi)$. In this case, the area element is $dA = \sin \phi \, d\phi \, d\theta$.

Denote the mapping (3.8.4) from spherical to Cartesian coordinates by $\boldsymbol{h}$, and let $\boldsymbol{g} = \boldsymbol{h}^{-1}$. Let $f(\mathbf{x}) = \frac{3}{4\pi} \mathbb{1}(\mathbf{x} \in B_3)$ be the p.d.f. of the uniform distribution $\mathcal{U}(B_3)$. Using the change of variables formula (A.1), we immediately obtain the following result. In this proposition, $\mathbf{X} \in B_3$ is a random point in the ball B_3, and (D, Φ, Θ) is the point in spherical coordinates, i.e.,

$$(D \cos \Theta \sin \Phi, \, D \sin \Theta \sin \Phi, \, D \cos \Phi).$$

Proposition 3.8.4 *Let $C \subset B_3$ and $f(\mathbf{x})$ be the p.d.f. of the uniform distribution $\mathcal{U}(B_3)$. Then*

$$\int_C f(\mathbf{x}) \, d\mathbf{x} = \int \frac{3}{4\pi} \, d\mathbf{x} = \int_{gC} 3r^2 \, dr \, \frac{1}{2} \sin \phi \, d\phi \, \frac{1}{2\pi} \, d\theta. \tag{3.8.6}$$

As a result, (D, Φ, Θ) are independent random variables, $D \overset{\mathcal{D}}{=} U^{1/3}$, Φ has a p.d.f. $\frac{1}{2} \sin \phi$, and Θ is uniformly distributed on $[0, 2\pi)$. We have $\Phi \overset{\mathcal{D}}{=} \arccos(1 - 2U)$, where $U \sim \mathcal{U}[0, 1)$.

Proof Formula (3.8.6) follows from (A.1). Hence, we see that D has a p.d.f. $3r^2 \mathbb{1}(r \in (0, 1])$. Conversely, $\mathbb{P}(U^{1/3} \le x) = x^3$ for $x \in (0, 1]$, which has a p.d.f. $3r^2$. The c.d.f. F of Φ is

$$F(x) = \frac{1}{2} \int_0^x \sin \phi \, d\phi = \frac{1}{2}(1 - \cos x),$$

so $F^{-1}(y) = \arccos(1 - 2y)$. Hence $\Phi \overset{\mathcal{D}}{=} \arccos(1 - 2U)$, where $U \sim \mathcal{U}[0, 1)$. $\quad\square$

For the sphere S_2, the surface area is $2\pi^{3/2}/(\pi^{1/2}/2) = 4\pi$ since $\Gamma(3/2) = \sqrt{\pi}/2$. Thus, the uniform distribution $\sigma_2(\cdot)$ has a p.d.f. $f(\mathbf{x}) = \frac{1}{4\pi}$, $\mathbf{x} \in S_2$. Here, the *spherical coordinates* are

$$x = \cos \theta \sin \phi, \qquad y = \sin \theta \sin \phi, \tag{3.8.7}$$

where $\theta \in [0, 2\pi)$ and $\phi \in [0, \pi)$. Notice that for a point $(x, y, z) \in S_2$,

$$z^2 = 1 - x^2 - y^2 = 1 - (\cos \theta \sin \phi)^2 - (\sin \theta \sin \phi)^2 = (\cos \phi)^2.$$

Hence, $z = \cos \phi$. Denote the mapping (3.8.7) from spherical to Cartesian coordinates by $\mathbf{h}$. The matrix of derivatives is

$$J(\phi, \theta) = \begin{pmatrix} \frac{\cos \theta \sin \phi}{d\phi} & \frac{\sin \theta \sin \phi}{d\phi} & \frac{\cos \phi}{d\phi} \\ \frac{\cos \theta \sin \phi}{d\theta} & \frac{\sin \theta \sin \phi}{d\theta} & \frac{\cos \phi}{d\theta} \end{pmatrix} = \begin{pmatrix} \cos \theta \cos \phi & \sin \theta \cos \phi & -\sin \phi \\ -\sin \theta \sin \phi & \cos \theta \sin \phi & 0 \end{pmatrix}. \tag{3.8.8}$$

Noting that

$$\cos^2 \theta \cos^2 \phi + \sin^2 \theta \cos^2 \phi + \sin^2 \phi = \cos^2 \phi (\cos^2 \theta + \sin^2 \theta) + \sin^2 \phi = 1$$

we have

$$\det \boldsymbol{J}(\phi, \theta)\,\boldsymbol{J}^T(\phi, \theta) = \det \begin{pmatrix} 1 & 0 \\ 0 & \sin^2 \phi \end{pmatrix},$$

thus $\sqrt{|\det \boldsymbol{J}(\phi, \theta)\,\boldsymbol{J}^T(\phi, \theta)|} = \sin \phi$. Hence, the surface element is

$$dA = \sin \phi \, d\phi \, d\theta.$$

The uniform distribution σ_2 on the sphere S_2, by the change of variable formula (3.8.2), in spherical coordinates has the p.d.f.

$$f(\theta, \phi)\, d\theta \, d\phi = \frac{1}{4\pi}\, d\mathbf{x} = \frac{1}{4\pi} \sin \phi \, d\theta \, d\phi = \frac{1}{2\pi}\, d\theta \, \frac{1}{2} \sin \phi \, d\phi, \qquad (3.8.9)$$

where $\theta \in [0, 2\pi)$ and $\phi \in [0, \pi)$. Let (Φ, Θ) be a random point on S_2 in spherical coordinates. From (3.8.9), we see that Θ and Φ are independent. The marginal p.d.f. of Θ is $1/(2\pi)$, and Φ is $\frac{1}{2} \sin \phi$.

Remark 3.8.5 An old question was how to simulate Φ. Note that a naïve (but incorrect) approach would be to simulate $\Phi \sim \mathcal{U}[0, \pi)$. $\blacksquare$

The following fact serves as the foundation for the so-called Marsaglia algorithm, a technique for generating random points on the sphere S_2.

Proposition 3.8.6 *Suppose that Z_1, Z_2, Z_3 are i.i.d. standard normal $\mathcal{N}(0, 1)$ random variables, and let*

$$X_1 = \frac{Z_1}{\sqrt{Z_1^2 + Z_2^2 + Z_3^2}}, \qquad X_2 = \frac{Z_2}{\sqrt{Z_1^2 + Z_2^2 + Z_3^2}}, \qquad X_3 = \frac{Z_3}{\sqrt{Z_1^2 + Z_2^2 + Z_3^2}}.$$

Then (X_1, X_2, X_3) is a random point on the sphere S_2.

Proof Consider the mapping $\boldsymbol{g}_1$ that transforms a point $\mathbf{z} \in \mathbb{R}^3 - \{\mathbf{0}\}$ from Euclidean to spherical coordinates:

$$\mathbb{R}^3 - \{\mathbf{0}\} \ni \mathbf{z} \to \boldsymbol{y} = (r, \theta, \phi) \in (0, \infty) \times [0, 2\pi) \times [0, \pi), \qquad (3.8.10)$$

where

$$\mathbf{z} = (r \cos \theta \sin \phi, r \sin \theta \sin \phi, r \cos \phi).$$

The Jacobian of $\boldsymbol{h}_1 = \boldsymbol{g}_1^{-1}$ is $J_{\boldsymbol{h}_1}(\boldsymbol{y}) = r^2 \sin \phi$. The joint p.d.f. of (Z_1, Z_2, Z_3) is

$$\frac{1}{(2\pi)^{3/2}} \exp\left(-\frac{z_1^2 + z_2^2 + z_3^2}{2} \right).$$

Applying the change of variables formula (A.1), we get

$$\frac{1}{(2\pi)^{3/2}} \exp\left(-\frac{z_1^2 + z_2^2 + z_3^2}{2}\right) dz_1 dz_2 dz_3$$

$$= \frac{1}{(2\pi)^{3/2}} \exp\left(-\frac{r^2}{2}\right) r^2\, dr\, \sin\phi\, d\phi\, d\theta.$$

Let C be defined as

$$C = \{(z_1, z_2, z_3) \in \mathbb{R}^3 : \left(\frac{z_1}{\sqrt{z_1^2 + z_2^2 + z_3^2}}, \frac{z_2}{\sqrt{z_1^2 + z_2^2 + z_3^2}}, \frac{z_3}{\sqrt{z_1^2 + z_2^2 + z_3^2}}\right) \in E\},$$

where E is a subset of S_2. In spherical coordinates, we have

$$z_1 = r\cos\theta\sin\phi, \quad z_2 = r\sin\theta\sin\phi, \quad z_3 = r\cos\phi,$$

and the variables

$$x_1 = \frac{z_1}{\sqrt{z_1^2 + z_2^2 + z_3^2}} = \cos\theta\sin\phi, \quad x_2 = \frac{z_2}{\sqrt{z_1^2 + z_2^2 + z_3^2}} = \sin\theta\sin\phi,$$

and

$$x_3 = \frac{z_3}{\sqrt{z_1^2 + z_2^2 + z_3^2}} = \cos\phi$$

are independent of r. Using the change of variables formula (A.1), we then have

$$\int_E \frac{1}{(2\pi)^{3/2}} \exp\left(-\frac{x_1^2 + x_2^2 + x_3^2}{2}\right) dA$$

$$= \int_C \frac{1}{(2\pi)^{3/2}} \exp\left(-\frac{r^2}{2}\right) r^2\, dr\, \sin\phi\, d\phi\, d\theta$$

$$= \int_0^\infty r^2 \exp\left(-\frac{r^2}{2}\right) dr \int_{C'} \frac{1}{(2\pi)^{3/2}} \sin\phi\, d\phi\, d\theta,$$

where $C = (0, \infty) \times C'$ and $C' \subset C$. Let $g_2 : S_2 \to [0, 2\pi) \times [0, \pi)$ map from Euclidean to spherical coordinates. Using the fact that

$$\int_0^\infty r^2 e^{-r^2/2}\, dr = \sqrt{\frac{\pi}{2}},$$

we get

$$\int_C \frac{1}{(2\pi)^{3/2}} \exp\left(-\frac{r^2}{2}\right) r^2\, dr \sin\phi\, d\phi\, d\theta = \frac{1}{4\pi} \int_{g_2(E)} \sin\phi\, dr d\theta d\phi$$

$$= \frac{1}{4\pi} \int_E dA,$$

which completes the proof because $\frac{1}{4\pi} dA$ is the p.d.f. of σ_2. $\qquad\square$

Remark 3.8.7 Suppose that $\mathbf{X}$ is a random point on S_2 and let $\mathbf{P}$ be an orthonormal matrix ($\mathbf{P}^{-1} = \mathbf{P}^T$). Then for $\mathbf{Z} = (Z_1, Z_2, Z_3)^T$, where Z_1, Z_2, Z_3 are i.i.d. standard normal variables, we have $\mathbf{PZ} \overset{\mathcal{D}}{=} \mathbf{Z}$. Therefore, a random point $\mathbf{X}$ on S_2 satisfies $\mathbf{X} \overset{\mathcal{D}}{=} \mathbf{PX}$. $\qquad\blacksquare$

3.8.2   Spherical Coordinates

In general, a point $\mathbf{x} = (x_1, \ldots, x_d) \in \mathbb{R}^d \setminus \{\mathbf{0}\}$ in spherical coordinates is expressed by

$$x_1 = r \cos\phi_1,$$

$$x_2 = r \sin\phi_1 \cos\phi_2,$$

$$x_3 = r \sin\phi_1 \sin\phi_2 \cos\phi_3,$$

$$\vdots$$

$$x_{d-1} = r \sin\phi_1 \cdots \sin\phi_{d-2} \cos\theta,$$

$$x_d = r \sin\phi_1 \cdots \sin\phi_{d-2} \sin\theta,$$

where $r > 0, \phi_1, \ldots, \phi_{d-2} \in [0, \pi)$, and $\theta \in [0, 2\pi)$. The Jacobian is

$$J(r, \phi_1, \ldots, \phi_{d-1}, \theta) = r^{d-1} \sin^{d-2}\phi_1 \sin^{d-3}\phi_2 \cdots \sin\phi_{d-2}.$$

Then the volume element is

$$dV = r^{d-1}\, dr\, \sin^{d-2}\phi_1\, d\phi_1\, \sin^{d-3}\phi_2\, d\phi_2 \cdots \sin\phi_{d-2}\, d\phi_{d-2}\, d\theta.$$

Similarly, the surface area on S_{d-1} is

$$dA = \sin^{d-2}\phi_1\, d\phi_1\, \sin^{d-3}\phi_2\, d\phi_2 \cdots \sin\phi_{d-2}\, d\phi_{d-2}\, d\theta.$$

It is computed from $\sqrt{|\det(\boldsymbol{J}(t)\boldsymbol{J}^T(t))|}$, where $\boldsymbol{J}(t)$ was given in (3.8.1) and $t = (\phi_1, \phi_2, \ldots, \phi_{d-2}, \theta) \in \mathbb{R}^{d-1}$, thus $\boldsymbol{J}(t)$ is the $(d-1) \times d$ matrix of derivatives. For $d = 3$ we have $t = (\phi, \theta)$, the matrix $\boldsymbol{J}(t)$ was given in (3.8.8), yielding $|\det(\boldsymbol{J}(\phi, \theta)\boldsymbol{J}^T(\phi, \theta))| = \sin\phi$.

We need the following counterpart of Lemma 3.7.1 for general dimension. To this end, we prove the following lemma.

Lemma 3.8.8 *For the ball $B_d \subset \mathbb{R}^d$ we have*

$$\mathrm{Vol}(B_d) = \frac{1}{d} \left(\int_0^\pi \sin^{d-2}\phi_1 \, d\phi_1 \right) \cdots \left(\int_0^\pi \sin\phi_{d-1} \, d\phi_{d-2} \right) \left(\int_0^{2\pi} d\phi_{d-1} \right)$$

$$= \frac{\pi^{d/2}}{\Gamma(\frac{d}{2} + 1)}$$

$$\mathrm{Area}(S_{d-1}) = \left(\int_0^\pi \sin^{d-2}\phi_1 \, d\phi_1 \right) \cdots \left(\int_0^\pi \sin\phi_{d-1} \, d\phi_{d-2} \right) \left(\int_0^{2\pi} d\phi_{d-1} \right)$$

$$= \frac{d\pi^{d/2}}{\Gamma(\frac{d}{2} + 1)}.$$

Corollary 3.8.9 *The inverse of the volume of B_d can be expressed as:*

$$\frac{1}{\mathrm{Vol}(B_d)} = dc_1 \cdots c_{d-2} \frac{1}{2\pi},$$

where $c_j = 1/(I_j)$ and $I_j = \int_0^\pi \sin^j x \, dx$.

Proof It is known that $I_1 = 2$ and, for $n \geq 2$,

$$I_n = \int_0^\pi \sin^n x \, dx = \begin{cases} 2\dfrac{n-1}{n}\dfrac{n-3}{n-2} \cdots \dfrac{2}{3} & \text{if } n \text{ is odd} \\ \pi\dfrac{n-1}{n}\dfrac{n-3}{n-2} \cdots \dfrac{1}{2} & \text{if } n \text{ is even} \end{cases} = \begin{cases} 2\dfrac{(n-1)!!}{n!!} & \text{if } n \text{ is odd,} \\ \pi\dfrac{(n-1)!!}{n!!} & \text{if } n \text{ is even.} \end{cases}$$

Suppose that $d \geq 2$ is even. Then for $d = 2k$,

$$\mathrm{Vol}(B_d) = \frac{1}{d} \left[\pi \frac{(d-3)!!}{(d-2)!!} \right] \left[2\frac{(d-4)!!}{(d-3)!!} \right] \cdots [2][2\pi]$$

$$= \frac{1}{d}\pi^{k-1} \frac{1}{(d-2)!!} 2^{k-1}(2\pi) = \frac{2^k \pi^k}{d!!}.$$

Since $d = 2k$, we have $(2k-2)!! = 2^{k-1}(k-1)!$. Hence, $\mathrm{Vol}(B_d) = \pi^{d/2}/k!$. Suppose now $d = 2k + 1$. Then

$$\mathrm{Vol}(B_d) = \frac{1}{d} \left[2\frac{(d-3)!!}{(d-2)!!} \right] \left[\pi \frac{(d-4)!!}{(d-3)!!} \right] \cdots [2][2\pi] = \frac{1}{d}\pi^k \frac{1}{(d-2)!!} 2^{k+1}.$$

Hence,

$$\mathrm{Vol}(B_d) = \frac{2(2\pi)^k}{d!!}.$$

The proof is completed using

$$(2k-1)!! = \pi^{-1/2} 2^k \Gamma\left(k + \frac{1}{2}\right).$$

$\square$

Lemma 3.8.10 *Let $(D, \Phi_1, \ldots, \Phi_{d-2}, \Theta)$ be the spherical coordinates of a point $(Z_1, \ldots, Z_d) \in \mathbb{R}^d$, where Z_i are i.i.d. standard normal random variables. Then $(D, \Phi_1, \ldots, \Phi_{d-2}, \Theta)$ are independent random variables, with D^2 being chi-square distributed, χ_d^2, with d degrees of freedom, i.e.,*

$$f_{\chi_d^2}(r) = \frac{1}{2^{d/2}\Gamma(d/2)} r^{d/2-1} e^{-r/2}, \qquad r > 0.$$

Furthermore, Φ_i have p.d.f. $c_{d-i-1} \sin^{d-i-1} x$, $x \in [0, \pi)$, where $(c_i)^{-1} = \int_0^\pi \sin^i x \, dx$, and $\Theta \sim \mathcal{U}[0, 2\pi)$.

Proof From formula (A.1), we see that the joint p.d.f of the point $(Z_1, \ldots, Z_d)$ in generalized spherical coordinates is

$$\frac{1}{(2\pi)^{d/2}} \exp\left(-\frac{z_1^2 + \cdots + z_d^2}{2}\right)$$

$$= \frac{1}{(2\pi)^{d/2}} e^{-r^2/2} r^{d-1} \sin^{d-2}\phi_1 \cdots \sin\phi_{d-2} \, dr \, d\phi_1 \cdots d\phi_{d-2} \, d\theta.$$

We have

$$\int_0^\infty e^{-r^2/2} r^{d-1} = 2^{d/2-1}\Gamma(d/2),$$

and so

$$g_D(r) = \frac{1}{2^{d/2-1}\Gamma(d/2)} e^{-r^2/2} r^{d-1}$$

Table 3.1 Values of $c_i^{-1} = \int_0^\pi \sin^i \phi \, d\phi$ for $i = 2, \ldots, 8$

i	2	3	4	5	6	7	8
c_i	$2/\pi$	$3/4$	$8/(3\pi)$	$15/16$	$16/(5\pi)$	$35/32$	$128/(35\pi)$

is a p.d.f. of D. Let $g_{\Phi_j}(\phi_j) = c_{d-j-1} \sin^{d-j-1} \phi_j$ where $c_j^{-1} = \int_0^\pi \sin^j \phi_j \, d\phi_j$ is the p.d.f. of Φ_j and $g_\Theta(\theta) = 1/(2\pi)$. Then

$$\frac{1}{(2\pi)^{d/2}} \exp\left(-\frac{z_1^2 + \cdots + z_d^2}{2}\right)$$

$$= \left[\frac{1}{(2\pi)^{d/2}} 2^{d/2-1} \Gamma(d/2) c_1^{-1} \cdots c_{d-2}^{-1} 2\pi\right] g_D(r) g_{\Phi_1}(\phi_1) \cdots g_{\Phi_{d-2}}(\phi_{d-2}) g_\Theta(\theta).$$

Using Lemma 3.8.8, we now see that the expression in brackets equals 1 because $\Gamma(d/2)/\Gamma(d/2 + 1) = 2/d$. To complete the proof, we note that

$$\frac{\mathbb{P}(D^2 \le x)}{dx} = \frac{d}{dx} \int_0^{\sqrt{x}} \frac{1}{2^{d/2-1}\Gamma(d/2)} e^{-r^2/2} r^{d-1} \, dr = \frac{1}{2^{d/2}\Gamma(d/2)} e^{-x/2} x^{d/2-1},$$

which is the p.d.f of the chi-square distribution with d degrees of freedom. $\square$

This means that $\mathbf{Z}$ can be represented by a random variable D and an independent random point on S^{d-1}, where D^2 is χ_d^2 distributed. In Table 3.1 we show values of c_i^{-1} for $i = 2, \ldots, 8$.

Proposition 3.8.6 can be generalized to allow random points to be generated on S_{d-1}. We leave the proof of this result and the next proposition to the reader.

Proposition 3.8.11 *Let* $\mathbf{Z} \sim \mathcal{N}(0, \mathbf{Id})$ *and* $\mathbf{X} = \mathbf{Z}/||\mathbf{Z}||$, *where* $||\mathbf{Z}||^2 = \sum_{j=1}^d Z_j^2$. *Then* $\mathbf{X}$ *is a random point on* S_{d-1}.

Knowing how to generate a random point on the sphere S_{d-1}, one can ask how to obtain a random point inside the ball B_d. The following extends Lemma 3.7.1 for $d = 2$, as demonstrated in Sect. 3.7.

Lemma 3.8.12 *Let* $\mathbf{X}$ *be a random point uniformly distributed within the ball* B_d, *and let* $(D, \Phi_1, \ldots, \Phi_{d-2}, \Theta)$ *be its representation in generalized spherical coordinates. Then* $D, \Phi_1, \ldots, \Phi_{d-2}, \Theta$ *are independent random variables, where* D *has the p.d.f.*

$$f(r) = d \, r^{d-1}, \qquad 0 < r \le 1,$$

and $(\Phi_1, \ldots, \Phi_{d-2}, \Theta)$ *is a random point on the* $(d-1)$-*sphere* S_{d-1}.

Proof Consider a mapping $\mathbf{g} : \mathbb{R}^d \ni \mathbf{x} \to (r, \phi_1, \ldots, \phi_{d-1}, \theta)$ and then apply the change of variables formula (A.1). $\square$

Proposition 3.8.13 *Let $\mathbf{X}$ be a random point on the $(d-1)$-sphere uniformly distributed $\mathcal{U}(S_{d-1})$ (in Cartesian coordinates). Then $\mathbf{Y} = U^{1/d}\mathbf{X} \in B_d$ is uniformly distributed $\mathcal{U}(B_d)$, where U and $\mathbf{X}$ are independent.*

Proof We have

$$\frac{d}{2\pi c_1 \cdots c_{d-2}} = \mathrm{Vol}(B_d),$$

where $c_j^{-1} = \int_0^\pi \sin^j x \, dx$. Let $B_d \ni \mathbf{y} \to (r, \phi_1, \ldots, \phi_{d-2}, \theta)$ be a mapping $\mathbf{g}$ from Cartesian to spherical coordinates in $\mathbb{R}^d$. Let $\mathbf{h} = \mathbf{g}^{-1}$. Then the Jacobian is

$$r^{d-1} \sin^{d-2}(\phi_1) \cdots \sin \phi_{d-2}.$$

If $\mathbf{Y}$ in new coordinates $(D, \mathbf{X})$ is uniformly distributed in B_d, then its p.d.f. is

$$\frac{1}{\mathrm{Vol}(B_d)} r^{d-1} \sin^{d-2}(\phi_1) \cdots \sin \phi_{d-2} dr \, d\phi_1 \cdots d\phi_{d-2}$$

$$= dr^{d-1} \frac{1}{c_{d-2}} \sin^{d-2}(\phi_1) \cdots \frac{1}{c_1} \sin \phi_{d-2} \frac{1}{2\pi}.$$

Since $U^{1/d}$ has the p.d.f. dr^{d-1} for $r \in (0, 1]$, the proof is complete. $\qquad\square$

3.8.3 Ellipsoids

Before we discuss how to generate a random point within a d-ellipsoid, we note the following fact.

Lemma 3.8.14 *Consider a random point $\mathbf{X}$ on a subset $B \subset \mathbb{R}^d$ that is uniformly distributed as $\mathcal{U}(B)$, where $0 < \mathrm{Vol}(B) < \infty$. Let $\mathbf{T}$ be a $d \times d$ non-singular matrix, with $\mathbf{X}$ considered as a column vector. Then $\mathbf{TX}$ is a random point in $\mathbf{T}B$ with a uniform distribution having p.d.f. $\frac{1}{|\det \mathbf{T}|\mathrm{Vol}(B)} \mathbb{1}(\mathbf{x} \in \mathbf{T}B)$.*

Proof Let $f(\mathbf{x}) = 1/\mathrm{Vol}(B)$ be the p.d.f. of $\mathbf{X}$, and set $\mathbf{y} = \mathbf{Tx}$. Then $\mathbf{x} = \mathbf{T}^{-1}\mathbf{y}$, and the Jacobian of this transformation is $\det \mathbf{T}^{-1} = 1/\det(\mathbf{T})$. From the change of variables formula (A.1), $\mathbf{TX}$ has p.d.f. $1/(|\det(\mathbf{T})|\mathrm{Vol}(B))\mathbb{1}(\mathbf{y} \in \mathbf{T}B)$. $\qquad\square$

Consider a hyper-ellipsoid $\mathcal{E}_d = \{\mathbf{x} \in \mathbb{R}^d : \mathbf{x}^T\mathbf{A}^{-1}\mathbf{x} \leq 1\}$, where $\mathbf{A}$ is a positive-definite matrix, which is diagonalizable. For some orthogonal matrix $\mathbf{P}$ and vector $\mathbf{a}$, we have $\mathbf{A} = \mathbf{P}\boldsymbol{\Lambda}\mathbf{P}^T$, where $\boldsymbol{\Lambda} = \mathbf{diag}(\mathbf{a}^2)$ and $\mathbf{a}^2 = (a_1^2, \ldots, a_d^2)$. We aim to show that $\mathcal{E}_d = \mathbf{L}B_d$, where

$$\mathbf{L} = \mathbf{Pdiag}(\mathbf{a}).$$

In other words, for $B_d \ni \mathbf{x} \to \mathbf{y} = \mathbf{L}\mathbf{x}$, we have $\mathbf{x} = \mathbf{L}^{-1}\mathbf{y}$. For $\mathbf{x} \in B_d$, we have

$$1 \geq \mathbf{x}^T\mathbf{x} = (\mathbf{L}^{-1}\mathbf{y})^T (\mathbf{L}^{-1}\mathbf{y}) = \mathbf{y}^T(\mathbf{L}\mathbf{L}^T)^{-1}\mathbf{y},$$

or, substituting $\mathbf{A} = \mathbf{L}\mathbf{L}^T$, we obtain $\mathbf{y}^T\mathbf{A}^{-1}\mathbf{y} \leq 1$. Hence, $\mathbf{y} = \mathbf{L}\mathbf{x}$ belongs to

$$\mathbf{y}^T\mathbf{A}^{-1}\mathbf{y} \leq 1.$$

In particular, setting $\mathbf{P} = \mathbf{Id}$, we get

$$\sum_i^d \frac{y_i^2}{a_i^2} \leq 1,$$

which represents a hyper-ellipsoid with principal axes parallel to the coordinate axes. Any ellipsoid not centered at the origin can be obtained from this by rotation with a general orthonormal matrix $\mathbf{P}$ and a shift.

Therefore, using Lemma 3.8.14, we present the following algorithm for generating random points within an ellipsoid $\mathcal{E}_d$.

Generating a Random Point $\mathbf{Y} \sim \mathcal{U}(\mathcal{E}_d)$ The following Algorithm 23 generates a random point uniformly distributed within the ellipsoid

$$\mathcal{E}_d = \{\mathbf{x} \in \mathbb{R}^d : \mathbf{x}^T\mathbf{A}^{-1}\mathbf{x} \leq 1\},$$

where $\mathbf{A}$ is a symmetric positive-definite matrix. The method transforms a random point from the unit ball B_d using a linear mapping derived from $\mathbf{A}$. To accomplish this, we first diagonalize $\mathbf{A}$ as $\mathbf{A} = \mathbf{P}\mathbf{diag}(a)\mathbf{diag}(a)\mathbf{P}^T$, where $\mathbf{P}$ is an orthonormal matrix, and then return $\mathbf{L}\mathbf{X}$, where $\mathbf{X}$ is sampled uniformly $\mathcal{U}(B_d)$.

Algorithm 23 Generating a random point within ellipsoid $\mathbf{x}^T\mathbf{A}^{-1}\mathbf{x} = 1$

1: Diagonalize $\mathbf{A} = \mathbf{P}\mathbf{diag}(a)\mathbf{diag}(a)\mathbf{P}^T$
2: Set $\mathbf{L} = \mathbf{P}\mathbf{diag}(a)$
3: Generate $\mathbf{X} \sim \mathcal{U}(B_d)$
4: Set $\mathbf{Y} = \mathbf{L}\mathbf{X}$
5: **return Y**

Example 3.8.15 Consider a two-dimensional ellipse with principal axes aligned with the coordinate axes, represented by $(x/a)^2 + (y/b)^2 \leq 1$. This corresponds to a diagonal matrix $\mathbf{A} = \begin{pmatrix} a^2 & 0 \\ 0 & b^2 \end{pmatrix}$. Since $\mathbf{A}$ is diagonal, we have $\mathbf{P} = \mathbf{Id}$ and $\mathbf{L} = \begin{pmatrix} a & 0 \\ 0 & b \end{pmatrix}$. A point uniformly distributed within the ellipse can be generated as follows:

Algorithm 24 Generating a random point within a two-dimensional ellipse

---
1: Sample i.i.d. $U_1, U_2 \sim \mathcal{U}[0, 1)$
2: Set $D = \sqrt{U_2}$ ▷ Radial distance for uniform distribution within the unit ball
3: Set $\mathbf{x} = (x, y)^T$ with $x = D\cos(2\pi U_1)$ and $y = D\sin(2\pi U_1)$
4: **return** $y = \mathbf{L}\mathbf{x} = (ax, by)$

Figure 3.7 shows $n = 1000$ points $\mathbf{x}_i$, $i = 1, \dots, n$, from the unit ball B_2, and the corresponding transformed points $\mathbf{y}_i = \mathbf{L}\mathbf{x}_i$, $i = 1, \dots, n$, with $a = 5$ and $b = 1$. $\Diamond$

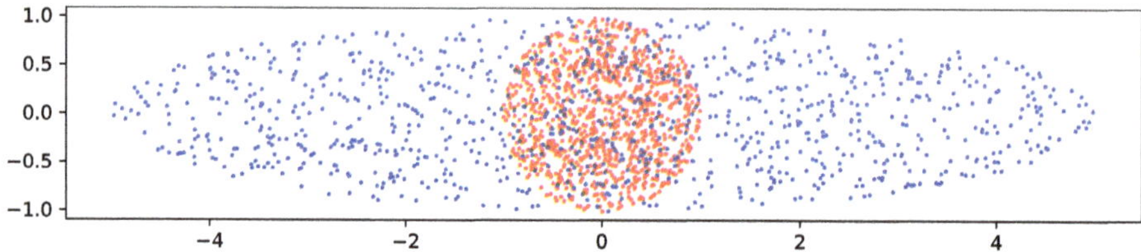

Fig. 3.7 (*red*) points $\mathbf{x}_i$, chosen uniformly from within the unit ball B_2, and (*blue*) points transformed to $\mathbf{y}_i = \mathbf{L}\mathbf{x}_i$, $i = 1, \dots, 1000$

Next, we discuss how to generate a random point **on** an ellipse.[1] Following the approach for generating random points within an ellipsoid does not yield a uniform distribution when generating points **on** the boundary. To see why, let us consider the example of a two-dimensional ellipse with principal axes aligned with the coordinate axes, i.e., $(x/5)^2 + y^2 = 1$. This corresponds to a diagonal matrix $\mathbf{A} = \begin{pmatrix} 5^2 & 0 \\ 0 & 1 \end{pmatrix}$. Let $S_1 = \{\mathbf{x}^T\mathbf{x} = 1\}$ denote the unit circle, and let the transformation be defined by $\mathbf{L} = \begin{pmatrix} 5 & 0 \\ 0 & 1 \end{pmatrix}$.

In Fig. 3.8a, we observe a regular grid on the circle S_1 (red circles) and its transformation to the ellipse $(x/5)^2 + y^2 = 1$ (blue circles), revealing that distances between points are not uniform. Let $\sigma_1(\cdot)$ represent the uniform distribution on S_1.

Although $\mathbf{L}$ maps a random element $\mathbf{X} \in B_2$ to a random element $\mathbf{L}\mathbf{X} \in \mathcal{E}_2$ within the ellipse, in Fig. 3.8b we observe that, after mapping with $x := 5x$ and $y := y$, the points on the circle transform to an uneven grid on $(x/5)^2 + y^2 = 1$. This demonstrates that a naïve mapping does not yield a uniform distribution on the ellipse.

---

[1] To generate a point on a 2-dimensional ellipsoid, see https://math.stackexchange.com/questions/973101/how-to-generate-points-uniformly-distributed-on-the-surface-of-an-ellipsoid.

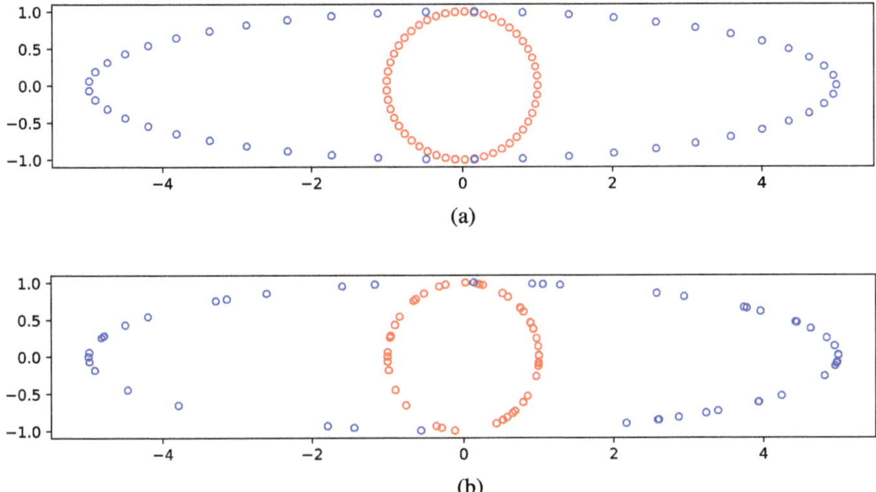

Fig. 3.8 Points on a circle S_1, in a regular grid (*top row*) and chosen uniformly (*bottom row*), transformed to an ellipse $(x/5)^2 + y^2 = 1$. (**a**) Regular grid $\mathbf{x}_i, i = 1, \ldots, 50$, on the circle S_1 (*red* points) and transformed points $\mathbf{y}_i = \mathbf{L}\mathbf{x}_i, i = 1, \ldots, 50$ (*blue* points). (**b**) Uniformly chosen points $\mathbf{x}_i, i = 1, \ldots, 50$, from S_1 (*red* points) and transformed points $\mathbf{y}_i = \mathbf{L}\mathbf{x}_i, i = 1, \ldots, 50$ (*blue* points)

To generate a uniform point on the ellipse, consider the ellipse defined by

$$\left(\frac{x}{a}\right)^2 + \left(\frac{y}{b}\right)^2 = 1.$$

Its parametrization in terms of the angle θ is given by

$$x = a\cos\theta,$$
$$y = b\sin\theta.$$

Thus, $E_1 = \{(a\cos\theta, b\sin\theta) : \theta \in [0, 2\pi]\}$, and let C be an arc on this ellipse from point $A = (a, 0)$ to point $B = (a\cos\theta, b\sin\theta)$. The matrix of derivatives is $\mathbf{J}(\theta) = (-a\sin\theta, b\cos\theta)$, thus

$$\sqrt{g(\theta)} = \sqrt{|\mathbf{J}(\theta)\mathbf{J}(\theta)^T|} = \sqrt{a^2\sin^2\theta + b^2\cos^2\theta}$$

$$= \sqrt{a^2\sin^2\theta + b^2\cos^2\theta} = b\sqrt{1 - \frac{b^2-a^2}{b^2}\sin^2\theta}.$$

We denote by $L(\theta)$ the distance between $A = (a, 0)$ and $B = (a\cos\theta, b\sin\theta)$. Then,

$$\int_C 1\, dA = L(\theta) = b \int_0^\theta \sqrt{1 - \frac{b^2 - a^2}{b^2} \sin^2 s}\, ds.$$

The integral above is an example of an incomplete elliptic integral of the second kind:

$$E(\theta, m) = \int_0^\theta \sqrt{1 - m\sin^2 s}\, ds,$$

where $L(\theta) = bE(\theta, m)$ and $m = \frac{b^2 - a^2}{b^2}$. Thus, to sample a uniform random point on E_1, we need the distribution of the angle Θ, which ensures that $(a\cos\Theta, b\sin\Theta)$ is uniformly distributed on E_1. Its cumulative distribution function is given by

$$F(\theta) = \frac{L(\theta)}{L(2\pi)} = \frac{E(\theta, m)}{E(2\pi, m)},$$

so we have the representation $\Theta \overset{\mathcal{D}}{=} F^{-1}(U)$, where $U \sim \mathcal{U}[0, 1)$. Unfortunately, elliptic integrals cannot be expressed in terms of elementary functions, so the solution $u = L(x)$ must be computed using a numerical method.

The Incomplete Elliptic Integral and Its Inverse in Python SciPy (short for *Scientific Python*) is a popular library that includes functionality for computing the incomplete elliptic integral of the second kind, $E(\theta, m)$, via `scipy.special.ellipeinc`.[2] Additionally, the pynverse library[3] is widely used for numerically finding the roots of equations of the form $y = f(x)$ for a broad class of bounded functions f. This library minimizes $(f(x) - y)^2$ using Brent's method,[4] a hybrid approach that combines bisection, secant, and inverse quadratic interpolation methods. Python Listing 3.2 provides an example for sampling a single point uniformly from an ellipse with parameters $a = 5$ and $b = 1$ (giving $m = 24$).

Python Listing 3.2 sampling a point from an ellipse uniformly at random

```python
from pynverse import inversefunc
from scipy.special import ellipeinc
from numpy.random import default_rng, PCG64
```

---

[2] https://docs.scipy.org/doc/scipy/reference/generated/scipy.special.ellipeinc.html.

[3] https://pypi.org/project/pynverse/.

[4] https://en.wikipedia.org/wiki/Brent%27s_method.

```
m = 24
rng = default_rng(PCG64())
U = rng.uniform()
F_fun = (lambda x: ellipeinc(x,m)/ellipeinc(2*np.pi,m))
theta = inversefunc(F_fun , y_values=U)


x=a*np.cos(theta)
y=b*np.sin(theta)
```

Example 3.8.16 We continue the example with an ellipse $(x/5)^2 + y^2 = 1$. We have $a = 5, b = 1$ thus $m = 24$. The c.d.f. $F(\theta)$ is shown in Fig. 3.9. To sample uniformly from the ellipse, we generate i.i.d. samples $U_1, \ldots, U_n$ from the uniform $\mathcal{U}(0, 1]$ distribution and compute the angles $\theta_i = F^{-1}(U_i)$ for $i = 1, \ldots, n$. In Fig. 3.10, $n = 500$ such samples from ellipse are presented. In Fig. 3.11, $n = 100$ equally spaced points $(U_1, \ldots, U_n$ form a regular grid, and the corresponding angles $\theta_i = F^{-1}(U_i), i = 1, \ldots, n$ are computed) are shown.

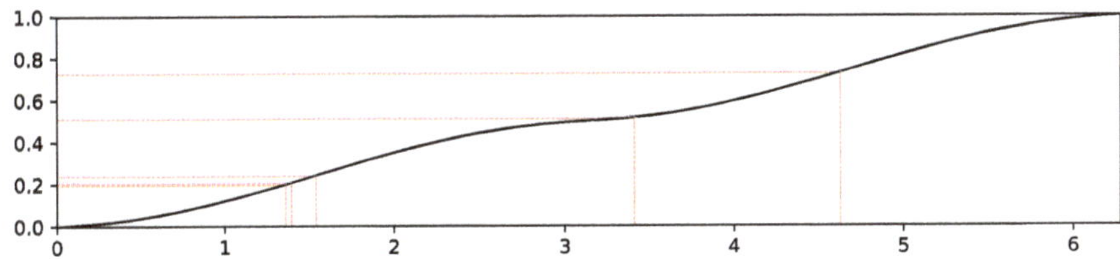

Fig. 3.9 A c.d.f. F for an ellipse $(x/5)^2 + y^2 = 1$ together with 5 uniformly chosen $U_1, \ldots, U_5$ and numerically computed $\theta_i = F^{-1}(U_i), i = 1, \ldots, 5$

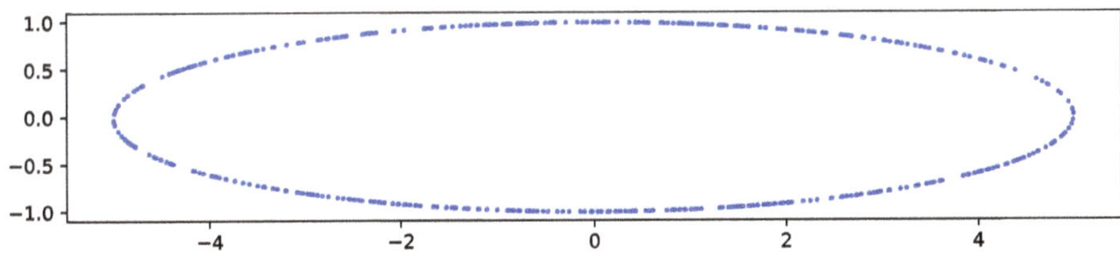

Fig. 3.10 500 points from ellipse $(x/5)^2 + y^2 = 1$

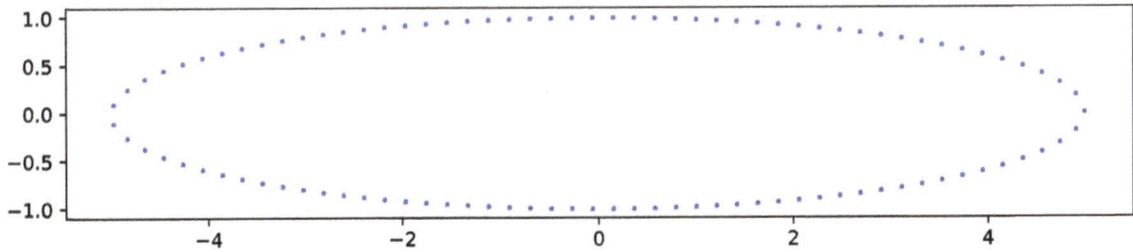

Fig. 3.11 100 equally distanced points from ellipse $(x/5)^2 + y^2 = 1$

Figures related to the ellipse $(x/5)^2 + y^2 = 1$, i.e., Figs. 3.7, 3.8, 3.9, 3.10 and 3.11 were generated by `ch3_ellipse_points.py`. $\Diamond$

3.8.4 Numerically Sampled Random Point on an Ellipsoid

Sampling a random point on a 2-dimensional surface in $\mathbb{R}^3$ can be challenging. For example, consider an ellipsoid E_2 given by

$$\frac{x^2}{a^2} + \frac{y^2}{b^2} + \frac{z^2}{c^2} = 1.$$

In this case, we can compute

$$g(\phi, \theta) = a^2 b^2 \cos^4 \theta \cos^2 \phi \sin^2 \phi + 2a^2 b^2 \cos^2 \theta \cos^2 \phi \sin^2 \theta \sin^2 \phi$$

$$+ a^2 b^2 \cos^2 \phi \sin^4 \theta \sin^2 \phi + b^2 c^2 \cos^2 \theta \sin^4 \phi + a^2 c^2 \sin^2 \theta \sin^4 \phi$$

$$= \sin^2 \phi \left[c^2 (a^2 - b^2) \cos^2 \theta (\cos^2(\phi) - 1) + a^2 ((b^2 - c^2) \cos^2 \phi + c^2) \right].$$

This implies that we should sample a random point $\mathbf{X}$ with p.d.f.

$$\frac{\sqrt{g(\phi, \theta)}}{\int_0^\pi \int_0^{2\pi} \sqrt{g(\phi, \theta)} \, d\phi \, d\theta},$$

which can be complex as it defines a 2-dimensional density where coordinates are not independent (unless for special cases of a, b, c). For example, for $a = b = 1$, we have

$$\sqrt{g(\phi, \theta)} = \sqrt{(1 - c^2) \cos^2 \phi + c^2} \sin \phi.$$

For general values of a, b, c, we propose the following method to sample a point from a surface C parametrized by $x_1(\mathbf{t}), x_2(\mathbf{t}), x_3(\mathbf{t})$, where $\mathbf{t} \in [k_0, k_1] \times [l_0, l_1]$. For an ellipsoid given by $x^2/a^2 + y^2/b^2 + z^2/c^2 = 1$, a general spherical parametrization is as follows:

$$x(\mathbf{t}) = a \cos \theta \sin \phi,$$

$$y(\mathbf{t}) = b \sin \theta \sin \phi, \qquad\qquad (3.8.11)$$

$$z(\mathbf{t}) = c \cos \phi,$$

where $\mathbf{t} = (\theta, \phi)$ with $\theta \in [0, 2\pi)$ and $\phi \in [0, \pi)$.

This numerical method approximates the formula (3.8.2) by subdividing the surface into small parallelograms. The approach is as follows: divide the surface C into small patches, approximate their areas, then sample one of these patches proportionally to its area. After selecting a patch, sample within it. This approach is analogous to the line segment sampling in Example 3.8.3.

For the rectangle $[k_0, k_1] \times [l_0, l_1]$, construct a regular grid by dividing $[k_0, k_1]$ into n_k intervals and $[l_0, l_1]$ into n_l intervals. Thus, the grid points are

$$\mathbf{p}_{i,j} = (k_0 + \kappa i, l_0 + \ell j), \quad i = 0, \dots, n_k, \ j = 0, \dots, n_l,$$

where

$$\kappa = \frac{k_1 - k_0}{n_k}, \qquad \ell = \frac{l_1 - l_0}{n_l}.$$

The parameters n_k and n_l define the precision of the grid. The rectangle

$$\mathrm{Rect}_{i,j} = [\mathbf{p}_{i,j}, \mathbf{p}_{i+1,j}, \mathbf{p}_{i,j+1}, \mathbf{p}_{i+1,j+1}]$$

is mapped to a surface patch $C_{i,j} \subseteq C$. We use the convention that $n_k + 1 = 0$ and $n_l + 1 = 0$. Each $\mathrm{Rect}_{i,j}$ corresponds to a disjoint surface patch $C_{i,j}$, whose area is given by

$$\mathrm{Area}\,(C_{i,j}) = \int_{C_{i,j}} dA = \int_{\mathrm{Rect}_{i,j}} \sqrt{g(\mathbf{t})}\, d\mathbf{t}$$

$$= \int_{k_0+\kappa i}^{k_0+\kappa(i+1)} \int_{l_0+\ell j}^{l_0+\ell(j+1)} \sqrt{g(k, l)}\, dl\, dk.$$

For cases where $i = n_k$, compute $\int_0^\kappa \dots dl$; similarly, handle cases where $j = n_l$. The area of $C_{i,j}$ can be approximated by the area of the parallelogram $\mathrm{Paral}_{i,j}$, spanned by points $\mathbf{q}_{i,j}^{(1)}, \mathbf{q}_{i,j}^{(2)}, \mathbf{q}_{i,j}^{(3)}$, which are following

$$\mathbf{q}_{i,j}^{(1)} = (x(k_0 + \kappa i, l_0 + \ell j),\ y(k_0 + \kappa i, l_0 + \ell j),\ z(k_0 + \kappa i, l_0 + \ell j)),$$

$$\mathbf{q}_{i,j}^{(2)} = (x(k_0 + \kappa(i + 1), l_0 + \ell j),\ y(k_0 + \kappa(i + 1), l_0 + \ell j),$$

$$z(k_0 + \ell(i + 1), l_0 + \ell j)),$$

$$\mathbf{q}_{i,j}^{(3)} = (x(k_0 + \kappa i, l_0 + \ell(j + 1)),\ y(k_0 + \kappa i, l_0 + \ell(j + 1)),$$

$$z(k_0 + \kappa i, l_0 + \ell(j + 1))),$$

see Fig. 3.12. Note that although points $\mathbf{q}_{i,j}^{(1)}, \mathbf{q}_{i,j}^{(2)}, \mathbf{q}_{i,j}^{(3)}$ lie on C, this is not necessarily the case for the point

$$(x(k_0 + \kappa(i+1), l_0 + \ell(j+1)), \; y(k_0 + \kappa(i+1), l_0 + \ell(j+1)),$$

$$z(k_0 + \kappa(i+1), l_0 + \ell(j+1)))).$$

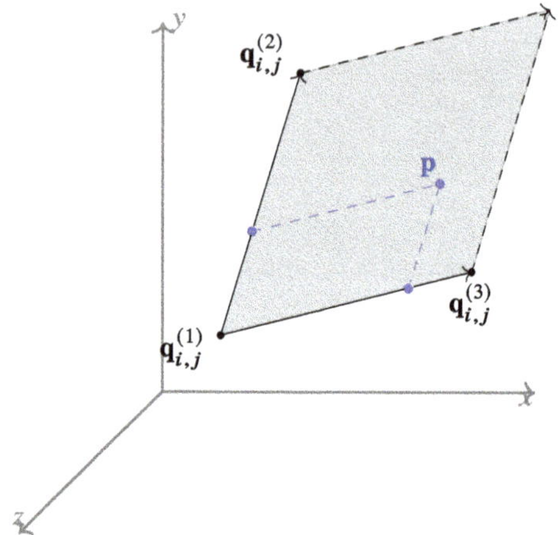

Fig. 3.12 A parallelogram span over $\mathbf{q}_{i,j}^{(1)} = (1, 1, 1)$, $\mathbf{q}_{i,j}^{(2)} = (3, 5, 4)$, $\mathbf{q}_{i,j}^{(3)} = (6, 4, 7)$. Points chosen uniformly on $(\mathbf{q}_{i,j}^{(1)}, \mathbf{q}_{i,j}^{(2)})$ and on $(\mathbf{q}_{i,j}^{(1)}, \mathbf{q}_{i,j}^{(3)})$ and resulting random point in a parallelogram depicted

We approximate Area $(C_{i,j})$ by the area of the parallelogram $\mathtt{Paral}_{i,j}$. With this approximation, we define a categorical distribution over the patches $C_{i,j}$.

$$\mathbb{P}(X = C_{i,j}) = \frac{\text{Area}\,(\mathtt{Paral}_{i,j})}{\sum_{i',j'} \text{Area}\,(\mathtt{Paral}_{i',j'})}. \tag{3.8.12}$$

Once a parallelogram X is sampled, we sample a uniform point $\mathbf{p}$ within it (by simply sampling uniformly along corresponding vectors—see Fig. 3.12) and then chose the closest to it (not depicted).

Example 3.8.17 We simulated uniformly distributed points on the ellipsoid

$$\left(\frac{x}{2}\right)^2 + \left(\frac{y}{3}\right)^2 + \left(\frac{z}{5}\right)^2 = 1.$$

Using $n_k = n_l = 120$, the grid comprised 14,400 segments, meaning the ellipsoid's surface was divided into so many sections. The resulting points are shown in Fig. 3.13.

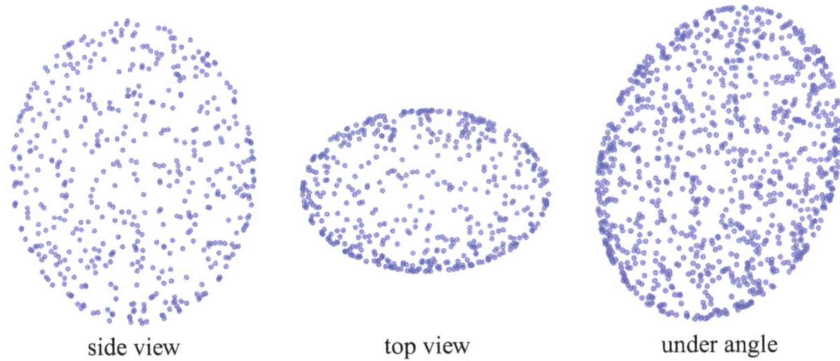

Fig. 3.13 Uniformly distributed points on the ellipsoid $x^2/2^2 + y^2/3^2 + z^2/5^2 = 1$. In the side and top views, points on the hidden side are not depicted

All images in this example, along with the surface area approximation computed by the method described below, were produced by `ch3_ellipsoid3D.py`. In Fig. 3.14 the heatmap of areas of `Paral`$_{i,j}$ (i.e., approximations to the areas of $C_{i,j}$) are depicted, and similarly in Fig. 3.15 for a sphere with radius 2.

Heatmaps from Figs. 3.14 and 3.15 are shown as regular 3D surfaces in Fig. 3.16.

If we uniformly sample the angle $\theta \sim \mathcal{U}[0, 2\pi)$ and $\phi \sim \mathcal{U}[0, \pi)$, the points on an ellipsoid, computed using the parametrization in (3.8.11), will not be uniformly distributed, as illustrated in Fig. 3.17.

The exact surface area of the ellipsoid is given by

$$\text{Area}\,(C) = \int_0^\pi \int_0^{2\pi} \sqrt{g(\phi, \theta)}\, d\phi\, d\theta,$$

where $g(\phi, \theta)$ is defined in (3.8.11). Using elliptic functions, we find that the exact surface area of an ellipsoid with $a = 2$, $b = 3$, and $c = 5$ is

$$\text{Area}\,(C) = 134.7751.$$

Additionally, in `ch3_surface_area_ellipsoid.py` we provide code that performs numerical integration using the `scipy.integrate.dblquad` function and the direct formula (3.8.11) for $g(\phi, \theta)$; it returns the same result. From the simulation results, the total area of the approximating parallelograms (which provides an approximation to Area$\,(C)$) was **134.7399**, showing close agreement. Notably, the approximated area is slightly less than the actual area, as expected, since the parallelograms are contained within the ellipsoid.

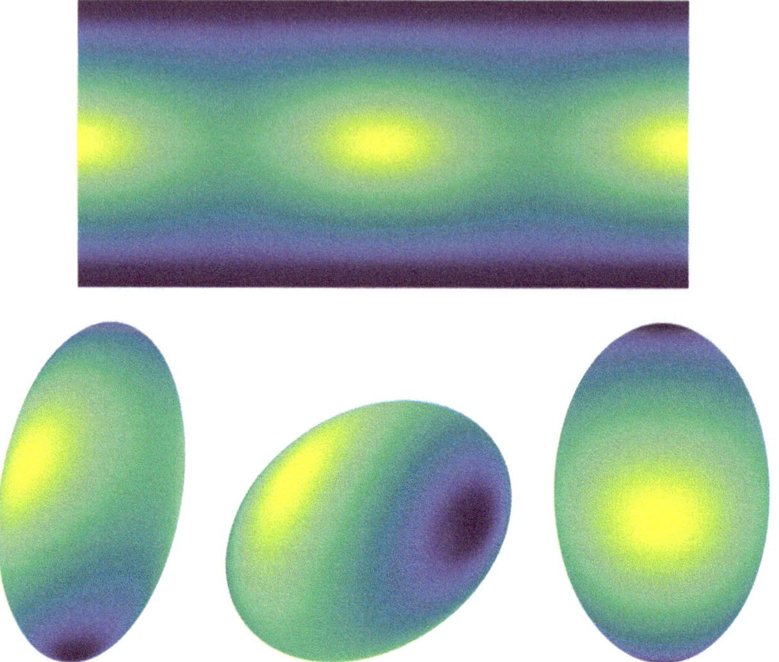

Fig. 3.14 Ellipsoid $x^2/2^2 + y^2/3^2 + z^2/5^2 = 1$. The top image shows a heatmap of the distribution (3.8.12), where $\theta \in [0, 2\pi)$ is on the horizontal axis and $\phi \in [0, \pi)$ is on the vertical axis. Bright yellow corresponds to higher values, and dark blue to lower values. The remaining images depict several views of the ellipsoid colored by the same distribution

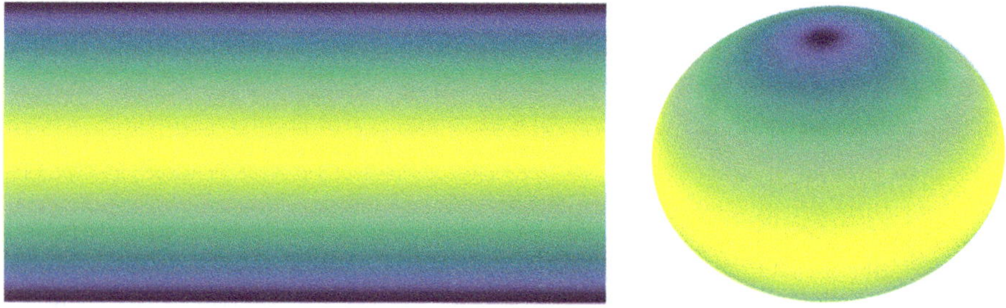

Fig. 3.15 Sphere $x^2/2^2 + y^2/2^2 + z^2/2^2 = 1$. See description in Fig. 3.14

According to Knud Thomsen,[5] an alternative approximation for the surface area of an ellipsoid is given by

$$\text{Area}\,(C) \approx \tilde{S} := 4\pi \left(\frac{(ab)^p + (bc)^p + (ac)^p}{3} \right)^{\frac{1}{p}},$$

<hr>

[5] http://www.numericana.com/answer/ellipsoid.htm#thomsen.

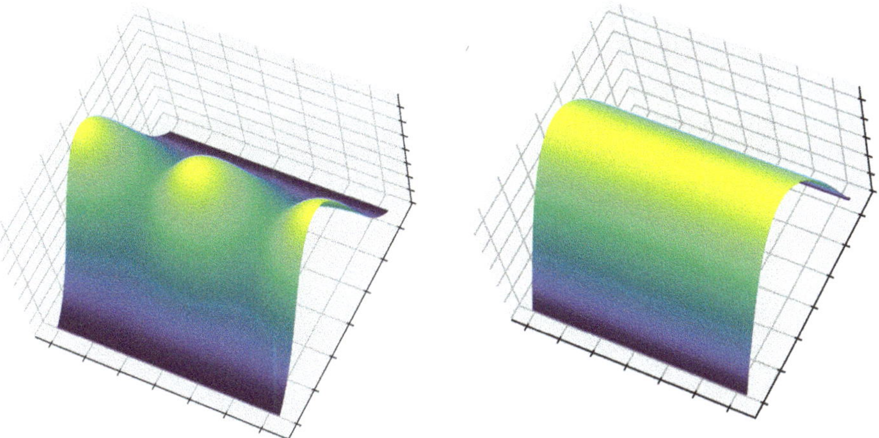

Fig. 3.16 Heatmaps from Figs. 3.14 (left) and 3.15 (right) presented as 3D surface plots

where $p = 1.6075$. This approximation is guaranteed to have a relative error of at most 1.41544%. For $a = 2$, $b = 3$, and $c = 5$, this gives $\tilde{S} = 134.8149868$, which is a close match to the simulation-based approximation, though the simulation error is smaller. $\qquad\Diamond$

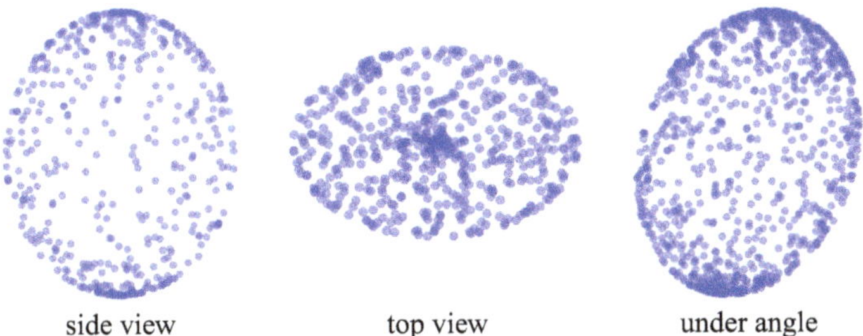

Fig. 3.17 Non-uniformly (incorrectly) sampled points on an ellipsoid $x^2/2^2 + y^2/3^2 + z^2/5^2 = 1$: Angle θ sampled uniformly from $(0, 2\pi)$ and angle ϕ sampled uniformly from $(0, \pi)$; point (x, y, z) computed using parametrization (3.8.7). In the side and top views, points on the hidden side are not depicted

Remark 3.8.18 We considered a rectangle with points $\mathbf{p}_{i,j}$, $\mathbf{p}_{i+1,j}$, $\mathbf{p}_{i,j+1}$, $\mathbf{p}_{i+1,j+1}$ which was transformed into a parallelogram. Instead of this, we can construct two triangles based on these points: the first triangle on $\mathbf{p}_{i,j}$, $\mathbf{p}_{i+1,j}$, $\mathbf{p}_{i,j+1}$ and the second one on $\mathbf{p}_{i+1,j}$, $\mathbf{p}_{i,j+1}$, $\mathbf{p}_{i+1,j+1}$. These triangles would then be transformed into two triangles. Note that, although the triangles in (ϕ, θ) space lie on the same hyperplane, the transformed triangles do not. We can proceed with a similar approach, creating distributions on these transformed triangles, which would provide slightly better accuracy for the same values of n_k and n_l. ∎

Remark 3.8.19 (Example from [2]). A set $K \subset \mathbb{R}^d$ is called star-shaped if there exists an $s_0 \in K$ such that for all $s \in K$, the line segment from s_0 to s lies in K. If s_0 is the center, we have a star-shaped set at the origin. Examples include various types of generalized hyper-ellipsoids, such as a fermatoid:

$$\sum_{j=1}^{d} \left(\frac{x_j}{a_j} \right)^{2m} = 1,$$

where $m = 1, 2, \ldots$.

Suppose that $K \subset \mathbb{R}^d$ is a bounded star-shaped geometrical object centered at the origin, parameterized as

$$(\rho(\mathbf{x}), \mathbf{x}), \qquad \mathbf{x} \in S_{d-1},$$

where $\mathbf{x} = (\phi_1, \ldots, \phi_{d-2}, \theta)$, and $\rho(\mathbf{x})$ is the radial function defined by

$$\rho(\mathbf{x}) = \max\{\lambda > 0 : \lambda \mathbf{x} \in K\}.$$

We aim to compute $\mathrm{Vol}(K)$. Switching to spherical coordinates, we find:

$$\int_{\mathbb{R}^d} \mathbb{1}_K(\mathbf{x})\, d\mathbf{x} \tag{3.8.13}$$

$$= \int_0^{\infty} \int_{S_{d-1}} \mathbb{1}_K(r\mathbf{x}) r^{d-1} \sin^{d-2}\phi_1 \ldots \sin\phi_{d-2}\, d\phi_1 \ldots d\phi_{d-2}\, d\theta\, dr\, d\mathbf{x}$$

$$= \int_{S_{d-1}} \int_0^{\rho(\mathbf{x})} r^{d-1}\, dr\, d\mathbf{x} = \frac{1}{d} \int_{S_{d-1}} \rho(\mathbf{x})^d\, d\mathbf{x}$$

$$= \frac{\mathrm{Area}(S_{d-1})}{d} \frac{1}{\mathrm{Area}(S_{d-1})} \int_{S_{d-1}} \rho(\mathbf{x})^d\, d\mathbf{x}$$

$$= \mathrm{Vol}(B_d) \frac{1}{\mathrm{Area}(S_{d-1})} \int_{S_{d-1}} \rho(\mathbf{x})^d\, d\mathbf{x} \tag{3.8.14}$$

$$= \frac{\pi^{d/2}}{\Gamma\left(\frac{d}{2}+1\right)} \frac{1}{\mathrm{Area}(S_{d-1})} \int_{S_{d-1}} \rho(\mathbf{x})^d\, d\mathbf{x},$$

where we used $\mathrm{Area}(S_{d-1}) = d\,\mathrm{Vol}(B_d)$ in Eq. (3.8.14). Let $V \sim \mathcal{U}(S_{d-1})$ be a random point on S_{d-1} and $V_1, \ldots, V_R$ be replications of V. Then

$$\hat{V}_R = \frac{\pi^{d/2}}{\Gamma\left(\frac{d}{2}+1\right)} \frac{1}{R} \sum_{j=1}^{R} \rho(V_j)^d$$

is a Monte Carlo estimator of $\mathrm{Vol}(K)$. In [2], a Monte Carlo estimation of the volume of fermatoids and hyper-ellipsoids was conducted using this estimator $\hat{V}_R$. The authors reported an experiment in which they first generated a uniformly distributed set of points on the unit sphere's surface. They then used this sample to send out random rays from the origin to intersect with the ellipsoid's surface, where $\rho(V_j)$ represents the length of the ray intercepting the surface of K for each j.   ■

Remark 3.8.20   Notice the following fact. If $\mathbf{X}$ is a random point on the surface C, and if $\mathbf{P}$ is an orthogonal matrix and $\boldsymbol{m}$ is a vector in $\mathbb{R}^d$, then $\mathbf{P}\mathbf{X} + \boldsymbol{m}$ is a random point on the surface $\mathbf{P}C + \boldsymbol{m}$. This observation explains why our method applies to the general form of an ellipsoid $\mathbf{x}^T \mathbf{A} \mathbf{x} = 1$, where $\mathbf{A}$ is a positive-definite matrix that can be diagonalized.   ■

3.8.5   Example: Average Area of a Shadow of a Convex 3D Body

Consider a bounded convex polyhedron $B \subset \mathbb{R}^3$ with faces $F_1, \ldots, F_m$. For example, a cube has six square faces $F_1, \ldots, F_6$. Imagine that the sun is infinitely far away, so its rays fall parallel onto the polyhedron. The figure casts a shadow on the "ground"—meaning a point (x, y, z) on the surface of B is projected onto $(x, y, 0)$. This projected shadow, a 2D figure, is denoted $\mathrm{shadow}(B)$.

Let F be a face of B (e.g., a square or triangle), which is a 2D figure placed in 3D space. Assume that the z-axis is perpendicular to the "ground," and that the face has a rotation with azimuth θ and altitude ϕ. Note that only the altitude ϕ affects the area of the projected shadow, as the azimuth θ merely rotates the face without altering the shadow's area; see Fig. 3.18 (left).
The area of the shadow is $|\cos(\phi)|$ times the area of face F, i.e.,

$$\mathrm{Area}\,(\mathrm{shadow}(F)) = \mathrm{Area}\,(F) \cdot |\cos \phi|.$$

To demonstrate this, consider first F as a triangle. Since changing the azimuth θ does not affect the area of the shadow, we may assume that one side (edge) of the triangle is parallel to the ground. The area of the shadow is then equal to the base of the triangle on the ground multiplied by the height scaled by $|\cos \phi|$.

Now consider a bounded convex polyhedron consisting of faces $F_1, \ldots, F_m$. Every ray from the sun intersects exactly two faces, except for vertices and edges, which contribute zero to the projected shadow area. This implies that the area of the shadow of B is half the sum of the shadow areas of all faces, i.e.,

$$\mathrm{Area}\,(\mathrm{shadow}(B)) = \frac{1}{2} \sum_{i=1}^{m} \mathrm{Area}\,(\mathrm{shadow}(F_i)) = \frac{1}{2} \sum_{i=1}^{m} \mathrm{Area}\,(F_i) \cdot |\cos \phi|.$$

$$(3.8.15)$$

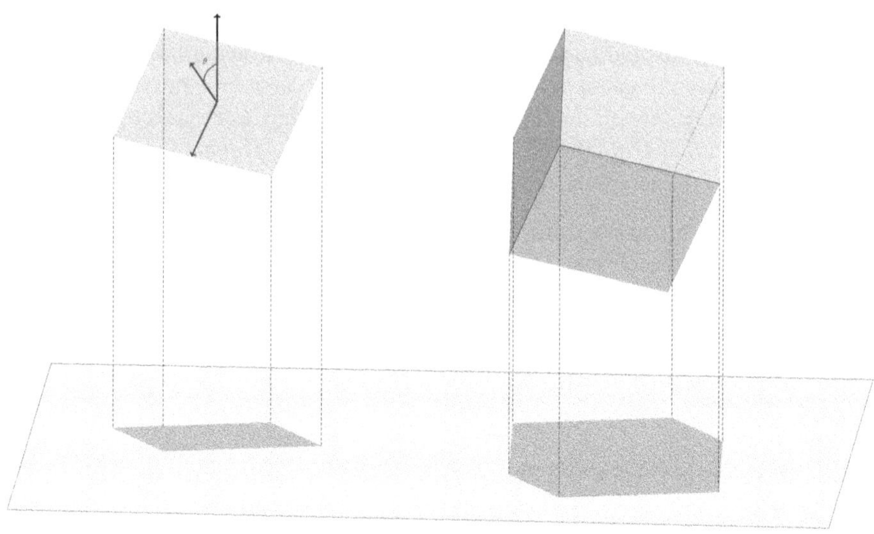

Fig. 3.18 Shadow of a square (left) and shadow of a cube (right)

Average Shadow Consider now a *randomly* rotated face. To define a randomly rotated face, we must specify what is meant by "randomly rotated." We can do this by rotating a normal vector attached to the center of the face, selecting a point on the unit sphere uniformly at random, and aligning the vector to this point. Recall the spherical coordinates (3.8.7) for (x, y, z):

$$x = \cos\theta \sin\phi, \qquad\qquad (3.8.16)$$
$$y = \sin\theta \sin\phi,$$
$$z = \cos\phi.$$

If we take $\Theta = 2\pi U_1$ and $\Phi = \arccos(1 - 2U_2)$ with U_1, U_2 independent $\mathcal{U}[0, 1)$ random variables, then the corresponding (x, y, z) is chosen randomly from the unit sphere. In other words, Θ has a uniform distribution on $(0, 2\pi)$ and Φ has the density $\frac{1}{2}\sin\phi$ on $(0, \pi)$. Thus, the average shadow of a face F is given by:

$$\mathbb{E}[\text{Area}\,(\text{shadow}(F))] = \text{Area}\,(F) \int_0^{2\pi} \frac{1}{2\pi} \int_0^{\pi} |\cos\phi| \frac{1}{2} \sin\phi \, d\theta \, d\phi$$
$$= \frac{\text{Area}\,(F)}{2} \int_0^{\pi} |\cos\phi| \sin\phi \, d\phi$$

$$= \frac{\text{Area}\,(F)}{4} \left[2 \int_0^{\pi/2} \sin(2\phi)\, d\phi \right]$$

$$= \frac{\text{Area}\,(F)}{2}.$$

Using (3.8.15), we have a simple formula for the average shadow of a convex polyhedron B:

$$\mathbb{E}(\text{Area}\,(\texttt{shadow}(B))) = \frac{1}{2} \sum_{i=1}^m \mathbb{E}[\text{Area}\,(\texttt{shadow}(F_i))]$$

$$= \frac{1}{2} \sum_{i=1}^m \frac{\text{Area}\,(F_i)}{2} = \frac{\text{Area}\,(\texttt{surface}(B))}{4}. \tag{3.8.17}$$

In particular, if B is a cube with edge length s, its surface area is Area $(\texttt{surface}(B)) = 6s^2$ and thus $\mathbb{E}(\text{Area}\,(\texttt{shadow}(B))) = \frac{3}{2}s^2$.

The formula (3.8.17) holds for any bounded convex body B, which can be shown by approximating B by a sequence of bounded convex polyhedrons B_n. Note that this result agrees for a ball: a ball of radius r has surface area $4\pi r^2$, so its expected shadow area is $4\pi r^2/4 = \pi r^2$, which corresponds to the area of a circle. In this case, the shadow remains a circle of fixed area, independent of the ball's orientation.

3.9 Simulating Bivariate Random Vectors: Conditional Inverse Transform and Copulas

We present two methods for simulating bivariate random vectors: the conditional inverse transform method and the copula-based approach.

3.9.1 Conditional Inverse Transform Method

We focus here on two-dimensional random variables; however, the method extends naturally to higher dimensions. Consider the joint random variable (X, Y) with the joint probability density function (p.d.f.) $f(x, y)$. Assume that X takes values in $[x_0, x_1)$ and Y takes values in $[y_0, y_1)$, i.e.,

$$\int_{x_0}^{x_1} \int_{y_0}^{y_1} f(x, y)\, dx\, dy = 1.$$

The marginal densities and cumulative distribution functions (c.d.f.s) are given by:

$$f_X(x) = \int_{y_0}^{y_1} f(x, y)\, dy, \qquad F_X(x) = \int_{x_0}^{x} f_X(s)\, ds = \int_{x_0}^{x} \left(\int_{y_0}^{y_1} f(s, y)\, dy \right) ds,$$

$$f_Y(y) = \int_{x_0}^{x_1} f(x, y)\, dx, \qquad F_Y(y) = \int_{y_0}^{y} f_Y(t)\, dt = \int_{y_0}^{y} \left(\int_{x_0}^{x_1} f(x, t)\, dx \right) dt.$$

The conditional c.d.f. of Y given $X = \mathsf{x}$ is defined as (provided $f_X(\mathsf{x}) \neq 0$):

$$F_{Y|X=\mathsf{x}}(y) = \frac{F_{X=\mathsf{x}, Y}(\mathsf{x}, y)}{f_X(\mathsf{x})},$$

where

$$F_{X=\mathsf{x}, Y}(\mathsf{x}, y) = \int_{y_0}^{y} f(\mathsf{x}, u)\, du.$$

To compute $F_{Y|X=\mathsf{x}}(y)$, we find the partial derivative of $F(x, y)$ with respect to x at $x = \mathsf{x}$, denoted $F'_x(\mathsf{x}, y)$.

If X and Y are independent, they can be sampled independently using their marginal distributions. Otherwise, we first sample one marginal (e.g., X) and then generate the conditional distribution of the other variable (e.g., Y) given the sampled value. The procedure is as follows:

- Compute the marginal p.d.f. $f_X(x) = \int_{y_0}^{y_1} f(x, y)\, dy$ and its c.d.f. $F_X(x) = \int_{x_0}^{x} f_X(s)\, ds$.
- Sample from this marginal using the inverse transform method (ITM): generate $U_1 \sim \mathcal{U}[0, 1)$ and set $X = F_X^{-1}(U_1)$.
- Conditional sampling: Given $X = \mathsf{x}$, generate Y from the conditional c.d.f. $F_{Y|X=\mathsf{x}}(y) = F'_x(\mathsf{x}, y)$. This can be done using ITM: generate $U_2 \sim \mathcal{U}(0, 1)$, independent of U_1, and set $Y = F_{Y|X=\mathsf{x}}^{-1}(U_2)$.
- Output the pair (X, Y).

Example 3.9.1 Let the random variable (X, Y) have the joint p.d.f.

$$f(x, y) = x \exp(-x(y + 1)), \quad \text{for } x, y \geq 0.$$

The marginal p.d.f.s are:

$$f_X(x) = e^{-x}, \qquad f_Y(y) = \frac{1}{(y + 1)^2}.$$

Thus, $X \sim \mathrm{Exp}(1)$ and $Y \sim \mathrm{Par}(1)$. Here, $\mathrm{Par}(\alpha)$ denotes standard Pareto distribution with shape parameter $\alpha > 0$. We sample from $\mathrm{Exp}(1)$ by setting $X = -\log(U_1)$ for $U_1 \sim \mathcal{U}[0, 1)$. Given $X = \mathsf{x}$, the conditional distribution is:

$$f_{Y|X=\mathsf{x}}(y) = \frac{f(\mathsf{x}, y)}{f_X(\mathsf{x})} = \mathsf{x}\exp(-\mathsf{x}y),$$

which follows an exponential distribution with parameter x. To sample from this distribution, we use:

$$Y = -\frac{\log(U_2)}{\mathsf{x}},$$

where $U_2 \sim \mathcal{U}[0, 1)$, independent of U_1.

A sample of size 100 from this distribution, generated using the described method (the implementation is available in `ch3_copula_Exp1_Par1_marginals.py`), is shown in Fig. 3.19, with y-values truncated at 10 for clarity. $\Diamond$

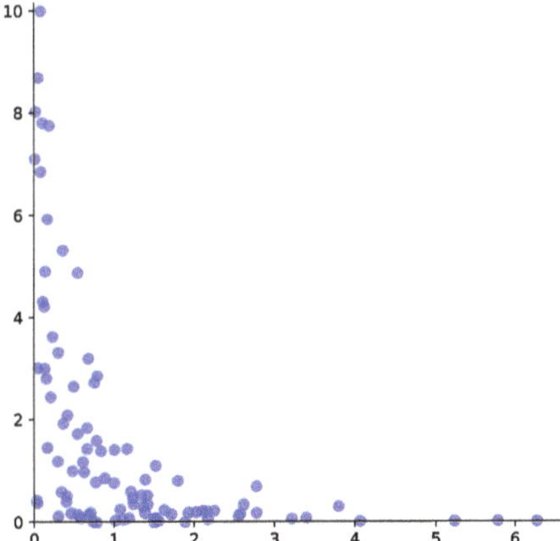

Fig. 3.19 Scatter plot of 100 samples of (X, Y) from the density $f(x, y) = x\exp(-x(y+1))$, with y-values truncated at 10 for clarity

3.9.2 Bivariate Copulas

A copula $C(u_1, u_2)$ is defined as a bivariate c.d.f. with uniformly distributed $\mathcal{U}[0, 1)$ marginal c.d.f.s. Suppose that (X, Y) is a random vector with joint c.d.f. $F_{X,Y}(t_1, t_2)$. The marginal distributions are $F_X(t_1) = F_{X,Y}(t_1, \infty)$ and $F_Y(t_2) = F_{X,Y}(\infty, t_2)$. Since X and Y are not assumed to be independent, knowing the marginal c.d.f.s F_X and F_Y is insufficient to describe the dependence structure, the dependence is captured by the copula.

Formally, a bivariate **copula** is any two-dimensional c.d.f. with uniform $\mathcal{U}[0, 1)$ marginals, denoted by $C(u_1, u_2)$. A function $C : [0, 1] \times [0, 1] \to [0, 1]$ is a copula if it satisfies the following conditions:

1. $\lim_{u_i \to 0} C(u_1, u_2) = 0$ for $i = 1, 2$,
2. $\lim_{u_1 \to 1} C(u_1, u_2) = u_2$ and $\lim_{u_2 \to 1} C(u_1, u_2) = u_1$,
3. $C(v_1, v_2) - C(u_1, v_2) - C(v_1, u_2) + C(u_1, u_2) \geq 0$ for all $u_1 \leq v_1$ and $u_2 \leq v_2$.

The following result can be established. A generalization for d-dimensional random vectors is known as **Sklar's theorem**.

Theorem 3.9.2 *For any c.d.f. $F_{X,Y}(t_1, t_2)$ of a random vector (X, Y), there exists a copula $C(u_1, u_2)$ such that*

$$F_{X,Y}(t_1, t_2) = C(F_X(t_1), F_Y(t_2)), \qquad -\infty < t_1, t_2 < \infty. \tag{3.9.1}$$

*If the marginal distributions $F_X(t)$ and $F_Y(t)$ **are continuous, then the copula C is unique.** Conversely, if C is a copula and F_X and F_Y are c.d.f.s, then the function*

$$F_{X,Y}(t_1, t_2) = C(F_X(t_1), F_Y(t_2))$$

is a two-dimensional c.d.f. with marginals F_X and F_Y.

From (3.9.1), if (X, Y) has a joint density $f_{X,Y}$, we obtain:

$$f_{X,Y}(t_1, t_2) = f_X(t_1) f_Y(t_2) c(F_X(t_1), F_Y(t_2)), \tag{3.9.2}$$

where $c(u_1, u_2)$ is the joint p.d.f. of the copula C, and f_X and f_Y are the marginal p.d.f.s. This expression allows us to write:

$$\mathbb{E}(k(X)k(Y)) = \int \int k_1(t_1) k_2(t_2) f_X(t_1) f_Y(t_2) c(F_X(t_1), F_Y(t_2)) \, dt_1 \, dt_2.$$

Since the marginal c.d.f.s are continuous, copulas cannot have atoms, ensuring that $C(v_1, v_2)$ is continuous in both v_1 and v_2. It can be shown (see Embrechts et al. [48]) that the copula can be decomposed as:

$$C(v_1, v_2) = p C_{\mathrm{s}}(v_1, v_2) + (1 - p) C_{ac}(v_1, v_2),$$

where $0 \leq p \leq 1$, C_{s} is a singular c.d.f., and C_{ac} is an absolutely continuous part with a p.d.f. We focus on the case where C has a joint p.d.f. $c(v_1, v_2)$, which provides the joint distribution of (X, Y) as the triplet (F_X, F_Y, C).

Recall the concept of a *generalized inverse $F^{\leftarrow}(t)$* as defined in (3.1.1) in Sect. 3.1. We can express the copula C of a given joint distribution $F_{X,Y}$ as follows.

Proposition 3.9.3 *Let $F_{X,Y}$ be a joint c.d.f. with continuous marginal c.d.f.s F_X and F_Y. Then the unique copula associated with $F_{X,Y}$ is given by:*

$$C_{X,Y}(u_1, u_2) = F_{X,Y}(F_X^{\leftarrow}(u_1), F_Y^{\leftarrow}(u_2)). \tag{3.9.3}$$

The following proposition generalizes the well-known property that if $F(t)$ is the continuous c.d.f. of a random variable X, then $F(X) \sim \mathcal{U}[0, 1)$.

Proposition 3.9.4 *If the marginal c.d.f.s F_X and F_Y are continuous, then the random vector $(F_X(X), F_Y(Y))$ follows the copula $C_{X,Y}(u_1, u_2)$ given in (3.9.3), which is uniquely determined by the joint distribution of (X, Y).*

From now on, we assume that the marginal distributions of any joint distribution are continuous. In survival analysis, survival functions are used instead of c.d.f.s. Thus, we introduce the corresponding **tail copula**:

$$C^*(u_1, t_2) = 1 - u_1 - u_2 + C(u_1, u_2),$$

which represents the probability $\mathbb{P}(U_1 > u_1, U_2 > u_2)$.

Lemma 3.9.5 *For a bivariate survival function*

$$S_{X,Y}(t_1, t_2) = \mathbb{P}(T_X > t_1, T_Y > u_2)$$

with marginal survival functions S_X and S_Y, we have:

$$S_{X,Y}(t_1, t_2) = C^*_{X,Y}(1 - S_X(t_1), 1 - S_Y(t_2)).$$

General Procedure for Sampling from a Copula C Let $(V_1, V_2) \sim C$ be a random vector sampled from a copula C. Assume that the partial derivative $C'_{v_1}(V_1, v_2)$ and its inverse with respect to v_2 can be computed in closed form. In this case, the following procedure can be used to simulate the copula.

First, simulate V_1, say $V_1 = v_1$, where $V_1 \sim \mathcal{U}[0, 1)$. Since the marginal distributions have a p.d.f. equal to 1, the conditional p.d.f. of V_2 given $V_1 = \mathsf{v}_1$ is the partial derivative of $C(v_1, v_2)$ with respect to v_1, evaluated at $v_1 = \mathsf{v}_1$, denoted by $C'_{v_1}(\mathsf{v}_1, v_2)$. The procedure is summarized in Algorithm 25.

Algorithm 25 Inverse Transform Method (ITM) for sampling from a copula

---

Require: Generalized inverse of the partial derivative of the copula C.
1: Sample i.i.d. $U_2, V_1 \sim \mathcal{U}[0, 1)$, denote $U_2 = \mathsf{u}_2$ and $V_1 = \mathsf{v}_1$
2: Set $V_2 = (C'_{v_1})^{\leftarrow}(\mathsf{v}_1, \mathsf{u}_2)$
3: **return** (V_1, V_2)

Note that, in some cases, calculating the partial derivative C'_{v_1} or its generalized inverse $(C'_{v_1})^{\leftarrow}(v_1, u_2)$ can be challenging or infeasible in closed form. Therefore, copula-specific simulation methods are often employed.

Fréchet-Hoeffding Bounds The following inequality characterizes the range of dependence structures achievable with copulas and has practical applications.

Theorem 3.9.6 (Fréchet-Hoeffding Bounds) *For a two-dimensional c.d.f. $F(x, y)$ of the random vector (Y_1, Y_2) with the same marginal distribution $F(x) = \mathbb{P}(Y_1 \leq x) = \mathbb{P}(Y_2 \leq x)$,*

$$(F(x) + F(y) - 1)_+ \leq F(x, y) \leq \min(F(x), F(y)). \qquad (3.9.4)$$

The lower bound is attained for $\mathbb{P}(F^{\leftarrow}(U) \leq x, F^{\leftarrow}(1 - U) \leq y)$.

We leave the proof of the Fréchet-Hoeffding bounds to the reader as an exercise (see Exercise 3.T.36).

The remark regarding the lower bound can be justified as follows. Using Lemma 3.1.1(a), we obtain:

$$\mathbb{P}(F^{\leftarrow}(U) \leq x, F^{\leftarrow}(1 - U) \leq y) = \mathbb{P}(U \leq F(x), 1 - U \leq F(y))$$

$$\leq F(x) + F(y) - 1,$$

provided that $F(x) + F(y) - 1 \geq 0$.

Remark 3.9.7 The inequality (3.9.4) can be interpreted in terms of stochastic ordering for random vectors $\mathbf{X}_1 = (X_{11}, X_{12})$ and $\mathbf{X}_2 = (X_{21}, X_{22})$, with joint c.d.f.s $F_1(x, y)$ and $F_2(x, y)$, respectively.

- We say that $\mathbf{X}_1 \leq_{\text{lo}} \mathbf{X}_2$ if $F_1(x, y) \geq F_2(x, y)$ for all x, y.
- Another stochastic order is defined based on the upper tails of the c.d.f., where $\bar{F}_i(x, y) = 1 + F_i(x, y) - F_i(x, \infty) - F_i(\infty, y)$, representing upper orthants. We say that $F_1 \leq_{\text{uo}} F_2$ if $\bar{F}_1(x, y) \leq \bar{F}_2(x, y)$ for all x, y.

Let U_1, U_2 be i.i.d. uniform $\mathcal{U}[0, 1)$ random variables. If $F_2(x, y) = F(x, y)$ and $F_1(x, y)$ is the c.d.f. of $(F^{\leftarrow}(U_1), F^{\leftarrow}(1 - U_2))$, then, in terms of stochastic ordering, $F_1 \geq_{\text{lo}} F_2$. Since $F_1(x, y)$ and $F_2(x, y)$ share the same marginal c.d.f. $F(x)$, it follows that $F_1 \leq_{\text{uo}} F_2$.

According to Müller and Stoyan [159], Theorem 3.3.16 states that $\mathbf{X}_1 \leq_{\text{uo}} \mathbf{X}_2$ if and only if, for every non-decreasing univariate function k_i (under mild integrability conditions),

$$\mathbb{E}k_1(X_{11})k_2(X_{12}) \leq \mathbb{E}k_1(X_{21})k_2(X_{22}).$$

Thus, for all (X_1^*, X_2^*) with the same marginals $X_1^* \sim X$ and $X_2^* \sim X$,

$$\mathrm{Corr}\,(F^{\leftarrow}(U), F^{\leftarrow}(1-U)) \leq \mathrm{Corr}\,(X_1^*, X_2^*).$$

We will revisit this result in the context of antithetic sampling; see Sect. 5.1.2. ∎

3.9.3 Some Classical Copulas

We denote a random vector with c.d.f. $C(u_1, u_2)$ by (V_1, V_2). Common examples of copulas include:

- **Fréchet upper bound copula:**

$$C_U(u_1, u_2) = \min(u_1, u_2), \qquad 0 \leq u_1, u_2 \leq 1;$$

  simulated by $V_1 = U$ and $V_2 = U$, where $U \sim \mathcal{U}[0, 1)$.
- **Fréchet lower bound copula:**

$$C_L(u_1, u_2) = \max(0, u_1 + u_2 - 1), \qquad 0 \leq u_1, u_2 \leq 1;$$

  simulated by $V_1 = U$ and $V_2 = 1 - U$, where $U \sim \mathcal{U}[0, 1)$.
- **Independence copula:**

$$C_I(u_1, u_2) = u_1 u_2, \qquad 0 \leq u_1, u_2 \leq 1;$$

  simulated by $V_1 = U_1$ and $V_2 = U_2$, where U_1, U_2 are independent and uniformly distributed on $\mathcal{U}[0, 1)$.

Remark 3.9.8 The Fréchet upper bound corresponds to perfect positive dependence, with mass concentrated along the diagonal $u_1 = u_2$, while the Fréchet lower bound corresponds to perfect negative dependence along the diagonal $u_1 = 1 - u_2$. The independence copula C_I represents independent uniform distributions. A convex combination of C_U, C_L, and C_I also defines a valid copula. Fréchet introduced a new copula defined by

$$C_{\alpha,\beta}(u_1, u_2) = \alpha C_U(u_1, u_2) + (1 - \alpha - \beta)C_I(u_1, u_2) + \beta C_L(u_1, u_2), \qquad (3.9.5)$$

where $\alpha, \beta \geq 0$ and $\alpha + \beta \leq 1$. ∎

Example 3.9.9 (Normal Copula) We will consider the unit variance and general variance cases separately.

Bivariate Normal with Unit Variance Consider a bivariate normal distribution (X, Y) with mean zero and correlation matrix:

$$\Sigma = \begin{pmatrix} 1 & \rho \\ \rho & 1 \end{pmatrix},$$

Here, $\mathbb{V}\mathrm{ar}(X) = \mathbb{V}\mathrm{ar}(Y) = 1$, and ρ is the correlation between X and Y. The p.d.f. of this distribution is given by:

$$\phi_{\Sigma}(t_1, t_2) = \frac{1}{2\pi\sqrt{1 - \rho^2}} \exp\left(-\frac{z}{2(1 - \rho^2)}\right),$$

where

$$z = t_1^2 - 2\rho t_1 t_2 + t_2^2.$$

Denote the corresponding bivariate normal c.d.f. by Φ_{Σ}. By Proposition 3.9.3, the Gaussian (normal) copula is given by:

$$C(u_1, u_2) = \Phi_{\Sigma}(\Phi^{-1}(u_1), \Phi^{-1}(u_2)).$$

This can be expressed as:

$$C(u_1, u_2) = \int_{-\infty}^{\Phi^{-1}(u_1)} \int_{-\infty}^{\Phi^{-1}(u_2)} \frac{1}{2\pi\sqrt{1 - \rho^2}} \exp\left(-\frac{x^2 + y^2 - 2\rho xy}{2(1 - \rho^2)}\right) dx\, dy.$$

To simulate (V_1, V_2), we apply Proposition 3.9.4:

$$(V_1, V_2) = (\Phi(X), \Phi(Y)), \tag{3.9.6}$$

where $(X, Y) \sim \mathcal{N}(0, \Sigma)$. Recall a standard method to generate (X, Y):

- Generate independent standard normal variables $Z_1, Z_2 \sim \mathcal{N}(0, 1)$.
- Set: $X = Z_1, \quad Y = \rho Z_1 + \sqrt{1 - \rho^2} Z_2$.

Bivariate Normal with General Variance Now consider a bivariate normal vector (X', Y') with an arbitrary covariance matrix:

$$\Sigma' = \begin{pmatrix} \sigma_X^2 & \rho\sigma_X\sigma_Y \\ \rho\sigma_X\sigma_Y & \sigma_X^2 \end{pmatrix}.$$

Here, $\rho = \mathrm{Corr}(X', Y')$. To sample (X', Y'), we first generate $(X, Y) \sim \mathcal{N}(\mathbf{0}, \Sigma)$ and scale it as follows:

$$X' = \sigma_X X, \quad Y' = \sigma_Y Y.$$

Thus, to generate a bivariate normal copula with covariance matrix Σ', we sample (X', Y') from $\mathcal{N}(\mathbf{0}, \Sigma')$ and transform it:

$$(V_1', V_2') = \left(\Phi\left(\frac{X'}{\sigma_X}\right), \, \Phi\left(\frac{Y'}{\sigma_Y}\right) \right).$$

This transformation ensures that the dependence structure remains the same regardless of the individual variances.

Limiting Behavior   It can be shown that as $\rho \to 0$, the limiting copula is C_I; as $\rho \to 1$, it converges to C_U; and as $\rho \to -1$, it converges to C_L.

Figure 3.20 shows scatter plots of 200 simulated replications of normal copulas with $\rho = 0.9$ and $\rho = -0.8$. The script `ch3_copulas.py` shows samples from the normal, Clayton, and Frank copulas. $\Diamond$

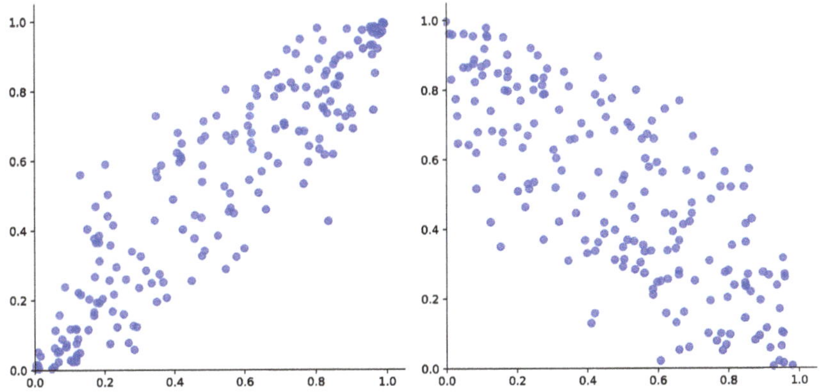

Fig. 3.20 Simulated normal copula: $\rho = 0.9$ (left), $\rho = -0.8$ (right)

Example 3.9.10 (Impact of Dependence Structure on Joint Distribution) In Example 3.9.1, we considered a joint distribution (X, Y) with the density:

$$f(x, y) = x \exp(-x(y + 1)), \quad x, y \geq 0,$$

where the marginal distributions were exponential Exp(1) for X and Pareto Par(1) for Y. A scatter plot of 100 samples from this distribution was previously shown in Fig. 3.19.

To further illustrate the role of the dependence structure in shaping a joint distribution, we now compare this original sample to (X', Y'), where the marginals remain the same (i.e., $X' \sim \text{Exp}(1)$ and $Y' \sim \text{Par}(1)$), but the dependence is altered using a normal copula. We generate (X', Y') as follows:

- First, sample (V_1, V_2) from a bivariate normal copula with a given correlation ρ, as described in (3.9.6).
- Then, transform these copula samples into the desired marginals:

$$
X' = F^{-1}_{\text{Exp}(1)}(V_1) = \log(1 - V_1),
$$

$$
Y' = F^{-1}_{\text{Par}(1)}(V_2) = \frac{1}{1 - V_2} - 1. \tag{3.9.7}
$$

Here, $F^{-1}_{\text{Exp}(1)}$ and $F^{-1}_{\text{Par}(1)}(X)$ denote the inverse cumulative distribution functions of the Exp(1) and Par(1) distributions, respectively. Thus, X' follows an exponential Exp(1) distribution, and Y' follows a Pareto Par(1) distribution (see Exercise 3.T.37).

Figure 3.21 presents scatter plots of 100 samples of (X', Y') generated using a normal copula with $\rho = -0.5$ (left panel) and with $\rho = 0.8$ (right panel). The implementation is available in `ch3_copula_Exp1_Par1_marginals.py`. Comparing these plots to Fig. 3.19, we observe that:

- For $\rho = -0.5$, there is a moderate negative dependence between X' and Y', meaning large values of X' tend to correspond to smaller values of Y', and vice versa.
- For $\rho = 0.8$, a strong positive dependence is imposed, so larger values of X' are more likely to be associated with larger values of Y'.
- In contrast, the original distribution in Fig. 3.19 has a different intrinsic dependence structure, as dictated by its joint density function.

These simulations demonstrate how different dependence structures can significantly affect the shape of a joint distribution, even when the marginals remain unchanged. $\Diamond$

3.10 Can We See the "Heaviness" of a Distribution?

In this section, we provide sample simulations based on the algorithms discussed in this chapter. Additionally, we aim to visually illustrate the behavior of light and heavy-tailed random variables. The plots were generated using the script `ch3_heavy_vs_light_tail.py`.

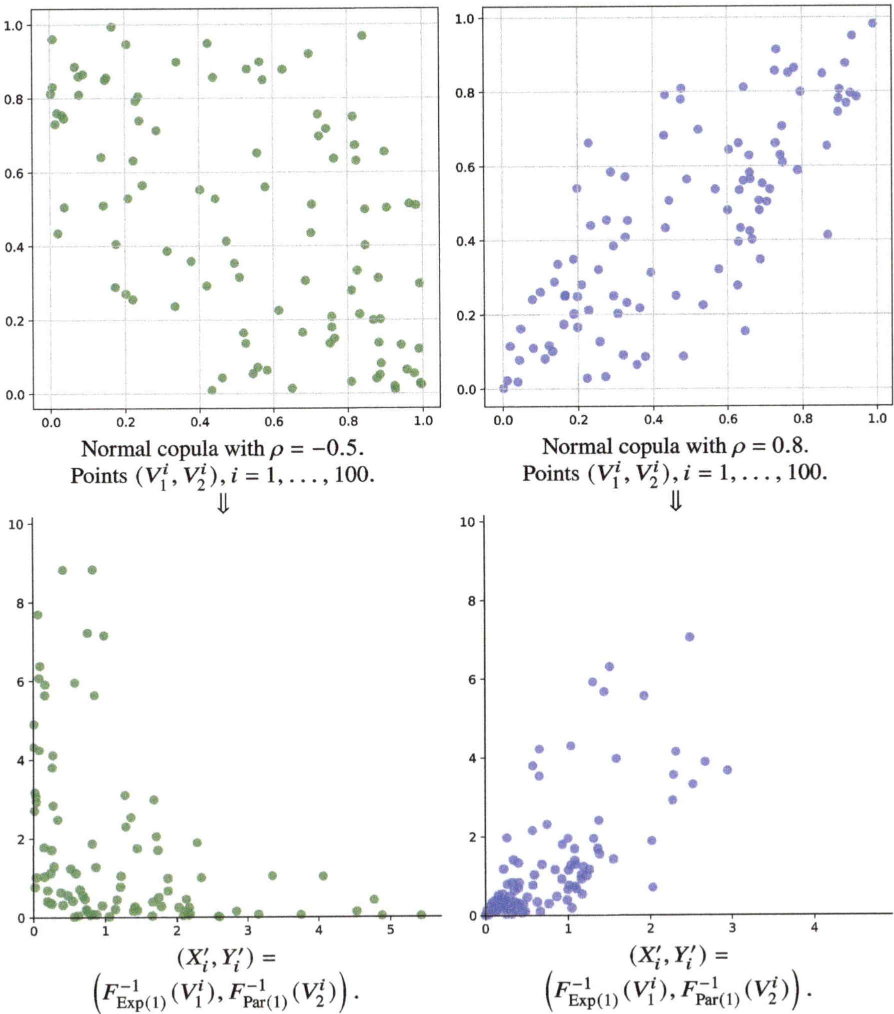

Fig. 3.21 Scatter plot of 100 samples of (X', Y') with Exp(1) and Par(1) marginals, generated using a normal copula with $\rho = -0.5$ and $\rho = 0.8$

Suppose we generate a sequence $X_1, \ldots, X_R$ of i.i.d. random variables and define

$$S_n = \sum_{i=1}^{n} X_i, \quad n = 0, 1, \ldots, R, \quad \text{with } S_0 = 0.$$

We will consider several distributions of X, all with the same mean value $\mathbb{E}X = 1$.

In Fig. 3.22, we display two panels:

(a) The running averages S_n/n for $n = 1, \ldots, 5000$ for three different distributions:

- $X_i \sim \mathcal{N}(1, 1)$ (green curve).
- $X_i \sim \mathrm{Exp}(1)$ (blue curve),
- $X_i = 0.2\, Y_i$ with $Y_i \sim \mathrm{Pareto}(1.2)$ (red curve),

Since the Pareto variables have mean 5, scaling by 0.2 ensures that all three distributions have mean 1. Notice that the heavy tail of the Pareto distribution causes markedly slower convergence of S_n/n.

(b) The time series of the individual X_i values used in (a):

- $X_1, \ldots, X_{5000}$ are i.i.d. $\mathcal{N}(1, 1)$,
- $X_{5001}, \ldots, X_{10000}$ are i.i.d. $\mathrm{Exp}(1)$,
- $X_{10001}, \ldots, X_{15000}$ are i.i.d. $0.2 \cdot \mathrm{Pareto}(1.2)$.

Observe the distinct characteristics of each segment: although each segment has a mean of 1, the spread of values increases significantly from the first segment (normal distribution) to the second (exponential distribution) and most dramatically in the third (Pareto distribution) due to its heavier tail. For clarity, the y-axis is clipped at 25 (the maximum observed value was 5.21 for $\mathcal{N}(1, 1)$, 7.93 for $\mathrm{Exp}(1)$ and 591.29 for $0.2 \cdot \mathrm{Pareto}(1.2)$).

In Fig. 3.23, we similarly display two panels:

(a) The running averages S_n/n for $n = 1, \ldots, 5000$ for three different distributions:

- $X_i \sim \mathrm{LN}(-1/2, 1)$ (green curve).
- $X_i \sim \mathrm{W}(1/2, \sqrt{2})$ (blue curve),
- $X_i = 0.2\, Y_i$ with $Y_i \sim \mathrm{Pareto}(1.2)$ (red curve)

(b) The time series of the individual X_i values used in (a):

- $X_1, \ldots, X_{5000}$ are i.i.d. $\mathrm{LN}(-1/2, 1)$,
- $X_{5001}, \ldots, X_{10000}$ are i.i.d. $\mathrm{W}(1/2, \sqrt{2})$,
- $X_{10001}, \ldots, X_{15000}$ are i.i.d. $0.2 \cdot \mathrm{Pareto}(1.2)$.

Note that the Pareto distribution is exactly the same as in the previous case—however, the randomness was different. Observe the distinct characteristics of each segment: all three distributions are heavy-tailed, however, we can see differences in samples. Note that the Pareto distribution has infinite variance. As previously, the y-axis is clipped at 25 (the maximum observed value was 28.2 for $\mathrm{LN}(-1/2, 1)$, 27.82 for $\mathrm{W}(1/2, \sqrt{2})$ and 388.32 for $0.2 \cdot \mathrm{Pareto}(1.2)$).

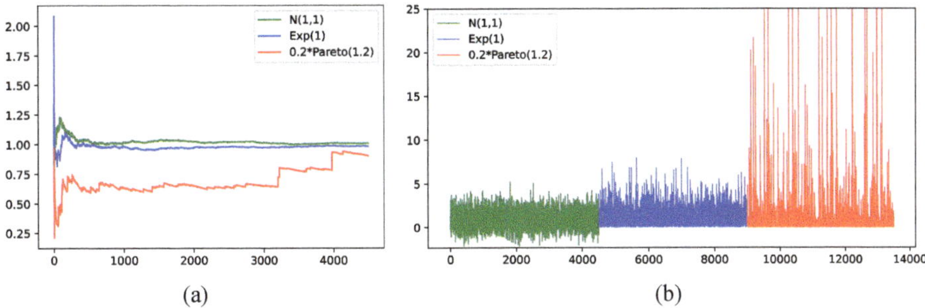

(a)   (b)

Fig. 3.22 (**a**) Running averages S_n/n for $n = 1, \ldots, 5000$ with $S_n = \sum_{i=1}^{n} X_i$, where the X_i are i.i.d. with $\mathcal{N}(1, 1)$ (*green*) , Exp(1) (*blue*), and 0.2·Pareto(1.2) (*red*) distributions. (**b**) The corresponding time series of X_i: $X_1, \ldots, X_{5000}$ from $\mathcal{N}(1, 1)$, $X_{5001}, \ldots, X_{10000}$ from Exp(1), and $X_{10001}, \ldots, X_{15000}$ from 0.2·Pareto(1.2); the y-axis is clipped at 25

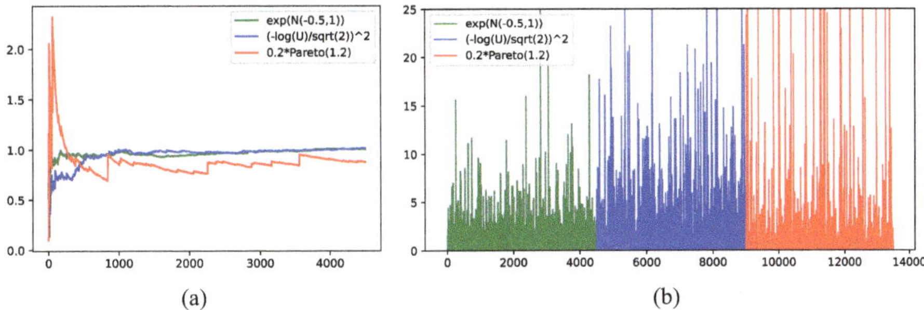

(a)   (b)

Fig. 3.23 (**a**) Running averages S_n/n for $n = 1, \ldots, 5000$ with $S_n = \sum_{i=1}^{n} X_i$, where the X_i are i.i.d. with LN$(-1/2, 1)$ (*green*) , Weibull W$(1/2, \sqrt{2})$ (*blue*), and 0.2·Pareto(1.2) (*red*) distributions. (**b**) The corresponding time series of X_i: $X_1, \ldots, X_{5000}$ from LN$(-1/2, 1)$, $X_{5001}, \ldots, X_{10000}$ from Weibull W$(1/2, \sqrt{2})$, and $X_{10001}, \ldots, X_{15000}$ from 0.2·Pareto(1.2); the y-axis is clipped at 25

3.11 Bibliographical Notes

The topic of generating random variables with given distributions has been extensively studied. For an elementary introduction, we refer to Ross [182], while a more advanced treatment can be found in Asmussen and Glynn [8], Fishman [55], and Dagpunar [29]. A concise yet insightful perspective is also provided in Madras [138].

The inverse transform method is one of the fundamental techniques for generating random variables and is well described in Knuth [107] and Devroye [33]. Properties of generalized inverse functions, essential for this method, can be found in Embrechts and Hofert [47]. The application of the inverse transform method to geometric and other lattice distributions is covered in the statistical literature, particularly in Ripley [175].

The acceptance-rejection method, introduced by John von Neumann, is a versatile approach when direct inversion is impractical. A comprehensive discussion of this method and its applications can be found in Devroye [33] and Ripley [175]. The fundamental theoretical aspects of the method, including its efficiency and its relationship to importance sampling, are discussed in Fishman [55]. For details on its applications, particularly in Bayesian statistics, see Robert and Casella [176].

The composition method, which expresses a target distribution as a mixture of simpler distributions, is treated in detail in Devroye [33] and Dagpunar [28]. An example of its use in modeling non-central chi-squared distributions is provided in Madras [138].

The generation of normally distributed random numbers has a long history, with ad hoc methods such as the Box-Muller transformation being widely used. The original formulation of the Box-Muller method can be found in Box and Muller [18]. A refinement of the Box-Muller method was proposed by Marsaglia and Bray [143]. Additionally, the Ziggurat algorithm, a widely used approach for generating Gaussian and other distributions, is discussed in Marsaglia and Tsang [144].

There is a substantial body of literature on generating random elements uniformly distributed on various geometric objects, though relevant results are often scattered across different sources. A comprehensive reference for the general theory of area measures on manifolds, including hyper-spheres and hyper-ellipsoids, is found in Chapter 5 of Munkres [160]. A fundamental tool for these constructions, spherical coordinates, is well explained in Blumenson [17]. Another method for generating random points on a sphere is presented by Marsaglia [140]. Additional methods not discussed in this chapter can be found in Harman and Lacko [79].

Section 3.8.5 on the average area of a shadow of a convex 3D body was inspired by the video "*A tale of two problem solvers | Average cube shadow area*" by 3Blue1Brown, available on YouTube (https://www.youtube.com/watch?v=ltLUadnCyi0).

For copulas, refer to the book by Nelsen [164], the book by Joe [91], and the paper by Embrechts et al. [48].

3.12 Exercises

Theoretical Exercises

3.T.1 Demonstrate properties (P2), (P5) and (P6) of the generalized inverse $F^{\leftarrow}(t)$ from Sect. 3.1 (ITM method; exponential and Pareto distribution), using diagrams similar to those included in that section.

3.T.2 Let $x_1 < x_2 < \cdots < x_n$ and $p_i > 0$, $\sum_{i=1}^{n} p_i = 1$. Suppose the c.d.f. is given as:

$$F(x) = \begin{cases} 0, & x < x_1, \\ p_1 + \cdots + p_i, & x_i \le x < x_{i+1}, \quad i = 1, \ldots, n-1, \\ 1, & x_n \le x. \end{cases}$$

Write down $F^{\leftarrow}(t)$.

3.T.3 Suppose $X_1, \ldots, X_R$ are i.i.d. and let

$$\hat{F}_R(x) = \frac{1}{R} \sum_{i=1}^{R} \mathbb{1}(X_i \le x)$$

be the empirical distribution function. Find $\hat{F}_R^{\leftarrow}(u)$.

3.T.4 Let X be a random variable with a c.d.f. F, and consider a conditional random variable $Y \overset{\mathcal{D}}{=} [X \mid X \in (a, b)]$, where $\mathbb{P}(X \in (a, b)) > 0$ (or $F(b) > F(a)$).

(a) Show that $Y \sim G$, where

$$G(x) = \begin{cases} 0, & x < a, \\ \frac{F(x)-F(a)}{F(b)-F(a)}, & a \le x < b, \\ 1, & x \ge b. \end{cases}$$

(b) Let

$$V = F(a) + (F(b) - F(a))U, \quad \text{where } U \sim \mathcal{U}[0, 1).$$

Show that $Y \overset{\mathcal{D}}{=} F^{\leftarrow}(V)$ has the required c.d.f. G.

3.T.5 Let $X \sim F$. Show that a random variable distributed as the overshoot $Y \overset{\mathcal{D}}{=} (X - a \mid X > a)$ can be generated as

$$F^{\leftarrow}(U\overline{F}(a) + F(a)) - a.$$

Here $U \sim \mathcal{U}[0, 1)$ and $\overline{F}(a) < 1$.

3.T.6 Modify Algorithm 10 (ITR) to allow generating lattice random variables on $\mathbb{Z}$ with probability function $(p_i, \ i \in \mathbb{Z})$.

3.T.7 Let $U_1, U_2, \ldots$ be i.i.d. $\mathcal{U}[0, 1)$ distributed random variables. Define

$$N = \min\{n : n \ge 2, \ U_n > U_{n-1}\}.$$

(a) Show that $\mathbb{E}N = e$ (hence, e can be estimated by simulating uniform random variables and stopping the first time one exceeds its immediate predecessor; the procedure needs to be repeated several times to estimate $\mathbb{E}N$).

(b) Compute $\mathbb{V}\text{ar}N$.

Hint: Compute $\mathbb{P}(N > n)$ and then the distribution $\mathbb{P}(N = n) = \mathbb{P}(N > n - 1) - \mathbb{P}(N > n)$.

3.T.8 The c.d.f. is given as:

$$
F(x) = \begin{cases}
0, & x < 0, \\[2mm]
\frac{1}{2}x, & x \in [0, 1), \\[2mm]
\frac{1}{2}, & x \in [1, 2), \\[2mm]
\frac{3}{4}, & x \in [2, 3), \\[2mm]
\frac{3}{8} + \frac{1}{8}x, & x \in [3, 5), \\[2mm]
1, & x \geq 5.
\end{cases}
$$

Write a procedure to generate $Y \sim F$ using the composition method.

3.T.9 Write a procedure to generate a random number X with a triangular distribution $\text{Tri}(a, b)$, having density:

$$
f(x) = \begin{cases}
c(x - a), & a < x < \frac{a+b}{2}, \\[2mm]
c(b - x), & \frac{a+b}{2} < x < b,
\end{cases}
$$

where the normalizing constant is $c = \frac{4}{(b-a)^2}$.
Hint: First consider the density of $U_1 + U_2$.

3.T.10 Suppose $U \sim \mathcal{U}[0, 1)$. Demonstrate that:

$$
X = \log\left(\frac{U}{1 - U}\right)
$$

has a logistic distribution with c.d.f. $F(x) = \frac{1}{1+e^{-x}}$.

3.T.11 Show the procedure for generating a random number X with density

$$
f(x) = n(1 - x)^{n-1}.
$$

Hint: You can compute the density function of $\min(U_1, \ldots, U_n)$.

3.T.12 Recall that the density of Gamma(α, λ) is given by

$$f(x) = \frac{\lambda^\alpha}{\Gamma(\alpha)} x^{\alpha-1} e^{-\lambda x}, \quad x \geq 0.$$

Let $X \sim$ Gamma(a, λ) and $Y \sim$ Gamma(b, λ), assuming X and Y are independent. Show that:

(a) $X + Y \sim$ Gamma$(a + b, \lambda)$.
(b) Let Z_i, $i = 1, \ldots, d$, be i.i.d. $\mathcal{N}(0, 1)$ random variables. Show that

$$\chi_d^2 = Z_1^2 + \cdots + Z_d^2$$

has a Gamma $\left(\frac{d}{2}, \frac{1}{2}\right)$ distribution.

(c) Let $X' \sim$ Gamma$(a, 1)$. Show that $X = \frac{X'}{\lambda}$ has a Gamma(a, λ) distribution.

Note that (c) implies that to sample $X \sim$ Gamma(a, λ), it suffices to sample $X' \sim$ Gamma$(a, 1)$ and take $X = \frac{X'}{\lambda}$.

3.T.13 Show two procedures for generating a random number with density function

$$f(x) = \sum_{j=1}^{N} a_j x^j, \qquad 0 \leq x \leq 1,$$

where $a_j > 0$. What condition must the coefficients (a_j) satisfy?
Hint: In one method, apply the composition method.

3.T.14 A random variable Z has a Cauchy distribution if it has the density $f(z) = \frac{1}{\pi(1+z^2)}$, $z \in \mathbb{R}$.

(a) Let $U \sim \mathcal{U}(0, 1)$. Show that $Z = \tan(\pi(U - 1/2))$ has a Cauchy distribution.
(b) Let X and Y be independent standard normal random variables. Show that $Z = \frac{X}{Y}$ has a Cauchy distribution.

3.T.15 For a non-negative random variable Y with c.d.f. F, define the cumulative hazard rate function $H(t)$ as:

$$H(t) = -\log(1 - F(t)).$$

(a) Show that if Y has a density f, then $H(t)$ can be expressed as:

$$H(t) = \int_0^t h(s)\,ds,$$

$$h(t) = \frac{f(t)}{1 - F(t)},$$

for all t where f is continuous. The function $h(t)$ is called the hazard rate function and represents $\mathbb{P}(t < Y \le t + \Delta t \mid Y > t) = h(t)\Delta t$.

(b) Show that one can sample Y with a cumulative hazard rate in the following way:

(i) Sample $X \sim \mathrm{Exp}(1)$ (a standard exponential random variable),
(ii) Return $Y = H^{\leftarrow}(X)$, where $H^{\leftarrow}(x) = \inf\{y : H(y) \ge x\}$.

3.T.16 A random variable X has an Erlang distribution $\mathrm{Erl}(n, \lambda)$ if its density is

$$f(x) = \frac{\lambda^n}{\Gamma(n)} x^{n-1} e^{-\lambda x}, \quad x \ge 0.$$

Show that

$$\bar{F}(x) = 1 - F(x) = \mathbb{P}(X > x) = e^{-\lambda x}$$

$$\times \left(1 + \lambda x + \frac{(\lambda x)^2}{2!} + \cdots + \frac{(\lambda x)^{n-1}}{(n-1)!}\right).$$

3.T.17 Define the Weibull distribution $W(\alpha, c)$ by:

$$F(x) = \begin{cases} 0, & x < 0, \\ 1 - \exp(-cx^\alpha), & x \ge 0. \end{cases}$$

Assuming $c > 0, \alpha > 0$, show that

$$X = \left(\frac{-\log U}{c}\right)^{1/\alpha}$$

has a $W(\alpha, c)$ distribution, where $U \sim \mathcal{U}[0, 1)$.

3.T.18 A random variable Y has a discrete Pareto distribution $\mathrm{DPar}(\alpha)$ if it has the probability function

$$p_k = \frac{1}{k^\alpha} - \frac{1}{(1+k)^\alpha}, \quad k = 1, 2, \ldots.$$

(a) Show that $Y = \lceil X \rceil$ is $\mathrm{DPar}(\alpha)$ if $X \sim \mathrm{Par}(\alpha)$.
(b) Show that $Y = \lfloor U^{-1/\alpha} \rfloor$ has a $\mathrm{DPar}(\alpha)$ distribution.
(c) Show that $\mathbb{P}(Y = k) \sim \alpha k^{-\alpha-1}$ as $k \to \infty$.

3.T.19 Write down an algorithm and compute the acceptance probability in the AR-d method for generating a random number X with the probability function $(p_k, \ k = 1, 2, \ldots)$, where

$$p_k = \frac{90}{\pi^4} \frac{1}{k^4}.$$

3.T.20 Show how to generate a random variable with a Laplace distribution (or bilateral exponential distribution) with density:

$$f(x) = \begin{cases} e^{2x}, & -\infty < x < 0, \\[2mm] e^{-2x}, & 0 \le x < \infty. \end{cases}$$

3.T.21 Recall that $(a, b, 0)$ is a class of probability functions on $\{0, 1, \ldots\}$ following the recurrence

$$p_k = p_{k-1}\left(a + \frac{b}{k}\right), \quad k = 1, 2, \ldots.$$

(a) Show that binomial, Poisson, and negative binomial distributions belong to this class.

(b) Consider the similar class $(a, b, 1)$ of probability functions on $\{1, 2, \ldots\}$ fulfilling the same recurrence, but for $k = 2, 3, \ldots$. Show that truncated binomial, truncated Poisson, and truncated negative binomial distributions, as well as the logarithmic distribution with probability function

$$p_k = \frac{p^k}{-k\log(1-p)}, \quad k = 1, 2, \ldots,$$

belong to this class.

For a distribution p_k on $\{0, 1, \ldots\}$, we define a truncated distribution as:

$$p'_k = \frac{p_k}{c}, \quad k = 1, 2, \ldots, \quad \text{where } c = \sum_{k \ge 1} p_k.$$

3.T.22 In Example 3.5.4, we demonstrated how to generate random numbers with a normal distribution $\mathcal{N}(0, 1)$ using the AR-c method with $g(y) = e^{-1}\mathbf{1}(y \ge 0)$. Present a version of the algorithm in which the distribution $\mathrm{Exp}(\lambda)$ is used for some $\lambda > 0$, i.e.,

$$g_2(y) = \lambda e^{-\lambda y}\mathbf{1}(y \ge 0).$$

For which λ is the acceptance probability maximized?

3.T.23 Show (and justify) the AR-c method for generating random numbers with a normal distribution $X \sim \mathcal{N}(0, 1)$ using a Cauchy-distributed Y. Show the optimal c.

Can we simulate Cauchy-distributed random numbers using a normally distributed Y with the AR-c method?

3.T.24 What is the distribution of X, the output of Algorithm 26?

Algorithm 26

1: Generate Y
2: **repeat**
3:      Generate $U_1 \sim \mathcal{U}[0, 1)$ and set $Y = \left\lfloor -\frac{\log U_1}{\log 2} \right\rfloor$
4:      Generate $U_2 \sim \mathcal{U}[0, 2)$
5: **until** $U_2 \leq \frac{2^{Y+1}}{4Y!}$
6: **return** $X = Y$

3.T.25 What is the distribution of X, the output of Algorithm 27?

Algorithm 27

1: **repeat**
2:      Generate Y uniformly on $\{1, 2, \ldots, n\}$
3:      Generate $U \sim \mathcal{U}[0, 1)$
4: **until** $U \leq \frac{Y}{n}$
5: **return** $X = Y$

3.T.26 What is the distribution of X, the output of Algorithm 28?

Algorithm 28

1: **repeat**
2:      Generate $U_1 \sim \mathcal{U}[0, 1)$
3:      Set

$$Y = j \quad \text{if} \quad \frac{(j-1)j}{n(n+1)} \leq U_1 \leq \frac{j(j+1)}{n(n+1)}.$$

         (Y takes values $1, 2, \ldots, n$).
4:      Generate $U \sim \mathcal{U}[0, 1)$
5: **until** $U \cdot Y \leq 1$
6: **return** $X = Y$

3.T.27 Consider the density function

$$f(x) = Cxe^{-x^4}, \qquad x \geq 0,$$

where $C = \frac{4}{\sqrt{\pi}}$. Show two different algorithms for generating random numbers with this distribution.

3.T.28 Show how to sample from the densities $g_1(x)$ and $g_2(x)$ from Example 3.5.5 .

3.T.29 Suppose the two-dimensional distribution of the random vector $\mathbf{X} = (X_1, X_2)$ is given by

$$f(x_1, x_2) = ce^{-|\mathbf{x}|^4} = ce^{-(x_1^2 + x_2^2)^2}, \qquad (x_1, x_2) \in \mathbb{R}^2,$$

where $c = \frac{\pi^{3/2}}{2}$.

(a) Express this density in polar coordinates (D, Θ). Show that D and Θ are independent. Describe how to generate $\mathbf{X} \sim f$ using polar coordinates.

(b) Show how to apply the AR-c algorithm. Is it necessary to know the exact value of c?

3.T.30 Let η be a random variable with distribution $\mathbb{P}(\eta = -1) = \mathbb{P}(\eta = 1) = \frac{1}{2}$, independent of X, where X has the half-normal distribution with the p.d.f.

$$f(x) = \frac{2}{\sqrt{2\pi}} e^{-\frac{x^2}{2}}, \qquad 0 < x < \infty.$$

Show that ηX is normally distributed $\mathcal{N}(0, 1)$.

3.T.31 Show two algorithms for generating a random point with the uniform distribution $\mathcal{U}[B]$, where $B = \{x \in \mathbb{R}^2 : |x| \le 1\}$.

Hint: In one of the algorithms, use Lemma 3.7.5 on the polar characterization of standard normal variables.

3.T.32 Justify the following methods for generating the multinomial distribution $M(n, \mathbf{a})$, where $\mathbf{a} = (a_1, \ldots, a_n)$.

- The first method generalizes the ad hoc methods for the binomial distribution from Sect. 3.3.1. It involves generating independent random vectors with the multinomial distribution $M(1, \mathbf{a})$, taking values $(0, \ldots, 1, \ldots, 0)$ (where 1 is in the i-th position) with probability a_i, and then summing up these vectors. Specifically:

$$v = \min\{j \in \{1, \ldots, k\} : \sum_{i=1}^{j} a_i \ge U\},$$

where $U \sim \mathcal{U}[0, 1)$, assigns $\mathbf{X}$ a value of 1 at the v-th coordinate and 0 elsewhere.

- The second method uses the unique property of the multinomial distribution $\mathbf{X} \sim M(n, a_1, \ldots, a_n)$ of the vector $\mathbf{X} = (X_1, \ldots, X_d)$:

  - The distribution of X_{j+1} conditional on $X_1 = k_1, \ldots, X_j = k_j$ is binomial $B(n - k_1 - \ldots - k_j, a_{j+1}/(a_1 + \cdots + a_j))$.
  - X_1 has the binomial distribution $B(n, a_1)$.

3.T.33 [Flury [59]] Let $\mathbf{X} \in \mathbb{R}^n$ be a random vector and $Y \in \mathbb{R}$ a random variable.

(i) If $\mathbf{X}$ has density $g(\mathbf{x})$ and, for $c > 1$, the random variable $(Y \mid \mathbf{X} = \mathbf{x})$ has distribution $\mathcal{U}(0, cg(\mathbf{x}))$, then the random vector $(\mathbf{X}, Y) \in \mathbb{R}^{n+1}$ is uniformly distributed $\mathcal{U}(\mathcal{A})$, where

$$\mathcal{A} = \{(\mathbf{x}, y) : \mathbf{x} \in \mathbb{R}^n, y \in \mathbb{R}, 0 < y < cg(\mathbf{x})\}.$$

(ii) If the $(n + 1)$-dimensional random vector $(\mathbf{X}, Y)$ is uniformly distributed $\mathcal{U}(\mathcal{B})$, where

$$\mathcal{B} = \{(\mathbf{x}, y) : \mathbf{x} \in \mathbb{R}^n, y \in \mathbb{R}, 0 < y < f(\mathbf{x})\},$$

then $\mathbf{X}$ has the density function $f(\mathbf{x})$.

(iii) Using (i), (ii), and the result from Sect. 3.4, prove Proposition 3.5.3.

3.T.34 Using the result of Exercise 3.T.33, propose an algorithm to generate a random variable $X \sim \text{Beta}(\alpha, \beta)$.

3.T.35 In actuarial mathematics, the *Makeham distribution* describes the future lifetime T_x of a life at age x:

$$\mathbb{P}(T_x > t) = e^{-At - m(c^{t+x} - c^x)}.$$

Write pseudocode for generating random numbers from Makeham(A, m, c, x) with $A, m > 0$ and $c > 1$. For a specific example, take $A = 5 \cdot 10^{-4}$, $m = 7.5858 \cdot 10^{-5}$, and $c = \log 1.09144$ (based on Danish G82 mortality tables for males).

Hint: The tail distribution function can be factorized as $e^{-At - m(c^{t+x} - c^x)} = e^{-At} e^{-m(c^{t+x} - c^x)}$. Consider how to generate a random number with the tail distribution $e^{-m(c^{t+x} - c^x)}$.

3.T.36 [Asmussen and Glynn [8]] For a two-dimensional c.d.f. $F(x, y)$ of (Y_1, Y_2) with the same marginal distribution $F(x) = \mathbb{P}(Y_1 \le x) = \mathbb{P}(Y_2 \le x)$, the so-called Fréchet-Hoeffding bounds are given by:

$$(F(x) + F(y) - 1)_+ \le F(x, y) \le \min(F(x), F(y)).$$

(i) Show that the lower bound is attained for

$$\mathbb{P}(F^{\leftarrow}(U) \le x, F^{\leftarrow}(1 - U) \le y),$$

where $U \sim \mathcal{U}[0, 1)$, and prove that

$$\mathbb{P}(Y_1 > x, Y_2 > y) = 1 + F(x, y) - F(x) - F(y).$$

From these results, demonstrate (5.1.16).

(ii) Show that the following are valid c.d.f.s:

$$F_{\mathrm{L}}(x, y) = (F(x) + F(y) - 1)_+, \qquad F_{\mathrm{U}}(x, y) = \min(F(x), F(y)).$$

(iii) Prove the Fréchet-Hoeffding bounds.

Hint: Set $A = \{Y_1 \le x\}$ and $B = \{Y_2 \le y\}$. Use Boole's inequality:

$$\mathbb{P}(A \cup B) \le \mathbb{P}(A) + \mathbb{P}(B),$$

and the fact that

$$\mathbb{P}(A \cap B) \le \min(\mathbb{P}(A), \mathbb{P}(B)).$$

3.T.37 Let (V_1, V_2) be a random vector with marginals

$$V_i \sim \mathcal{U}[0, 1), \quad i = 1, 2.$$

Define the random vector (X', Y') (cf. with (3.9.7)) by

$$X' = -\log V_1, \qquad Y' = \frac{1}{1 - V_2} - 1.$$

Show that then, the marginal distributions of (X', Y') are

$$X' \sim \mathrm{Exp}(1) \quad \text{and} \quad Y' \sim \mathrm{Par}(1).$$

3.T.38 Show that the *Clayton copula* with parameter $\theta > 0$ can be sampled as follows: Draw independent $X \sim \mathrm{Gamma}(1/\theta, 1)$ and $U_1, U_2 \sim \mathcal{U}[0, 1)$, then substitute:

$$(V_1, V_2) = \left(\left(1 - \frac{\log U_1}{X} \right)^{-1/\theta}, \left(1 - \frac{\log U_2}{X} \right)^{-1/\theta} \right).$$

Lab Exercises

3.L.1 Consider the uniform distribution on the sphere S_2. In spherical coordinates (Θ, Φ), its p.d.f. is:

$$f(\theta, \phi) = \frac{1}{4\pi} \sin \phi, \qquad \theta \in (0, 2\pi), \ \phi \in (0, \pi).$$

Hence, the marginal p.d.f.s of Θ and Φ are $\frac{1}{2\pi}$ and $\frac{1}{2} \sin \phi$, respectively. Simulate 100 such random points and plot them.

For comparison, simulate points where both Θ and Φ have uniform distributions on $(0, 2\pi)$ and $(0, \pi)$, respectively. Plot the points and comment on the differences.

Note: This task is conceptually similar to the procedure shown, e.g., in Fig. 3.17, but applied to a sphere instead of an ellipsoid.

3.L.2 Consider a geometric distribution with success parameter $p = 1/100$. Simulate 1000 such numbers using two methods: (1) the Inverse Transform Method (ITM), (2) directly by computing $\lfloor \log U / \log p \rfloor$ for $U \sim \mathcal{U}[0, 1)$. Compare the running times of these methods.

3.L.3 Implement the Acceptance-Rejection method to generate random variables following a standard normal distribution. Use Exp(1) as the proposal distribution. Compare this method with the Box-Muller method in terms of performance (time/steps taken) and the quality of the generated numbers (e.g., using a chi-square test).

3.L.4 Design an Acceptance-Rejection method to generate random variables with the density provided in Example 3.6.2:

$$f(x) = C \sinh(\sqrt{xc})e^{-bx}, \qquad x \geq 0,$$

where $b, c > 0$ are fixed parameters and C is a normalizing constant. Consider an exponential distribution as the proposal distribution. Implement the method for $b = 4$ and $c = 2$.

Recall that $\sinh(t) = \frac{e^t - e^{-t}}{2}$.

3.L.5 Consider the set $\mathcal{S}_n$ of permutations of $\{1, \ldots, n\}$. Let $\sigma_{\mathrm{id}} = (1, 2, \ldots, n)$ denote the identity permutation. For $\sigma, \sigma' \in \mathcal{S}_n$, define a distance function:

$$d(\sigma, \sigma') = \frac{1}{n} \sum_{i=1}^{n} |\sigma(i) - \sigma'(i)|.$$

Explain why this is a valid distance function. Define a distribution on permutations:

$$\pi(\sigma) = \frac{1}{C} \lambda^{d(\sigma, \sigma_{\mathrm{id}})}, \quad \text{where } \lambda \in (0, 1).$$

In this distribution, permutations "close" to the identity permutation are more likely.

Implement a sampling procedure for $\lambda = 0.75$ and $n = 52$ using the Acceptance-Rejection method with a uniform proposal distribution. Simulate 1000 permutations from π and estimate the acceptance probability of the AR algorithm.

Discuss the feasibility of running the algorithm for smaller values of λ, e.g., $\lambda = 0.6$ or $\lambda = 0.5$.

Note: The distribution π (often with the Kendall τ distance) is called the *Mallows distribution*, used in ranking and sorting problems.

In Exercise 7.L.2, the reader will implement the Metropolis algorithm for this distribution, applicable for a wider range of λ.

3.L.6 Consider Exercise 3.L.5, but replace $d(\sigma, \sigma_{\mathrm{id}})$ with the number of fixed points:

$$\pi(\sigma) = \frac{1}{C}\lambda^{\mathrm{FP}(\sigma)}, \quad \text{where } \mathrm{FP}(\sigma) = \#\{i : \sigma(i) = i\}.$$

Implement the Acceptance-Rejection method for $n = 52$ and $\lambda = 0.75$. Can it be applied for $\lambda = 2$? What about the efficiency?

In Chap. 7, you will implement the Metropolis algorithm for this distribution (see Exercise 7.L.3).

3.L.7 Generate and plot 1000 points from a 2D normal distribution with mean (0,0) and covariance matrix:

$$\Sigma = \begin{pmatrix} 1 & 0.6 \\ 0.6 & 2 \end{pmatrix}.$$

Use the Cholesky decomposition (you may use a built-in procedure for this, but you are allowed to sample only i.i.d. standard normal random variables). Analyze the empirical correlation of the generated random vectors and compare it with the theoretical correlation.

3.L.8 Consider a 2D random walk. Initially, we are at (0,0). Each step has a 2D normal distribution with parameters from Exercise 3.L.7.

- Generate 1000 steps and plot the path of the random walk.
- Generate 1000 such random walks and plot their final positions.
- Scale the final positions by dividing both coordinates by $\sqrt{1000}$. Compare the plotted points with those from Exercise 3.L.7. Verify if the distribution of the endpoints matches the expected bivariate normal distribution.

3.L.9 The Erdős-Rényi graph $G(n, p)$ is a random graph constructed as follows: given n vertices, an edge between any pair of vertices exists with probability p (independently of other edges).

- The degree of a vertex in such a graph is a random variable. It is known that the limiting distribution (as $n \to \infty$) is Poiss(λ) with $\lambda = np$. Verify this via simulations. For $n = 100$ and $p = 0.1$, simulate $G(n, p)$ 200 times. Estimate the degree distribution (accumulated over 200 simulations) and compare it with the Poisson distribution with $\lambda = np = 0.1$. Plot the observed and expected distributions and perform a chi-square test (report the final p-value).
- It is known that for $p < \frac{\log n}{n}$, the graph is likely to be disconnected, whereas for $p > \frac{\log n}{n}$, the graph is likely to be connected (and as p increases, the probability of the graph being connected approaches 1).

Estimate the probability of the graph being connected for $p = 0.1$ and $p = 0.01$. Verify the results against the theory. Note that for $n = 100$, $\frac{\log n}{n} \approx 0.04605$.

3.L.10 Repeat the experiment from Sect. 3.10, simulating independent copies $X_1, \ldots, X_{15000}$, where:

- $X_1, \ldots, X_{5000}$ are i.i.d. Weibull $W(2, c)$, with c chosen such that $\mathbb{E}[X_1] = 1$,
- $X_{5001}, \ldots, X_{10000}$ are i.i.d. Weibull $W(1/2, c)$, with c chosen such that $\mathbb{E}[X_{5001}] = 1$,
- $X_{10001}, \ldots, X_{15000}$ are i.i.d. log-normal $LN(0, b)$, with b chosen such that $\mathbb{E}[X_{10001}] = 1$.

Comment on the results.

3.L.11 Simulate $n = 1000$ 2D normally distributed points $(X_1^1, X_2^1), \ldots, (X_1^n, X_2^n)$ with mean $(0, 0)$ and covariance matrix:

$$\Sigma = \begin{pmatrix} 1 & 0.7 \\ -0.7 & 1 \end{pmatrix}.$$

Plot the points (X_1^i, X_2^i) together with the transformed points $(U_i, U_i') = (F(X_1^i), F(X_2^i))$, where F is the cumulative distribution function (c.d.f.) of a standard normal $\mathcal{N}(0, 1)$ distribution. In other words, (U_i, U_i') are sampled from a Gaussian copula.

Verify the uniformity of the sequences (U_i) and (U_i') by creating histograms and performing a goodness-of-fit test.

How should the task be modified when $\mathbb{V}\mathrm{ar} X_2 = 4$?

3.L.12 Suppose $K \subset \mathbb{R}^3$ is defined as $K = E_1 \cup E_2$, where:

$$E_1 = \left\{ (x, y, z) : \frac{x^2}{a^2} + y^2 + z^2 = 1 \right\},$$

$$E_2 = \left\{ (x, y, z) : x^2 + y^2 + \frac{z^2}{c^2} = 1 \right\}.$$

Choose $a = 5$ and $c = 5$.

Estimate $\mathrm{Vol}(K)$ using a Monte Carlo method. Analyze two possible approaches:

- Crude Monte Carlo: Use a cuboid $[-5, 5]^3$ containing K to estimate the volume. Generate random points uniformly in the cuboid, and compute the proportion of points inside K.
- Star-shaped region method: Use the method described in Remark 3.8.19 to estimate the volume by sampling points efficiently in a region tailored to the geometry of K.

Chapter 4
Simulation Output Analysis: Independent Replications

In Monte Carlo simulations, the results are inherently stochastic. For instance, when estimating a quantity such as π using the sample average of simulated random variables, the raw numerical outcome is incomplete without understanding its precision. The Central Limit Theorem (CLT) provides a framework to establish confidence intervals (or confidence regions for d-dimensional quantities), making simulation results meaningful. In this chapter, we establish the so-called *fundamental relationship* between the required precision, the number of replications, and variability.

Suppose we aim to estimate a quantity I that can be expressed as:

$$I = \mathbb{E}Y,$$

where Y is a random variable (a random number) with a prescribed distribution, satisfying $\mathbb{E}Y^2 < \infty$. The variance of Y is denoted by σ_Y^2. Additionally, we assume that Y can be simulated.

Note that for a given I, there are many random variables Y such that $I = \mathbb{E}Y$. This chapter and the next aim to explore how to choose the "best" Y for a given problem and how to determine the number of replications required for a prescribed accuracy.

Suppose we simulate a sample $Y_1, Y_2, \ldots, Y_R$ of i.i.d. random variables. This chapter studies the relationships between the error b, the number of replications R, and estimators of I. We denote estimators by a hat, e.g., $\hat{I}$ represents an estimator of I.

Two types of errors are considered:

- **Absolute error:** $|\hat{I} - I|$, which measures the magnitude of the error.
- **Relative error:** $|\hat{I}/I - 1| = |\hat{I} - I|/I$, which evaluates the error relative to the true value.

© The Author(s), under exclusive license to Springer Nature Switzerland AG 2025 187
P. Lorek, T. Rolski, *Lectures on Monte Carlo Theory*, Probability Theory
and Stochastic Modelling 108, https://doi.org/10.1007/978-3-032-01190-9_4

We aim to bound these errors probabilistically by specifying $\varepsilon, \delta \in (0, 1)$ such that:

$$\text{Absolute error:} \quad \mathbb{P}(|\hat{I} - I| \geq \varepsilon) \leq \delta, \tag{E1}$$

$$\text{Relative error:} \quad \mathbb{P}(|\hat{I}/I - 1| \geq \varepsilon) = \mathbb{P}(|\hat{I} - I| \geq \varepsilon I) \leq \delta. \tag{E2}$$

Equivalently, these conditions can be expressed as:

$$\text{Absolute error:} \quad \mathbb{P}(|\hat{I} - I| \leq \varepsilon) \geq 1 - \delta, \tag{E1'}$$

$$\text{Relative error:} \quad \mathbb{P}(|\hat{I}/I - 1| \leq \varepsilon) = \mathbb{P}(|\hat{I} - I| \leq \varepsilon I) \geq 1 - \delta. \tag{E2'}$$

Depending on the problem, we may be given ε and δ and tasked with constructing $\hat{I}$ such that conditions (E1) (or equivalently (E1')) for absolute error, or (E2) (or equivalently (E2')) for relative error, hold.

This chapter primarily focuses on absolute error; relative error is revisited in Sect. 4.4. The relative error is also discussed in the context of *approximate counting* in Sect. 6.5. In this context, using terminology from the literature, $\hat{I}$ is referred to as an (ε, δ)-approximation scheme for I if condition (E2) holds.

4.1 Estimation of I

4.1.1 One-dimensional Case

Assume we want to compute I, which can be expressed as:

$$I = \mathbb{E}Y,$$

for some random variable Y satisfying $\mathbb{E}Y^2 < \infty$. In this subsection, we assume that $Y_1, \ldots, Y_R$ are independent replications of Y, and we consider an *estimator* of I in the form of a sample mean:

$$\hat{Y}_R = \frac{1}{R} \sum_{j=1}^{R} Y_j. \tag{1.1}$$

Often, Y is expressed as a function of another random variable X, namely $Y = k(X)$, which leads to:

$$I = \mathbb{E}Y = \mathbb{E}k(X).$$

In this case, we require that $\mathbb{V}\mathrm{ar}\,k(X) < \infty$. The estimator (1.1) remains valid, with $Y_j = k(X_j)$ for i.i.d. $X_1, \ldots, X_R$. Using elementary probability theory, we note that $\hat{Y}_R$ is *unbiased*, i.e.,

$$\mathbb{E}\hat{Y}_R = I \, ,$$

and *strongly consistent*, meaning that:

$$\hat{Y}_R \to I,$$

with probability one as $R \to \infty$.

Strong consistency implies (more importantly for us) *weak consistency*, which means that for any $b > 0$, we have:

$$\mathbb{P}(|\hat{Y}_R - I| \geq b) \to 0.$$

Note that $|\hat{Y}_R - I| \leq b$ means that the absolute error does not exceed b. This implies that for any $b > 0$ and $0 < \alpha < 1$, there exists R_0 (we are typically interested in the smallest possible value of R_0) such that for $R \geq R_0$:

$$\mathbb{P}(\hat{Y}_R - b < I \leq \hat{Y}_R + b) \geq 1 - \alpha. \tag{1.2}$$

Thus, with probability at least $1 - \alpha$, the quantity I lies within the random interval $(\hat{Y}_R - b, \hat{Y}_R + b)$. This is why α is called the *significance level*, or equivalently, the *confidence level*. Alternatively, we may seek to determine the error b for a given number of iterations R and a confidence level α, or, for a given R and error $b = b_R$, we may determine the corresponding confidence level α. Chebyshev's inequality provides a rough estimate for this relationship.

Remark 4.1.1 Let b_R denote a prescribed error for R replications. Using Chebyshev's inequality, we can bound the probability of exceeding the error b_R:

$$\mathbb{P}(|\hat{Y}_R - I| \geq b_R) \leq \frac{\mathbb{V}\mathrm{ar}\hat{Y}_R}{b_R^2}.$$

If we want the error to be at level α, then:

$$\alpha = \frac{\mathbb{V}\mathrm{ar}\hat{Y}_R}{b_R^2} = \frac{\sigma_Y^2}{Rb_R^2},$$

and solving for b_R, we find:

$$b_R = \frac{1}{\alpha^{1/2}} \frac{\sigma_Y}{\sqrt{R}}, \tag{1.3}$$

which shows that $b_R = O(R^{-1/2})$. $\blacksquare$

Now let us derive a more precise relationship between R, b, and α than that provided by (1.2). Since R is usually large in Monte Carlo methods, we may apply the *central limit theorem* (CLT), which states that if $Y_1, \ldots, Y_R$ are independent and identically distributed random variables with $\sigma_Y = \sqrt{\mathbb{V}\mathrm{ar} Y_j} < \infty$, then as $R \to \infty$:

$$\mathbb{P}\left(\sum_{j=1}^{R} \frac{Y_j - I}{\sigma_Y \sqrt{R}} \leq x\right) \to \Phi(x), \tag{1.4}$$

where $\Phi(x)$ is the cumulative distribution function of the standard normal $\mathcal{N}(0, 1)$ random variable.

Let $z_{1-\alpha/2}$ denote the $1 - \alpha/2$–quantile of the standard normal distribution, i.e.,

$$z_{1-\alpha/2} = \Phi^{-1}\left(1 - \frac{\alpha}{2}\right).$$

From this, we obtain:

$$\lim_{R \to \infty} \mathbb{P}\left(-z_{1-\alpha/2} < \sum_{j=1}^{R} \frac{Y_j - I}{\sigma_Y \sqrt{R}} \leq z_{1-\alpha/2}\right) = 1 - \alpha,$$

which implies:

$$\mathbb{P}\left(-z_{1-\alpha/2} < \sum_{j=1}^{R} \frac{Y_j - I}{\sigma_Y \sqrt{R}} \leq z_{1-\alpha/2}\right) \approx 1 - \alpha,$$

or equivalently:

$$\mathbb{P}\left(\hat{Y}_R - z_{1-\alpha/2} \frac{\sigma_Y}{\sqrt{R}} \leq I \leq \hat{Y}_R + z_{1-\alpha/2} \frac{\sigma_Y}{\sqrt{R}}\right) \approx 1 - \alpha, \tag{1.5}$$

providing the *fundamental relationship* between R, α, and σ_Y^2:

$$b = z_{1-\alpha/2} \frac{\sigma_Y}{\sqrt{R}}. \tag{1.6}$$

Conversely, to achieve a prescribed error b, i.e.,

$$\mathbb{P}(\hat{Y}_R - b \leq I \leq \hat{Y}_R + b) \approx 1 - \alpha,$$

we require:

$$R = z_{1-\alpha/2}^2 \frac{\sigma_Y^2}{b^2} \tag{1.7}$$

replications.

Remark 4.1.2 Note that the CLT (1.4) can also be expressed as:

$$\sqrt{R}(\hat{Y}_R - I) \xrightarrow{\mathcal{D}} \mathcal{N}(0, \sigma_Y^2).$$

■

In practice, we often do not know the value of σ_Y^2. In such cases, we replace it with the sample variance, defined as:

$$\hat{S}_R^2 = \frac{1}{R-1} \sum_{j=1}^{R} (Y_j - \hat{Y}_R)^2.$$

From the theory of statistics, it is known that the estimator $\hat{S}_R^2$ is unbiased, meaning $\mathbb{E}\hat{S}_R^2 = \sigma_Y^2$, and it is strongly consistent. From probability theory, we also know the following holds:

$$\mathbb{P}\left(\sum_{j=1}^{R} \frac{Y_j - I}{\hat{S}_R \sqrt{R}} \leq x \right) \to \Phi(x), \tag{1.8}$$

where $\Phi(x)$ is the cumulative distribution function of the standard normal distribution. Consequently,

$$\mathbb{P}\left(-z_{1-\alpha/2} < \sum_{j=1}^{R} \frac{Y_j - I}{\hat{S}_R \sqrt{R}} \leq z_{1-\alpha/2} \right) \approx 1 - \alpha,$$

and after rearranging, we obtain the equivalent form:

$$\mathbb{P}\left(\hat{Y}_R - z_{1-\alpha/2} \frac{\hat{S}_R}{\sqrt{R}} < I \leq \hat{Y}_R + z_{1-\alpha/2} \frac{\hat{S}_R}{\sqrt{R}} \right) \approx 1 - \alpha.$$

In other words, the *confidence interval at confidence level* $1 - \alpha$, or in short, the $1 - \alpha$-confidence interval, is given by:

$$\left(\hat{Y}_R - z_{1-\alpha/2} \frac{\hat{S}_R}{\sqrt{R}}, \ \hat{Y}_R + z_{1-\alpha/2} \frac{\hat{S}_R}{\sqrt{R}} \right). \tag{1.9}$$

Note that whether the interval is considered open (as written here) or closed is irrelevant, as the distribution function $\Phi(x)$ is continuous.

Pilot and Actual Simulations Throughout this book, we frequently make recommendations regarding simulation procedures. In particular, we run preliminary, so-called pilot simulations to determine the necessary number of replications, as well as required constants (e.g., variance) and other parameters. The main part, referred to as the actual simulation, is then conducted based on the values obtained from the pilot simulations. This approach is commonly known as a two-stage procedure, which has been extensively studied for constructing confidence intervals while accounting for dependency; see e.g., Nakayama [162].

We have the following lemma, which will be needed for the next remark:

Lemma 4.1.3 *Let (X_n) and (Y_n) be two sequences of random variables such that $Y_n \xrightarrow{\mathcal{D}} Y$ and $X_n \to 1$ with probability 1. Then $X_n Y_n \xrightarrow{\mathcal{D}} Y$.*

Sketch of Proof Since $Y_n \xrightarrow{\mathcal{D}} Y$, the sequence Y_n converges to Y in distribution. Additionally, $X_n \to 1$ almost surely, meaning X_n gets arbitrarily close to 1 with probability 1. The product $X_n Y_n$ combines these two properties. The almost sure convergence of X_n to 1 ensures that X_n does not alter the limiting distribution of Y_n. Hence, $X_n Y_n \xrightarrow{\mathcal{D}} Y$. $\square$

Remark 4.1.4 In Monte Carlo methods, we often estimate σ_Y using the sample variance $\hat{S}_R$. Consequently, confidence intervals rely on $\sum_{j=1}^{R} \frac{Y_j - I}{\hat{S}_R \sqrt{R}}$. The random variable $\sum_{j=1}^{R} \frac{Y_j - I}{\sigma_Y \sqrt{R}}$ has an approximately standard normal distribution, whereas $\sum_{j=1}^{R} \frac{Y_j - I}{\hat{S}_R \sqrt{R}}$ follows approximately a Student's t-distribution with $R - 1$ degrees of freedom. For large R, the t-distribution converges to the standard normal distribution, that is why (1.8) holds. This practical equivalence is an important aspect of Monte Carlo methods, as R is typically large. The approximation can be justified using Lemma 4.1.3. Specifically, we rewrite:

$$\sum_{j=1}^{R} \frac{Y_j - I}{\hat{S}_R \sqrt{R}} = \frac{\sigma_Y}{\hat{S}_R} \sum_{j=1}^{R} \frac{Y_j - I}{\sigma_Y \sqrt{R}},$$

and note that due to the strong consistency of $\hat{S}_R$:

$$\frac{\sigma_Y}{\hat{S}_R} \to 1$$

with probability 1. By applying Lemma 4.1.3, we conclude that:

$$\sum_{j=1}^{R} \frac{Y_j - I}{\hat{S}_R \sqrt{R}} \xrightarrow{\mathcal{D}} \mathcal{N}(0, 1).$$

Thus, for large R, the confidence intervals derived from either distribution are nearly identical in practice. ∎

In the theory of Monte Carlo methods, a confidence level of $\alpha = 0.05$ is commonly used, corresponding to $z_{1-0.025} = 1.96$. This confidence level will be adopted in most cases throughout the book. For a higher confidence level, such as 99%, the corresponding quantile is $z_{1-0.005} = 2.58$. For a confidence level of $\alpha = 0.05$, the fundamental relation is given by:

$$b = \frac{1.96\sigma_Y}{\sqrt{R}}, \tag{1.10}$$

which implies that to achieve a prescribed accuracy b, the required number of replications is:

$$R = \frac{1.96^2\sigma_Y^2}{b^2}.$$

As previously mentioned, the value of σ_Y^2 is often unknown in practice. To address this, we first estimate σ_Y^2 using m independent replications to obtain $\hat{S}_m^2$. Using this estimate, we calculate the required R from formula (1.10). Next, we estimate I using $\hat{Y}_R$, and independently estimate σ_Y^2 (using the R determined from the pilot simulation) with $\hat{S}_R^2$. The resulting confidence interval is given in (1.9).

Sample Mean and Sample Variance Suppose $Y_1, \ldots, Y_R$ are outcomes of an experiment. The *sample mean* is given by:

$$\hat{Y}_R = \frac{\sum_{i=1}^{R} Y_i}{R}.$$

There are two possible definitions for the *sample variance*:

- *Unbiased sample variance:*

$$\hat{S}_R^2 = \frac{\sum_{i=1}^{R}(Y_i - \hat{Y}_R)^2}{R-1}.$$

- *Standard (biased) sample variance:*

$$\hat{\sigma}_R^2 = \frac{\sum_{i=1}^{R}(Y_i - \hat{Y}_R)^2}{R}.$$

Since in simulations R is typically large, the difference between the two definitions is negligible. Efficient computation is often essential when updating statistics with new data points. The following recursive formulas allow for such updates:

$$\hat{Y}_{R+1} = \hat{Y}_R + \frac{Y_{R+1} - \hat{Y}_R}{R+1}, \tag{1.11}$$

$$\hat{\sigma}_{R+1}^2 = \hat{\sigma}_R^2 + (\hat{Y}_R)^2 - (\hat{Y}_{R+1})^2 + \frac{(Y_{R+1})^2 - \hat{\sigma}_R^2 - (\hat{Y}_R)^2}{R+1}. \tag{1.12}$$

Remark 4.1.5 We have two formulas for the error: the one given in (1.3)—denoted by b^{Ch}, and the one given in (1.10)—denoted by b^{CLT}. Their ratio is given by:

$$\frac{b^{\mathrm{CLT}}}{b^{\mathrm{Ch}}} = z_{1-\alpha/2}\alpha^{1/2} = 43.82\% \quad \text{(for } \alpha = 0.05\text{)}.$$

Consequently, comparing the required number of replications calculated using b^{CLT} (denoted R^{CLT}) and b^{Ch} (denoted R^{Ch}), we have:

$$\frac{R^{\mathrm{CLT}}}{R^{\mathrm{Ch}}} = z_{1-\alpha/2}^2 \alpha = 19.20\% \quad \text{(for } \alpha = 0.05\text{)}.$$

Thus, using the formula derived from the CLT results in approximately 19.20% fewer replications compared to the formula derived from Chebyshev's inequality, while maintaining the same significance level ($\alpha = 0.05$). ∎

Below, we present two examples for estimating π, illustrating the impact of choosing different random variables Y. It is important to emphasize that estimating the number of required replications depends on knowing the variance, which is typically unknown in practice. However, such examples are useful for understanding key concepts and selecting good estimators. In practice, the variance is either estimated from pilot simulations or bounded (see Remark 4.1.12).

Example 4.1.6 (Asmussen and Glynn [8]) Assume we want to compute $I = \pi = 3.1415\ldots$ using Monte Carlo methods with an error $b = 0.5$ (achieving an estimate accurate to the first leading digit). Let $Y = 4\mathbb{1}(U_1^2 + U_2^2 \leq 1)$, where U_1, U_2 are i.i.d. random variables with a standard uniform distribution $\mathcal{U}[0, 1)$. Given $U_1, \ldots, U_{2R}$, the estimator is:

$$\hat{Y}_R = \frac{1}{R}\sum_{j=1}^{R} Y_j, \quad \text{where } Y_j = 4\mathbb{1}(U_{2j-1}^2 + U_{2j}^2 \leq 1).$$

Note that $\mathbb{1}(U_1^2 + U_2^2 \leq 1)$ is a Bernoulli random variable with mean $\pi/4$, thus

$$\mathbb{E}Y = \pi,$$

and

$$\mathbb{V}\mathrm{ar}Y = 16(\pi/4)(1 - \pi/4) = 2.6968.$$

To achieve the desired accuracy of the first leading digit, we need approximately:

$$R \approx 1.96^2 \cdot 2.6968/0.5^2 \approx 42 \text{ replications.}$$

For the second leading digit, we require:

$$R \approx 1.96^2 \cdot 2.6968/0.05^2 \approx 4144 \text{ replications, and so on.}$$

$\Diamond$

Example 4.1.7 Again, we aim to estimate $I = \pi = 3.1415\ldots$. The area of the circle with radius 1 is π. Consider the quarter-circle $\{(x, y) : x^2 + y^2 \le 1, 0 \le x, y \le 1\}$ inside the square $[0, 1] \times [0, 1]$. Its area is:

$$\frac{\pi}{4} = \int_0^1 \sqrt{1 - x^2}\, dx.$$

This suggests taking $Y = 4\sqrt{1 - U^2}$, where $U \sim \mathcal{U}[0, 1)$. Then:

$$\mathbb{E}Y = \int_0^1 4\sqrt{1 - x^2}\, dx = \pi.$$

The corresponding estimator is:

$$\hat{Y}_R = \frac{1}{R} \sum_{j=1}^R Y_j, \quad \text{where } Y_j = 4\sqrt{1 - U_j^2},$$

and $U_j, j = 1, \ldots, R$ are i.i.d. $\mathcal{U}[0, 1)$. We compute:

$$\mathbb{V}\mathrm{ar}Y = \int_0^1 16(1 - x^2)dx - (\mathbb{E}Y)^2 = \frac{32}{3} - \pi^2 = 0.797.$$

For the first leading digit, we need approximately:

$$R \approx 1.96^2 \cdot 0.797/0.5^2 \approx 12 \text{ replications.}$$

For the second leading digit, we require:

$$R \approx 1.96^2 \cdot 0.797/0.05^2 \approx 1224 \text{ replications, and so on.}$$

$\Diamond$

Note that the variance of the estimator for π in Example 4.1.7 is approximately 3.38 times smaller than the variance in Example 4.1.6.

Computing Integrals with Monte Carlo Methods Suppose the quantity I can be expressed as:

$$I = \int_B k(x)\, dx,$$

where B is a bounded subset of $\mathbb{R}^d$. We assume that $\mathrm{Vol}\,(B) = 1$ (if not, the problem can be rescaled accordingly). Typically, B is either $[0,\,1]$ or $[0,\,1]^d$.

Consider $Y = k(V)$, where V has the uniform distribution on B, so that $\mathbb{E}Y = I$. Let $V_1, \ldots, V_R$ be independent copies of V with the uniform distribution $\mathcal{U}(B)$. The estimator:

$$\hat{Y}_R^{\mathrm{CMC}} = \frac{1}{R} \sum_{j=1}^{R} k(V_j)$$

is called a *crude Monte Carlo estimator* of I. We will often refer to it as the CMC estimator.

More generally, the quantity I can be expressed as:

$$I = \mathbb{E}Y = \mathbb{E}k(X) = \int_B k(x) f(x)\, dx, \tag{1.13}$$

where f is a density of X on B and $Y = k(X)$. In this case, the CMC estimator of I is:

$$\hat{Y}_R^{\mathrm{CMC}} = \frac{1}{R} \sum_{j=1}^{R} k(X_j),$$

where $X_1, \ldots, X_R$ are i.i.d. random variables with density f.

Remark 4.1.8 The term *crude Monte Carlo estimator* does not uniquely define a single estimator for a given I. There are multiple ways to choose Y, provided that $\mathbb{E}Y = I$. Taking the sample mean of independent replications of Y will always yield a valid CMC estimator. For instance, in Examples 4.1.6 and 4.1.7, we presented two different CMC estimators for $I = \pi$, both valid under this framework. ∎

Example 4.1.9 To illustrate this non-uniqueness, consider estimating $I = \int_0^\infty x^2 e^{-x}\, dx$. This integral can be rewritten as:

$$I = \int_0^\infty x^2 e^{-x}\, dx = \int_0^\infty k_1(x) f_1(x)\, dx = \int_0^\infty x \left(x e^{-x}\right) dx$$

$$= \int_0^\infty k_2(x) f_2(x)\, dx,$$

where $k_1(x) = x^2$, $f_1(x) = e^{-x}$, and $k_2(x) = x$, $f_2(x) = xe^{-x}$. Thus, we can construct two CMC estimators:

$$\hat{Y}_R^{\mathrm{CMC_1}} = \frac{1}{R}\sum_{j=1}^R X_j^2, \quad X_j \sim f_1, \qquad \hat{Y}_R^{\mathrm{CMC_2}} = \frac{1}{R}\sum_{j=1}^R X_j', \quad X_j' \sim f_2,$$

where $\hat{Y}_R^{\mathrm{CMC_1}}$ is the average of X_j^2 for $X_j \sim \mathrm{Exp}(1)$, and $\hat{Y}_R^{\mathrm{CMC_2}}$ is the average of X_j' for $X_j' \sim \mathrm{Gamma}(2, 1)$. $\Diamond$

Example 4.1.10 (Madras [138]) Let $k : [0, 1] \to \mathbb{R}$ be an integrable function satisfying: $\int_0^1 k^2(x)\, dx < \infty$. Define $Y = k(U)$, where $U \sim \mathcal{U}[0, 1)$. Then:

$$\mathbb{E}Y = \int_0^1 k(x)\, dx, \quad \mathbb{E}Y^2 = \int_0^1 k^2(x)\, dx,$$

and the variance is:

$$\sigma_Y^2 = \int_0^1 k^2(x)\, dx - \left(\int_0^1 k(x)\, dx\right)^2.$$

For example, let:

$$k(x) = \frac{e^x - 1}{e - 1}.$$

Then:

$$\int_0^1 k(x)\, dx = \frac{1}{e - 1}\left(\int_0^1 e^x\, dx - 1\right) = \frac{e - 2}{e - 1} \approx 0.418,$$

$$\int_0^1 k^2(x)\, dx = 0.2567.$$

From the above, we compute $\sigma_Y^2 = 0.0820$, so the required number of replications for accuracy b is: $R = \frac{0.3280}{b^2}$. $\Diamond$

Example 4.1.11 (The Hit-or-miss Method) The hit-or-miss (HoM) method offers an alternative approach for estimating $I = \int_B k(x)\, dx$. Assume that $B = [0, 1]$ and $0 \le k(x) \le 1$. Visualize $k(x)$ as a curve within the rectangle $[0, 1] \times [0, 1]$, splitting the square into two regions: one above and one below the curve. The region below the curve corresponds to the desired integral. The idea is to simulate random points (U_1, U_2) within the square and count how many fall under the curve $k(x)$. This forms the basis of the hit-or-miss estimator. We hit the target if $k(U_1) > U_2$; see Fig. 4.1 for an illustration. Formally, we define:

$$Y' = \mathbb{1}(k(U_1) > U_2).$$

We can write:

$$I = \mathbb{E}Y' = \int_0^1 \int_0^1 \mathbb{1}(k(u_1) > u_2)\, du_2\, du_1 = \int_0^1 k(u_1)\, du_1.$$

Given uniform random variables $U_1, \ldots, U_{2R}$, we define $Y'_j = \mathbb{1}(k(U_{2j-1}) > U_{2j})$ for $j = 1, \ldots, R$. The hit-or-miss estimator is then:

$$\hat{Y}_R^{\mathrm{HoM}} = \frac{1}{R} \sum_{j=1}^R Y'_j.$$

Returning to Example 4.1.10 (with $k(x) = (e^x - 1)/(e - 1)$), we can compute the variance $\mathbb{V}\mathrm{ar}Y' = I(1 - I) = 0.2433$ yielding $R = \frac{0.9346}{b^2}$, which shows that to achieve the same accuracy, three times more replications are required compared to the CMC method. Furthermore, simulating Y' requires twice as many random numbers as Y from Example 4.1.10. However, the simplicity of the hit-or-miss method often makes it straightforward to implement. $\Diamond$

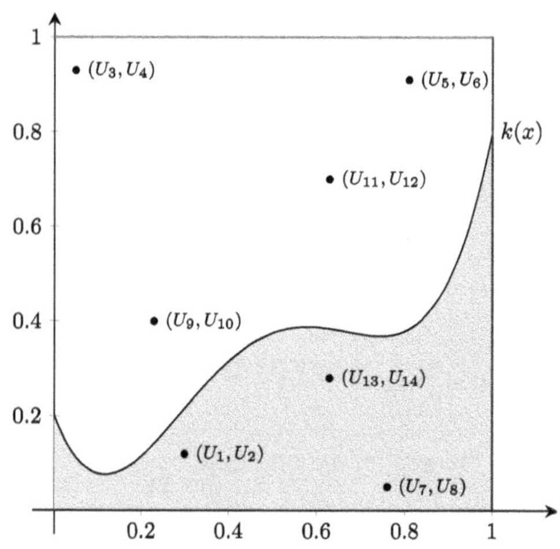

Fig. 4.1 Example: estimating $\int_0^1 k(x)$ with a hit-or-miss estimator using $U_1, \ldots, U_{2R}$ random numbers, $R = 7$. In this case $Y_1 = Y_4 = Y_7 = 1$, $Y_2 = Y_3 = Y_5 = Y_6 = 0$ and $\hat{Y}_R^{\mathrm{HoM}} = \frac{3}{7}$

Remark 4.1.12 For simplicity, in Examples 4.1.7, 4.1.10, and 4.1.11, we used the exact value of I to compute the variances, even though our goal is to estimate I. As mentioned, in practice we may estimate it from pilot simulation. Alternatively, an upper bound on the variance suffices to determine the number of replications R

required for a given significance level. For estimating $I = \int_B k(x) f(x) \, dx$, where f is a density of B, we may rewrite $I = \mathbb{E}k(X) = \mathbb{E}Y$, where $Y = k(X)$ and X has density f. In general, we have

$$\mathbb{V}\mathrm{ar}Y = \int_B k^2(x) f(x) \, dx - \left(\int_B k(x) f(x) \, dx \right)^2.$$

In Example 4.1.10 we had $B = [0, 1]$, thus the variance simplifies to

$$\mathbb{V}\mathrm{ar}Y = \int_0^1 k^2(x) \, dx - \left(\int_0^1 k(x) \, dx \right)^2,$$

so, using $k(x) = \frac{e^x - 1}{e - 1}$, we have bounds:

$$\int_0^1 k(x) \, dx = \int_0^1 \frac{e^x - 1}{e - 1} \, dx \geq \int_0^1 \frac{2.71^x - 1}{1.72} \, dx =: k_{\mathrm{down}},$$

and:

$$\int_0^1 k^2(x) \, dx = \int_0^1 \frac{(e^2)^x - e^x + 1}{(e - 1)^2} \, dx \leq \int_0^1 \frac{(2.72^2)^x - 2.71^x + 1}{(1.71)^2} \, dx =: k_{\mathrm{up}}^2.$$

Evaluating these integrals gives:

$$k_{\mathrm{down}} \geq 0.4157, \quad k_{\mathrm{up}}^2 \leq 0.2622.$$

Thus:

$$\mathbb{V}\mathrm{ar}Y \leq 0.2622 - (0.4157)^2 = 0.0894,$$

which is close to the exact value 0.0820. Similarly, in Example 4.1.7, the variance:

$$\mathbb{V}\mathrm{ar}Y = 16 \left(\frac{\pi}{4} \right) \left(1 - \frac{\pi}{4} \right),$$

can be bounded by:

$$\mathbb{V}\mathrm{ar}Y \leq 16 \left(\frac{3.15}{4} \right) \left(1 - \frac{3.14}{4} \right) = 2.7090,$$

which is again close to the actual value 2.6968.

In practice, only an upper bound on $\mathbb{V}\mathrm{ar}Y$ is necessary, with the penalty being a slight increase in the number of replications required to achieve the desired significance level. ∎

As illustrated, the goal is to find estimators with small variance. However, it is also essential to consider the time required for each replication. For instance, computing Y' takes approximately twice as long as computing Y in Example 4.1.10.

Simulation Budget When comparing multiple experiments, such as estimating $I = \mathbb{E}Y = \mathbb{E}Y'$, the time required for each replication must also be considered. Let t and t' represent the average times for generating one replication of Y and Y', respectively. Suppose the total available simulation time is C. We can perform C/t and C/t' replications, respectively. Using formula (1.10), the achievable accuracies are $z_{1-\alpha/2}\frac{\sigma_Y}{\sqrt{C/t}}$ and $z_{1-\alpha/2}\frac{\sigma_{Y'}}{\sqrt{C/t'}}$. We compare the two experiments by comparing $\sqrt{t}\sigma_Y$ and $\sqrt{t'}\sigma_{Y'}$. The quantities $t\sigma_Y^2$ and $t'\sigma_{Y'}^2$ are referred to as the *efficiency* of an estimator. For a more detailed treatment of estimator efficiency considering simulation budget constraints, see Glynn and Whitt [67].

4.1.2 Multidimensional Case

Before proceeding to Monte Carlo estimators, we will first a) recall several facts about the multivariate normal distribution, and b) discuss the shapes of confidence regions for a multivariate normal random variable.

4.1.2.1 Several Facts About the Multivariate Normal Distribution

Let $\mathbb{R}^d \ni \mathbf{X} \sim \mathcal{N}(\boldsymbol{\mu}, \boldsymbol{\Sigma})$. In the case $d = 1$, that is $X \sim \mathcal{N}(\mu, \sigma^2)$, we can write $X \overset{\mathcal{D}}{=} \sigma Z + \mu$, where $Z \sim \mathcal{N}(0, 1)$. We now extend this representation to the case $d > 1$.

Some Facts from Matrix Theory We begin by reviewing some essential matrix theory and establishing the necessary terminology. Consider a $d \times d$ matrix $\mathbf{S}$. The matrix $\mathbf{S}$ is said to be positive-definite if it is symmetric and satisfies $\mathbf{x}^T\mathbf{S}\mathbf{x} > 0$ for all nonzero vectors $\mathbf{x}$. Henceforth, we assume that $\mathbf{S}$ is positive-definite.

A matrix $\mathbf{S}$ is **diagonalizable** if it can be written in the form $\mathbf{S} = \mathbf{F}\mathbf{D}\mathbf{F}^{-1}$, where $\mathbf{D}$ is a diagonal matrix. Let λ_i denote the eigenvalues of $\mathbf{S}$ and $\mathbf{f}_i$ the corresponding eigenvectors for $i = 1, \ldots, d$. These satisfy the equation $\mathbf{S}\mathbf{f}_i = \lambda_i\mathbf{f}_i$. Assuming the eigenvectors $\mathbf{f}_i$ are linearly independent, we construct the matrix $\mathbf{F} = [\mathbf{f}_1, \ldots, \mathbf{f}_d]$ (with columns as eigenvectors) and the diagonal matrix $\boldsymbol{\Lambda} = \mathrm{diag}(\lambda_1, \ldots, \lambda_d)$ (with eigenvalues on the diagonal). In matrix notation, this gives:

$$\mathbf{S}\mathbf{F} = \mathbf{F}\boldsymbol{\Lambda}, \quad \text{and thus } \mathbf{S} = \mathbf{F}\boldsymbol{\Lambda}\mathbf{F}^{-1}.$$

For symmetric positive-definite matrices, all eigenvalues are strictly positive. Furthermore, we may assume without loss of generality that the eigenvectors are normalized to unit length, i.e., $\mathbf{f}_i^T\mathbf{f}_i = 1$. In this case, the eigenvectors are orthonormal, meaning $\mathbf{f}_i^T \mathbf{f}_j = 0$ for $i \neq j$. Consequently, $\mathbf{F}$ is an orthonormal matrix, and it follows that $\mathbf{F}^{-1} = \mathbf{F}^T$. Thus, the diagonalization of $\mathbf{S}$ simplifies to:

$$\mathbf{S} = \mathbf{F}\boldsymbol{\Lambda}\mathbf{F}^T = \sum_{i=1}^{d} \lambda_i \mathbf{f}_i \mathbf{f}_i^T.$$

Decompositions of $\mathbf{S}$ In this context, we will use two types of decompositions:

(D1) **Cholesky decomposition:** For any positive-definite matrix $\mathbf{S}$, there exists a decomposition $\mathbf{S} = \mathbf{A}\mathbf{A}^T$. In particular, the Cholesky decomposition $\mathbf{S} = \mathbf{L}\mathbf{L}^T$ uses a lower triangular matrix $\mathbf{L}$. This decomposition is not unique, and other matrices $\mathbf{A}$ may satisfy $\mathbf{S} = \mathbf{A}\mathbf{A}^T$.

(D2) **Square root of $\mathbf{S}$**: The matrix square root $\mathbf{S}^{1/2}$ is the unique positive-definite matrix such that $\mathbf{S} = \mathbf{S}^{1/2}\mathbf{S}^{1/2}$. If $\mathbf{S}$ can be diagonalized as $\mathbf{S} = \mathbf{F}\boldsymbol{\Lambda}\mathbf{F}^T$, then:

$$\mathbf{S}^{1/2} = \mathbf{F}\boldsymbol{\Lambda}^{1/2}\mathbf{F}^T,$$

where the square root of the diagonal matrix is given by

$$\boldsymbol{\Lambda}^{1/2} = \mathrm{diag}(\lambda_1^{1/2}, \ldots, \lambda_d^{1/2}).$$

While the type of decomposition does not affect the distributional properties of a multivariate normal random variable, it becomes crucial when studying properties such as confidence ellipsoids, which may depend on the decomposition method used.

Square Root of $\boldsymbol{\Sigma}$ Assume $\boldsymbol{\Sigma}$ is a nonsingular $d \times d$ covariance matrix. If $\boldsymbol{\Sigma}$ can be diagonalized as $\boldsymbol{\Sigma} = \mathbf{F}\boldsymbol{\Lambda}\mathbf{F}^{-1}$, where $\mathbf{F}$ consists of eigenvectors of $\boldsymbol{\Sigma}$ and $\boldsymbol{\Lambda}$ is a diagonal matrix with eigenvalues as entries, then:

$$\boldsymbol{\Sigma} = \mathbf{F}\boldsymbol{\Lambda}\mathbf{F}^T = \mathbf{F}\boldsymbol{\Lambda}^{1/2}\boldsymbol{\Lambda}^{1/2}\mathbf{F}^T = \boldsymbol{\Sigma}^{1/2}\boldsymbol{\Sigma}^{1/2}.$$

Here, $\boldsymbol{\Sigma}^{1/2} = \mathbf{F}\boldsymbol{\Lambda}^{1/2}\mathbf{F}^T$ is the square root of $\boldsymbol{\Sigma}$, satisfying $\boldsymbol{\Sigma} = \boldsymbol{\Sigma}^{1/2}\boldsymbol{\Sigma}^{1/2}$.

Proposition 4.1.13 *Consider a d-dimensional random vector normally distributed $\mathbf{X} \sim \mathcal{N}(\boldsymbol{\mu}, \boldsymbol{\Sigma})$, where $\boldsymbol{\Sigma}$ is non-singular and $\boldsymbol{\Sigma} = \mathbf{A}\mathbf{A}^T$. Then*

(a) We have

$$\mathbf{Z} = \mathbf{A}^{-1}(\mathbf{X} - \boldsymbol{\mu}) \sim \mathcal{N}(\mathbf{0}, \mathbf{Id}_d),$$

where $\mathbf{Id}_d$ is a d-dimensional identity matrix.

(b) $(\mathbf{X} - \boldsymbol{\mu})^T \boldsymbol{\Sigma}^{-1}(\mathbf{X} - \boldsymbol{\mu})$ has the chi-square distribution (χ_d^2-distribution) with d degrees of freedom.

Proof

(a) This follows from Proposition 3.7.8.
(b) From part a) we have $\mathbf{Z} = \mathbf{A}^{-1}(\mathbf{X} - \boldsymbol{\mu}) \sim \mathcal{N}(\mathbf{0}, \mathbf{Id}_d)$. Recall from matrix algebra: $(\mathbf{A}^T)^{-1} = (\mathbf{A}^{-1})^T$, hence

$$\boldsymbol{\Sigma}^{-1} = (\mathbf{A}\mathbf{A}^T)^{-1} = (\mathbf{A}^T)^{-1}\mathbf{A}^{-1} = (\mathbf{A}^{-1})^T\mathbf{A}^{-1}.$$

Let $\mathbf{Z} = (Z_1, \ldots, Z_d)$. We have

$$(\mathbf{X} - \boldsymbol{\mu})^T \boldsymbol{\Sigma}^{-1}(\mathbf{X} - \boldsymbol{\mu}) = (\mathbf{X} - \boldsymbol{\mu})^T (\mathbf{A}^{-1})^T \mathbf{A}^{-1}(\mathbf{X} - \boldsymbol{\mu})$$

$$= (\mathbf{A}^{-1}(\mathbf{X} - \boldsymbol{\mu}))^T \mathbf{A}^{-1}(\mathbf{X} - \boldsymbol{\mu}) = \mathbf{Z}^T\mathbf{Z} = \sum_{i=1}^{d} Z_i^2.$$

By (a) we have $\mathbf{Z} \sim \mathcal{N}(\mathbf{0}, \mathbf{Id}_d)$, i.e., $Z_i, i = 1, \ldots, d$ are i.i.d. random variables $\mathcal{N}(0, 1)$. Thus, $(\mathbf{X} - \boldsymbol{\mu})^T \boldsymbol{\Sigma}^{-1}(\mathbf{X} - \boldsymbol{\mu})$ is a sum of squares of d independent standard normal random variable, i.e., it has the chi-square distribution with d degrees of freedom.

$\square$

For a general distribution F, a quantity q_p is the p-th quantile of F if

$$F(q_p-) \le p \le F(q_p).$$

In case of a continuous F, we have $p = F(q_p)$. We will denote by $q_{1-\alpha}(\chi_d^2)$ the $1 - \alpha$ quantile of the chi-square with d degrees of freedom. Proposition 4.1.13 implies that for $\mathbf{X} \sim \mathcal{N}(\boldsymbol{\mu}, \boldsymbol{\Sigma})$ we have that

$$\mathbb{P}\left((\mathbf{X} - \boldsymbol{\mu})^T \boldsymbol{\Sigma}^{-1}(\mathbf{X} - \boldsymbol{\mu}) \le q_{1-\alpha}(\chi_d^2)\right) \approx 1 - \alpha.$$

4.1.2.2 Shapes of Confidence Regions for Multivariate Normal Random Variables

For a given random variable $\mathbf{X} \in \mathbb{R}^d$, a confidence region $C \subset \mathbb{R}^d$ at level $1 - \alpha$ is any region (of any shape) such that:

$$\mathbb{P}(\mathbf{X} \in C) = 1 - \alpha.$$

In this section, we discuss confidence regions for multivariate normal random variables ($d \ge 2$) in the forms of hyper-ellipsoids, hyper-parallelograms, and hyper-rectangles.

Ellipsoid Assume $\mathbf{X} \sim \mathcal{N}(\mathbf{0}, \boldsymbol{\Sigma})$. From Proposition 4.1.13, we know:

$$\mathbb{P}\left(\mathbf{X}^T \boldsymbol{\Sigma}^{-1}\mathbf{X} \le q_{1-\alpha}(\chi_d^2)\right) = 1 - \alpha.$$

Thus, we can define the confidence region C as:

$$C = \left\{\mathbf{x} : \mathbf{x}^T \boldsymbol{\Sigma}^{-1}\mathbf{x} \le q_{1-\alpha}(\chi_d^2)\right\}.$$

If the covariance matrix $\boldsymbol{\Sigma}$ can be diagonalized as $\boldsymbol{\Sigma} = \mathbf{F}\boldsymbol{\Lambda}\mathbf{F}^T$, where $\boldsymbol{\Lambda} = \mathrm{diag}(\lambda_1, \ldots, \lambda_d)$ contains the eigenvalues of $\boldsymbol{\Sigma}$ and $\mathbf{F} = [\mathbf{f}_1, \ldots, \mathbf{f}_d]$ contains the eigenvectors as columns, the inverse of $\boldsymbol{\Sigma}$ can be expressed as:

$$\boldsymbol{\Sigma}^{-1} = \sum_{i=1}^{d} \frac{1}{\lambda_i} \mathbf{f}_i \mathbf{f}_i^T. \tag{1.14}$$

The boundary of the confidence region ∂C is given by:

$$\partial C = \left\{\mathbf{x} : \mathbf{x}^T \boldsymbol{\Sigma}^{-1}\mathbf{x} = q_{1-\alpha}(\chi_d^2)\right\}.$$

Using (1.14), we obtain:

$$q_{1-\alpha}(\chi_d^2) = \mathbf{x}^T \left(\sum_{i=1}^{d} \frac{1}{\lambda_i} \mathbf{f}_i \mathbf{f}_i^T\right)\mathbf{x} = \sum_{i=1}^{d} \frac{1}{\lambda_i} \left[\mathbf{f}_i^T \mathbf{x}\right]^2. \tag{1.15}$$

Substituting $t_i = \mathbf{f}_i^T \mathbf{x}$, the boundary of the confidence region takes the form:

$$\sum_{i=1}^{d} \frac{t_i^2}{\lambda_i} = q_{1-\alpha}(\chi_d^2),$$

which is a standard representation of a hyper-ellipsoid.

Thus, $C \equiv C(\boldsymbol{\Sigma})$ is an ellipsoid centered at $\mathbf{0}$ with principal axes $\sqrt{q_{1-\alpha}(\chi_d^2)\lambda_i}\mathbf{f}_i$, $i = 1, \ldots, d$. The length of each axis is $\sqrt{q_{1-\alpha}(\chi_d^2)\lambda_i}$.

For $\mathbf{X}' = \mathbf{X} + \boldsymbol{\mu}$ (i.e., $\mathbf{X}' \sim \mathcal{N}(\boldsymbol{\mu}, \boldsymbol{\Sigma})$), the confidence region becomes an ellipsoid centered at $\boldsymbol{\mu}$ with the same axes:

$$C' \equiv C'(\boldsymbol{\mu}, \boldsymbol{\Sigma}) = C(\boldsymbol{\Sigma}) + \boldsymbol{\mu}.$$

Parallelogram Assume $\mathbf{X} \sim \mathcal{N}(\boldsymbol{\mu}, \boldsymbol{\Sigma})$. Then:

$$\boldsymbol{\Sigma}^{-1/2}(\mathbf{X} - \boldsymbol{\mu}) \sim \mathcal{N}(\mathbf{0}, \mathbf{Id}_d).$$

Define a hypercube $\mathcal{H} = (-z, z]^d$ such that:

$$\mathbb{P}\left(\boldsymbol{\Sigma}^{-1/2}(\mathbf{X} - \boldsymbol{\mu}) \in \mathcal{H}\right) = 1 - \alpha. \tag{1.16}$$

Since the coordinates are independent, we have:

$$\mathbb{P}\left(\boldsymbol{\Sigma}^{-1/2}(\mathbf{X} - \boldsymbol{\mu}) \in \mathcal{H}\right) = \prod_{i=1}^{d} \mathbb{P}\left(-z \leq \left(\boldsymbol{\Sigma}^{-1/2}(\mathbf{X} - \boldsymbol{\mu})\right)_i \leq z\right)$$

$$= (2\Phi(z) - 1)^d = 1 - \alpha.$$

Solving for z, we obtain:

$$z = \Phi^{-1}\left(\frac{\sqrt[d]{1 - \alpha} + 1}{2}\right). \tag{1.17}$$

Transforming (1.16), we find:

$$\mathbb{P}\left(\mathbf{X} \in \boldsymbol{\Sigma}^{1/2}\mathcal{H} + \boldsymbol{\mu}\right) = 1 - \alpha.$$

Thus, the confidence region is:

$$C = \boldsymbol{\Sigma}^{1/2}\mathcal{H} + \boldsymbol{\mu}, \tag{1.18}$$

which is a parallelogram, with vertices given by $\boldsymbol{\Sigma}^{1/2}\mathbf{v} + \boldsymbol{\mu}$ for $\mathbf{v} = (\pm z, \pm z, \ldots, \pm z)$.

Rectangular Confidence Region Consider $\mathbf{X} \sim \mathcal{N}(\mathbf{0}, \boldsymbol{\Sigma})$. We now construct a rectangular region $C \subset \mathbb{R}^d$ such that $\mathbb{P}(\mathbf{X} \in C) \geq 1 - \alpha$. The region is defined as:

$$C = [-c_1, c_1] \times \ldots \times [-c_d, c_d],$$

where $c_1, \ldots, c_d > 0$. We have

$$\mathbb{P}\left(|X_1| \leq c_1, \ldots, |X_d| \leq c_d\right) \geq \prod_{i=1}^{d} \mathbb{P}\left(|X_i| \leq c_i\right) \tag{1.19}$$

$$= \prod_{i=1}^{d} \mathbb{P}\left(\frac{c_i}{\sigma_i} \leq \frac{X_i}{\sigma_i} \leq \frac{c_i}{\sigma_i}\right) = \prod_{i=1}^{d}\left(2\Phi\left(\frac{c_i}{\sigma_i}\right) - 1\right),$$

where (1.19) is the Šidák inequality [192]. To ensure the confidence level $1 - \alpha$, we set:

$$2\Phi\left(\frac{c_i}{\sigma_i}\right) - 1 = \sqrt[d]{1 - \alpha}, \quad i = 1, \ldots, d,$$

leading to:

$$c_i = \sigma_i \Phi^{-1}\left(\frac{1 + \sqrt[d]{1 - \alpha}}{2}\right).$$

Example 4.1.14 Consider a bivariate normal random variable $\mathbf{X} \sim \mathcal{N}(\mathbf{0}, \boldsymbol{\Sigma})$ with the covariance matrix:

$$\boldsymbol{\Sigma} = \begin{pmatrix} 1 & 2 \\ 2 & 6 \end{pmatrix}.$$

We aim to find confidence regions with a confidence level $1 - \alpha = 0.95$, i.e., C such that:

$$\mathbb{P}(\mathbf{X} \in C) = 0.95.$$

All confidence regions for this example presented below are generated/computed by `ch4_bivariate_normal_conf_regions.py`.

Ellipsoid To use (1.15), we decompose $\boldsymbol{\Sigma} = \mathbf{F}\boldsymbol{\Lambda}\mathbf{F}^{-1}$:

$$\mathbf{F} = \begin{bmatrix} \mathbf{f}_1 & \mathbf{f}_2 \end{bmatrix}, \quad \mathbf{f}_1 = \begin{pmatrix} -0.3310 \\ -0.9436 \end{pmatrix}, \quad \mathbf{f}_2 = \begin{pmatrix} -0.9436 \\ 0.3310 \end{pmatrix},$$

$$\boldsymbol{\Lambda} = \begin{pmatrix} \lambda_1 & 0 \\ 0 & \lambda_2 \end{pmatrix}, \quad \lambda_1 = \frac{7 + \sqrt{41}}{2} = 6.7015, \quad \lambda_2 = \frac{7 - \sqrt{41}}{2} = 0.2984.$$

Using $q_{1-\alpha}(\chi_d^2) = q_{0.95}(\chi_2^2) = 5.991$, the confidence region C^{el} is an ellipsoid centered at $\mathbf{0} = (0, 0)$ with principal axes:

- $\sqrt{q_{1-\alpha}(\chi_d^2)\lambda_1}\mathbf{f}_1$, of length $\sqrt{q_{1-\alpha}(\chi_d^2)\lambda_1} = 6.3365$,
- $\sqrt{q_{1-\alpha}(\chi_d^2)\lambda_2}\mathbf{f}_2$, of length $\sqrt{q_{1-\alpha}(\chi_d^2)\lambda_2} = 1.3372$.

The ellipse, along with its principal axes and 5000 sampled points, is depicted in Fig. 4.2.

Parallelogram First, to find the hypercube $\mathcal{H}$ (which, for $d = 2$, is a square), we need to compute z from (1.17):

$$z = \Phi^{-1}\left(\frac{\sqrt{0.95}+1}{2}\right) = 2.2365. \tag{1.20}$$

The corners of the hypercube $\mathcal{H}$ (which is a square for $d = 2$) are:

$$\mathbf{v}_1 = \begin{pmatrix} -z \\ -z \end{pmatrix}, \ \mathbf{v}_2 = \begin{pmatrix} -z \\ z \end{pmatrix}, \ \mathbf{v}_3 = \begin{pmatrix} z \\ z \end{pmatrix}, \ \mathbf{v}_4 = \begin{pmatrix} z \\ -z \end{pmatrix}. \tag{1.21}$$

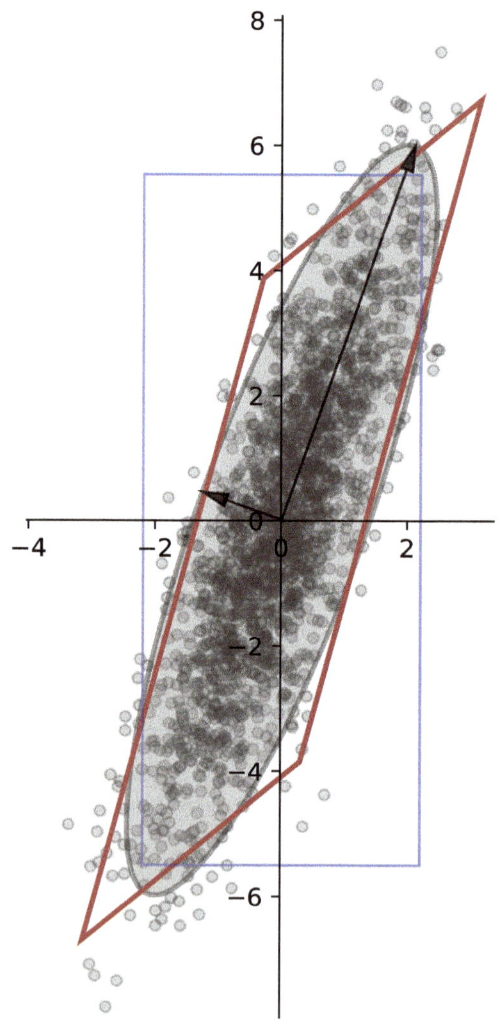

Fig. 4.2 Example 4.1.14: 5000 sampled points (*gray*) from a bivariate normal, confidence ellipse (*gray*) with principal axes (**black**) at confidence level $1 - \alpha = 0.95$; confidence parallelogram (*brown*) at confidence level $1 - \alpha = 0.95$; rectangular confidence region (*blue*) at confidence level $\geq 1 - \alpha = 0.95$ (see text)

Transforming these via $\Sigma^{1/2}$, the corners of the parallelogram C^{par} are $\mathbf{w}_i = \Sigma^{1/2}\mathbf{v}_i$, $i = 1, 2, 3, 4$, i.e.,

$$\mathbf{w}_1 = \begin{pmatrix} -3.1374 \\ -6.6984 \end{pmatrix}, \quad \mathbf{w}_2 = \begin{pmatrix} -0.2839 \\ 3.8798 \end{pmatrix}, \quad \mathbf{w}_3 = \begin{pmatrix} 3.1605 \\ 6.7333 \end{pmatrix}, \quad \mathbf{w}_4 = \begin{pmatrix} 0.3070 \\ -3.8449 \end{pmatrix}.$$

The confidence parallelogram is depicted in Fig. 4.2.

Rectangular Confidence Region For the rectangular region, note that $\sigma_1 = 1$ and $\sigma_2 = \sqrt{6}$. Using $z = \Phi^{-1}\left(\frac{\sqrt{0.95}+1}{2}\right) = 2.2365$, the bounds are:

$$c_1 = \sigma_1 \cdot 2.2365 = 2.2365, \quad c_2 = \sigma_2 \cdot 2.2365 = 5.4782.$$

This provides a conservative rectangular region:

$$C^{\text{rec}} = [-2.2365, 2.2365] \times [-5.4782, 5.4782],$$

ensuring $\mathbb{P}(\mathbf{X} \in C^{\text{rec}}) \geq 0.95$.

Double-check and Verifying with Exact Values To verify the correctness of the computed confidence regions, recall the general formula for the density of $\mathcal{N}(\boldsymbol{\mu}, \boldsymbol{\Sigma})$:

$$f(\mathbf{x}) = \frac{1}{(2\pi)^{d/2}\det(\boldsymbol{\Sigma})^{1/2}} \exp\left(-\frac{1}{2}(\mathbf{x} - \boldsymbol{\mu})^T \boldsymbol{\Sigma}^{-1}(\mathbf{x} - \boldsymbol{\mu})\right).$$

In our example, this becomes:

$$f(x, y) = \frac{\sqrt{2}}{4\pi} \exp\left(-\frac{3}{2}x^2 + xy - \frac{1}{4}y^2\right). \tag{1.22}$$

Using this density, one can directly verify that:

$$\int_{C^{\text{el}}} f(x, y)\, dx\, dy = \int_{C^{\text{par}}} f(x, y)\, dx\, dy = 0.95,$$

for the computed ellipsoid and parallelogram as confidence regions.

For the rectangular confidence region, the approximate region provides coverage ≥ 0.95, as expected, with:

$$\int_{C^{\text{rec}}} f(x, y)\, dx\, dy = \int_{-2.2365}^{2.2365} \int_{-5.4782}^{5.4782} f(x, y)\, dx\, dy \approx 0.9598.$$

To compare, we may find the rectangular confidence region C^{exactRec} with exact confidence level $1 - \alpha = 0.95$ by solving:

$$\int_{-z\sigma_1}^{z\sigma_1} \int_{-z\sigma_2}^{z\sigma_2} f(x, y)\, dx\, dy = 0.95,$$

yielding $z = 2.146$, thus

$$C^{\text{exactRec}} = [-2.146, 2.146] \times [-5.258, 5.258].$$

4.1.2.3   Confidence Regions for Estimator $\hat{\mathbf{Y}}_R$ of I

In this section, we aim to estimate:

$$\mathbf{I} = (I^{(1)}, \ldots, I^{(d)})^T = \mathbb{E}\mathbf{Y} = \mathbb{E}(Y^{(1)}, \ldots, Y^{(d)})^T,$$

for some random vector $\mathbf{Y}$. Assume that $\mathbf{Y}_1, \ldots, \mathbf{Y}_R$ are independent replications of $\mathbf{Y}$, and we consider an estimator in the form:

$$\hat{\mathbf{Y}}_R = \frac{1}{R} \sum_{j=1}^{R} \mathbf{Y}_j, \qquad \mathbf{Y}_j = \begin{pmatrix} Y_j^{(1)} \\ \vdots \\ Y_j^{(d)} \end{pmatrix}, \qquad j = 1, \ldots, R.$$

Let $\boldsymbol{\Sigma}_{\mathbf{Y}}$ denote the covariance matrix of $\mathbf{Y}$, i.e., $\boldsymbol{\Sigma}_{\mathbf{Y}} = \mathbb{E}[(\mathbf{Y} - \mathbf{I})(\mathbf{Y} - \mathbf{I})^T]$. The multivariate version of the Central Limit Theorem states:

$$\sqrt{R}(\hat{\mathbf{Y}}_R - \mathbf{I}) \xrightarrow{\mathcal{D}} \mathcal{N}(\mathbf{0}, \boldsymbol{\Sigma}_{\mathbf{Y}}), \qquad (1.23)$$

which implies that $\hat{\mathbf{Y}}_R$ has an approximate distribution:

$$\mathcal{N}\left(\mathbf{I}, \frac{1}{R}\boldsymbol{\Sigma}_{\mathbf{Y}}\right).$$

Ellipsoid Confidence Region   The normalized eigenvectors $\mathbf{f}_i, i = 1, \ldots, d$, of $\boldsymbol{\Sigma}_{\mathbf{Y}}$ and $\boldsymbol{\Sigma}_{\mathbf{Y}}/R$ are the same. Denote by λ_i' the eigenvalue of $\boldsymbol{\Sigma}_{\mathbf{Y}}/R$ and by λ_i the eigenvalue of $\boldsymbol{\Sigma}_{\mathbf{Y}}$ corresponding to $\mathbf{f}_i$. We have:

$$\lambda_i' = \frac{\lambda_i}{R}, \quad i = 1, \ldots, d.$$

Thus, following (1.15), the confidence ellipsoid at level $1 - \alpha$ for $\mathbf{I}$ is an ellipsoid centered at $\hat{\mathbf{Y}}_R$ with axes:

$$\sqrt{\frac{q_{1-\alpha}(\chi_d^2)\lambda_i}{R}}\,\mathbf{f}_i,\ i = 1, \ldots, d. \tag{1.24}$$

In practice, we estimate $\boldsymbol{\Sigma}_\mathbf{Y}$ using the sample covariance matrix $\hat{\mathbf{S}}_R^2$:

$$\hat{\mathbf{S}}_R^2(i, j) = \frac{1}{R-1} \sum_{k=1}^{R} (Y_k^{(i)} - \hat{Y}_R^{(i)})(Y_k^{(j)} - \hat{Y}_R^{(j)}),$$

where $\hat{Y}_R^{(s)} = \frac{1}{R} \sum_{k=1}^{R} Y_k^{(s)}$. Equivalently, in matrix form:

$$\hat{\mathbf{S}}_R^2 = \frac{1}{R-1} \sum_{k=1}^{R} (\mathbf{Y}_k - \hat{\mathbf{Y}}_R)(\mathbf{Y}_k - \hat{\mathbf{Y}}_R)^T. \tag{1.25}$$

The procedure for constructing the **estimated confidence ellipsoid** from replications $\mathbf{Y}_1, \ldots, \mathbf{Y}_R$ is as follows:

1. Compute $\hat{\mathbf{Y}}_R$.
2. For the sample covariance matrix $\hat{\mathbf{S}}_R^2$, calculate eigenvalues λ_i and eigenvectors $\mathbf{f}_i$, for $i = 1, \ldots, d$.
3. Determine the ellipsoid C with axes given in (1.24).
4. The desired confidence ellipsoid is $\hat{\mathbf{Y}}_R + C$.

Remark 4.1.15 When the unknown covariance matrix $\boldsymbol{\Sigma}_\mathbf{Y}$ is replaced with the sample covariance matrix $\hat{\mathbf{S}}_R^2$, the statistic $R(\hat{\mathbf{Y}}_R - \mathbf{I})^T (\hat{\mathbf{S}}_R^2)^{-1}(\hat{\mathbf{Y}}_R - \mathbf{I})$ follows Hotelling's T^2-distribution. For a large number of replications R (as in our applications), this statistic is approximately chi-square distributed. The resulting confidence regions are indistinguishable in practice, which is why we use the chi-square distribution, consistent with Remark 4.1.4. ∎

Example 4.1.16 Instead of simulating data, here we use a real dataset SOCR Data Dinov 020108 HeightsWeights. [1] The dataset contains 25,000 samples $\mathbf{Y}_i = (Y_i^{(1)}, Y_i^{(2)})^T, i = 1, \ldots, R$, where $Y_i^{(1)}$ represents heights (in inches) and $Y_i^{(2)}$ represents weights (in pounds) of 18-year-old children. For this example, we consider the first $R = 1000$ entries of the dataset. The goal is to approximate $\mathbf{I} = (I^{(1)}, I^{(2)})^T$, where $I^{(1)}$ is the mean height, and $I^{(2)}$ is the mean weight of an 18-year-old child.

---

[1] Available at https://wiki.socr.umich.edu/index.php/SOCR_Data_Dinov_020108_HeightsWeights.

- **Confidence intervals for heights and weights.** Computing independently for the two variables, we obtain:

$$\hat{Y}_R^{(1)} = 127.4817 \quad \hat{S}_R^{(1)} = 11.8022,$$
$$\hat{Y}_R^{(2)} = 68.047 \quad\quad \hat{S}_R^{(2)} = 1.9361.$$

The corresponding confidence intervals (1.9) at confidence level $1 - \alpha = 0.95$ (where $z_{1-\alpha/2} = 1.96$) are:

$$\text{Height}: \ \mathbb{P}(126.7502 \le I^{(1)} \le 128.2132) = 95\%,$$
$$\text{Weight}: \ \mathbb{P}(67.927 \le I^{(2)} \le 68.167) \quad\quad = 95\%.$$

- **Confidence ellipse.** Estimating the covariance matrix, we find:

$$\hat{\mathbf{S}}_R^2 = \begin{pmatrix} 139.2911 & 11.4039 \\ 11.4039 & 3.7484 \end{pmatrix}, \quad (\hat{\mathbf{S}}_R^2)^{-1} = \begin{pmatrix} 0.0095 & -0.0290 \\ -0.0290 & 0.3552 \end{pmatrix}.$$

The eigenvalues and eigenvectors of $\hat{\mathbf{S}}_R^2$ are:

$$\lambda_1 = 140.2439, \quad \mathbf{f}_1 = (0.9965, 0.0832)^T,$$
$$\lambda_2 = 2.7856, \quad\quad \mathbf{f}_2 = (-0.0832, 0.9965)^T.$$

For $\alpha = 0.05$, we have $q_{1-\alpha}(\chi_d^2) = 5.991$. The principal axes of the $1 - \alpha$ confidence ellipse are:

$$\mathbf{g}_1 = \sqrt{\tfrac{5.991\lambda_1}{R}} \cdot \mathbf{f}_1 = \sqrt{0.8402} \cdot \mathbf{f}_1 = (0.9134, 0.7631)^T,$$
$$\mathbf{g}_2 = \sqrt{\tfrac{5.991\lambda_2}{R}} \cdot \mathbf{f}_2 = \sqrt{0.0167} \cdot \mathbf{f}_2 = (-0.0107, 0.1289)^T.$$

- **Confidence parallelogram.** To compute the parallelogram, we calculate:

$$\hat{\mathbf{S}}_R = \begin{pmatrix} 11.7719 & 0.8438 \\ 0.8438 & 1.7425 \end{pmatrix}.$$

For $1 - \alpha = 0.95$, we have $z = 2.2365$ as computed earlier. The vertices $\mathbf{v}_1, \ldots, \mathbf{v}_4$ of $\mathcal{H}$ are given in (1.21). The parallelogram corners are:

$$\mathbf{w}_i = \frac{1}{\sqrt{R}}\hat{\mathbf{S}}_R\mathbf{v}_i + \hat{\mathbf{Y}}_R, \quad i = 1, \ldots, 4,$$

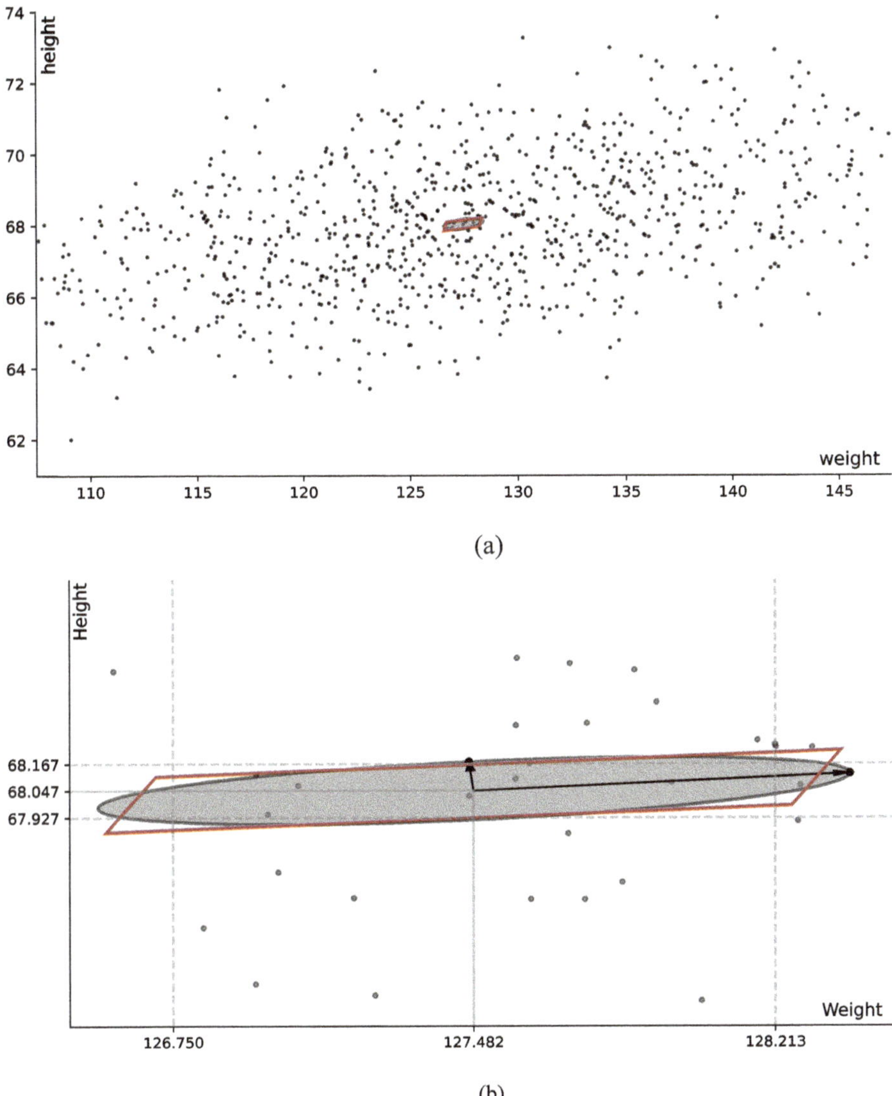

(a)

(b)

Fig. 4.3 First $R = 1000$ data points from dataset SOCR Data Dinov 020108 HeightsWeights together with computed confidence ellipse and confidence parallelogram, all at the significance level $1 - \alpha = 0.95$. (**a**) Data points (weight, height) and the confidence ellipse and parallelogram. (**b**) Confidence ellipse (and its principal axes) and parallelogram enlarged. One-dimensional confidence intervals for $I^{(1)}$ and $I^{(2)}$ depicted

resulting in:

$$\mathbf{w}_1 = \begin{pmatrix} 126.5895 \\ 67.8641 \end{pmatrix}, \ \mathbf{w}_2 = \begin{pmatrix} 126.7089 \\ 68.1105 \end{pmatrix}, \ \mathbf{w}_3 = \begin{pmatrix} 128.3740 \\ 68.2299 \end{pmatrix}, \ \mathbf{w}_4 = \begin{pmatrix} 128.2546 \\ 67.9834 \end{pmatrix}.$$

The data points, together with the computed confidence ellipse and confidence parallelogram, are presented in Fig. 4.3.

All plots and computations for this example are provided in `ch4_human_height_weight_conf_regions.py`.

◊

4.2 Estimation of $k(I)$

In this section, we focus on approximating

$$z = k(I),$$

(or $z = k(\mathbf{I})$ in the multidimensional case), where k is a prescribed function, $I = \mathbb{E}Y$, and we know how to simulate Y. We assume that $k : \mathbb{R}^d \to \mathbb{R}$ is continuous in a neighborhood of I. For a given strongly consistent estimator $\hat{X}_R$ of I, we define the estimator of z as

$$\hat{Z}_R = k(\hat{X}_R).$$

In the previous section, we considered the case where the estimator is the sample mean of independent replications, i.e., $\hat{X}_R = \hat{Y}_R = \frac{1}{R}\sum_{j=1}^{R} Y_j$, where Y_j are i.i.d. replications of Y. In this case, $\hat{Z}_R = k(\hat{Y}_R)$. Throughout this section, $\hat{Y}_R$ will always denote the sample mean of R i.i.d. replications of Y, where $I = \mathbb{E}Y$, and $\hat{Y}_R$ is an unbiased estimator of I. However, we will also consider other strongly consistent estimators of I, which we denote generically by $\hat{X}_R$. Note that when $k(x) = x$ and $\hat{X}_R = \hat{Y}_R$, we revert to the case of Sect. 4.1, where the goal is to approximate $z = I = \mathbb{E}Y$.

Before discussing how to simulate a function of a parameter I, let us introduce an important constant that will appear frequently in these lecture notes. Suppose $\hat{Z}_R$ is an estimator of z. Then, provided the limit exists, the quantity

$$\varsigma^2 = \lim_{R \to \infty} R \mathbb{V}\mathrm{ar}\hat{Z}_R$$

is called the *asymptotic variance*. In some contexts, this is referred to as the *time average variance constant* (TAVC). Note that when $\hat{Z}_R = \hat{Y}_R$ and $k(x) = x$ (i.e., $\hat{Z}_R = \hat{Y}_R$ and $z = I$), the asymptotic variance simplifies to $\varsigma^2 = \mathbb{V}\mathrm{ar}Y$.

Recall that an estimator $\hat{X}_R$ is strongly consistent for I if $\hat{X}_R \to I$ with probability 1 as $R \to \infty$, while it is unbiased if $\mathbb{E}\hat{X}_R = I$. For an example of a consistent but biased estimator, see Exercise 4.T.10.

4.2.1 The Case $\hat{X}_R = \hat{Y}_R$

In this section, we consider an estimator of $k(I)$ of the form $\hat{Z}_R = k(\hat{Y}_R)$, where $\hat{Y}_R$ is the sample mean of R independent replications of Y. We will separately address the one-dimensional and multidimensional cases.

One-dimensional Case We aim to compute $z = k(I)$, where $I = \mathbb{E}Y$ and we know how to simulate Y. Assuming that k is continuous in a neighborhood of I, a natural estimator is given by:

$$\hat{Z}_R = k(\hat{Y}_R) \,, \tag{2.1}$$

where $\hat{Y}_R = \frac{1}{R}\sum_{j=1}^{R} Y_j$ and $Y_1, Y_2, \ldots$ are independent replications of Y. This estimator is strongly consistent, but it is generally not unbiased. Assuming $\sigma_Y^2 = \mathbb{V}\mathrm{ar}Y < \infty$, we have:

$$\sqrt{R}(\hat{Y}_R - I) \overset{\mathcal{D}}{\to} \mathcal{N}(0, \sigma_Y^2) \,. \tag{2.2}$$

In applications, once an estimator is fixed, it is essential to compute or estimate its variance (or asymptotic variance). Only then, using the *fundamental relation* (1.10), can we provide a meaningful confidence interval. The asymptotic variance of the estimator $\hat{Z}_R$ in (2.1) can be obtained from the following result.

Proposition 4.2.1 *Let $Y_1, Y_2, \ldots$ be independent replications of Y, and let $\hat{Y}_R = \frac{1}{R}\sum_{j=1}^{R} Y_j$ be a strongly consistent estimator of $\mathbb{E}Y = I$. Suppose $k(x)$ is continuously differentiable in a neighborhood of I. Then, for the estimator $\hat{Z}_R = k(\hat{Y}_R)$ of $z = k(I)$, as $R \to \infty$,*

$$\sqrt{R}(k(\hat{Y}_R) - k(I)) \overset{\mathcal{D}}{\to} \mathcal{N}(0, \varsigma^2),$$

where

$$\varsigma^2 = k'(I)^2 \sigma_Y^2 = \lim_{R\to\infty} R\,\mathbb{V}\mathrm{ar}\hat{Z}_R.$$

Proof We use a proof technique based on the Taylor expansion, commonly referred to as the *delta method* (see van der Vaart [207]). The Taylor expansion yields

$$k(x) - k(I) = k'(I)(x - I) + o(x - I),$$

hence:

$$\sqrt{R}(k(\hat{Y}_R) - k(I)) = k'(I)\sqrt{R}(\hat{Y}_R - I) + \sqrt{R}\, o(\hat{Y}_R - I).$$

Since:

$$\sqrt{R}\, o(\hat{Y}_R - I) = \sqrt{R}(\hat{Y}_R - I)o(1) \xrightarrow{\text{Pr}} 0$$

(details can be found in van der Vaart [207]), using (2.2), we obtain:

$$\sqrt{R}(k(\hat{Y}_R) - k(I)) \xrightarrow{\mathcal{D}} \mathcal{N}(0, k'(I)^2\sigma_Y^2).$$

$\square$

Thus, in the above setting, the asymptotic variance is:

$$\varsigma^2 = k'(I)^2\sigma_Y^2.$$

We can now write the confidence interval at the confidence level $1 - \alpha$ as:

$$\left(k(\hat{Y}_R) - \frac{z_{1-\alpha/2}\sigma_Y k'(I)}{\sqrt{R}}, k(\hat{Y}_R) + \frac{z_{1-\alpha/2}\sigma_Y k'(I)}{\sqrt{R}}\right).$$

In practice, however, σ_Y^2 and I are rarely known. Typically, as mentioned earlier, m pilot simulations are used first to compute R using the fundamental relationship (1.6). The final confidence interval is then:

$$\left(k(\hat{Y}_R) - \frac{z_{1-\alpha/2}\hat{S}_R k'(\hat{Y}_R)}{\sqrt{R}}, k(\hat{Y}_R) + \frac{z_{1-\alpha/2}\hat{S}_R k'(\hat{Y}_R)}{\sqrt{R}}\right). \tag{2.3}$$

Example 4.2.2 In life insurance mathematics, the net premium for whole life assurance is given by:

$$\Pi = \frac{\mathbb{E}e^{-rT}}{\mathbb{E}\int_0^T e^{-rt}\, dt},$$

where $T = T_x$ is the future lifetime of the insured aged x-years, and r is the interest rate. Note that $\int_0^T e^{-rt}\, dt = (1 - e^{-rT})/r$. Let:

$$Y = e^{-rT}, \quad I = \mathbb{E}Y, \quad \sigma_Y^2 = \mathbb{V}\text{ar}Y, \quad \hat{Y}_R = \frac{1}{R}\sum_{j=1}^{R} Y_j.$$

We can rewrite Π as:

$$\Pi = r\frac{\mathbb{E}e^{-rT}}{1 - \mathbb{E}e^{-rT}} = k(I),$$

where $k(I) = \frac{rI}{1-I}$ and $k'(x) = \frac{r}{(1-x)^2}$. Using (2.3), the confidence interval for the confidence level $1 - \alpha$ is:

$$\left(k(\hat{Y}_R) - \frac{z_{1-\alpha/2}r\hat{S}_R}{(1 - \hat{Y}_R)^2\sqrt{R}}, k(\hat{Y}_R) + \frac{z_{1-\alpha/2}r\hat{S}_R}{(1 - \hat{Y}_R)^2\sqrt{R}}\right).$$

$\Diamond$

Multidimensional Case We now extend Proposition 4.2.1 to the multidimensional case. Suppose we aim to estimate $z = k(I^{(1)}, \ldots, I^{(d)})$, where $(I^{(1)}, \ldots, I^{(d)}) = (\mathbb{E}Y^{(1)}, \ldots, \mathbb{E}Y^{(d)})$, and we can simulate the vector $\mathbf{Y} = (Y^{(1)}, \ldots, Y^{(d)})^T$.

To simplify the notation, let:

$$\mathbf{Y} = \begin{pmatrix} Y^{(1)} \\ \vdots \\ Y^{(d)} \end{pmatrix}, \quad \mathbf{I} = \begin{pmatrix} I^{(1)} \\ \vdots \\ I^{(d)} \end{pmatrix},$$

and define an **unbiased estimator** of $\mathbf{I}$:

$$\hat{\mathbf{Y}}_R = \frac{1}{R}\sum_{j=1}^{R}\mathbf{Y}_j,$$

where $\mathbf{Y}_1 = (Y_1^{(1)}, \ldots, Y_1^{(d)})^T, \ldots, \mathbf{Y}_R = (Y_R^{(1)}, \ldots, Y_R^{(d)})^T$ are independent replications of $\mathbf{Y}$.

For any $\mathbf{x} = (x_1, \ldots, x_d)^T$, we denote $k(\mathbf{x})$ by $k(x_1, \ldots, x_d)$. The natural estimator of z is $\hat{Z}_R = k(\hat{\mathbf{Y}}_R)$. We denote the gradient of k at $\mathbf{I}$ by $\nabla k(\mathbf{I}) = (k^{(1)}(\mathbf{x}), \ldots, k^{(d)}(\mathbf{x}))^T$, where $k^{(i)}(\mathbf{x}) = \frac{\partial k(\mathbf{x})}{\partial x_i}$.

Notation In the one-dimensional case, for $I = \mathbb{E}Y$ and replications $Y_1, \ldots, Y_R$, the sample mean: $\hat{Y}_R = \frac{1}{R}\sum_{j=1}^{R}Y_j$ is an estimator of I, and the sample variance:

$$\hat{S}_R^2 = \frac{1}{R-1}\sum_{j=1}^{R}(Y_j - \hat{Y}_R)^2$$

is an estimator of $\sigma_Y^2 = \mathbb{V}\mathrm{ar}Y$. For the multidimensional case, suppose $\mathbf{I} = \mathbb{E}\mathbf{Y}$, where $\mathbf{Y} = (Y^{(1)}, \ldots, Y^{(d)})^T \in \mathbb{R}^d$. The sample mean is:

$$\hat{\mathbf{Y}}_R = \frac{1}{R} \sum_{j=1}^{R} \mathbf{Y}_j = (\hat{Y}_R^{(1)}, \ldots, \hat{Y}_R^{(d)})^T.$$

The covariance matrix of $\mathbf{Y}$ is:

$$\boldsymbol{\Sigma}_\mathbf{Y} = \mathbb{E}\left[(\mathbf{Y} - \mathbf{I})(\mathbf{Y} - \mathbf{I})^T\right],$$

and the sample covariance matrix is:

$$\hat{\mathbf{S}}_R^2 = \frac{1}{R-1} \sum_{j=1}^{R} (\mathbf{Y}_j - \hat{\mathbf{Y}}_R)(\mathbf{Y}_j - \hat{\mathbf{Y}}_R)^T,$$

which is an unbiased estimator of $\boldsymbol{\Sigma}_\mathbf{Y}$.

Two-Dimensional Case ($d = 2$) Let:

$$\mathbf{Y} = \begin{pmatrix} Y^{(1)} \\ Y^{(2)} \end{pmatrix}, \quad \mathbf{I} = \begin{pmatrix} I^{(1)} \\ I^{(2)} \end{pmatrix}, \quad \hat{\mathbf{Y}}_R = \begin{pmatrix} \hat{Y}_R^{(1)} \\ \hat{Y}_R^{(2)} \end{pmatrix} = \frac{1}{R} \sum_{j=1}^{R} \mathbf{Y}_j.$$

For a continuously differentiable function $k(y_1, y_2)$ at $\mathbf{I}^T = (I^{(1)}, I^{(2)})$, the Taylor expansion yields:

$$k(y_1, y_2) = k(I^{(1)}, I^{(2)}) + k'_{y_1}(I^{(1)}, I^{(2)})(y_1 - I^{(1)})$$

$$+ k'_{y_2}(I^{(1)}, I^{(2)})(y_2 - I^{(2)}) + o\left(\max(|y_1 - I^{(1)}|, |y_2 - I^{(2)}|)\right).$$

$$(2.4)$$

Setting $y_1 = \hat{Y}_R^{(1)}$ and $y_2 = \hat{Y}_R^{(2)}$, and simplifying:

$$k(\hat{\mathbf{Y}}_R) - k(\mathbf{I}) = \left(k'_{y_1}(I^{(1)}, I^{(2)})(\hat{Y}_R^{(1)} - I^{(1)}) + k'_{y_2}(I^{(1)}, I^{(2)})(\hat{Y}_R^{(2)} - I^{(2)})\right)$$

$$+ o\left(\max|\hat{Y}_R^{(1)} - I^{(1)}|, |\hat{Y}_R^{(2)} - I^{(2)}|\right).$$

Recall that for a function $k(y_1, y_2)$ we define

$$\nabla k(y_1, y_2) = (k'_{y_1}(y_1, y_2), k'_{y_2}(y_1, y_2)),$$

a gradient row vector. Rewrite

$$\left(k'_{y_1}(I^{(1)}, I^{(2)})(\hat{Y}_R^{(1)} - I^{(1)}) + k'_{y_2}(I^{(1)}, I^{(2)})(\hat{Y}_R^{(2)} - I^{(2)})\right)$$

$$= \frac{1}{R} \sum_{j=1}^{R} \nabla k(\mathbf{I}) \begin{pmatrix} Y_j^{(1)} - I^{(1)} \\ Y_j^{(2)} - I^{(2)} \end{pmatrix} = \frac{1}{R} \sum_{j=1}^{R} W_j,$$

where

$$W_j = k'_{y_1}(\mathbf{I})(Y_j^{(1)} - I^{(1)}) + k'_{y_2}(\mathbf{I})(Y_j^{(2)} - I^{(2)}) \ .$$

Random variables $W_1, \ldots, W_R$ are i.i.d., $\mathbb{E}W_j = 0$ and

$$\mathbb{V}\mathrm{ar}W_j = (k'_{y_1}(\mathbf{I}))^2 \mathbf{\Sigma_Y}(1, 1) + (k'_{y_2}(\mathbf{I}))^2 \mathbf{\Sigma_Y}(2, 2) + 2k'_{y_1}(\mathbf{I})k'_{y_2}(\mathbf{I})\mathbf{\Sigma_Y}(1, 2) \ .$$

Similarly as in Proposition 4.2.1 we have that

$$\sqrt{R} \ o\left(\max |\hat{Y}_R^{(1)} - I^{(1)}|, |\hat{Y}_R^{(2)} - I^{(2)}|\right)$$

goes in probability to 0. Hence

$$\sqrt{R}(k(\hat{\mathbf{Y}}_R) - k(\mathbf{I}))$$

$$= \sqrt{R}\frac{1}{R} \sum_{j=1}^{R} W_j + \sqrt{R} \ o\left(\max |\hat{Y}_R^{(1)} - I^{(1)}|, |\hat{Y}_R^{(2)} - I^{(2)}|\right) \ .$$

Letting $R \to \infty$ we can state the following proposition:

Proposition 4.2.3 *Let $\mathbf{Y}_1, \mathbf{Y}_2, \ldots$ be independent replications of $\mathbf{Y}$, and let $k(y_1, y_2)$ be continuously differentiable in a neighborhood of $(I^{(1)}, I^{(2)})$. Then, for the estimator $k(\hat{\mathbf{Y}}_R)$ of $k(I^{(1)}, I^{(2)})$, we have as $R \to \infty$:*

$$\sqrt{R}(k(\hat{\mathbf{Y}}_R) - k(\mathbf{I})) \overset{\mathcal{D}}{\to} \mathcal{N}(0, \varsigma^2), \tag{2.5}$$

where

$$\varsigma^2 = (k'_{y_1}(\mathbf{I}))^2 \mathbf{\Sigma_Y}(1, 1) + (k'_{y_2}(\mathbf{I}))^2 \mathbf{\Sigma_Y}(2, 2) + 2k'_{y_1}(\mathbf{I})k'_{y_2}(\mathbf{I})\mathbf{\Sigma_Y}(1, 2). \tag{2.6}$$

Hence, we have the asymptotic variance

$$\varsigma^2 = \lim_{R \to \infty} R \ \mathbb{V}\mathrm{ar}(k(\hat{\mathbf{Y}}_R) - k(\mathbf{I})).$$

In practice, we estimate the covariance matrix $\hat{\mathbf{S}}_R^2$ from simulations and then estimate the asymptotic variance by $\hat{\varsigma}_R^2 =$

$$(k'_{y_1}(\hat{\mathbf{Y}}_R))^2 \hat{\mathbf{S}}_R^2(1, 1) + (k'_{y_2}(\hat{\mathbf{Y}}_R))^2 \hat{\mathbf{S}}_R^2(2, 2) + 2k'_{y_1}(\hat{\mathbf{Y}}_R)k'_{y_2}(\hat{\mathbf{Y}}_R)\hat{\mathbf{S}}_R^2(1, 2) . \tag{2.7}$$

The confidence interval at confidence level $1 - \alpha$ is as follows:

$$\left(k(\hat{\mathbf{Y}}_R) - \frac{z_{1-\alpha/2}\sqrt{\hat{\varsigma}_R^2}}{\sqrt{R}}, k(\hat{\mathbf{Y}}_R) + \frac{z_{1-\alpha/2}\sqrt{\hat{\varsigma}_R^2}}{\sqrt{R}} \right) . \tag{2.8}$$

A Study of the Bias of $k(\hat{Y}_R)$ It is important to note that $k(\hat{\mathbf{Y}}_R)$ is generally not unbiased. To explore its bias (albeit heuristically), we rely on the second-order Taylor expansion of k around $(I^{(1)}, I^{(2)})$. This requires the additional assumption that the second derivative of k is continuous in a neighborhood of $(I^{(1)}, I^{(2)})$. We have

$$k(y_1, y_2) =$$

$$k(I^{(1)}, I^{(2)})$$

$$+ k'_{y_1}(I^{(1)}, I^{(2)})(y_1 - I^{(1)}) + k'_{y_2}(I^{(1)}, I^{(2)})(y_2 - I^{(2)})$$

$$+ \frac{1}{2}\left[k''_{y_1 y_1}(I^{(1)}, I^{(2)})(y_1 - I^{(1)})^2 + 2k''_{y_1 y_2}(I^{(1)}, I^{(2)})(y_1 - Y^{(1)})(y_2 - I^{(2)}) \right.$$

$$\left. + k''_{y_2 y_2}(I^{(1)}, I^{(2)})(y_2 - I^{(2)})^2 \right] + o\left(\max(y_1 - I^{(1)})^2, (y_2 - I^{(2)})^2 \right) . \tag{2.9}$$

Now set $y_i = \hat{Y}_R^{(i)}, i = 1, 2$ and take the expectation of both sides. Notice that

$$\mathbb{E}[k'_{y_1}(I^{(1)}, I^{(2)})(\hat{Y}_R^{(1)} - I^{(1)}) + k'_{y_2}(I^{(1)}, I^{(2)})(\hat{Y}_R^{(2)} - I^{(2)})] = 0$$

and

$$\mathbb{E}\frac{1}{2}\left[k''_{y_1 y_1}(I^{(1)}, I^{(2)})(\hat{Y}_R^{(1)} - I^{(1)})^2 + 2k''_{y_1 y_2}(I^{(1)}, I^{(2)})(\hat{Y}_R^{(1)} - I^{(1)})(\hat{Y}_R^{(2)} - I^{(2)}) \right.$$

$$\left. + k''_{y_2 y_2}(I^{(1)}, I^{(2)})(\hat{Y}_R^{(2)} - I^{(2)})^2 \right]$$

$$= \frac{1}{2R}\left[k''_{y_1 y_1}(I^{(1)}, I^{(2)})\mathbf{\Sigma_Y}(1, 1)) + 2k''_{y_1 y_2}(I^{(1)}, I^{(2)})\mathbf{\Sigma_Y}(1, 2) \right.$$

$$\left. + k''_{y_2 y_2}(I^{(1)}, I^{(2)})\mathbf{\Sigma_Y}(2, 2) \right] .$$

Recall that $\mathbf{\Sigma_Y}$ is the covariance matrix of $\mathbf{Y}$. Concluding we have

$$
\begin{aligned}
\mathbb{E}k(\hat{\mathbf{Y}}_R) &- k(\mathbf{I}) \\
&= \frac{1}{2R}\Big(k''_{y_1 y_1}(I^{(1)}, I^{(2)})\mathbf{\Sigma_Y}(1,1) + 2k''_{y_1 y_2}(I^{(1)}, I^{(2)})\mathbf{\Sigma_Y}(1,2) \\
&\qquad + k''_{y_2 y_2}(I^{(1)}, I^{(2)})\mathbf{\Sigma_Y}(2,2)\Big) + o(1/R) \\
&= R^{-1}\beta_R + o(1/R),
\end{aligned}
\tag{2.10}
$$

where

$$
\beta = \frac{1}{2}\Big[k''_{y_1 y_1}(I^{(1)}, I^{(2)})\mathbf{\Sigma_Y}(1,1) + k''_{y_2 y_2}(I^{(1)}, I^{(2)})\mathbf{\Sigma_Y}(2,2) \tag{2.11}
$$

$$
+ 2k''_{y_1 y_2}(I^{(1)}, I^{(2)})\mathbf{\Sigma_Y}(1,2)\Big].
$$

A formal proof of (2.10) will require some regularity assumptions. Note that

$$
\lim_{R\to\infty} R\left(\mathbb{E}k(\mathbf{Y}) - k(\mathbf{I})\right) = \beta.
$$

Again, in practice we estimate the covariance matrix with $\hat{\mathbf{S}}_R^2$, asymptotic variance with $\hat{\varsigma}^2$ in (2.7) and the asymptotic bias with

$$
\hat{\beta}_R = \frac{1}{2}\Big[k''_{y_1 y_1}(\hat{Y}^{(1)}, \hat{Y}^{(2)})\hat{\mathbf{S}}_R^2(1,1) + k''_{y_2 y_2}(\hat{Y}^{(1)}, \hat{Y}^{(2)})\hat{\mathbf{S}}_R^2(2,2)
$$

$$
+ 2k''_{y_1 y_2}(\hat{Y}^{(1)}, \hat{Y}^{(2)})\hat{\mathbf{S}}_R^2(1,2)\Big].
$$

We can thus add the correction and estimate $k(\mathbf{I})$ by $k(\hat{\mathbf{Y}}_R) - R^{-1}\hat{\beta}_R$ and as a corrected confidence interval, take

$$
\left(k(\hat{\mathbf{Y}}_R) - R^{-1}\hat{\beta}_R - \frac{z_{1-\alpha/2}\sqrt{\hat{\varsigma}_R^2}}{\sqrt{R}}, k(\hat{\mathbf{Y}}_R) - R^{-1}\hat{\beta}_R + \frac{z_{1-\alpha/2}\sqrt{\hat{\varsigma}_R^2}}{\sqrt{R}}\right).
$$

Remark 4.2.4 As noted in [8], bias corrections are often implemented due to a widely held *belief* that they improve accuracy in small-sample scenarios. However, in our applications, the number of replications R is typically (very) large, making the difference between bias-corrected and uncorrected estimators negligible.　■

Example 4.2.5 (Estimating Ratios of Expectations) Assume we want to estimate the ratio $z = I^{(1)}/I^{(2)}$, where there exists a random vector $\mathbf{Y} = (Y^{(1)}, Y^{(2)})^T$ (assuming the distribution of $Y^{(2)}$ has no mass at 0) such that $I^{(1)} = \mathbb{E}Y^{(1)}$ and

$I^{(2)} = \mathbb{E}Y^{(2)}$. We do not assume that $Y^{(1)}$ is independent of $Y^{(2)}$. Having R independent replications $(Y_1^{(1)}, Y_1^{(2)})^T, \ldots, (Y_R^{(1)}, Y_R^{(2)})^T$, a natural estimator is:

$$\hat{Z}_R = \frac{\hat{Y}_R^{(1)}}{\hat{Y}_R^{(2)}}, \qquad \text{where} \quad \hat{Y}_R^{(1)} = \frac{1}{R}\sum_{i=1}^{R} Y_i^{(1)}, \quad \hat{Y}_R^{(2)} = \frac{1}{R}\sum_{i=1}^{R} Y_i^{(2)}.$$

We compute the asymptotic variance of $\hat{Z}_R$ using Proposition 4.2.3. Setting $k(y_1, y_2) = y_1/y_2$, we estimate $z = k(I^{(1)}, I^{(2)})$, where $(I^{(1)}, I^{(2)}) = (\mathbb{E}Y^{(1)}, \mathbb{E}Y^{(2)})$. The partial derivatives of $k(y_1, y_2)$ are:

$$k'_{y_1}(y_1, y_2) = \frac{1}{y_2}, \qquad k'_{y_2}(y_1, y_2) = -\frac{y_1}{y_2^2},$$

and taking $(y_1, y_2) = (I^{(1)}, I^{(2)})$, we obtain:

$$\varsigma^2 = \frac{1}{(I^{(2)})^2}\boldsymbol{\Sigma}_{\mathbf{Y}}(1, 1) - 2\frac{1}{I^{(2)}}\frac{I^{(1)}}{(I^{(2)})^2}\boldsymbol{\Sigma}_{\mathbf{Y}}(1, 2) + \frac{(I^{(1)})^2}{(I^{(2)})^4}\boldsymbol{\Sigma}_{\mathbf{Y}}(2, 2)$$

$$= \frac{(I^{(1)})^2}{(I^{(2)})^2}\left[\frac{\boldsymbol{\Sigma}_{\mathbf{Y}}(1, 1)}{(I^{(1)})^2} - 2\frac{\boldsymbol{\Sigma}_{\mathbf{Y}}(1, 2)}{I^{(1)}\,I^{(2)}} + \frac{\boldsymbol{\Sigma}_{\mathbf{Y}}(2, 2)}{(I^{(2)})^2}\right]. \tag{2.12}$$

Alternatively, this can be reformulated in a more convenient form:

$$\varsigma^2 = \frac{\boldsymbol{\Sigma}_{\mathbf{Y}}(1, 1) - 2z\boldsymbol{\Sigma}_{\mathbf{Y}}(1, 2) + z^2\boldsymbol{\Sigma}_{\mathbf{Y}}(2, 2)}{(I^{(2)})^2}$$

$$= \frac{\mathbb{E}\left(Y^{(1)} - \mathbb{E}Y^{(1)} - z(Y^{(2)} - \mathbb{E}Y^{(2)})\right)^2}{(I^{(2)})^2}$$

$$= \frac{\mathbb{E}\left(Y^{(1)} - zY^{(2)}\right)^2}{(I^{(2)})^2}.$$

Thus, using Proposition 4.2.3, we approximate the variance of $\hat{Z}_R$ as:

$$\mathbb{V}\mathrm{ar}\hat{Z}_R = \frac{\varsigma^2}{R} + o\left(R^{-1}\right) = \frac{1}{R}\frac{\mathbb{E}\left(Y^{(1)} - zY^{(2)}\right)^2}{\left(\mathbb{E}Y^{(2)}\right)^2} + o\left(R^{-1}\right).$$

This variance can be estimated by:

$$\widehat{\mathbb{V}\mathrm{ar}}(\hat{Z}_R) = \frac{1}{R^2(\hat{Y}_R^{(2)})^2}\sum_{i=1}^{R}\left(Y_i^{(1)} - \hat{Z}_R Y_i^{(2)}\right)^2. \tag{2.13}$$

Finally, the confidence interval can be written as:

$$\left(\hat{Z}_R - z_{1-\alpha/2}\sqrt{\widehat{\mathbb{V}\mathrm{ar}}(\hat{Z}_R)}, \ \hat{Z}_R + z_{1-\alpha/2}\sqrt{\widehat{\mathbb{V}\mathrm{ar}}(\hat{Z}_R)} \right).$$

$\Diamond$

Example 4.2.6 Continuing Example 4.2.5, the second derivatives are:

$$k''_{y_1 y_1}(y_1, y_2) = 0, \quad k''_{y_1 y_2}(y_1, y_2) = -\frac{1}{y_2^2}, \quad k''_{y_2 y_2}(y_1, y_2) = \frac{2y_1}{y_2^3}.$$

Thus, the asymptotic bias of $k(\hat{\mathbf{Y}}_R)$ is:

$$\beta = \frac{I^{(1)}\boldsymbol{\Sigma}_{\mathbf{Y}}(2, 2)}{(I^{(2)})^3} - \frac{\boldsymbol{\Sigma}_{\mathbf{Y}}(1, 2)}{(I^{(2)})^2}.$$

Using (2.12), we estimate this bias as:

$$\hat{\beta}_R = \frac{\hat{Y}_R^{(1)}\hat{\mathbf{S}}_R^2(2, 2)}{(\hat{Y}_R^{(2)})^3} - \frac{\hat{\mathbf{S}}_R^2(1, 2)}{(\hat{Y}_R^{(2)})^2}.$$

The corrected confidence interval becomes:

$$\left(\hat{Z}_R - R^{-1}\hat{\beta}_R - z_{1-\alpha/2}\sqrt{\widehat{\mathbb{V}\mathrm{ar}}(\hat{Z}_R)}, \ \hat{Z}_R - R^{-1}\hat{\beta}_R + z_{1-\alpha/2}\sqrt{\widehat{\mathbb{V}\mathrm{ar}}(\hat{Z}_R)} \right).$$

$\Diamond$

Example 4.2.7 Assume that a random variable X takes values in E. For a given function k and a set $A \subseteq E$, we are interested in estimating the conditional expectation:

$$I = \mathbb{E}(k(X) \mid X \in A).$$

Let $w(X) = \mathbb{1}(X \in A)$. Note that this can be rewritten as:

$$\mathbb{E}(k(X) \mid X \in A) = \frac{\mathbb{E}(k(X)w(X))}{\mathbb{E}w(X)}.$$

Given i.i.d. replications $X_1, \ldots, X_R$, such that $w(X_i) = 1$ for at least one $i \in \{1, \ldots, R\}$, the estimator of I is:

$$\hat{Y}_R = \frac{\frac{1}{R}\sum_{i=1}^{R} k(X_i)w(X_i)}{\frac{1}{R}\sum_{i=1}^{R} w(X_i)} = \frac{\sum_{i=1}^{R} k(X_i)w(X_i)}{\sum_{i=1}^{R} w(X_i)} = \frac{1}{R_A}\sum_{i=1}^{R} k(X_i)w(X_i),$$

where $R_A = \sum_{i=1}^{R} w(X_i)$. We define $\hat{\mu}_A = \frac{1}{R} \sum_{i=1}^{R} w(X_i) = \frac{R_A}{R}$. Using Eq. (2.13), the variance of $\hat{Y}_R$ is given by:

$$\widehat{\mathbb{V}\text{ar}}(\hat{Y}_R) = \frac{1}{R^2 \left(\frac{1}{R} \sum_{i=1}^{R} w(X_i) \right)^2} \sum_{i=1}^{R} \left(k(X_i)w(X_i) - \hat{Y}_R w(X_i) \right)^2$$

$$= \frac{1}{R_A^2} \sum_{i=1}^{R} w(X_i)\left(k(X_i) - \hat{Y}_R \right)^2. \tag{2.14}$$

Thus, the error is:

$$b = z_{1-\alpha/2} \sqrt{\widehat{\mathbb{V}\text{ar}}(\hat{Y}_R)}.$$

Denoting $\sum_{i=1}^{R} w(X_i)\left(k(X_i) - \hat{Y}_R \right)^2$ by γ, we can write the error more compactly as:

$$b = z_{1-\alpha/2} \frac{\sqrt{\gamma}}{R}.$$

◇

General Case: $d \geq 2$ Recall that $\nabla k(\mathbf{I})$ denotes the gradient of k, i.e., the vector $(k^{(1)}(\mathbf{I}), \ldots, k^{(d)}(\mathbf{I}))$, where $k^{(i)}(\mathbf{x}) = \frac{\partial k(\mathbf{x})}{\partial x_i}$. The following proposition is a key result for analyzing the asymptotic behavior of $k(\hat{\mathbf{Y}}_R)$. Its proof can be found in van der Vaart [207].

Proposition 4.2.8 *If $k : \mathbb{R}^d \to \mathbb{R}$ is continuously differentiable in a neighborhood of $\mathbf{I}$ and*

$$\sqrt{R}(\hat{\mathbf{Y}}_R - \mathbf{I}) \xrightarrow{\mathcal{D}} \mathcal{N}(\mathbf{0}, \mathbf{\Sigma})$$

for some finite covariance matrix $\mathbf{\Sigma}$, then:

$$\sqrt{R}\left(k(\hat{\mathbf{Y}}_R) - k(\mathbf{I}) \right) \xrightarrow{\mathcal{D}} \mathcal{N}\left(\mathbf{0}, \nabla k(\mathbf{I})\mathbf{\Sigma}(\nabla k(\mathbf{I}))^{\mathsf{T}}\right).$$

In particular, if $\mathbf{Y}$ has a finite covariance matrix $\mathbf{\Sigma}_{\mathbf{Y}} = \left(\mathbf{\Sigma}_{\mathbf{Y}}(i, j)\right)_{i, j=1, \ldots, d}$ and $\hat{\mathbf{Y}}_R = \frac{1}{R} \sum_{j=1}^{R} \mathbf{Y}_j$, where $\mathbf{Y}_1, \ldots, \mathbf{Y}_R$ are independent replications of $\mathbf{Y}$, then a confidence interval at the confidence level $\alpha = 0.05$ for the estimator $\hat{Z}_R = k(\hat{\mathbf{Y}}_R)$ is given by:

$$k(\hat{\mathbf{Y}}_R) \pm 1.96 \frac{\varsigma}{\sqrt{R}},$$

where

$$\varsigma^2 = \nabla k(\mathbf{I})\boldsymbol{\Sigma}_{\mathbf{Y}}(\nabla k(\mathbf{I}))^{\mathrm{T}} = \sum_{i,j=1}^{d} k^{(i)}(\mathbf{I})k^{(j)}(\mathbf{I})\boldsymbol{\Sigma}_{\mathbf{Y}}(i,j). \tag{2.15}$$

Unfortunately, the formula for ς^2 involves the unknown parameters $\mathbf{I}$ and $\boldsymbol{\Sigma}_{\mathbf{Y}}$. Therefore, these must first be estimated using a pilot simulation. The resulting confidence interval is:

$$k(\hat{\mathbf{Y}}_R) \pm 1.96\frac{\hat{\varsigma}}{\sqrt{R}},$$

where the estimated asymptotic variance is:

$$\hat{\varsigma}^2 = (\nabla k(\hat{\mathbf{Y}}_R))\hat{\mathbf{S}}_R^2(\nabla k(\hat{\mathbf{Y}}_R))^{\mathrm{T}} = \sum_{i,j=1}^{d} k^{(i)}(\hat{\mathbf{Y}}_R)k^{(j)}(\hat{\mathbf{Y}}_R)\hat{\mathbf{S}}_R^2(i,j). \tag{2.16}$$

4.2.2 General Consistent $\hat{X}_R$

Let us return to the one-dimensional case. Our goal is to estimate $z = k(I)$ using the estimator $\hat{Z}_R = k(\hat{X}_R)$, where $\hat{X}_R$ is a strongly consistent estimator of I. The bias of the estimator $\hat{X}_R$ is defined as:

$$\delta_R^X = \mathbb{E}\hat{X}_R - I.$$

In the specific case where $\hat{X}_R = \hat{Y}_R$ (recall that $\hat{Y}_R = \sum_{i=1}^{R} Y_i/R$, based on independent replications of Y), the bias is clearly 0. It is important to distinguish the concept of bias from the asymptotic bias of order R^{-1}, which is addressed in Definition (A2') below.

The proof of Proposition 4.2.1 demonstrates that the assumption of independent replications is unnecessarily strong. Of course, under this assumption, $k(\hat{Y}_R)$ is consistent. Now, consider another estimator $\hat{X}_R$ which is a strongly consistent estimator of I. Then, $\hat{Z}_R = k(\hat{X}_R)$ is also a consistent estimator of $z = k(I)$.

The critical point is the convergence behavior, which should resemble that in (2.2). For this, we introduce the following assumptions:

(A1) $\sqrt{R}(\hat{X}_R - \mathbb{E}\hat{X}_R) \xrightarrow{\mathcal{D}} \mathcal{N}(0,\sigma^2)$, where $0 < \sigma^2 < \infty$ as $R \to \infty$.
(A2) $\sqrt{R}(\mathbb{E}\hat{X}_R - I) \to 0$ as $R \to \infty$.

Using the properties of convergence of random variables, assumptions (A1) and (A2) imply:

$$\sqrt{R}(\hat{X}_R - I) \xrightarrow{\mathcal{D}} \mathcal{N}(0, \sigma^2).$$

Summarizing, we have the following result: *Consider the setup of Proposition 4.2.1, where $\hat{Y}_R$ is replaced by a consistent estimator $\hat{X}_R$ of I, satisfying assumptions (A1) and (A2). Then, as $R \to \infty$:*

$$\sqrt{R}(k(\hat{X}_R) - k(I)) \xrightarrow{\mathcal{D}} \mathcal{N}(0, \varsigma^2),$$

where:

$$\varsigma^2 = k'(I)^2 \sigma^2 = \lim_{R \to \infty} R \mathbb{V}\mathrm{ar} \hat{Z}_R.$$

Asymptotic Bias of the Estimator $\hat{Z}_R$ of $k(I)$ We now discuss the asymptotic bias of the estimator $\hat{Z}_R$ for $k(I)$. Instead of assumption (A2), we now adopt the following stronger assumption:

(A2') For some finite $\beta^X \neq 0$, $R(\mathbb{E}\hat{X}_R - I) \to \beta^X$ as $R \to \infty$.

This means that β^X represents the R^{-1}-order asymptotic bias of the estimator $\hat{X}_R$ for I. Now, we aim to find the R^{-1}-order asymptotic bias β^Z of the estimator $\hat{Z}_R = k(\hat{X}_R)$ for $z = k(I)$. Assume the function k is twice continuously differentiable in a neighborhood of I, and that assumptions (A1) and (A2') hold. Then:

$$\mathbb{E}(\hat{Z}_R - z) = \mathbb{E}(k(\hat{X}_R) - k(I)) \sim \frac{1}{R}\left(\beta^X k'(I) + \frac{\sigma^2 k''(I)}{2}\right), \qquad (2.17)$$

implying that the R^{-1}-order asymptotic bias of $\hat{Z}_R$ is:

$$\beta^Z = \beta^X k'(I) + \frac{\sigma^2 k''(I)}{2}.$$

Sketch of Proof Using Taylor's expansion:

$$k(\hat{X}_R) - k(I) = k'(I)(\hat{X}_R - I) + \frac{k''(I)}{2}(\hat{X}_R - I)^2 + o((\hat{X}_R - I)^2). \qquad (2.18)$$

Multiplying both sides of (2.18) by R, taking expectations, and leveraging the following facts:

- $Rk'(I)(\mathbb{E}\hat{X}_R - I) \to \beta^X k'(I)$,
- $R\frac{k''(I)}{2}\mathbb{E}(\hat{X}_R - I)^2 \to \frac{k''(I)}{2}\sigma^2$,
- $R\mathbb{E}[o((\hat{X}_R - I)^2)] \to 0$,

we obtain the desired result in (2.17).

Providing a formal proof would require additional assumptions and the use of properties like uniform integrability, which are beyond the current scope. $\square$

Bias Correction   For biased estimation of $z = k(I)$ using $\hat{Z}_R$, one can include a correction term:

$$k(\hat{X}_R) - \frac{\beta^Z}{R}.$$

If the bias β^Z is analytically computable, a common rule of thumb is to subtract an estimate of the bias from the original estimator. However, if the bias cannot be computed and is of order $R^{-1/2}$ or larger, the standard deviation $\varsigma/\sqrt{R}$ will dominate the bias. In such cases, confidence intervals centered on the biased estimator will lead to incorrect coverage probabilities. Following [8], the general principle is: *The standard deviation should dominate the bias for confidence intervals to remain meaningful.* To illustrate, note that:

$$\sqrt{R}\left(k(\hat{X}_R) - \frac{\beta^Z}{R}\right) \xrightarrow{\mathcal{D}} \mathcal{N}(0, \varsigma^2).$$

Thus, the $1 - \alpha$ confidence interval becomes:

$$k(\hat{X}_R) - \frac{\beta^Z}{R} \pm z_{1-\alpha/2}\frac{\varsigma}{\sqrt{R}}.$$

Example 4.2.9 Consider $k(y) = y^2$ and $\hat{Z}_R = k(\hat{Y}_R)$, where $\hat{Y}_R = \sum_{j=1}^{R} Y_j/R$, and $Y_1, \ldots, Y_R$ are i.i.d. random variables. Compute the asymptotic bias β of order R^{-1}. We can compute β directly from the identity:

$$\mathbb{E}(\hat{Y}_R)^2 = \tfrac{1}{R^2}\left(R\mathbb{E}Y^2 + (R^2 - R)(\mathbb{E}Y)^2\right)$$

$$= (\mathbb{E}Y)^2 + R^{-1}\sigma_Y^2 = k(I) + R^{-1}\sigma_Y^2.$$

This result is consistent with the method using the Taylor expansion in (2.18). Expanding $k(\hat{Y}_R)$ around $I = \mathbb{E}Y$, we have:

$$k(\hat{Y}_R) = k(I) + k'(I)(\hat{Y}_R - I) + \frac{k''(I)}{2}(\hat{Y}_R - I)^2 + o((\hat{Y}_R - I)^2).$$

Taking expectations on both sides (noting that $\mathbb{E}(\hat{Y}_R - I) = 0$, since $\hat{Y}_R$ is unbiased):

$$\mathbb{E}k(\hat{Y}_R) = k(I) + \frac{k''(I)}{2}\mathbb{E}(\hat{Y}_R - I)^2 + o(1/R).$$

Using $k(y) = y^2$ (so $k''(y) = 2$) and the fact that $\sigma_Y^2 = \mathbb{V}\mathrm{ar}Y = \mathbb{E}(\hat{Y}_R - I)^2$, we obtain:

$$\mathbb{E}(\hat{Y}_R)^2 = (\mathbb{E}Y)^2 + R^{-1}\sigma_Y^2.$$

In this example, the asymptotic bias of order R^{-1} for $\hat{Y}_R$ is zero because $\hat{Y}_R$ is unbiased. $\qquad\qquad\qquad\qquad\qquad\qquad\qquad\qquad\qquad\qquad\qquad\qquad\qquad\quad \Diamond$

4.2.3 Mean Square Error, Bias, and Variance

Finally, we comment on a measure of error encountered during computations. Suppose $\hat{X}_R$ is an estimator of I. The *mean square error* (MSE) is defined as:

$$\mathrm{MSE}(\hat{X}_R) = \mathbb{E}(\hat{X}_R - I)^2.$$

Using the identity $(a + b)^2 = a^2 + 2ab + b^2$ and the fact that

$$\mathbb{E}[(\hat{X}_R - \mathbb{E}\hat{X}_R)(\mathbb{E}\hat{X}_R - I)] = 0,$$

we have

$$\mathrm{MSE}(\hat{X}_R) = \mathbb{E}(\hat{X}_R - \mathbb{E}\hat{X}_R + \mathbb{E}\hat{X}_R - I)^2 = \mathbb{E}(\hat{X}_R - \mathbb{E}\hat{X}_R)^2 + (\mathbb{E}\hat{X}_R - I)^2. \tag{2.19}$$

The first term, $\mathbb{E}(\hat{X}_R - \mathbb{E}\hat{X}_R)^2$, represents the variance of $\hat{X}_R$, while the second term, $(\mathbb{E}\hat{X}_R - I)^2$, represents the squared bias. Equation (2.19) is often called the *bias-variance decomposition* of MSE.

For large R, recall that:

$$\lim_{R\to\infty} R\mathbb{E}(\hat{X}_R - \mathbb{E}\hat{X}_R)^2 = \varsigma^2 \quad \text{and} \quad \lim_{R\to\infty} R(\mathbb{E}\hat{X}_R - I) = \beta,$$

provided these limits exist. Combining these results, the MSE for large R can be approximated as:

$$\mathrm{MSE}(\hat{X}_R) \approx \frac{\varsigma^2}{R} + \frac{\beta^2}{R^2}.$$

4.3 Estimators of a Quantile

Consider a random variable X with c.d.f. F and density function $f(x)$. A widely used characteristic of a distribution is the p-quantile ($p \in (0, 1)$), defined as

$$q_p = F^{\leftarrow}(p) = \inf\{x : F(x) \geq p\}.$$

In financial mathematics, this is often denoted by $q_p = \text{VaR}_X(p)$, called the *Value-at-Risk*.

Suppose we want to estimate q_p based on a sample $X_1, \ldots, X_R$. Consider the empirical c.d.f. $\hat{F}_R(x)$ and its corresponding p-th sample quantile, defined as $\hat{F}_R^{\leftarrow}(p)$. Using the order statistics $X_{(1)}, \ldots, X_{(R)}$, it can be shown that

$$\hat{F}_R^{\leftarrow}(p) = \begin{cases} X_{(\lfloor Rp \rfloor)} & \text{if } pR \text{ is an integer,} \\[2mm] X_{(\lfloor Rp \rfloor + 1)} & \text{otherwise.} \end{cases} \tag{3.1}$$

The reader is invited to verify this in Exercise 4.T.12.

To simplify the analysis, we define an estimator of q_p as

$$\hat{q}_p = X_{(\lfloor Rp \rfloor)}.$$

Using Lemma 4.3.3 below, we have

$$|\hat{F}_R^{\leftarrow}(p) - \hat{q}_p| \leq X_{(\lfloor Rp \rfloor + 1)} - X_{(\lfloor Rp \rfloor)} \to 0, \quad \text{as } R \to \infty.$$

This alternative definition avoids the need to handle the cases of $X_{(\lfloor Rp \rfloor)}$ and $X_{(\lfloor Rp \rfloor + 1)}$ separately, while preserving the same asymptotic behavior. It turns out that $\hat{q}_p$ is a strongly consistent estimator. Additionally, the normal approximation of the empirical quantile can be described by the following result:

Proposition 4.3.1 *Assume that f is continuous and positive in a neighborhood of q_p. Then:*

(i) $\hat{q}_p$ is a strongly consistent estimator of q_p.
(ii) The asymptotic distribution is given by

$$\sqrt{R}(\hat{q}_p - q_p) \xrightarrow{\mathcal{D}} \mathcal{N}(0, \varsigma^2), \quad \text{where } \varsigma^2 = \frac{p(1-p)}{f(q_p)}. \tag{3.2}$$

To outline the proof of Proposition 4.3.1, we first recall some properties of the Beta distribution, which play a key role in analyzing order statistics. These are summarized in the lemma below; see Johnson et al. [93] for details.

Lemma 4.3.2 *Suppose $W \sim \mathrm{Beta}(n, m)$, where $n, m \in \{1, 2, \ldots\}$. Let $U_1, U_2, \ldots,$ U_{n+m} be i.i.d. random variables uniformly distributed on $\mathcal{U}[0, 1)$, and let $U_{(1)}, \ldots,$ $U_{(n+m)}$ denote their order statistics. Then:*

(i) $U_{(n)} \sim \mathrm{Beta}(n, m)$.

(ii) *Let* $X \sim \mathrm{Erl}(n, 1)$ *and* $Y \sim \mathrm{Erl}(m, 1)$ *be independent. Then* $X/(X + Y) \sim$ $\mathrm{Beta}(n, m)$.

(iii) *As a consequence of (ii), for i.i.d.* $\mathrm{Exp}(1)$ *random variables* $\tau_1, \ldots, \tau_{n+m}$,

$$\frac{\tau_1 + \cdots + \tau_n}{\tau_1 + \cdots + \tau_{n+m}} \sim \mathrm{Beta}(n, m).$$

From this, we can establish the following convergence result for order statistics:

Lemma 4.3.3 *Consider the setup of Lemma 4.3.2, and suppose $s = o(r)$ and $p \in (0, 1)$. Then, as $r \to \infty$,*

$$U_{\lfloor rp \rfloor + s} \to p \quad \text{in probability (and almost surely).}$$

Sketch of Proof From property (iii) of Lemma 4.3.2 (taking $n = \lfloor pr \rfloor + s$ and $n + m = r$) and the law of large numbers, we have

$$U_{\lfloor pr \rfloor + s} \overset{\mathcal{D}}{=} \frac{\tau_1 + \cdots + \tau_{\lfloor pr \rfloor + s}}{\tau_1 + \cdots + \tau_r}$$

$$= \frac{\tau_1 + \cdots + \tau_{\lfloor pr \rfloor + s}}{\lfloor pr \rfloor + s} \cdot \frac{\lfloor pr \rfloor + s}{\tau_1 + \cdots + \tau_r}.$$

Using the strong law of large numbers, the first fraction converges to 1, while the second fraction converges to p. Thus, $U_{\lfloor pr \rfloor + s} \to p$ almost surely. $\qquad\square$

Proof of Proposition 4.3.1 Let $X_1, \ldots, X_R$ be i.i.d. random variables with general distribution F. We can express them as $F^{\leftarrow}(U_1), \ldots, F^{\leftarrow}(U_R)$, where $U_1, \ldots, U_R$ are i.i.d. uniformly distributed random variables on $\mathcal{U}[0, 1)$.

(i) We begin by noting that $X_{\lfloor Rp \rfloor} \overset{\mathcal{D}}{=} F^{\leftarrow}(U_{\lfloor Rp \rfloor})$. To complete the proof, observe that $U_{\lfloor pR \rfloor} \leq p$ and that $U_{\lfloor pR \rfloor} \to p$ as $R \to \infty$. Additionally, since $F^{\leftarrow}(x)$ is left-continuous, it follows that

$$F^{\leftarrow}(U_{\lfloor pR \rfloor}) \to F^{\leftarrow}(p) \quad \text{as } R \to \infty.$$

(ii) The intuition behind Eq. (3.2) is as follows:

- First, consider a sample $U_1, \ldots, U_R$ of uniformly distributed random variables on $\mathcal{U}[0, 1)$. We then focus on the $\lfloor Rp \rfloor$-th order statistic. Given that $U_{\lfloor Rp \rfloor} = u$, there are $\lfloor Rp \rfloor - 1$ values to the left of it and $R - \lfloor Rp \rfloor - 1$ values to the right. Thus, $U_{\lfloor Rp \rfloor}$ has the p.d.f.

$$\frac{R!}{(\lfloor pR \rfloor - 1)!(R - \lfloor pR \rfloor - 1)!} u^{\lfloor pR \rfloor - 1}(1 - u)^{R - \lfloor pR \rfloor} du,$$

which is the p.d.f. of a Beta($\lfloor pR \rfloor$, $R - \lfloor pR \rfloor$) distribution.

- From Lemma 4.3.2, we know that

$$U_{\lfloor Rp \rfloor} \overset{\mathcal{D}}{=} \frac{\tau_1 + \cdots + \tau_{\lfloor pR \rfloor}}{\tau_1 + \cdots + \tau_R},$$

where $\tau_1, \ldots, \tau_R$ are i.i.d. exponentially distributed random variables with Exp(1) distribution. This is also Beta($\lfloor pR \rfloor$, $R - \lfloor pR \rfloor$).

- We can further show that, as $R \to \infty$,

$$\sqrt{R}(U_{\lfloor pR \rfloor} - p) \overset{\mathcal{D}}{\to} \mathcal{N}(0, \varsigma^2),$$

where $\varsigma^2 = p(1 - p)$. Specifically,

$$\begin{aligned}
U_{\lfloor Rp \rfloor} &= \frac{\xi_1 + \cdots + \xi_{\lfloor Rp \rfloor}}{\xi_1 + \cdots + \xi_R} - p \\
&= \frac{(1 - p) \sum_{i=1}^{\lfloor Rp \rfloor} \xi_i - p \sum_{i=\lfloor Rp \rfloor + 1}^{R} \xi_i}{\xi_1 + \cdots + \xi_R} - p \\
&= \frac{(1 - p) \sum_{i=1}^{\lfloor Rp \rfloor} \xi_i - p \sum_{i=\lfloor Rp \rfloor + 1}^{R} \xi_i / R}{(\xi_1 + \cdots + \xi_R)/R} - p.
\end{aligned}$$

Simplifying further, and using the law of large numbers, we arrive at the expression

$$\frac{(1 - p) \sum_{i=1}^{\lfloor pR \rfloor} \xi_i - p \sum_{i=\lfloor pR \rfloor + 1}^{R} \xi_i}{\sqrt{R}}.$$

The reader is invited to complete the argument, showing that this converges in distribution to $\mathcal{N}(0, p(1 - p))$.

- Finally, let $X_1, \ldots, X_R$ be i.i.d. random variables, where X has a general distribution F with density function f, which is continuous and positive around q_p. Consider the transformation $F^{-1}(U_1), \ldots, F^{-1}(U_R)$. From Proposition 3.1.2 (a), we know that these transformed random variables have the c.d.f. F. Therefore, $F^{\leftarrow}(U_{(\lfloor pR \rfloor)})$ has the same distribution as $X_{(\lfloor pR \rfloor)}$.

- From Taylor's expansion, we have

$$F^{\leftarrow}(p + x) = F^{\leftarrow}(p) + \frac{1}{f(F^{\leftarrow}(p))} x + o(x).$$

Therefore,

$$F^{\leftarrow}\left(U_{(\lfloor pR \rfloor)}\right) = F^{\leftarrow}\left(p + U_{(\lfloor pR \rfloor)} - p\right)$$

$$= F^{\leftarrow}\left(p + \frac{1}{\sqrt{R}} \frac{\sqrt{R}(U_{(\lfloor pR \rfloor)} - p)}{\sqrt{p(1-p)}} \sqrt{p(1-p)}\right)$$

$$= F^{\leftarrow}(p) + \frac{1}{f(F^{\leftarrow}(p))} \frac{1}{\sqrt{R}} \sqrt{p(1-p)} N_R,$$

where N_R is approximately normal with $\mathcal{N}(0, 1)$ as $R \to \infty$.

$\square$

4.3.1 Estimation of $1/f(q_p)$

To estimate $1/f(q_p)$, we assume that f is the p.d.f. of $X \sim F$, which is continuous in the neighborhood of the quantile q_p. Let $Q(p) = F^{-1}(p)$. From calculus, we know that

$$Q'(p) = \frac{1}{f(Q(p))}.$$

Using the finite difference approximation:

$$Q'(p) = \lim_{\delta \to 0} \frac{F^{-1}(p+\delta) - F^{-1}(p-\delta)}{2\delta},$$

a natural estimator of $Q'(p)$ is

$$\hat{Q}'_{p,R} = \frac{R}{2m}\left(X_{(\lfloor pR \rfloor + m)} - X_{(\lfloor pR \rfloor - m + 1)}\right), \tag{3.3}$$

where $m = m_R$ is a suitably chosen parameter.

Proposition 4.3.4 *Suppose that f has a continuous derivative in the neighborhood of q_p and that for $R \to \infty$ we have $m_R \to \infty$ such that $m_R = o(R)$. Then, $\hat{Q}'_{p,R}$ is a strongly consistent estimator of $Q'(p)$.*

Sketch of Proof From the Taylor expansion, we have

$$F^{-1}(p+\Delta) - F^{-1}(p) = \frac{1}{f(q_p)}\Delta + o(\Delta).$$

Similarly,

$$Q'(p) = \lim_{\Delta_1, \Delta_2 \to 0} \frac{F^{-1}(p + \Delta_1) - F^{-1}(p + \Delta_2)}{\Delta_1 - \Delta_2}. \tag{3.4}$$

From Lemma 4.3.3, we know that, with probability 1,

$$\Delta_1 = U_{\lfloor pR \rfloor + s} - p \to 0, \quad \Delta_2 = U_{\lfloor pR \rfloor - s + 1} - p \to 0$$

whenever $s = o(R)$. Thus, a good estimator of $Q'(p)$ is

$$
\begin{aligned}
\hat{Q}'_{p,R} &= \frac{F^{-1}(p + \Delta_1) - F^{-1}(p + \Delta_2)}{\Delta_1 - \Delta_2} \\
&= \frac{q_p + \frac{1}{f(q_p)}(U_{\lfloor pR \rfloor + s} - p) - \left(q_p + \frac{1}{f(q_p)}(U_{\lfloor pR \rfloor - s} - p) \right)}{\Delta_1 - \Delta_2} \\
&= \frac{\frac{1}{f(q_p)}(U_{\lfloor pR \rfloor + s} - U_{\lfloor pR \rfloor - s}) + o(U_{\lfloor pR \rfloor + s} - p, U_{\lfloor pR \rfloor - s} - p)}{\Delta_1 - \Delta_2}.
\end{aligned}
$$

Using Lemma 4.3.3, we observe that

$$U_{\lfloor pR \rfloor + s} - U_{\lfloor pR \rfloor - s} \overset{\mathcal{D}}{=} \frac{\sum_{i=\lfloor pR \rfloor - s + 1}^{\lfloor pR \rfloor + s} \tau_i}{\sum_{i=1}^{R} \tau_i},$$

where τ_i are i.i.d. exponentially distributed random variables. By the law of large numbers, for $R \to \infty$ and $s = o(R)$, we have

$$\Delta_1 - \Delta_2 = U_{\lfloor pR \rfloor + s} - U_{\lfloor pR \rfloor - s} \sim \frac{2s}{R}.$$

Thus, the estimator $\hat{Q}'_{p,R}$ defined in (3.3) is consistent. This completes the proof. $\qquad \square$

Choosing s_R A key consideration is the choice of s_R. One recommendation is $s_R = cR^{1/2}$. However, Bloch and Gastwirth [16] demonstrated that, under the additional assumption that the density f has the first three derivatives around q_p, the asymptotic mean square error (MSE) can be minimized. Specifically, if $s_R = cR^{4/5}$ for some constant c, the MSE is approximately

$$\mathbb{E}(\hat{Q}'_{p,R} - Q'(p))^2 \sim \frac{5}{4} \cdot 9^{-1/5}[Q'(p)]^{4/5}[Q''(p)]^{3/5} R^{-2/5}.$$

4.4   More on Absolute vs Relative Errors

In this section, we explore additional problems related to errors encountered during simulations and their probabilistic bounds. We utilize tools such as Chebyshev's inequality, the Central Limit Theorem, and introduce a powerful tool known as the *Chernoff bound*.

4.4.1   Chernoff Bound

We begin with the following proposition.

Proposition 4.4.1 (Chernoff Bound) *Consider $S_R = Y_1 + \cdots + Y_R$, where Y_i, $i = 1, \ldots, R$, are i.i.d. indicator random variables with $\mathbb{P}(Y_i = 1) = p$ and $\mathbb{P}(Y_i = 0) = 1 - p$. Then for any $0 < \varepsilon < 1$, the following bounds hold:*

$$\textit{Upper tail:} \quad \mathbb{P}(S_R \geq (1 + \varepsilon)Rp) \leq \exp\left(-\frac{\varepsilon^2 Rp}{2 + \varepsilon}\right) \leq \exp\left(-\frac{\varepsilon^2 Rp}{3}\right),$$

$$\tag{4.1}$$

$$\textit{Lower tail:} \quad \mathbb{P}(S_R \leq (1 - \varepsilon)Rp) \leq \exp\left(-\frac{\varepsilon^2 Rp}{2}\right). \tag{4.2}$$

As a result,

$$\mathbb{P}(|S_R - Rp| \geq \varepsilon Rp) \leq 2\exp\left(-\frac{\varepsilon^2 Rp}{3}\right), \tag{4.3}$$

or equivalently, for $\hat{Y}_R = \frac{1}{R}\sum_{i=1}^{R} Y_i$,

$$\mathbb{P}(|\hat{Y}_R - p| \geq \varepsilon p) \leq 2\exp\left(-\frac{\varepsilon^2 Rp}{3}\right). \tag{4.4}$$

Sketch of Proof Using Markov's inequality (Exercise 2.T.1), for any $x > 0$ and $s > 0$, we have

$$\mathbb{P}(S_R \geq x) = \mathbb{P}(\exp(s S_R) \geq \exp(sx)) \leq \frac{\mathbb{E}[\exp(s S_R)]}{\exp(sx)}.$$

Now calculate $\mathbb{E}[\exp(s S_R)]$:

$$\mathbb{E}[\exp(s S_R)] = \prod_{i=1}^{R} \mathbb{E}[\exp(s Y_i)] = (p \exp(s) + 1 - p)^R$$

$$= (1 + p(\exp(s) - 1))^R$$

$$\leq \exp\left(Rp(\exp(s) - 1)\right),$$

where the last inequality follows from $1 + t \leq \exp(t)$ for all t. Substituting $x = (1 + \varepsilon)Rp$, we get

$$\mathbb{P}(S_R \geq (1 + \varepsilon)Rp) \leq \exp\left(Rp(\exp(s) - 1 - s(1 + \varepsilon))\right).$$

The bound holds for any $s > 0$. Choosing $s = \ln(1 + \varepsilon)$ (which minimizes the right-hand side), we obtain

$$\exp\left(Rp(\exp(s) - 1 - s(1 + \varepsilon))\right) = \exp\left(-Rp\left((1 + \varepsilon)\ln(1 + \varepsilon) - \varepsilon\right)\right).$$

Using the inequality $\ln(1 + \varepsilon) > \frac{2\varepsilon}{2+\varepsilon}$, we have

$$(1 + \varepsilon)\log(1 + \varepsilon) - \varepsilon \geq \frac{\varepsilon + \varepsilon^2 - \varepsilon - \varepsilon^2/2}{1 + \varepsilon/2} = \frac{\varepsilon^2/2}{1 + \varepsilon/2} = \frac{\varepsilon^2}{2 + \varepsilon}.$$

Thus, we obtain the upper tail bound in (4.1).

For the lower tail bound in (4.2), apply Markov's inequality with $\exp(-s S_R)$:

$$\mathbb{P}(S_R \leq x) = \mathbb{P}(\exp(-s S_R) \geq \exp(-sx)) \leq \frac{\mathbb{E}[\exp(-s S_R)]}{\exp(-sx)}.$$

Proceeding analogously by choosing $s = \ln(1 - \varepsilon)$ and using the inequality $\ln(1 - \varepsilon) \geq -\varepsilon + \frac{\varepsilon^2}{2}$ for $\varepsilon \in (0, 1)$, the lower tail bound follows. $\quad\square$

The following corollary directly follows from (4.3):

Corollary 4.4.2 *Let $Y_1, \ldots, Y_R$ be i.i.d. indicator random variables with $\mathbb{P}(Y_i = 1) = p$ and $\mathbb{P}(Y_i = 0) = 1 - p$. Define $\hat{Y}_R = \frac{1}{R}\sum_{i=1}^{R} Y_i$. If $\varepsilon, \delta \in (0, 1)$ and $R \geq \frac{3\ln(2/\delta)}{\varepsilon^2 p}$, then*

$$\mathbb{P}\left(|\hat{Y}_R - p| \geq \varepsilon p\right) \leq \delta.$$

Example 4.4.3 Consider a sequence of random bits $B_1, \ldots, B_R$, which are indicator random variables with $p = 1/2$. We will compare bounds derived from Chebyshev and Chernoff inequalities, as well as the bound obtained from the Central Limit Theorem.

- *Chebyshev inequality*—see (2.7.2). We have

$$\mathbb{P}\left(|B_1 + \cdots + B_R - R/2| \geq \frac{R}{4}\right) \leq \frac{R\mathbb{V}\mathrm{ar}\,B_1}{R^2/4^2} = \frac{4}{R}. \tag{4.5}$$

- *Chernoff inequality.* Using (4.3), we get

$$\mathbb{P}\left(|B_1 + \cdots + B_R - R/2| \geq \varepsilon\frac{R}{2}\right) \leq 2\exp\left(-\frac{\varepsilon^2 R}{6}\right).$$

Taking $\varepsilon = 1/2$, we have

$$\mathbb{P}\left(|B_1 + \cdots + B_R - R/2| \geq \frac{R}{4}\right) \leq 2\exp\left(-\frac{R}{24}\right), \tag{4.6}$$

which—for larger R—provides a significant improvement over the result derived from Chebyshev inequality. For example, when $R = 10^3$, we have $4/R = 0.004$ compared to $2\exp(-41.6) = 1.71 \times 10^{-18}$.

Taking $\varepsilon = \sqrt{6\log R/R}$, we get

$$\mathbb{P}\left(|B_1 + \cdots + B_R - R/2| \geq \frac{R}{2}\sqrt{6\log R/R}\right) \leq 2\exp\left(-\frac{\frac{6\log R}{R}\cdot R}{6}\right) = \frac{2}{R},$$

and taking twice as large $\varepsilon = 2\sqrt{6\log R/R}$ we have

$$\mathbb{P}\left(|B_1 + \cdots + B_R - R/2| \geq \frac{R}{2}2\sqrt{6\log R/R}\right)$$

$$\leq 2\exp\left(-\frac{4\cdot 6\frac{\log R}{R}\cdot R}{6}\right) = \frac{2}{R^4}.$$

- Using the Central Limit Theorem:

$$\mathbb{P}\left(|B_1 + \cdots + B_R - R/2| \geq \frac{R}{4}\right) = \mathbb{P}\left(\frac{|B_1 + \cdots + B_R - R/2|}{\sqrt{R}/2} \geq \frac{R/4}{\sqrt{R}/2}\right)$$

$$\approx 2\mathbb{P}(Z \geq \sqrt{R}/2),$$

where $Z \sim \mathcal{N}(0, 1)$. Using the asymptotics:

$$1 - \Phi(x) \sim \frac{\exp(-x^2/2)}{\sqrt{2\pi}x}, \quad \text{as } x \to \infty,$$

we obtain an asymptotic approximation:

$$\mathbb{P}\left(|B_1 + \cdots + B_R - R/2| \geq \frac{R}{4}\right) \approx \sqrt{\frac{8}{\pi R}}\,\exp(-R/8). \tag{4.7}$$

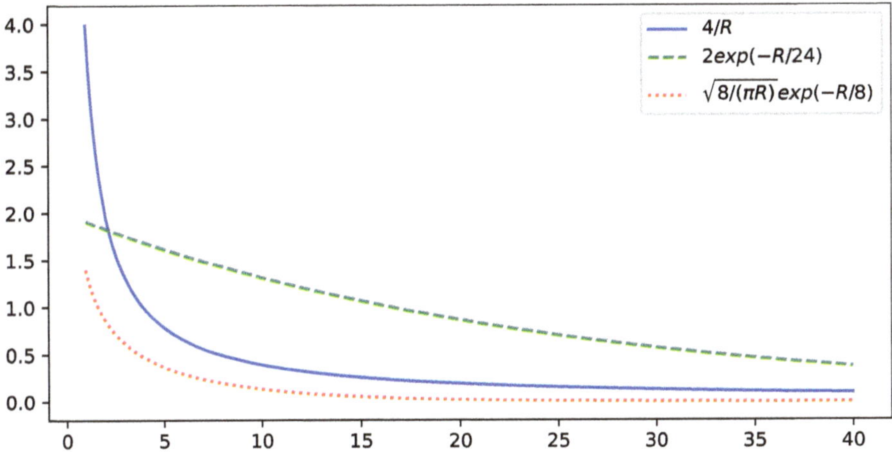

Fig. 4.4 $\mathbb{P}\left(|B_1 + \cdots + B_R - R/2| \geq \frac{R}{4}\right)$: Comparison of Chebyshev bound (4.5), Chernoff bound (4.6) and CLT approximation (4.7)

Comparing the bounds/approximations for $\mathbb{P}\left(|B_1 + \cdots + B_R - R/2| \geq \frac{R}{4}\right)$, we observe that the Central Limit Theorem gives the best result, the Chernoff bound comes second, and the Chebyshev bound provides the least precise result. The bounds for $R = 1, \ldots, 40$ are depicted in Fig. 4.4. The figure is generated using the script `ch4_binomial_chebyshev_chernoff_clt.py`. For much larger R, the differences become even more pronounced. $\qquad\qquad\qquad\diamond$

4.4.2   Discussion on Errors

Consider the following scenario. Assume that $\mathbb{P}(Y = 1) = p = 0.01 = 1 - \mathbb{P}(Y = 0)$, and we aim to estimate $I = \mathbb{E}Y$ using i.i.d. random variables $Y_1, \ldots, Y_R$. Suppose we flip a coin just once, i.e., $R = 1$. Then, $\hat{Y}_R = Y_1$, and

$$\mathbb{P}(|\hat{Y}_1 - I| \leq 0.01) = \mathbb{P}(|Y_1 - I| \leq 0.01) = \mathbb{P}(Y_1 = 0) = 0.99,$$

which satisfies (E1) with $\varepsilon = \delta = 0.01$. Clearly, such an estimator is not very informative. Note that at the same time, $|\hat{Y}_1 - I|$ is either 0.01 or 0.99, and in any case:

$$\mathbb{P}(|\hat{Y}_1 - I| \le \hat{Y}_1) = 1,$$

which indicates that the *relative error* is huge. Thus, (E2) and (E2') hold with $\varepsilon = 1$ and any $\delta \in (0, 1)$:

$$\mathbb{P}(|\hat{Y}_1 - I| \ge \hat{Y}_1) \le \delta, \qquad \mathbb{P}(|\hat{Y}_1 - I| \le \hat{Y}_1) \ge 1 - \delta.$$

This shows that the number of replications required to achieve a small *absolute error* may seem counter-intuitive. From the fundamental relationship (1.6) and (1.7), we need—using the current notation—

$$R = z_{1-\delta/2}^2 \frac{\mathbb{V}\mathrm{ar}Y_1}{\varepsilon^2} = z_{1-\delta/2}^2 \frac{I(1-I)}{\varepsilon^2}$$

replications to ensure that $\mathbb{P}(|\hat{Y}_R - I| \ge \varepsilon) \le \delta$. For $\varepsilon = \delta = 0.01$, we compute:

$$I = 0.1 \quad R = \lceil 2.57^2 \cdot 0.1 \cdot 0.9/0.01^2 \rceil = 5945,$$

$$I = 0.01 \quad R = \lceil 2.57^2 \cdot 0.01 \cdot 0.99/0.01^2 \rceil = 654,$$

$$I = 0.001 \quad R = \lceil 2.57^2 \cdot 0.001 \cdot 0.999/0.01^2 \rceil = 66.$$

This shows that the smaller the p, the smaller the number of replications required to achieve a small *absolute error*. Roughly speaking, if we estimate I, which is close to 0, using an estimator that equals 0, the *absolute error* is small. In such cases, it is better to consider the *relative error*. We aim to find R such that:

$$\mathbb{P}(|\hat{Y}_R - I| \ge \varepsilon I) \le \delta.$$

Using the Chernoff bound (see Corollary 4.4.2), for i.i.d. indicator random variables $Y_1, \ldots, Y_R$ with $I = \mathbb{E}Y_i$ and $\hat{Y}_R = \frac{1}{R} \sum_{i=1}^{R} Y_i$, if

$$R \ge \frac{3 \ln(2/\delta)}{\varepsilon^2 I},$$

then:

$$\mathbb{P}\left(\left|\hat{Y}_R - I\right| \ge \varepsilon I\right) \le \delta, \quad \text{or equivalently,} \quad \mathbb{P}\left(\left|\hat{Y}_R - I\right| \le \varepsilon I\right) \ge 1 - \delta.$$

$$(4.8)$$

For $\delta = 0.01$ and $\varepsilon = 0.01$, this requires:

$$R \geq \frac{3 \ln \frac{2}{0.99}}{0.01^2 I},$$

which gives:

$$I = 0.1 \quad R \geq 1.59 \cdot 10^6,$$

$$I = 0.01 \quad R \geq 1.59 \cdot 10^7,$$

$$I = 0.001 \quad R \geq 1.59 \cdot 10^8.$$

Remark 4.4.4 In the above computations, we used the exact value of I. In practice, approximate bounds on I would yield similar results. ∎

Influence of $\mathbb{V}\mathrm{ar}(\hat{I})$ **on Absolute and Relative Errors** The variance of the estimator $\hat{I}$ plays a fundamental role in controlling both absolute and relative errors. The absolute error is defined as:

$$\mathbb{P}(|\hat{I} - I| \geq \varepsilon) \leq \delta,$$

and the relative error is defined as:

$$\mathbb{P}\left(\left|\frac{\hat{I}}{I} - 1\right| \geq \varepsilon\right) \leq \delta.$$

Using the Central Limit Theorem (CLT), for a sufficiently large number of replications R, the estimator $\hat{I}$ is approximately normally distributed:

$$\hat{I} \sim \mathcal{N}\left(I, \frac{\mathbb{V}\mathrm{ar}(Y)}{R}\right),$$

where Y represents the underlying random variable being sampled, and $\mathbb{V}\mathrm{ar}(Y)$ is its variance.

To achieve a small absolute error, the number of replications R must satisfy:

$$R = \frac{z_{1-\delta/2}^2 \, \mathbb{V}\mathrm{ar}(Y)}{\varepsilon^2}.$$

This formula shows that reducing the variance $\mathbb{V}\mathrm{ar}(Y)$ is essential to minimize computational cost for a given absolute error threshold ε.

Similarly, to achieve a small relative error, the number of replications R must satisfy:

$$R = \frac{z_{1-\delta/2}^2 \mathbb{V}\mathrm{ar}(Y)}{\varepsilon^2 I^2}.$$

Here, the dependence on I^2 indicates that achieving a small relative error becomes increasingly expensive when I is small. This underscores the importance of variance reduction techniques, such as stratification or importance sampling, in efficiently controlling the relative error.

In general, a small variance $\mathbb{V}\mathrm{ar}(\hat{I})$ is beneficial for both absolute and relative errors. For absolute error, it directly reduces the probability of exceeding the error threshold ε. For relative error, it mitigates the amplified impact of small I, making the estimator more efficient in terms of the required replications R.

The above formulas serve as the foundation for understanding the influence of variance on error control in simulations, regardless of whether I represents a probability or another parameter of interest.

4.4.3 Absolute vs Relative Error for Estimating π

The number π is neither too small nor too large, making it reasonable to consider both absolute and relative errors. In Example 4.1.6, we considered the following unbiased estimator of π (for given i.i.d. $U_1, \ldots, U_{2R^{\mathrm{abs}}}$ from the $\mathcal{U}[0, 1)$ distribution):

$$\hat{Y}_{R^{\mathrm{abs}}} = \frac{1}{R^{\mathrm{abs}}} \sum_{j=1}^{R^{\mathrm{abs}}} Y_j, \quad \text{where } Y_j = 4\mathbb{1}(U_{2j-1}^2 + U_{2j}^2 \le 1).$$

Since $\mathbb{E}Y_1 = \pi$ and $\mathbb{V}\mathrm{ar}Y_1 = 16(\pi/4)(1 - \pi/4) = 2.6968$, we can compute the number of replications required to achieve a small absolute error. Fixing $\varepsilon = 0.01$ and $\delta = 0.01$, to satisfy:

$$\mathbb{P}(|\hat{Y}_{R^{\mathrm{abs}}} - \pi| \le 0.01) \ge 0.99,$$

we need:

$$R^{\mathrm{abs}} = z_{1-\delta/2}^2 \frac{\mathbb{V}\mathrm{ar}Y_1}{\varepsilon^2} = 2.57^2 \cdot \frac{2.6968}{0.01^2} \approx 1.78 \cdot 10^5$$

replications. To have a small relative error, we use the Chernoff bound from Proposition 4.4.1. Taking $\varepsilon = \delta = 0.01$, to ensure:

$$\mathbb{P}(|\hat{Y}_{R^{\text{rel},1}} - \pi| \le 0.01\pi) \ge 0.99,$$

we require:

$$R^{\text{rel},1} \ge \frac{3 \ln \frac{2}{0.01}}{0.01^2 \pi} \approx 5 \cdot 10^4$$

replications.

Comparing Relative and Absolute Errors Instead of comparing absolute and relative errors directly for the same ε and δ, we "reduce" the relative error bound to match the absolute one by setting $\varepsilon' = 0.01/\pi$. Then, to achieve:

$$\mathbb{P}(|\hat{Y}_{R^{\text{rel},2}} - \pi| \le \varepsilon'\pi) = \mathbb{P}\left(|\hat{Y}_{R^{\text{rel},2}} - \pi| \le \frac{0.01}{\pi} \cdot \pi\right)$$

$$= \mathbb{P}(|\hat{Y}_{R^{\text{rel},2}} - \pi| \le 0.01) \ge 0.99,$$

we require:

$$R^{\text{rel},2} \ge \frac{3 \ln \frac{2}{0.01}}{(\varepsilon')^2 \pi} \ge \frac{3 \ln \frac{2}{0.01}}{\frac{0.01^2}{\pi^2} \cdot \pi} = \frac{3\pi \ln \frac{2}{0.01}}{0.01^2} \approx 5 \cdot 10^5.$$

For a fixed $\delta = 0.01$, Fig. 4.5 shows the number of replications required—expressed as functions of ε—to achieve a given absolute error as well as both types of relative error.
The figure is generated by `ch4_pi_required_R_absolute_relative.py`.

4.5 Conclusions

A fundamental relationship between the absolute error b and the number of replications R was established:

$$b = \frac{z_{1-\alpha/2} D}{\sqrt{R}}, \tag{5.1}$$

where the constant D depends on the variability of the experiment under consideration. For independent replications, the following cases apply:

- If $I = \mathbb{E}Y$, then $D = \sigma_Y$.
- If $z = k(I)$, where $I = \mathbb{E}Y$, then $D = k'(I)\sigma_Y$.

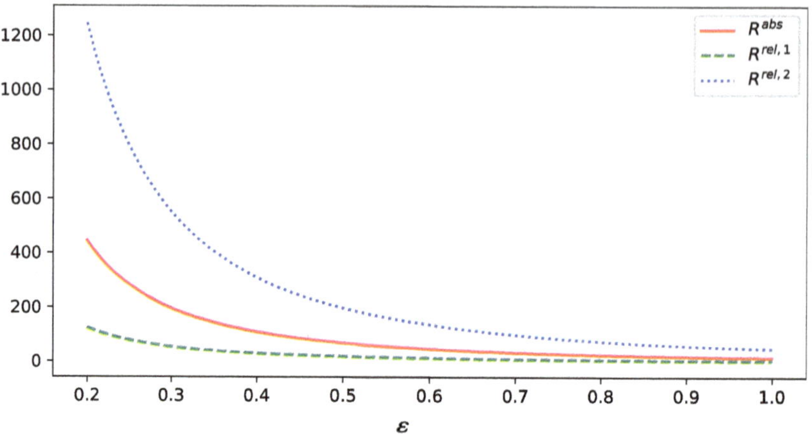

Fig. 4.5 Number of replications required to estimate π within absolute and relative error bounds at a confidence level of $\delta = 0.01$, shown as a function of ε

- If $z = k(\mathbf{I})$, where $\mathbf{I} = \mathbb{E}\mathbf{Y}$, then

$$
D = \sqrt{\sum_{i,j=1}^{d} k^{(i)}(\mathbf{I})k^{(j)}(\mathbf{I})\mathbf{\Sigma}_{\mathbf{Y}}(i,j)}.
$$

In practice, this relationship (5.1) is used to design simulations with a prescribed error by determining the required number of replications R. Typically, I and σ_Y (or $\mathbf{I}$ and $\mathbf{\Sigma}$) are estimated through a pilot simulation of size m. Subsequently, R replications are simulated to compute $\hat{Y}_R$ and $\hat{S}_R^2$ (or $\hat{\mathbf{Y}}_R$ and $\hat{\mathbf{\Sigma}}_R$), leading to:

- A confidence interval for I:

$$
\hat{Y}_R \pm \frac{z_{1-\alpha/2}\hat{S}_R}{\sqrt{R}}.
$$

- A confidence interval for $k(I)$:

$$
k(\hat{Y}_R) \pm \frac{z_{1-\alpha/2}k^{'}(\hat{Y}_R)\hat{S}_R}{\sqrt{R}}.
$$

- A confidence interval for $k(\mathbf{I})$:

$$
k(\hat{\mathbf{Y}}_R) \pm \frac{z_{1-\alpha/2}\sqrt{\sum_{i,j=1}^{d} k^{(i)}(\hat{\mathbf{Y}}_R)k^{(j)}(\hat{\mathbf{Y}}_R)\hat{\mathbf{S}}^2(i,j)}}{\sqrt{R}}.
$$

In summary, the general approach to obtaining a confidence interval for estimating a quantity is to leverage a suitable version of the central limit theorem. This ensures robust estimation with well-defined probabilistic guarantees.

4.6 Bibliographical Notes

For an introduction to statistical methods in simulation, we refer to Ross [182], while more advanced treatments can be found in Asmussen and Glynn [8], and Glasserman [63]. The fundamental relationship between the required precision, the number of replications, and variability is a key topic in these works.

The Central Limit Theorem (CLT) plays a crucial role in deriving confidence intervals and understanding simulation errors. Proofs of the CLT for quantile estimation, particularly results related to (3.2), can be found in Serfling [191], specifically Sections 2.3.3 and 2.5. Properties of the estimator for $1/f(q_p)$ follow from results by Bloch and Gastwirth [16]. A broader perspective on quantile estimation via Monte Carlo methods is surveyed by Dong and Nakayama [40]. A short introduction to *delta method* can be found e.g., in van der Vaart [207]. A detailed treatment of estimator efficiency (also considering simulation budget constraints), see Glynn and Whitt [67].

Estimating the required number of replications in simulations often involves a two-stage procedure, where an initial pilot simulation is conducted to estimate variance. The construction of confidence intervals in such two-stage procedures, while accounting for dependency between the stages, has been extensively studied; see e.g., Nakayama [162].

In multivariate settings, simulation analysis often requires tools from matrix algebra, particularly eigen-decomposition and diagonalization techniques. These concepts are well-covered in standard texts such as Johnson and Horn [92].

The theory of rectangular confidence regions was first developed by Šidák [192], whose results remain fundamental for constructing high-dimensional confidence regions in simulation studies.

The problem of estimating ratios of expectations is discussed in Glasserman [63], particularly in the context of financial simulations. The impact of asymptotic bias in simulations of functions of parameters of interest is examined in depth in Asmussen and Glynn [8].

In many applications, pilot simulations are used to determine the number of required replications, as well as to estimate key constants such as variance. The importance of pilot and adaptive simulation techniques is emphasized throughout the literature on simulation efficiency and variance reduction.

4.7 Exercises

Theoretical Exercises

4.T.1 How can we simulate

$$I = \int_{-\infty}^{\infty} g(x)e^{-a|x|}\, dx,$$

where $a > 0$, assuming the integral is absolutely convergent?

4.T.2 [Asmussen and Glynn [8], p.71] We aim to compute I using the Monte Carlo method, where $I = I'I''$, and we know that $I' = \mathbb{E}Y'$ and $I'' = \mathbb{E}Y''$. Let $Y'_1, \ldots, Y'_R$ be R independent replications of Y', and independently, $Y''_1, \ldots, Y''_R$ be R independent replications of Y''. Consider two estimators for I:

$$\hat{Z}_1 = \left(\frac{1}{R} \sum_{i=1}^{R} Y'_i \right) \left(\frac{1}{R} \sum_{i=1}^{R} Y''_i \right),$$

$$\hat{Z}_2 = \frac{1}{R} \sum_{i=1}^{R} Y'_i Y''_i.$$

Show that both estimators are unbiased and that $\hat{Z}_1$ has smaller variance. *Hint:* First, show that for two independent random variables X, Y with finite second moments:

$$\mathbb{V}\mathrm{ar}(XY) = \mathbb{V}\mathrm{ar}X\,\mathbb{V}\mathrm{ar}Y + (\mathbb{E}X)^2\mathbb{V}\mathrm{ar}Y + (\mathbb{E}Y)^2\mathbb{V}\mathrm{ar}X.$$

4.T.3 (Buffon's needle and the number π) Suppose that we drop a short needle on ruled paper. What is the probability p that the needle crosses one of the lines? Let d be the distance between the lines, and l the length of the needle. It is known that:

$$p = \frac{2l}{\pi d}.$$

Suppose we run R replications of the needle drop experiment and obtain $Y_1, \ldots, Y_R$, where $Y_j = 1$ if the needle crosses a line, and $Y_j = 0$ otherwise. Then:

$$\hat{p} = \frac{1}{R} \sum_{j=1}^{R} Y_j$$

is an estimator of p, from which we can obtain the following estimator of π:

$$\hat{\pi} = \frac{2l}{\hat{p}} = k(\hat{p}).$$

(a) Estimate the required number of experiments to compute $k(p)$ with a given precision $b = 0.01$ and confidence level $\alpha = 0.05$.
(b) For $\alpha = 0.05$, estimate the number of replications needed to obtain π up to ν significant digits. (The number of significant digits ν is defined such that the relative error is less than $10^{-\nu}$.)

4.T.4 Estimate the number of replications needed to simulate:

$$I = \mathbb{E}e^{\mu+\sigma Z},$$

where $Z \sim \mathcal{N}(0, 1)$. Recall that the random variable $Y = e^{\mu+\sigma Z}$ has a log-normal distribution $LN(\mu, \sigma^2)$.

Consider the Monte Carlo estimator:

$$\hat{Y}_R = \frac{1}{R} \sum_{j=1}^{R} Y_j$$

for I. The goal is to control the relative error:

$$\frac{\hat{Y}_R^{CMC} - I}{I}.$$

From the Central Limit Theorem:

$$\mathbb{P}\left(-\frac{1.96\sqrt{\mathbb{V}\mathrm{ar}Y}}{\sqrt{RI}} \leq \frac{\hat{Y}_R^{CMC} - I}{I} \leq \frac{1.96\sqrt{\mathbb{V}\mathrm{ar}Y}}{\sqrt{RI}}\right) \geq 0.95.$$

Suppose $\sigma = 1$. Using known formulas for the log-normal distribution, find the number of replications needed to achieve a relative error of 10% with significance level $\alpha = 0.05$.

What about $\sigma = 5$? Consider relative errors of 10% and 50%.

Hint: Using the formulas for the log-normal distribution (recalled below), compute:

$$\frac{1.96\sqrt{\mathbb{V}\mathrm{ar}Y}}{\sqrt{RI}}.$$

We have:

$$\mathbb{E}Y^n = e^{n\mu + n^2\sigma^2/2},$$

and:

$$\sqrt{\mathbb{V}\mathrm{ar}Y} = e^{\mu+\sigma^2/2}\sqrt{e^{\sigma^2} - 1}.$$

4.T.5 Suppose we aim to estimate:

$$\mathbf{I} = (\mathbb{E}(X^{(1)} + X^{(3)}), \mathbb{E}(X^{(2)} + X^{(3)})),$$

where $X^{(1)}$, $X^{(2)}$, $X^{(3)}$ are i.i.d. with exponential distribution $\mathrm{Exp}(1/\theta)$. Let:

$$\mathbf{Y} = (X^{(1)}, X^{(2)}, X^{(3)})^T.$$

Define an estimator of $\mathbf{I}$ as $\hat{\mathbf{Y}}_R$, where $\mathbf{Y}_1, \mathbf{Y}_2, \ldots$ are copies of $\mathbf{Y}$. Write down the formula for an elliptical confidence region at significance level $1 - \alpha$.

4.T.6 For i.i.d. random variables $Z_1, \ldots, Z_n$ normally distributed $\mathcal{N}(0, \sigma^2)$, the random variable:

$$X^2 = \frac{(n-1)\hat{S}_n^2}{\sigma^2}$$

has a chi-square distribution with $n-1$ degrees of freedom. Its mean is $n-1$, and its variance is $2(n-1)$.

Hence, for $Y_1, \ldots, Y_n$, i.i.d. standard normal $\mathcal{N}(0, \sigma^2)$:

$$\frac{(n-1)\hat{S}_n^2/\sigma^2 - n + 1}{\sqrt{2n-2}} \xrightarrow{\mathcal{D}} \mathcal{N}(0, 1), \quad \text{as } n \to \infty,$$

(see [207]).

For generally distributed r.v.s Y_j with finite fourth moment:

$$\sqrt{n}\left(\hat{S}_n^2 - \sigma^2\right) \xrightarrow{\mathcal{D}} \mathcal{N}(0, \mu_4 - \sigma^4),$$

where $\mu_4 = \mathbb{E}(Y - \mathbb{E}Y)^4$.

Suppose $\alpha = 0.05$. Estimate the number of replications needed to estimate σ^2 with 10% accuracy, i.e.:

$$\mathbb{P}\left(\left|\frac{\hat{S}_R - \sigma^2}{\sigma^2}\right| \le 0.1\right) \ge 0.95.$$

Consider both variants: (1) Normal distribution, (2) General distribution. *Hint:* The formula for the number of replications involves $\gamma_2 = \mu_4/\sigma^4$, the kurtosis. For standard normal $\mathcal{N}(0, 1)$, $\gamma_2 = 3$.

4.T.7 In Example 4.1.9, two CMC estimators for:

$$I = \int_0^\infty x^2 e^{-x}\, dx,$$

namely $\hat{Y}_R^{\mathrm{CMC1}}$ and $\hat{Y}_R^{\mathrm{CMC2}}$, were introduced. Compute the variances of both estimators and determine which performs better.

4.T.8 The formula for the level net premium for term assurance for time n, when the insured sum is 1, is:

$$\Pi = \frac{\mathbb{E} e^{-r T_x} \mathbb{1}(T_x \le n)}{\mathbb{E} \int_0^{\min(T_x, n)} e^{-rt}\, dt}.$$

Perform an error analysis and estimate the number of replications required to compute Π with ν significant digits.

4.T.9 Let $(B_i : i = 1, \ldots)$ be a sequence of i.i.d. random variables assuming values 1 with probability p and 0 otherwise. Consider an estimator of p in the form:

$$\hat{B}_R = \frac{\sum_{i=1}^R B_i}{R}.$$

We aim to control the relative error:

$$\left| \frac{\hat{B}_R - p}{p} \right| \le b_r.$$

Find the required number of replications $R(p)$ to achieve the error b_r with probability 0.95. Show the asymptotics as $p \downarrow 0$.

4.T.10 Let $X \sim \mathcal{U}[0, \theta)$, where $\theta > 0$ is an unknown parameter. Show that the estimator:

$$\hat{\theta}_R = \max(X_1, \ldots, X_R)$$

is consistent but biased. Perform an error analysis and compute the number of replications required to achieve a prescribed accuracy $b = 0.1$.

4.T.11 Consider $k(y) = y^3$ and $\hat{Z}_R = k(\hat{Y}_R)$, where:

$$\hat{Y}_R = \frac{\sum_{j=1}^R Y_j}{R},$$

and $Y_1, \ldots, Y_R$ are i.i.d. Compute the asymptotic bias β of order R^{-1} directly and using Taylor's formula.

4.T.12 For a sequence $x_1, \ldots, x_R$, let $x_{(1)} \leq \ldots \leq x_{(R)}$ denote the sequence in ascending order. Define the empirical c.d.f.:

$$\hat{F}_R(t) = \frac{1}{R} \sum_{i=1}^{R} \mathbb{1}(x_i \leq t).$$

Show that for $\frac{i-1}{R} < \alpha \leq \frac{i}{R}$ $(i = 1, \ldots, R)$, we have:

$$\hat{F}_R^{\leftarrow}(\alpha) = x_{(i)}.$$

In other words, $\hat{F}_R^{\leftarrow}(\alpha) = \lceil \alpha R \rceil$.

4.T.13 Verify recursive formulas for the sample mean and variance (1.11) and (1.12).

Lab Exercises

4.L.1 Consider the Dirichlet integral:

$$\int_0^{\infty} \frac{\sin x}{x}\, dx.$$

It is known that it equals $\pi/2$. Can we simulate this integral? One approach would be:

$$\int_0^{\infty} \frac{\sin x}{x}\, dx = \int_0^1 \frac{\sin x}{x}\, dx + \int_1^{\infty} \frac{\sin x}{x}\, dx.$$

The first integral on the r.h.s. is approximately 0.30117. Think about an efficient way to simulate this integral.

However, the second integral is not absolutely convergent. To explore this issue, consider the identity:

$$\int_1^{\infty} \frac{\sin x}{x}\, dx = \int_0^1 y \sin(1/y)\, dy.$$

Assume everything is correct (although it is not), and proceed to simulate this integral using the Crude Monte Carlo method:

$$\frac{1}{R} \sum_{j=1}^{R} U_j \sin(1/U_j),$$

where $U_j \sim \mathcal{U}[0, 1)$. Explain the problem.

4.L.2 Estimate the value of the integral $I = \int_0^1 e^{-x^2}\, dx$ using two different estimators:

- Crude Monte Carlo (CMC) estimator:

$$\hat{Y}_R^{\text{CMC}} = \frac{1}{R} \sum_{i=1}^{R} e^{-U_i^2},$$

  where $U_1, \ldots, U_R$ are i.i.d. $\mathcal{U}[0, 1)$ random variables.
- Hit-or-Miss (HoM) estimator:

$$\hat{Y}_R^{\text{HoM}} = \frac{1}{R} \sum_{i=1}^{R} Y_i, \quad Y_i = \mathbb{1}\left(e^{-U_{2i}^2} > U_{2i-1}\right),$$

  where $U_1, \ldots, U_{2R}$ are i.i.d. $\mathcal{U}[0, 1)$ random variables.

Take $R = 10^4$. Compute both estimators, their variances, and confidence intervals. Which estimator is more reliable?

4.L.3 Simulate $R = 1000$ random vectors $X_i = (X_i^{(1)}, X_i^{(2)})$ from a bivariate normal distribution with mean $(0, 0)$ and covariance matrix:

$$\Sigma = \begin{pmatrix} 1 & 0.7 \\ 0.7 & 1 \end{pmatrix}.$$

Estimate the mean $\hat{\mu}$ and covariance matrix $\hat{\Sigma}$ from the sample. Following Sect. 4.1.2.2, find and draw the following confidence regions at the 95% level for the mean vector:

(a) Elliptical confidence region,
(b) Parallelogram,
(c) Rectangular confidence region.

4.L.4 Estimate $k(I) = e^I$, where $I = \mathbb{E}Y$ and Y follows a standard normal distribution with mean 1.5 and variance 0.25. Simulate $R = 10^3$ i.i.d. samples $Y_1, \ldots, Y_{1000}$ from the distribution of Y. Estimate the mean I and compute $k(I)$. Calculate the asymptotic variance using Proposition 4.2.1 and provide a 95% confidence interval for $k(I)$.

4.L.5 Continue Exercise 4.T.5. Estimate:

$$\mathbf{I} = (\mathbb{E}(X^{(1)} + X^{(3)}), \mathbb{E}(X^{(2)} + X^{(3)})),$$

where $X^{(1)}, X^{(2)}, X^{(3)}$ are i.i.d. with exponential distribution $\text{Exp}(1/\theta)$, and $\theta = 2$. In this case, $\mathbf{I} = (4, 4)$. Let:

$$\mathbf{Y} = (X^{(1)}, X^{(2)}, X^{(3)})^T.$$

Sample $\mathbf{Y}_i$, $i = 1, \ldots, R$, with $R = 1000$, and estimate $\mathbf{I}$ via:

$$\hat{\mathbf{I}}_R = \left(\frac{1}{R} \sum_{i=1}^{R} (X_i^{(1)} + X_i^{(3)}), \ \frac{1}{R} \sum_{i=1}^{R} (X_i^{(2)} + X_i^{(3)}) \right).$$

- Estimate the covariance matrix from the sample. Provide an elliptical confidence region at the 95% level. Visualize this elliptical confidence region.
- Repeat the procedure 1000 times and check how often the true parameter $\mathbf{I} = (4, 4)$ is within the region.

Chapter 5
Variance Reduction Techniques

In this chapter, we explore various methods for approximating $I = \mathbb{E}Y$ and constructing estimators with reduced variance compared to the crude Monte Carlo (CMC) estimator $\hat{Y}_R^{\text{CMC}}$. The CMC estimator is defined as the arithmetic mean:

$$\hat{Y}_R^{\text{CMC}} = \frac{1}{R} \sum_{j=1}^{R} Y_j,$$

where $Y_1, \ldots, Y_R$ are i.i.d. replications of Y. The variance of $\hat{Y}_R^{\text{CMC}}$ is given by $\mathbb{V}\mathrm{ar}Y/R$. The starting point is the fundamental relation (1.6) between absolute error (relative error) and the number of replications R under a prescribed confidence level α from which we can figure out the half-length of confidence interval $\frac{z_{1-\alpha/2}\sigma_Y}{\sqrt{R}}$ (or $\frac{z_{1-\alpha/2}\sigma_Y}{\sqrt{R}I}$ respectively). So to reduce the half-length of confidence interval we need to look for estimators with smaller variance.

In this section our goal is to identify alternative estimators of I that achieve a smaller variance than the crude Monte Carlo approach.

In addition to reducing variance, practical considerations such as the ease of generating replications play an important. In some cases, relaxing the assumption of independent replications can also prove advantageous.

This chapter introduces and examines the following variance reduction techniques:

1. stratified sampling (str),
2. antithetic variates (anti),
3. common random numbers (CRN),
4. control variates (CV),
5. conditional Monte Carlo (cond),
6. importance sampling (IS) and the cross-entropy method.

© The Author(s), under exclusive license to Springer Nature Switzerland AG 2025
P. Lorek, T. Rolski, *Lectures on Monte Carlo Theory*, Probability Theory
and Stochastic Modelling 108, https://doi.org/10.1007/978-3-032-01190-9_5

Let $I = \mathbb{E}Y$. In Chap. 4, we considered estimators of the form $\hat{Y}_R$, based on i.i.d. replications of Y. In this chapter, we aim to derive new estimators of I that provide tighter confidence bounds for the same number of replications R. These new estimators, denoted $\hat{Y}_R^{\text{new}}$, are often constructed using non-i.i.d. samples $Y_1^*, \ldots, Y_R^*$. The loss of independence or identical distribution is a characteristic feature of several techniques discussed here.

To ensure the validity of $\hat{Y}_R^{\text{new}}$, the following properties must hold:

- **Consistency:** $\hat{Y}_R^{\text{new}} \to I$ with probability 1 as $R \to \infty$.
- **Central Limit Theorem (CLT):**

$$\sqrt{R}(\hat{Y}_R^{\text{new}} - I) \xrightarrow{\mathcal{D}} \mathcal{N}(0, (\varsigma^{\text{new}})^2).$$

Then, the confidence interval for I is stated as:

$$\mathbb{P}\left(\hat{Y}_R^{\text{new}} - z_{1-\alpha/2}\frac{\varsigma^{\text{new}}}{\sqrt{R}} \leq I \leq \hat{Y}_R^{\text{new}} + z_{1-\alpha/2}\frac{\varsigma^{\text{new}}}{\sqrt{R}} \right) \approx 1 - \alpha.$$

Informally, we can write a confidence interval for I:

$$\mathbb{P}\left(\hat{Y}_R^{\text{new}} - z_{1-\alpha/2}\sqrt{\mathbf{Var}(\hat{Y}_R^{\text{new}})} \leq I \leq \hat{Y}_R^{\text{new}} + z_{1-\alpha/2}\sqrt{\mathbf{Var}(\hat{Y}_R^{\text{new}})} \right) \approx 1 - \alpha.$$

The term $(\varsigma^{\text{new}})^2$, referred to as the *time average variance constant* (TAVC) or *asymptotic variance*, is defined as:

$$(\varsigma^{\text{new}})^2 = \lim_{R \to \infty} R\mathbf{Var}\hat{Y}_R^{\text{new}}.$$

Note that in the confidence interval, we used the approximation " $\approx 1 - \alpha$". This approximation, resulting from the CLT, may pose issues for small sample sizes R. However, in our applications, R is usually relatively large, so this approximation is not a significant concern.

A primary objective of *variance reduction* techniques is to achieve $(\varsigma^{\text{new}})^2 < \mathbf{Var}Y$, enabling the construction of tighter confidence bounds. While the example above considers confidence bounds for the absolute error, reducing the variance also yields tighter confidence intervals for the relative error. To see this, recall from Chap. 4 (Eq. (E2)) that the relative error satisfies:

$$\mathbb{P}\left(\left| \frac{\hat{Y}_R^{\text{new}}}{I} - 1 \right| \geq \varepsilon \right) = \mathbb{P}\left(\left| \hat{Y}_R^{\text{new}} - I \right| \geq \varepsilon I \right) \leq \frac{(\varsigma^{\text{new}})^2}{\varepsilon^2 I^2 R}.$$

Thus, the smaller the $(\varsigma^{\text{new}})^2$, the tighter the confidence interval for the relative error.

5.1 Survey of Variance Reduction Methods

5.1.1 Stratified Sampling

As usual, we want to compute $I = \mathbb{E}Y$. In general, by *strata*, we mean a partition of the sample space Ω into disjoint events $S^1, \ldots, S^m$. There are several ways to introduce the strata. One way to define strata is as follows. Assume that Y takes values in a space E (e.g., E could be $\mathbb{R}, \mathbb{R}^d$). Let $A^1, \ldots, A^m$ be disjoint subsets of E such that $\mathbb{P}(Y \in \bigcup_{j=1}^m A^j) = 1$. This partition of E defines events $S^j = \{Y \in A^j\}$, which are strata; an illustration for an example with $E = [0, 1]$ is presented in Fig. 5.1. For simplicity, we may slightly abuse the notation and refer to A^j as a stratum.

In some cases, strata may also be defined indirectly by considering a transformation of Y. For example, if $Y = f(U)$, where f is 1-1, U is a random variable uniformly distributed on $[0, 1)$, and if $B^1, \ldots, B^m$ is a partition of $[0, 1)$, the strata can be defined as $S^j = \{U : U \in f^{-1}(B^j)\}$. In the special case where $E = [0, 1]$, as illustrated in Fig. 5.1, the sets $S^1, \ldots, S^m$ and $B^1, \ldots, B^m$ coincide because U directly partitions $[0, 1]$ into the strata.

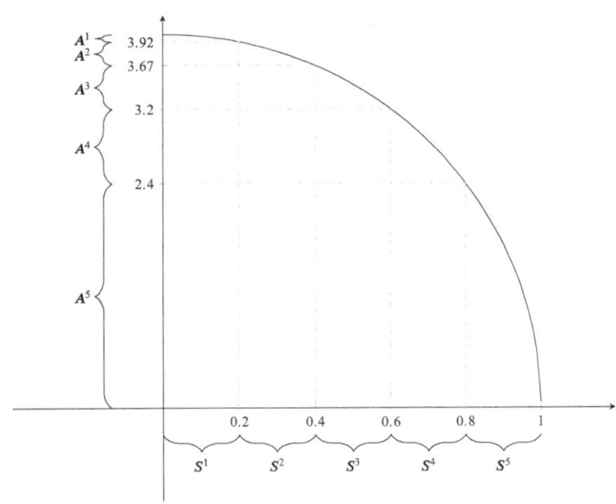

Fig. 5.1 Example $I = \pi = EY$: $Y = 4\sqrt{1 - U^2}$ with $U \sim \mathcal{U}[0, 1)$ with 5 strata $S^j = (0.2(j - 1), 0.2j)$ and corresponding sets/split $A^j, j = 1, \ldots, 5$ depicted

The key idea of stratified sampling is that, with an appropriate choice of strata and the number of replications within each stratum, it is possible to reduce the variance of the estimator.

Define:

- $p_j = \mathbb{P}(Y \in A^j)$: the probability of Y being in the j-th stratum (assumed to be known),
- $I^j = \mathbb{E}[Y \mid Y \in A^j]$: the conditional expectation within the j-th stratum (typically unknown, as we need to estimate it).

From the law of total probability, we have:

$$I = \mathbb{E}Y = p_1 I^1 + \cdots + p_m I^m.$$

The stratified sampling method relies on sampling within each individual stratum, using the fact that:

$$Y^j \overset{\mathcal{D}}{=} (Y \mid Y \in A^j), \tag{5.1.1}$$

Let $\sigma_j^2 = \mathbb{V}\mathrm{ar}Y^j$ denote the variance within the j-th stratum, which is equivalent to $\mathbb{V}\mathrm{ar}(Y \mid J = j)$, where J represents the random choice of stratum j, i.e., $\mathbb{P}(J = j) = p_j$.

We will need the following fact from basic probability theory.

Proposition 5.1.1 (Law of Total Variance) . *Suppose Y and J are two random variables. If $\mathbb{V}\mathrm{ar}Y < \infty$, then:*

$$\mathbb{V}\mathrm{ar}Y = \mathbb{E}[\mathbb{V}\mathrm{ar}(Y \mid J)] + \mathbb{V}\mathrm{ar}[\mathbb{E}(Y \mid J)]. \tag{5.1.2}$$

Proof The law of total variance can be derived as follows:

$$\mathbb{E}\mathbb{V}\mathrm{ar}(Y|J) + \mathbb{V}\mathrm{ar}(\mathbb{E}[Y|J])$$

$$= \mathbb{E}\left[\mathbb{E}[Y^2|J] - (\mathbb{E}[Y|J])^2\right] + \left[\mathbb{E}(\mathbb{E}[Y|J])^2 - (\mathbb{E}(\mathbb{E}[Y|J]))^2\right]$$

$$= \mathbb{E}Y^2 - (\mathbb{E}Y)^2 = \mathbb{V}\mathrm{ar}Y.$$

$\square$

The variance $\mathbb{V}\mathrm{ar}Y$ can be decomposed into **within-stratum** and **between-stratum** components as follows:

Corollary 5.1.2

$$\mathbb{V}\mathrm{ar}Y = \sum_{j=1}^{m} p_j \sigma_j^2 + \sum_{j=1}^{m} p_j (I^j - I)^2. \tag{5.1.3}$$

Proof To derive (5.1.3), define a random variable $J(\omega)$ such that $J(\omega) = j$ if and only if $\omega \in S^j$, where S^j are the strata. Using the law of total variance (Proposition 5.1.1), we write:

$$\mathbb{V}\mathrm{ar}Y = \mathbb{E}[\mathbb{V}\mathrm{ar}(Y \mid J)] + \mathbb{V}\mathrm{ar}[\mathbb{E}(Y \mid J)].$$

The two terms in this decomposition can be calculated as follows:

- **Within-stratum variance:**

$$\mathbb{E}[\mathbb{V}\mathrm{ar}(Y \mid J)] = \sum_{j=1}^{m} \mathbb{P}(J = j)\mathbb{V}\mathrm{ar}(Y \mid J = j) = \sum_{j=1}^{m} p_j \sigma_j^2$$

- **Between-stratum variance:**

$$\mathbb{V}\mathrm{ar}[\mathbb{E}(Y \mid J)] = \mathbb{E}\big[(\mathbb{E}(Y \mid J) - \mathbb{E}[\mathbb{E}(Y \mid J)])^2\big] = \mathbb{E}\big[(\mathbb{E}(Y \mid J) - \mathbb{E}Y)^2\big]$$

$$= \sum_{j=1}^{m} \mathbb{P}(J = j)(\mathbb{E}(Y \mid J = j) - I)^2 = \sum_{j=1}^{m} p_j (I^j - I)^2,$$

where $I^j = \mathbb{E}(Y \mid J = j)$ and $I = \mathbb{E}Y$.

Combining these results, we obtain (5.1.3). $\square$

Let R denote the total number of replications, which is distributed among the strata such that there are R_j replications in the j-th stratum. Specifically, let $Y_1^j, \ldots, Y_{R_j}^j$ be i.i.d. replications of Y^j within the j-th stratum. We assume that the replications are independent both within and across the strata. Naturally, the total number of replications satisfies:

$$R = R_1 + \cdots + R_m.$$

Define the sample mean within the j-th stratum as:

$$\hat{Y}_{R_j}^j = \frac{1}{R_j} \sum_{i=1}^{R_j} Y_i^j,$$

which serves as an estimator of $I^j = \mathbb{E}[Y \mid Y \in A^j]$. The variance within the j-th stratum is denoted by $\sigma_j^2 = \mathbb{V}\mathrm{ar}Y^j$. Note that $\hat{Y}_{R_j}^j$ is an unbiased estimator of I^j, and its variance is:

$$\mathbb{V}\mathrm{ar}\hat{Y}_{R_j}^j = \frac{\sigma_j^2}{R_j}.$$

Using these stratum-specific estimators, we define the **stratified estimator**:

$$\hat{Y}_R^{\mathrm{str}} = p_1 \hat{Y}_{R_1}^1 + \cdots + p_m \hat{Y}_{R_m}^m. \tag{5.1.4}$$

Assuming that it is possible to simulate individual Y^j, we verify that the stratified estimator is unbiased. Specifically:

$$\mathbb{E}\hat{Y}_R^{\text{str}} = p_1\mathbb{E}\hat{Y}_{R_1}^1 + \cdots + p_m\mathbb{E}\hat{Y}_{R_m}^m$$

$$= \sum_{j=1}^{m} p_j \frac{\mathbb{E}[Y\mathbb{1}_{\{Y\in A^j\}}]}{p_j} = \mathbb{E}Y = I.$$

Furthermore, the variance of the stratified estimator is given by:

$$\text{Var}\hat{Y}_R^{\text{str}} = \sum_{j=1}^{m} p_j^2 \text{Var}\hat{Y}_{R_j}^j = \sum_{j=1}^{m} \frac{p_j^2\sigma_j^2}{R_j}. \tag{5.1.5}$$

It is convenient to use the following notation: for $\mathbf{x} = (x_1, \ldots, x_m)$, where $x_j > 0$ and $\sum_{j=1}^{m} x_j = 1$, we define:

$$\sigma_{\text{str}}^2(\mathbf{x}) := \sum_{j=1}^{m} \frac{p_j^2}{x_j}\sigma_j^2. \tag{5.1.6}$$

We can rewrite the right-hand side of (5.1.5) as:

$$\text{Var}\hat{Y}_R^{\text{str}} = \frac{1}{R}\left(p_1^2 \frac{R}{R_1}\sigma_1^2 + \cdots + p_m^2 \frac{R}{R_m}\sigma_m^2\right)$$

$$= \frac{1}{R}\left(\frac{p_1^2}{R_1/R}\sigma_1^2 + \cdots + \frac{p_m^2}{R_m/R}\sigma_m^2\right)$$

$$= \frac{1}{R}\sigma_{\text{str}}^2((R_1/R, \ldots, R_m/R)). \tag{5.1.7}$$

Next, consider the case where $R \to \infty$ in such a way that:

$$\gamma_{j,R} = R_j/R \quad \text{converges to (say)} \quad \gamma_j \quad (j = 1, \ldots, m). \tag{5.1.8}$$

Clearly, $\boldsymbol{\gamma} = (\gamma_1, \ldots, \gamma_m)$ is a probability distribution. In this case:

$$\sigma_{\text{str}}^2((R_1/R, \ldots, R_m/R)) \to \sigma_{\text{str}}^2(\boldsymbol{\gamma}).$$

The following result is key for analyzing the behavior of the stratified estimator:

Proposition 5.1.3 *If $R \to \infty$ such that (5.1.8) holds, then:*

$$\sqrt{R}(\hat{Y}_R^{\text{str}} - I) \xrightarrow{\mathcal{D}} \mathcal{N}(0, \varsigma^2),$$

where $\varsigma^2 = \sigma_{\text{str}}^2(\boldsymbol{\gamma})$.

Proof We rely on the independence of replications within and across strata. Then:

$$\sqrt{R}\left(\sum_{j=1}^{m} p_j \hat{Y}_{R_j}^{j} - I\right) = \sqrt{R}\left(\sum_{j=1}^{m} p_j \hat{Y}_{R_j}^{j} - \sum_{j=1}^{m} p_j I^j\right)$$

$$= \sum_{j=1}^{m} p_j \sigma_j \sqrt{\frac{R}{R_j}}\left[\frac{\sqrt{R_j}}{\sigma_j}\left(\hat{Y}_{R_j}^{j} - I^j\right)\right]$$

$$= \sum_{j=1}^{m} \frac{p_j \sigma_j}{\sqrt{\gamma_{j,R}}}\left[\frac{\sqrt{R_j}}{\sigma_j}\left(\hat{Y}_{R_j}^{j} - I^j\right)\right].$$

Since:

$$\frac{\sqrt{R_j}}{\sigma_j}\left(\hat{Y}_{R_j}^{j} - I^j\right) \overset{\mathcal{D}}{\to} N_j,$$

where $N_1, \ldots, N_m$ are i.i.d. $\mathcal{N}(0, 1)$, it follows that:

$$\sqrt{R}(\hat{Y}_R^{\mathrm{str}} - I) \overset{\mathcal{D}}{\to} \mathcal{N}\left(0, \sum_{j=1}^{m} \frac{p_j^2}{\gamma_j} \sigma_j^2\right).$$

Thus, $\varsigma^2 = \sigma_{\mathrm{str}}^2(\boldsymbol{\gamma})$. $\square$

At this point, we can ask two important questions:

- Given strata (i.e., knowing p_j and σ_j^2) and a fixed total number of replications R, how should the number of replications R_j in each stratum be allocated to minimize the variance $\mathbb{V}\mathrm{ar}\hat{Y}_R^{\mathrm{str}}$?
- How can we choose strata to minimize the variance? (Addressing this question depends strongly on the specific problem.)

Regarding the first question, we will focus on two common allocation strategies: *proportional allocation* and *optimal allocation*.

5.1.1.1   Choosing the Number of Replications in Strata

Proportional Allocation A straightforward approach is to set $R_j = Rp_j$, for $j = 1, \ldots, m$. As the following proposition shows, this allocation results in a variance that is never worse—and in practice often better—than the variance of the crude Monte Carlo (CMC) estimator $\hat{Y}_R^{\mathrm{CMC}}$. Let $\hat{Y}_R^{\mathrm{pa}}$ denote the stratified estimator $\hat{Y}_R^{\mathrm{str}}$ with $R_j = Rp_j$. For simplicity, we ignore rounding issues; in practice, one typically uses $R_j = \lceil Rp_j \rceil$.

Proposition 5.1.4

$$\mathbb{V}ar(\hat{Y}_R^{\mathrm{CMC}}) = \mathbb{V}ar(\hat{Y}_R^{\mathrm{pa}}) + \frac{1}{R} \sum_{j=1}^{m} p_j (I^j - I)^2 \geq \mathbb{V}ar(\hat{Y}_R^{\mathrm{pa}}).$$

Proof For $\mathbb{V}ar(\hat{Y}_R^{\mathrm{pa}})$, note that $R_j / R = p_j$, so from (5.1.5), we have:

$$\mathbb{V}ar(\hat{Y}_R^{\mathrm{pa}}) = \frac{1}{R} \sum_{j=1}^{m} p_j \sigma_j^2.$$

Using the *within-stratum and between-stratum variance decomposition* of $\mathbb{V}ar Y$ from (5.1.3), we have:

$$\mathbb{V}ar Y = \sum_{j=1}^{m} p_j \sigma_j^2 + \sum_{j=1}^{m} p_j (I^j - I)^2,$$

which implies:

$$\frac{1}{R} \mathbb{V}ar Y = \frac{1}{R} \sum_{j=1}^{m} p_j \sigma_j^2 + \frac{1}{R} \sum_{j=1}^{m} p_j (I^j - I)^2.$$

Thus:

$$\mathbb{V}ar(\hat{Y}_R^{\mathrm{CMC}}) = \mathbb{V}ar(\hat{Y}_R^{\mathrm{pa}}) + \frac{1}{R} \sum_{j=1}^{m} p_j (I^j - I)^2.$$

$\square$

Optimal Allocation While proportional allocation improves over the crude Monte Carlo estimator, it is not always optimal. The following theorem identifies the allocation that minimizes the variance.

Theorem 5.1.5 *Let R be fixed, with $R_1 + \cdots + R_m = R$. The variance:*

$$\mathbb{V}ar(\hat{Y}_R^{\mathrm{str}}) = \sum_{j=1}^{m} \frac{p_j^2}{R_j} \sigma_j^2$$

is minimized when:

$$R_j = \frac{p_j \sigma_j}{\sum_{i=1}^{m} p_i \sigma_i} R. \tag{5.1.9}$$

Proof Define $D = \sum_{j=1}^{m} p_j \sigma_j$. Let $\hat{Y}_R^{\text{opt}}$ be the stratified estimator with $R_j = \frac{p_j \sigma_j}{D} R$. Substituting this into (5.1.5), we have:

$$\text{Var}(\hat{Y}_R^{\text{opt}}) = \frac{1}{R} \sum_{j=1}^{m} \frac{p_j^2}{p_j \sigma_j / D} \sigma_j^2 = \frac{1}{R} D^2.$$

Now, let $\hat{Y}_R^{\text{str}}$ be a stratified estimator with $R_j = x_j R$, where $\mathbf{x} = (x_1, \ldots, x_m)$ satisfies $\sum_{j=1}^{m} x_j = 1$ and $x_j > 0$. Using the Cauchy-Schwarz inequality:

$$\left(\sum_{j=1}^{m} w_j z_j \right)^2 \leq \sum_{k=1}^{m} w_k^2 \sum_{j=1}^{m} z_j^2,$$

set $w_k = \sqrt{x_k}$ and $z_j = \frac{p_j \sigma_j}{\sqrt{x_j}}$. Then:

$$\text{Var}(\hat{Y}_R^{\text{opt}}) = \frac{1}{R} D^2 = \frac{1}{R} \left(\sum_{j=1}^{m} p_j \sigma_j \right)^2 = \frac{1}{R} \left(\sum_{j=1}^{m} \sqrt{x_j} \frac{p_j \sigma_j}{\sqrt{x_j}} \right)^2$$

$$\leq \frac{1}{R} \sum_{k=1}^{m} x_k \sum_{j=1}^{m} \frac{p_j^2 \sigma_j^2}{x_j} = \frac{1}{R} \sum_{j=1}^{m} \frac{p_j^2 \sigma_j^2}{x_j} = \text{Var}(\hat{Y}_R^{\text{opt}}).$$

Thus, the variance is minimized for $\mathbf{x} = \boldsymbol{\gamma}^* = \left(\frac{p_1 \sigma_1}{D}, \ldots, \frac{p_m \sigma_m}{D} \right)$. $\square$

The allocation in (5.1.9) is called *optimal allocation*. In practice, the variances σ_j^2 are usually unknown. A common approach is to first estimate them using R' simulations with proportional allocation $R'_j = p_j R'$, obtaining estimators $\hat{S}_j'^2$. Then, for the main simulations, we use:

$$R_j = \frac{p_j \hat{S}_j'}{\sum_{k=1}^{m} p_k \hat{S}_k'} R$$

replications in the j-th stratum. Finally, the variance of the stratified estimator is estimated as:

$$\widehat{\text{Var}}(\hat{Y}_R^{\text{str}}) = \sum_{j=1}^{m} \frac{p_j^2}{R_j} \hat{S}_j^2, \tag{5.1.10}$$

and the confidence interval for I at level α is:

$$\left(\hat{Y}_R^{\text{str}} - z_{1-\alpha/2}\sqrt{\widehat{\mathbb{V}\text{ar}}(\hat{Y}_R^{\text{str}})},\ \hat{Y}_R^{\text{str}} + z_{1-\alpha/2}\sqrt{\widehat{\mathbb{V}\text{ar}}(\hat{Y}_R^{\text{str}})} \right).$$

Modifications Instead of a single random variable Y, we can consider a pair (Y, X) and define strata in the following way. Let $A^1, \ldots, A^m$ be disjoint subsets such that $\mathbb{P}(X \in \bigcup_{j=1}^m A^j) = 1$. Then the j-th stratum is defined as $S^j = \{X \in A^j\}$, and

$$I = \mathbb{E}Y = \sum_{j=1}^m p_j \mathbb{E}[Y|X \in A^j],$$

where $p_j = \mathbb{P}(X \in A^j)$. To construct the stratified estimator, we redefine:

$$Y^j \overset{\mathcal{D}}{=} (Y|X \in A^j).$$

In this case, we leave it as an exercise for the reader to write down the stratified estimator. For example, if $Y = g(U)$, where $U \sim \mathcal{U}[0, 1)$, we can define strata $S^1, \ldots, S^m$ either for Y or directly for $X = U$. Another example is when $Y = g(Z)$, where $Z \sim \mathcal{N}(0, 1)$. In this case, strata for Z will be shown later in Example 5.1.9. Note that in the general case, strata S^j ($j = 1, \ldots, m$) form a partition of the sample space Ω. In the most general sense, strata partition Ω into events $S^1, \ldots, S^m$.

Example 5.1.6 (Stratified Sampling for Arbitrary One-dimensional Distributions) We discuss methods for simulating a random variable with a prescribed distribution using stratified sampling concepts. This method applies when the ITM method (Sect. 3.1) is applicable, i.e., when the cumulative distribution function F has a computable generalized inverse $F^{\leftarrow}$.

Recall that

$$V_i^j = \frac{j-1}{m} + \frac{U_i}{m}, \qquad i = 1, \ldots, R_j,$$

where $U_1, \ldots, U_{R_j}$ are i.i.d. random variables uniformly distributed on $\mathcal{U}[0, 1)$. These variables are uniformly distributed within the interval $((j-1)/m, j/m]$. Setting $R_j = 1$, the variables $V^1, \ldots, V^m$ create a stratified sample from the uniform distribution.

Now, consider how to perform stratified simulations for a random variable Y with c.d.f. F. For simplicity, assume F is strictly increasing. While our considerations focus on $\mathbb{R}$, they can be adapted to distributions on subsets of $\mathbb{R}$. Let $p_1, \ldots, p_m$ be fixed strata probabilities such that $p_j > 0$ and $\sum_{j=1}^m p_j = 1$. We divide $\mathbb{R}$ into m intervals:

$$A^1 = (a_0, a_1],\ A^2 = (a_1, a_2], \ldots,\ A^m = (a_{m-1}, a_m],$$

where $a_0 = -\infty$ and $a_m = \infty$, with

$$\mathbb{P}(Y \in (a_{j-1}, a_j]) = p_j.$$

The endpoints a_j are determined as follows:

$$a_0 = -\infty,$$
$$a_1 = F^{\leftarrow}(p_1),$$
$$\vdots$$
$$a_m = F^{\leftarrow}(1).$$

If the support E of F is a proper subset of $\mathbb{R}$, the values of a_j should be adjusted accordingly.

To simulate $Y^j \overset{\mathcal{D}}{=} (Y|Y \in A^j)$, we assume the ability to compute the generalized inverse $F^{\leftarrow}$. With $p_j = \mathbb{P}(A^j)$, we use the following result, which follows directly from Exercise 3.T.4:

Proposition 5.1.7 *Let $V^j = a_{j-1} + (a_j - a_{j-1})U$, where $U \sim \mathcal{U}[0, 1)$. Then*

$$F^{\leftarrow}(V^j) \overset{\mathcal{D}}{=} (Y|Y \in A^j).$$

$\Diamond$

Example 5.1.8 In Example 4.1.7, we estimated π using a CMC estimator:

$$\hat{Y}_R^{\mathrm{CMC}} = \frac{1}{R} \sum_{j=1}^{R} Y_j, \quad \text{where } Y_j = 4\sqrt{1 - U_j^2}, \ U_j \sim \mathcal{U}[0, 1) \text{ i.i.d.}$$

We now apply stratified sampling to improve the variance of this estimate. Define m strata:

$$B^1 = (0, 1/m], \ B^2 = (1/m, 2/m], \ldots, \ B^m = ((m-1)/m, 1].$$

The strata are equally probable, i.e., $\mathbb{P}(U \in B^j) = p_j = 1/m$ for $j = 1, \ldots, m$.

Simulating $Y^j \overset{\mathcal{D}}{=} (Y|Y \in B_j)$ is straightforward:

$$Y^j = 4\sqrt{1 - (V^j)^2}, \quad \text{where } V^j = \frac{j-1}{m} + \frac{1}{m}U, \ U \sim \mathcal{U}[0, 1).$$

This example is implemented in `ch5_strat_sampling_pi.py`.

Pilot Simulations for Variance Estimation To apply the optimal allocation scheme, we estimate the variances in each stratum using R' pilot simulations. Specifically:

- Simulate $R'_j = R'/m$ random points in each stratum j.
- Compute the sample variance for the j-th stratum:

$$\hat{Y}'_{R'_j} = \frac{1}{R'_j} \sum_{i=1}^{R'_j} Y_i^j, \quad \hat{S}'^2_j = \frac{1}{R'_j - 1} \sum_{i=1}^{R'_j} \left(Y_i^j - \hat{Y}'_{R'_j} \right)^2.$$

Optimal Allocation for Actual Simulations Using the estimated variances, the optimal number of replications for R total simulations is:

$$R_j = \left\lceil \frac{\hat{S}'_j}{\sum_{i=1}^{m} \hat{S}'_i} R \right\rceil.$$

For $R' = 100$ and $m = 10$, Table 5.1 shows the pilot estimates of standard deviations $\hat{S}'_j$ and the resulting optimal R_j for $R = 200$.

Simulation Results We conducted simulations for $R \in \{200, 10^4\}$ replications and $m \in \{5, 10, 20\}$ strata. The results are summarized in Table 5.2. For each case, the variance $\mathbb{V}\mathrm{ar}(\hat{Y}_R^{\mathrm{str}})$ of the stratified estimator was estimated using:

$$\widehat{\mathbb{V}\mathrm{ar}}(\hat{Y}_R^{\mathrm{str}}) = \sum_{j=1}^{m} \frac{p_j^2}{R_j} \hat{S}_j^2.$$

The error $\hat{b} = 1.96\sqrt{\widehat{\mathbb{V}\mathrm{ar}}(\hat{Y}_R^{\mathrm{str}})}$ was then computed.

For $R = 200$ replications, the stratified estimator consistently produces lower standard errors and absolute errors than the CMC estimator, regardless of the number of strata. For $R = 10^4$, increasing the number of strata further reduces the error, demonstrating that with an appropriate choice of m, stratified sampling reliably outperforms CMC.

Visualization Figure 5.2 shows the simulated points for $R = 200$, $m = 10$ strata. For the CMC estimator, points $U_i \sim \mathcal{U}[0, 1)$ were simulated, and $Y_i = 4\sqrt{1 - U_i^2}$

Table 5.1 Estimating standard deviations in $m = 10$ stratas using $R' = 100$ simulations. Resulting R_j, $j = 1, \ldots, m$, for $R = 200$ displayed

Stratum j	1	2	3	4	5	6	7	8	9	10
$\hat{S}'_j$	0.006	0.018	0.031	0.034	0.059	0.078	0.112	0.097	0.169	0.453
R_j	1	3	5	6	11	14	21	18	31	90

Table 5.2 Simulation results for estimating π using the CMC and stratified estimators with $R \in \{200, 10^4\}$ replications. The rows displaying the smallest error $\hat{b}$ and the absolute error $|\hat{Y}_R - \pi|$ are highlighted in **bold**

$R = 200$				
	CMC	Stratified estimator		
		Number of strata m		
		5	10	20
$\hat{Y}_R$	3.19479	3.14373	3.12173 4	3.15037
$\hat{b}$	0.12178	0.02834	0.01435	**0.00793**
$|\hat{Y}_R - \pi|$	0.05319	**0.00214**	0.01986	0.00878
$R = 10^4$				
	CMC	Stratified estimator		
		Number of strata m		
		5	10	20
$\hat{Y}_R$	3.14022	3.14251	3.14230	3.14182
$\hat{b}$	0.01765	0.00400	0.00209	**0.00110**
$|\hat{Y}_R - \pi|$	0.00137	0.0092	0.00071	**0.00023**

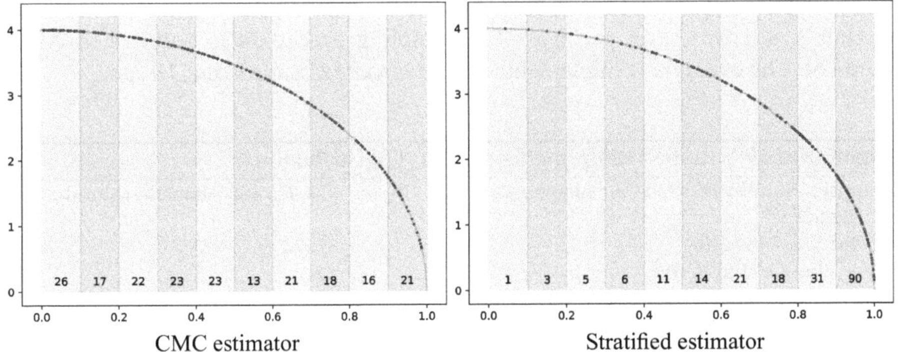

Fig. 5.2 Simulation with $R = 200$ and $m = 10$ strata. Left: CMC estimator. Right: Stratified estimator with optimal allocation. Numbers above the x-axis indicate the number of replications within each stratum

was computed. For the stratified estimator, R_j points were simulated in each stratum using V^j.

Example 5.1.9 In Fig. 5.3, $m = 10$ equally probable strata for the $\mathcal{N}(0, 1)$ distribution are shown. The strata are defined as:

$$A^j = \left(\Phi^{-1}\left(\frac{j-1}{m}\right), \Phi^{-1}\left(\frac{j}{m}\right) \right],$$

where Φ^{-1} denotes the quantile function (inverse c.d.f.) of the standard normal distribution. We use the convention $\Phi^{-1}(0) = -\infty$ and $\Phi^{-1}(1) = +\infty$, so that:

$$A^1 = \left(-\infty, \Phi^{-1}\left(\frac{1}{m}\right)\right], \quad A^m = \left(\Phi^{-1}\left(\frac{m-1}{m}\right), +\infty\right).$$

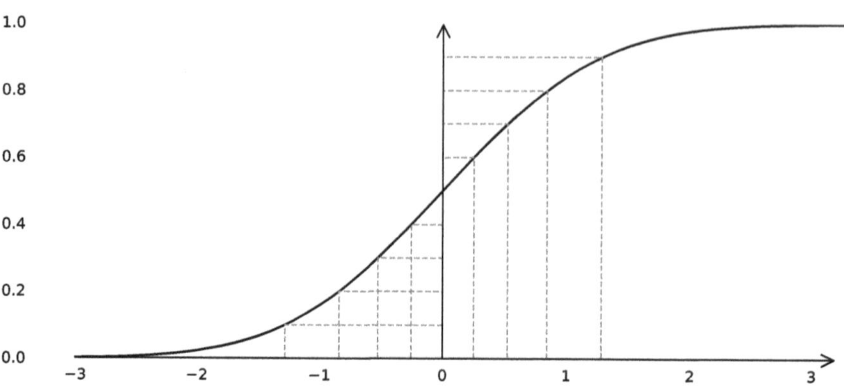

Fig. 5.3 Standard normal c.d.f. with 10 equally probable strata

The numerical evaluation of Φ^{-1} was performed using the `ppf` function from Python's `scipy.stats` package. The sampling procedure is outlined in Algorithm 29. The example is implemented in `ch5_strat_sampling_1D.py`.

Algorithm 29 Stratified sampling from $\mathcal{N}(0, 1)$ distribution.

Require: Number of strata: m; sample size: $R = R_1 + \cdots + R_m$ (R_j simulations in the j-th stratum)
1: **for** $j = 1$ to m **do**
2:      **for** $i = 1$ to R_j **do**
3:          Sample $U \sim \mathcal{U}[0, 1)$
4:          Compute $V = (j - 1)/m + U/m$
5:          Compute $Y_i^j = \Phi^{-1}(V)$
6:      **end for**
7: **end for**
8: **return** $(Y_1^1, Y_2^1, \ldots, Y_{R_1}^1, Y_1^2, \ldots, Y_{R_m}^m)$

In Fig. 5.4, the results of a stratified sampling experiment are presented. The experiment involved $R = 500$ replications from the $\mathcal{N}(0, 1)$ distribution. Two approaches were compared:

- **Regular sampling:** 500 independent samples $U_1, \ldots, U_{500} \sim \mathcal{U}[0, 1)$ were drawn, and the transform $X_i = \Phi^{-1}(U_i)$ was applied. The resulting histogram is shown on the right-hand side.
- **Stratified sampling:** Using Algorithm 29, $m = 100$ strata were used, with $R_j = 5$ samples drawn in each stratum ($j = 1, \ldots, 100$). The resulting histogram is shown on the left-hand side.

The same set of random values $(U_1, \ldots, U_{500})$ was used for both methods, and both histograms were generated using 50 bins. As observed, the stratified sampling method produced a **more accurate** sample—its sample mean and variance are

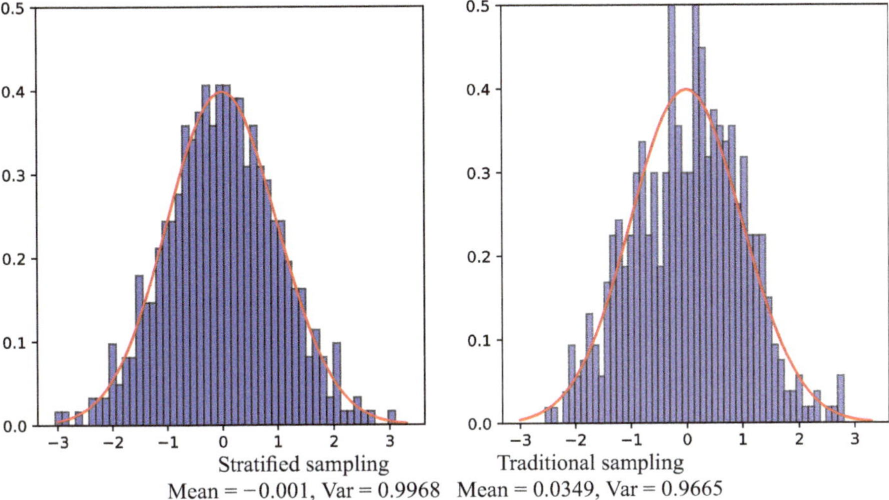

Fig. 5.4 Comparison of stratified sampling (left, $m = 100$ strata, $R_j = 5$ for $j = 1, \ldots, m$) with regular sampling (right) for $R = 500$ observations. In both cases, the same randomness was used. Histograms were constructed with 50 bins

notably closer to the expected values of 0 and 1, respectively, compared to the traditional method.

Example 5.1.10 (Probability of Network Failure) Consider a connection network represented as a connected graph consisting of nodes, where two nodes, s (*source*) and t (*target*), are distinguished. Edges connect the nodes, numbered $1, \ldots, n$. Each edge can either be operating (0) or failed (1). When an edge is in a failed state, we refer to it as an *interruption*. Removing an edge results in a subgraph that may or may not remain connected.

The state of the network is described by a vector $\mathbf{e} = (e_1, \ldots, e_n)$, where $e_i \in \{0, 1\}$. The entire network is considered operational (0) if there exists a path (connection) between s and t. Conversely, it is considered failed (1) if the removal of failed edges disconnects the graph. The state space is $\mathcal{S} = \{0, 1\}^n$. Formally, the network state is characterized by a function $k : \mathcal{S} \to \{0, 1\}$, where $k(\mathbf{e}) = 0$ if the network is operational and $k(\mathbf{e}) = 1$ if it has failed.

The failure or operational status of the network depends on the subset B of failed edges, where $e_j = 1$ for $j \in B$, and $e_j = 0$ otherwise. Below are two specific examples:

- *Serial system:* Fails if at least one edge fails. Thus, $k(\mathbf{e}) = 1$ iff $|\mathbf{e}| \geq 1$.
- *Parallel system:* Fails only if all edges fail. Thus, $k(\mathbf{e}) = 1$ iff $|\mathbf{e}| = n$.

Assume each edge fails independently with probability q. The state of the system is described by a random vector X, which takes values in $\mathcal{S}$. The probability distribution of X is given by:

$$\mathbb{P}(X = \mathbf{e}) = p(\mathbf{e}) = q^{\sum_{j=1}^{n} e_j} (1 - q)^{n - \sum_{j=1}^{n} e_j}, \quad \text{for } \mathbf{e} \in \mathcal{S}.$$

Our goal is to compute the probability that the system fails:

$$I = \mathbb{E}[k(X)] = \sum_{\mathbf{e} \in \mathcal{S}} k(\mathbf{e}) p(\mathbf{e}).$$

This probability I can be estimated using the crude Monte Carlo (CMC) method:

$$\hat{Y}_R^{\mathrm{CMC}} = \frac{1}{R} \sum_{i=1}^{R} k(X_i),$$

where $X_1, \ldots, X_R$ are i.i.d. replications of X. Since $k(X)$ is a Bernoulli random variable, we have:

$$I = \mathbb{P}(k(X) = 1) = \mathbb{E}[k(X)], \quad \text{and} \quad \mathrm{Var}[k(X)] = I(1 - I).$$

Thus, the variance of the CMC estimator is:

$$\mathrm{Var}(\hat{Y}_R^{\mathrm{CMC}}) = \frac{I(1 - I)}{R}.$$

We suggest performing a simulation for a specific system from the book by Madras [138]. The network consists of 11 nodes connected by 22 edges, as depicted in Fig. 5.5.

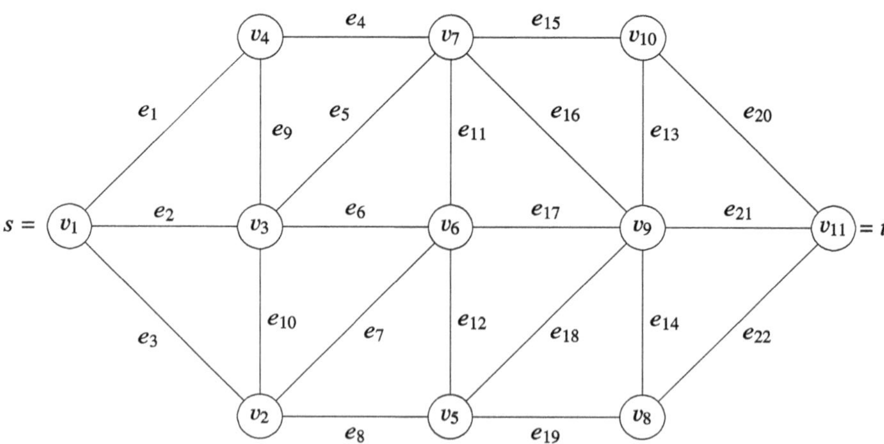

Fig. 5.5 A network with $L = 11$ nodes and $n = 22$ edges considered in Madras [138]

Typically, the probability I of network failure is very small; hence, estimating this probability accurately requires a large number of replications. This necessity motivates the construction of improved estimators for I.

Stratification Recall that for a given subset $B \subset S$, the number of failed edges is denoted by $|B|$. A natural choice for stratification is to consider the following strata:

$$A^j = \{B \subset S : |B| = j\}, \quad j = 0, \ldots, n,$$

where the j-th stratum consists of all edge failure states B such that $|B| = j$.

The probability of X belonging to stratum A^j is:

$$p_j = \mathbb{P}(X \in A^j) = \mathbb{P}(\text{exactly } j \text{ edges fail}) = \binom{n}{j} q^j (1-q)^{n-j}. \qquad (5.1.11)$$

Within each stratum A^j, each configuration $B \subset S$ has the same probability $q^j(1-q)^{n-j}$. Furthermore:

$$\mathbb{P}(X = B \mid X \in A^j) = \frac{1}{\binom{n}{j}}, \quad \text{for all } B \in A^j.$$

Sampling from stratum A^j is straightforward: it suffices to choose $n - j$ working edges uniformly at random out of the n total edges (without replacement). Then, we use these edges to build the adjacency matrix step by step.

The stratified estimation of network failure probability using proportional allocation proceeds as follows. For each stratum j:

- Simulate the network with j failed edges by choosing $n - j$ working edges uniformly (without replacement) from the set of n edges.
- Build the adjacency matrix $\mathbf{A}$ step by step by adding the $n - j$ working edges.
- Check whether a path exists between nodes s and t.
- Repeat the above steps $R_j = \lceil p_j R \rceil$ times, recording the fraction of failures as $\hat{Y}^j_{R_j}$.

The final stratified estimator is:

$$\hat{Y}^{\text{str}}_R = \sum_{j=0}^{n} p_j \hat{Y}^j_{R_j}.$$

Storing the network structure in an adjacency matrix $\mathbf{A}$ is convenient. The matrix $\mathbf{A}$ is a binary matrix of size $L \times L$, where L is the number of vertices. For example, in the network shown in Fig. 5.5, the edges are:

$$\text{edges} = [\ [v_1, v_2], [v_1, v_3], [v_1, v_4], [v_2, v_3], [v_2, v_5], [v_2, v_6],$$

$$[v_3, v_6], [v_3, v_7], [v_4, v_7], [v_5, v_6], [v_5, v_8], [v_5, v_9],$$

$$[v_6, v_7], [v_6, v_9], [v_7, v_9], [v_7, v_{10}], [v_8, v_9], [v_8, v_{11}],$$

$$[v_9, v_{11}], [v_9, v_{10}], [v_{10}, v_{11}]\ ].$$

For each edge $(v_l, v_r) \in \text{edges}$, we set $\mathbf{A}[v_l, v_r] = \mathbf{A}[v_r, v_l] = 1$. To sample X uniformly in the j-th stratum, we select $n - j$ working edges (chosen uniformly at random without replacement from edges) and add them step by step to $\mathbf{A}$. The detailed procedure is outlined in Algorithm 30.

Algorithm 30 Probability of a network failure: Stratified sampling with proportional allocation

---

Require: List of edges: $\text{edges} = [[v_l^1, v_r^1], [v_l^2, v_r^2], \ldots [v_l^n, v_r^n]]$, edge failure probability q, number of replications R, and number of vertices L.

1: **for** $j = 0$ to n **do**   ▷ nr of strata = nr of edges + 1
2:     Compute $R_j = \lceil p_j R \rceil$, where p_j is given in (5.1.11).
3: **end for**
4: **for** $j = 0$ to n **do**
5:     **for** $i = 1$ to R_j **do**
6:         Construct $L \times L$ adjacency matrix $\mathbf{A}$ with all entries initialized to 0.
7:         Sample $n - j$ working edges uniformly without replacement from edges.
8:         **for** each sampled edge (v_l, v_r) **do**
9:             Set $\mathbf{A}[v_l, v_r] = \mathbf{A}[v_r, v_l] = 1$.   ▷ Add edge $[v_l, v_r]$ to $\mathbf{A}$
10:         **end for**
11:         $Y_i^j = 1$ if there is no path from s to t in $\mathbf{A}$.
12:     **end for**
13:     Compute $\hat{Y}_{R_j}^j = \frac{1}{R_j} \sum_{i=1}^{R_j} Y_i^j$.
14: **end for**
15: **return** $\hat{Y}_R^{\text{str}} = \sum_{j=0}^{n} p_j \hat{Y}_{R_j}^j$.

In line 11 of Algorithm 30, we need to determine whether a path exists between s and t. For small networks, this can be achieved by computing powers of the adjacency matrix $\mathbf{A}^k$ for $k = 1, 2, \ldots, n$ and checking if $\mathbf{A}^k(s, t) > 0$. If this condition holds, a path exists; otherwise, it does not. However, for larger networks, this approach is computationally infeasible, see e.g., Fig. 5.6. For larger networks, where the number of nodes significantly exceeds the number of edges, representing the network as an adjacency matrix is inefficient. In such cases, we recommend using the **Union-Find** algorithm for connectivity checks.

Union-Find Algorithm   In Example 5.1.10, after sampling a graph (by randomly—in a usual or stratified way—removing edges), we need to check whether there is a path from the source s to the target t. This should be done in an efficient manner. A larger network is presented in Fig. 5.6. The Union-Find algorithm is a very useful tool for such a task.

Fig. 5.6 Is there a path from source s to target t?

It allows us to *quickly* answer the question of whether there is a path between any two vertices v and w. Before providing the details of the algorithm, let us briefly describe its functionality. In Fig. 5.7, we have a sample graph G with 10 vertices. The main goal of the algorithm is to answer questions such as *"Is there a path from v_8 to v_9?"*. Additionally, it can partition all vertices into sets corresponding to the connected components of the graph. For the example shown in Fig. 5.7, the algorithm would return three sets: $A_1 = \{v_1, v_2, v_7\}$, $A_2 = \{v_6\}$, and $A_3 = \{v_3, v_4, v_5, v_8, v_9, v_{10}\}$. This example is implemented in `ch5_union_find.py`.

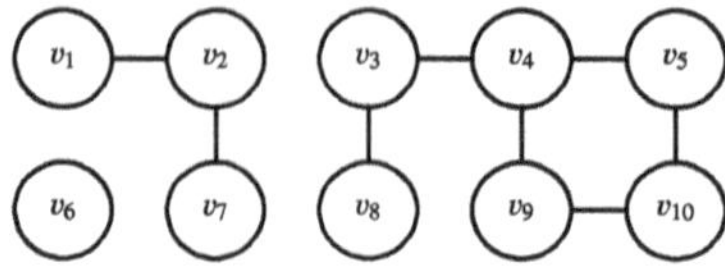

Fig. 5.7 A graph with 10 vertices with three sets corresponding to disjoint connected components: $A_1 = \{v_1, v_2, v_7\}$, $A_2 = \{v_6\}$, $A_3 = \{v_3, v_4, v_5, v_8, v_9, v_{10}\}$

We will use a simple table representation of a parenthood structure. The sets will have *representatives*, which will be one of the vertices. For example, if v_i and v_j share the same representative w, then v_i and v_j belong to the same set.

Initialization Step Initially, each vertex is its own set. In other words, each vertex is its own representative. This initial step is illustrated in Fig. 5.8 , where the parent of each vertex is stored in the `Arr` array.

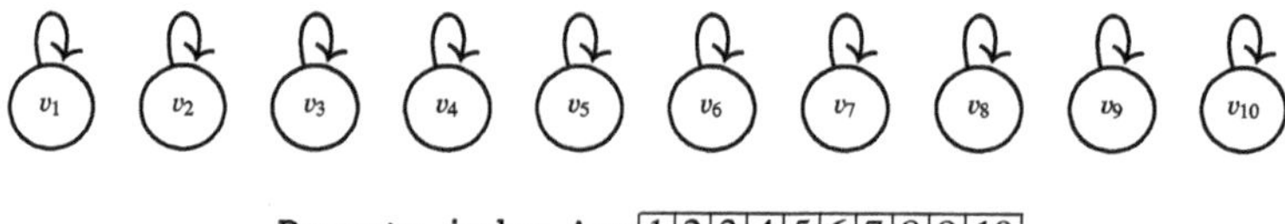

Parents index Arr	1	2	3	4	5	6	7	8	9	10
Vertex index:	1	2	3	4	5	6	7	8	9	10

Fig. 5.8 Union-Find: Initialization step

Union The procedure $\mathrm{Union}(i, j)$ assigns the same representative to both v_i and v_j. It determines the representatives of the sets to which v_i and v_j belong—using the `Find` operation—and places one of the trees (representing a set) under the root node of the other tree/set. For example, $\mathrm{Union}(1,2)$ sets `Arr[2]=1`, as shown in Fig. 5.9. A subsequent call to $\mathrm{Union}(2,7)$ is illustrated in Fig. 5.10. Performing all the remaining unions for the example in Fig. 5.7, i.e., $\mathrm{Union}(3,4)$, $\mathrm{Union}(3,8)$, $\mathrm{Union}(4,5)$, $\mathrm{Union}(4,9)$, $\mathrm{Union}(5,10)$, and $\mathrm{Union}(9,10)$, results in the structure depicted in Fig. 5.11. The procedure `Union` is detailed in Fig. 5.12, Algorithm 31.

Find The procedure $\mathrm{Find}(i)$ identifies the parent of i. The simplest implementation recursively traverses the parent array `Arr` until it reaches a node that is its own parent. This approach is described in the procedure `Find_slow`, provided in Fig. 5.12, Algorithm 33. However, in practice, a relatively simple modification is often used to improve efficiency by **compressing the height** of the tree. This is achieved by incorporating a small caching mechanism during the traversal. The optimized `Find` procedure is detailed in Fig. 5.12, Algorithm 32.

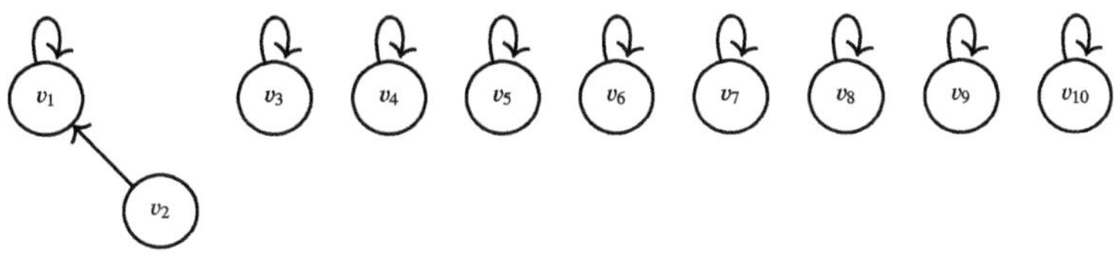

Parents index Arr: | 1 | 1 | 3 | 4 | 5 | 6 | 7 | 8 | 9 | 10 |
Vertex index: | 1 | 2 | 3 | 4 | 5 | 6 | 7 | 8 | 9 | 10 |

Fig. 5.9 Union-Find: Union(1, 2)

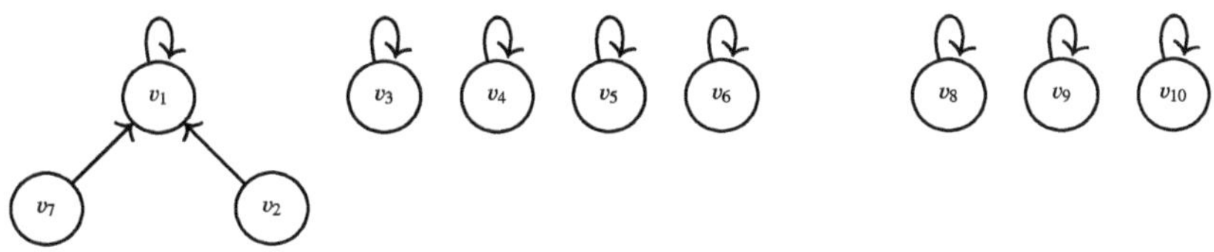

Parents index Arr: | 1 | 1 | 3 | 4 | 5 | 6 | 1 | 8 | 9 | 10 |
Vertex index: | 1 | 2 | 3 | 4 | 5 | 6 | 7 | 8 | 9 | 10 |

Fig. 5.10 Union-Find: Union(2, 7)

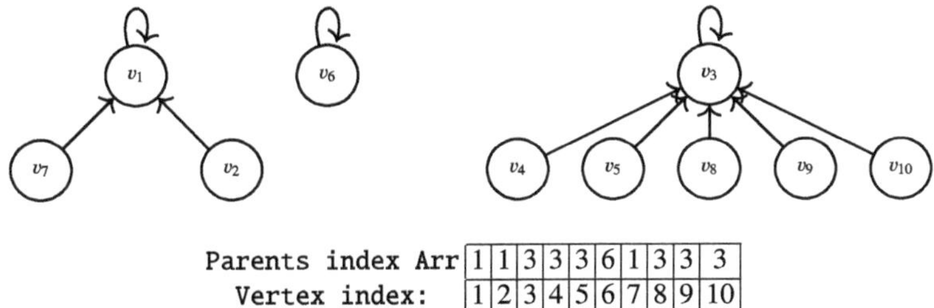

Parents index Arr: | 1 | 1 | 3 | 3 | 3 | 6 | 1 | 3 | 3 | 3 |
Vertex index: | 1 | 2 | 3 | 4 | 5 | 6 | 7 | 8 | 9 | 10 |

Fig. 5.11 Union-Find: Union(3,4), Union(3,8), Union(4,5), Union(4,9), Union(5,10), Union(9,10)

Summing up: given a graph $G = (V, E)$ (e.g., a network presented in Fig. 5.5, a large network shown in Fig. 5.6, or the small example in Fig. 5.7), we must initialize $\text{Arr}[i] = i$ for $i = 1, \ldots, |V|$, then perform $\text{Union}(i, j)$ for all pairs (i, j) such that $(v_i, v_j) \in E$. Finally, we check whether $\text{Find}(s) == \text{Find}(t)$ (indicating that there is a path from s to t) or not (indicating that no such path exists). We emphasize that the Union-Find algorithm is a rich topic with numerous enhancements and variations. For instance, beyond *path compression*, there are other optimizations, such as *"union by rank"*, *"union by size"*, or tailored approaches like *"iterative find with path compression"* and *"linking by rank with path halving or splitting"*. In certain scenarios, a pointer-based approach (instead of an array-based one) may also provide additional advantages.

Algorithm 31 Union(*i*, *j*)

```
int repr_i = Find(i)
int repr_j = Find(j)
Arr[repr_j]=repr_i
```

Algorithm 32 Find(*i*)

---

if Arr[*i*]==*i* **then**
    return *i*
else
        ▷ Recursively find the representative:
    result=Find(Arr[*i*])
        ▷ Cache the result, move *i* node:
        ▷ directly under the repr.
    Arr[*i*]=result
    return result
end if

Algorithm 33 Find_slow(*i*)

---

if Arr[*i*]==*i* **then**
    return *i*
else
        ▷ Recursively find the repr.
    return Find_slow(Arr[*i*])
end if

Fig. 5.12 Union(*i*, *j*) and Find(*i*) algorithms

The following two examples are based on the idea of stratification and yield a super-canonical rate (faster than $O(R^{-1/2})$) of convergence.

Example 5.1.11 (Asmussen and Glynn [8]) Let us consider computing the integral

$$I = \int_0^1 g(x)\, dx \tag{5.1.12}$$

and set $Y = g(U)$. The crude Monte Carlo estimator $\hat{Y}_R^{\mathrm{CMC}}$ has a convergence rate of $O(R^{-1/2})$. We now consider a specific stratified estimator. Take $X = U$ and introduce strata in the following way: $S^j = \{U \in \left(\frac{j-1}{R}, \frac{j}{R} \right]\}$, $j = 1, \ldots, R$, (note that we construct as many strata as the number of replications). In this case, we can easily simulate $Y^j \overset{\mathcal{D}}{=} (Y \mid U \in (j-1/R, j/R])$. Namely, the random variable

$$V^j = \frac{j-1}{R} + \frac{U}{R}$$

has a uniform distribution on $\left(\frac{j-1}{R}, \frac{j}{R} \right]$, and $Y^j \overset{\mathcal{D}}{=} F^{-1}(V^j)$, where $Y \sim F$. Moreover, $p_j = 1/R$. Using proportional allocation with $R_j = 1$, the resulting stratified estimator of I is:

$$\hat{Y}_R^{\text{pa}} = \frac{1}{R} \sum_{j=1}^{R} g(V^j).$$

Let us see how the stratified estimator reduces the variance in computing (5.1.12), where g is a continuously differentiable function with

$$M = \sup_{x \in [0,1]} |g'(x)| < \infty. \tag{5.1.13}$$

Note that

$$Y^j \overset{\mathcal{D}}{=} \left(g(U) \,\Big|\, \frac{j-1}{R} \leq U < \frac{j}{R} \right) \overset{\mathcal{D}}{=} g\left(\frac{j-1}{R} + \frac{U}{R} \right).$$

Plugging $m = R$ and $p_j = 1/R$ into (5.1.5), we obtain the following variance of this proportional estimator $\hat{Y}_R^{\text{pa}}$:

$$\text{Var}\hat{Y}_R^{\text{pa}} = \frac{1}{R} \sum_{j=1}^{R} \sigma_j^2,$$

where

$$\sigma_j^2 = \text{Var}\left(g\left(\frac{j-1}{R} + \frac{1}{R}U \right) \right).$$

Using the mean value theorem:

$$g\left(\frac{j-1}{R} + \frac{U}{R} \right) = g\left(\frac{j-1}{R} \right) + g'(\theta_j)\frac{U}{R},$$

where θ_j is a point in the interval $\left(\frac{j-1}{R}, \frac{j-1+U}{R} \right)$, we have:

$$\sigma_j^2 = \text{Var}\left(g\left(\frac{j-1}{R} + \frac{U}{R} \right) \right) = \frac{\text{Var}(g'(\theta_j)U)}{R^2} \leq \frac{\mathbb{E}\left(g'(\theta_j)^2 U^2 \right)}{R^2}.$$

Using (5.1.13):

$$\sigma_j^2 \leq \frac{\mathbb{E}\left(g'(\theta_j)^2 U^2 \right)}{R^2} \leq \frac{M^2}{R^2}.$$

Thus, the variance of the stratified estimator with proportional allocation is of order:

$$\operatorname{Var}\hat{Y}_R^{\mathrm{pa}} = \frac{1}{R}\sum_{j=1}^{R}\sigma_j^2 \leq \frac{1}{R}\sum_{j=1}^{R}\frac{M^2}{R^2} = \frac{M^2}{R^2} = O(R^{-2}).$$

Using the Chebyshev inequality (see Remark 4.1.1), we have:

$$\mathbb{P}(|\hat{Y}_R^{\mathrm{pa}} - I| \geq b_R) \leq \frac{\operatorname{Var}\hat{Y}_R^{\mathrm{pa}}}{b_R^2} \leq \frac{O(R^{-2})}{b_R^2}.$$

From the basic relationship $\frac{O(R^{-2})}{b_R^2} = \alpha$, we obtain:

$$b_R = O(R^{-1}),$$

which is better than the canonical Monte Carlo convergence rate $O(R^{-1/2})$. $\Diamond$

Example 5.1.12 (Asmussen and Glynn [8])   As a curiosity, we now present another estimator that is not stratified but draws inspiration from the previous Example 5.1.11. This estimator achieves a better convergence rate than the canonical rate of $R^{-1/2}$. Again, we aim to estimate $I = \int_0^1 g(x)\,dx$, and similarly as before we assume that g is differentiable with

$$M = \sup_{x\in[0,1]} g'(x) < \infty.$$

Consider the estimator:

$$\hat{Y}_R = \frac{1}{R}\sum_{j=1}^{R} g\left(\frac{j-1+U_j}{R}\right),$$

where $U_j \sim \mathcal{U}(0,1)$. This estimator is unbiased since:

$$\mathbb{E}[\hat{Y}_R] = \frac{1}{R}\sum_{j=1}^{R}\int_0^1 g\left(\frac{j-1+u}{R}\right)du$$

$$= \sum_{j=1}^{R}\int_{(j-1)/R}^{j/R} g(u)\,du = \int_0^1 g(u)\,du.$$

Now, let us examine its variance. Using the mean value theorem, we have:

$$g\left(\frac{j-1}{R} + \frac{U_j}{R}\right) = g\left(\frac{j-1}{R}\right) + g'(\theta_j)\frac{U_j}{R},$$

where $\theta_j \in \left(\frac{j-1}{R}, \frac{j}{R}\right)$. Since $g\left(\frac{j-1}{R}\right)$ is deterministic, the variance becomes:

$$\mathbb{V}\mathrm{ar}(\hat{Y}_R) = \frac{1}{R^2} \sum_{j=1}^{R} \mathbb{V}\mathrm{ar}\left[g\left(\frac{j-1+U_j}{R}\right)\right]$$

$$= \frac{1}{R^2} \sum_{j=1}^{R} \mathbb{V}\mathrm{ar}\left(\frac{g'(\theta_j)U_j}{R}\right)$$

$$= \frac{1}{R^4} \sum_{j=1}^{R} \mathbb{V}\mathrm{ar}\left(g'(\theta_j)U_j\right).$$

Applying the variance bound and properties of expectations:

$$\mathbb{V}\mathrm{ar}(\hat{Y}_R) \leq \frac{1}{R^4} \sum_{j=1}^{R} \mathbb{E}\left[\left(g'(\theta_j)U_j\right)^2\right]$$

$$\leq \frac{1}{R^4} \sum_{j=1}^{R} \left(g'(\theta_j)\right)^2 \mathbb{E}[U_j^2]$$

$$= \frac{1}{R^4} \sum_{j=1}^{R} \left(g'(\theta_j)\right)^2 \cdot \frac{1}{3}$$

$$= \frac{1}{3R^3} \sum_{j=1}^{R} \left(g'(\theta_j)\right)^2 \leq \frac{M^2}{3R^3}.$$

By Chebyshev's inequality and standard relationships, we obtain:

$$b_R = O(1/R^{3/2}),$$

demonstrating the improved convergence rate compared to the canonical $O(1/R^{1/2})$. $\qquad\qquad\Diamond$

5.1.2   Antithetic Variates

We will consider simulation methods for $I = \mathbb{E}Y$, where replications are not independent. We say that random variables Y_1^* and Y_2^* are **negatively correlated** if

$$\mathrm{Corr}\left(Y_1^*, Y_2^*\right) = \frac{\mathrm{Cov}\left(Y_1^*, Y_2^*\right)}{\sqrt{\mathbb{V}\mathrm{ar}Y_1^* \mathbb{V}\mathrm{ar}Y_2^*}} < 0.$$

As usual, we aim to estimate $I = \mathbb{E}Y$. Consider R replications $Y_1^*, Y_2^*, \ldots, Y_R^*$ of a random variable Y, assuming that R is even and:

- Pairs (Y_{2i-1}^*, Y_{2i}^*), $i = 1, \ldots, R/2$, are i.i.d.
- Random variables $(Y_j^*)_{j=1}^R$ have the same marginal distribution as Y.

Thus, (Y_1^*, Y_2^*) is a vector of dependent random variables such that $Y \overset{\mathcal{D}}{=} Y_1^* \overset{\mathcal{D}}{=} Y_2^*$, and $Y_1^*, Y_2^*, \ldots$ are replications of Y (which are not independent). Recall that the crude Monte Carlo (CMC) estimator,

$$\hat{Y}_R^{\text{CMC}} = \frac{1}{R} \sum_{j=1}^R Y_j,$$

(where $Y_1, \ldots, Y_R$ are independent replications of Y) has variance $\mathbb{V}\text{ar}Y/R$. The estimator

$$\hat{Y}_R^* = \frac{1}{R} \sum_{j=1}^R Y_j^* = \frac{1}{R} \sum_{j=1}^{R/2} (Y_{2j-1}^* + Y_{2j}^*) \tag{5.1.14}$$

is also an estimator of I, with variance:

$$\mathbb{V}\text{ar}(\hat{Y}_R^*) = \frac{\frac{R}{2}\mathbb{V}\text{ar}(Y_1^* + Y_2^*)}{R^2}$$

$$= \frac{1}{2R}\left(2\mathbb{V}\text{ar}(Y) + 2\text{Cov}(Y_1^*, Y_2^*)\right)$$

$$= \frac{1}{R}\mathbb{V}\text{ar}(Y)(1 + \text{Corr}(Y_1^*, Y_2^*)) = \mathbb{V}\text{ar}(\hat{Y}_R^{\text{CMC}})(1 + \text{Corr}(Y_1^*, Y_2^*)).$$

$$\tag{5.1.15}$$

Thus, the variance is reduced if the correlation $\text{Corr}(Y_1^*, Y_2^*)$ is negative. We have:

$$\sqrt{R}(\hat{Y}_R^* - I) \overset{\mathcal{D}}{\to} \varsigma Z, \qquad \varsigma = \sqrt{\frac{\mathbb{V}\text{ar}(Y_1^* + Y_2^*)}{2}},$$

where $Z \sim \mathcal{N}(0, 1)$. This implies that the confidence interval is $\hat{Y}_R^* \pm z_{1-\alpha/2}\varsigma$, with:

$$\varsigma = \sigma_Y(1 + \text{Corr}(Y_1^*, Y_2^*))^{1/2}.$$

We say that two functions $k_1(x_1, \ldots, x_n)$ and $k_2(x_1, \ldots, x_n)$ are **simultaneously monotone** if, for every $i = 1, \ldots, n$, the one-variable functions $f_i^{(1)}(x_i) = k_1(x_1, \ldots, x_i, \ldots, x_n)$ and $f_i^{(2)}(x_i) = k_2(x_1, \ldots, x_i, \ldots, x_n)$ are either both non-decreasing or both non-increasing.

We have the following result.

Theorem 5.1.13 *If $X_1, \ldots, X_n$ are i.i.d. random variables and the functions $k_1(x_1, \ldots, x_n)$ and $k_2(x_1, \ldots, x_n)$ are simultaneously monotone, then:*

$$\mathbb{E}[k_1(X_1, \ldots, X_n)k_2(X_1, \ldots, X_n)] \geq \mathbb{E}[k_1(X_1, \ldots, X_n)]\mathbb{E}[k_2(X_1, \ldots, X_n)],$$

provided the expectations are finite.

Proof The proof proceeds by induction on the number of variables $n = 1, 2, \ldots$.

For $n = 1$, suppose k_1 and k_2 are either both non-decreasing or both non-increasing. Then:

$$(k_1(x) - k_1(y))(k_2(x) - k_2(y)) \geq 0,$$

which implies:

$$\mathbb{E}[(k_1(X) - k_1(Y))(k_2(X) - k_2(Y))] \geq 0.$$

Expanding this yields:

$$\mathbb{E}[k_1(X)k_2(X)] - \mathbb{E}[k_1(X)k_2(Y)] - \mathbb{E}[k_1(Y)k_2(X)] + \mathbb{E}[k_1(Y)k_2(Y)] \geq 0,$$

which for independent X and Y is

$$\mathbb{E}[k_1(X)k_2(X)] - \mathbb{E}[k_1(X)]\mathbb{E}[k_2(Y)] - \mathbb{E}[k_1(Y)]\mathbb{E}[k_2(X)] + \mathbb{E}[k_1(Y)k_2(Y)] \geq 0.$$

Moreover, if X and Y are identically distributed as X_1, we conclude:
If X and Y are independent and identically distributed as X_1, we conclude:

$$\mathbb{E}[k_1(X_1)k_2(X_1)] \geq \mathbb{E}[k_1(X_1)]\mathbb{E}[k_2(X_1)].$$

For $n > 1$, assume the theorem holds for $n - 1$. Using this assumption for all fixed x_n, we have:

$$\mathbb{E}[k_1(X_1, \ldots, X_{n-1}, x_n)k_2(X_1, \ldots, X_{n-1}, x_n)]$$

$$\geq \mathbb{E}[k_1(X_1, \ldots, X_{n-1}, x_n)]\mathbb{E}[k_2(X_1, \ldots, X_{n-1}, x_n)],$$

Integrating over x_n, the result follows. $\square$

Corollary 5.1.14 *If a function $k(x_1, \ldots, x_n)$, $0 \leq x_1, \ldots, x_n \leq 1$, is monotone (increasing in some variables and decreasing in others), then for i.i.d. random variables $U_1, \ldots, U_n$ uniformly distributed as $\mathcal{U}(0, 1)$, we have:*

$$\mathrm{Cov}\left(k(U_1, \ldots, U_n), k(1 - U_1, \ldots, 1 - U_n)\right) \leq 0.$$

Proof of Corollary 5.1.14 Let $k(x_1, \ldots, x_n)$ be monotone, increasing in some variables and decreasing in others. Define:

$$k_1(U_1, \ldots, U_k) = k(U_1, \ldots, U_n), \quad k_2(U_1, \ldots, U_k) = -k(1 - U_1, \ldots, 1 - U_n).$$

Note that $U_1, \ldots, U_k \sim \mathcal{U}(0, 1)$ are i.i.d. random variables, and k_1 and k_2 are simultaneously monotone. This is because flipping $U_i \to 1 - U_i$ reverses the monotonicity of k, and the negative sign in k_2 restores the simultaneous monotonicity condition. Thus, the assumptions of Theorem 5.1.13 are satisfied. Applying the theorem, we have:

$$\mathbb{E}[k_1(U_1, \ldots, U_k)k_2(U_1, \ldots, U_n)] \geq \mathbb{E}[k_1(U_1, \ldots, U_n)]\mathbb{E}[k_2(U_1, \ldots, U_n)].$$

Substituting k_1 and k_2 we get:

$$\mathbb{E}[k(U_1, \ldots, U_n)(-k(1 - U_1, \ldots, 1 - U_n))]$$

$$\geq \mathbb{E}[k(U_1, \ldots, U_n)]\mathbb{E}[-k(1 - U_1, \ldots, 1 - U_n)],$$

i.e.,

$$\mathbb{E}[k(U_1, \ldots, U_n)k(1 - U_1, \ldots, 1 - U_n)]$$

$$\leq \mathbb{E}[k(U_1, \ldots, U_n)]\mathbb{E}[k(1 - U_1, \ldots, 1 - U_n)].$$

This inequality implies:

$$\mathrm{Cov}\,(k(U_1, \ldots, U_n), k(1 - U_1, \ldots, 1 - U_n))$$

$$= \mathbb{E}[k(U_1, \ldots, U_n)k(1 - U_1, \ldots, 1 - U_n)]$$

$$- \mathbb{E}[k(U_1, \ldots, U_n)]\mathbb{E}[k(1 - U_1, \ldots, 1 - U_n)] \leq 0.$$

Thus, the random variables $k(U_1, \ldots, U_n)$ and $k(1 - U_1, \ldots, 1 - U_n)$ are negatively correlated. $\square$

In other words, the random variables $k(U_1, \ldots, U_k)$ and $k(1 - U_1, \ldots, 1 - U_k)$ are negatively correlated. Moreover, note that U and $1 - U$ have the same uniform distribution $\mathcal{U}[0, 1)$.

Assume that for some function $k : [0, 1]^k \to \mathbb{R}$, we have $Y_1^{\mathrm{anti}} \stackrel{\mathcal{D}}{=} k(U_1, \ldots, U_n)$ and that k fulfills the assumptions of Corollary 5.1.14. Then $Y_2^{\mathrm{anti}} = k(1 - U_1, \ldots, 1 - U_n)$, and the random variables Y_1^{anti}, Y_2^{anti} are negatively correlated. We further define $Y_3^{\mathrm{anti}} = k(U_{n+1}, \ldots, U_{2n})$, $Y_4^{\mathrm{anti}} = k(1 - U_{n+1}, \ldots, 1 - U_{2n})$, etc. Note that the sequence $Y_1^{\text{anti}}, \ldots, Y_R^{\text{anti}}$, defined this way, satisfies our postulates. Then:

$$\hat{Y}_R^{\text{anti}} = \frac{1}{R} \sum_{j=1}^{R} Y_j^{\text{anti}} = \frac{1}{R} \sum_{j=1}^{R/2} (Y_{2j-1}^{\text{anti}} + Y_{2j}^{\text{anti}})$$

is called the *antithetic estimator*.

Remark 5.1.15 (Antithetic Variates: ITM Method) In some cases, Y is sampled from a c.d.f. F directly via the ITM method, i.e., $Y = F^{\leftarrow}(U)$. Note that $F^{\leftarrow}(\cdot)$ is monotone, thus the assumptions of Corollary 5.1.14 are fulfilled. This means that we may take the antithetic estimator (using i.i.d. $U_1, \ldots, U_{R/2}$):

$$\hat{Y}_R^{\text{anti}} = \frac{\sum_{j=1}^{R} Y_j^{\text{anti}}}{R}, \quad Y_{2i-1}^{\text{anti}} = F^{\leftarrow}(U_i), \; Y_{2i}^{\text{anti}} = F^{\leftarrow}(1 - U_i),$$

and we surely have that:

$$\mathbb{V}\text{ar}(\hat{Y}_R^{\text{anti}}) \leq \mathbb{V}\text{ar}(\hat{Y}_R^{\text{CMC}}).$$

Suppose now we aim at estimating:

$$I = \mathbb{E}[g(Y)].$$

If Y is sampled via the ITM method, i.e., $Y = F^{\leftarrow}(U)$, the standard CMC estimator is:

$$\hat{Y}_R = \frac{1}{R} \sum_{i=1}^{R} Y_i, \quad Y_i = g(F^{\leftarrow}(U)).$$

If g is non-decreasing, so is $k = g \circ F^{\leftarrow}$. Thus, the antithetic estimator:

$$\hat{Y}_R^{\text{anti}} = \frac{\sum_{j=1}^{R} Y_j^{\text{anti}}}{R}, \quad Y_{2i-1}^{\text{anti}} = g\left(F^{\leftarrow}(U_i)\right), \; Y_{2i}^{\text{anti}} = g\left(F^{\leftarrow}(1 - U_i)\right)$$

again has surely no larger variance. ∎

Example 5.1.16 We continue the task of estimating π. Recall that in Example 4.1.6, we presented a crude Monte Carlo (CMC) estimator with variance $\frac{2.6968}{R}$. In Example 4.1.7, we showed another estimator with variance $\frac{0.797}{R}$. Additionally, Example 4.1.7 included a stratified estimator, although its exact variance was not computed and was instead estimated via simulations. Here, we introduce the antithetic estimator.

As in Example 4.1.7, consider the CMC estimator:

$$\hat{Y}_R = \frac{1}{R} \sum_{j=1}^{R} Y_j, \quad \text{where } Y_j = k(U_j) = 4\sqrt{1 - U_j^2},$$

and U_j, $j = 1, \ldots, R$, are i.i.d. $\mathcal{U}[0, 1)$ random variables. Assume R is even. The antithetic estimator is defined as:

$$\hat{Y}_R^{\text{anti}} = \frac{1}{R} \sum_{i=1}^{R} Y_i^{\text{anti}}, \quad Y_{2i-1}^{\text{anti}} = k(U_i), \quad Y_{2i}^{\text{anti}} = k(1 - U_i).$$

From (5.1.15), the variance of the antithetic estimator is:

$$\mathbb{V}\text{ar}(\hat{Y}_R^{\text{anti}}) = \mathbb{V}\text{ar}(\hat{Y}_R^{\text{CMC}})(1 + \text{Corr}\,(k(U), k(1 - U))).$$

Now, compute the covariance $\text{Cov}\,(k(U), k(1 - U))$:

$$\text{Cov}\,(k(U), k(1 - U)) = \mathbb{E}[k(U)k(1 - U)] - (\mathbb{E}[k(U)])^2$$

$$= \int_0^1 16\sqrt{(1 - x^2)(1 - (1 - x)^2)}\, dx - \pi^2$$

$$= -0.5776.$$

Note that, as stated in Remark 5.1.15, the covariance is guaranteed to be non-positive. The correlation $\text{Corr}\,(k(U), k(1 - U))$ is:

$$\text{Corr}\,(k(U), k(1 - U)) = \frac{\text{Cov}\,(k(U), k(1 - U))}{\mathbb{V}\text{ar}k(U)} = \frac{-0.5776}{0.797} = -0.7246.$$

Finally, the variance of the antithetic estimator is:

$$\mathbb{V}\text{ar}(\hat{Y}_R^{\text{anti}}) = \frac{0.797}{R}(1 - 0.7246) = \frac{0.219}{R},$$

i.e., ~ 3.03 times smaller than the variance of the CMC estimator. $\Diamond$

Remark 5.1.17 Sometimes we aim to compute $I = \mathbb{E}g(Z)$, where Z is a standard normal random variable $\mathcal{N}(0, 1)$, and g is a monotone function. Suppose we simulate Z using the inverse transform method (ITM), where $Z = \Phi^{-1}(U)$ for $U \sim \mathcal{U}[0, 1)$. Note that $\Phi(1 - U) = -Z$, meaning $Z \overset{\mathcal{D}}{=} -Z$. Using Theorem 5.1.13 with $k_1(x) = g(x)$ and $k_2(x) = -g(-x)$, we obtain:

$$\mathbb{E}[g(Z)g(-Z)] \leq (\mathbb{E}[g(Z)])^2.$$

Hence, the correlation between $g(Z)$ and $g(-Z)$ satisfies:

$$\mathrm{Corr}\,(g(Z), g(-Z)) = \frac{\mathbb{E}[g(Z)g(-Z)] - (\mathbb{E}[g(Z)])^2}{\mathbb{V}\mathrm{ar}[g(Z)]} \le 0.$$

Now, consider a sample where $Z_1, Z_2, \ldots, Z_{R/2}$ are i.i.d. standard normal random variables, and construct the sequence:

$$Y_{2i-1}^{\mathrm{anti}} = g(Z_i), \quad Y_{2i}^{\mathrm{anti}} = g(-Z_i), \quad i = 1, \ldots, R/2.$$

Here, $Y_1^{\mathrm{anti}}, \ldots, Y_R^{\mathrm{anti}}$ are dependent replications, constructed such that the correlation between $g(Z)$ and $g(-Z)$ reduces the variance. Importantly, this approach avoids the computational cost of evaluating the inverse cumulative distribution function $\Phi^{-1}(\cdot)$, as we can directly generate $Z \sim \mathcal{N}(0, 1)$ using standard methods. The corresponding antithetic estimator is:

$$\hat{Y}_R^{\mathrm{anti}} = \frac{1}{R} \sum_{j=1}^{R} Y_j^{\mathrm{anti}}.$$

This estimator reduces the variance since its variance is given by:

$$\mathbb{V}\mathrm{ar}[\hat{Y}_R^{\mathrm{anti}}] = \frac{\mathbb{V}\mathrm{ar}[g(Z)](1 + \mathrm{Corr}\,(g(Z), g(-Z)))}{R}.$$

Because $\mathrm{Corr}\,(g(Z), g(-Z)) \le 0$, the antithetic estimator achieves variance reduction compared to the standard CMC estimator.   ∎

Remark 5.1.18 Antithetic estimators have some limitations in variance reduction. Consider a pair of non-negative random variables (Y_1, Y_2) with the same marginal c.d.f. F. Define a pair of antithetic random variables:

$$Y_1^{\mathrm{anti}} = F^{\leftarrow}(U), \quad Y_2^{\mathrm{anti}} = F^{\leftarrow}(1 - U),$$

where $U \sim \mathcal{U}[0, 1)$. Notice that Y_i and Y_i^{anti} share the same marginal c.d.f. F. Recall that:

$$\mathbb{P}(Y_1 > x, Y_2 > y) = 1 + \mathbb{P}(Y_1 \le x, Y_2 \le y) - F(x) - F(y).$$

From the Fréchet-Hoeffding lower bound (see Theorem 3.9.6), we have:

$$\mathbb{P}(Y_1 > y_1, Y_2 > y_2) = 1 + \mathbb{P}(Y_1 \le x, Y_2 \le y) - F(x) - F(y)$$

$$\ge 1 + \mathbb{P}(Y_1^{\mathrm{anti}} \le x, Y_2^{\mathrm{anti}} \le y) - F(x) - F(y) = \mathbb{P}(Y_1^{\mathrm{anti}} > x, Y_2^{\mathrm{anti}} > y).$$

Hence:

$$
\mathbb{E}(Y_1 Y_2) = \int_0^\infty \int_0^\infty \mathbb{P}(Y_1 > y_1, Y_2 > y_2)\, dy_1\, dy_2
$$

$$
\geq \int_0^\infty \int_0^\infty \mathbb{P}(Y_1^{\mathrm{anti}} > y_1, Y_2^{\mathrm{anti}} > y_2)\, dy_1\, dy_2
$$

$$
= \mathbb{E}\left(F^{\leftarrow}(U) F^{\leftarrow}(1 - U)\right). \tag{5.1.16}
$$

In terms of correlations:

$$
\mathrm{Corr}\,(Y_1, Y_2) = \frac{\mathbb{E}(Y_1 Y_2) - (\mathbb{E}Y)^2}{\mathbb{V}\mathrm{ar}\,Y} \geq \mathrm{Corr}\,(Y_1^{\mathrm{anti}}, Y_2^{\mathrm{anti}}),
$$

and so the use of an antithetic estimator for $I = \mathbb{E}Y$ reduces variance by a factor of $1 + \mathrm{Corr}\,(F^{\leftarrow}(U), F^{\leftarrow}(1-U))$. For example, in the case of the standard exponential distribution, this factor is 0.335; see Exercise 5.T.4. Thus, among all estimators $\hat{Y}_R^*$ of the form (5.1.14) with dependent pairs Y_i^*, Y_{i+1}^*, the optimal one is for antithetic $Y_i^* = Y_i^{\mathrm{anti}}, Y_{i+1} = Y_{i+1}^{\mathrm{anti}}$ pairs. $\blacksquare$

One may question the optimality of the antithetic method and its limitations for estimating $I = \mathbb{E}k(X)$ within a specific class of replications:

$$
X_1^*, X_2^*, \ldots, X_R^*, \quad \text{where } X_1^* \overset{\mathcal{D}}{=} X,\ X \sim F,
$$

and pairs (X_{2i-1}^*, X_{2i}^*) are independent, as assumed earlier in this section. The estimator in this setting is:

$$
\hat{Y}_R^* = \frac{1}{R} \sum_{j=1}^{R} Y_j^*, \quad Y_j^* = k(X_j^*).
$$

In the class of replications $X_1^*, X_2^*, \ldots, X_R^*$, there exists an antithetic construction where:

$$
X_{2i-1}^{\mathrm{anti}} = F^{\leftarrow}(U_{2i-1}), \quad X_{2i}^{\mathrm{anti}} = F^{\leftarrow}(1 - U_{2i}),
$$

with $U_i \sim \mathcal{U}[0, 1)$, i.i.d. Thus, the antithetic estimator reduces variance for $I = \mathbb{E}k(X)$ by a factor:

$$
1 + \mathrm{Corr}\,(k(F^{\leftarrow}(U)), k(F^{\leftarrow}(1 - U))),
$$

whenever k is monotone non-decreasing (since $g(f(x))$ is non-increasing if $g(x)$ is non-increasing and $f(x)$ is non-decreasing). For example, if $k(x) = x$, then for the standard exponential distribution, this factor is 0.335; see Exercise 5.T.4.

Example 5.1.19 Consider the following scheduling problem. Assume we have N tasks to process, with task processing times being i.i.d. random variables having a c.d.f. F. There are c available servers. We analyze the following scheduling rules:

- Longest Remaining Processing Time (LRPT),
- Shortest Remaining Processing Time (SRPT),
- First Come First Served (FCFS),
- Alternating (ALT): The i-th task is assigned to processor $((i-1) \bmod c) + 1$.

In the case of LRPT and SRPT, decisions are made upon the completion of tasks, and tasks are started immediately after the previous one is completed. Preemptions are allowed at task completion times, meaning tasks may be exchanged. If C_j denotes the completion time of the j-th task, then $C = C_N$ is called the *makespan*.

Our goal is to compute the mean makespans $\mathbb{E}C^{\mathrm{SRPT}}$, $\mathbb{E}C^{\mathrm{LRPTF}}$, $\mathbb{E}C^{\mathrm{FCFS}}$, and $\mathbb{E}C^{\mathrm{ALT}}$, where C is the makespan under the respective scheduling rule.

Illustration Let us illustrate the rules with a simple example. Suppose $N = 5$, $c = 2$, and the task processing times are $3, 1, 2, 4, 5$.

- **SRPT:** Start with processing times 1 and 2. After one time step, the remaining processing times are $3, 0, 1, 4, 5$. Thus, we now have $N = 4$ and $c = 2$. Continuing, we find $C^{\mathrm{SRPT}} = 9$.
- **LRPT:** Start with processing times 4 and 5. After 4 time units, the remaining processing times are $3, 1, 2, 0, 1$. Thus, we now have $N = 4$ and $c = 2$. Continuing, we find $C^{\mathrm{LRPT}} = 8$.

Analysis It is an interesting question which scheduling rule is optimal. However, we focus here on estimating the mean makespan $I = \mathbb{E}C$.

Assume that processing times are i.i.d. $\mathrm{Exp}(1)$ random variables. The ALT scheduling rule is easy to analyze. Assuming N is even, we have:

$$\mathbb{E}C^{\mathrm{ALT}} = \mathbb{E}\max(X_1, X_2),$$

where:

$$X_1 = \sum_{i=1}^{N/2} \xi_{1i}, \quad X_2 = \sum_{i=1}^{N/2} \xi_{2i},$$

and ξ_{ij} are i.i.d. $\mathrm{Exp}(1)$. Unfortunately, we lack theoretical formulas for LRPT, SRPT, and even FCFS.

We perform simulations for $N = 10$ tasks and $c = 2$ processors. In each case, the makespan C is represented as $C = k(U_1, \ldots, U_{10})$, where k depends on the scheduling rule. While we do not have explicit formulas for k under LRPT, SRPT, and FCFS, we heuristically assume that k is non-decreasing.

Simulation Results In Table 5.3, we present simulation results for $\mathbb{E}C$ using crude Monte Carlo (CMC) and antithetic methods, both with $R = 1000$ replications.

Table 5.3 Estimating $\mathbb{E}C$ via CMC and antithetic methods; $N = 10$, number of replications $R = 1000$

	Discipline	$\hat{I}$	$\hat{S}$	$\hat{b}$
CMC	FCFS	5.4982	1.7938	0.1112
	SRPT	5.9252	1.9460	0.1206
	LRPT	5.0678	1.5833	0.0981
Ant	FCFS	5.4528	0.8251	0.0511
	SRPT	5.9347	1.0872	0.0674
	LRPT	5.0941	0.7232	0.0448

Table 5.4 Estimating C via MCM and antithetic methods; $N = 10$, number of replications $R = 10^5$

	Discipline	$\hat{I}$	$\hat{S}$	$\hat{b}$
CMC	FCFS	5.4939	1.7971	0.0111
	SRPTF	5.9014	1.9166	0.0119
	LRPTF	5.0571	1.5956	0.0099
Ant	FCFS	5.4955	0.8745	0.0054
	SRPTF	5.9023	0.9818	0.0061
	LRPTF	5.0535	0.6829	0.0042

Implementation of this example is available in `ch5_scheduling_means.py`. Here:

- $\hat{Y}_R$ denotes the simulated $\mathbb{E}C$,
- $\hat{S}$ is the standard deviation $\sqrt{\mathbb{V}\text{ar}(C)}$,
- $\hat{b}$ is half the confidence interval width at confidence level $\alpha = 0.05$ (i.e., the error).

For comparison, Table 5.4 presents simulation results with $R = 10^5$ replications.

$\Diamond$

5.1.3 Common Random Numbers

Assume we want to estimate a difference $I = I_1 - I_2 = \mathbb{E}k_1(Y) - \mathbb{E}k_2(Y)$, where Y is a random variable. This approach can also be extended to random vectors $\mathbf{Y}$. We can proceed in (at least) two ways:

- **Method 1: Crude Monte Carlo (CMC).** Using $2R$ independent replications of Y—denoted $Y_1^{(1)}, \ldots, Y_R^{(1)}, Y_1^{(2)}, \ldots, Y_R^{(2)}$—we separately estimate I_1 and I_2 and then compute their difference. The CMC estimator is:

$$\hat{Y}_R^{\text{CMC}} = \frac{1}{R} \sum_{i=1}^{R} k_1(Y_i^{(1)}) - \frac{1}{R} \sum_{i=1}^{R} k_2(Y_i^{(2)}).$$

The variance of this estimator is:

$$\mathrm{Var}(\hat{Y}_R^{\mathrm{CMC}}) = \frac{1}{R}\left(\mathrm{Var}k_1(Y) + \mathrm{Var}k_2(Y)\right).$$

This method is not always optimal, especially when the difference $I_1 - I_2$ is very small.

- **Method 2: Common Random Numbers (CRN).** Writing $I = \mathbb{E}(k_1(Y) - k_2(Y))$, we can estimate this directly using R independent replications $Y_1, \ldots, Y_R$ of Y. The CRN estimator is:

$$\hat{Y}_R^{\mathrm{CRN}} = \frac{1}{R}\sum_{i=1}^{R}(k_1(Y_i) - k_2(Y_i)).$$

The variance of this estimator is:

$$\mathrm{Var}(\hat{Y}_R^{\mathrm{CRN}}) = \frac{1}{R}\left(\mathrm{Var}k_1(Y) + \mathrm{Var}k_2(Y) - 2\mathrm{Cov}\left(k_1(Y), k_2(Y)\right)\right).$$

Thus, if $\mathrm{Cov}\left(k_1(Y), k_2(Y)\right) > 0$, the variance is reduced. This often occurs when k_1 and k_2 are identically monotone (i.e., both increasing or both decreasing).

Example 5.1.20 We continue with a series of examples estimating π. We may decompose:

$$\pi = 2\pi - \pi = \int_0^1 \frac{8}{1+x^2}\,dx - \int_0^1 4\sqrt{1-x^2}\,dx.$$

This can be reformulated as estimating $I = \mathbb{E}k_1(U) - \mathbb{E}k_2(U)$, where $U \sim \mathcal{U}[0,1)$, $k_1(x) = \frac{8}{1+x^2}$, and $k_2(x) = 4\sqrt{1-x^2}$.

- **CMC Estimator.** Take $2R$ independent replications

$$U_1^{(1)}, \ldots, U_R^{(1)}, U_1^{(2)}, \ldots, U_R^{(2)}$$

of $U \sim \mathcal{U}[0,1)$ and set:

$$\hat{Y}_R^{\mathrm{CMC}} = \frac{1}{R}\sum_{i=1}^{R}\frac{8}{1+(U_i^{(1)})^2} - \frac{1}{R}\sum_{i=1}^{R}4\sqrt{1-(U_i^{(2)})^2}.$$

Using `Maple`, we compute:

$$\mathrm{Var}(k_1(U)) = \int_0^1 \frac{64}{(1+x^2)^2}\,dx - (2\pi)^2 = 1.6543,$$

and $\mathbb{V}\mathrm{ar}(k_2(U)) = 0.797$ (as computed in Example 4.1.7). Hence:

$$\mathbb{V}\mathrm{ar}(\hat{Y}_R^{\text{CMC}}) = \frac{1}{R}\left(\mathbb{V}\mathrm{ar}k_1(U) + \mathbb{V}\mathrm{ar}k_2(U)\right) = \frac{2.4513}{R}.$$

- **CRN Estimator.** Take R independent replications $U_1, \ldots, U_R$ of $U \sim \mathcal{U}[0, 1)$ and set:

$$\hat{Y}_R^{\text{CRN}} = \frac{1}{R}\sum_{i=1}^{R}\left[\frac{8}{1 + U_i^2} - 4\sqrt{1 - U_i^2}\right].$$

Computing the covariance:

$$\mathrm{Cov}\,(k_1(U), k_2(U)) = \mathbb{E}(k_1(U)k_2(U)) - \mathbb{E}(k_1(U))\mathbb{E}(k_2(U))$$

$$= 16\pi(\sqrt{2} - 1) - 2\pi^2 = 1.0814.$$

Hence:

$$\mathbb{V}\mathrm{ar}(\hat{Y}_R^{\text{CRN}}) = \tfrac{1}{R}\left(\mathbb{V}\mathrm{ar}(k_1(U)) + \mathbb{V}\mathrm{ar}(k_2(U)) - 2\mathrm{Cov}\,(k_1(U)k_2(U))\right)$$

$$= \frac{2.4513 - 2 \cdot 1.0814}{R} = \frac{0.2885}{R}.$$

Thus, using the CRN method, the variance is reduced $2.4513/0.2885 = 8.4967$ times. $\diamond$

Example 5.1.21 We now continue Example 5.1.19. This time, we ask: what is the difference $C^{\text{SRPT}} - C^{\text{LRPT}}$? Specifically, we aim to estimate $I = \mathbb{E}(C^{\text{SRPT}} - C^{\text{LRPT}})$. While it is possible to subtract the estimates from Example 5.1.19, the resulting variance would be the sum of the corresponding variances, which would lead to a larger error for the same number of replications. Instead, we use the CRN method for this estimation.

Table 5.5 presents the simulation results for $I = \mathbb{E}(C^{\text{SRPT}}) - \mathbb{E}(C^{\text{LRPT}})$ using both CRN and CMC methods for $N = 10$ tasks and $R \in \{1000, 10^5\}$ replications. As before, $\hat{S}$ represents the standard deviation $\sqrt{\mathbb{V}\mathrm{ar}(C^{\text{SRPT}} - C^{\text{LRPT}})}$,

Table 5.5 Estimating I using CRN and CMC methods; $N = 10$, number of replications $R \in \{1000, 10^5\}$. $\hat{I}$ denotes the respective estimator, $\hat{Y}_R^{\text{CMC}}$ or $\hat{Y}_R^{\text{CRN}}$

	$R = 1000$			$R = 10^5$		
	$\hat{I}$	$\hat{S}$	$\hat{b}$	$\hat{I}$	$\hat{S}$	$\hat{b}$
CRN	0.8739	0.5039	0.0312	0.8668	0.5136	0.0032
CMC	0.8101	2.5060	0.1553	0.8649	2.4906	0.0154

and $\hat{b}$ denotes half of the confidence interval at a significance level of $\alpha = 0.05$. Implementation is available in `ch5_scheduling_diff.py`.

From Table 5.5, we observe that the variance is significantly reduced when using the CRN method compared to the CMC method; for example, for $R = 1000$, the variance is reduced by a factor of approximately $2.5793^2/0.5431^2 = 22.5$.   ◊

5.1.4 *Ratios of Expectations*

Following Example 4.2.5, consider the problem of estimating

$$I = \frac{I^{(1)}}{I^{(2)}},$$

where independent replications of the random vector $(Y^{(1)}, Y^{(2)})$ are available:

$$(Y_1^{(1)}, Y_1^{(2)}), \ldots, (Y_R^{(1)}, Y_R^{(2)}),$$

and

$$I^{(1)} = \mathbb{E}Y^{(1)}, \quad I^{(2)} = \mathbb{E}Y^{(2)}. \tag{5.1.17}$$

Till now, we assumed no specific dependence structure for $(Y^{(1)}, Y^{(2)})$ beyond the fixed first moments defined in (5.1.17). Define the estimator:

$$\hat{Y}_R^{\mathrm{RoE}} = \frac{\sum_{i=1}^R Y_i^{(1)}}{\sum_{i=1}^R Y_i^{(2)}}.$$

The key question is how to choose the joint distribution of $(Y^{(1)}, Y^{(2)})$ to minimize the asymptotic variance of this estimator.

In Example 4.2.5 (*Estimating ratios of expectations*), we derived a formula for the asymptotic variance of this estimator (see formula (2.12)). Using those results, we can write:

$$\varsigma^{\mathrm{RoE}} = \frac{I^{(1)}}{I^{(2)}} \left[\frac{\Sigma_{\mathbf{Y}}(1,1)}{(I^{(1)})^2} - \frac{\Sigma_{\mathbf{Y}}(1,2)}{I^{(1)}I^{(2)}} + \frac{\Sigma_{\mathbf{Y}}(2,2)}{(I^{(2)})^2} \right]^{1/2},$$

$$\varsigma^{\mathrm{CMC}} = \frac{I^{(1)}}{I^{(2)}} \left[\frac{\mathbb{V}\mathrm{ar}Y^{(1)}}{(I^{(1)})^2} + \frac{\mathbb{V}\mathrm{ar}Y^{(2)}}{(I^{(2)})^2} \right]^{1/2}.$$

The crude Monte Carlo estimator corresponds to the case where $\boldsymbol{\Sigma}_{\mathbf{Y}}(1, 2) = 0$. Define:

$$\hat{Y}_R^{\text{CMC}} = \frac{\sum_{i=1}^{R} X_i^{(1)}}{\sum_{i=1}^{R} X_i^{(2)}},$$

where $X_1^{(1)}, \ldots, X_R^{(1)}$ are independent replications of $Y^{(1)}$, and $X_1^{(2)}, \ldots, X_R^{(2)}$ are independent replications of $Y^{(2)}$. Additionally, $X_i^{(1)}$ and $X_i^{(2)}$ are assumed to be independent for each i, with $I^{(1)} = \mathbb{E}Y^{(1)}$ and $I^{(2)} = \mathbb{E}Y^{(2)}$. The variance of $\hat{Y}_R^{\text{RoE}}$ is smaller than or equal to that of $\hat{Y}_R^{\text{CMC}}$ for large R if and only if:

$$\frac{I^{(1)}}{I^{(2)}} \left(\boldsymbol{\Sigma}_{\mathbf{Y}}(2, 2) - \boldsymbol{\Sigma}_{\mathbf{X}}(2, 2)\right) + \frac{1}{I^{(1)}/I^{(2)}} \left(\boldsymbol{\Sigma}_{\mathbf{Y}}(1, 1) - \boldsymbol{\Sigma}_{\mathbf{X}}(1, 1)\right) \leq 2\boldsymbol{\Sigma}_{\mathbf{Y}}(1, 2).$$

For example, if $\mathbb{V}\text{ar}X^{(1)} = \mathbb{V}\text{ar}Y^{(1)}$ and $\mathbb{V}\text{ar}X^{(2)} = \mathbb{V}\text{ar}Y^{(2)}$, then the variance of $\hat{Y}_R^{\text{RoE}}$ is smaller than or equal to that of the crude estimator if and only if $\boldsymbol{\Sigma}_{\mathbf{Y}}(1, 2) > 0$, i.e., $\text{Cov}\,(Y^{(1)}, Y^{(2)}) > 0$.

In practice, as demonstrated in the following example, we can estimate the variance directly using:

$$\widehat{\mathbb{V}\text{ar}}(\hat{Y}_R) = \frac{1}{R^2(\hat{Y}_R^{(2)})^2} \sum_{i=1}^{R} \left(Y_i^{(1)} - \hat{Y}_R Y_i^{(2)}\right)^2.$$

Example 5.1.22 We aim to estimate the ratio:

$$I = \frac{I^{(1)}}{I^{(2)}} = \frac{32 \int_0^1 \arctan(x)/(1 + x^2)\, dx}{\int_0^1 e^x\, dx}. \tag{5.1.18}$$

The exact value is:

$$I = \frac{\pi^2}{e - 1} = 5.7438.$$

For the CMC estimator, we take i.i.d. random variables $U_1^{(1)}, \ldots, U_R^{(1)}$ and $U_1^{(2)}, \ldots, U_R^{(2)}$ and compute:

$$\hat{Y}_R^{\text{CMC}} = \frac{\sum_{i=1}^{R} X_i^{(1)}}{\sum_{i=1}^{R} X_i^{(2)}}, \qquad X_i^{(1)} = \frac{32 \arctan(U_i^{(1)})}{1 + (U_i^{(1)})^2}, \qquad X_i^{(2)} = e^{U_i^{(2)}}.$$

Table 5.6 Simulation results for estimation of a ratio I given in (5.1.18) with the CMC and the RoE estimators. Estimated variances and errors at confidence level 95% are provided. The exact value of z in this example is known: $z = 5.7438$

	Number of replications $R =$			
	10	100	1000	100,000
$\hat{Y}_R^{\mathrm{CMC}}$	4.74973	6.37042	5.68613	5.74376
$\hat{Y}_R^{\mathrm{RoE}}$	5.53775	5.64188	5.71912	5.74874
$\lvert \hat{Y}_R^{\mathrm{CMC}} - I \rvert$	0.99415	0.62654	0.05775	0.00012
$\lvert \hat{Y}_R^{\mathrm{RoE}} - I \rvert$	0.99415	0.62654	0.05775	0.00012
$\widetilde{\mathrm{Var}}(\hat{Y}_R^{\mathrm{CMC}})$	0.69098	0.08901	0.00800	0.00078
$\widetilde{\mathrm{Var}}(\hat{Y}_R^{\mathrm{RoE}})$	0.07144	0.01710	0.00165	0.00016
b^{CMC}	1.62926	0.58475	0.17536	0.05460
b^{RoE}	0.52389	0.25633	0.07971	0.02468
$\dfrac{\widetilde{\mathrm{Var}}(\hat{Y}_R^{\mathrm{CMC}})}{\widetilde{\mathrm{Var}}(\hat{Y}_R^{\mathrm{RoE}})}$	9.67161	5.20412	4.83952	4.89384

For the RoE estimator, we similarly take:

$$\hat{Y}_R^{\mathrm{RoE}} = \frac{\sum_{i=1}^{R} Y_i^{(1)}}{\sum_{i=1}^{R} Y_i^{(2)}}, \quad Y_i^{(1)} = \frac{32 \arctan(U_i^{(1)})}{1 + (U_i^{(1)})^2}, \quad Y_i^{(2)} = e^{U_i^{(1)}},$$

i.e., we use the same randomness for both $Y^{(1)}$ and $Y^{(2)}$.

In Table 5.6, we present simulation results for $R \in \{10, 100, 1000, 100000\}$. The RoE method reduces the variance approximately 5 times for large $R \geq 100$ and around 7 times for $R = 10$.

This example is implemented in `ch5_ratios_of_expectations.py`. $\qquad \Diamond$

5.1.5 Control Variates

Recall, to estimate I, we find a random variable Y such that $I = \mathbb{E}Y$. The CMC estimator is given by: $\hat{Y}_R^{\mathrm{CMC}} = \frac{1}{R} \sum_{j=1}^{R} Y_j$, where $Y_1, \ldots, Y_R$ are independent replications of Y. Its variance is $\mathrm{Var}(\hat{Y}_R^{\mathrm{CMC}}) = \frac{\mathrm{Var}Y}{R}$.

Now suppose we simultaneously simulate another random variable X, i.e., we have independent replications of pairs $(Y_1, X_1), \ldots, (Y_R, X_R)$, for which we know the exact value of $\mathbb{E}X$. Define:

$$\hat{Y}_R^{\mathrm{CV}} = \hat{Y}_R^{\mathrm{CMC}} + c(\hat{X}_R - \mathbb{E}X),$$

where $\hat{X}_R = \frac{1}{R} \sum_{j=1}^{R} X_j$. Clearly, $\mathbb{E}\hat{Y}_R^{\mathrm{CV}} = I$, and its variance is:

$$\mathrm{Var}(\hat{Y}_R^{\mathrm{CV}}) = \frac{\mathrm{Var}(Y + cX)}{R} = \frac{\mathrm{Var}Y + c^2 \mathrm{Var}X + 2c\mathrm{Cov}\,(Y, X)}{R}.$$

This is a quadratic function of c, minimized when:

$$c = -\frac{\mathrm{Cov}\,(Y,\,X)}{\mathbb{V}\mathrm{ar}X}.$$

Substituting the optimal c, the variance becomes:

$$\mathrm{Var}(\hat{Y}_R^{CV}) = \frac{1}{R}\left(\mathbb{V}\mathrm{ar}Y + \frac{(\mathrm{Cov}\,(Y,\,X))^2}{(\mathbb{V}\mathrm{ar}X)^2}\cdot\mathbb{V}\mathrm{ar}X - \frac{2(\mathrm{Cov}\,(Y,\,X))^2}{\mathbb{V}\mathrm{ar}X}\right)$$

$$= \frac{\mathbb{V}\mathrm{ar}Y}{R}\left(1 - \frac{(\mathrm{Cov}\,(Y,\,X))^2}{\mathbb{V}\mathrm{ar}X\mathbb{V}\mathrm{ar}Y}\right) = \frac{\mathbb{V}\mathrm{ar}Y}{R}(1 - \rho^2)$$

$$= \mathbb{V}\mathrm{ar}\hat{Y}_R^{CMC}(1 - \rho^2),$$

where $\rho = \mathrm{Corr}\,(Y,\,X)$. We call this the *control variate estimator (CV)*. The variance is reduced by a factor of $1 - \rho^2$. To achieve significant reduction, Y and X must be highly correlated, and we must be able to simulate the pair $(Y,\,X)$.

Let $Y^{CV} = Y + c(X - \mathbb{E}X)$, and let $Y_1^{CV}, \ldots, Y_R^{CV}$ be independent replications of Y^{CV}, where c is optimal. Then:

Proposition 5.1.23 *The estimator $\hat{Y}_R^{CV}$ is unbiased and strongly consistent, and:*

$$\sqrt{R}(\hat{Y}_R^{CV} - I) \xrightarrow{\mathcal{D}} \mathcal{N}(0,\,\varsigma^2),$$

where $\varsigma^2 = \mathbb{V}\mathrm{ar}Y(1 - \rho^2)$.

In practice, we do not know $\mathrm{Cov}\,(Y,\,X)$ or $\mathbb{V}\mathrm{ar}(X)$, so we estimate these quantities using:

$$\hat{c} = -\frac{\hat{S}_{YX}^2}{\hat{S}_X^2},$$

where:

$$\hat{S}_X^2 = \frac{1}{R-1}\sum_{j=1}^{R}(X_j - \hat{X}_R)^2, \quad \hat{S}_{YX}^2 = \frac{1}{R-1}\sum_{j=1}^{R}(Y_j - \hat{Y}_R^{CMC})(X_j - \hat{X}_R).$$

Thus, the CLT for $\hat{Y}_R^{CV}$ becomes:

$$\sqrt{R}(\hat{Y}_R^{CV} - I) \xrightarrow{\mathcal{D}} \mathcal{N}(0,\,\varsigma^2),$$

where:

$$\varsigma^2 = \hat{S}_X^2 (1 - \hat{S}_{YX}^2).$$

Hence, half the length of the confidence interval is $z_{1-\alpha/2}\varsigma/\sqrt{R}$.

In the next two examples, the results (on the resulting variances) require the knowledge of π. In practice, we find ρ by simulations. We treat presented examples as toy ones.

Example 5.1.24 In Example 4.1.6, we estimated π using the CMC method with $Y = 4\ \mathbb{1}(U_1^2 + U_2^2 \le 1)$. Recall that $\mathrm{Var}Y = 2.6968$. As a control variate, take $X = \mathbb{1}(U_1 + U_2 \le 1)$, which is strongly correlated with Y and has $\mathbb{E}X = 1/2$. We denote the corresponding control variate estimator by $\hat{Y}_R^{\mathrm{cv},1}$. We have: $\mathrm{Var}X = \frac{1}{2}(1 - \frac{1}{2}) = \frac{1}{4}$ and

$$\mathbb{E}XY = 4 \int_0^1 \int_0^1 \mathbb{1}(u_1^2 + u_2^2 \le 1)\mathbb{1}(u_1 + u_2 \le 1)\,du_1\,du_2 = 2.$$

Thus:

$$1 - \rho^2 = 1 - \frac{(\mathbb{E}XY - \mathbb{E}X\mathbb{E}Y)^2}{\mathrm{Var}X\mathrm{Var}Y} = 1 - \frac{\left(2 - \frac{1}{2}\pi\right)^2}{\pi\left(1 - \frac{1}{4}\pi\right)} = 2 - \frac{4}{\pi} = 0.727,$$

and

$$\mathrm{Var}\hat{Y}_R^{\mathrm{cv},1} = \frac{\mathrm{Var}Y}{R}(1 - \rho^2) = \frac{2.6968 \cdot 0.727}{R} = \frac{1.959}{R}.$$

Now, let us take another control variate $X = \mathbb{1}(U_1 + U_2 > \sqrt{2})$. Denote this estimator by $\hat{Y}_R^{\mathrm{cv},2}$. Then:

$$\mathbb{E}X = \frac{(2 - \sqrt{2})^2}{2}, \quad \mathrm{Var}X = (\sqrt{2} - 1)(2 - \sqrt{2})^2,$$

and

$$\mathbb{E}XY = 4 \int_0^1 \int_0^1 \mathbb{1}(u_1^2 + u_2^2 \le 1)\mathbb{1}(u_1 + u_2 \ge 2)\,du_1\,du_2 = 0.$$

Thus:

$$1 - \rho^2 = 1 - \frac{(0 - \mathbb{E}X\mathbb{E}Y)^2}{\mathbb{V}\mathrm{ar}X\,\mathbb{V}\mathrm{ar}Y} = 1 - \frac{\left(\frac{(2-\sqrt{2})^2\pi}{2}\right)^2}{(\sqrt{2}-1)(2-\sqrt{2})^2 \cdot 4\pi\left(1 - \frac{\pi}{4}\right)}$$

$$= \frac{8(\sqrt{2}-1) - \pi}{2(\sqrt{2}-1)(4-\pi)} = 0.242$$

and

$$\mathbb{V}\mathrm{ar}\hat{Y}_R^{\mathrm{cv},2} = \frac{\mathbb{V}\mathrm{ar}Y}{R}(1 - \rho^2) = \frac{2.6968 \cdot 0.242}{R} = \frac{0.6527}{R}.$$

◇

Example 5.1.25 We want to estimate $\int_0^1 g(x)\,dx$. We can take $Y = g(U)$ (where $U \sim \mathcal{U}[0, 1)$) and $X = f(U)$ as a control variate, provided we know $\int_0^1 f(x)\,dx$. To be more specific: in Example 4.1.7, we estimated π using the CMC method with $Y = 4\sqrt{1 - U^2}$. Recall that we had $\mathbb{V}\mathrm{ar}Y = 32/3 - \pi^2 = 0.797$.

As a control variate, let us take $X = 1 - U$, which is strongly correlated with Y and satisfies $\mathbb{E}X = 1/2$. Denote the control variate estimator by $\hat{Y}_R^{\mathrm{cv},3}$. We compute:

$$\mathbb{E}XY = \int_0^1 4\sqrt{1 - u^2}(1 - u)\,du = \pi - \frac{4}{3}.$$

Of course, we have $\mathbb{V}\mathrm{ar}X = \frac{1}{12}$, and as computed in Example 4.1.7, $\mathbb{V}\mathrm{ar}Y = \frac{32}{3} - \pi^2$. Thus, we find:

$$1 - \rho^2 = 1 - \frac{(\mathbb{E}XY - \mathbb{E}X\mathbb{E}Y)^2}{\mathbb{V}\mathrm{ar}X\,\mathbb{V}\mathrm{ar}Y} = 1 - \frac{(8 - 3\pi)^2}{32 - \pi^2} = 0.151.$$

Finally, the variance of the control variate estimator is:

$$\mathbb{V}\mathrm{ar}(\hat{Y}_R^{\mathrm{cv},3}) = \frac{\mathbb{V}\mathrm{ar}Y}{R}(1 - \rho^2) = \frac{0.797 \cdot 0.151}{R} = \frac{0.119}{R}.$$

◇

Note that if we take (see Exercise 5.T.12) as control variate X or $a + bX$, then the reduction will be the same.

Example 5.1.26 (Control Variate: Cauchy) The goal is to estimate the following integral:

$$I = \int_2^\infty \frac{1}{\pi(1 + x^2)}\,dx.$$

We recognize that $f(x) = \frac{1}{\pi(1+x^2)}$ is the probability density function of the Cauchy distribution ($x \in \mathbb{R}$), and denote the corresponding random variable by X. Therefore, we can rewrite the integral as:

$$I = \mathbb{P}(X > 2).$$

The CMC estimator is given by:

$$\hat{Y}_R^{\mathrm{CMC}} = \frac{1}{R}\sum_{i=1}^{R} Y_i = \frac{1}{R}\sum_{i=1}^{R} \mathbf{1}(X_i > 2),$$

where the X_i are i.i.d. random variables with the Cauchy distribution. This is easy to simulate: $X_i = \tan(\pi(U_i - 0.5))$, where U_i are i.i.d. $\mathcal{U}[0, 1)$.

In the calculations below, we "cheat" slightly by using the exact value of I (i.e., the value we are trying to estimate), which is $I = \frac{1}{2} - \frac{\arctan(2)}{\pi} = 0.147583$. This allows us to compare the methods. In practical applications, we would use bounds on I to control variances, as discussed in Remark 4.1.12.

Note that Y_i is a Bernoulli random variable with parameter I, so its variance is:

$$\mathbb{V}\mathrm{ar}(\hat{Y}_R^{\mathrm{CMC}}) = \frac{I(1 - I)}{R} = \frac{0.1258}{R}. \tag{5.1.19}$$

Reformulation of the Problem We can rewrite I as:

$$I = \int_2^{\infty} \frac{1}{\pi(1 + x^2)}\,dx = \frac{1}{2} - \int_0^2 \frac{1}{\pi(1 + x^2)}\,dx,$$

and reformulate the problem. We now need to estimate:

$$I_2 = \int_0^2 \frac{1}{\pi(1 + x^2)}\,dx = \int_0^2 \frac{2}{\pi(1 + x^2)} \cdot \frac{1}{2}\,dx.$$

The CMC Estimator for the Reformulated Problem The CMC estimator for the reformulated problem requires the simulation of:

$$Y = \frac{2}{\pi(1 + V^2)} := k_Y(V),$$

where $V \sim \mathcal{U}[0, 2)$. Therefore, the CMC estimator is:

$$\hat{Y}_R^{\mathrm{CMC}} = \frac{1}{R}\sum_{i=1}^{R} Y_i = \frac{1}{R}\sum_{i=1}^{R} \frac{2}{\pi(1 + V_i^2)}.$$

The variance of this estimator is computed as follows (using exact calculations performed in `Maple`):

$$\mathbb{E}Y = \int_0^2 \frac{2}{\pi(1+x^2)} \cdot \frac{1}{2} = \frac{\arctan(2)}{\pi},$$

$$\mathbb{E}Y^2 = \int_0^2 \left(\frac{2}{\pi(1+x^2)} \right)^2 \cdot \frac{1}{2} = \frac{1}{5} \cdot \frac{5\arctan(2)+2}{\pi^2},$$

and thus the variance is:

$$\mathrm{Var}Y = \frac{\arctan(2)+2/5}{\pi^2} - \frac{(\arctan(2))^2}{\pi^2}$$

$$= \frac{\arctan(2)(1 - \arctan(2)) + 2/5}{\pi^2} = 0.0285087854.$$

Finally, the variance of the CMC estimator is:

$$\mathrm{Var}(\hat{Y}_R^{\mathrm{CMC}}) = \frac{0.0285087854}{R}. \tag{5.1.20}$$

Notice that thanks to the reformulation of the problem, the variance of the new CMC estimator is approximately four times smaller, as compared to (5.1.19).

Let us define $X = k_X(V)$, and if we know the exact value of $\mathbb{E}X$, we can use a control variate estimator:

$$\hat{Y}_R^{\mathrm{CV}} = \hat{Y}_R^{\mathrm{CMC}} + c(\hat{X}_R - \mathbb{E}X), \quad \text{where} \quad \hat{X}_R = \frac{1}{R} \sum_{i=1}^R X_i.$$

Recall that the smallest variance is achieved when $c = -\frac{\mathrm{Cov}(X,Y)}{\mathrm{Var}X}$. In this case, the variance of the control variate estimator is:

$$\mathrm{Var}(\hat{Y}_R^{\mathrm{CV}}) = \mathrm{Var}(\hat{Y}_R^{\mathrm{CMC}})(1 - \rho^2),$$

where $\rho = \mathrm{Corr}(X, Y)$.

The Control Variates Estimator for the Reformulated Problem: First-degree Polynomial We recall that $Y = \frac{2}{\pi(1+V^2)} = k_Y(V)$, where $V \sim \mathcal{U}[0, 2)$. As a control variate, let us take:

$$X = \frac{2}{\pi} - \frac{4}{5\pi} V =: k_X^{(1)}(V).$$

In Fig. 5.13, we can see that $k_X^{(1)}(v)$ for $v \in [0, 2]$ is "close" to $k_Y(v)$.

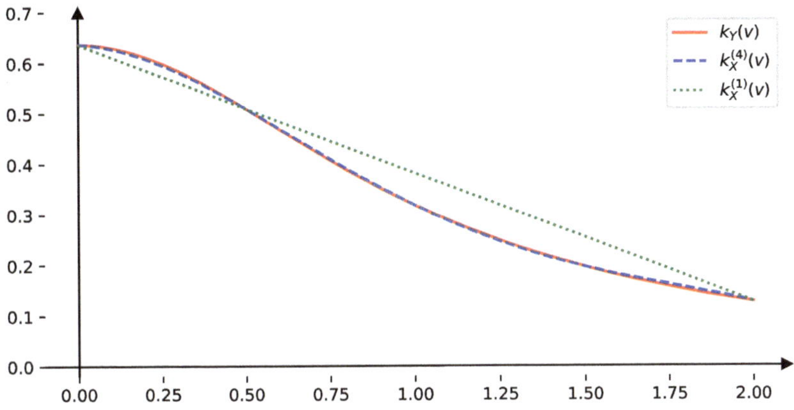

Fig. 5.13 Function $k_Y(v) = 2/(\pi(1+v^2))$ and fitted polynomials $k_X^{(1)}(v)$ (first degree) and $k_X^{(4)}(v)$ (fourth degree). See description in the text

We have the following values:

$$
\mathbb{E}X \;=\; \int_0^2 \left(\frac{2}{\pi} - \frac{4}{5\pi} x \right) \cdot \frac{1}{2} dx \;=\; \frac{6}{5\pi},
$$

$$
\mathbb{E}X^2 \;=\; \int_0^2 \left(\frac{2}{\pi} - \frac{4}{5\pi} x \right)^2 \cdot \frac{1}{2} dx \;=\; \frac{124}{75\pi^2},
$$

$$
\mathbb{V}\mathrm{ar}X \;=\; \frac{124}{75\pi^2} - \left(\frac{6}{5\pi} \right)^2 \;=\; \frac{16}{75\pi^2} = 0.02161518583,
$$

$$
\mathbb{E}XY \;=\; \int_0^2 \left(\frac{2}{\pi} - \frac{4}{5\pi} x \right) \cdot \frac{2}{\pi(1+x^2)} \cdot \frac{1}{2} dx = 0.1591271754 \quad (\text{using Maple}).
$$

The correlation coefficient is:

$$
\rho = \frac{\mathbb{E}XY - \mathbb{E}X\mathbb{E}Y}{\sqrt{\mathbb{V}\mathrm{ar}X\mathbb{V}\mathrm{ar}Y}} = 0.9875200579.
$$

Finally, we have $1 - \rho^2 = 0.0248041347$, which means that we have reduced the variance by a factor of approximately **40**.

The Control Variates Estimator for the Reformulated Problem: Fourth-degree Polynomial This time let us take $X = k_X(V)$, where $k_X^{(4)}(v)$ is a fourth degree polynomial, such that $k_X^{(4)}(v) = k_Y(v)$ for $v \in \{0, 1/2, 1, 3/2, 2\}$. In `ch5_ContrVariates_Cauchy.py`, we implement the Vandermonde-matrix approach to compute these polynomial coefficients and also generate Fig. 5.13. We have

$$k_X(x) = -0.0979415x^4 + 0.470119x^3 - 0.661105x^2$$

$$-0.0293825x + 0.63662.$$

In Fig. 5.13 we can see that $k_X(v)$ is "very close" to $k_Y(v)$. One can calculate:

$$\mathbb{E}X \ = 0.352589, \quad \mathbb{E}X^2 = 0.152314,$$

$$\mathbb{V}\text{ar}X = 0.027995, \quad \mathbb{E}XY = 0.152506.$$

Finally, $1 - \rho^2 = 0.0000859848$ and we reduced the variance:

$$(0.0000859848)^{-1} \approx \mathbf{11630} \text{ times.}$$

Let us finally compare the control variates estimator (with $k_X(v)$ being a fourth-degree polynomial) with the CMC estimator. We have:

$$\mathbb{V}\text{ar}(Y_R^{\text{CV}}) = (1 - \rho^2)\mathbb{V}\text{ar}(Y_R^{\text{CMC}})$$

$$= \frac{0.0000859848e \cdot 0.0285087854}{R} = \frac{0.00000245\,1322211}{R},$$

where the control variate was a fourth degree polynomial and $\mathbb{V}\text{ar}(Y_R^{\text{CMC}})$ was given in (5.1.20), i.e., after reformulating the problem. Before reformulation we had (i.e. for $Y_i = \mathbf{1}(X_i > 2)$)—see (5.1.19)—$\mathbb{V}\text{ar}(Y_R^{\text{CMC}}) = \frac{0.1258022581}{R}$. Thus "reformulating + control variates" reduced the variance:

$$\frac{0.1258022581}{0.00000245\,1322211} \approx 51320$$

times. $\Diamond$

Example 5.1.27 (Control Variate: $P(Z > 4)$ **for** $Z \sim \mathcal{N}(0, 1)$**)** Our goal is to estimate $I = P(Z > 4)$, where $Z \sim \mathcal{N}(0, 1)$. We write $I = \mathbb{E}Y$ with $Y = \mathbf{1}(Z > 4)$. A CMC estimator for this probability is $\hat{Y}_R^{\text{CMC}} = \frac{1}{R}\sum_{i=1}^{R} \mathbf{1}(Z_i > 4)$, where Z_i are independent standard normal samples. We generate Z using ITM method, i.e., $Z = \Phi^{-1}(U)$ for $U \sim \mathcal{U}$. Then

$$Y = \mathbf{1}(\Phi^{-1}(U) > 4) = \mathbf{1}(U > \Phi(4)).$$

Since $\Phi(4)$ is close to 1, we choose a control variate that also depends on the upper tail of U. Let

$$X = \mathbf{1}(U > 0.9999).$$

Clearly,

$$\mathbb{E}X = 1 - 0.9999 = 0.0001, \quad \mathbb{V}\mathrm{ar}X = 0.9999 \times 0.0001 \approx 1 \times 10^{-4}.$$

In computations we will use exact values

$$\mathbb{E}Y = 3.17 \times 10^{-5}, \quad \mathbb{V}\mathrm{ar}Y = 3.17 \times 10^{-5}(1 - 3.17 \times 10^{-5}) \approx 3.17 \times 10^{-5}.$$

Since

$$\Phi(4) \approx 0.9999683 > 0.9999,$$

we have

$$\mathbb{E}[XY] = \mathbb{P}(U > \Phi(4), U > 0.9999) = \mathbb{P}(U > \Phi(4)) = 3.17 \times 10^{-5}.$$

Thus, the covariance is

$$\mathrm{Cov}\,(X, Y) = \mathbb{E}[XY] - \mathbb{E}X\mathbb{E}Y = 3.17 \times 10^{-5} - 3.17 \times 10^{-5} \times 0.0001 \approx 3.17 \times 10^{-5},$$

and consequently the squared correlation is

$$\rho^2 = \frac{(3.17 \times 10^{-5})^2}{3.17 \times 10^{-5} \cdot 10^{-4}} \approx 0.317.$$

Hence, the variance of the control variate estimator is:

$$\mathbb{V}\mathrm{ar}(\hat{Y}_R^{CV}) = \mathbb{V}\mathrm{ar}(\hat{Y}_R^{CMC})(1 - \rho^2) = 0.683 \cdot \mathbb{V}\mathrm{ar}(\hat{Y}_R^{CMC}).$$

In other words, we have reduced the variance by about 31.7%, i.e., a factor of roughly 1.46.

Simulation Results In this example, we used the exact value of $P(Z > 4)$ for illustrative computations. In practice, this probability is unknown and is the very quantity we wish to estimate. We computed it analytically here solely to confirm the theoretical correlation and the resulting variance reduction. In a real-world scenario, you would select $X = \mathbf{1}(U > 0.9999)$ based on the intuition that both X and $Y = \mathbf{1}(Z > 4)$ "activate" only when U is extremely close to 1, ensuring a strong relationship. After a pilot simulation to estimate c, actual runs would show empirically that the variance is indeed reduced, even without knowing $P(Z > 4)$ in advance.

We performed a pilot simulation with $R'_{\mathrm{pilot}} = 10^6$ samples to estimate the coefficient c, and then ran a main simulation with $R = 10^7$ samples. Using the control variate $X = \mathbf{1}(U > 0.9999)$, we observed results close to the theoretical prediction, with estimated $\rho^2 = 0.308$ with variance reduced by a factor of

Table 5.7 Numerical results demonstrating variance reduction with the chosen control variate. The actual value of $P(Z > 4)$ is approximately 3.167×10^{-5}

	$\hat{Y}_R$	$\mathbb{V}\mathrm{ar}(\hat{Y}_R)$	$\left(\mathbb{V}\mathrm{ar}(\hat{Y}_R)\right)^{1/2}$	$\lvert \hat{Y}_R - \mathbb{P}(Z > 4) \rvert$
CMC	3.0400×10^{-5}	3.0399×10^{-12}	1.7435×10^{-6}	1.2712×10^{-6}
CV	3.0799×10^{-5}	2.1037×10^{-12}	1.4504×10^{-6}	8.7238×10^{-7}

$2.1037/3.0399 = 0.692$, as can be seen in Table 5.7. Implementation of this example is available in `ch5_ContrVariates_normal_1D.py`.

$\diamond$

Remark 5.1.28 In the previous Example 5.1.27, the CMC estimator was $\hat{Y}_R^{\mathrm{CMC}} = \frac{1}{R} \sum_{i=1}^{R} \mathbb{1}(Z_i > 4)$, where $Z_i \sim \mathcal{N}(0, 1)$. Note that since $\mathbb{P}(\mathbb{1}(Z_i > 4) = 1) = \mathbb{P}(Z_i > 4) = I$ is small, R must be large for the estimator to be meaningful. The control variate $X = \mathbb{1}(U > 0.9999)$, while showing some improvement, faces the same problem—it is most often 0, requiring a large R for the CV estimator to be meaningful. For such small I, other estimators, such as importance sampling, are better suited—see Example 5.1.36, where the variance was reduced by a factor of 879.478. ∎

5.1.6 Conditional Monte Carlo

Again, our starting point is an estimation of $I = \mathbb{E}[Y]$. Recall that the crude Monte Carlo estimator is given by $\hat{Y}_R^{\mathrm{CMC}} = \frac{1}{R} \sum_{j=1}^{R} Y_j$, where $Y_1, \ldots, Y_R$ are independent replications of Y. Now, assume we also have X. For simplicity, assume that the random vector (Y, X) has a joint density $f_{Y,X}(y, x)$. The marginal p.d.f. of X is then given by

$$f_X(x) = \int f_{Y,X}(y, x)\, dy.$$

Let

$$c(x) = \mathbb{E}[Y \mid X = x] = \frac{\int y f_{Y,X}(y, x)\, dy}{f_X(x)}.$$

From the law of total probability, we have

$$I = \mathbb{E}[Y] = \mathbb{E}[c(X)].$$

We then define the *conditional estimator (cond)* as

$$\hat{Y}_R^{\text{cond}} = \frac{1}{R} \sum_{j=1}^{R} Y_j^{\text{cond}},$$

where $Y_1^{\text{cond}}, \ldots, Y_R^{\text{cond}}$ are independent replications of $Y^{\text{cond}} = c(X)$, i.e., we sample X_j from f_X and compute $Y_j^{\text{cond}} = c(X_j)$. The estimator $\hat{Y}_R^{\text{cond}}$ is unbiased. We will show that its variance is not greater than the one of the crude Monte Carlo estimator.

From Jensen's inequality, we have

$$\mathbb{E}[Y^2 | X = x] \geq (\mathbb{E}[Y | X = x])^2.$$

If X has a density f_X, then

$$\begin{aligned}
\mathbb{V}\text{ar}Y &= \int \mathbb{E}[Y^2 | X = x] f_X(x)\, dx - I^2 \\
&\geq \int (\mathbb{E}[Y | X = x])^2 f_X(x)\, dx - I^2 \\
&= \int c^2(x) f_X(x)\, dx - I^2 \\
&= \mathbb{V}\text{ar}(c(X)).
\end{aligned}$$

Another way is to use the *law of total variance*, which follows from formula (5.1.2):

$$\text{Var}(Y) = \mathbb{E}[\text{Var}(Y | X)] + \text{Var}[\mathbb{E}[Y | X]].$$

As an immediate consequence, we have

$$\text{Var}(Y) \geq \text{Var}[\mathbb{E}[Y | X]] = \mathbb{V}\text{ar}(c(X)).$$

Example 5.1.29 We continue from Example 4.1.6. Recall that we estimated π with $Y = 41(U_1^2 + U_2^2 \leq 1)$. Taking $X = U_1$, we have

$$c(x) = \mathbb{E}[Y | X = x] = 4\mathbb{P}(x^2 + U_2^2 \leq 1) = 4\mathbb{P}(U_2 \leq \sqrt{1 - x^2}) = 4\sqrt{1 - x^2}.$$

The variance of $\hat{Y}_R^{\text{cond}} = \frac{1}{R} \sum_{i=1}^{R} 4\sqrt{1 - U_i^2}$ is

$$\text{Var}(\hat{Y}_R^{\text{cond}}) = \frac{1}{R} \left(\int_0^1 16(1 - x^2)\, dx - \pi^2 \right) = \frac{0.797}{R}.$$

Note that in Example 4.1.6, for the crude Monte Carlo method, we computed that $\text{Var}(\hat{Y}_R^{\text{CMC}}) = 2.6968/R$, so the variance is reduced by a factor of 3.38 compared to the crude Monte Carlo estimator. We will continue to analyze this estimator in Example 5.1.35.   ◊

Remark 5.1.30 In Examples 4.1.6 and 4.1.7, we showed CMC estimators of π. The first used $Y^a = 41(U_1^2 + U_2^2 \leq 1)$, whereas the latter used $Y^b = 4\sqrt{1 - U^2}$. As shown, the latter has a variance reduced by a factor of 3.38. Note that the CMC estimator for Y^a with $X = U_1$ yields Y^b.   ∎

Example 5.1.31 Let $I = \mathbb{P}(\xi_1 + \xi_2 > y)$, where ξ_1, ξ_2 are i.i.d. random variables with a cumulative distribution F, which is known, but for which we do not have an analytic formula for F^{*2}. By $\overline{F}$, we denote the tail of F, i.e., $\overline{F}(x) = 1 - F(x)$. We can define

$$Y = \mathbf{1}(\xi_1 + \xi_2 > y)$$

and $X = \xi_1$. Then,

$$c(x) = \mathbb{E}[Y|X = x] = \mathbb{P}(\xi_1 + \xi_2 > y|\xi_1 = x) = \mathbb{P}(\xi_2 > y - x) = \overline{F}(y - x),$$

and

$$Y^{\text{cond}} = \overline{F}(y - \xi_1).$$

Let $X_1, \ldots, X_R$ be i.i.d. random variables with c.d.f. F. Then,

$$\hat{Y}_R^{\text{cond}} = \frac{1}{R} \sum_{j=1}^{R} \overline{F}(y - X_j)$$

is an unbiased estimator of I with no larger variance than in the CMC estimator. ◊

Example 5.1.32 In some sense, we continue from Example 5.1.31. Let

$$S_n = \xi_1 + \cdots + \xi_n,$$

where $\xi_1, \xi_2, \ldots, \xi_n$ are i.i.d. random variables with density function f. The conditional estimator can also be used for estimating the density function $f^{*n}(x)$ of the cumulative distribution function $F^{*n}(x)$, provided f is known. The conditional estimator for $f^{*n}(y)$ is defined by:

$$Y^{\text{cond}} = f(y - S_{n-1}).$$

◊

5.1.7 *Importance Sampling*

Assume we want to estimate I, which can be written as

$$I = \int_E k(x) f(x)\, dx = \mathbb{E}k(X),$$

where E is a subset of $\mathbb{R}$ or $\mathbb{R}^d$, f is a p.d.f. of $X \in E$, and the function $k : \mathbb{R}^d \to \mathbb{R}$ satisfies $\int_E k(x)^2 f(x)\, dx < \infty$. Thus, X is defined on a probability space with values in E. Recall that the crude Monte Carlo estimator for I is:

$$\hat{Y}_R^{\text{CMC}} = \frac{1}{R} \sum_{j=1}^{R} Y_j,$$

where $Y_1 = k(X_1), \ldots, Y_R = k(X_R)$ are independent replications of $Y = k(X)$. Let $\tilde{f}(x)$ (often called the *proposal distribution*) be a density function on E, which is assumed to be strictly positive on E (actually on $\{x : f(x) > 0\}$), and let $\tilde{X}$ be a random variable with density function $\tilde{f}(x)$. Then,

$$I = \int_E k(x) \frac{f(x)}{\tilde{f}(x)} \tilde{f}(x)\, dx = \mathbb{E}k_1(\tilde{X}),$$

where $k_1(x) = \frac{k(x) f(x)}{\tilde{f}(x)}$. Now we define the *importance sampling estimator*:

$$\hat{Y}_R^{\text{IS}} = \frac{1}{R} \sum_{j=1}^{R} k(\tilde{X}_j) \frac{f(\tilde{X}_j)}{\tilde{f}(\tilde{X}_j)},$$

where $\tilde{X}_1, \ldots, \tilde{X}_R$ are independent replications of $\tilde{X}$. Note that we can rewrite the estimator as:

$$\hat{Y}_R^{\text{IS}} = \frac{1}{R} \sum_{j=1}^{R} Y_j^{\text{IS}}, \quad \text{where} \quad Y_j^{\text{IS}} = k_1(\tilde{X}_j) = k(\tilde{X}_j) \frac{f(\tilde{X}_j)}{\tilde{f}(\tilde{X}_j)}.$$

In other words, $\hat{Y}_R^{\text{IS}}$ is a CMC estimator using independent replications $Y_j^{\text{IS}}, j = 1, \ldots, R$. Note that $\hat{Y}_R^{\text{IS}}$ is unbiased since

$$\mathbb{E}k_1(\tilde{X}) = \int_E k(x) \frac{f(x)}{\tilde{f}(x)} \tilde{f}(x)\, dx = I,$$

with variance

$$\mathrm{Var}(\hat{Y}_R^{\mathrm{IS}}) = \frac{1}{R}\mathrm{Var}\left[k(\tilde{X})\frac{f(\tilde{X})}{\tilde{f}(\tilde{X})}\right] = \frac{1}{R}\left(\int_E \left(k(x)\frac{f(x)}{\tilde{f}(x)}\right)^2 \tilde{f}(x)\,dx - I^2\right).$$

It is important to note that $\mathrm{Var}\left[k(\tilde{X})\frac{f(\tilde{X})}{\tilde{f}(\tilde{X})}\right]$ need not be finite; see Example 5.1.35 (ii). Define

$$\varsigma^2 = \left(\int_E \left(k(x)\frac{f(x)}{\tilde{f}(x)}\right)^2 \tilde{f}(x)\,dx - I^2\right).$$

Let us summarize the properties of importance sampling estimators.

Proposition 5.1.33 *The estimator $\hat{Y}_R^{\mathrm{IS}}$ is unbiased and strongly consistent, and*

$$\sqrt{R}(\hat{Y}_R^{\mathrm{IS}} - I) \xrightarrow{\mathcal{D}} \mathcal{N}(0, \varsigma^2).$$

Example 5.1.34 In Example 5.1.26 we were estimating $I = \mathbb{P}(X > 2)$, where X has a Cauchy distribution. Recall that

$$I = \int_2^\infty \frac{1}{\pi(1 + x^2)}\,dx = \int_{-\infty}^\infty k(x)f(x)\,dx,$$

where $k(x) = \mathbf{1}(x \geq 2)$ and $f(x) = \frac{1}{\pi(1+x^2)}$ ($x \in \mathbb{R}$) is the corresponding density. Recall also that the actual value of I is $\frac{1}{2} - \frac{\arctan(2)}{\pi} = 0.147583$ (we will use this for comparison). The CMC estimator (using replications of $Y = \mathbf{1}(X > 2)$) has variance $\mathbb{V}\mathrm{ar}(\hat{Y}_R^{\mathrm{CMC}}) = \frac{0.1258022581}{R}$, as earlier given in (5.1.19). Now, define the importance sampling estimator using the p.d.f.

$$\tilde{f}(x) = \frac{2}{x^2}\mathbf{1}(x > 2).$$

We compute

$$I = \int_2^\infty \frac{\frac{1}{\pi(1+x^2)}}{\frac{2}{x^2}}\frac{2}{x^2}\,dx = \int_2^\infty \frac{1}{2\pi\left(1 + \frac{1}{x^2}\right)}\frac{2}{x^2}\,dx = \int_2^\infty k_1(x)\tilde{f}(x)\,dx,$$

where $k_1(x) = \dfrac{1}{2\pi\left(1+\frac{1}{x^2}\right)}$. A random variable $\tilde{X}$ with density $\tilde{f}$ can be easily simulated since $\tilde{X} \sim \frac{2}{U}$. Thus,

$$
\hat{Y}_R^{\mathrm{IS}} = \frac{1}{R}\sum_{i=1}^{R} Y_i^{\mathrm{IS}} = \frac{1}{R}\sum_{i=1}^{R}\frac{1}{2\pi\left(1+\frac{1}{\tilde{X}_i^2}\right)}, \quad \text{where} \quad \tilde{X}_i = \frac{2}{U_i}
$$

and $U_i \sim \mathcal{U}[0, 1)$ are independent replications of U. We can compute theoretically $\mathbb{V}\mathrm{ar}(\hat{Y}_R^{\mathrm{IS}}) = \frac{1}{R}\mathbb{V}\mathrm{ar}(Y_1^{\mathrm{IS}})$ using

$$
\mathbb{E}((Y_1^{\mathrm{IS}})^2) = \mathbb{E}\left(k_1(\tilde{X})^2\right) = \int_2^{\infty}\left(\frac{1}{2\pi\left(1+\frac{1}{x^2}\right)}\right)^2 \frac{2}{x^2}\,dx = 0.02187645.
$$

Thus,

$$
\mathbb{V}\mathrm{ar}(\hat{Y}_R^{\mathrm{IS}}) = \frac{1}{R}\left(\mathbb{E}((Y_1^{\mathrm{IS}})^2) - \mathbb{E}((Y_1^{\mathrm{IS}}))^2\right) = \frac{9.55 \times 10^{-5}}{R},
$$

and the variance reduction is

$$
\frac{\mathbb{V}\mathrm{ar}(\hat{Y}_R^{\mathrm{CMC}})}{\mathbb{V}\mathrm{ar}(\hat{Y}_R^{\mathrm{IS}})} = \frac{0.1258022581}{9.55 \times 10^{-5}} \approx 1317 \text{ times.}
$$

$\Diamond$

As we have seen, by choosing $\tilde{f}(x)$ appropriately, we can significantly reduce the variance. However, this is not always the case, as the following example shows.

Example 5.1.35 The formula gives the function of the quarter circle

$$
k(x) = 4\sqrt{1 - x^2}, \qquad x \in [0, 1]
$$

and of course

$$
\int_0^1 4\sqrt{1 - x^2}\,dx = \pi.
$$

In this example $Y = k(U)$, where U is uniformly distributed on $E = [0, 1]$. We consider the following situations.

(i) In this case, we have p.d.f. $f(x) = \mathbb{1}_{[0,1]}(x)$ and then the CMC estimator is following

$$
\hat{Y}_R^{\mathrm{CMC}} = \frac{1}{R}\sum_{j=1}^{n} 4\sqrt{1 - U_j^2}. \tag{5.1.21}
$$

Note that we considered this estimator in Example 5.1.29 (where it was presented as a conditional Monte Carlo estimator), wherein we computed its variance

$$\mathbb{V}\mathrm{ar}(\hat{Y}_R^{\mathrm{CMC}}) = \frac{1}{R}\left(\int_0^1 16(1-x^2)\,dx - \pi^2\right) = \frac{0.797}{R}.$$

(ii) Now let us take a proposal p.d.f. $\tilde{f}_1(x) = 2x$. Then we have the following importance sampling estimator

$$\hat{Y}_R^{\mathrm{IS}} = \frac{1}{R}\sum_{i=1}^R \frac{4\sqrt{1-\tilde{X}_i^2}}{2\tilde{X}_i},$$

where $\tilde{X}_1, \ldots, \tilde{X}_R$ are independent replications of a random variable $\tilde{X}$ with p.d.f. $2x$, i.e.,

$$\hat{Y}_R^{\mathrm{IS}} = \frac{1}{R}\sum_{i=1}^R Y_i^{\mathrm{IS}}, \quad \text{where } Y_i^{\mathrm{IS}} = k_1(\tilde{X}_i)$$

and $k_1(x) = k(x)\frac{f(x)}{\tilde{f}_1(x)} = \frac{4\sqrt{1-x^2}}{2x}$.

Its variance is

$$\mathbb{V}\mathrm{ar}(\hat{Y}_R^{\mathrm{IS}}) = \frac{\mathbb{V}\mathrm{ar}k_1(\tilde{X})}{R},$$

where

$$\mathbb{V}\mathrm{ar}k_1(\tilde{X}) = \mathbb{E}\left(\frac{4\sqrt{1-\tilde{X}^2}}{2\tilde{X}}\right)^2 - I^2 = \int_0^1 \frac{16(1-x^2)}{4x^2}\cdot 2x\,dx - I^2$$

$$= 8\int_0^1 \frac{1-x^2}{x}\,dx - \pi^2 = \infty.$$

(iii) Now let us take another density

$$\tilde{f}_2(x) = (4-2x)/3$$

for which the importance estimator has a smaller variance than for the estimator given in (5.1.21). We have

$$\hat{Y}_R^{\mathrm{IS}} = \frac{1}{R}\sum_{i=1}^R \frac{4\sqrt{1-\tilde{X}_i^2}}{(4-2\tilde{X}_i)/3},$$

i.e.,

$$\hat{Y}_R^{\text{IS}} = \frac{1}{R}\sum_{i=1}^{R} Y_i^{\text{IS}}, \quad \text{where } Y_i^{\text{IS}} = k_2(\tilde{X}_i)$$

and $k_2(x) = k(x)\frac{f(x)}{\tilde{f}(x)} = \frac{4\sqrt{1-x^2}}{(4-2x)/3}$. Hence we have

$$\text{Var}\, k_2(\tilde{X}) = \int_0^1 \frac{16(1-x^2)}{(4-2x)/3} - \pi^2 = 0.224,$$

thus this importance sampling estimator with proposal p.d.f. $\tilde{f}_2$ has around 3.5 times smaller variance than the CMC estimator.

$$\diamond$$

Example 5.1.36 Suppose we want to estimate $I = \mathbb{P}(X > 4)$, where $X \sim \mathcal{N}(0, 1)$. Since $I = \int_{-\infty}^{\infty} \mathbb{1}(x > 4)f(x)\,dx$ (where f is the p.d.f. of X), the CMC estimator is $\hat{Y}_R^{\text{CMC}} = \frac{1}{R}\sum_{i=1}^{R} Y_i^{\text{CMC}}$, where $Y_i^{\text{CMC}} = \mathbb{1}(X_i > 4)$ for i.i.d. $X_i \sim f$. The actual value of I is approximately 3.1671×10^{-5}, and similarly, $\text{Var}(Y_1^{\text{CMC}}) = I(1-I) \approx 3.1670 \times 10^{-5}$. Note that the actual value of I is very small, and we are estimating it with an estimator that is also close to zero. In such cases, we should focus more on relative error (see Sect. 4.4). But recall that a smaller variance also decreases the relative error.

The challenge here is that many of the X_i's will be less than or equal to 4, requiring a large R for the CMC estimator to produce a reasonable result. One approach to address this is to sample from $\mathcal{N}(4, 1)$, so that, on average, half of the $\tilde{X}_i$'s will be less than or equal to 4, and the other half will be greater than 4. Let $\tilde{f}$ be the p.d.f. of $\mathcal{N}(4, 1)$. Then:

$$I = \int_{-\infty}^{\infty} \mathbb{1}(x > 4)f(x)\,dx = \int_{-\infty}^{\infty} \frac{\mathbb{1}(x > 4)f(x)}{\tilde{f}(x)}\tilde{f}(x)\,dx.$$

The IS estimator is:

$$\hat{Y}_R^{\text{IS}} = \frac{1}{R}\sum_{i=1}^{R} Y_i^{\text{IS}}, \quad \text{where } Y_i^{\text{IS}} = \frac{1}{R}\sum_{i=1}^{R} \frac{\mathbb{1}(\tilde{X}_i > 4)f(\tilde{X}_i)}{\tilde{f}(\tilde{X}_i)},$$

and $\tilde{X}_i \sim \tilde{f}$. We may compute that:

$$\text{Var}(Y_1^{\text{IS}}) = \int_4^{\infty} \left(\frac{f(x)}{\tilde{f}(x)}\right)^2 \tilde{f}(x)\,dx - I^2 = 3.601 \times 10^{-8},$$

and thus the variance was reduced by a factor of:

$$\frac{\mathrm{Var}(\hat{Y}_R^{\mathrm{CMC}})}{\mathrm{Var}(\hat{Y}_R^{\mathrm{IS}})} = 879.478.$$

One might ask whether the proposal p.d.f. used is optimal within the class of distributions $\tilde{f}_\theta \sim \mathcal{N}(\theta, 1)$. Is $\theta = 4$ "the best" choice? The answer is **no**, and later we will see how we can choose a "better" θ using the cross-entropy method. $\Diamond$

How to Choose $\tilde{f}$? Let $I = \int_E k(x) f(x) \, dx$. There is no one-size-fits-all recipe for choosing a proposal distribution. However, there are several key points worth remembering while searching for an appropriate proposal distribution. Assume for a moment that the function k is non-negative and consider a new p.d.f.

$$g^*(x) = \frac{k(x) f(x)}{\int_E k(x) f(x) \, dx} = \frac{k(x) f(x)}{I}, \tag{5.1.22}$$

which we call the *optimal proposal distribution*. Then, the variance of the importance sampling estimator is 0 since

$$\mathrm{Var} Y^{\mathrm{IS}} = \int_E \left(\frac{k(x) f(x)}{k(x) f(x)/I} \right)^2 \frac{k(x) f(x)}{I} \, dx - I^2 = 0.$$

Of course, such an estimator does not make practical sense because we need I to define it (and we would need its exact value to sample from g^*, e.g., via ITM). Nevertheless, this remark suggests how to choose $\tilde{f}$. Namely, it should be a density having a shape "similar" to $k(x) f(x)$. In Exercise 5.T.8, we propose to characterize g^* as the solution to a variational problem. It is worth mentioning that the cross-entropy method (which will be introduced in Sect. 5.1.8) aims to find a distribution (within a given class) that is in a sense "close" to g^*.

Recall estimating π in Example 5.1.35: in Fig. 5.14, we can see a comparison of $k(x) f(x)$ (which is simply $k(x)$, since in this example $f(x) = \mathbf{1}_{[0,1]}(x)$) with the densities $\tilde{f}_1$ and $\tilde{f}_2$.

Out of $\tilde{f}_1$ and $\tilde{f}_2$, the function $\tilde{f}_2$ has a shape that is "closer" (by eye inspection) to $k(x) f(x)$ than $\tilde{f}_1$. As expected, $\tilde{f}_2$ resulted in a smaller variance for the IS estimator.

Example 5.1.37 We continue estimating π as in Example 5.1.35. Following the suggestion that the shape of $k(x) f(x) = k(x) = 4\sqrt{1 - x^2}$ should be "similar" to the proposal distribution, let us consider a function $f_3(x) = a_4 x^4 + a_3 x^3 + a_2 x^2 + a_1 x + a_0$, which is a polynomial of degree 4 having the same values as $k(x) f(x)$ at

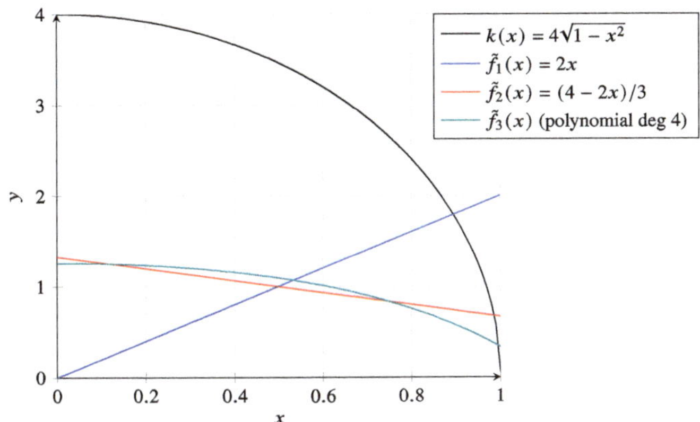

Fig. 5.14 Estimating $k(x) = 4\sqrt{1 - x^2}$ using importance sampling with densities $\tilde{f}_1(x) = 2x$ and $\tilde{f}_2(x) = (4 - 2x)/3$

the points $0, 1/5, 2/5, 3/5, 4/5$. After determining the coefficients we compute that $c = \int_0^1 f_3(x)\,dx = 3.17177$. Taking $\tilde{f}_3(x) = \frac{f_3(x)}{c}$, we obtain:

$$\tilde{f}_3(x) = -0.6598816639x^4 + 0.5249798009x^3 - 0.8092835203x^2$$

$$+ 0.01873718398x + 1.261123964.$$

Define $k_3(x) = \frac{k(x)f(x)}{\tilde{f}_3(x)}$. One may compute that $\mathbb{V}\mathrm{ar}(k_3(\tilde{X})) = 0.0308$, which means that the IS estimator with the proposal distribution $\tilde{f}_3$ has a variance that is $\frac{0.797}{0.0308} = 25.87$ times smaller than the variance of the CMC estimator. The density $\tilde{f}_3(x)$ is depicted in Fig. 5.14. $\diamond$

Example 5.1.38 (Asmussen and Glynn [8]) Let $\mathbb{R}^d \ni \mathbf{X} \sim \mathcal{N}(\mathbf{0}, \boldsymbol{\Sigma})$ and $A \subset \mathbb{R}^d$. We want to estimate $I = \mathbb{P}(\mathbf{X} \in A)$. An interesting example is

$$\mathbb{P}(X_1 > x_1, \ldots, X_d > x_d),$$

where $A = \prod_{j=1}^d (x_j, \infty)$. The method we are discussing is important when I is small, particularly when $\mathbf{0} \notin A$. Assuming that $\boldsymbol{\Sigma}$ is non-singular, the density of $\mathcal{N}(\mathbf{a}, \boldsymbol{\Sigma})$ exists and is given by

$$\phi_{\mathbf{a}, \boldsymbol{\Sigma}}(\mathbf{x}) = \frac{1}{\sqrt{(2\pi)^d |\det \boldsymbol{\Sigma}|}} e^{-\frac{1}{2}(\mathbf{x}-\mathbf{a})^T \boldsymbol{\Sigma}^{-1}(\mathbf{x}-\mathbf{a})}.$$

Thus,

$$I = \int_{\mathbb{R}^d} \mathbb{1}(\mathbf{x} \in A) \phi_{\mathbf{0}, \boldsymbol{\Sigma}}(\mathbf{x})\,d\mathbf{x}.$$

According to the previous suggestion, the density should be proportional to

$$\mathbb{1}(\mathbf{x} \in A)\phi_{\mathbf{0},\boldsymbol{\Sigma}}(\mathbf{x}).$$

We look for the maximizer

$$\mathbf{x}^* = \arg \sup_{\mathbf{x} \in A} \phi_{\mathbf{0},\boldsymbol{\Sigma}}(\mathbf{x}).$$

Now we search for a density $\tilde{f}(\mathbf{x})$ that is concentrated around $\mathbf{x}^*$ and for which the ratio

$$\frac{\phi_{\mathbf{0},\boldsymbol{\Sigma}}(\mathbf{x})}{\tilde{f}(\mathbf{x})}$$

is "simple", i.e., computable. A possible choice is

$$\tilde{f}(\mathbf{x}) = \phi_{\mathbf{x}^*,\boldsymbol{\Sigma}}(\mathbf{x}),$$

and then we have

$$L(\mathbf{x}) = \frac{\phi_{\mathbf{0},\boldsymbol{\Sigma}}(\mathbf{x})}{\phi_{\mathbf{x}^*,\boldsymbol{\Sigma}}(\mathbf{x})} = e^{-(\mathbf{x}^*)^T \boldsymbol{\Sigma}^{-1}\mathbf{x}+\frac{1}{2}(\mathbf{x}^*)^T \boldsymbol{\Sigma}^{-1}\mathbf{x}^*}. \tag{5.1.23}$$

Thus, the importance sampling estimator is

$$\hat{Y}_R^{\mathrm{IS}} = \frac{1}{R} \sum_{j=1}^{R} Y_j^{\mathrm{IS}},$$

where

$$Y_j^{\mathrm{IS}} = e^{-(\mathbf{x}^*)^T \boldsymbol{\Sigma}^{-1}\tilde{\mathbf{X}}_j+\frac{1}{2}(\mathbf{x}^*)^T \boldsymbol{\Sigma}^{-1}\mathbf{x}^*}\mathbb{1}(\tilde{\mathbf{X}}_j \in A),$$

and $\hat{\mathbf{X}}_j \sim \phi_{\mathbf{x}^*,\boldsymbol{\Sigma}}$ are independent replications.

Simulation Results To be more specific, consider the estimation of $I = \mathbb{P}(X_1 > 2, X_2 > 2)$ for a two-dimensional $\mathbf{X} \sim \mathcal{N}(\mathbf{0}, \boldsymbol{\Sigma})$ with

$$\boldsymbol{\Sigma} = \begin{pmatrix} 1 & \sqrt{2} \\ \sqrt{2} & 4 \end{pmatrix},$$

i.e., with density

$$\varphi_{\mathbf{0},\boldsymbol{\Sigma}}((x_1, x_2)) = \frac{\sqrt{2}}{4\pi} \exp\left(-x_1^2 + \frac{\sqrt{2}}{2}x_1 x_2 - \frac{1}{4}x_2^2\right).$$

One can numerically find that

$$\arg\max_{x_1>2,\,x_2>2} \varphi_{\mathbf{0},\boldsymbol{\Sigma}}((x_1,x_2)) = (2,2.83) =: (x_1^*,x_2^*)$$

(for example, $\varphi_{\mathbf{0},\boldsymbol{\Sigma}}((2,2)) = 0.01283 < 0.01523 = \varphi_{\mathbf{0},\boldsymbol{\Sigma}}((2,2.83)))$.

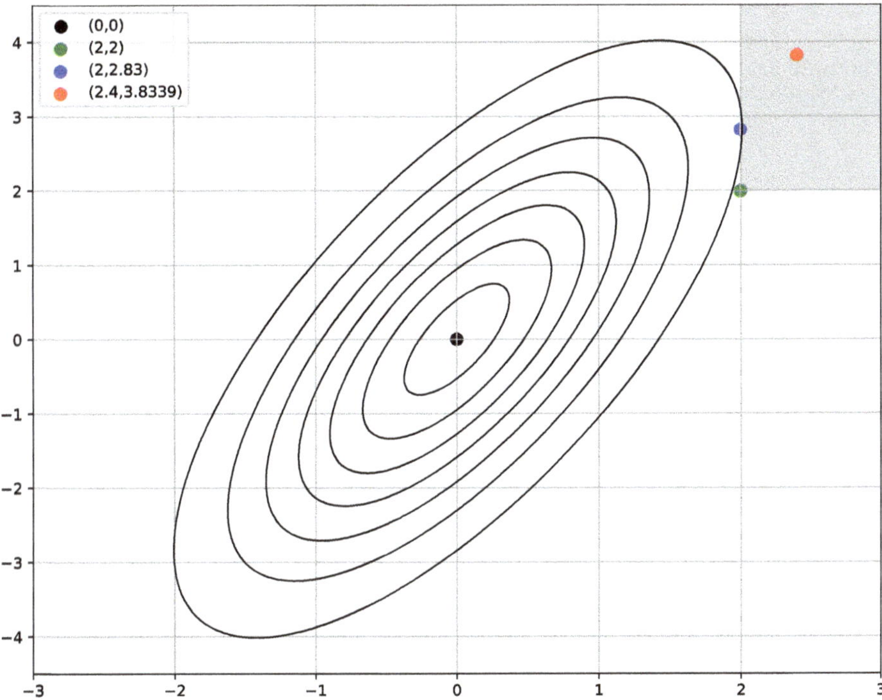

Fig. 5.15 Contours of 2D normal distribution considered in Example 5.1.38 together with several points used in IS estimator. Area $(x_1 > 2, x_2 > 2)$ shaded

Running $R = 10^5$ simulations, we compared the estimators, ranked from worst to best: i) CMC; ii) IS with $\mathbf{x}^* = (2,2)$; iii) IS with $\mathbf{x}^* = (2,2.83)$ (found using the method described above); iv) IS/CE with $\mathbf{x}^* = (2.4,3.8339)$ (found using the cross-entropy method, see Example 5.1.48 later on). By "worst/best," we refer to both the variance of the estimator and its estimated value, which we can compare to the true value $I = 0.01862$. The results are presented in Table 5.8, and the density of X and the points $\mathbf{x}^*$ are shown in Fig. 5.15. This example is implemented in `ch5_IS_CE_normal_2D.py`. $\diamondsuit$

Table 5.8 Simulation results for estimating $I = \mathbb{P}(X_1 > 2, X_2 > 2)$ for $\mathbf{X} = (X_1, X_2)$ normally distributed using importance sampling with proposal distribution $\phi_{\mathbf{x}^*, \mathbf{\Sigma}}$, see description in text

Estimator	Value of $\hat{Y}_R$	Difference $|\hat{Y}_R - I|$	$\mathbb{V}\mathrm{ar}\hat{Y}_R$
CMC	0.01832	$2.982 \cdot 10^{-4}$	$1.798 \cdot 10^{-7}$
IS $\mathbf{x}^* = (2, 2)$	0.01883	$2.150 \cdot 10^{-4}$	$1.901 \cdot 10^{-8}$
IS $\mathbf{x}^* = (2, 2.83)$	0.01853	$8.050 \cdot 10^{-5}$	$1.010 \cdot 10^{-8}$
IS/CE $\mathbf{x}^* = (2.4, 3.8339)$	0.01860	$1.167 \cdot 10^{-5}$	$9.504 \cdot 10^{-9}$

Example 5.1.39 Consider $X = (X_1, \ldots, X_n)$, where X_i are i.i.d. random variables with $\mathbb{P}(X_i = 0) = \mathbb{P}(X_i = 1) = \frac{1}{2}$. Let $\mu = \mathbb{E}(X) = 0.5$. The goal is to estimate

$$I(n) = \mathbb{P}\left(S_n > n(\mu + \varepsilon)\right), \quad \text{for } \varepsilon \in (0, 1/2).$$

The Crude Monte Carlo (CMC) estimator is:

$$\hat{Y}_R^{\mathrm{CMC}} = \frac{1}{R} \sum_{i=1}^{R} Y_i, \quad \text{where } Y_i = \mathbb{1}(S_n^{(i)} > n(\mu + \varepsilon)),$$

and $S_n^{(i)} = \sum_{j=1}^{n} X_j^{(i)}$ is the sum for the i-th simulation.

Let $\mathbf{e} = (e_1, \ldots, e_n)$, where $e_i \in \{0, 1\}$. Then, $p(\mathbf{e})$ denotes the probability that X takes a specific value $\mathbf{e}$, and since the X_i's are i.i.d., we have

$$\mathbb{P}(X = \mathbf{e}) = p(\mathbf{e}) = \left(\frac{1}{2}\right)^n.$$

For large ε, the probability that $S_n^{(i)} > n(\mu + \varepsilon)$ becomes very small, so large R would be needed for the CMC estimator to give reasonable results. The idea is to choose a proposal distribution that shifts the probability mass to the region where the indicator $\mathbb{1}(S_n^{(i)} > n(\mu + \varepsilon))$ is more likely to be 1.

A natural choice for the proposal distribution is to flip a biased coin, where the probability of heads is $\mu + \varepsilon$. Thus, let $\tilde{X} = (\tilde{X}_1, \ldots, \tilde{X}_n)$, where $\mathbb{P}(\tilde{X}_i = 1) = \mu + \varepsilon$. In other words, the proposal distribution $\tilde{p}$ is:

$$\tilde{p}(\mathbf{e}) = \beta^{\sum_{j=1}^{n} e_j} (1 - \beta)^{n - \sum_{j=1}^{n} e_j}, \quad \text{where } \beta = \mu + \varepsilon.$$

Let $\tilde{S}_n^{(i)} = \sum_{j=1}^{n} \tilde{X}_j^{(i)}$, where $\tilde{X}^{(i)} = (\tilde{X}_1^{(i)}, \ldots, \tilde{X}_n^{(i)})$ are the results of the i-th simulation. Thus, the Importance Sampling estimator is:

$$\hat{Y}_R^{\mathrm{IS}} = \frac{1}{R} \sum_{i=1}^{R} \frac{\mathbb{1}(\tilde{S}_n^{(i)} > n(\mu + \varepsilon))}{\tilde{p}(\tilde{X}^{(i)})} = \frac{1}{R} \sum_{i=1}^{R} \frac{\mathbb{1}(\tilde{S}_n^{(i)} > n(\mu + \varepsilon))}{\beta^{\tilde{S}_n^{(i)}} (1 - \beta)^{n - \tilde{S}_n^{(i)}}}.$$

Table 5.9 Comparison of CMC and IS estimators for different parameters using $R = 1000$ replications

n	ε	Exact $I(n)$	Estimator	Mean	Variance
50	0.25	0.0001529320	CMC	0.0000000	0.00000000
			IS	0.0001707	0.00000008
50	0.10	0.0594602263	CMC	0.0600000	0.05645650
			IS	0.0591562	0.00717780

For $n = 50$ and $\varepsilon \in \{0.25, 0.1\}$, we summarize the results of the CMC and IS estimators in Table 5.9. Note that for $\varepsilon = 0.25$, the event of interest is extremely rare under the original distribution, and the CMC estimator never observed a single occurrence in $R = 1000$ trials, resulting in an estimated probability of zero and consequently a zero variance. In contrast, the IS estimator provided a non-zero estimate close to the exact value with a small variance, demonstrating the effectiveness of importance sampling in handling rare events. The example is implemented in `ch5_IS_coin_flipping.py`. $\Diamond$

Example 5.1.40   We continue Example 5.1.10 (*Probability of network failure*). Recall that a state is described by $\mathbf{e} = (e_1, \ldots, e_n)$, where $e_i = 1$ means the edge is broken/removed and $e_i = 0$ means the edge is working. Thus, we have a random variable X, describing the network with the distribution:

$$\mathbb{P}(X = \mathbf{e}) = p(\mathbf{e}) = q^{\sum_{j=1}^{n} e_j} (1 - q)^{n - \sum_{j=1}^{n} e_j}.$$

We will analyze an importance sampling estimator with $\tilde{p}(\mathbf{e}) = \beta^{|\mathbf{e}|}(1 - \beta)^{n - |\mathbf{e}|}$, where we remove each edge with probability β. We have:

$$I = \sum_{\mathbf{e} \in S} k(\mathbf{e}) p(\mathbf{e}) = \sum_{\mathbf{e} \in S} \frac{k(\mathbf{e}) p(\mathbf{e})}{\tilde{p}(\mathbf{e})} \tilde{p}(\mathbf{e}).$$

Note that $k(\mathbf{e}) = 1$ if and only if the network fails, otherwise $k(\mathbf{e}) = 0$. Thus:

$$I = \sum_{\mathbf{e}:k(\mathbf{e})=1} \frac{p(\mathbf{e})}{\tilde{p}(\mathbf{e})} \tilde{p}(\mathbf{e}),$$

where

$$\frac{p(\mathbf{e})}{\tilde{p}(\mathbf{e})} = \frac{q^{|\mathbf{e}|}(1 - q)^{n - |\mathbf{e}|}}{\beta^{|\mathbf{e}|}(1 - \beta)^{n - |\mathbf{e}|}} = \left(\frac{1 - q}{1 - \beta}\right)^n \left(\frac{q(1 - \beta)}{\beta(1 - q)}\right)^{|\mathbf{e}|},$$

with $|\mathbf{e}| = \sum_{j=1}^{n} e_j$. We assume $\beta \geq q$, which implies:

$$\frac{q(1 - \beta)}{\beta(1 - q)} \leq 1.$$

Let d be the smallest number of failed edges that cause the entire network to fail. For example, for a serial network, $d = 1$, while for a parallel network, $d = n$. For the network in Fig. 5.5, we have $d = 3$. Thus, if $k(\mathbf{e}) = 1$, then $|\mathbf{e}| \geq d$. Assuming $|\mathbf{e}| \geq d$, we have:

$$\left(\frac{1 - q}{1 - \beta}\right)^n \left(\frac{q(1 - \beta)}{\beta(1 - q)}\right)^{|\mathbf{e}|} \leq \left(\frac{1 - q}{1 - \beta}\right)^n \left(\frac{q(1 - \beta)}{\beta(1 - q)}\right)^d = \frac{(1 - q)^{n-d} q^d}{\beta^d (1 - \beta)^{n-d}}.$$

Now, we want to minimize the right-hand side of the above equation for $\beta \geq q$. The problem is equivalent to finding the maximum of $\beta^d (1 - \beta)^{n-d}$. By setting the derivative equal to zero, we get:

$$d\beta^{d-1}(1 - \beta)^{n-d} - (n - d)\beta^d (1 - \beta)^{n-d-1} = 0,$$

which simplifies to $\beta = \frac{d}{n}$. Thus, for $|\mathbf{e}| \geq d$, we have:

$$\left(\frac{1 - q}{1 - \frac{d}{n}}\right)^n \left(\frac{q(1 - \frac{d}{n})}{\frac{d}{n}(1 - q)}\right)^d = \delta.$$

If $\frac{d}{n} \geq q$, then $\delta \leq 1$. Now, considering the second moment:

$$\tilde{\mathbb{E}}k^2(X)\frac{p^2(X)}{\tilde{p}^2(X)} = \sum_{\{\mathbf{e}:\, k(\mathbf{e})=1\}} \frac{p^2(\mathbf{e})}{\tilde{p}^2(\mathbf{e})}\tilde{p}(\mathbf{e}) = \sum_{\{\mathbf{e}:\, \kappa(\mathbf{e})=1\}} \frac{p(\mathbf{e})}{\tilde{p}(\mathbf{e})}p(\mathbf{e})$$

$$= \sum_{\{\mathbf{e}:\, k(\mathbf{e})=1\}} \left(\frac{1 - q}{1 - \beta}\right)^n \left(\frac{q(1 - \beta)}{\beta(1 - q)}\right)^{|\mathbf{e}|} p(\mathbf{e})$$

$$\leq \delta \sum_{\{\mathbf{e}:\, |\mathbf{e}| \geq d\}} p(\mathbf{e}) = \delta I.$$

Since I is usually small, we approximate the variance of the estimator:

$$\mathbf{Var}\hat{Y}_R = \frac{I(1 - I)}{R} \approx \frac{I}{R}.$$

On the other hand, the variance of the IS estimator is:

$$\mathbf{Var}\hat{Y}_R^{\text{IS}} \sim \frac{\delta}{R}I.$$

This shows that the IS estimator has $1/\delta$ times smaller variance than the CMC estimator.

For the network in Fig. 5.5, with $q = 0.01$ and $d = 3$, we have $\beta = \frac{3}{22}$, and thus $\delta = 0.0053$. Using the IS estimator, we need approximately $\sqrt{1/\delta} \approx 14$ times fewer replications than with the CMC estimator. $\qquad\qquad\qquad\qquad\diamond$

Self-normalized Importance Sampling Recall that we want to estimate $I = \mathbb{E}k(X) = \int k(x)f(x)\,dx$. Sometimes we can sample from $f(x)$, but we know it only up to a constant (e.g., it has the desired shape, etc.). This means $f(x) = cf_0(x)$, where we can compute $f_0(x)$, but we do not know c (it is determined by $\int f(x) = 1$, but we may be unable to compute it). Suppose we have another density $\tilde{f}(x)$, and similarly, we can sample from this density, but we know it only up to a constant. Say $\tilde{f}(x) = \tilde{c}\tilde{f}_0(x)$, where we can compute $\tilde{f}_0(x)$, but we do not know $\tilde{c}$. If we knew the ratio $c/\tilde{c}$ (which is rarely the case), we could proceed with the importance sampling estimator as previously. Otherwise, the idea is as follows. Let us define $w(x) = \frac{f_0(x)}{\tilde{f}_0(x)}$ and compute:

$$I = \int k(x)f(x)\,dx = \int \frac{k(x)f(x)}{\tilde{f}(x)}\tilde{f}(x) = \frac{\int k(x)\frac{f(x)}{\tilde{f}(x)}\tilde{f}(x)}{\int \frac{f(x)}{\tilde{f}(x)}\tilde{f}(x)}.$$

Thus, we have:

$$I = \frac{\int k(x)w(x)\tilde{f}(x)}{\int w(x)\tilde{f}(x)} = \frac{\mathbb{E}(k(\tilde{X})w(\tilde{X}))}{\mathbb{E}(w(\tilde{X}))},$$

where $\tilde{X}$ is a random variable with density $\tilde{f}(x)$. As a result, we have the following estimator for I:

$$\hat{Y}_R = \frac{\sum_{i=1}^{R} k(\tilde{X}_i)w(\tilde{X}_i)}{\sum_{i=1}^{R} w(\tilde{X}_i)}.$$

To provide a confidence interval for I, we need an estimator $\widehat{\mathbb{V}\mathrm{ar}}(\hat{Y}_R)$ for $\mathbb{V}\mathrm{ar}(\hat{Y}_R)$. Using the delta method and the formula (2.13), we get:

$$\widehat{\mathbb{V}\mathrm{ar}}(\hat{Y}_R) = \frac{1}{R^2\left(\frac{1}{R}\sum_{i=1}^{R} w(X_i)\right)^2}\sum_{i=1}^{R}(k(X_i)w(X_i) - \hat{Y}_R w(X_i))^2$$

$$= \frac{1}{\left(\sum_{i=1}^{R} w(X_i)\right)^2}\sum_{i=1}^{R}(w(X_i))^2(k(X_i) - \hat{Y}_R)^2$$

$$= \sum_{i=1}^{R}(w'(X_i))^2(k(X_i) - \hat{Y}_R)^2,$$

where $w'(X_i) = w(X_i)/\sum_{j=1}^{R} w(X_j)$ is the so-called i-th normalized weight. Finally, the error of the estimator is

$$b = z_{1-\alpha/2}\sqrt{\widehat{\mathbb{V}\mathrm{ar}}(\hat{Y}_R)}.$$

Remark 5.1.41 Note that above we applied the delta method to $w(x) = \frac{f_0(x)}{\tilde{f}_0(x)}$. This is an extension of Example 4.2.7, where we applied the method to $w(x) = \mathbb{1}(x \in A)$. ∎

Likelihood Ratio Function Consider random variables $X_1, X_2, \ldots$ under two probabilities: $\mathbb{P}$ and $\tilde{\mathbb{P}}$ such that $X_1, X_2, \ldots$ are i.i.d. random variables with density f or $\tilde{f}$, respectively.

Proposition 5.1.42 *Let $\tilde{f}(x) > 0$ whenever $f(x) > 0$. Then we have:*

$$\mathbb{P}(d\omega) = L_n(\omega)\tilde{\mathbb{P}}(d\omega),$$

where

$$L_n = \prod_{j=1}^{n} \frac{f(X_j)}{\tilde{f}(X_j)}. \tag{5.1.24}$$

For a random variable Y

$$I = \mathbb{E}Y = \tilde{\mathbb{E}}[Y L_n].$$

provided Y is integrable.

Proof Let $A = \{X_1 \in B_1, \ldots, X_n \in B_n\}$. Then:

$$\mathbb{P}(A) = \mathbb{P}(X_1 \in B_1, \ldots, X_n \in B_n) = \int_{B_1} \cdots \int_{B_n} f(x_1) \cdots f(x_n) \, dx_1 \ldots dx_n.$$

This is equivalent to:

$$\int_{B_1} \cdots \int_{B_n} \frac{f(x_1) \cdots f(x_n)}{\tilde{f}(x_1) \cdots \tilde{f}(x_n)} \tilde{f}(x_1) \cdots \tilde{f}(x_n) \, dx_1 \ldots dx_n,$$

which means:

$$\mathbb{E}\mathbb{1}(A) = \tilde{\mathbb{E}}[L_n \mathbb{1}_A].$$

Thus, approximating Y with simple functions, we obtain:

$$I = \tilde{\mathbb{E}}[Y L_n].$$

$\square$

Note that L, defined in (5.1.24), is called a *likelihood ratio*.

Example 5.1.43 We now calculate the likelihood ratio for the case where f is the density of the standard normal distribution $\mathcal{N}(0, 1)$, while $\tilde{f}$ is the density of a normal distribution $\mathcal{N}(\mu, 1)$. Then:

$$L_n = \exp\left(-\mu \sum_{j=1}^{n} X_j + \frac{\mu^2}{2} n\right).$$

On the other hand, let us define $\widetilde{\mathbb{P}}$ on $\mathcal{F}_n$ as:

$$d\tilde{\mathbb{P}} = L_n^{-1} d\mathbb{P}.$$

Then, with the new probability $\tilde{\mathbb{P}}$, the variables $X_1, X_2, \ldots$ are i.i.d. with distribution $\mathcal{N}(\mu, 1)$. This is an example of exponential change of measure. $\qquad\qquad \Diamond$

Exponential Change of Measure; Rare Event Simulation Suppose that $X_1, X_2, \ldots$ are i.i.d. random variables with p.d.f. $f(x)$. Let $\mu = \mathbb{E}X_1$. For a given $\varepsilon \in (0, 1)$ and integer n, our goal is to estimate

$$I(n) = \mathbb{P}(S_n > n(\mu + \varepsilon)).$$

For $\theta \geq 0$, we consider the distribution $f_\theta(x) = \frac{e^{\theta x} f(x)}{M(\theta)}$, where $M(\theta) = \mathbb{E}e^{\theta X}$ is the moment generating function of X. We also define the logarithmic moment generating function as

$$\kappa(\theta) = \log M(\theta).$$

Under the probability $\mathbb{P}_\theta$, random variables $X_1, X_2, \ldots$ are i.i.d. with p.d.f. $f_\theta(x)$. In other words, $X_1, X_2, \ldots$ are the same *random variables*, but with a different underlying distribution. Note that $f_0(x) = f(x)$, so $\mathbb{P}_0 = \mathbb{P}$. By $\mathbb{E}_\theta g(X_1, \ldots, X_n)$ and $\mathbb{V}\mathrm{ar}_\theta g(X_1, \ldots, X_n)$, we denote the expectation and variance of $g(X_1, \ldots, X_n)$ under the probability $\mathbb{P}_\theta$, i.e., for X_i with density $f_\theta(x)$, $i = 1, \ldots, n$.

Let us compute the moment generating function $\mathbb{E}_\theta e^{sX}$ of X:

$$\mathbb{E}_\theta e^{sX} = \int e^{sx} f_\theta(x)\, dx$$

$$= \frac{M(s + \theta)}{M(\theta)}. \qquad\qquad (5.1.25)$$

The first moment is $\mathbb{E}_\theta X_1 = \frac{M'(\theta)}{M(\theta)} = \kappa'(\theta)$.

Let $\mathbf{X} = (X_1, \ldots, X_n)$. For a Borel subset $B \subset \mathbb{R}^d$

$$\mathbb{P}(\mathbf{X} \in B) = \int \ldots \int \mathbb{1}_{\{\mathbf{x} \in B\}} f(x_1) \ldots f(x_n)\, dx_1 \ldots dx_n \qquad (5.1.26)$$

$$= \int \ldots \int \mathbb{1}_{\{\mathbf{x} \in B\}} \frac{f(x_1) \ldots f(x_n)}{f_\theta(x_1) \ldots f_\theta(x_n)} f_\theta(x_1) \ldots f_\theta(x_n)\, dx_1 \ldots dx_n$$

$$= \int \ldots \int \mathbb{1}_{\{\mathbf{x} \in B\}} M^n(\theta) e^{-\theta(x_1 + \cdots + x_n)} f_\theta(x_1) \ldots f_\theta(x_n)\, dx_1 \ldots dx_n$$

$$= \int \ldots \int \mathbb{1}_{\{\mathbf{x} \in B\}} e^{-\theta(x_1 + \cdots + x_n) + n\kappa(\theta)} f_\theta(x_1) \ldots f_\theta(x_n)\, dx_1 \ldots dx_n$$

$$= \mathbb{E}_\theta \left[\mathbb{1}_{\{\mathbf{X} \in B\}} e^{-\theta S_n + n\kappa(\theta)} \right].$$

In the sequel, let $A = \{S_n > n(\mu + \varepsilon)\}$. Note that for any $\theta \geq 0$,

$$I(n) = \mathbb{P}(A) = \mathbb{P}(S_n > n(\mu + \varepsilon)) = \mathbb{E}_\theta[\mathbb{1}_{\{S_n > n(\mu+\varepsilon)\}} e^{-\theta S_n + n\kappa(\theta)}].$$

Consider the random variable

$$Y^\theta(n) = \mathbb{1}_A e^{-\theta S_n + n\kappa(\theta)}$$

and let $Y_1^\theta(n), \ldots Y_R^\theta(n)$ of be i.i.d. replications of $Y^\theta(n)$. Denote their mean by

$$\hat{Y}_R^\theta(n) = \frac{1}{R} \sum_{i=1}^{R} Y_i^\theta(n).$$

Of course $\mathbb{E}\hat{Y}_R^\theta(n) = \mathbb{E}Y^\theta(n) = I(n)$ for any θ and $n \geq 0$. Classically—see (1.5)—we have the following **confidence interval for** $I(n)$ at level α:

$$\mathbb{P}\left(I(n) \in \left(\hat{Y}_R^\theta(n) - z_{1-\alpha/2} \frac{\sqrt{\mathbb{V}\mathrm{ar}_\theta Y^\theta(n)}}{\sqrt{R}}, \hat{Y}_R^\theta(n) + z_{1-\alpha/2} \frac{\sqrt{\mathbb{V}\mathrm{ar}_\theta Y^\theta(n)}}{\sqrt{R}}\right)\right)$$

$$\approx 1 - \alpha.$$

Notice that $I(n) \to 0$, which means that we are dealing with a *rare event simulation*. In this theory, we focus on the **relative error**

$$\frac{\hat{Y}_R^\theta(n) - I(n)}{I(n)}.$$

From the Central Limit Theorem (see Exercise 4.T.4) we have the following **confidence interval for relative error** at level α

$$\mathbb{P}\left(\frac{\hat{Y}_R^{\theta}(n) - I(n)}{I(n)} \in \left(-\frac{z_{1-\alpha/2}}{\sqrt{R}}\sqrt{\frac{\mathbb{V}\mathrm{ar}_\theta Y^\theta(n)}{I^2(n)}}, \frac{z_{1-\alpha/2}}{\sqrt{R}}\sqrt{\frac{\mathbb{V}\mathrm{ar}_\theta Y^\theta(n)}{I^2(n)}}\right)\right) \approx 1 - \alpha.$$

$$(5.1.27)$$

We approximate the confidence interval by

$$\left(-\frac{z_{1-\alpha/2}}{\sqrt{R}}\sqrt{\frac{\widehat{\mathbb{V}\mathrm{ar}}Y^\theta(n)}{(\hat{Y}_R^\theta(n))^2}}, \frac{z_{1-\alpha/2}}{\sqrt{R}}\sqrt{\frac{\widehat{\mathbb{V}\mathrm{ar}}Y^\theta(n)}{(\hat{Y}_R^\theta(n))^2}}\right), \tag{5.1.28}$$

i.e., we estimate $I(n)$ via the sample mean $\hat{Y}_R^\theta$ and $\mathbb{V}\mathrm{ar}_\theta Y^\theta(n)$ via the sample variance

$$\widehat{\mathbb{V}\mathrm{ar}}Y^\theta(n) = \frac{1}{R}\sum_{i=1}^{R}(Y_i^\theta(n) - \hat{Y}_R^\theta(n))^2.$$

As noted earlier, $\mathbb{E}\hat{Y}_R^\theta(n) = I(n)$ independently from θ, however, the variance $\mathbb{V}\mathrm{ar}_\theta Y^\theta(n)$ does depend on θ. We would like to find a "good" θ. A desired property for the estimator $\hat{Y}_R^\theta(n)$ is that it has a *bounded relative error*, that is

$$\limsup_{n\to\infty} \frac{\mathbb{V}\mathrm{ar}_\theta Y^\theta(n)}{I^2(n)} < \infty.$$

In other words, that the length of the confidence interval for the relative error does not grow to infinity as $n \to \infty$. However, in the following theorem, we are able to prove a slightly weaker property, namely that the estimator is *logarithmically effective*, i.e., that for every $0 < \epsilon < 2$ we have

$$\limsup_{n\to\infty} \frac{\mathbb{V}\mathrm{ar}_\theta Y^\theta(n)}{I^{2-\epsilon}(n)} < \infty.$$

We need to recall a known fact: the moment generating function is log-convex, thus $\kappa'(\theta)$ is strictly increasing. Furthermore $\kappa'(0) = \mu$. We now choose $\theta = \theta_0$ fulfilling

$$\mathbb{E}_{\theta_0} X_1 = \kappa'(\theta_0) = \mu + \varepsilon.$$

The situation is depicted in Fig. 5.16.

In the following theorem we demonstrate optimal properties of the estimator $\hat{Y}_R^{\theta_0}$ defined by $Y^{\theta_0}(n) = \mathbb{1}_{\{S_n > n(\mu+\varepsilon)\}} e^{-\theta_0 S_n + n\kappa(\theta_0)}$, under probability $\mathbb{P}_{\theta_0}$.

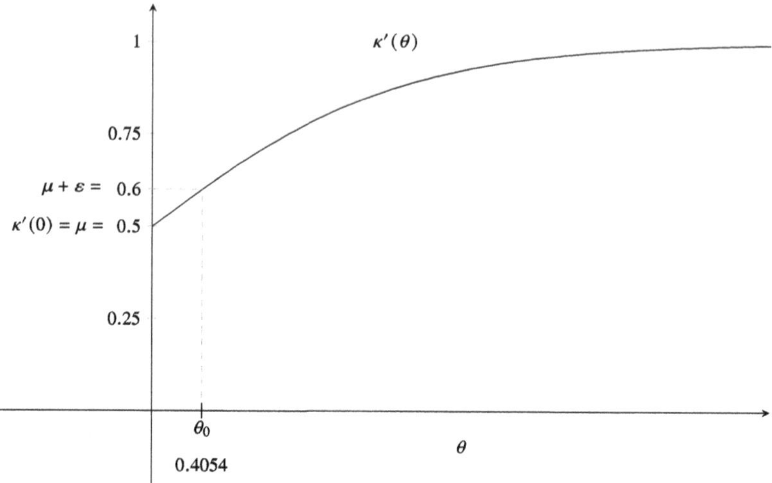

Fig. 5.16 Derivative of the logarithmic moment genarating function κ for X with distribution $\mathbb{P}(X_1 = 0) = \mathbb{P}(X_1 = 1) = 1/2$ and solution $\kappa'(\theta_0) = \mu + \varepsilon$. See Example 5.1.46

Theorem 5.1.44 *Assume that* $\kappa''(\theta_0) < \infty$. *The estimator* $\hat{Y}_R^{\theta_0}(n)$ *is logarithmically effective.*

Proof Set $\gamma = \theta_0(\mu + \varepsilon) - \kappa(\theta_0)$. We will prove the following facts:

$$\mathbb{V}\mathrm{ar}_{\theta_0} Y^{\theta_0}(n) \leq e^{-2n\gamma}, \tag{5.1.29}$$

and

$$\liminf_{n \to \infty} e^{n\gamma + \theta_0\sqrt{n}} \mathbb{E}_{\theta_0}[\mathbb{1}_{\{S_n > n(\mu+\varepsilon)\}} e^{-\theta_0 S_n + n\kappa(\theta_0)}] = c > 0. \tag{5.1.30}$$

Hence, for some $\delta \in (0, c)$ and n_0

$$I(n) = \mathbb{E}_{\theta_0}[\mathbb{1}_{\{S_n > n(\mu+\varepsilon)\}} e^{-\theta_0 S_n + n\kappa(\theta_0)}] \geq (c - \delta)e^{-n\gamma - \theta_0\sqrt{n}}, \qquad n \geq n_0.$$

Thus

$$\limsup_{n \to \infty} \frac{\mathbb{V}\mathrm{ar}_{\theta_0} Y^{\theta_0}(n)}{(I(n))^{2-\varepsilon}} \leq \limsup_{n \to \infty} \frac{e^{-n\gamma}}{((c - \delta)e^{-n\gamma - \theta_0\sqrt{n}})^{2-\varepsilon}}$$

$$= \limsup_{n \to \infty} e^{-\varepsilon\gamma n + (\varepsilon - 2)\sqrt{n}\theta_0} = 0,$$

which means that $\hat{Y}_R^{\theta_0}(n)$ is logarithmically bounded.

To complete the proof we must verify (5.1.29) and (5.1.30).

We have $\mathbb{V}\mathrm{ar}_{\theta_0} Y^{\theta_0}(n) \le \mathbb{E}_{\theta_0}[(Y^{\theta_0}(n))^2]$ and

$$\mathbb{E}_{\theta_0}[(Y^{\theta_0}(n))^2] = \mathbb{E}_{\theta_0}[\mathbb{1}_{\{S_n > n(\mu+\varepsilon)\}} e^{-2(\theta_0 S_n + n\kappa(\theta_0))}].$$

Now we use the Central Limit Theorem on $S_n = \sum_{j=1}^{n} X_j$, where $X_1, X_2, \ldots$ are i.i.d. with logarithmic moment generating function $\log M(s + \theta_0) - \log M(\theta_0)$,

$$\mathbb{1}_{\{S_n > n(\mu+\varepsilon)\}} e^{-2(\theta_0 S_n - n\kappa(\theta_0))} = \mathbb{1}_{\{S_n > n(\mu+\varepsilon)\}} e^{-2(\theta_0 (S_n - n(\mu+\varepsilon)) + \theta_0 n(\mu+\varepsilon) + n\kappa(\theta_0))}$$

$$\le e^{-2(\theta_0 \cdot 0 + n(\theta_0(\mu+\varepsilon) + \kappa(\theta_0)))}$$

$$= e^{-2n\gamma}.$$

To demonstrate (5.1.30) we have to observe that under probability $\mathbb{P}_{\theta_0}$ the Central Limit Theorem holds in the form:

$$\frac{S_n - n(\mu + \varepsilon)}{\sqrt{n}} \overset{\mathcal{D}}{\to} \mathcal{N}(0, \kappa''(\theta_0)),$$

because X_i has the logarithmic moment generating function $\log M(s + \theta_0) - \log M(\theta_0)$ and hence

$$\mathbb{E}_{\theta_0} X_i = \kappa'(\theta_0) = \mu + \varepsilon \qquad \mathbb{V}\mathrm{ar}_{\theta_0} X_i = \kappa''(\theta_0).$$

Now, as $n \to \infty$ we have

$$e^{n\gamma + \theta_0\sqrt{n}} \mathbb{E}_{\theta_0}[\mathbb{1}_{\{S_n > n(\mu+\varepsilon)\}} e^{-(\theta_0 S_n + n\kappa(\theta_0))}]$$

$$= e^{n\gamma + \theta_0\sqrt{n}} e^{n\gamma} \mathbb{E}_{\theta_0}[\mathbb{1}_{\{S_n > n(\mu+\varepsilon)\}} e^{-(\theta_0(S_n - n(\mu+\varepsilon)))}]$$

$$\ge e^{n\gamma + \theta_0\sqrt{n}} e^{n\gamma + \theta_0\sqrt{n}} \mathbb{P}_{\theta_0}\left(\frac{S_n - n(\mu + \varepsilon)}{\sqrt{n}} \in (0, 1)\right)$$

$$\to \mathbb{P}(\sqrt{\kappa''(\theta_0)} Z \in (0, 1)) = c > 0,$$

where $Z \sim \mathcal{N}(0, 1)$. $\qquad\qquad\square$

Remark 5.1.45 It can be proven that the estimator $\hat{Y}_R^{\theta_0}(n)$ from Theorem 5.1.44 is the only one that is logarithmically efficient in the class of all importance sampling estimators (not necessarily of the form $f_\theta(x) = e^{\theta x} f(x)$) for the problem $I(n) = \mathbb{P}(X_1 + \cdots + X_n > n(\mu + \epsilon))$. $\qquad\blacksquare$

The presented theory works also for discrete random variables $X_1, X_2, \ldots$ on the probability space with probability function $(p_j)_{j \ge 0}$ (i.e., $\mathbb{P}(X_i = j) = $

$p_j, j = 0, 1, \ldots$). Let $\mu = \mathbb{E}X$. The moment generating function is then $M(s) = \sum_{j=0}^{\infty} p_j e^{sj}$, the goal is again to estimate (recall, $S_n = X_1 + \cdots, X_n$)

$$I(n) = \mathbb{P}(S_n > n(\mu + \varepsilon)).$$

Random variables $X_1, X_2, \ldots$ under probability $\mathbb{P}^\theta$ are i.i.d. with distribution function

$$p_j^\theta = e^{j\theta} p_j, \quad j = 0, 1, \ldots, \quad \theta \geq 0.$$

(Note that $p_j^0 \equiv p_j$). Similarly as in (5.1.27) one may show that

$$\mathbb{P}(S_n > n(\mu + \varepsilon) = \mathbb{E}_\theta \mathbb{1}_{S_n > n(\mu+\varepsilon)} e^{-\theta S_n + n\kappa(\theta)}.$$

Everything else remains unchanged.

Example 5.1.46 Consider $X_1, X_2, \ldots$ i.i.d. random variables, $\mathbb{P}(X_i = 0) = \mathbb{P}(X_i = 1) = 1/2$, thus $\mu = 0.5$. The goal is to estimate

$$I(n) = \mathbb{P}(S_n > n(\mu + \varepsilon)).$$

In Example 5.1.39 we considered importance sampling for $I(n)$. We have $M(s) = (1 + e^s)/2$ and $\kappa(s) = \log M(s)$. Under $\mathbb{P}_\theta$, the moment generating function is

$$\mathbb{E}_\theta e^{sX} = M(s + \theta)/M(\theta) = \frac{1}{1 + e^\theta} + \frac{e^\theta}{1 + e^\theta} e^s.$$

To compute θ_0 notice that $\kappa'(s) = M'(s)/M(s)$, thus we need to solve

$$\kappa'(\theta_0) = \mu + \varepsilon \quad \Leftrightarrow \quad \frac{e^{\theta_0}}{1 + e^{\theta_0}} = 1/2 + \varepsilon \quad \Leftrightarrow \quad e^{\theta_0} = \frac{1 + 2\varepsilon}{1 - 2\varepsilon},$$

$$\theta_0 = \log(1 + 2\varepsilon) - \log(1 - 2\varepsilon).$$

Thus, under $\mathbb{P}_{\theta_0}$, X_i takes value 0 with probability $\frac{1}{2} - \varepsilon$ and 1 with probability $\frac{1}{2} + \varepsilon$.

For $\varepsilon = 0.1$ we performed simulations for $n = 2, \ldots, 50$, for each n we used $R = 1000$ replications to compute $\hat{Y}_R^\theta(n)$ and its variance. For this case we have $\theta_0 = 0.40546$, we also considered $\theta \in \{0.1, 0.2, 1.4, 1.6\}$. The derivative of the logarithmic moment generating function, i.e., $\kappa'(\theta)$, as well as the solution $\kappa'(\theta_0) = \mu + \varepsilon$, are depicted in Fig. 5.16.

Note that one can also numerically compute $I(n)$ for this case, simply

$$I(n) = \mathbb{P}\left(S_n > n(\mu + \varepsilon)\right) = \mathbb{P}\left(S_n > 0.6n\right) = \frac{1}{2^n} \sum_{k=\lfloor 0.6n \rfloor + 1}^{n} \binom{n}{k}.$$

We will use true values of $I(n)$ for a comparison with estimators. Moreover, for a comparison we also estimated $I(n)$ for each n using CMC estimator, i.e., we simply simulated $X_1^i, \ldots, X_n^i$ i.i.d. with p.d.f. $f(x)$ for $i = 1, \ldots, R$ and computed

$$Y_i(n) = \mathbb{1}\left(X_1^i + \cdots + X_n^i > 0.6n\right),$$

$$\hat{Y}_R^{\mathrm{CMC}}(n) = \frac{1}{R} \sum_{i=1}^{R} Y_i(n), \quad \widehat{\mathrm{Var}}Y^{\mathrm{CMC}}(n) = \frac{1}{R} \sum_{i=1}^{R} (Y_i(n) - \hat{Y}_R^{\mathrm{CMC}}(n))^2.$$

The confidence intervals for relative errors are computed using (5.1.28) (where in the case of CMC we plug in $\widehat{\mathrm{Var}}Y^{\mathrm{CMC}}(n)$ and $\hat{Y}_R^{\mathrm{CMC}}(n)$ correspondingly). We considered CMC (i.e., $\theta = 0$), optimal $\theta_0 = 0.405$ and $\theta = 1.4$. The results are collected in Fig. 5.17.

As we can see in Fig. 5.17a, all estimators estimate the true value of $I(n)$ pretty well at first glance (with $\theta = 1.4$ being noticeably worse). However, in Fig. 5.17b (plotted on a log scale), we can see that the ratio

$$\frac{\widehat{\mathrm{Var}}Y^{\theta}(n)}{(\hat{Y}_R^{\theta}(n))^2}$$

is the smallest (best) for the optimal $\theta = \theta_0$, worse for CMC, and the worst for $\theta = 1.4$. This implies that the relative errors—which, in this example, we could calculate since we actually know the true values of $I(n)$—(depicted in Fig. 5.18a) are (most often) smallest for θ_0 and largest for $\theta = 1.4$ (this is more evident for larger n). In turn, the confidence intervals for the relative error are narrowest for the optimal θ_0, as can be seen in Fig. 5.18b.

Simulations confirm that using the method with $\theta = \theta_0$ yields the best results. We can see (Fig. 5.17b) that the ratio

$$\frac{\widehat{\mathrm{Var}}_{\theta_0}[Y^{\theta_0}(n)]}{(I_{\theta_0}(n))^2}$$

is constant (or grows very slowly) for θ_0, and we can conjecture that it grows linearly for other θ's. Note that in this example we were able to compute the true value of $I(n)$; in most examples, we would be able to provide only confidence intervals for the relative error. This example is implemented in `ch5_rare_event_expon_change_measure.py`. $\Diamond$

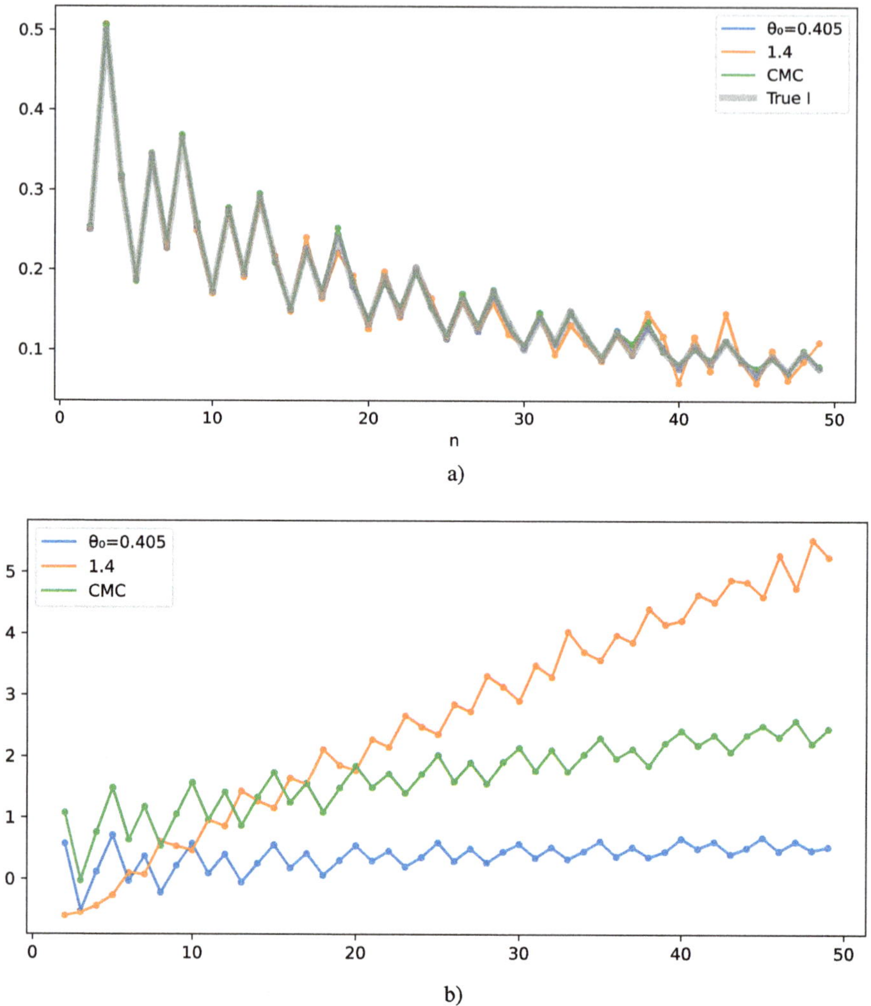

Fig. 5.17 (**a**) and (**b**): Example 5.1.46 for $\varepsilon = 0.1, n = 2, 3, \ldots, 50, \theta \in \{0 \text{ (CMC)}, \theta_0 = 0.405, 1.4\}$. For each n we performed $R = 1000$ replications. (**a**) Estimations $\hat{Y}_R^\theta(n)$ (*green*: value of $\hat{Y}_R^{\mathrm{CMC}}$, *gray*: true value of I). (**b**) Log ratio $\log \frac{\widehat{\mathrm{Var}} Y^\theta(n)}{(\hat{Y}_R^\theta(n))^2}$ (*green*: ratio for CMC estimator)

5.1.8 The Cross-entropy Method

The goal is to estimate

$$I = \mathbb{E}_f k(\mathbf{X}) = \int k(\mathbf{x}) f(\mathbf{x}) \, d\mathbf{x},$$

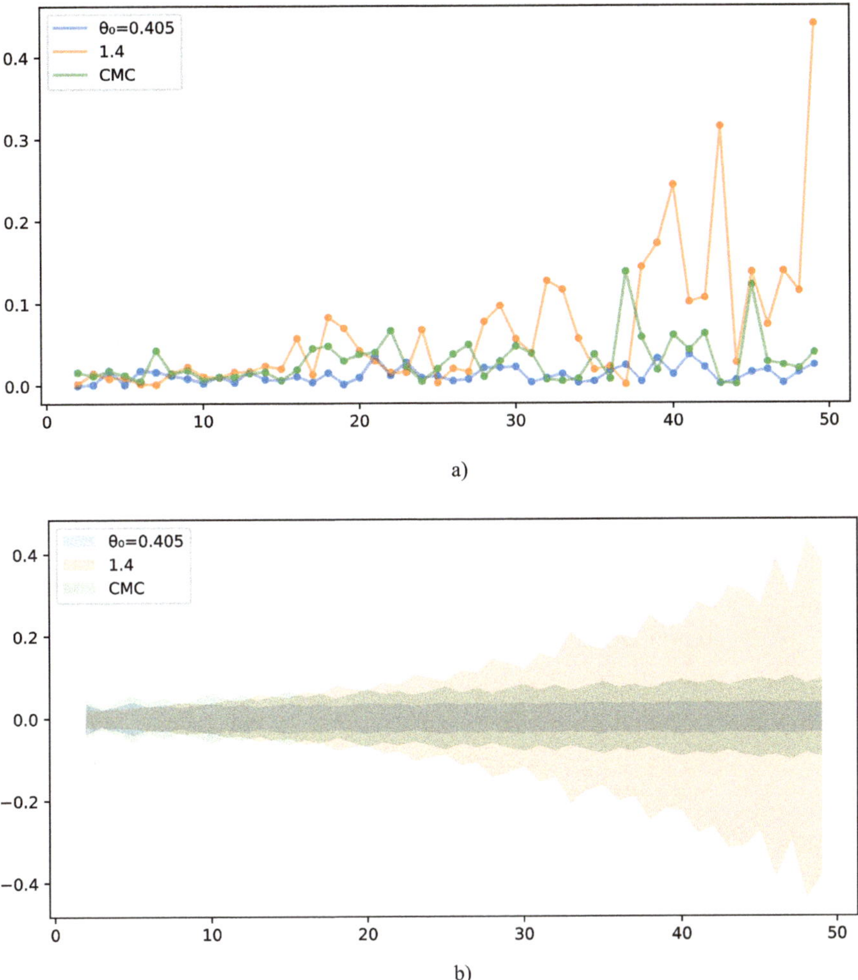

a)

b)

Fig. 5.18 (**a**) and (**b**) Example 5.1.46: corresponding absolute errors and confidence intervals. (**a**) Corresponding absolute relative errors $\left|\frac{\hat{Y}_R^\theta(n)-I(n)}{I(n)}\right|$. (**b**) Corresponding confidence intervals for relative errors. Colors overlap

where $\mathbf{X} = (X_1,\ldots,X_d)$ has a p.d.f. f. Whenever we write $\mathbb{E}_f(\cdot)$ or $\mathbb{P}_f(\cdot)$, we mean computing the quantities for $\mathbf{X}$ having p.d.f. f.

In the current notation, the optimal importance sampling distribution (5.1.22) is

$$g^*(\mathbf{x}) = \frac{k(\mathbf{x})f(\mathbf{x})}{\int_E k(\mathbf{x})f(\mathbf{x})\,d\mathbf{x}} = \frac{k(\mathbf{x})f(\mathbf{x})}{I}.$$

As already mentioned, the use of g^* is impractical, but we may try to find a proposal distribution that is "close" to it. The whole point of the cross-entropy method is to find "best" proposal distributions within a prespecified class of distributions $\{\tilde{f}_\theta(\mathbf{x})\}$ parametrized by $\theta \in \Theta$. In previous importance sampling examples, we chose proposal distributions $\tilde{f}_\theta$ with some fixed θ, see Examples 5.1.36 and 5.1.38. For this, we can use the so-called *Kullback-Leibler divergence* (KL) between g^* and $\tilde{f}_\theta$:

$$\mathcal{D}(g^*, \tilde{f}_\theta) = \int g^*(\mathbf{x}) \log g^*(\mathbf{x}) \, d\mathbf{x} - \int g^*(\mathbf{x}) \log \tilde{f}_\theta(\mathbf{x}) \, d\mathbf{x},$$

which can be rewritten as

$$\mathcal{D}(g^*, \tilde{f}_\theta) = \mathbb{E}_{g^*} \log \left(\frac{g^*(\mathbf{X})}{\tilde{f}_\theta(\mathbf{X})} \right).$$

Note that $\mathcal{D}$ is called a divergence, not a distance, since it is not symmetric, i.e., $\mathcal{D}(g^*, \tilde{f}_\theta) \neq \mathcal{D}(\tilde{f}_\theta, g^*)$ (unless $\tilde{f}_\theta \equiv g^*$). The heuristic behind the cross-entropy method is the following: try to find θ such that the KL divergence between g^* and $\tilde{f}_\theta$ is small. Ideally:

$$\theta^* = \operatorname*{argmin}_\theta \mathcal{D}(g^*, \tilde{f}_\theta) = \operatorname*{argmax}_\theta \int g^*(\mathbf{x}) \log \tilde{f}_\theta(\mathbf{x}) \, d\mathbf{x}. \tag{5.1.31}$$

Substituting the optimal g^*, we have (note that I has no influence on argmax):

$$\theta^* = \operatorname*{argmax}_\theta \int \frac{k(\mathbf{x}) f(\mathbf{x})}{I} \log \tilde{f}_\theta(\mathbf{x}) \, d\mathbf{x}$$

$$= \operatorname*{argmax}_\theta \int k(\mathbf{x}) \log \tilde{f}_\theta(\mathbf{x}) f(\mathbf{x}) \, d\mathbf{x}$$

$$= \operatorname*{argmax}_\theta \mathbb{E}_f \left[k(\mathbf{X}) \log \tilde{f}_\theta(\mathbf{X}) \right]. \tag{5.1.32}$$

Finding the exact θ^* from (5.1.32) may be challenging. In practice, we sample $\mathbf{X}_1, \ldots, \mathbf{X}_M$ from f and estimate θ^* via

$$\hat{\theta}^* = \operatorname*{argmax}_\theta \frac{1}{M} \sum_{i=1}^{M} k(\mathbf{X}_i) \log \tilde{f}_\theta(\mathbf{X}_i). \tag{5.1.33}$$

Note that gradient methods may be applied to find or estimate $\hat{\theta}^*$.

Below we consider two examples in which θ^* can be found explicitly.

Example 5.1.47 We continue Example 5.1.36, where we estimated $I = \mathbb{P}(X > 4)$ for $X \sim \mathcal{N}(0, 1)$ using the importance sampling with the $\mathcal{N}(4, 1)$ proposal distribution, i.e, $\tilde{f}(x) = \phi_{4,1}(x)$. Denote the corresponding IS estimator by $\hat{Y}^{IS}_{R,\theta=4}$, for which we showed that $\text{Var}Y^{IS}_{\theta=4} = 3.601 \cdot 10^{-8}$ (we also showed that $\text{Var}Y^{CMC} = 3.1670 \cdot 10^{-5}$).

Now consider a family $\{\tilde{f}_\theta\}$ of normal distributions with mean $\theta \in \mathbb{R}$ and variance 1. We have $f(x) = \frac{e^{-x^2/2}}{\sqrt{2\pi}}$ and $\tilde{f}_\theta(x) = \frac{e^{-(x-\theta)^2/2}}{\sqrt{2\pi}}$. Let us compute θ^* from (5.1.32), namely:

$$\theta^* = \underset{\theta}{\text{argmax}}\, \mathbb{E}_f \left[\mathbb{1}(X \geq 4) \log \tilde{f}_\theta(x) \right]$$

$$= \underset{\theta}{\text{argmax}} \int_4^\infty \log\left(\frac{e^{-(x-\theta)^2/2}}{\sqrt{2\pi}} \right) \frac{e^{-x^2/2}}{\sqrt{2\pi}}\, dx$$

$$= \underset{\theta}{\text{argmin}} \int_4^\infty (x-\theta)^2 e^{-x^2/2}\, dx.$$

In Exercise 5.T.25, the reader is asked to show that $\theta^* = \frac{\sqrt{2}e^{-8}}{\sqrt{\pi}(1-\text{erf}(2\sqrt{2}))} = 4.2256$. Hence, we can compute that $\text{Var}Y^{IS}_{\theta=4.2256} = 4.5162 \cdot 10^{-9}$, which is $\text{Var}Y^{IS}_{\theta=4}/\text{Var}Y^{IS}_{\theta=4.2256} = 7.97$ times smaller than $\text{Var}Y^{IS}_{\theta=4}$. Finding

$$\theta^* = \underset{\theta}{\text{argmin}}\, \mathbb{E}_f \left[\mathbb{1}(X > \gamma) \log \tilde{f}_\theta(x) \right]$$

numerically is implemented in `ch5_cross_entropy_normal_1D.py`.   ◇

Example 5.1.48 We continue estimating $I = \mathbb{P}(X_1 > 2, X_2 > 2)$ for $\mathbf{X} = (X_1, X_2) \sim \mathcal{N}(\mathbf{0}, \boldsymbol{\Sigma})$ considered in Example 5.1.38, i.e., for $\boldsymbol{\Sigma} = \begin{pmatrix} 1 & \sqrt{2} \\ \sqrt{2} & 4 \end{pmatrix}$. In our current notation, the parameter $\boldsymbol{\theta} = (\mu_1, \mu_2)$ is the mean of the 2-dimensional normal distribution. Let $f_{(0,0)} = f$ denote the p.d.f. of $\mathcal{N}(\mathbf{0}, \boldsymbol{\Sigma})$ and let $\tilde{f}_{(\mu_1,\mu_2)}$ denote the p.d.f. of $\mathcal{N}((\mu_1, \mu_2), \boldsymbol{\Sigma})$. We may rewrite

$$I = \mathbb{P}(X_1 > 2, X_2 > 2)$$

$$= \int_2^\infty \int_2^\infty f_{(0,0)}(x_1, x_2)\, dx_1\, dx_2$$

$$= \int_{-\infty}^\infty \int_{-\infty}^\infty k(x_1, x_2) f_{(0,0)}(x_1, x_2)\, dx_1\, dx_2,$$

where $k(\mathbf{x}) = \mathbb{1}(x_1 > 2)\mathbb{1}(x_2 > 2)$. Now we find $\boldsymbol{\theta}^* = (\mu_1^*, \mu_2^*)$ from (5.1.32), i.e.,

$$(\mu_1^*, \mu_2^*) = \underset{\mu_1,\mu_2}{\text{argmax}} \int_2^\infty \int_2^\infty \log(\tilde{f}_{(\mu_1,\mu_2)}(x_1, x_2)) f_{(0,0)}(x_1, x_2)\, dx_1\, dx_2.$$

Substituting the formulas for $\tilde{f}_{(\mu_1,\mu_2)}(x_1, x_2)$ and $f_{(0,0)}(x_1, x_2)$, we differentiate the r.h.s. with respect to μ_1 and μ_2, and (setting the derivatives to zero) find that

$$(\mu_1^*, \mu_2^*) = (2.4, 3.8339).$$

We confirmed numerically that the IS estimator with proposal distribution $\tilde{f}_{(\mu_1^*,\mu_2^*)}$ yields the best results, see Table 5.8 (the density of $\mathbf{X}$ is depicted in Fig. 5.15). $\Diamond$

5.1.8.1 Does the Cross-entropy Method Always Yield a Better IS Estimator?

The answer is **no**. Although the cross-entropy method is often used to improve IS estimators over the standard CMC method, it does not guarantee improvement over other choices within the same IS family. In particular, the variance may increase— or become infinite. Although the optimal density g^* provided in (5.1.22) has zero variance, taking the "ideal" θ^* which minimizes the KL divergence between g^* and $\tilde{f}_\theta$—i.e., θ^* from (5.1.31)—does not necessarily imply that the IS estimator with proposal distribution $\tilde{f}_{\theta^*}$ will have smaller variance than the one using some other parameter θ_0. In fact, it may even happen that the variance of the IS estimator with proposal distribution $\tilde{f}_{\theta^*}$ is infinite, as the following example shows:

Example 5.1.49 (Infinite Variance with Optimal CE Estimator) In Example 5.1.36, we considered the problem of estimating $I = \mathbb{P}(X > 4)$ for $X \sim \mathcal{N}(0, 1)$ using an IS estimator with proposal density $\tilde{f} \sim \mathcal{N}(4, 1)$. In Example 5.1.47, we reconsidered this problem using the cross-entropy method, with a family of proposal distributions $f_\theta \sim \mathcal{N}(\theta, 1)$. Solving (5.1.31) yielded $\theta = 4.2256$, which indeed improved (decreased) the variance compared to the CMC estimator—recall, we computed:

$$\mathbb{V}\mathrm{ar}Y^{\mathrm{CMC}} = 3.1670\cdot 10^{-5}, \quad \mathbb{V}\mathrm{ar}Y_{\theta=4}^{\mathrm{IS}} = 3.601\cdot 10^{-8}, \quad \mathbb{V}\mathrm{ar}Y_{\theta=4.2256}^{\mathrm{IS}} = 4.5162\cdot 10^{-9}.$$

Now, let us reconsider the cross-entropy method, but take $\theta = (\mu, \sigma^2)$ and consider a family of proposal distributions

$$f_\theta \equiv f_{(\mu,\sigma^2)} \sim \mathcal{N}(\mu, \sigma^2).$$

Rewriting (5.1.31) gives:

$$\theta^* = \operatorname*{argmax}_{\mu,\sigma^2} \int_4^\infty \log f_{(\mu,\sigma^2)}(x) f_{0,1}(x)\, dx$$

$$= \operatorname*{argmin}_{\mu,\sigma^2} \int_4^\infty \frac{(x-\mu)^2}{2\sigma^2} e^{-x^2/2}\, dx.$$

Solving this numerically (implemented in `ch5_cross_entropy_normal_1D_optimal_mu_s2.py`) yields

$$\mu^* = 4.2256, \quad (\sigma^2)^* = 0.046671.$$

However, any IS estimator with proposal distribution $\mathcal{N}(\mu, \sigma^2)$ for $\sigma^2 < 1/2$ has infinite variance. The second moment of such an estimator is given by:

$$\mathbb{E}\left(Y^{\mathrm{IS}}_{\mu,\sigma^2}\right)^2 = \int_4^\infty \left(\frac{f_{0,1}(x)}{f_{\mu,\sigma^2}(x)}\right)^2 f_{\mu,\sigma^2}(x)\, dx = \int_4^\infty \frac{f_{0,1}^2(x)}{f_{\mu,\sigma^2}(x)}\, dx$$

$$= \frac{\sigma}{\sqrt{2\pi}} \int_4^\infty \exp\left(\frac{(x-\mu)^2}{2\sigma^2} - x^2\right) dx$$

$$= \frac{\sigma}{\sqrt{2\pi}} \exp\left(\frac{\mu^2}{2\sigma^2}\right) \int_4^\infty \exp\left(\left(\frac{1}{2\sigma^2} - 1\right)x^2 - \frac{\mu}{\sigma^2}x\right) dx = \infty$$

$$\tag{5.1.34}$$

for $\sigma^2 < 1/2$. Note that this is also infinite if $\mu < 0$.

By the way, solving

$$\boldsymbol{\theta}^{*\prime} = \operatorname*{argmax}_{\mu,\sigma^2,\ \sigma^2 \geq 1/2} \int_4^\infty \log f_{(\mu,\sigma^2)}(x) f_{0,1}(x)\, dx$$

yields

$$\mu^{*\prime} = 4.2256, \quad (\sigma^2)^{*\prime} = 0.5.$$

Plugging this into (5.1.34) gives the second moment:

$$\mathbb{E}\left(Y^{\mathrm{IS}}_{\mu^{*\prime},\sigma^2=1/2}\right)^2 = \frac{1}{\sqrt{4\pi}} \exp\left(\mu^{*\prime} - \mu^{*\prime}\frac{1}{2\mu^{*\prime}}\right) \overset{\mu^{*\prime}=4.2256}{=} 3.9524 \cdot 10^{-9},$$

yielding the final variance

$$\mathbb{V}\mathrm{ar} Y^{\mathrm{IS}}_{\mu^{*\prime},\sigma^2=1/2} = 2.9494 \cdot 10^{-9},$$

which is the best among all considered options. $\Diamond$

General Result: Finite Variance Under Multivariate Gaussian Proposals In the previous Example 5.1.49, we showed that optimizing the cross-entropy loss without constraints may lead to importance sampling (IS) estimators with infinite variance, even though the chosen proposal minimizes KL divergence from the optimal density. A similar phenomenon occurs in higher dimensions. For symmetric positive definite matrices $\boldsymbol{\Sigma}$ and $\boldsymbol{\Sigma}'$, we write $\boldsymbol{\Sigma}' \succ \boldsymbol{\Sigma}$ if $\boldsymbol{\Sigma}' - \boldsymbol{\Sigma}$ is positive definite. The following theorem provides a sufficient condition for finite variance when the proposal distribution is multivariate normal:

Theorem 5.1.50 *Assume* $\mathbf{X} \sim \mathcal{N}(\boldsymbol{\mu}, \boldsymbol{\Sigma})$ *is a d-dimensional Gaussian vector and we estimate* $I = \mathbb{P}(g(\mathbf{X}) > \gamma)$ *using importance sampling with proposal* $f_{\boldsymbol{\theta}} = \mathcal{N}(\boldsymbol{\mu}', \boldsymbol{\Sigma}')$. *If* $\boldsymbol{\Sigma}' \succ \frac{1}{2}\boldsymbol{\Sigma}$, *then the IS estimator*

$$Y_{\boldsymbol{\theta}}^{\mathrm{IS}} = \mathbb{1}(g(\mathbf{X}) > \gamma) \cdot \frac{f_0(\mathbf{X})}{f_{\boldsymbol{\theta}}(\mathbf{X})}, \quad \mathbf{X} \sim f_{\boldsymbol{\theta}},$$

has finite variance, regardless of the choice of $\boldsymbol{\mu}'$ *or the shape of the rare event set* $\{\mathbf{x} : g(\mathbf{x}) > \gamma\}$.

Proof We study the second moment:

$$\mathbb{E}\left(Y_{\boldsymbol{\theta}}^{\mathrm{IS}}\right)^2 = \int_{\mathbb{R}^d} \mathbb{1}(g(\mathbf{x}) > \gamma) \cdot \frac{f_0(\mathbf{x})^2}{f_{\boldsymbol{\theta}}(\mathbf{x})} \, d\mathbf{x}.$$

Substituting the Gaussian densities gives:

$$\frac{f_0(\mathbf{x})^2}{f_{\boldsymbol{\theta}}(\mathbf{x})} = C \cdot \exp\left(-(\mathbf{x} - \boldsymbol{\mu})^T \boldsymbol{\Sigma}^{-1}(\mathbf{x} - \boldsymbol{\mu}) + \frac{1}{2}(\mathbf{x} - \boldsymbol{\mu}')^T \boldsymbol{\Sigma}'^{-1}(\mathbf{x} - \boldsymbol{\mu}')\right),$$

where $C > 0$ is a constant. Expanding the exponent,

$$-(\mathbf{x} - \boldsymbol{\mu})^T \boldsymbol{\Sigma}^{-1}(\mathbf{x} - \boldsymbol{\mu}) + \frac{1}{2}(\mathbf{x} - \boldsymbol{\mu}')^T \boldsymbol{\Sigma}'^{-1}(\mathbf{x} - \boldsymbol{\mu}')$$

$$= \mathbf{x}^T \left(\frac{1}{2}\boldsymbol{\Sigma}'^{-1} - \boldsymbol{\Sigma}^{-1}\right) \mathbf{x} + (\text{lower order terms}).$$

The assumption $\boldsymbol{\Sigma}' \succ \boldsymbol{\Sigma}$ implies $\frac{1}{2}\boldsymbol{\Sigma}'^{-1} - \boldsymbol{\Sigma}^{-1}$ is negative definite, so the integrand decays exponentially in all directions as $\|\mathbf{x}\| \to \infty$. Thus, the second moment is finite for any choice of $\boldsymbol{\mu}'$ and any measurable set $\{g(\mathbf{x}) > \gamma\}$. $\qquad\square$

Remark 5.1.51 In practice, we optimize the Kullback-Leibler divergence over the parameters of the proposal distribution $f_{\boldsymbol{\theta}} = \mathcal{N}(\boldsymbol{\mu}', \boldsymbol{\Sigma}')$ under constraints to ensure finite variance of the resulting IS estimator.

In optimization, it is often convenient to parameterize $\boldsymbol{\Sigma}'$ via its Cholesky factor:

$$\boldsymbol{\Sigma}' = \mathbf{L}\mathbf{L}^T + \left(\tfrac{1}{2} + \varepsilon\right)\boldsymbol{\Sigma},$$

where $\mathbf{L}$ is a lower-triangular matrix and $\varepsilon > 0$ is fixed. This parameterization automatically ensures that $\boldsymbol{\Sigma}' \succ \boldsymbol{\Sigma}$, and the resulting optimization problem can be efficiently solved numerically. $\qquad\blacksquare$

5.1.8.2   The Cross-entropy Method for Rare Event Simulation

We now study the cross-entropy method in the context of rare event simulation. By a *rare event*, we mean that its probability is small, say less than 10^{-5}. Here we have the following setup. Let f denote the original distribution of $\mathbf{X}$, and let $\{\tilde{f}_\theta(\mathbf{x})\} \equiv \{f_\theta(x)\}$ (i.e., from now on we omit the tildes) be a parametric family of proposal distributions for importance sampling, parametrized by $\theta \in \Theta$. Our goal is to find or construct $\theta' \in \Theta$ that enables efficient estimation of rare event probabilities under the original distribution f.

Moreover, for short, we will write $\mathbb{E}_\theta(\cdot)$ or $\mathbb{P}_\theta(\cdot)$ for computing the quantities having p.d.f f_θ. Relaxing the method so that $\mathbf{X}$ has a discrete distribution is straightforward.

Let h be a real-valued function (called the **performance function**) on $\mathbb{R}^d$. We are interested in the probability that $h(\mathbf{X}) \geq \gamma$ for some fixed $\gamma \in \mathbb{R}$ under f, i.e.,

$$I = \mathbb{P}_f(h(\mathbf{X}) \geq \gamma) = \mathbb{E}_f \mathbb{1}(h(\mathbf{X}) \geq \gamma) = \int \underbrace{\mathbb{1}(h(\mathbf{x}) \geq \gamma)\, f(\mathbf{x})}_{k(\mathbf{x})}\, d\mathbf{x}.$$

In this case, the optimal parameter θ^* from (5.1.32) is given by

$$\theta^* = \underset{\theta}{\operatorname{argmax}}\, \mathbb{E}_f \mathbb{1}(h(\mathbf{X}) \geq \gamma) \log f_\theta(\mathbf{X}). \tag{5.1.35}$$

To estimate θ^* from (5.1.35), we could proceed with (5.1.33), i.e., we would estimate it by sampling i.i.d. $\mathbf{X}_1, \ldots, \mathbf{X}_M$ from f and computing

$$\hat{\theta}^* = \underset{\theta}{\operatorname{argmax}}\, \frac{1}{M} \sum_{i=1}^{M} \mathbb{1}(h(\mathbf{X}_i) \geq \gamma) \log f_\theta(\mathbf{X}_i).$$

However, we still have a problem: in the case of a rare event, $I = \mathbb{P}_f(h(\mathbf{X}) > \gamma)$ is very small, thus the value of M would have to be very large so that $\hat{\theta}^*$ closely approximates θ^*. As a rule of thumb, by "very small", we mean that $I \leq 10^{-5}$. In other words, we may proceed with the above recipe if $I > 10^{-5}$, otherwise, we apply the so-called **adaptive** or **multi-level** approach, to be described below.

First let us proceed with estimating θ^* from (5.1.35) in the following way. We will use an importance sampling method again—still from the $\{f_\theta\}$ family—with another parameter θ', namely from p.d.f. $f_{\theta'}$. Let us rewrite the expectation in (5.1.35)

$$\mathbb{E}_f \left[\mathbb{1}(h(\mathbf{X}) \geq \gamma) \log f_\theta(\mathbf{X}) \right]$$

$$= \int \mathbb{1}(h(\mathbf{x}) \geq \gamma) \log f_\theta(\mathbf{x})\, f(\mathbf{x})\, d\mathbf{x}$$

$$= \int \mathbb{1}(h(\mathbf{x}) \geq \gamma) \frac{f(\mathbf{x})}{f_{\theta'}(\mathbf{x})} \log f_\theta(\mathbf{x}) f_{\theta'}(\mathbf{x}) \, d\mathbf{x}$$

$$= \mathbb{E}_{\theta'} \left[\mathbb{1}(h(\mathbf{X}) \geq \gamma) L(\mathbf{X}, f, f_{\theta'}) \log f_\theta(\mathbf{X}) \right],$$

where L is the *likelihood ratio* defined as

$$L(\mathbf{x}, f, f_{\theta'}) = \frac{f(\mathbf{x})}{f_{\theta'}(\mathbf{x})}. \tag{5.1.36}$$

The optimal parameter can thus be expressed as

$$\boldsymbol{\theta}^* = \underset{\boldsymbol{\theta}}{\operatorname{argmax}} \, \mathbb{E}_{\theta'} \left[\mathbb{1}(h(\mathbf{X}) \geq \gamma) L(\mathbf{X}, f, f_{\theta'}) \log f_\theta(\mathbf{X}) \right], \tag{5.1.37}$$

which may be estimated by sampling M independent replications $\mathbf{X}_1, \ldots, \mathbf{X}_M$ from $\mathbf{X} \sim f_{\theta'}$ and solving

$$\hat{\boldsymbol{\theta}}^* = \underset{\boldsymbol{\theta}}{\operatorname{argmax}} \, \frac{1}{M} \sum_{i=1}^{M} \mathbb{1}(h(\mathbf{X}_i) \geq \gamma) L(\mathbf{X}_i, f, f_{\theta'}) \log f_\theta(\mathbf{X}_i). \tag{5.1.38}$$

The key idea is to find a parameter $\boldsymbol{\theta}'$ such that many of the samples $\mathbf{X}_i$, $i = 1, \ldots, M$, drawn from $f_{\theta'}$, satisfy $\mathbb{1}(h(\mathbf{X}_i) \geq \gamma) = 1$. This typically ensures that $\hat{\boldsymbol{\theta}}^*$ provides a sufficiently accurate approximation of $\boldsymbol{\theta}^*$ even for moderately small M. Moreover, often the r.h.s. of (5.1.38) is convex, and the solution can be obtained by solving the equations

$$\frac{1}{M} \sum_{i=1}^{M} \mathbb{1}(h(\mathbf{X}_i) \geq \gamma) L(\mathbf{X}_i, f, f_{\theta'}) \nabla \log f_\theta(\mathbf{X}_i) = \mathbf{0},$$

which can be analytically solved, e.g., when f_θ belongs to a *natural exponential family*, see [185]. Alternatively, for more complex cases where no closed-form solution exists, gradient-based numerical methods can be employed to approximate the solution efficiently on a computer.

How to Choose $\boldsymbol{\theta}'$ Adaptively?   Roughly speaking, we will be estimating "easier" events, namely events $I_t = \mathbb{P}(h(\mathbf{X}) \geq \gamma_t)$. We will construct a sequence of parameters $(\boldsymbol{\theta}'_t)$ and a sequence of corresponding *levels* (γ_t), $t = 0, 1, \ldots$, in the following way (starting with some fixed $\boldsymbol{\theta}_0$):

- We fix some not too small ρ, e.g., $\rho = 0.01$ and set $\boldsymbol{\theta}'_0 = \boldsymbol{\theta}_0$.
- Set $\gamma_1 = \operatorname{argmax}_r \{ \mathbb{P}_{\theta'_0}(h(\mathbf{X}) \geq r) \geq \rho \}$, i.e., quantile $q_{1-\rho}$ of $h(\mathbf{X})$.
- Then let $\boldsymbol{\theta}'_1$ be a set of optimal parameters for estimating I_1, i.e., obtained by solving (5.1.37).
- Now set $\gamma_2 = \operatorname{argmax}_r \{ \mathbb{P}_{\theta'_1}(h(\mathbf{X}) \geq r) \geq \rho \}$ and so on.

The whole procedure is depicted in Algorithm 34.

Algorithm 34 The cross-entropy method for rare event simulation: deterministic version

---

Require: Family of distributions $\{f_\theta\}$, $\theta \in \Theta$, fixed θ_0, real-valued function s, parameters $\gamma \in \mathbb{R}$, $\rho \in (0, 1)$, sample size R

1: Define $\theta'_0 = \theta_0$, set $t = 0$
2: **repeat**
3:     $t = t + 1$
4:     **Update** γ_t: Set

$$\gamma_t = \underset{r}{\mathrm{argmax}}\{\mathbb{P}_{\theta'_{t-1}}(h(\mathbf{X}) \geq r) \geq \rho\}. \tag{5.1.39}$$

5:     **Update** θ'_t: For current θ'_{t-1} and γ_t compute

$$\theta'_t = \underset{\theta}{\mathrm{argmax}}\, \mathbb{E}_{\theta'_{t-1}} \mathbb{1}(h(\mathbf{X}) \geq \gamma_t) L(\mathbf{X}, f, f_{\theta'_{t-1}}) \log f_\theta(\mathbf{X}). \tag{5.1.40}$$

6: **until** $\gamma_t < \gamma$
7: Set $T = t$, sample $\mathbf{X}_1, \ldots, \mathbf{X}_R$ from f_{θ_T} and compute

$$\hat{Y}_R^{\mathrm{CE}} = \frac{1}{R} \sum_{i=1}^{R} \mathbb{1}(h(\mathbf{X}_i) > \gamma) L(\mathbf{X}_i, f, f_{\theta'_T}) \tag{5.1.41}$$

8: **return** $\hat{Y}_R^{\mathrm{CE}}$

Algorithm 34 is called a *deterministic version* (of CE for rare event estimation), it requires finding exact γ_t in (5.1.39) and it requires solving (5.1.40). In practice, instead of finding exact γ_t, we estimate it from simulations by computing $(1 - \rho)$ sample quantile from performances of the sample (see e.g., Exercise 4.T.12). Similarly, instead of finding exact θ'_t, we solve system of equations resulting from simulations. This more practical procedure, called a *stochastic version*, is provided in Algorithm 35, where computing γ_t in (5.1.39) is replaced with (5.1.42) and updating θ'_t in (5.1.40) is replaced with (5.1.43).

Note that this way the problem of having a very large M is mitigated. We are estimating events which now have a prescribed probability ρ, usually small, e.g., 0.01, but not very small, e.g., 10^{-6}. The number of replications R for a final estimation of I is usually larger—exactly as, e.g., in usual CMC methods—note that it is used once.

Algorithm 35 The cross-entropy method for rare event simulation: stochastic version

---

Require: Family of distributions $\{f_\theta\}$, $\theta \in \Theta$, fixed θ_0, real-valued function h, parameters $\gamma \in \mathbb{R}$, $\rho \in (0, 1)$, $\alpha \in (0, 1]$, sample sizes M, R.

1: Define $\theta'_0 = \theta_0$, set $t = 0$
2: **repeat**
3:     $t = t + 1$
4:     **Update** γ_t: generate samples $\mathbf{X}_1, \ldots, \mathbf{X}_M$ from density $f_{\theta_{t-1}}(\cdot)$,
        compute their performances $\eta_1 = h(\mathbf{X}_i), \ldots, \eta_M = h(\mathbf{X}_M)$,
        sort them $\eta_{(1)} \leq \cdots \leq \eta_{(M)}$ and compute $(1 - \rho)$ sample quantile

$$\gamma_t = \eta_{(\lceil(1-\rho)M\rceil)} \tag{5.1.42}$$

5:     **Update** θ'_t: using **the same** samples $\mathbf{X}_1, \ldots, \mathbf{X}_M$
        solve

$$\theta'_t = \underset{\theta}{\mathrm{argmax}} \; \frac{1}{M} \sum_{i=1}^{M} \mathbb{1}(h(\mathbf{X}_i) \geq \gamma_t) L(\mathbf{X}_i, f, f_{\theta'_{t-1}}) \log f_\theta(\mathbf{X}_i) \tag{5.1.43}$$

    In practice we often apply so-called **smoothing** with $\alpha \in (0, 1)$

$$\theta'_t := \alpha \theta'_t + (1 - \alpha)\theta'_{t-1},$$

6: **until** $\gamma_t < \gamma$
7: Set $T = t$, sample $\mathbf{X}_1, \ldots, \mathbf{X}_R$ from f_{θ_T} and compute

$$\hat{Y}_R^{\mathrm{CE}} = \frac{1}{R} \sum_{i=1}^{R} \mathbb{1}(h(\mathbf{X}_i) > \gamma) L(\mathbf{X}_i, f, f_{\theta'_T})$$

8: **return** $\hat{Y}_R^{\mathrm{CE}}$

Remark 5.1.52 In Algorithms 34 and 35, the iteration stops once the level γ_t reaches or exceeds the target threshold γ, and the final proposal parameter θ_T is used to estimate the rare-event probability $\mathbb{P}(h(\mathbf{X}) \geq \gamma)$. However, observe that θ_T was optimized for level γ_T, which may be strictly greater than γ. Therefore, it may be beneficial to perform one additional refinement step: generate new samples from f_{θ_T} and re-optimize θ using the original threshold γ in place of γ_T in the objective function (5.1.38). Formally, we solve

$$\theta_{T+1} = \underset{\theta \in \Theta}{\mathrm{argmax}} \; \frac{1}{M} \sum_{i=1}^{M} \mathbb{1}(h(\mathbf{X}_i) \geq \gamma) \, L(\mathbf{X}_i, f, f_{\theta_T}) \log f_\theta(\mathbf{X}_i),$$

where $\mathbf{X}_1, \ldots, \mathbf{X}_M \sim f_{\theta_T}$. This final adjustment may improve the quality of the estimator, especially when γ_T overshoots γ significantly. ∎

Updating θ_t' in (5.1.43) for d-dimensional Normal Distribution   Assume that f_θ is a p.d.f. of $\mathcal{N}(\theta, \boldsymbol{\Sigma})$ (where $\boldsymbol{\Sigma}$ is fixed and θ is the parameter). That is, we have:

$$f_\theta(\mathbf{x}) = \frac{1}{\sqrt{(2\pi)^d |\boldsymbol{\Sigma}|}} \exp\left(-\frac{1}{2}(\mathbf{x} - \theta)^T \boldsymbol{\Sigma}^{-1}(\mathbf{x} - \theta)\right).$$

Denote the weights by

$$w_{i,t-1} = L(\mathbf{X}_i, f, f_{\theta_{t-1}'}),$$

where $t - 1$ indicates that the weights are computed using the parameters from the previous iteration. Note that:

$$w_{i,t-1} \log f_\theta(\mathbf{X}_i) = w_{i,t-1}\left[-\frac{d}{2}\log(2\pi) - \frac{1}{2}\log|\boldsymbol{\Sigma}|\right]$$
$$- w_{i,t-1}\left[\frac{1}{2}(\mathbf{X}_i - \theta)^T \boldsymbol{\Sigma}^{-1}(\mathbf{X}_i - \theta)\right].$$

Finding the argmax of (5.1.43) with respect to θ is equivalent to finding

$$\underset{\theta}{\text{argmin}} \sum_{i=1}^{M} \mathbb{1}(h(\mathbf{X}_i) \geq \gamma_t) w_{i,t-1}(\mathbf{X}_i - \theta)^T \boldsymbol{\Sigma}^{-1}(\mathbf{X}_i - \theta),$$

which can be done by taking the derivative of the expression inside the above argmin and setting it to $\mathbf{0}$:

$$\frac{\partial}{\partial \theta}\left(\sum_{i=1}^{M} \mathbb{1}(h(\mathbf{X}_i) \geq \gamma_t) w_{i,t-1}(\mathbf{X}_i - \theta)^T \boldsymbol{\Sigma}^{-1}(\mathbf{X}_i - \theta)\right) = \mathbf{0},$$

which yields:

$$\theta_t' = \frac{\sum_{i=1}^{M} \mathbb{1}(h(\mathbf{X}_i) \geq \gamma_t) w_{i,t-1}\mathbf{X}_i}{\sum_{i=1}^{M} \mathbb{1}(h(\mathbf{X}_i) \geq \gamma_t) w_{i,t-1}},$$

i.e., a weighted average. Component-wise, defining $\theta_t' = (\theta_t'(1), \ldots, \theta_t'(d))$ and $\mathbf{X}_i = (X_i(1), \ldots, X_i(d))$, we have:

$$\theta_t'(j) = \frac{\sum_{i=1}^{M} \mathbb{1}(h(\mathbf{X}_i) \geq \gamma_t) w_{i,t-1} X_i(j)}{\sum_{i=1}^{M} \mathbb{1}(h(\mathbf{X}_i) \geq \gamma_t) w_{i,t-1}}. \tag{5.1.44}$$

In Exercise 5.T.26, the reader is asked to derive update formulas for the case where both $\boldsymbol{\theta}$ (mean) and $\boldsymbol{\Sigma}$ (covariance matrix) are treated as parameters of the proposal distribution $\mathcal{N}(\boldsymbol{\theta}, \boldsymbol{\Sigma})$. This extends the formulation discussed in this section to account for the joint optimization of both parameters.

Remark 5.1.53 It turns out that the formula for updating $\boldsymbol{\theta}$ in (5.1.43) becomes (5.1.44) not only for multivariate normal distributions, but for a wider class. Namely, when $f_{\boldsymbol{\theta}}$ belongs to a *natural exponential family*, i.e., it is of the form:

$$f_{\boldsymbol{\theta}}(\mathbf{x}) = \prod_{i=1}^{d} f_{\theta_i}^{(i)}(x_i),$$

where each $\{f_{\theta_i}^{(i)}, \theta_i \in \Theta_i\}$ is a 1-parameter exponential family. Recall that $f_{\theta_i}^{(i)}$ is a 1-parameter exponential family if it is of the form:

$$f_{\theta_i}^{(i)}(x_i) = m(x_i)\exp(\kappa(\theta_i)T(x_i) - A(\theta_i)),$$

where $\kappa, A : \Theta \to \mathbb{R}$ and functions $T, m : \mathbb{R}^d \to \mathbb{R}$, for more details see [185]. ∎

Smoothed Updating Often, a smoothed updating rule is used. Having $\boldsymbol{\theta}_t'$ from the current step and $\boldsymbol{\theta}_{t-1}'$, we update:

$$\boldsymbol{\theta}_t' := \alpha\boldsymbol{\theta}_t' + (1-\alpha)\boldsymbol{\theta}_{t-1}',$$

where $\alpha \in (0, 1)$ is a smoothing parameter (by multiplying a scalar by a vector or a matrix, we mean multiplying each entry of the latter). This update is especially relevant for discrete optimization (see Sect. 7.4). The authors in [31] empirically found that α between 0.4 and 0.9 gives the best results. Note that smoothing is possible if we assume that Θ is convex, which is the case in all our examples.

Example 5.1.54 We will continue estimating $I = \mathbb{P}(X > 4)$ for $X \sim \mathcal{N}(0, 1)$, again considering the family of proposal distributions $f_\theta(x) \equiv \phi_{\theta,1}(x)$, i.e., the normal distribution with mean θ and variance 1. Recall, in Example 5.1.47, we computed that the exact optimal parameter is $\theta^* = 4.2256$. We implemented Algorithm 35 with $M = 10^4$, $\rho = 0.55$, and a smoothing parameter $\alpha = 0.4$. The algorithm stopped after $t = 46$ steps: γ_t ranged from $\gamma_1 = -0.114$ to $\gamma_{47} = 3.975$, while θ_t' was converging to the optimal θ^*, namely from $\theta_0' = 0$, $\theta_1 = 0.286$ to $\theta_{46} = 4.152 =: \theta^{\mathrm{CE}}$. The sequences (γ_t) and (θ_t') are depicted in Fig. 5.19.

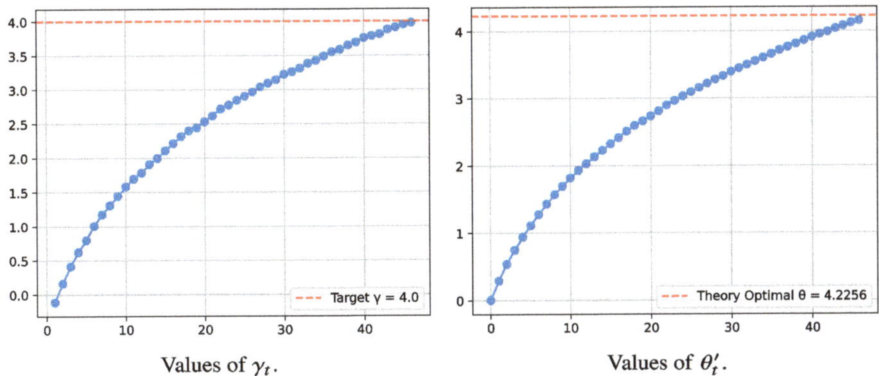

Values of γ_t.   Values of θ_t'.

Fig. 5.19 Sequences (γ_t) and (θ_t') for $t = 1, \ldots, 46$ computed when running Algorithm 35 for Example 5.1.54

Table 5.10 Numerical results for estimation of $\mathbb{P}(X > 4)$ (which is, approximately, 3.167×10^{-5}) using CMC and importance sampling estimators. See description in text

	$\hat{Y}_R$	$\mathbb{V}\mathrm{ar}(\hat{Y}_R)$	$|\hat{Y}_R - \mathbb{P}(X > 4)|$
CMC	0	0	3.16712×10^{-5}
IS $\theta = 4$	3.2566×10^{-5}	4.69327×10^{-13}	8.94758×10^{-7}
IS $\theta^{\mathrm{CE}} = 4.1521$	3.23104×10^{-5}	4.52492×10^{-13}	6.39163×10^{-7}
IS $\theta^* = 4.2256$	3.10587×10^{-5}	4.41932×10^{-13}	6.12528×10^{-7}

We also performed an estimation of $I = \mathbb{P}(X > 4)$ with CMC as well as using the importance sampling method with $\theta = 4$, $\theta^{\mathrm{CE}} = 4.152$ (found by the simulated cross-entropy method) and $\theta^* = 4.2256$ ("ideal" cross entropy—as computed in Example 5.1.47). The results are presented in Table 5.10. $\diamond$

5.1.9 Reducing Variance for Estimating π: A Summary

In Table 5.11, we compare different methods of variance reduction based on examples computing π.

5.2 Examples of Variance Reduction for Problems in Finance and Actuarial Science

In this section, we present selected examples of variance reduction methods applied to problems in financial engineering. To set the stage, we begin with a brief introduction to basic stochastic processes. Classical financial mathematics often

Table 5.11 Comparison of methods for estimating $\pi = 3.1415$. Exact values of variances provided for all estimators except stratified ones—for those the variance was approximated from simulations

Example	Estimator		Var($\hat{Y}_{10^4}$)
4.1.6	CMC	$Y = 4\mathbb{1}(U_1^2 + U_2^2)$	$2.6968 \cdot 10^{-4}$
5.1.24	CV	$Y = 4\mathbb{1}(U_1^2 + U_2^2),\ X = \mathbb{1}(U_1 + U_2 \leq 1)$	$1.959 \cdot 10^{-4}$
4.1.7	CMC	$Y = 4\sqrt{1 - U^2}$	$7.970 \cdot 10^{-5}$
5.1.24	CV	$Y = 4\mathbb{1}(U_1^2 + U_2^2),\ X = \mathbb{1}(U_1 + U_2 \geq \sqrt{2})$	$6.526 \cdot 10^{-5}$
5.1.20	CRN	$Y = 8/(1 + U^2) - 4\sqrt{1 - U^2}$	$2.885 \cdot 10^{-5}$
5.1.16	IS	$Y = 4\sqrt{1 - U^2},$ proposal $\tilde{f}(x) = (4 - 2x)/3$	$2.24 \cdot 10^{-5}$
5.1.16	anti	$Y_1 = 4\sqrt{1 - U^2},\ Y_2 = 4\sqrt{1 - (1 - U)^2}$	$2.19 \cdot 10^{-5}$
5.1.24	CV	$Y = 4\sqrt{1 - U^2},\ X = 1 - U$	$1.119 \cdot 10^{-5}$
5.1.8	str	$m = 10$ strata, optimal allocation	$1.12 \cdot 10^{-6}$
5.1.8	str	$m = 20$ strata, optimal allocation	$2.96 \cdot 10^{-7}$

relies on the concept of Brownian motion, with particular emphasis on geometric Brownian motion as a fundamental process.

A stochastic process is defined as a collection of random variables $(X(t) : t \in \mathcal{T})$ indexed by a parameter space $\mathcal{T}$. Typically, $\mathcal{T}$ is one of the following:

$$\mathcal{T} = [0, T], \quad T < \infty, \quad T = \infty, \quad \text{or} \quad \mathcal{T} = [a, b].$$

If $\mathcal{T}$ is a subset of the integers, the process is referred to as a sequence or a chain of random variables. In this chapter, we focus on the case $\mathcal{T} = [0, T]$, where a stochastic process represents a random evolution over time $t \in [0, T]$.

Each random variable in the process is formally denoted $X(t) = X(t, \omega)$, where $\omega \in \Omega$. A function $t \mapsto X(t, \omega)$, viewed as a function of t for fixed ω, is called a trajectory or sample path of the process. Just as the cumulative distribution function (c.d.f.) characterizes a random variable in probability theory, the finite-dimensional distributions of the vectors $X(t_1), \ldots, X(t_n)$ for all $n = 1, 2, \ldots$ and $0 \leq t_1 < \cdots < t_n \leq T$ are required to define a stochastic process $(X(t) : t \in [0, T])$.

5.2.1 *Brownian Motion and Related Stochastic Processes*

5.2.1.1 Brownian Motion

Definition 5.2.1 *A stochastic process $(B(t) : t \in [0, T])$ is called a standard Brownian motion if it satisfies the following conditions:*

(i) $B(0) = 0$,
(ii) the increments $B(t_1), B(t_2) - B(t_1), \ldots, B(t_n) - B(t_{n-1})$ are independent for all n and $0 \leq t_1 < \cdots < t_n \leq T$,

(iii) $B(t_2) - B(t_1) \sim \mathcal{N}(0, t_2 - t_1)$,
(iv) with probability 1, the function $t \mapsto B(t)$ is continuous.

A stochastic process satisfying conditions (ii) and (iii) is referred to as a process with *independent and stationary increments*. Notably, Brownian motion is *self-similar*, meaning $B(at) \stackrel{\mathcal{D}}{=} \sqrt{a}\, B(t)$.

A widely studied class of stochastic processes $(X(t))$ is a Gaussian processes $(X(t) : t \in [0, T])$ if every vector $(X(t_1), \ldots, X(t_n))$ has a multivariate normal distribution $\mathcal{N}(\mathbf{0}, \Sigma_{t_1,\ldots,t_n})$ for all $n = 1, 2, \ldots$ and $0 \le t_1 < \cdots < t_n$. The following proposition is straightforward to demonstrate:

Proposition 5.2.2 *A stochastic process with almost surely continuous trajectories,* $(B(t) : t \in [0, T])$, *is a standard Brownian motion if and only if:*

$$\mathrm{Cov}\,(B(s), B(t)) = \min(s, t), \quad \textit{for all } 0 \le s \le t \le T.$$

In Exercise 5.T.27, the reader is asked to show that Brownian motion is a Gaussian process.

This property provides an immediate method for simulating Brownian motion. Without loss of generality, we assume $T = 1$. It is not feasible to simulate a continuous trajectory $B(t)$ for $t \in [0, 1]$. Instead, we approximate it by simulating $B(t_0), B(t_1), \ldots, B(t_N)$ on a grid (t_j), where:

$$0 = t_0 < t_1 < \cdots < t_N = 1.$$

Typically, the grid has a uniform spacing $\Delta = 1/N$, with:

$$t_0 = 0, \quad t_1 = \Delta, \quad t_2 = 2\Delta, \quad \ldots, \quad t_N = N\Delta = 1.$$

The parameter Δ is referred to as the *mesh size* of the grid. Recall that $B(0) = 0$. Algorithm 36 utilizes properties (ii) and (iii) of Brownian motion to generate the process on this grid.

Algorithm 36 Algorithm BM; simulating $B(t)$ for $t \in [0, 1]$.

1: Generate $Z_1, \ldots, Z_N$ i.i.d. $\sim \mathcal{N}(0, 1)$
2: Set $B(0) = 0$
3: **for** $i = 1$ to N **do**
4:     Compute $B(i\Delta) = B((i - 1)\Delta) + \sqrt{\Delta}\, Z_i$
5: **end for**

Let $\Delta = 1/n$. Algorithm 36 can be compactly rewritten as

$$(B(1/n), B(2/n), \ldots, B(1)) \stackrel{\mathcal{D}}{=} \frac{1}{\sqrt{n}}(Z_1, Z_1 + Z_2, \ldots, Z_1 + \cdots + Z_n), \qquad (5.2.1)$$

or equivalently

$$B(k/n) = \frac{1}{\sqrt{n}} \sum_{i=1}^{k} Z_i, \tag{5.2.2}$$

where $Z_1, \ldots, Z_n$ are i.i.d. standard normal random variables.

A more general process is the *Brownian motion with drift μ* and *diffusion coefficient σ*, given by:

$$X(t) = \sigma B(t) + \mu t, \qquad \mu \in \mathbb{R},\ \sigma > 0.$$

It can be shown that this process also has independent and stationary increments. An adaptation of Algorithm 36 to generate such a process is left as an exercise for the reader.

Geometric Brownian Motion A geometric Brownian motion is a stochastic process $(S(t) : t \in [0, 1])$, where $S(t) = S(0) \exp(X(t))$, and $X(t)$ is a Brownian motion with drift μ and diffusion coefficient σ. Assume the initial value $S(0) = x$. To simulate a geometric Brownian motion on a grid with mesh Δ, define $Y_i = S(i\Delta)/S((i-1)\Delta)$. Then $Y_i \sim LN(\mu, \sigma)$ follows a log-normal distribution, i.e., $Y_i \overset{\mathcal{D}}{=} \exp(\sigma\sqrt{\Delta}Z_i + \Delta\mu)$, where $Z_i \sim \mathcal{N}(0, 1)$. The simulation procedure is outlined in Algorithm 37.

Algorithm 37 Generating geometric Brownian motion $S(t) : 0 \le t \le 1$

1: Generate $Y_1, \ldots, Y_N$ i.i.d. $\sim LN(\mu\Delta, \sigma\sqrt{\Delta})$
2: $S(0) = x$
3: **for** $i = 1$ to N **do**
4:      $S(i\Delta) = S((i-1)\Delta)Y_i$
5: **end for**

Brownian Bridge The standard *Brownian bridge* is a Brownian motion $(X(t), t \in [0, 1])$ conditioned on $B(1) = 0$. An equivalent definition is given by $X(t) = B(t) - tB(1)$. It is a straightforward exercise to show that $\mathrm{Cov}\,(X(s), X(t)) = s(1 - t)$ for $s \le t$. We will revisit this result in a more general context in Proposition 5.2.3 below.

This concept generalizes to a stochastic process $(X(t))$ over an arbitrary interval $t \in [a, b]$, tied down at a to x and at b to y. Formally, it can be shown that the process $(X^{a,b,x,y}(t) : t \in [a, b])$ is defined as:

$$X^{a,b,x,y}(t) = B\left(\frac{t-a}{b-a}\right) - \frac{t-a}{b-a}B(1) + \frac{b-t}{b-a}x + \frac{t-a}{b-a}y.$$

Proposition 5.2.3 *Consider a centered (mean-zero) Gaussian process $(Y(t) : t \in [a, b])$, where $0 \leq a < b \leq 1$, with covariance function $\Sigma_Y(s, t) = \min(s, t)$.*

(i) For the random vector $(Y(a), Y(s), Y(t), Y(b))$, where $a \leq s, t \leq b$, the conditional distribution of $(Y(s), Y(t))$ given $(Y(a) = x, Y(b) = y)$ is multivariate normal with mean vector and covariance matrix:

$$\mu = \left(\frac{(b-s)x + (s-a)y}{b-a}, \frac{(b-t)x + (t-a)y}{b-a} \right),$$

$$\Gamma(s, t) = \frac{(s-a)(b-t)}{(b-a)^2}.$$

(ii) The stochastic process $(Y(t) : t \in [a, b])$, conditioned on $Y(a) = x$ and $Y(b) = y$, is $(X^{a,b,x,y}(t) : t \in [a, b])$.

Proof For $a \leq s \leq t \leq b$, the covariance matrix of the vector $(Y(s), Y(t), Y(a), Y(b))$ is:

$$\begin{pmatrix} s & s & a & a \\ s & t & a & t \\ \hline a & a & a & a \\ s & t & a & b \end{pmatrix} = \begin{pmatrix} \Sigma_{11} & \Sigma_{12} \\ \Sigma_{21} & \Sigma_{22} \end{pmatrix}.$$

The conditional covariance matrix of $(Y(s), Y(t))|(Y(a) = x, Y(b) = y)$ is given by:

$$\Sigma_{11} - \Sigma_{12}\Sigma_{22}^{-1}\Sigma_{21},$$

and the conditional mean vector is:

$$\Sigma_{12}\Sigma_{22}^{-1}(x, y)^T.$$

After some algebra, part (i) is proved.

For part (ii), verification involves checking that:

$$\mathbb{E}X^{a,b,x,y}(t) = \frac{b-t}{b-a}x + \frac{t-a}{b-a}y,$$

and:

$$\text{Cov}\,(X^{a,b,x,y}(s), X^{a,b,x,y}(t)) = \frac{(s-a)(b-t)}{(b-a)^2},$$

which completes the proof. $\qquad\square$

For future use, we note the following result.

Corollary 5.2.4 *Consider a centered (mean-zero) Gaussian process $(Y(t) : t \in [a, b])$, where $0 \le a < b \le 1$, with covariance function $\Sigma_Y(s, t) = \min(s, t)$.*

(i) For $s < t < u$, the conditional vector $(Y(t)|Y(s) = x, Y(u) = y)$ is normally distributed:

$$\mathcal{N}\left(\frac{(u - t)x + (t - s)y}{u - s}, \frac{(u - t)(t - s)}{u - s}\right).$$

(ii) In particular, $Y(t + h)$, under the conditions $Y(t) = x$ and $Y(t + 2h) = y$, is normally distributed:

$$\mathcal{N}\left(\frac{x + y}{2}, \frac{h}{2}\right).$$

Generating Brownian Motion via the Bisection Method Typically, stochastic processes are simulated on a grid (t_j), where $t_j = j/2^k$ for $j = 0, \ldots, 2^k$. For such grid we may rewrite (5.2.1) as

$$(B(1/2^k), B(2/2^k), \ldots, B(1)) \overset{\mathcal{D}}{=} 2^{-k/2}(Z_1, Z_1 + Z_2, \ldots, Z_1 + \cdots + Z_{2^k}),$$

where $Z_1, \ldots, Z_{2^k}$ are i.i.d. standard normal random variables.

We now present an alternative method for simulating

$$(B(1/2^k), B(2/2^k), \ldots, B(1)),$$

which can be useful in specific cases, such as for stratified sampling. This method utilizes a reordering of the random variables $Z_1, \ldots, Z_{2^k}$. Specifically, we first use Z_{2^k}, followed by $Z_{1/2 \cdot 2^k}$, then $Z_{1/4 \cdot 2^k}$ and $Z_{3/4 \cdot 2^k}$, and so on.

The procedure begins with $B(1) \overset{\mathcal{D}}{=} Z_{2^k}$. Using Corollary 5.2.4 with $s = 0$, $u = 1$, and $y = 0$ (and $x = 0$ by definition), we find that:

$$(B(1/2)|B(1) = y) \sim \mathcal{N}(y/2, 1/4).$$

Equivalently:

$$(B(1/2)|B(1) = y) \overset{\mathcal{D}}{=} \frac{1}{2}Z_{1/2 \cdot 2^k} + \frac{y}{2}.$$

The simulation proceeds by generating $B(1/4)$ and $B(3/4)$, and so on, using a recursive approach. The procedure is summarized in Algorithm 38.

Connecting the points $B((j-1)/2^k)$ and $B(j/2^k)$ linearly results in a continuous path, approximating the Brownian motion.

Table 5.12 Simulated values of $(B(0), B(1/4), B(2/4), B(3/4), B(1))$ using Algorithm 38. Here $k = 2$, and the simulated $(Z_{2^{k-2}}, Z_{2^{k-1}}, Z_{3 \cdot 2^{k-2}}, Z_{2^k}) = (Z_1, Z_2, Z_3, Z_4) = (0.3919, -1.4309, -0.8694, 0.8812)$

$i \backslash t$	0	$\frac{1}{4}$	$\frac{2}{4}$	$\frac{3}{4}$	$\frac{4}{4}$
$i = 0$					$B(1) = Z_4$ $= 0.8812$
$i = 1$			$\frac{B(0)+B(1)}{2} + \frac{Z_2}{2}$ $= -0.2748$		
$i = 2$		$\frac{B(0)+B(1/2)}{2} + \frac{Z_1}{\sqrt{8}}$ $= 0.0011$		$\frac{B(1/2)+B(1)}{2} + \frac{Z_3}{\sqrt{8}}$ $= -0.0042$	

Algorithm 38 Generating Brownian motion at $0, 1/2^k, \ldots, (2^k - 1)/2^k, 1$ via the bisection method

---

Require: k
1: Generate $Z_1, \ldots, Z_{2^k}$, i.i.d. $\mathcal{N}(0, 1)$ random variables
2: Set $B(0) = 0$ and $B(1) = Z_{2^k}$
3: **for** $i = 1$ to k **do**
4:     **for** $j = 0$ to $2^{i-1} - 1$ **do**
5:         $B\left(\frac{2j+1}{2^i}\right) = \frac{B\left(\frac{j}{2^{i-1}}\right)+B\left(\frac{j+1}{2^{i-1}}\right)}{2} + 2^{-(i+1)/2} Z_{2^{k-i}+j2^{k-i+1}}$
6:     **end for**
7: **end for**
8: **return** $(B(0), B(1/2^k), B(2/2^k), \ldots, B(1))$

The simulation results for $k = 2$ are presented in Table 5.12. For sampled

$$(Z_{2^{k-2}}, Z_{2^{k-1}}, Z_{3 \cdot 2^{k-2}}, Z_{2^k}) = (Z_1, Z_2, Z_3, Z_4)$$
$$= (0.3919, -1.4309, -0.8694, 0.8812)$$

the corresponding $B(1/4), B(2/4), B(3/4), B(4/4)$ are computed.

The values from Table 5.12 are depicted in Fig. 5.20 as red dots. Both the values in Table 5.12 and the plots in Fig. 5.20 are obtained using the bisection method implemented in the script `ch5_BrownianMotion_bisection.py`.

Generating Conditional Brownian Motion The bisection method can be easily adapted to **simulate a Brownian motion conditioned** on fixed values $B(0) = b_0 = 0, B(1/2^s) = b_1, B(2/2^s) = b_2, \ldots, B(1) = b_{2^s}$, where $s < k$. For $k = 15$ and $s = 2$, Fig. 5.20a shows a realization of a Brownian motion conditioned on fixed values

$$B(1/4) = 0.0011, \quad B(2/4) = -0.2748, \quad B(3/4) = -0.0042, \quad B(1) = 0.8812.$$

Note that these values are taken from Table 5.12, thus they result from $(Z_{2^{k-2}}, Z_{2^{k-1}}, Z_{3 \cdot 2^{k-2}}, Z_{2^k}) = (0.3919, -1.4309, -0.8694, 0.8812)$. (In other words, conditioning on fixed $B(1/2^s), \ldots, B(1)$ is the same as conditioning on the corresponding Z's). Figure 5.20b presents five different realizations with the same $B(1/4), B(2/4), B(3/4), B(4/4)$.

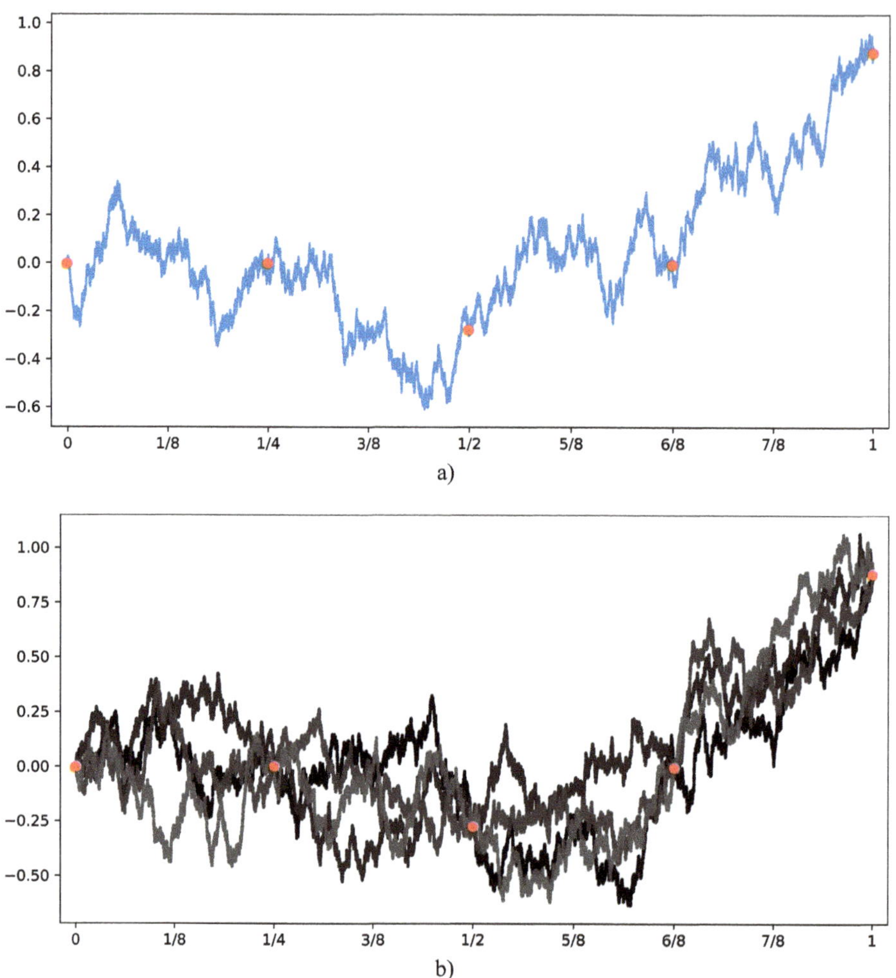

Fig. 5.20 Simulations of a Brownian motion for $k = 15$ with the same $B(1), B(1/4), B(1/2), B(3/4), B(1)$

The procedure to simulate a conditional Brownian motion is straightforward. In Algorithm 38, replace the loop range in line 3 from "$i = 1$ to k" to "$i = s + 1$ to k". The modified procedure is provided in Algorithm 39. Note that in line 6, some Z_j are redundant, and the algorithm could be optimized for efficiency.

Algorithm 39 Generating a Brownian motion conditioned on fixed values at $t = 0, 1/2^s, \ldots, 1$ via the bisection method

---

Require: $k, s, \{b_0, b_1, \ldots, b_{2^s}\}$
1: Generate $Z_1, \ldots, Z_{2^k}$, i.i.d. $\mathcal{N}(0, 1)$ random variables
2: Set $B(0) = b_0$, $B(1) = b_{2^s}$
3: **for** $i = 1$ to s **do** $\qquad\qquad\qquad\qquad\qquad\qquad$ $\triangleright$ Assign fixed values
4: $\quad$ Set $B(j/2^s) = b_j$ for $j = 1, \ldots, 2^s$
5: **end for**
6: **for** $i = s + 1$ to k **do**
7: $\quad$ **for** $j = 0$ to $2^{i-1} - 1$ **do**
8: $\qquad B\left(\frac{2j+1}{2^i}\right) = \frac{B\left(\frac{j}{2^{i-1}}\right) + B\left(\frac{j+1}{2^{i-1}}\right)}{2} + 2^{-(i+1)/2} Z_{2^{k-i} + j2^{k-i+1}}$
9: $\quad$ **end for**
10: **end for**
11: **return** $(B(0), B(1/2^k), B(2/2^k), \ldots, B(1))$

Notice that simulating a standard Brownian bridge is a straightforward modification of Algorithm 38—in line 2 we simply set $B(1) = 0$, as summarized in Algorithm 40. Ten sample simulations of a Brownian bridge are depicted in Fig. 5.21. This Brownian bridge example is implemented in `ch5_BrownianMotion_bridge.py`.

Algorithm 40 Generating Brownian bridge at $0, 1/2^k, \ldots, (2^k - 1)/2^k, 1$

Require: k
1: Apply Algorithm 36 with line 2 replaced with:
2: Set $B(0) = 0$ and $B(1) = 0$

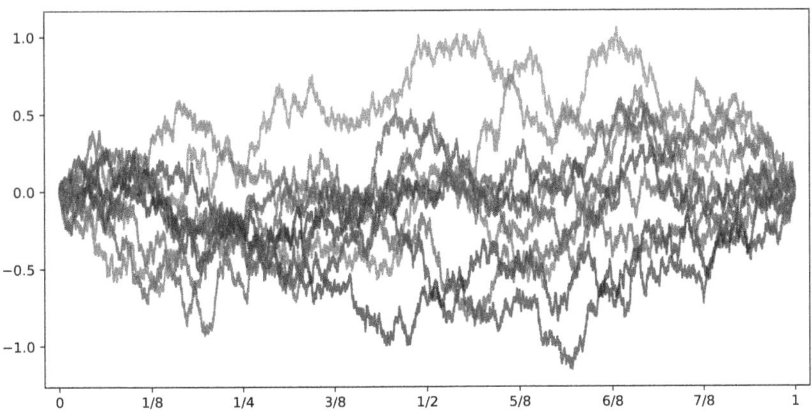

Fig. 5.21 Ten simulations of a Brownian bridge using Algorithm 40 with $k = 15$

5.2.2   *Information on Financial Contracts*

In mathematical finance, the price of a stock $S(t)$ is modeled as a function of time t, where $0 \leq t \leq T$. An option is a financial derivative that provides a payoff depending on the evolution of the stock price $S(t)$. Various types of options exist, and in this section, we focus on options that are functions of $S(t_1), \ldots, S(t_n)$, where $0 \leq t_1 < \cdots < t_n = T$. For simplicity, we assume $T = 1$. Examples of such options include:

- **European call option**:

$$f(S(1)) = (S(1) - K)_+,$$

  where K is called the *strike price*.
- **European put option**:

$$f(S(1)) = (K - S(1))_+.$$

- **Asian call option**:

$$f\left(S\left(\frac{1}{n}\right), \ldots, S\left(\frac{n-1}{n}\right), S(1)\right) = \left(\frac{1}{n}\sum_{j=1}^{n} S\left(\frac{j}{n}\right) - K\right)_+.$$

- **Up-and-out call option** (a modification of the European call option):

$$f(S(t_1), \ldots, S(t_{n-1}), S(1)) = (S(1) - K)_+ \mathbb{1}(S(t_1) < b, \ldots, S(t_{n-1}) < b),$$

  where b is a specified upper barrier.

It is also possible to consider d assets, $(S_1(t), \ldots, S_d(t))$ for $0 \leq t \leq 1$. For given weights $w_1, \ldots, w_d$, an example of a multi-asset option is:

- **European basket call option**:

$$f(S_1(1), \ldots, S_d(1)) = \left(\sum_{j=1}^{d} w_j S_j(1) - K\right)_+.$$

The evolution of stock prices (or asset prices) is typically modeled as a stochastic process. In this section, we focus on the *risk-neutral geometric Brownian motion*, denoted $\mathrm{GBM}(r, \sigma)$, which is defined by:

$$S(t) = S(0)\exp\left(\left(r - \frac{\sigma^2}{2}\right)t + \sigma B(t)\right), \qquad 0 \leq t \leq 1, \tag{5.2.3}$$

where $B(t)$ $(0 \leq t \leq 1)$ is a standard Brownian motion.

Here:

- r represents the interest rate.
- $\sigma > 0$ is a parameter known as *volatility*.

In this context, since we evaluate $S(t)$ at discrete time points (t_j), it suffices to understand the joint multidimensional distribution of $B(t_1), \ldots, B(t_n)$.

The price of an option for a single asset evolving according to $S(t)$ is given by:

$$e^{-r} \mathbb{E} f(S(t_1), \ldots, S(t_n)),$$

where f is the payoff function specific to the option under consideration.

Black-Scholes Model The Black-Scholes model is a foundational mathematical framework for modeling the dynamics of financial markets that include derivative instruments. The model assumes a market with two types of assets:

- A risky asset, such as a stock, denoted by $S(t)$.
- A risk-free asset, such as the money market, denoted by $R(t)$.

The dynamics of these assets are described by the following system of stochastic differential equations:

$$\begin{cases} dS(t) = \mu S(t)\,dt + \sigma S(t)\,dB(t), \\[2mm] dR(t) = r R(t)\,dt, \end{cases} \tag{5.2.4}$$

where:

- μ is the drift or growth rate of the risky asset.
- $\sigma > 0$ is the volatility of the risky asset.
- $r > 0$ is the risk-free interest rate.
- $B(t)$ is a standard Brownian motion.

The solution to the stochastic differential equation for $S(t)$ with the initial condition $S(0) = x$ is given by:

$$S(t) = x \exp\left(\left(\mu - \frac{\sigma^2}{2}\right) t + \sigma B(t)\right), \qquad t \geq 0.$$

However, in the pricing of financial derivatives, the theory of financial mathematics dictates the use of the *risk-neutral measure*. Under this measure, the price of the risky asset evolves according to a geometric Brownian motion with drift r (the risk-free rate) instead of μ. Thus, the dynamics of $S(t)$ are described by:

$$S(t) = S(0) \exp\left(\left(r - \frac{\sigma^2}{2}\right) t + \sigma B(t)\right), \qquad t \geq 0. \tag{5.2.5}$$

Log-normal Distributions and Black-Scholes Formula If $S(t)$ evolves as the geometric Brownian motion $\text{GBM}(\mu, \sigma)$, then:

$$S(1) \stackrel{\mathcal{D}}{=} S(0)\exp\left(\mu - \frac{\sigma^2}{2} + \sigma Z\right),$$

where $Z \sim \mathcal{N}(0, 1)$.

However, when computing the option price under the risk-neutral measure, with an interest rate r and volatility σ, we consider $\text{GBM}(r, \sigma)$. Let $\mu^* = r - \sigma^2/2$. Then:

$$S(1) \stackrel{\mathcal{D}}{=} S(0)\exp\left(\mu^* + \sigma Z\right).$$

For a European call option, the price is given by:

$$I = e^{-r}\mathbb{E}\left(S(1) - K\right)_+, \tag{5.2.6}$$

where $S(1) = S(0)\exp(\mu^* + \sigma B(1))$ and $B(1) \sim \mathcal{N}(0, 1)$. In this case, the exact Black-Scholes formula is:

$$\mathbb{E}\left(S(1) - K\right)_+ = S(0)\Phi(d_1) - Ke^{-r}\Phi(d_2), \tag{5.2.7}$$

where:

$$d_1 = \frac{1}{\sigma}\left[\log\left(\frac{S(0)}{K}\right) + r + \frac{\sigma^2}{2}\right], \quad d_2 = \frac{1}{\sigma}\left[\log\left(\frac{S(0)}{K}\right) + r - \frac{\sigma^2}{2}\right].$$

Estimating quantities from log-normal distributions can be challenging due to their heavy-tailed nature. Exercise 4.T.4 illustrates some of the difficulties that may arise in such cases.

In the following subsections, we will explore applications of the variance reduction techniques introduced earlier to problems in finance, particularly in option pricing and risk management.

5.2.3   Stratified Sampling

We will showcase several examples of stratified sampling applied to financial problems, emphasizing its effectiveness in variance reduction.

Example 5.2.5 (Stratified Sampling: European Call Option) Consider a European call option with a price given by the formula:

$$I = e^{-r}\mathbb{E}(S(1) - K)_+, \tag{5.2.8}$$

where $S(1) = S(0) \exp(\mu^* + \sigma B(1))$ and $B(1) \sim N(0, 1)$. The exact value of I can be calculated using the Black-Scholes formula provided in (5.2.7). We will estimate I via simulations and compare the results with the exact value. Fix the parameters as follows: $r = 0.05$, $\sigma = 0.25$ (thus $\mu^* = r - \sigma^2/2 = -0.0125$), $S(0) = 100$, and $K = 100$.

For the **Crude Monte Carlo** (CMC) estimator, we fix the number of simulations R, simulate $B^1(1), \ldots, B^R(1)$ i.i.d. $\sim N(0, 1)$, and compute:

$$S^i(1) = S(0) \exp(\mu^* + \sigma B^i(1)), \quad Y_i = (S^i(1) - K)_+.$$

The final estimator is:

$$\hat{Y}_R^{CMC} = \frac{1}{R} \sum_{i=1}^{R} Y_i.$$

The variance of the estimator is $\mathrm{Var}(\hat{Y}_R^{CMC}) = \mathrm{Var}Y/R$, which is estimated by:

$$\widehat{\mathrm{Var}}(\hat{Y}_R^{CMC}) = \frac{\hat{S}^2}{R},$$

where $\hat{S}^2$ is the sample variance of Y. The confidence interval error b is:

$$b = z_{1-\alpha/2} \sqrt{\widehat{\mathrm{Var}}(\hat{Y}_R^{CMC})} = z_{1-\alpha/2} \frac{\hat{S}}{\sqrt{R}},$$

which for $\alpha = 0.05$ is $b = 1.96 \cdot \frac{\hat{S}}{\sqrt{R}}$.

For the **stratified estimator**, we first perform a **pilot simulation** with a budget of $R' = 10{,}000$ total replications. Using m strata as presented in Example 5.1.9, we partition the range into strata $A_1, \ldots, A_m$, where $A_j = (a_{j-1}, a_j]$ with $a_0 = -\infty$, $a_m = \infty$, and $a_j = \Phi^{-1}(j/m)$ for $j = 1, \ldots, m - 1$ (ensuring $p_i = 1/m$). Within each stratum j, simulate $R'_j = R'/m$ points, denoted by $B_i^j(1)$ for $i = 1, \ldots, R'_j$, and calculate:

$$S_i^j(1) = S(0) \exp(\mu^* + \sigma B_i^j(1)), \quad Y_i^j = (S_i^j(1) - K)_+.$$

The stratum-wise mean and variance are:

$$\hat{Y}'_{R_j} = \frac{1}{R'_j} \sum_{i=1}^{R'_j} Y_i^j, \quad \hat{S}_j^2 = \frac{1}{R'_j} \sum_{i=1}^{R'_j} \left(Y_i^j - \hat{Y}'_{R_j}\right)^2.$$

Table 5.13 Simulation results for estimating the price I of the European call option with parameters $R = 10^5$, $S(0) = 100$, $r = 0.05$, $\sigma = 0.25$, $K = 100$. The table shows the estimates from the CMC and stratified estimators for strata $m \in \{5, 10, 50, 100, 200\}$, along with their absolute errors $|\hat{Y}_R - I|$ and respective errors $b = 1.96\sqrt{\widehat{\mathrm{Var}}(\hat{Y}_R)}$. The last row reports the ratio of the CMC variance to the stratified variance. The exact value of I is 12.3360

	CMC	Stratified estimator				
		Number of strata m				
		5	10	50	100	200
$\hat{Y}_R$	12.2799	12.3507	12.3278	12.3383	12.3357	12.3361
$|\hat{Y}_R - I|$	0.05607	0.0147	0.00819	0.00234	0.00030	0.00013
b	0.11463	0.0428	0.02450	0.00627	0.00345	0.00186
$\dfrac{\mathrm{Var}(\hat{Y}_R^{\mathrm{CMC}})}{\mathrm{Var}(\hat{Y}_R^{\mathrm{str}})}$	–	7.1562	21.9368	334.225	1105.62	3778.45

For the **actual simulation**, we determine (according to Theorem 5.1.5) the optimal allocation:

$$R_j = \left\lceil \frac{\hat{S}_j}{\sum_{k=1}^{m} \hat{S}_k} R \right\rceil .$$

Then simulate R_j points in each stratum, compute the means $\hat{Y}_{R_j}$, and variances $\hat{S}_j^2$ in each stratum. The final estimator and its variance are:

$$\hat{Y}_R^{\mathrm{str}} = \sum_{j=1}^{m} p_j \hat{Y}_{R_j} = \frac{1}{m} \sum_{j=1}^{m} \hat{Y}_{R_j}, \quad \widehat{\mathrm{Var}}(\hat{Y}_R^{\mathrm{str}}) = \sum_{j=1}^{m} \frac{p_j^2 \hat{S}_j^2}{R_j} .$$

We performed simulations using $R = 10^5$ replications and various numbers of strata, namely $m \in \{5, 10, 50, 100, 200\}$. Table 5.13 reports the estimated option prices. As shown, stratified sampling substantially reduces the error. In particular, for $m = 200$ strata, the variance is reduced by a factor of 3778, which corresponds to an approximate 60-fold reduction in the error b. The example is implemented in `ch5_european_option_stratified.py`.   ◊

Example 5.2.6 (Tail Estimation for a Collective Risk Model) In the collective risk model, the random sum $S = \sum_{j=1}^{N} X_j$ represents the aggregate claims generated by the portfolio for the period under study. Suppose that claims X_j have Erlang(4, 10/22) distribution and that the number of claims has Poisson(λ) distribution with parameter $\lambda = 3.3$. The question is: what is the probability that the amount of $u = 35$ suffices to cover all claims in the considered period? In other words, we are interested in estimating the ruin probability $I = \mathbb{P}(S > 32)$. For $N = 0$, we set $S = 0$. Note that the average total amount paid by the insurance company is $\mathbb{E}S = \mathbb{E}N \cdot \mathbb{E}X = 3.3 \cdot 4 \cdot \frac{22}{10} = 29.04$.

Table 5.14 Stratification of the collective risk model for $m = 7$ strata: stratum probabilities p_i and the corresponding allocations R_i for each stratum (with the last stratum corresponding to $N \geq 6$)

i	0	1	2	3	4	5	$(N \geq 6)$
p_i	0.0369	0.1217	0.2008	0.2209	0.1823	0.1203	0.1171
R_i	1	1218	2009	2210	1823	1203	1536

For the **CMC estimator**, we fix a number of replications R and then sample $N_1, \ldots, N_R$ i.i.d. Poisson(λ) random variables. Then, given $N_i = n_i$, we sample $X_1^i, \ldots, X_{n_i}^i$ i.i.d. Erlang(4, 10/22) random variables and set

$$\hat{Y}_R^{\text{CMC}} = \frac{1}{R} \sum_{i=1}^{R} Y_i, \quad Y_i = \mathbf{1}\left(\sum_{k=1}^{n_i} X_k^i > 32 \right).$$

Consider now the **stratified estimator** resulting from stratifying N. Take m strata: $A^k = \{N = k\}, k = 0, \ldots, m - 2$, and $A^{m-1} = \{N \geq m - 1\}$. Proportional allocation yields

$$R_j = \lceil p_j R \rceil, \quad j = 0, \ldots, m-2, \quad R_{m-1} = \left\lceil R \sum_{k=m-1}^{\infty} p_k \right\rceil, \quad \text{where } p_k = \frac{\lambda^k}{k!} e^{-\lambda}.$$

We need to estimate I in each stratum. Note that if $N = 0$, then $S = 0$ and $\mathbf{1}(S > 32) = 0$ for sure, so there is no need to estimate it. Therefore, we can take $R_0 = 1$, and then rescale the remaining R_i correspondingly. For example, for $R = 10000$ replications and $m = 7$ strata, the strata probabilities p_i and corresponding allocations R_i are shown in Table 5.14.

In each stratum $i = 1, \ldots, m - 1$, we need to simulate S. For $i < m - 1$, we simply simulate i i.i.d. Erlang random variables. Then we compute whether the sum of these variables exceeds 32. For a given j, we repeat the procedure n_j times, denoting simulated Erlang random variables by $X_k^{i,j}, j = 1, \ldots, m$ (stratum), i (iteration number), $k = 1, \ldots, n_j$. The estimator in each stratum is

$$\hat{Y}_j = \frac{1}{n_j} \sum_{i=1}^{n_j} Y_i^j, \quad \text{where } Y_i^j = \mathbf{1}\left(\sum_{k=1}^{j} X_k^{i,j} > 32 \right).$$

For stratum $m - 1$, we proceed as follows: we first simulate N conditioned on $N \geq m - 1$. This can be done using the ITM method. Let $F_{\text{Pois}(\lambda)}$ denote the cdf of the Poisson(λ) distribution. Set $\rho_{m-1} = \sum_{i=1}^{m-1} p_i$, simulate $U \sim \mathcal{U}[0, 1)$, and set

$$N = F_{\text{Pois}(\lambda)}^{-1} \left(\rho_{m-1} + (1 - \rho_{m-1})U \right).$$

Table 5.15 Simulation results for ruin probability $I = \mathbb{P}(S > 32)$ with parameters $R = 10^4$, $\lambda = 3.3$, claims distributed Erlang(4, 10/22). The values of CMC and stratified estimators (for number of strata $m \in \{3, 4, 5, 6, 7\}$), as well as errors $b = 1.96\sqrt{\widehat{\mathrm{Var}}(Y_R)}$ are presented. The last row reports the ratio of the CMC variance to the stratified variance

	CMC	Stratified estimator				
		Number of strata $m =$				
		3	4	5	6	7
$\hat{Y}_R$	0.39350	0.39794	0.38365	0.38346	0.39091	0.38935
b	0.00958	0.00879	0.00763	0.00622	0.00618	0.00619
$\dfrac{\mathrm{Var}(\hat{Y}_R^{\mathrm{CMC}})}{\mathrm{Var}(\hat{Y}_R^{\mathrm{str}})}$	–	1.18751	1.57588	2.36690	2.39724	2.39690

Given $N = n \geq m - 1$, we simulate n i.i.d. Erlang random variables and check whether the sum exceeds 32. Denote the corresponding estimator by $\hat{Y}_{m-1}$. For each stratum, we also calculate the variance of the estimator, denoting it by $\hat{S}_i^2$, $i = 0, \ldots, m - 1$.

Finally, our stratified estimator and its variance are given respectively by

$$\hat{Y}_R^{\mathrm{str}} = \sum_{j=0}^{m-1} p_j \hat{Y}_j, \qquad \widehat{\mathrm{Var}}(\hat{Y}_R^{\mathrm{str}}) = \sum_{j=0}^{m-1} \frac{p_j^2 \hat{S}_j^2}{R_i}.$$

In Table 5.15, simulation results for the above parameters ($R = 10^4$ replications) are presented. Using stratified sampling with proportional allocation reduces the variance by a factor of 2.4 for $m = 6, 7$ strata.

We leave it to the reader to apply the optimum allocation for this example (Exercise 5.L.16).

This example, including the values reported in Tables 5.14 and 5.15, is implemented in the script `ch5_collective_risk_Erlang_stratified.py`.

$\Diamond$

Stratified Sampling for Brownian Motion We consider two methods for stratified sampling of a Brownian motion at time instants $t_0 = 0, t_1 = 1/n, t_2 = 2/n, \ldots, t_n = 1$, where $n \geq 1$:

- **Method 1: Stratified sampling for a multivariate normal distribution.** A vector $\mathbf{B} = (B(t_1), B(t_2), \ldots, B(t_n))$ follows a multivariate normal distribution with covariance matrix given in Proposition 5.2.2, i.e., $\mathrm{Cov}\,(B(t_i), B(t_j)) = \min(t_i, t_j)$. Below, we demonstrate how to stratify and simulate a multivariate normal random variable with a general non-singular covariance matrix $\boldsymbol{\Sigma}$.
- **Method 2: Stratification in the bisection algorithm.** For time instants $t_1 = 1/2^k, t_2 = 2/2^k, \ldots, t_{2^k} = 1$ (i.e., $n = 2^k$), we use Algorithm 38. In this algorithm, $Z_1, \ldots, Z_{2^k}$ are i.i.d. standard normal random variables, which can be stratified partially or entirely.

5.2.3.1 Method 1: Stratified Sampling of a Multivariate Normal Distribution

Fix the number of strata m and a natural number n. Let

$$\mathbf{B} = (B(1/n),\ B(2/n),\ \ldots,\ B(1)),$$

which follows a multivariate normal distribution with covariance matrix:

$$\mathbf{\Sigma}(i, j) = \frac{1}{n}\min(i, j).$$

The covariance matrix can be factored as $\mathbf{\Sigma} = \mathbf{A}\mathbf{A}^T$, where:

$$\mathbf{A}(i, j) = \begin{cases} \frac{1}{\sqrt{n}}, & j \le i, \\ 0, & \text{otherwise.} \end{cases}$$

Let $\mathbf{Z} = (Z_1, \ldots, Z_n)^T$ be a standard multivariate normal vector. Define strata $A^1, \ldots, A^m$ as concentric rings (Fig. 5.22), where $\mathbb{P}(\mathbf{Z} \in A^i) = 1/m$. Sampling from A^i ensures $\mathbf{B}^i = \mathbf{A}\mathbf{Z}^i$ belongs to the corresponding stratum.

For the one-dimensional case ($n = 1$), the stratified sampling method for standard normal random variables was given in Example 5.1.9. For general n, the procedure is as follows:

1. Generate $\boldsymbol{\xi} = (\xi_1, \ldots, \xi_n)^T$, where $\xi_i \sim \mathcal{N}(0, 1)$ are i.i.d. Normalize to obtain $\mathbf{X} = (\xi_1/\|\boldsymbol{\xi}\|, \ldots, \xi_n/\|\boldsymbol{\xi}\|)^T$, a vector uniformly distributed on the unit sphere.
2. Sample $U \sim \mathcal{U}[0, 1)$ and compute $D^2 = F_{\chi_n^2}^{-1}\left(\frac{i-1}{m} + \frac{U}{m}\right)$, where $F_{\chi_n^2}$ is the c.d.f. of the χ_n^2 distribution.
3. Set $\mathbf{Z}^i = D\mathbf{X}$.
4. Return $\mathbf{B}^i = \mathbf{A}\mathbf{Z}^i$.

Note that D^2 is distributed $Z_1^2 + \cdots + Z_n^2$ for i.i.d. standard normal random variables Z_i, i.e., it has a χ_n^2 distribution (χ^2 with n degrees of freedom). Its p.d.f. and c.d.f. are

$$f_{\chi_n^2}(r) = \frac{1}{2^{n/2}\Gamma(n/2)} r^{n/2-1} e^{-r/2},$$

$$F_{\chi_n^2}(r) = \frac{1}{\Gamma(n/2)} \gamma_{n/2}(r/2),$$

where Γ is a gamma function, and γ is the so-called *incomplete gamma function*.[1] For $n = 2$ the random variable D has the so-called Rayleigh distribution.

---

[1] https://en.wikipedia.org/wiki/Incomplete_gamma_function.

Admittedly, there is no explicit formula for the inverse function of $F_{\chi_n^2}(r)$ for general n, but numerically this inverse is available in several libraries. [2]

Example 5.2.7 In Fig. 5.22, 5000 points are sampled within 4 strata using this method. The left plot shows points from a standard normal distribution, while the right plot illustrates points from a normal distribution with covariance matrix $\Sigma(i, j) = \min(i, j)/3$. This example is implemented in `ch5_strat_normal_3D.py`.

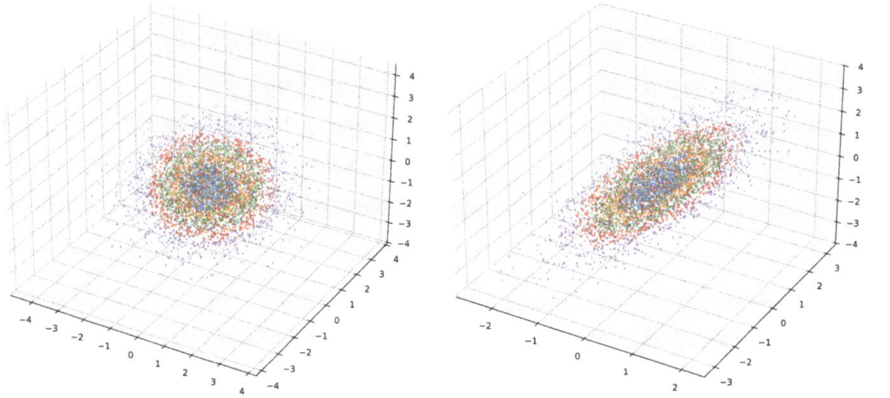

Fig. 5.22 5000 stratified samples from a 3-dimensional normal distribution. Left: standard normal. Right: covariance matrix $\Sigma(i, j) = \min(i, j)/3$

$\Diamond$

Example 5.2.8 (Asian Call Option: Method 1) Consider an Asian call option (with discounted payoff at time 1) with a price given by:

$$I = e^{-r}\mathbb{E}(A - K)_+, \tag{5.2.9}$$

where $A = \frac{S(1/2)+S(1)}{2}$, and $S(t) = S(0)\exp(\mu^* t + \sigma B(t))$. The parameters are:

$$r = 0.05, \quad \sigma = 0.25 \quad (\text{thus } \mu^* = r - \sigma^2/2 = -0.0125),$$

$$S(0) = 100, \quad K = 100.$$

Crude Monte Carlo Estimator Fix a number of simulations R and simulate $B^1(1), \ldots, B^R(1)$ i.i.d. $\sim \mathcal{N}(0, 1)$. Using Corollary 5.2.4, the conditional distribution of $B^k(1/2) \mid B^k(1) = y_k$ is $\mathcal{N}(y_k, 1/4)$. Generate $B^k(1/2)$ in this way for

---

[2] E.g. https://docs.scipy.org/doc/scipy/reference/generated/scipy.stats.chi2.html in `Python`.

$k = 1, \ldots, R$. Then compute:

$$S^k(t) = S(0) \exp(\mu^* t + \sigma B^k(t)), \quad k = 1, \ldots, R, \ t = 1/2, 1,$$

and set:

$$\hat{Y}_R^{\text{CMC}} = \frac{1}{R} \sum_{k=1}^{R} Y_k, \quad Y_k = \left(\frac{S^k(1/2) + S^k(1)}{2} - K \right)_+.$$

Stratified Estimator   Define strata A_i, $i = 1, \ldots, m$. The vector $(B(1/2), B(1))$ has a bivariate normal distribution $\mathcal{N}(\mathbf{0}, \boldsymbol{\Sigma})$, where:

$$\boldsymbol{\Sigma} = \begin{pmatrix} 1/2 & 1/2 \\ 1/2 & 1 \end{pmatrix}.$$

A natural stratification is based on confidence ellipses, where $p_i = \mathbb{P}(\mathbf{Z} \in A_i) = 1/m$ and $\mathbf{Z} = (Z_1, Z_2) \sim \mathcal{N}(\mathbf{0}, \boldsymbol{\Sigma})$. See Fig. 5.23 (right).

We need to simulate $\mathbf{Z}^i \overset{d}{=} (\mathbf{Z} \mid \mathbf{Z} \in A_i)$, where $\mathbf{Z} = (Z_1, Z_2) \sim \mathcal{N}(\mathbf{0}, \mathbf{I})$, i.e., Z_1, Z_2 are i.i.d. $\mathcal{N}(0, 1)$. Define strata for $\mathbf{Z} \sim \mathcal{N}(\mathbf{0}, \mathbf{I})$: the first stratum is $A_1 = \{(x, y) : x^2 + y^2 \le r_1^2\}$, where $P(\mathbf{Z} \in A_1) = 1/m$. The radius r_1 is determined accordingly. The second stratum is $A_2 = \{(x, y) : r_1^2 < x^2 + y^2 \le r_2^2\}$, such that $P(\mathbf{Z} \in A_2) = 1/m$, and so on.

The vector $\mathbf{Z}$ in polar coordinates is (D, Θ), where:

$$f(r, \theta) = g(r)h(\theta) = re^{-r^2/2} \cdot \frac{1}{2\pi}, \quad r \ge 0, \ \theta \in [0, 2\pi),$$

with $D \sim$ Rayleigh and $\Theta \sim \mathcal{U}[0, 2\pi)$, independent.

The c.d.f. of D is:

$$F_D(r) = 1 - e^{-r^2/2}, \quad \text{with inverse} \quad F_D^{-1}(t) = \sqrt{-2\ln(1 - t)}.$$

To simulate $\mathbf{Z}^i \overset{\mathcal{D}}{=} (\mathbf{Z} \mid \mathbf{Z} \in A_i)$, proceed as follows:

1. Simulate $\theta \sim \mathcal{U}[0, 2\pi)$.
2. Simulate $U \sim \mathcal{U}[0, 1)$, and set:

$$D = F_D^{-1} \left(\frac{i-1}{m} + \frac{1}{m} U \right).$$

3. Return $(D \cos(\theta), D \sin(\theta))$.

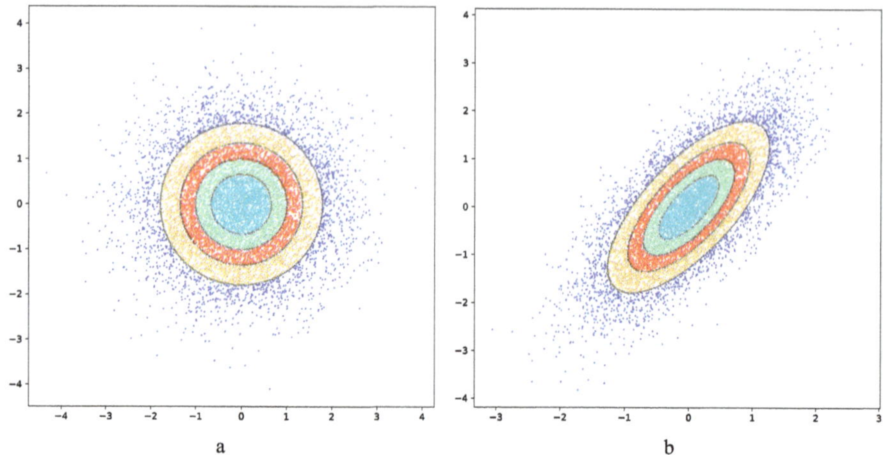

Fig. 5.23 5 strata: left $\mathbf{Z} = (Z_1, Z_2) \sim \mathcal{N}(\mathbf{0}, \boldsymbol{I})$, right $= \mathbf{X} = (X_1, X_2) \sim \mathcal{N}(\mathbf{0}, \boldsymbol{\Sigma})$, 20 000 points in each strata

See Fig. 5.23 (left) for simulations in 5 strata. To simulate $\mathbf{X}^i \stackrel{\mathcal{D}}{=} (\mathbf{X} \mid \mathbf{X} \in S_i)$, where $\mathbf{X} \sim \mathcal{N}(\mathbf{0}, \boldsymbol{\Sigma})$, perform a Cholesky decomposition of $\boldsymbol{\Sigma}$:

$$\boldsymbol{\Sigma} = \mathbf{A}^T \mathbf{A}, \quad \text{where } \mathbf{A} = \frac{1}{2}\begin{pmatrix} \sqrt{2} & \sqrt{2} \\ 0 & \sqrt{2} \end{pmatrix}.$$

Simulate $\mathbf{Z}^i \stackrel{\mathcal{D}}{=} (\mathbf{Z} \mid \mathbf{Z} \in A_i)$ as described previously, and set:

$$\mathbf{X}^i = \mathbf{A}\mathbf{Z}^i.$$

See Fig. 5.23 (right) for simulation results.

Having all the tools needed, we proceed with the stratified sampling as follows. First, we perform a **pilot simulation**: we set $R' = 10000$ and simulate $R_i' = 10000/m$ bivariate random variables $(B_j^i(1/2), B_j^i(1)), j = 1, \ldots, R_i$ in each stratum i. We then set

$$S_j^i(t) = S(0)\exp(\mu^* t + \sigma B_j^i(t)), \quad i = 1, \ldots, m, \; j = 1, \ldots, R_i', \; t \in \{1/2, 1\}.$$

See Fig. 5.24 for the resulting $(S_j^i(1/2), S_j^i(1))$.

Fig. 5.24 Points $(S^i_j(1/2), S^i_j(1))$ for $B^i_j(t)$, $i = 1, \ldots, 5$, $j = 1, \ldots, 20\,000$, $t \in \{1/2, 1\}$ simulated using stratified sampling (presented in Fig. 5.23)

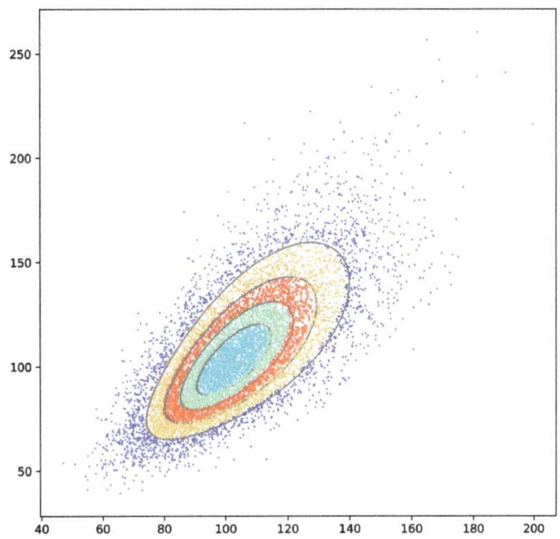

We compute

$$Y^i_j = \left(\frac{S^i_j(1/2) + S^i_j(1)}{2} - K \right)_+ ,$$

(the values of $(S^i_j(1/2) + S^i_j(1))/2$ are depicted in Fig. 5.25), estimators and variances in each stratum are given respectively by

$$\hat{Y}'_{R_i} = \frac{1}{R'_i} \sum_{j=1}^{R'_i} Y^i_j, \qquad \hat{S}^2_i = \frac{1}{R'_i} \sum_{j=1}^{R'_j} (Y^i_j - \hat{Y}'_{R_i}).$$

For the **actual simulation**, we determine (according to Theorem 5.1.5) the optimal allocation:

$$R_i = \left\lfloor \frac{\hat{S}_i}{\sum_{j=1}^m \hat{S}_j} R \right\rfloor ,$$

and in cases where $R_i = 0$, we update it to $R_i = 1$. We then perform the simulations again, this time simulating R_i points in the i-th stratum. Finally, as in Example 5.2.5, we compute the mean $\hat{Y}_{R_i}$ and the variance $\hat{S}^2_i$ in each stratum to obtain:

$$\hat{Y}^{\text{str}}_R = \frac{1}{m} \sum_{i=1}^m \hat{Y}_{R_i}, \qquad \widehat{\text{Var}}(\hat{Y}^{\text{str}}_R) = \frac{1}{m^2} \sum_{i=1}^m \frac{\hat{\sigma}^2_i}{R_i}.$$

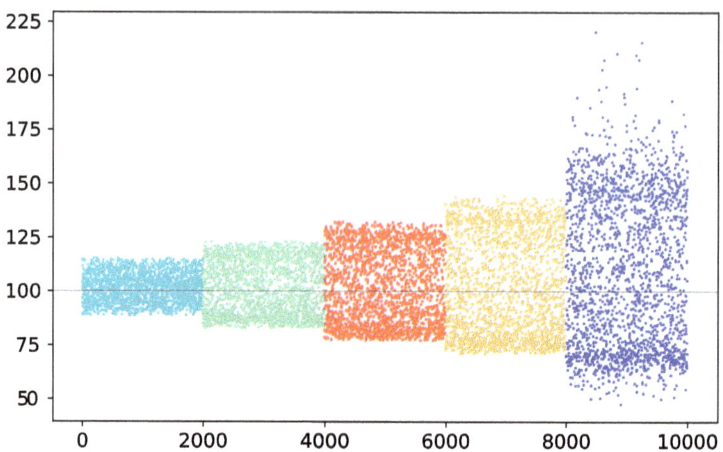

Fig. 5.25 Points (number of simulation, A), where $A = (S(1/2) + S(1))/2$ in pilot simulation. Colors correspond to strata. Gray line: threshold K.

The simulation results for $R = 10^5$, $S(0) = 100$, $r = 0.05$, $\sigma = 0.25$, $K \in \{100, 125, 150, 175\}$, and strata $m \in \{5, 10, 50, 100, 200\}$ are presented in Table 5.16. For larger values of K, the improvement is significant: e.g., for

Table 5.16 Simulation results for estimating Asian option I with $n = 2$ and parameters $R = 10^5, S(0) = 100, r = 0.05, \sigma = 0.25$ and $K \in \{100, 125, 150, 175\}$. The values of CMC and stratified estimators (for number of strata $m \in \{5, 10, 50, 100, 200\}$, as well as errors $b = 1.96\sqrt{\widehat{\mathrm{Var}}(Y_R)}$

		CMC	Stratified estimator				
$k \downarrow$			number of strata $m =$				
			5	10	50	100	200
100	$\hat{Y}$	9.6100	9.5829	9.6233	9.6142	9.5858	9.5609
	b	0.0875	0.0722	0.0713	0.0707	0.0705	0.0706
	$\dfrac{\mathrm{Var}(\hat{Y}^{\mathrm{CMC}})}{\mathrm{Var}(\hat{Y}^{\mathrm{str}})}$	–	1.4708	1.5071	1.5418	1.5215	1.5369
125	$\hat{Y}$	2.0629	2.0151	2.0149	2.0177	2.0202	2.0359
	b	0.0428	0.0232	0.0216	0.0209	0.0208	0.021
	$\dfrac{\mathrm{Var}(\hat{Y}^{\mathrm{CMC}})}{\mathrm{Var}(\hat{Y}^{\mathrm{str}})}$	–	3.3975	3.7877	4.0141	4.0964	4.0334
150	$\hat{Y}$	0.2988	0.3021	0.3041	0.3022	0.3034	0.2999
	b	0.0159	0.007	0.0051	0.0039	0.004	0.0038
	$\dfrac{\mathrm{Var}(\hat{Y}^{\mathrm{CMC}})}{\mathrm{Var}(\hat{Y}^{\mathrm{str}})}$	–	5.0682	10.023	17.0475	15.7522	18.5801
175	$\hat{Y}$	0.0363	0.0361	0.0353	0.0373	0.0361	0.0363
	b	0.0054	0.0024	0.0017	0.0008	0.0006	0.0006
	$\dfrac{\mathrm{Var}(\hat{Y}^{\mathrm{CMC}})}{\mathrm{Var}(\hat{Y}^{\mathrm{str}})}$	–	5.1083	10.1118	58.8544	74.3299	84.0150

$K = 175$ (see Fig. 5.25, where for this K only points from the last stratum give nonzero $(A - K)_+$), the variance for $m = 200$ is reduced by a factor of 84. This example, including the results in Table 5.16 and the plots in Figs. 5.23, 5.24, and 5.25, is implemented in `ch5_asian_option_stratified.py`.   ◇

5.2.3.2 Method 2: Stratified Sampling of a Brownian Motion Using Algorithm 38

Recall that Algorithm 38 simulates Brownian motion at time instants

$$1/2^k, 2/2^k, \ldots, 1$$

using i.i.d. standard normal random variables $Z_1, \ldots, Z_{2^k}$. Thus, we may stratify the simulation of each, or only some, of these random variables. Suppose we want to stratify the sampling of Z_j using equally probable m strata. Then the strata are:

$$A^1 = (a_0, a_1], \ A^2 = (a_1, a_2], \ldots, A^m = (a_{m-1}, a_m),$$

where $a_0 = -\infty$, $a_m = \infty$ and:

$$a_i = \Phi^{-1}\left(\frac{i}{m}\right), \quad i = 1, \ldots, m - 1.$$

The inverse of the cumulative distribution function of a standard normal random variable Φ^{-1} is readily available in most programming packages. To simulate a standard normal random variable from stratum A^j, we simulate $U \sim \mathcal{U}[0, 1)$ and then (see Proposition 5.1.7) compute:

$$\Phi^{-1}(V^j), \quad \text{where } V^j = a_{j-1} + (a_j - a_{j-1})U.$$

Consider the following case: we stratify Z_{2^k} and Z_{2^k-1}, which are used in Algorithm 38 to compute $B(1)$ and $B(1/2)$, respectively. We independently sample each of these random variables using stratification with m strata, resulting in m^2 strata for the bivariate random variable (Z_{2^k-1}, Z_{2^k}). These strata are defined as:

$$A^{i,j} = A^i \times A^j = (a_{i-1}, a_i] \times (a_{j-1}, a_j], \quad i, j = 1, \ldots, m.$$

From (Z_{2^k-1}, Z_{2^k}), we compute $(B(1/2), B(1))$ as follows:

$$B(1) = Z_{2^k},$$

$$B(1/2) = \frac{B(0)+B(1)}{2} + \frac{Z_{2^k-1}}{2} = \frac{Z_{2^k}+Z_{2^k-1}}{2}.$$

Note that:

$$B(1/2) \in \left(\frac{a_{i-1} + a_{j-1}}{2}, \frac{a_i + a_j}{2} \right).$$

In Fig. 5.26a, we present the following example:

- We fix $m = 5$.
- $R = 10$ realizations of stratified sampling of (Z_{2k-1}, Z_{2k}) from stratum $A^{4,2} = A^4 \times A^2$ are presented.
- The strata for Z_{2k-1} and Z_{2k} (note they are the same) are presented (shaded gray regions).

In Fig. 5.26b, we extend the example:

- $R = 10$ realizations of $(B(0), B(1/2^k), \ldots, B(1))$ with $k = 10$ are presented.
- $R = 10$ independent replications of $Z_1, \ldots, Z_{2k}$ were used, where (Z_{2k-1}, Z_{2k}) were stratified as in Fig. 5.26a.
- The strata for Z_{2k} (shaded gray region) as well as the "stratum" $C^{4,2}$ for $B(1/2)$ is presented (shaded red region).

Note that for the above example, there are $m^2 = 25$ strata in total. Similar realizations, but for stratum $A^2 \times A^9$ (one out of 10×10 strata), are presented in Fig. 5.27. The example is implemented in `ch5_BrownianMotion_bisection_stratified.py`. Default options return Fig. 5.26. To obtain Fig. 5.27 invoke:

```
ch5_BrownianMotion_bisection_stratified.py --nr_strata 10
--strata05 2 --strata1 9
```

5.2.4 Control Variates

Here we present an example of the control variate variance reduction method applied to a financial contract.

Example 5.2.9 (Control Variate: Asian Call Option) We are interested in estimating an Asian call option (with discounted payoff at time 1) with price given by the formula:

$$I = e^{-r} \mathbb{E}(A - K)_+,$$

where

$$A = \frac{1}{n} \sum_{i=1}^{n} S(i/n), \qquad S(t) = S(0) \exp(\mu^* t + \sigma B(t)).$$

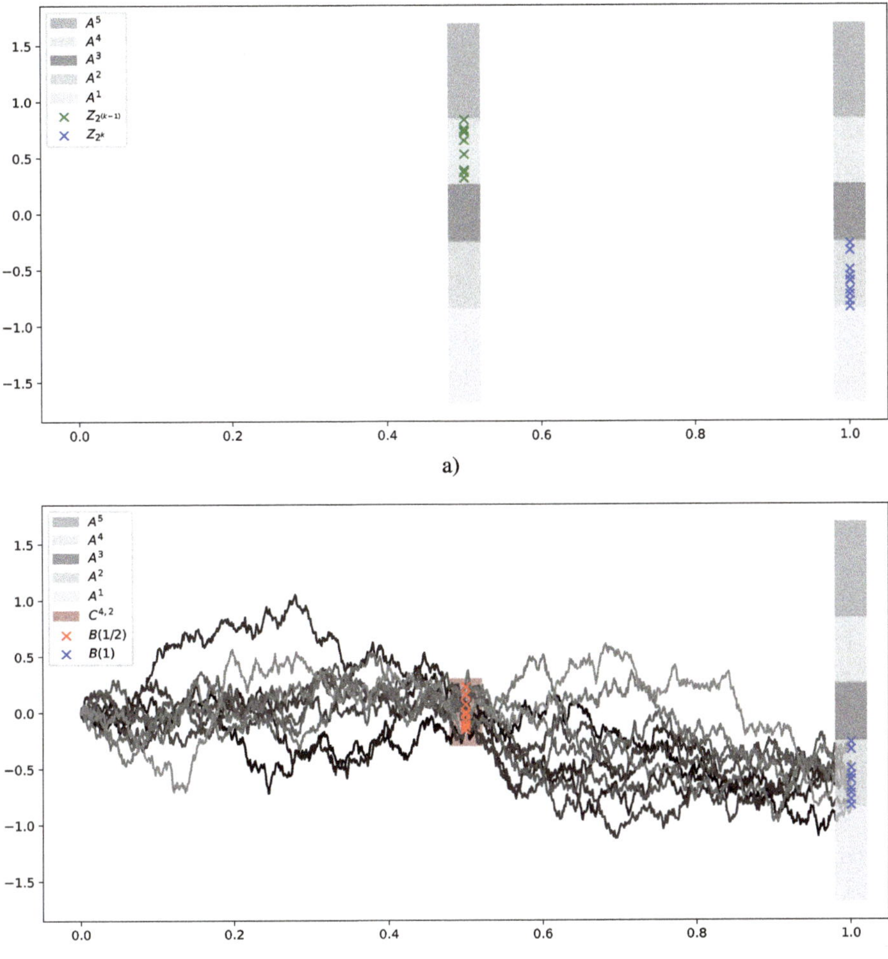

Fig. 5.26 Stratified sampling (one out of 5×5 strata) of a Brownian motion with $(Z_{2k-1}, Z_{2k}) \in A^4 \times A^2$. See description in text. (**a**) Stratified samples of $(Z_{2k-1}, Z_{2k}) \in A^4 \times A^2$ (one out of 5×5 strata). (**b**) Stratified samples of $(B(0), B(1/2^k), \ldots, B(1))$ with $k = 10$

Fix the parameters as in Example 5.2.8: $r = 0.05$, $\sigma = 0.25$ (thus $\mu^* = r - \sigma^2/2 = -0.0125$), $S(0) = 100$ and $K = 100$.

The main challenge in evaluating I analytically lies in the distribution of the arithmetic mean of $S(i/n), i = 1, \ldots, n$, which is intractable. However, the geometric average

$$G = \left(\prod_{i=1}^{n} S(i/n) \right)^{1/n}$$

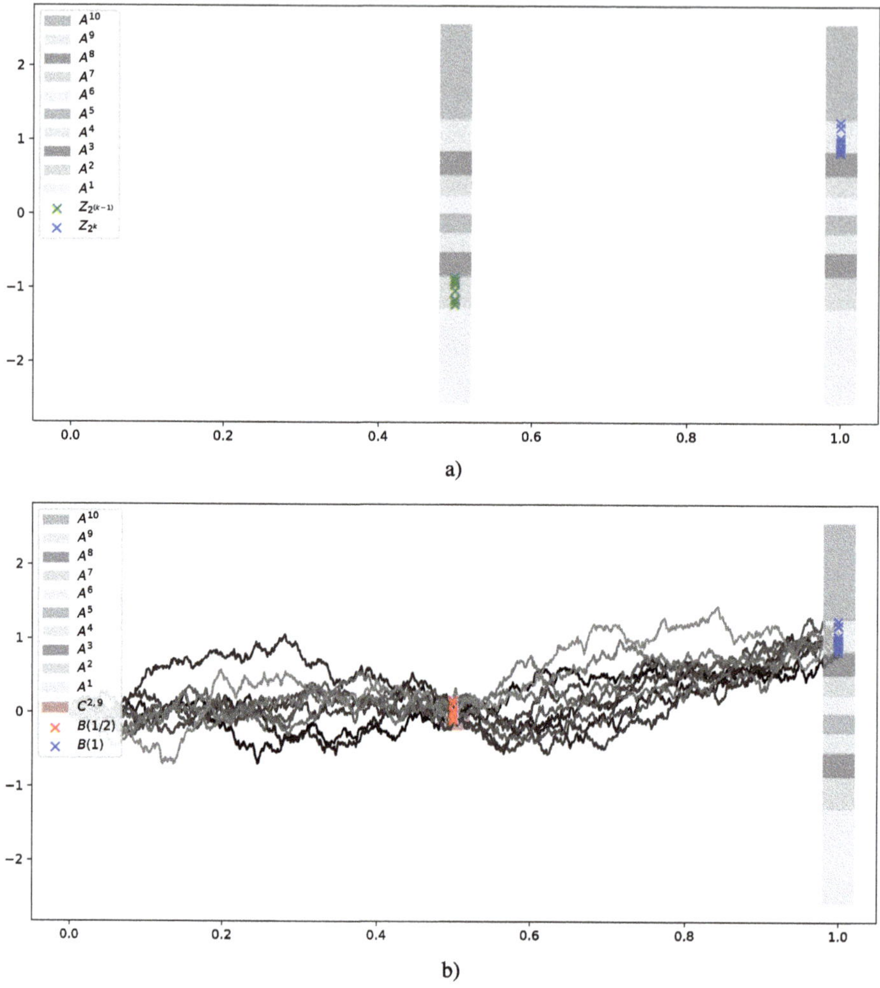

Fig. 5.27 Stratified sampling (one out of 10×10 strata) of a Brownian motion with $(Z_{2k-1}, Z_{2k}) \in A^2 \times A^9$. See description in text. (**a**) Stratified samples of $(Z_{2k-1}, Z_{2k}) \in A^2 \times A^9$ (one out of 10×10 strata). (**b**) Stratified samples of $(B(0), B(1/2^k), \ldots, B(1))$ with $k = 10$

is tractable:

$$G = S(0) \left(\prod_{i=1}^{n} \exp\left(\mu^* \frac{i}{n} + \sigma B\left(\frac{i}{n} \right) \right) \right)^{\frac{1}{n}}$$

$$= S(0) \left(\exp\left[\sum_{i=1}^{n} \left(\mu^* \frac{i}{n} + \sigma B\left(\frac{i}{n} \right) \right) \right] \right)^{\frac{1}{n}}$$

$$= S(0) \left(\exp \left[\mu^* \frac{n(n+1)}{2n} + \sum_{i=1}^{n} \sigma B \left(\frac{i}{n} \right) \right] \right)^{\frac{1}{n}}$$

$$= S(0) \exp \left[\mu^* \frac{n+1}{2n} + \frac{\sigma}{n} \sum_{i=1}^{n} B \left(\frac{i}{n} \right) \right].$$

The vector $\mathbf{B} = (B(1/n), \ldots, B(n/n))$ has a multivariate normal distribution $\mathcal{N}(\mathbf{0}, \boldsymbol{\Sigma})$ with $\Sigma(i, j) = \min(i/n, j/n)$. A linear transformation $\mathbf{LB}$ also has a multivariate normal distribution with mean $\mathbf{0}$ and the variance $\mathbf{L\Sigma L}^T$. Thus, for a sum—i.e., for $\mathbf{L} = (1, \ldots, 1)$ we have:

$$Y = \sum_{i=1}^{n} B \left(\frac{i}{n} \right),$$

whose variance is given by:

$$\sigma_Y^2 = (1, \ldots, 1) \boldsymbol{\Sigma} (1, \ldots, 1)^T = \frac{1}{n} \sum_{i=1}^{n} \sum_{j=1}^{n} \min(i, j) = \frac{1}{n} \sum_{i=1}^{n} i^2 = \frac{1}{6} (n+1)(2n+1).$$

In other words, $Y = \sigma_Y Z$, where $Z \sim \mathcal{N}(0, 1)$. We may express G as:

$$G = S(0) \exp \left[\mu^* \frac{n+1}{2n} + \frac{\sigma}{n} Y \right]$$

$$= S(0) \exp \left[\mu^* \frac{n+1}{2n} + \frac{\sigma \sigma_Y}{n} Z \right]$$

$$= S(0) \exp \left[\mu_N + \sigma_N Z \right]$$

with

$$\mu_N = \mu^* \frac{n+1}{2n} = (r - \sigma^2/2) \frac{n+1}{2n},$$

$$\sigma_N^2 = \frac{\sigma^2}{n^2} \frac{1}{n} \sum_{i=1}^{n} i^2 = \frac{\sigma^2}{6n^2} (n+1)(2n+1).$$

We can compute the exact value of $\mathbb{E}(G - K)_+$ from the Black-Scholes formula (5.2.7), and thus use:

$$W = e^{-r} \mathbb{E}(G - K)_+$$

as a control variate.

Table 5.17 Simulation results for the Asian option I with $n = 10$ using CMC, antithetic variates and control variates. Parameters: $R = 10^4, n = 10, S(0) = 100, r = 0.05, \sigma = 0.25$. The table shows the estimators and errors ($b = 1.96\hat{S}$) for various K

$K \downarrow$	$\hat{Y}^{\text{CMC}}$	Error $\hat{b}^{\text{CMC}}$	$\hat{Y}^{\text{CV}}$	Error $\hat{b}^{\text{CV}}$	$\hat{Y}^{\text{ant}}$	Error $\hat{b}^{\text{ant}}$
150	0.0919	2.4468	0.0586	0.3738	0.0518	0.0153
120	1.5706	10.2541	1.4219	0.5859	1.4258	0.0943
100	7.9348	21.6495	7.3853	0.7305	7.3243	0.2097
50	51.3216	30.6612	50.2194	0.9978	50.2376	0.3024
20	79.6775	30.3089	78.7550	0.9777	78.7681	0.3021

Let $Z_1, \ldots, Z_n$ be i.i.d. $\mathcal{N}(0, 1)$ random variables. Recall (5.2.2), which allows for a straightforward sampling of $B(k/n), k = 1, \ldots, n$:

$$B(k/n) = \frac{1}{\sqrt{n}} \sum_{i=1}^{k} Z_i.$$

In Algorithm 41 the summarized procedure on how to proceed with control variates for this problem is provided. Note that we first need to estimate the optimal c via $\hat{c}$ using R' pilot simulations.

In Table 5.17, we present simulation results for the control variate method described above for estimating $I = e^{-r}\mathbb{E}(A - K)_+$ with parameters $R' = 10^3$ (pilot simulations, used to estimate $\hat{c}$), $R = 10^4$ (actual simulations), $S(0) = 100, r = 0.05, \sigma = 0.25$, and $n = 10$. Note that for larger K, e.g., the error of control variate estimator is much smaller than those of CMC. The example (antithetic and control variates methods) is implemented in `ch5_asian_option_antithetic_control_variates.py`. $\qquad\qquad\diamondsuit$

5.2.5 Antithetic Variates

Here we present an example of the antithetic variates reduction method applied to a financial contract.

Example 5.2.10 (Antithetic Variates: Asian Call Option) We consider the same setup as in previous Example 5.2.9 (i.e., control variates for Asian call option). We are interested in estimating the price of an Asian call option (with discounted payoff at time 1) using the antithetic variates method for variance reduction. Recall the price is given by:

$$I = e^{-r}\mathbb{E}(A - K)_+,$$

Algorithm 41 Control variates for estimation of an Asian option

---

Require: Parameters $r, K, \sigma, S(0), n$

1: Simulate $Z_1^j, \ldots, Z_n^j$, i.i.d. $\mathcal{N}(0, 1)$ random variables.

2: Compute:

$$A_j = \frac{1}{n} \sum_{k=1}^{n} S^j(k/n), \quad Y_j^{\mathrm{CMC}} = e^{-r}(A_j - K)_+.$$

3: Compute:

$$G_j = \left(\prod_{k=1}^{n} S^j(k/n) \right)^{1/n}, \quad Y_j^G = e^{-r}(G_j - K)_+.$$

4: Set:

$$\hat{Y}^{\mathrm{CMC}} = \frac{1}{R} \sum_{j=1}^{R} Y_j^{\mathrm{CMC}}, \quad \hat{Y}^G = \frac{1}{R} \sum_{j=1}^{R} Y_j^G.$$

5: From the Black-Scholes formula (5.2.7), compute the exact value $\mathbb{E}(G - K)_+$, set $W = e^{-r}\mathbb{E}(G - K)_+$, and:

$$\hat{Y}^{\mathrm{CV}} = \hat{Y}^{\mathrm{CMC}} + \hat{c}(\hat{Y}^G - W),$$

where $\hat{c}$ is estimated from pilot simulations.

6: **return** $\hat{Y}^{\mathrm{CV}}$

---

where:

$$A = \frac{1}{n} \sum_{i=1}^{n} S(i/n), \qquad S(t) = S(0) \exp(\mu^* t + \sigma B(t)).$$

Refer to Example 5.2.9 for the Crude Monte Carlo (CMC) approach. For the antithetic variates method, we proceed as follows:

1. Generate $Z_1, Z_2, \ldots, Z_n$, i.i.d. $\mathcal{N}(0, 1)$ random variables.
2. Compute:

$$B(i/n) = \frac{1}{\sqrt{n}} \sum_{j=1}^{i} Z_j, \quad i = 1, \ldots, n.$$

3. Compute the antithetic realization:

$$B'(i/n) = \frac{1}{\sqrt{n}} \sum_{j=1}^{i} (-Z_j), \quad i = 1, \ldots, n.$$

4. Calculate A and A', the arithmetic averages for the original and antithetic paths, respectively:

$$A = \frac{1}{n}\sum_{i=1}^{n} S(i/n), \quad A' = \frac{1}{n}\sum_{i=1}^{n} S'(i/n),$$

where $S(i/n) = S(0)\exp(\mu^* i/n + \sigma B(i/n))$ and $S'(i/n) = S(0)\exp(\mu^* i/n + \sigma B'(i/n))$.

5. Compute the payoffs for both realizations:

$$Y = e^{-r}(A - K)_+, \quad Y' = e^{-r}(A' - K)_+.$$

6. The antithetic variates estimator is given by:

$$\hat{Y}_R^{\text{anti}} = \frac{1}{2R}\sum_{j=1}^{R}(Y_j + Y_j'),$$

where R is the number of independent simulations.

The simulation results for exactly the same parameters as in Example 5.2.9 are provided in Table 5.17. The results demonstrate that both the control variates and antithetic variates methods significantly reduce the error compared to the CMC estimator, with the antithetic variates method achieving smaller variance than the control variates method in all cases. Both variance reduction methods show particularly notable improvements for smaller K values. For example, for $K = 20$, the error of antithetic variates method was approximately 100 times smaller than those of CMC (thus, variance was reduced by a factor of 10^4.) Both, antithetic and control variates methods for this example are implemented in `ch5_asian_option_antithetic_control_variates.py`. $\Diamond$

5.2.6 Importance Sampling

Example 5.2.11 (Importance Sampling for a European Call Option) We consider the European call option with price given by:

$$I = e^{-r}\mathbb{E}(S(1) - K)_+$$

$$= e^{-r}\int_{-\infty}^{\infty}(S(0)e^{\mu^* + \sigma z} - K)_+\, \phi_{0,1}(z)\, dz, \qquad (5.2.10)$$

where $S(1) = S(0)\exp(\mu^* + \sigma B(1))$, $B(1) \sim \mathcal{N}(0, 1)$, and $\phi_{0,1}(z)$ is the p.d.f. of $\mathcal{N}(0, 1)$. Here, $\mu^* = r - \sigma^2/2$.

Crude Monte Carlo Estimator The standard CMC estimator for I is:

$$\hat{Y}_R^{\text{CMC}} = \frac{1}{R}\sum_{i=1}^{R} Y_i, \quad Y_i = e^{-r}(S(0)e^{\mu^*+\sigma Z_i} - K)_+,$$

where $Z_1, \ldots, Z_R \sim \mathcal{N}(0,1)$ are i.i.d. random variables.

Importance Sampling Estimator To reduce variance, we use importance sampling with a proposal distribution $\tilde{f}(z) = \phi_{x^*,1}(z)$, where $\phi_{x^*,1}(z)$ is the p.d.f. of $\mathcal{N}(x^*,1)$. The likelihood ratio is:

$$L(z) = \frac{\phi_{0,1}(z)}{\phi_{x^*,1}(z)} = e^{-x^*z+(x^*)^2/2}.$$

Rewriting I with this proposal distribution gives:

$$I = e^{-r}\mathbb{E}_{x^*}\left[(S(1) - K)_+ L(Z)\right],$$

where $\mathbb{E}_{x^*}$ denotes expectation under $\mathcal{N}(x^*,1)$. The IS estimator becomes:

$$\hat{Y}_R^{\text{IS}} = \frac{1}{R}\sum_{i=1}^{R} Y_i^{\text{IS}}, \quad Y_i^{\text{IS}} = e^{-r}(S(0)e^{\mu^*+\sigma Z_i} - K)_+ e^{-x^*Z_i+(x^*)^2/2},$$

where $Z_i \sim \mathcal{N}(x^*,1)$.

Methods for Choosing x^* We will consider two methods.

Heuristic Method The optimal proposal distribution—see (5.1.22)—is the following:

$$g^*(z) = \frac{(S(0)e^{\mu^*+\sigma z} - K)_+\phi_{0,1}(z)}{I}.$$

We approximate $g^*(z)$ by aligning its peak with the proposal $\phi_{x^*,1}(z)$, i.e., we choose x^* as

$$x^* = \arg\sup_z h(z), \quad \text{where } h(z) = (S(0)e^{\mu^*+\sigma z} - K)_+\phi_{0,1}(z).$$

Denoting $\frac{1}{\sigma}(\ln(K/S(0)) - \mu^*)$ by x_1, we may rewrite

$$h(z) = \begin{cases} (S(0)e^{\mu^*+\sigma z} - K)\phi_{0,1}(z) & \text{for } z \geq x_1, \\ 0 & \text{otherwise.} \end{cases}$$

It can be verified that the right-hand derivative of $h(z)$ at x_1 is positive, implying that the maximum of $h(z)$ occurs for some $z > x_1$. Setting $h'(z) = 0$ leads to the following equation for the *importance point*:

$$S(0)\sigma e^{\mu^* + \sigma z} - z\left(S(0)e^{\mu^* + \sigma z} - K\right) = 0, \tag{5.2.11}$$

which is solved numerically using root-finding methods like Newton-Raphson.

Cross-entropy Method The CE method minimizes the Kullback-Leibler divergence between $g^*(z)$ and $\phi_{x^*,1}(z)$, as provided in (5.1.32), which is equivalent to:

$$x_{\mathrm{CE}}^* = \underset{x}{\mathrm{argmax}}\left\{\mathbb{E}(S(0)e^{\mu^* + \sigma Z} - K)_+ \log \phi_{x,1}(Z)\right\}$$

$$= \underset{x}{\mathrm{argmin}}\left\{\mathbb{E}(S(0)e^{\mu^* + \sigma Z} - K)_+ e^{(Z-x)^2/2}\right\},$$

where $Z \sim \mathcal{N}(0, 1)$. Instead of solving this, we will estimate it from simulations. We proceed as in Eq. (5.1.33): we sample $Z_1, \ldots, Z_M$ i.i.d. $\mathcal{N}(0, 1)$ and compute

$$\hat{x}_{\mathrm{CE}}^* = \underset{x}{\mathrm{argmin}}\,\frac{1}{M}\sum_{i=1}^{M}(S(0)e^{\mu^* + \sigma Z_i} - K)_+ e^{(Z_i - x)^2/2}.$$

Differentiating and setting to zero gives the solution:

$$\hat{x}_{\mathrm{CE}}^* = \frac{\sum_{i=1}^{M} Z_i \cdot (S(0)e^{\mu^* + \sigma Z_i} - K)_+}{\sum_{i=1}^{M} (S(0)e^{\mu^* + \sigma Z_i} - K)_+}. \tag{5.2.12}$$

i.e., a weighted averages of payoffs.

Simulation Results We take parameters $S(0) = K = 100$, $r = 0.05$, $\sigma = 0.25$, and hence $\mu^* = 0.01875$.

- **Heuristic method**: we have $x_1 = -0.075$ and solving (5.2.11) numerically (the Newton-Raphson method is implemented in `ch5_european_option_IS_compute_heuristic_max_hz.py`), we find

$$x^* = 1.0331.$$

  The function $h(z)$ is depicted in Fig. 5.28.
- **Cross-entropy method**: Simulating $M = 1000$ i.i.d. $Z_i, i = 1, \ldots, M$, we obtain from (5.2.12) that

$$\hat{x}_{\mathrm{CE}}^* = 1.2316.$$

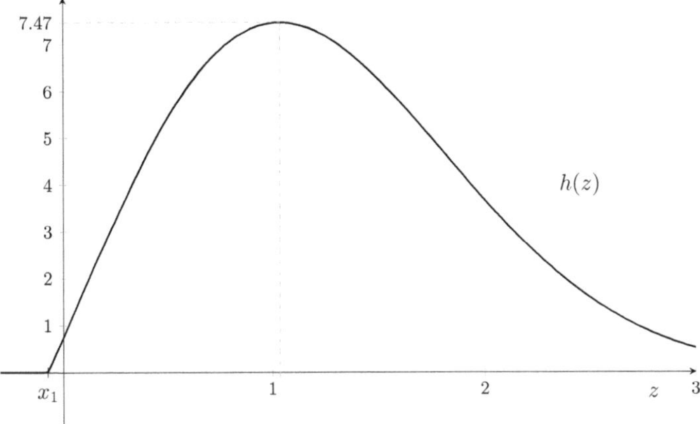

Fig. 5.28 Plot of $h(z) = (S(0)e^{\mu^*+\sigma z} - K)_+\phi_{0,1}(z)$ with $S(0) = K = 100, r = 0.05, \sigma = 0.25$, thus $\mu^* = 0.01875$

Table 5.18 Estimates for the European call option with parameters $S_0 = 100, K = 100, r = 0.05, \sigma = 0.25$. Comparison of CMC and IS methods with various x^*

	CMC	IS, $x^* = 1.0331$ (heuristic)	IS, $\hat{x}^*_{CE} = 1.2316$ (cross-entropy)
Estimate $\hat{Y}$	12.3905	12.3019	12.3057
Error b	0.3869	0.1245	0.1196
$\mathrm{Var}(\hat{Y}^{CMC})/\mathrm{Var}(\hat{Y}^{IS})$	–	9.6505	10.457

Thus, the final IS estimator is the following: we sample $\tilde{Z}_1, \ldots, \tilde{Z}_R$ i.i.d. with distribution $\mathcal{N}(x^*, 1)$ and compute

$$\hat{Y}_R^{IS,x^*} = \frac{1}{R}\sum_{i=1}^R Y_i^{IS,x^*}, \quad Y_i^{IS,x^*} = (S(0)e^{\mu^*+\sigma\tilde{Z}_i} - K)_+ e^{-x^*\tilde{Z}_i+(x^*)^2/2}.$$

We estimate I using $R = 10^4$ replications with CMC and IS estimators. The results are gathered in Table 5.18. The cross-entropy method for this example is implemented in `ch5_european_option_IS_CE.py`.

$\Diamond$

Conclusion The results show that importance sampling with either x^* or $\hat{x}^*_{CE}$ reduces the variance by approximately a factor of 9.6 compared to CMC. The cross-entropy method yields a slightly better x^*, resulting in the lowest variance.

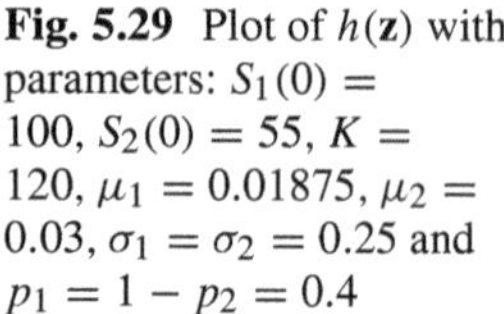

Fig. 5.29 Plot of $h(\mathbf{z})$ with parameters: $S_1(0) = 100$, $S_2(0) = 55$, $K = 120$, $\mu_1 = 0.01875$, $\mu_2 = 0.03$, $\sigma_1 = \sigma_2 = 0.25$ and $p_1 = 1 - p_2 = 0.4$

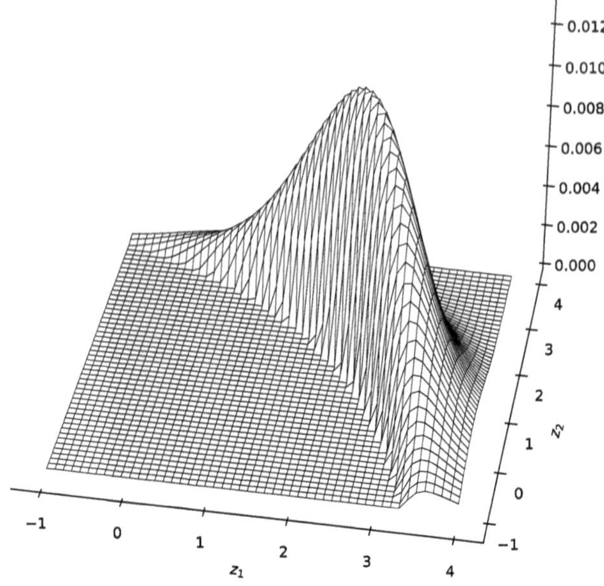

Example 5.2.12 For the *European basket call option*, which is a weighted sum of different assets that have been grouped together in a basket, we face the problem of estimating the option price

$$I = e^{-r}\mathbb{E}(D - K)_+,$$

where

$$D = p_1 S_1(0)e^{\mu_1^* + \sigma_1 Z_1} + \cdots + p_d S_d(0)e^{\mu_d^* + \sigma_d Z_d}$$

and $Z_1, \ldots, Z_d$ are i.i.d. standard normal random variables. To avoid going into details from financial mathematics, for simplicity, we suppose that the geometrical Brownian motions $S_1(t), \ldots, S_d(t)$ are independent and their parameters are chosen according to a risk neutral measure, which yields that the sought $I = e^{-r}\mathbb{E}(D-K)_+$ is the price of this option.

Consider

$$h(\mathbf{z}) = (p_1 S_1(0)e^{\mu_1 + \sigma_1 z_1} + \cdots + p_d S_d(0)e^{\mu_d + \sigma_d z_d} - K)_+ \phi_{\mathbf{0},\mathbf{I}}(\mathbf{z}),$$

where $\mathbf{z} = (z_1, \ldots, z_d)$. Based on intuitions from previous examples, we thus seek $\mathbf{z}^* = \arg\sup_{\mathbf{z}} h(\mathbf{z})$ and then the importance sampling estimator is given by

$$I = \mathbb{E}L(\tilde{\mathbf{Z}})(p_1 S_1(0)e^{\mu_1 + \sigma_1 \tilde{Z}_1} + \cdots + p_d S_d(0)e^{\mu_d + \sigma_d \tilde{Z}_d} - K)_+$$

where $\tilde{Z}_j$ ($j = 1, \ldots, d$) are independent, $\tilde{Z}_j \sim \mathcal{N}(z_j^*, 1)$ and $L(\mathbf{z})$ are given in (5.1.23). A plot of $h(\mathbf{z})$ for exemplary parameters is presented in Fig. 5.29. The plot of the function is implemented in `ch5_european_basket_option_hz_plot.py`. ◇

5.3 Bibliographical Notes

The topic of variance reduction in Monte Carlo methods has been extensively studied, with foundational treatments available in Asmussen and Glynn [8], Fishman [55], and Glasserman [63]. The book by Madras [138] serves as an accessible introduction to variance reduction techniques.

Stratified sampling is a well-known technique for variance reduction, and its theoretical properties are covered in standard references such as Fishman [55] and Asmussen and Glynn [8]. The mathematical justification behind variance decomposition and its applications in stratification are also discussed in Cochran [26]. An example of stratified sampling for computing π is presented in Madras [138].

The antithetic variates method, which relies on inducing negative correlation between paired samples, is discussed in detail in Asmussen and Glynn [8], see also Evans and Swartz [50]. The method's effectiveness in practical applications is illustrated in Hammersley and Handscomb [78] and Fishman [55]. Specific examples of its use for estimating π are explored in various Monte Carlo studies, including Marsaglia [141].

Common random numbers (CRN), which exploit shared randomness to reduce variance in comparative simulations, have been widely used in simulation studies.

Control variates (CV), another widely used technique, leverage auxiliary variables with known expectations to reduce estimator variance. The theoretical background and in-depth analysis for control variates can be found in Glynn and Szechtman [66], while applications in financial mathematics are covered in Glasserman [63].

Conditional Monte Carlo (CMC) is an approach that eliminates variance components through conditioning. The classical reference on CMC is Asmussen and Glynn [8], and further applications are explored in Hammersley and Handscomb [78].

Importance sampling (IS) is one of the most powerful variance reduction techniques and is widely used in rare-event simulations. The theoretical foundations of IS are well presented in Asmussen and Glynn [8] and in Fishman [55]. Its applications in financial risk management are covered in Glasserman [63], while heuristics for optimal importance sampling distributions are discussed in Rubinstein and Kroese [186]. For a broader overview of rare-event estimation, see Asmussen and Rubinstein [9], Heidelberger [81], or Juneja and Shahabuddin [95].

The cross-entropy method, an adaptive importance sampling approach, was introduced by Rubinstein [184]. A comprehensive treatment can be found in Rubinstein and Kroese [186], while a practical tutorial is provided by de Boer [31]. This method has gained popularity in optimization and rare-event probability estimation.

The construction of Brownian motion using the bisection method follows the approach detailed in Asmussen and Glynn [8], the method was also used in Knuth [106].

Certain topics, such as variance reduction techniques for quantile (or value-at-risk) estimation, have been omitted. For further details on this subject, refer to Chu and Nakayama [25], Dong and Nakayama [39], and Asmussen [6].

5.4 Exercises

Theoretical Exercises

5.T.1 (Stratified Sampling) Suppose we aim to compute $I = \mathbb{E}Y$ using stratified sampling with strata $S^1, \ldots, S^m$, defined by $A^1, \ldots, A^m$. Here, $S^j = \{Y \in A^j\}$, and we know $p_j = \mathbb{P}(Y \in A^j)$. Each replication in stratum j takes time t_j, and the total budget is constrained by $t_1 R_1 + \cdots + t_m R_m = C$. Let $\gamma_{j,R} = R_j/R$, which converges to γ_j as $R \to \infty$.

(a) Use C instead of R as the parameter for the estimator $\hat{Y}_C^{\mathrm{str}}$. Show that:

$$\sqrt{C}(\hat{Y}_C^{\mathrm{str}} - I) \xrightarrow{D} \mathcal{N}(0, \varsigma^2),$$

where:

$$\varsigma^2 = \sum_{j=1}^{m} \frac{p_j^2 \sigma_j^2}{\gamma_j} \sum_{k=1}^{m} t_k \gamma_k.$$

(b) Show that the variance ς^2 is minimized for $R_j \propto \frac{p_j \sigma_j}{\sqrt{t_j}} R$, i.e.,

$$R_j = \frac{1}{D} \frac{p_j \sigma_j}{\sqrt{t_j}} R, \quad D = \sum_{k=1}^{m} \frac{p_k \sigma_k}{\sqrt{t_k}}.$$

5.T.2 (Confidence Bounds with Stratified Sampling) Continuation of Exercise 5.T.1. Derive confidence bounds for I when stratified sampling is used with given times t_j $(j = 1, \ldots, m)$ and the total budget constraint C. Use the optimal allocation method.

5.T.3 Show that if X and Y have the same distribution, then:

$$\mathrm{Var}\left(\frac{X + Y}{2}\right) \leq \mathrm{Var}X.$$

5.T.4 (Antithetic Variates Method) Let F be the c.d.f. of the $\mathrm{Exp}(\lambda)$ distribution. Define $Z_1 = F^{-1}(U)$ and $Z_2 = F^{-1}(1 - U)$, where $U \sim \mathcal{U}[0, 1)$. Show that:

$$\mathbb{E}Z_i = \frac{1}{\lambda}, \quad \mathbb{E}Z_i^2 = \frac{2}{\lambda^2},$$

and:

$$\mathrm{Corr}(Z_1, Z_2) = -0.6449.$$

Hint: $\int_0^1 \log x \log(1 - x)\, dx = 0.3551$.

5.T.5 (Antithetic Variates Method) Use the antithetic variates method to estimate:

$$I = \int_0^1 e^x\, dx,$$

using $Y_{2j-1} = e^{U_j}$ and $Y_{2j} = e^{1-U_j}$. Determine how much the variance is reduced compared to the crude Monte Carlo (CMC) estimator.

5.T.6 Suppose $Z \sim \mathcal{N}(0, 1)$ and note that $Z \overset{d}{=} -Z$. Show that:

$$\Phi^{\leftarrow}(u) = -\Phi^{\leftarrow}(1 - u),$$

and conclude that $\Phi^{\leftarrow}(U)$ and $-\Phi^{\leftarrow}(1 - U)$, where $U \sim \mathcal{U}[0, 1)$, are antithetic random variables.

5.T.7 Consider estimating the probability $I = \mathbb{P}(X > 4)$, where $X \sim \mathcal{N}(0, 1)$, using Monte Carlo simulation. Construct an antithetic estimator by pairing samples $(X, -X)$. Show that the variance of the resulting antithetic estimator is almost the same as the CMC estimator, implying negligible variance reduction.

5.T.8 Let a p.d.f. $f(x)$ be given and consider a class of p.d.f.s $\widetilde{\mathcal{F}}$ with the same domain as f. Show that:

$$\frac{|k(x)|f(x)}{\int_E |k(x)|f(x)\, dx} = \underset{\tilde{f} \in \widetilde{\mathcal{F}}}{\mathrm{argmin}}\, \mathrm{Var}\left[k(\tilde{X})\frac{f(\tilde{X})}{\tilde{f}(\tilde{X})}\right].$$

Furthermore, if $k(x) \geq 0$, show that:

$$\min_{\tilde{f} \in \tilde{\mathcal{F}}} \mathbb{V}\mathrm{ar}\left[k(\tilde{X}) \frac{f(\tilde{X})}{\tilde{f}(\tilde{X})} \right] = 0,$$

and determine $g^*(x)$, the optimal solution.

5.T.9 (Importance sampling) The quarter-circle function is defined as:

$$k(x) = 4\sqrt{1 - x^2}, \quad x \in [0, 1].$$

We know:

$$\int_0^1 4\sqrt{1 - x^2}\, dx = \pi.$$

For the crude Monte Carlo estimator:

$$\hat{Y}_R^{\mathrm{CMC}} = \frac{1}{R} \sum_{j=1}^{n} 4\sqrt{1 - U_j^2},$$

consider a family of densities for importance sampling:

$$\tilde{f}(x) = ((2 - 2b)x + b)\mathbb{1}_{[0,1]}(x), \quad b \in [0, 2].$$

Derive the variance of the importance sampling estimator using $\tilde{f}(x)$. Plot the variance as a function of b, and for $b = 3/2$, compute the variance. Compare this to the variance of the importance sampling estimator in Example 5.1.35.

5.T.10 (Importance sampling) We can use the equality

$$\int_0^2 \frac{1}{1 + x^2}\, dx = \arctan(2)$$

to define a CMC estimator $\hat{Y}_R^{\mathrm{CMC}}$ for $\arctan(2)$, expressed as $\arctan(2) = \int_0^2 k(x) f(x)\, dx$, where $k(x) = 2/(1 + x^2)$ and $f(x) = 1/2$ for $x \in (0, 2)$. Consider the IS estimator $\hat{Y}_R^{\mathrm{IS}}$ with the proposal distribution $\tilde{f}(x)$ as a polynomial of degree 4. Proceed as follows:

- Let $h(x) = a_4 x^4 + a_3 x^3 + a_2 x^2 + a_1 x + a_0$. Determine coefficients such that $h(i/2) = k(i/2) f(i/2)$ for $i = 0, 1, 2, 4$.
- Normalize h to obtain $\tilde{f}(x) = h(x)/c$, where $c = \int_0^2 h(x)\, dx$.

Compute $\mathbb{V}\mathrm{ar}(\hat{Y}_R^{\mathrm{IS}})$ and the ratio $\mathbb{V}\mathrm{ar}(\hat{Y}_R^{\mathrm{CMC}})/\mathbb{V}\mathrm{ar}(\hat{Y}_R^{\mathrm{IS}})$, i.e., the variance reduction factor compared to the CMC estimator.

5.T.11 (Importance sampling) Let $X_1, \ldots, X_{100}$ be i.i.d. Bernoulli variables with $\mathbb{P}(X_i = 1) = 1 - \mathbb{P}(X_i = 0) = 1/2$. We aim to estimate

$$I = \mathbb{P}(X_1 + \cdots + X_{100} \geq 75).$$

(We know that $X_1 + \cdots + X_{100} \sim \mathrm{Bin}(100, 1/2)$ and $I = 2.818 \cdot 10^{-7}$ can be computed numerically.) The straightforward CMC estimator is as follows: repeat R independent simulations of flipping a coin 100 times. Denote the results of the i-th simulation by $X_1^{(i)}, \ldots, X_{100}^{(i)}$ and define

$$\hat{Y}_R^{\mathrm{CMC}} = \frac{1}{R} \sum_{i=1}^{R} Y_i, \qquad \text{where } Y_i = \mathbb{1}(X_1^{(i)} + \cdots + X_{100}^{(i)} \geq 75).$$

Construct an importance sampling estimator $\hat{Y}_R^{\mathrm{IS,p}}$, where the proposal distribution corresponds to flipping a coin with success probability $p \in (0, 1)$. Determine how many times the variance of $\hat{Y}_R^{\mathrm{IS,0.75}}$ is smaller than the variance of $\hat{Y}_R^{\mathrm{CMC}}$.

   Plot $\mathbb{V}\mathrm{ar}(\hat{Y}_R^{\mathrm{IS,p}})$ for $p \in (0.74, 0.76)$ and conclude that $p = 0.75$ is not the optimal choice. Compare this conclusion with the asymptotic result from Example 5.1.46 for $\varepsilon = 1/4$.

5.T.12 Assume we estimate I using the control variates method with independent replications $(Y_1, X_1), \ldots, (Y_R, X_R)$. Denote the resulting estimator by $\hat{Y}_R^{\mathrm{CV}}$. Consider another set of control variates with $X_i' = a + bX_i$, i.e., using independent replications $(Y_1, a+bX_1), \ldots, (Y_R, a+bX_R)$, with some fixed constants $a, b \in \mathbb{R}$. Denote the resulting estimator by $\hat{Y}_R^{'\mathrm{CV}}$. Show that $\mathbb{V}\mathrm{ar}(\hat{Y}_R^{\mathrm{CV}}) = \mathbb{V}\mathrm{ar}(\hat{Y}_R^{'\mathrm{CV}})$.

5.T.13 (Asmussen and Glynn [8]) Discuss reducing the variance for the estimation of

$$\int_0^\infty (x + 0.02x^2) \exp(0.1\sqrt{1 + \cos x} - x)\, dx$$

using Monte Carlo methods.

5.T.14 Suppose $X \sim F$, and the goal is to estimate the n-th convolution power of F at x, $I = F^{n*}(x)$. Let $S_n = X_1 + \cdots + X_n$, where X_j are independent copies of X. Consider $Y^{\mathrm{cond}} = \overline{F}(x - S_{n-1})$. For each replication of Y^{cond}, generate i.i.d. $X_1, \ldots, X_n$. Show that

$$\hat{Y}_R^{\mathrm{cond}} = \frac{1}{R} \sum_{j=1}^{R} \overline{F}(x - S_{n-1})$$

is an unbiased estimator of $\overline{F}^{n*}(x) = 1 - F^{*n}(x)$.

5.T.15 Estimate $\int_0^1 k(x)\,dx$, where $k(x) = 4\sqrt{1-x^2}$. Let $Y = k(U)$. Analyze using the control variate method with $X = h(U)$, where $h(x) = 4 - 4x$.

5.T.16 (Madras [138]) Consider estimating

$$\int_0^4 x^{1/2}e^{-x}\,dx$$

using Monte Carlo methods.

(a) Apply the importance sampling estimator with the density

$$\tilde{f}(x) = \frac{e^{-x}}{1 - e^{-4}}.$$

Provide a complete algorithm using R random numbers $U_1, \dots, U_R$.
(b) Assess whether it makes sense to use the antithetic variates method.
(c) Consider using other variance reduction techniques.

5.T.17 (Madras [138])

(a) We want to estimate $I = \int_0^\infty k(x)\,dx$, where

$$k(x) = \sin(x)(1+x)^{-2}e^{-x/5},$$

using the importance sampling estimator using $\tilde{X}$ exponentially distributed with a parameter λ. Show that for some values of λ the variance of the estimator is infinite.
(b) Show that if $k(x) = \sin(x)(1 + x)^{-2}$, then the variance of the importance sampling estimator with $\tilde{X}$ exponentially distributed is always infinite. Design a good method for estimating I.

5.T.18 Assume that the integral $I = \int_0^\infty k(x)\,dx$ is conditionally convergent (i.e., it is not absolutely convergent). Show that then any importance sampling estimator has infinite mean.

Hint: Let $\tilde{f}$ be an importance p.d.f. Thus, we have

$$\int_0^\infty \frac{k(x)}{\tilde{f}(x)}\,\tilde{f}(x)\,dx.$$

Denoting $\tilde{X} \sim \tilde{f}(x)$ and using importance sampling, $\hat{Y}_R^{\mathrm{IS}} = \frac{1}{R}\sum_{j=1}^R Y_j$, where i.i.d. $Y_j = \frac{k(\tilde{X}_j)}{\tilde{f}(\tilde{X}_j)}$. The mean of the estimator $\mathbb{E}\hat{Y}_R^{\mathrm{IS}}$ is finite if $\mathbb{E}|\hat{Y}_R^{\mathrm{IS}}| < \infty$, and this is clearly true if $\mathbb{E}|Y_j| < \infty$. However,

$$\mathbb{E}|Y_j| = \int_0^\infty \frac{|k(x)|}{\tilde{f}(x)}\,\tilde{f}(x)\,dx = \int_0^\infty |k(x)|\,dx = \infty.$$

5.T.19 We aim to estimate $I = \mathbb{E}Y$, where Y is a random variable with density

$$f(x) = \begin{cases} 2x & 0 \le x \le 1, \\ 0 & \text{otherwise.} \end{cases}$$

(a) Write an algorithm (and justify it) for generating a random variable Y.
(b) How many replications are needed to estimate I so that one digit after the decimal point is correct (i.e., error is $b = 0.01$) at significance level $\alpha = 0.05$? Consider

  (i) the crude Monte Carlo estimator,
 (ii) the antithetic estimator.

Hint: $B(3/2.3/2) = \pi/8$, $\text{Corr}(U^{1/2}, (1 - U)^{1/2}) = -0.156$ for $U \sim \mathcal{U}[0, 1)$.

5.T.20 (Importance sampling) Provide an importance sampling procedure for estimating

$$I = \int_0^1 e^{-x^2}\, dx$$

using proposal distribution $\tilde{f}(x) = re^{-x}$ (compute r so that $\tilde{f}$ is a density on $(0, 1)$). Use a computer to compute the appropriate integrals. How much is the variance reduced compared to CMC estimator?

5.T.21 Provide an importance sampling procedure for estimating

$$I = \int_0^1 x^{\alpha-1} e^{-x}\, dx$$

for $\alpha = 3/4$ taking the same $\tilde{X}$ as in Exercise 5.T.20. How much is the variance reduced compared to CMC estimator?

5.T.22 (Latin squares) When d is large, estimating

$$I = \int_{[0,1]^d} g(x)\, dx$$

using stratified sampling by stratifying along each coordinate becomes impractical, since it requires a large number of strata. Specifically, if we use n partitions along each axis, then the $[0, 1]^d$ cube will be divided into n^d

strata. Extending stratified sampling from one dimension to d dimensions would yield

$$
\mathbf{V} = \begin{pmatrix} V_{11} & \cdots & V_{1n} \\ \vdots & \ddots & \vdots \\ V_{d1} & \cdots & V_{dn} \end{pmatrix},
$$

where $V_{ij} = \frac{i-1}{n} + \frac{U_{ij}}{n}$. A more efficient alternative is to use the so-called Latin squares.

Let $U_i^{(j)}$ be i.i.d. $\mathcal{U}[0, 1)$ random variables, and let $\pi_i = (\pi_i(1), \ldots, \pi_i(n))$, for $i = 1, \ldots, d$, be independent random permutations of $(1, \ldots, n)$. For $j = 1, \ldots, n$, define

$$
V^{(j)} = (V_1^{(j)}, \ldots, V_d^{(j)}) \in [0, 1]^d, \quad \text{where} \quad V_i^{(j)} = \frac{\pi_i(j) - 1 + U_i^{(j)}}{n}.
$$

Show that the estimator

$$
\hat{Y}_n^{\mathrm{LS}} = \frac{1}{n} \sum_{j=1}^{n} g\left(V^{(j)}\right)
$$

is an unbiased estimator of I, and that

$$
\mathbb{Var}\left[\hat{Y}_n^{\mathrm{LS}}\right] = \frac{\sigma_Y^2}{n} + \frac{n-1}{n} \mathrm{Cov}\left(g(V^{(1)}), g(V^{(2)})\right),
$$

where $\sigma_Y^2 = \mathbb{Var}\left[g(U_1, \ldots, U_d)\right]$.

5.T.23 One can estimate π using the CMC method with $Y = 4\,\mathbb{1}(U_1^2 + U_2^2 \leq 1)$. As control variate take $X = \mathbb{1}(U_1 + U_2 \leq 1)$, which is strongly correlated with Y and $\mathbb{E}X = 1/2$. In this case $\rho^2 = 0.727$, however this result requires the knowledge of π. In practice we find ρ by simulations. Write in a pseudocode an algorithm for computing ρ^2.

5.T.24 For the network depicted in Fig. 5.5 (from Example 5.1.10) find a lower and upper bound for I. Analyze the precision of the simulation with the use of the crude estimator.

5.T.25 Define $h(\theta) = \int_4^\infty (x - \theta)^2 e^{-x^2/2}\, dx$. Show that

$$
\underset{\theta}{\mathrm{argmax}}\, h(\theta) = \frac{\sqrt{2}e^{-8}}{\sqrt{\pi}(1 - \mathrm{erf}(2\sqrt{2})}},
$$

which was needed in Example 5.1.47.

5.T.26 Consider the cross-entropy method described in Algorithm 35 for estimating $I = \mathbb{P}(h(\mathbf{X}) > \gamma)$, where $\mathbf{X}$ has a density f. Assume that the proposal distribution f_θ is a d-dimensional normal distribution $\mathcal{N}(\boldsymbol{\mu}, \boldsymbol{\Sigma})$, where both $\boldsymbol{\mu}$ (mean) and $\boldsymbol{\Sigma}$ (covariance matrix) are parameters to be updated. The p.d.f. is given by:

$$f_\theta(\mathbf{x}) = \frac{1}{\sqrt{(2\pi)^d |\boldsymbol{\Sigma}|}} \exp\left(-\frac{1}{2}(\mathbf{x} - \boldsymbol{\mu})^T \boldsymbol{\Sigma}^{-1}(\mathbf{x} - \boldsymbol{\mu}) \right).$$

Recall that the weights are defined as:

$$w_{i,t-1} = L(\mathbf{X}_i, f, f_{\theta'_{t-1}}).$$

Derive the update formulas for $\boldsymbol{\mu}$ and $\boldsymbol{\Sigma}$ by maximizing the cross-entropy objective:

$$\boldsymbol{\theta}'_t = (\boldsymbol{\mu}'_t, \boldsymbol{\Sigma}'_t) = \operatorname*{argmax}_{\theta, \Sigma} \frac{1}{M} \sum_{i=1}^{M} \mathbb{1}(h(\mathbf{X}_i) \geq \gamma_t) w_{i,t-1} \log f_\theta(\mathbf{X}_i).$$

Show that the solution is:

$$\boldsymbol{\mu}'_t = \frac{\sum_{i=1}^{M} \mathbb{1}(h(\mathbf{X}_i) \geq \gamma_t) w_{i,t-1} \mathbf{X}_i}{\sum_{i=1}^{M} \mathbb{1}(h(\mathbf{X}_i) \geq \gamma_t) w_{i,t-1}},$$

$$\boldsymbol{\Sigma}'_t = \frac{\sum_{i=1}^{M} \mathbb{1}(h(\mathbf{X}_i) \geq \gamma_t) w_{i,t-1}(\mathbf{X}_i - \boldsymbol{\mu}'_t)(\mathbf{X}_i - \boldsymbol{\mu}'_t)^T}{\sum_{i=1}^{M} \mathbb{1}(h(\mathbf{X}_i) \geq \gamma_t) w_{i,t-1}}.$$

Remark In Sect. 5.1.8 on page 5.1.8.2, we considered a situation where $\boldsymbol{\mu}$ was a parameter, and $\boldsymbol{\Sigma}$ was fixed. Note that the formula for $\boldsymbol{\mu}'_t$ is the same in both cases.

5.T.27 Show that a Brownian motion defined in Definition 5.2.1 is a centered Gaussian process, with covariance matrix $\operatorname{Cov}(B(s), B(t)) = \min(s, t)$.

5.T.28 Let $(S(t))$ be the geometric Brownian motion with drift μ and diffusion coefficient σ. Give a recursive procedure to simulate values of $(S(t))$ at $0 = t_0 < t_1 < \cdots < t_n$.

Lab Exercises

5.L.1 Implement estimating I from some previous theoretical exercises. In all cases use $R = 10^4$ replications.

(a) Exercise 5.T.5: Estimate $I = \int_0^1 e^x\, dx$ using CMC and antithetic variates taking $Y_{2j-1} = e^{U_j}$, $Y_{2j} = e^{1-U_j}$. Compare the results of the estimators.

(b) Exercise 5.T.13: implement CMC and a variance reduction method of your choice for estimating the integral.

(c) Exercise 5.T.15: implement the control variate method with $h(x) = 4 - 4x$. Compare the results with Example 5.1.25.

(d) Exercise 5.T.16: implement importance sampling with the provided proposal density. Compare the results with the CMC estimator.

(e) Exercise 5.T.17: implement importance sampling with the exponential proposal distribution (choose λ) for estimating I with k given in a).

(f) Exercise 5.T.19: compare CMC and the antithetic estimator.

(g) Exercise 5.T.20: implement the IS estimator with the suggested proposal distribution. Compare IS and CMC (based on the uniform proposal distribution).

(h) Exercise 5.T.23: estimate ρ^2 from simulations. Check the quality of estimation (comparing to the known $\rho^2 = 0.727$) using several values of R (number of replications).

5.L.2 Let $Z_1 = F^{-1}(U)$, $Z_2 = F^{-1}(1 - U)$ for $U \sim \mathcal{U}[0, 1]$. Estimate the correlation $\mathrm{Corr}(Z_1, Z_2)$ when F is:

(a) a Poisson distribution with mean λ,

(b) a geometric distribution $\mathrm{Geo}(p)$,

(c) a Pareto distribution $\mathrm{Par}(\alpha)$.

In each case plot the correlation as a function of a parameter.

5.L.3 Suppose

$$k(x) = \begin{cases} 10x & x \le \log 2, \\[2mm] x & x > \log 2. \end{cases}$$

Using $R = 10^4$ simulations estimate

$$I = \int_0^\infty k(x)e^{-x}\, dx.$$

- Define and implement CMC estimator (where $Y_i = k(X_i)$ and X_i are i.i.d. with $\mathrm{Exp}(1)$).
- Suppose $X \sim \mathrm{Exp}(1)$ and consider strata $S_1 = \{X \le \log 2\}$ and $S_2 = \{X > \log 2\}$. Let $X^{(1)} \overset{\mathcal{D}}{=} (X|X \le \log 2)$ and $Y^{(2)} \overset{\mathcal{D}}{=} (Y|Y > \log 2)$. Assume knowledge of approximate variances

$$\mathbb{V}\mathrm{ar} X^{(1)} \approx 0.3, \qquad \mathbb{V}\mathrm{ar} X^{(2)} \approx 1.0.$$

  Plan and write an algorithm for estimating I using stratified sampling with optimal allocation.

  Hint: Notice that $e^{-\log 2} = 1/2$.

Compare the efficiency of both estimators.

5.L.4 (Combination of stratified sampling and antithetic variates) In Example 5.1.8 we estimated π using stratification by defining m strata

$$B^1 = (0, 1/m], \ B^2 = (1/m, 2/m], \ldots, \ B^m = ((m-1)/m, 1].$$

and sampling $Y^j = 4\sqrt{1 - (V^j)^2}$, where

$$V^j = \frac{j-1}{m} + \frac{U}{m}.$$

Consider the following antithetic variate:

$$V'^j = \frac{j-1}{m} + \frac{1-U}{m}.$$

Show the formula for such estimator in the case of proportional allocation and provide a formula for its variance. Implement the case with $m = 10$ and $R = 200$ replications. Compare the pure stratified estimator with a combined one.

5.L.5 Estimate the integral

$$I = \int_0^1 \frac{e^{-x}}{x+1} \, dx.$$

(a) Implement the CMC estimator taking $Y_i = \frac{e^{-U_i}}{U_i+1}$ with i.i.d. $U_1, \ldots, U_R$ with uniform $\mathcal{U}[0, 1)$ distribution. Use $R = 10^4$ replications.
(b) Implement the IS estimator with proposal distribution the normal distribution truncated to $(0,1)$, i.e., $\tilde{f}(x) = \frac{1}{C}e^{-x^2}$ for $x \in (0, 1)$ (you can use numerical inversion of the standard normal c.d.f. available in programming packages). Compare it to the CMC estimator using the same number of replications.

5.L.6 Estimate the integral

$$I = \int_0^1 \int_0^1 e^{-(x^2+y^2)} \, dx \, dy$$

using

(a) The CMC estimator: simulate $R = 10^4$ pairs (X, Y) uniformly distributed over $[0, 1] \times [0, 1]$.
(b) Stratified sampling: Divide $[0, 1] \times [0, 1]$ into $m = 4 \times 4$ strata, and allocate samples optimally (based on a pilot simulation).

Compare the above methods—compare their variances, construct confidence intervals for the estimates and discuss the results.

5.L.7 Estimate $I = \mathbb{P}(X > 5)$ for $X \sim \mathcal{N}(0, 1)$ using the CMC estimator and the importance sampling method with $\tilde{X} \sim \mathcal{N}(5, 1)$. Perform $R = 10^4$ samples to estimate the tail probability using both estimators. Plot the estimated tail probabilities and confidence intervals. Discuss the variance reduction achieved by importance sampling.

5.L.8 Continue Example 5.2.11 with parameters given therein. Using the conditional MC method from Sect. 5.1.6 estimate (make a plot) of the p.d.f. of the random variable $\exp(-r)(S(1) - K)_+$.

5.L.9 Consider Example 5.1.38 with $\mathbf{X} = (X_1, X_2)$ having two-dimensional normal distribution with $\mathbb{V}\mathrm{ar}X_1 = 3, \mathbb{V}\mathrm{ar}X_2 = 1$ and correlation $\rho = -0.8$. Take $A = \{x_1 > a, x_2 > a\}$ with $a \in \{1, 3, 6, 9\}$.

(a) Estimate $I = \mathbb{P}(\mathbf{X} \in A)$ and the associated 95% confidence interval using the CMC method.

(b) For each A find the maximizer x^* and $I = \mathbb{P}(\mathbf{X} \in A)$ and the associated 95% confidence interval using the importance sampling method from the Example.

5.L.10 Suppose that $X \sim \mathrm{Par}(1.2)$. Consider the risk given by $Y = 0.2(X_1 + \cdots X_5)$, where $X_1, \ldots, X_5$ are i.i.d. Using the conditional MC method from Sect. 5.1.6 make a plot of $S(x) = \mathbb{P}(Y > y)$ and estimate $S(7)$.

5.L.11 Let $I = \mathbb{P}(L > x)$, where

$$L = \sum_{j=1}^{n} D_n X_n,$$

$X_1, \ldots, X_n$ are i.i.d. standard normal $\mathcal{N}(0,1)$ random variables, independent from $D_1, \ldots, D_n$. Let P be a r.v. with distribution Beta(1,19), i.e., with the density

$$f(p) = (1 - p)^{18}/19, \quad 0 < p < 1.$$

Conditioned on $P = p$ random variables $D_1, \ldots, D_n$ are independent Bernoulli random variables with success parameter p, independent from $X_1, \ldots, X_n$. Provide a simulation procedure for estimation of $I = \mathbb{P}(L > x)$ using the conditional Monte Carlo method, conditioning on $\sum_{j=1}^{n} D_j$. Take $n = 100$ and

$$x = 3\mathbb{E}L = 3n\mathbb{E}P\mathbb{E}X = 3 \cdot 100 \cdot 0.05 \cdot 3 = 45.$$

(Asmussen and Glynn [8], p. 147).

5.L.12 (Cross-entropy method) Consider estimating:

$$I = \mathbb{P}\left(\sum_{i=1}^{n} X_i \geq \gamma\right),$$

where $X_1, \ldots, X_n$ are i.i.d. standard normal variables $\mathcal{N}(0, 1)$, $n = 50$, and $\gamma = 20$ (for these parameters, the value of I is extremely small).
Note that I can be reformulated as:

$$I = \mathbb{P}(h(\mathbf{X}) \geq 0),$$

where $\mathbf{X}$ is an n-dimensional standard normal random variable $\mathcal{N}(\mathbf{0}, \mathbf{I}_n)$, and $h(x_1, \ldots, x_n) = \sum_{i=1}^{n} x_i$. Thus, this problem aligns with the setup described in Sect. 5.1.8.2 (The cross-entropy method for rare event simulation).
Implement the cross-entropy method, i.e., Algorithm 35, for estimating I, using parameters of your choice. As the proposal distribution, consider f_θ to be $\mathcal{N}(\boldsymbol{\mu}, \boldsymbol{\Sigma})$, where both the mean and covariance matrix $\boldsymbol{\theta} = (\boldsymbol{\mu}, \boldsymbol{\Sigma})$ are parameters. Note that the update formulas for this case are provided in Exercise 5.T.26.
Compare the results with crude Monte Carlo (CMC) estimations.

5.L.13 In Example 5.2.8, we estimated $I = e^{-r} E(A - K)_+$ using stratification with strata depicted in Fig. 5.23. Estimate I (with parameters used in the example) by stratifying the underlying Brownian motion via the bisection method, i.e., Algorithm 38. For the same number of total replications and the same number of strata, compare both methods (consider proportional and optimal allocation).

5.L.14 Consider a European call option with strike price $K = 100$, underlying asset price $S_0 = 100$, risk-free rate $r = 0.05$, volatility $\sigma = 0.2$. Estimate the stock price at $T = 1$ using CMC and control variates methods. Use the known expectation of the stock price $E S_T = S_0 e^{rT}$ as the control variate. Compare both methods, discuss the effectiveness of control variates in option pricing.

5.L.15 Make a simulation experiment exploiting the importance sampling method for computing I from Example 5.2.11, with the following data: $\sigma = 0.25$, $\mu^* = -0.0125$, $S(0) = K = 100$.

  (a) For computing I use the CMC and importance sampling method.
  (b) Combine the importance sampling methods with post-stratification. Namely. instead from drawing directly $\tilde{X} \sim \phi_{x^*, 1}$, we may make stratified sampling of Z according to the theory presented in Example 5.1.6 for a standard normal distribution (make 10 strata) and to the obtained results add the importance value x^*.

Compare the effectiveness of all three methods.

5.L.16 Recall the *tail estimation for a collective risk model* in Example 5.2.6, where aggregate claims $S = \sum_{j=1}^{N} X_j$ are generated by a portfolio. Here, $N \sim$ Poisson(λ) with $\lambda = 3.3$, and X_j are i.i.d. Erlang(4, 10/22). The goal is to estimate $I = P(S > 32)$.

Implement stratified sampling with *optimal allocation*. Perform $R' = 1000$ pilot simulations to estimate variances in each stratum, then allocate $R = 10^4$ replications optimally. Compare the results (estimates and errors) with those from proportional allocation and crude Monte Carlo (CMC).

5.L.17 Consider a European basket call option with strike price $K = 100$, with two stock prices, which are independent geometrical Brownian motions. Take $p_1 = 0.4$, $p_2 = 0.6$, $\sigma_1 = \sigma_2 = 0.25$ and $\mu_1 = -0.01$, $\mu_2 = -0.015$. Compute the price of this option by the CMC methods and compare to the one using the importance sampling method discussed in Example 5.2.12.

Chapter 6
Markov Chain Monte Carlo Methods

6.1 Introduction

In previous chapters (e.g., in variance reduction techniques), our goal was often to estimate

$$I = \mathbb{E}Y \quad \text{or} \quad I = \mathbb{E}f(Y),$$

where Y follows a predefined distribution, denoted in this chapter by π. Typically, it was assumed that we could sample independent replications $Y_1, \ldots, Y_R$ of Y, and that π was defined on a space such as $\mathbb{R}$, $\mathbb{R}^d$, or a finite state space S. For instance, consider estimating the number of fixed points in a random permutation. Recall that a random permutation of n elements is a random variable Y that takes each permutation with equal probability. In other words, the distribution of Y is uniform, i.e., $\pi_\sigma = \frac{1}{n!}$, $\forall \sigma \in S_n$. Let

$$f(\sigma) = \#\{i : \sigma(i) = i\}$$

represent the number of fixed points of a permutation σ, and assume $Y \sim \pi_\sigma$. Our goal is to estimate $I = \mathbb{E}f(Y)$. This can be achieved as follows: we generate R independent samples $Y_1, \ldots, Y_R$ of Y, and estimate I using the sample mean $\hat{Y}_R = \sum_{i=1}^{R} f(Y_i)/R$. A critical aspect here is the ability to generate uniformly distributed permutations. This can be done efficiently using the Fisher-Yates algorithm (Algorithm 1, page 26).

Now, let us complicate the scenario slightly. Suppose we still aim to estimate $I = \mathbb{E}f(Y)$, where $Y \sim \pi$, but the distribution π is no longer uniform. Instead, assume it takes the form:

$$\pi_\sigma = \frac{1}{z_\lambda} \exp(\lambda f(\sigma)), \tag{6.1.1}$$

where $\lambda \in \mathbb{R}$ is fixed and z_λ is the normalizing constant $z_\lambda = \sum_{\sigma' \in \mathcal{S}_n} \exp(\lambda f(\sigma'))$. The same estimation procedure (sampling independent replications $Y_1, \ldots, Y_R$ from π and averaging) could still be applied. However, generating samples from this distribution is no longer straightforward. The key challenges here are:

- The state space $\mathcal{S}$ is extremely large (e.g., $|\mathcal{S}| = n!$ for permutations or $|\mathcal{S}| = 2^d$ for binary spaces).
- The normalizing constant z_λ is computationally intractable.

In general, for distributions defined on large or infinite state spaces, exact sampling from π is often infeasible. Instead, we aim to construct a sequence of random variables $X_0, X_1, \ldots$ such that X_n converges, in some sense, to π. Remarkably, such approximate sampling methods are often sufficient for practical applications.

This is the focus of the current chapter. The sequence $X_0, X_1, \ldots$ is a **Markov chain**, and methods for constructing Markov chains that converge to a prescribed distribution π are called **Markov chain Monte Carlo (MCMC) methods**. These methods enable us to generate samples from a distribution close to π. In some specific cases, we will also be able to sample exactly from π. Such methods are referred to as **perfect sampling** or **perfect simulation**. Several algorithms exist for constructing Markov chains, with the **Metropolis–Hastings algorithm** being one of the most widely used. This algorithm serves as the foundation for many other MCMC variants, which will be discussed in the following sections.

6.2 Markov Chains

By a Markov chain, we mean a sequence of random variables $(X_n) \equiv (X_n)_{n \geq 0} = X_0, X_1, X_2, \ldots$, taking values in a *state space* $\mathcal{S}$. In this chapter, we assume $\mathcal{S} = \{s_1, \ldots, s_N\}$ is finite. A *transition matrix* (t.m.) $\mathbf{P} = (p_{ss'})_{s,s' \in \mathcal{S}}$ satisfies:

$$p_{ss'} \geq 0, \quad \text{for all } s, s' \in \mathcal{S},$$

and:

$$\sum_{s' \in \mathcal{S}} p_{ss'} = 1, \quad \text{for all } s \in \mathcal{S}.$$

Such a matrix is sometimes referred to as a *stochastic matrix*. Let $(\mu_s)_{s \in \mathcal{S}}$ (or shortly μ) be a probability distribution on $\mathcal{S}$. The sequence $(X_0, X_1, \ldots)$ is a *Markov chain* with *initial distribution* μ and transition matrix $\mathbf{P}$ if, for any $s_{i_0}, \ldots, s_{i_k} \in \mathcal{S}$:

$$\mathbb{P}(X_0 = s_{i_0}, X_1 = s_{i_1}, \ldots, X_k = s_{i_k}) = \mu_{s_{i_0}} p_{s_{i_0} s_{i_1}} \cdots p_{s_{i_{k-1}} s_{i_k}}.$$

The transition matrix $\mathbf{P}$ does not depend on the current step k, making the chain *time-homogeneous*. For cases where $\mathbf{P}$ depends on k, the chain is referred to as *time non-homogeneous*. Time non-homogeneous chains are discussed in Chap. 7.

We now state a few definitions.

- **Stationary distribution:** A probability distribution $(\pi_s)_{s \in \mathcal{S}}$ (denoted π) is stationary for (X_n) with t.m. $\mathbf{P}$ if:

$$\pi_s = \sum_{s \in \mathcal{S}} \pi_s \, p_{ss'}, \quad \text{for all } s' \in \mathcal{S}.$$

In matrix form:

$$\pi \mathbf{P} = \pi,$$

where $\pi = (\pi_{s_1}, \ldots, \pi_{s_N})$. Throughout this chapter, for a distribution v on $\mathcal{S}$, we represent it as a **row vector** $v = (v_{s_1}, \ldots, v_{s_N})$.
- **Transition probabilities:** $\mathbf{P}^n = (p_{ss'}^{(n)})_{s,s' \in \mathcal{S}}$, where $p_{ss'}^{(n)}$ denotes the probability of transitioning from state s to s' in n steps.
- **Communication:** Two states $s, s' \in \mathcal{S}$ are said to *communicate* if there exist $n_0, n_1 \geq 0$ such that $p_{ss'}^{(n_0)} > 0$ and $p_{s's}^{(n_1)} > 0$.
- **Irreducibility:** The matrix $\mathbf{P}$ is *irreducible* if all states in $\mathcal{S}$ communicate. Otherwise, $\mathbf{P}$ is *reducible*.
- **Periodicity:** The *period* of a state $s \in \mathcal{S}$ is defined as:

$$\gcd\{n \geq 1 : p_{ss}^{(n)} > 0\}.$$

The matrix $\mathbf{P}$ is *aperiodic* if all states $s \in \mathcal{S}$ have a period of 1.
- **Ergodicity:** The matrix $\mathbf{P}$ is *ergodic* (or *regular*) if there exists n_0 such that $\mathbf{P}^{n_0}$ has all entries strictly positive ($\mathbf{P}^{n_0} > \mathbf{0}$). One can show that $\mathbf{P}$ is regular if and only if $\mathbf{P}$ is irreducible and aperiodic. We say that (X_n) is ergodic.

Example 6.2.1 Consider two Markov chains on the state space $\mathcal{S} = \{1, 2, 3, 4\}$ with transition matrices:

$$\mathbf{P}_1 = \begin{pmatrix} 1/2 & 1/2 & 0 & 0 \\ 1/2 & 1/2 & 0 & 0 \\ 0 & 0 & 1/2 & 1/2 \\ 0 & 0 & 1/2 & 1/2 \end{pmatrix}, \quad \mathbf{P}_2 = \begin{pmatrix} 0 & 0 & 1/2 & 1/2 \\ 0 & 0 & 1/2 & 1/2 \\ 1/2 & 1/2 & 0 & 0 \\ 1/2 & 1/2 & 0 & 0 \end{pmatrix}.$$

The matrix $\mathbf{P}_1$ is reducible, while $\mathbf{P}_2$ is irreducible, but all states have a period of 2.

Distribution at Step n Denote the distribution of a chain (X_n) at step n by:

$$\boldsymbol{\mu}^n = (\mathbb{P}(X_n = s_1), \mathbb{P}(X_n = s_2), \dots, \mathbb{P}(X_n = s_N)).$$

Let $\boldsymbol{\mu}^0 \equiv \boldsymbol{\mu}$, which is the initial distribution. By induction:

$$\boldsymbol{\mu}^n = \boldsymbol{\mu}\mathbf{P}^n. \tag{6.2.1}$$

For future considerations, it is important to know the following proposition. We will use the terms Markov chain (X_n) and transition matrix (t.m.) $\mathbf{P}$ having a stationary distribution π interchangeably.

Proposition 6.2.2 (Ergodic Theorem) *If $\mathbf{P}$ is regular, then there exists a unique stationary distribution π, with all entries strictly positive. Furthermore:*

$$\mathbf{P}^n \to \boldsymbol{\pi}, \quad as\ n \to \infty,$$

where:

$$\mathbf{\Pi} = \begin{pmatrix} \boldsymbol{\pi} \\ \vdots \\ \boldsymbol{\pi} \end{pmatrix}.$$

Proposition 6.2.2 and Eq. (6.2.1) imply the following corollary.

Corollary 6.2.3 *Let (X_n) be an ergodic Markov chain, i.e., with a regular t.m. $\mathbf{P}$ and an arbitrary initial distribution μ. Then:*

$$\boldsymbol{\mu}\mathbf{P}^n \to \boldsymbol{\pi}, \quad as\ n \to \infty.$$

The distance between the distribution of X_n and the stationary distribution π is often measured using the *total variation distance*, which for two probability vectors $\boldsymbol{\mu} = (\mu_s)_{s \in \mathcal{S}}$ and $\boldsymbol{v} = (v_s)_{s \in \mathcal{S}}$ on $\mathcal{S}$ is defined as:

$$d_{\mathrm{TV}}(\boldsymbol{\mu}, \boldsymbol{v}) = \frac{1}{2} \sum_{s \in \mathcal{S}} |\mu_s - v_s|.$$

Corollary 6.2.3 implies that for an ergodic chain with any initial distribution:

$$\lim_{n \to \infty} d_{\mathrm{TV}}(\boldsymbol{\mu}\mathbf{P}^n, \boldsymbol{\pi}) = 0.$$

An important topic is the *rate of convergence* of a Markov chain to its stationary distribution. For instance, knowledge of (or a bound on) the *mixing time*:

$$\tau(\varepsilon) = \min_{n \in \mathbb{N}} \{d_{\mathrm{TV}}(\boldsymbol{\mu}\mathbf{P}^n, \boldsymbol{\pi}) \leq \varepsilon\} \tag{6.2.2}$$

is often desirable. However, the rate of convergence, while significant, is beyond the scope of this manuscript.

Next, we show a specific property of Markov chains that is useful for simulations.

Proposition 6.2.4 *For a given* $\mathbf{P}$ *and* μ, *there exist functions* $\varphi : \mathcal{S} \times [0, 1] \to \mathcal{S}$ *and* $\psi : [0, 1] \to \mathcal{S}$ *such that a sequence of random variables* (X_n) *defined recursively as:*

$$X_0 = \psi(U_0),$$

and:

$$X_{n+1} = \varphi(X_n, U_{n+1}), \quad n = 0, 1, \ldots, \tag{6.2.3}$$

where $U_0, U_1, \ldots$ *is an i.i.d. sequence of* $\mathcal{U}[0, 1]$ *uniformly distributed random variables, is a Markov chain with the initial distribution* μ *and the t.m.* $\mathbf{P}$.

Proof The following functions satisfy the required properties:

$$\varphi(s, u) = \begin{cases} s_1, & u \in [0, p_{ss_1}), \\[4pt] s_2, & u \in [p_{ss_1}, p_{ss_1} + p_{ss_2}), \\[4pt] \vdots & \vdots \\[4pt] s_m, & u \in [p_{ss_1} + \cdots + p_{s,s_{N-1}}, 1]. \end{cases} \tag{6.2.4}$$

and:

$$\psi(u) = \begin{cases} s_1, & u \in [0, \mu_{s_1}), \\[4pt] s_2, & u \in [\mu_{s_1}, \mu_{s_1} + \mu_{s_2}), \\[4pt] \vdots & \vdots \\[4pt] s_m, & u \in [\mu_{s_1} + \cdots + \mu_{s_{N-1}}, 1]. \end{cases}$$

$\square$

The function φ is called the **update function**, and ψ is called the **initialization function**. It is worth noting that this representation is not unique.

Given a transition matrix $\mathbf{P}$, one of the primary goals is determining its stationary distribution π. Instead of solving $\pi \mathbf{P} = \pi$, in some cases, one may use the following proposition, which will also be important later.

Proposition 6.2.5 *Let $(X_n)_{n\geq 0}$ be a Markov chain with the transition matrix* $\mathbf{P}$. *Assume there exist strictly positive numbers* w_s, $s \in S$, *such that:*

$$w_s p_{ss'} = w_{s'} p_{s's}, \qquad s, s' \in S. \tag{6.2.5}$$

Then:

$$\pi_s = \frac{w_s}{\sum_{s' \in S} w_{s'}}, \qquad s \in S,$$

is a stationary distribution of (X_n).

Proof Without loss of generality, assume that w_s in (6.2.5) sum up to 1. Then:

$$\sum_{s \in S} w_s p_{ss'} = \sum_{s \in S} w_{s'} p_{s's} = w_{s'} \sum_{s \in S} p_{s's} = w_{s'}.$$

$\square$

The condition (6.2.5) is called the **detailed balance condition (DBC)**. For stationary Markov chains (i.e., those with the initial distribution being the stationary distribution $\mu = \pi$), it is equivalent to reversibility.

A Markov chain $(X_n)_{n\geq 0}$ is called *(time) reversible* if, for $n = 0, 1, 2, \ldots$, we have:

$$(X_0, X_1, \ldots, X_n) \overset{\mathcal{D}}{=} (X_n, X_{n-1}, \ldots, X_0).$$

We leave the proof of the following corollary to the reader.

Corollary 6.2.6 *If the transition matrix* $\mathbf{P}$ *is symmetric, i.e.,* $p_{ss'} = p_{s's}$, *then the stationary distribution is uniform on* S, *i.e.,* $\pi_s = 1/|S|$, $s \in S$.

Later in this chapter, we will address a specific class of Markov chains whose state space has a graph structure. Therefore, we now provide a brief introduction to this concept.

Basic Concepts of Graphs A *graph G* consists of vertices V and edges E. We define $m = |E|$ and $M = |V|$. The edge collection can be identified with a subset of unordered pairs $\{v, v'\}$, where $v \neq v', v, v' \in V$. Two vertices v, v' are called *neighbors* if there is an edge connecting v with v'. The number of neighbors of a vertex v is denoted by $\deg(v)$. The maximum degree is denoted by $\Delta = \max_{v \in V} \deg(v)$. A graph is *connected* if it is possible to traverse from one vertex to any other via edges, not necessarily in a single step. A graph is *bipartite* if and only if it has no cycles with an odd number of edges. A *path* between two vertices in the graph is a sequence of edges connecting them. If two vertices are connected by a single edge (a path of length 1), they are called *adjacent*, which is another term for being neighbors.

Example 6.2.7 (Simple Random Walk on a Graph G) For a given graph $G = (V, E)$, consider a simple (symmetric) random walk on $\mathcal{S} = V$. The walker moves from the state v to any of its neighbors with equal probability:

$$
q_{vv'} = \begin{cases} \frac{1}{\deg(v)}, & \text{if } v' \text{ is a neighbor of } v, \\[2mm] 0, & \text{otherwise.} \end{cases} \tag{6.2.6}
$$

The transition matrix of the simple random walk on the graph $G = (V, E)$ is denoted by $\mathbf{Q} = (q_{vv'})_{v,v'\in V}$. It is easy to verify that if $w_v = \deg(v)$, then the equality (6.2.5) holds, and the stationary distribution is:

$$
\pi_v = \frac{\deg(v)}{\sum_{v'\in V} \deg(v')} = \frac{\deg(v)}{2|E|}, \quad v \in V. \tag{6.2.7}
$$

We encourage the reader to consider when the matrix $\mathbf{Q}$ is regular and when it is irreducible. The concepts of irreducibility and aperiodicity are crucial for Markov chain Monte Carlo methods. $\Diamond$

Example 6.2.8 (Simple Random Walk on a Hypercube) Consider the set of vertices $V = \{0, 1\}^d$. Its elements are d-dimensional 0-1 vectors, i.e., $v = (e_1, \ldots, e_d)$, where $e_j \in \{0, 1\}$. Neighboring vertices are at a Hamming distance of 1, i.e., $v_1 = (e_{11}, \ldots, e_{1d})$ and $v_2 = (e_{21}, \ldots, e_{2d})$ are neighbors if $\sum_{j=1}^{d} |e_{1j} - e_{2j}| = 1$. Thus, each vertex has d outgoing edges, i.e., $\deg(v) = d$. In one step, the chain moves from $v = (e_1, \ldots, e_i, \ldots, e_d)$ to a neighbor $v' = (e_1, \ldots, 1 - e_i, \ldots, e_d)$ with probability $1/d$. The stationary distribution is, therefore, the uniform distribution on V. $\Diamond$

As mentioned earlier, the aperiodicity of a simple random walk is not always evident. Consider, for example, a simple graph with vertices $V = \{1, \ldots, n\}$ and edges connecting neighboring vertices i and $i + 1$ ($i = 1, \ldots, n - 1$). The Markov chain with t.m. $\mathbf{Q}$ on this graph is periodic with a period of 2. To eliminate periodicity, the chain can be modified to allow staying at each state with some probability $\beta > 0$, which may be arbitrary and can depend on the graph.

Example 6.2.9 (The Simple Random Walk on a Graph G with Self-loops) Given $0 < \beta < 1$ (which may depend on the graph), the transition probabilities $q_{vv'}$ defined in (6.2.6) are redefined as follows:

$$
q_{vv'} = \begin{cases} \beta, & \text{if } v = v', \\[2mm] \frac{1-\beta}{\deg(v)}, & \text{if } v' \text{ is a neighbor of } v, \\[2mm] 0, & \text{otherwise.} \end{cases}
$$

Note that this modification ensures the chain is aperiodic while retaining the same stationary distribution π as given in (6.2.7). $\Diamond$

The examples above share a common characteristic: in one step, the chain transitions to the nearest neighbor. Therefore, instead of considering the full graph structure (V, E), it suffices to provide a list of neighbors $\mathcal{N}(v)$ for each state $v \in V$. We encourage the reader to explore the equivalence between describing all vertices and defining the set of neighbors for each state. Formally, a family $\{\mathcal{N}(v) : v \in V\}$ of subsets of V is called a *neighborhood system*. If $v \notin \mathcal{N}(v)$ for all $v \in V$, the system satisfies the basic neighborhood condition. Additionally, if for any $v, v' \in V$ $(v \neq v')$, there exists a path $v_1, \ldots, v_l$ such that $v_1 \in \mathcal{N}(v)$, $v_2 \in \mathcal{N}(v_1), \ldots, v' \in \mathcal{N}(v_l)$, the neighborhood system is said to be *communicated*.

We illustrate this concept in the next example (another simple random walk with self-loops). A more detailed example, the hard-core model, is presented later in Sect. 6.3.2.2.

Example 6.2.10 (Matchings in a Graph) Consider a complete graph $G = (V, E)$, where E contains all edges between every pair of vertices. Assume that $|V| = 2n$. A **perfect matching** $\mathcal{M} \subset E$ is a subset of edges such that every vertex in V is included in exactly one edge in $\mathcal{M}$. In Fig. 6.1, all possible matchings for 4 vertices are presented.

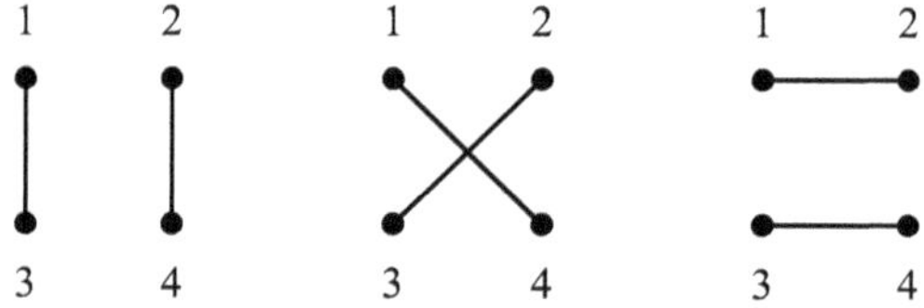

Fig. 6.1 All possible matchings for 4 vertices

We now consider a new graph, with vertices Λ—the set of all matchings— and edges defined by the following neighborhood structure: a matching $\mathcal{M}'$ is a neighbor of matching $\mathcal{M}$ (i.e., $\mathcal{M}' \in \mathcal{N}_\Lambda(\mathcal{M})$) if it differs from $\mathcal{M}$ by swapping two edges. Specifically, $\mathcal{M}$ contains edges (i_1, i_2), (j_1, j_2), and $\mathcal{M}'$ contains edges (i_1, j_1), (i_2, j_2) or edges (i_1, j_2), (i_2, j_1).

Thus, one should not confuse the neighborhood $\mathcal{N}_G(v)$, induced by the original graph G (in a complete graph, every vertex is a neighbor of every other), with the neighborhood $\mathcal{N}_\Lambda(\mathcal{M})$ on matchings. Note that each matching $\mathcal{M}$ has $2\binom{n}{2}$ neighbors (as we can choose a pair of edges in $\binom{n}{2}$ ways and swap them in two ways). Therefore, the degree of a "vertex" $\mathcal{M}$ in this new graph is $2\binom{n}{2}$.

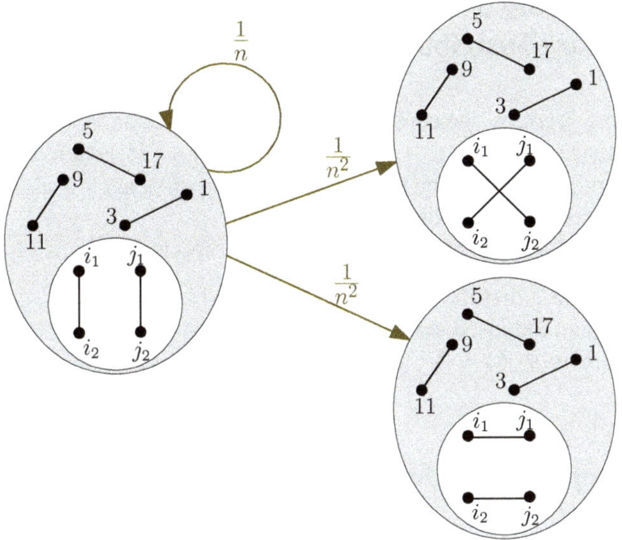

Fig. 6.2 Sample possible transitions from a fixed (left) matching to two other matchings (right) with a self-loop

We consider a simple random walk on this graph with a self-loop probability of $\beta = 1/n$. The transition probabilities for neighboring matchings are given by:

$$q_{M,M'} = (1 - \beta)\frac{1}{2\binom{n}{2}} = \left(\frac{n-1}{n}\right)\frac{1}{n(n-1)} = \frac{1}{n^2}.$$

Summarizing, the transition probabilities are:

$$q_{M,M'} = \begin{cases} \frac{1}{n}, & \text{if } M' = M, \\[2mm] \frac{1}{n^2}, & \text{if } M' \in \mathcal{N}_\Lambda(M), \\[2mm] 0, & \text{otherwise.} \end{cases}$$

See Fig. 6.2 for sample transitions.   ◇

6.3   Constructing a Markov Chain with Prescribed Stationary Distribution π (aka MCMC Methods)

In the following sections, we consider a probability function π, typically defined on a large state space $S = \{s_1, \ldots, s_N\}$ (e.g., $N = |S| = 2^d$, where d can be several hundreds). This probability function assigns mass π_s to each $s \in S$. In this context, π is referred to as the **target distribution**.

The **Markov chain Monte Carlo** (MCMC) methods enable us to:

(A1) Sample from a distribution "close" to π (and in some cases, exactly from π, providing an unbiased sample).

(A2) Approximate $\pi(A) = \sum_{s \in A} \pi_s$ for a specific subset $A \subseteq S$.

6.3.1   Metropolis and Metropolis-Hastings Algorithms

Let π be our target distribution on S. We choose (that is, it is a parameter of the algorithm) a transition matrix $\mathbf{Q} = (q_{ss'})_{s,s' \in S}$, which will play the role of a *candidate-generating matrix* (often $(q_{s\cdot})_{s \in S}$ are called *proposal distributions*). Notice that the distribution π is not the stationary one of $\mathbf{Q}$. Define now the transition matrix $\mathbf{P}$ generating the Markov chain (Y_n):

$$
p_{ss'} = \begin{cases} q_{ss'} \min(1, \pi_{s'}/\pi_s), & \text{if } s \neq s', \\ 1 - \sum_{k \neq i} q_{ss'} \min(1, \pi_{s'}/\pi_s), & \text{if } s = s'. \end{cases} \tag{6.3.1}
$$

We leave it to the reader to check that $\mathbf{P} = (p_{ss'})$ is a transition matrix. The next theorem is the basis of the Metropolis algorithm.

Proposition 6.3.1 (Metropolis)   *Assume that $\mathbf{Q}$ is a symmetric regular transition matrix and that π is a strictly positive probability function on S. Then $\mathbf{P}$ is a regular matrix with the stationary distribution π.*

Proof We will show that the DBC (6.2.5) holds. It holds, of course, for $s = s'$. For $s \neq s'$, from formula (6.3.1) we observe that $q_{ss'} > 0$ if and only if $p_{ss'} > 0$. Since $\mathbf{Q}$ is regular, so is $\mathbf{P}$. Let us check (6.2.5) with $w_s = \pi_s$. We have

$$
\pi_s p_{ss'} = \pi_s q_{ss'} \min(1, \pi_{s'}/\pi_s) = q_{ss'} \min(\pi_s, \pi_{s'}) = q_{s's} \min(\pi_{s'}, \pi_s) = \pi_{s'} p_{s's}.
$$

$$\square$$

The Metropolis algorithm is provided in Algorithm 42.

Algorithm 42 The Metropolis algorithm for sampling for a distribution π on $\mathcal{S}$ using a symmetric regular matrix $\mathbf{Q}$ (one step)

1: Assume that $Y_k = s$
2: Generate X according to a distribution $\mathbf{q}_s = (q_{ss_1}, \ldots, q_{ss_N})$
3: Set $\alpha = \min(1, \pi_X/\pi_s)$
4: Generate $U \sim \mathcal{U}[0, 1)$
5: **if** $U \le \alpha$ **then**
6: $\quad$ $Y_{k+1} = X$
7: **else**
8: $\quad$ $Y_{k+1} = Y_k$
9: **end if**
10: **return** Y_{k+1}

In the sequel, we will look for t.m.s in the form

$$p_{ss'} = q_{ss'}\alpha_{ss'}, \qquad \text{for } s \ne s'. \tag{6.3.2}$$

We want to eliminate the assumption that $\mathbf{Q}$ is symmetric. Consider postulates:

(I) $\mathbf{P}$ is a regular transition matrix,
(II) π is the stationary distribution for $\mathbf{P}$,
(III) the matrix $\mathbf{P}$ is a transition matrix of a reversible Markov chain, i.e., fulfilling the DBC (6.2.5).

Algorithm 43 The Metropolis-Hastings algorithm for sampling for a distribution π on $\mathcal{S}$ using a regular matrix $\mathbf{Q}$ (one step)

1: Assume that $Y_k = s$
2: Generate X according to a distribution $\mathbf{q}_s = (q_{ss_1}, \ldots, q_{ss_N})$

3: Set $\alpha = \min\left(1, \dfrac{\pi_X q_{Xs}}{\pi_s q_{sX}}\right)$

4: Generate $U \sim \mathcal{U}[0, 1)$
5: **if** $U \le \alpha$ **then**
6: $\quad$ $Y_{k+1} = X$
7: **else**
8: $\quad$ $Y_{k+1} = Y_k$
9: **end if**
10: **return** Y_{k+1}

Note that if $q_{ss'} = \pi_{s'}$ and $\alpha_i = 1$ $(s, s' \in \mathcal{S})$, then we have independent replications of (Y_j) with the distribution π. A generalization of the Metropolis algorithm not requiring that $\mathbf{Q}$ is symmetric is stated in the following proposition.

Proposition 6.3.2 (Metropolis-Hastings) *Let $\mathbf{Q}$ be a regular transition matrix such that $q_{ss'} > 0$ if and only if $q_{s's} > 0$. If*

$$\alpha_{ss'} = \min\left(1, \frac{\pi_{s'} q_{s's}}{\pi_s q_{ss'}}\right), \qquad s \neq s'$$

then $\mathbf{P}$ fulfills (I)–(III).

Proof Of course, $\mathbf{P}$ is a stochastic matrix. We need to check that the DBC (6.2.5) is fulfilled with $w_s = \pi_s$. We have

$$\pi_s q_{ss'} \min\left(1, \frac{\pi_{s'} q_{s's}}{\pi_s q_{ss'}}\right) = \min\left(\pi_s q_{ss'}, \pi_{s'} q_{s's}\right) = \pi_{s'} q_{s's} \min\left(1, \frac{\pi_s q_{ss'}}{\pi_{s'} q_{s's}}\right).$$

$\square$

The Metropolis-Hastings algorithm is provided in Algorithm 43.

Remark 6.3.3 The general scheme is as follows. For a candidate-generating matrix $\mathbf{Q}$, we look for the so-called *probabilities of acceptance* $\alpha_{ss'}$. The problem is that $\mathbf{P}$ defined by $p_{ij} = \alpha_{ss'} q_{ss'}$ must be a transition matrix. So far, we have had

$$\alpha_{ss'} = \min\left(1, \frac{\pi_{s'}}{\pi_s}\right) \tag{6.3.3}$$

or

$$\alpha_{ss'} = \min\left(1, \frac{\pi_{s'} q_{s's}}{\pi_{s'} q_{ss'}}\right). \tag{6.3.4}$$

Let

$$\alpha_{ss'} = \frac{t'_{ss'}}{1 + t_{ss'}}$$

for some symmetric matrix $(t'_{ss'})$ and

$$t_{ss'} = \frac{\pi_s q_{ss'}}{\pi_{s'} q_{s's}}.$$

To have $\alpha_{ss'} \in [0, 1]$, the following condition must be met:

$$t'_{ss'} \leq 1 + \min(t_{ss'}, t_{s's}).$$

In the case of equality, we have (6.3.4). If $q_{ss'}$ is constant, then we have (6.3.3). Another case is the so-called *Barker algorithm*:

$$\alpha_{ss'} = \frac{\pi_{s'} q_{ss'}}{\pi_{s'} q_{s's} + \pi_s q_{ss'}}.$$

In particular, if $q_{ss'} = 1/|\mathcal{S}|$, then we have

$$\alpha_{ss'} = \frac{\pi_{s'}}{\pi_s + \pi_{s'}}.$$

∎

Metropolis Algorithm on Graph G Consider a graph G from Example 6.2.7. Recall that $\Delta = \max_{v \in V} \deg(v)$ and $\mathcal{N}(v)$ is the set of all neighbors of a vertex $v \in V$. Let us define a Markov chain on $\mathcal{S} = V$ walking on "neighboring" vertices with the following transition matrix $\mathbf{Q}$:

$$q_{vv'} = \begin{cases} \frac{1}{K}, & v' \in \mathcal{N}(v), \\ 0, & v \neq v' \text{ and } v' \notin \mathcal{N}(v), \\ 1 - \frac{\deg(v)}{K}, & v = v', \end{cases} \qquad (6.3.5)$$

where $K \geq \Delta$. We assume that a Markov chain with this transition matrix is irreducible and aperiodic. This assumption follows from the neighborhood structure. Note that $v' \in \mathcal{N}(v)$ if and only if $v \in \mathcal{N}(v')$. This, combined with (6.3.5), implies that $\mathbf{Q}$ is symmetric. It can be shown that the DBC is fulfilled. Corollary 6.2.6 implies that the uniform distribution on V, i.e., $\mathcal{U}(V)$, is the stationary distribution for $\mathbf{Q}$. It is also straightforward to verify that the chain is reversible.

Now, on a state space $\mathcal{S} = V$, we want to construct a Markov chain with stationary distribution $\pi_v = b_v/B$, where $b_v > 0$ and $B = \sum_{v \in V} b_v$. In applications, the state space V is often enormous, and there is no effective way of computing B, but it is easy to compute $b_v/b_{v'}$. We can then use the following lemma.

Lemma 6.3.4   *Consider a graph* $G = (V, E)$ *from Example 6.2.7 with maximal degree* Δ*, and let* $\mathcal{N}(v)$ *be the set of all neighbors of* $v \in V$*. Define a Markov chain on* V *(a random walk on vertices) with the transition matrix* $\mathbf{P}$ *having entries*

$$p_{vv'} = \begin{cases} \frac{1}{K} \min(1, \pi_{v'}/\pi_v), & \text{if } v' \in \mathcal{N}(v), \\ 0, & \text{if } v \neq v' \text{ and } v' \notin \mathcal{N}(v), \\ 1 - \sum_{w \neq v} p_{vw}, & \text{if } v = v', \end{cases}$$

where $K \geq \Delta$*. If* $\mathbf{Q}$ *defined in (6.3.5) is regular, then so is* $\mathbf{P}$*. Moreover,* π *is its stationary distribution.*

Proof We will show that the DBC is fulfilled, and the assertion will then follow from Proposition 6.2.5. For $v \neq v'$, if $\pi_v \leq \pi_{v'}$, then $p_{vv'} = 1$ and $p_{v'v} = \pi_v/\pi_{v'}$. This implies that the DBC holds. A similar argument holds for the case $\pi_v \geq \pi_{v'}$.

$\square$

6.3.2   The Gibbs Sampler

We now consider a Markov chain evolving on configurations $\mathcal{S} = C^V$, where V and C are finite sets. In principle, $C = \{1, \ldots, r\}$; however, sometimes, for example, $C = \{0, 1\}$ (e.g., the hard-core model) or $C = \{-1, 1\}$ (e.g., the Ising model). We identify V with the set of vertices, and the distributions we consider here depend on a graph $G = (V, E)$. A state, which we denote by $\xi \in \mathcal{S}$, is called a *configuration*. It assigns values from C to each vertex in V, i.e., it is a function $\xi : V \to C$. Note that distributions on $\mathcal{S}$ depend on the graph G.

Let $H : C^V \to \mathbb{R}$ be an *energy* of the configuration. For example:

$$H(\xi) = \sum_{v \in V} U(\xi(v)),$$

or

$$H(\xi) = \sum_{(v,v'): v' \in \mathcal{N}_G(v)} W(v, v') + \sum_{v \in V} U(\xi(v)),$$

for some functions W on $V \times V$ and U on $\mathcal{S}$, where $\mathcal{N}_G(v)$ is the set of neighbors defined by edges E. On the space of configurations C^V, we define the **Gibbs distribution** by a probability function:

$$\pi_\xi = \frac{1}{z(T)} e^{-H(\xi)/T}, \tag{6.3.6}$$

where $T \neq 0$ (often $T > 0$ is interpreted as *temperature*), and $z(T)$ is called the *partition function*:

$$z(T) = \sum_{\xi \in \mathcal{S}} e^{-H(\xi)/T}.$$

Depending on the functions U and W, we may have $\pi_\xi = 0$ for some configurations.

Given a distribution π on a space of configurations, the Gibbs sampler (in statistical physics, it is called *Glauber dynamics*) is a Markov chain that has a stationary distribution π. The Gibbs sampler is particularly useful for the Gibbs distributions (6.3.6). It allows samples to be generated from (6.3.6) without requiring the computation of the partition function $z(T)$, which is often infeasible in practice.

As mentioned, we can assign values from C to each vertex in V. Fix some configuration ξ and $v \in V$. Denote by ξ_{-v} the set $\{\xi(v') : v' \in V, v' \neq v\}$, i.e., the fragment of configuration ξ that assigns $\xi(w')$ to all vertices $w' \in V$ except v. Let Z be a random element with distribution π, and consider the probability function:

$$p_{\xi,v}(c) = \mathbb{P}(Z(v) = c \mid Z_{-v} = \xi_{-v}), \quad c \in C. \tag{6.3.7}$$

We call the probability function $p_{\xi,v}(c)$, $c \in C$, the conditional distribution of π at v, given that the remaining vertices take values according to ξ_{-v}. One step of the **Gibbs sampler**, assuming that at step k, we are at $X_k = \xi$, is provided in Algorithm 44.

Algorithm 44 Gibbs sampler: transition from step k to step $k + 1$

1: $X_k = \xi$
2: Pick $v \in V$ uniformly at random
3: Set $X_{k+1}(w) = X_k(w)$ for all $w \neq v$
4: Pick $X_{k+1}(v)$ according to the conditional π distribution of the value at v, given that all other vertices take values according to ξ_{-v}
5: **return** X_{k+1}

The sequence of random elements (X_n) is a Markov chain on $\{\xi : \pi_\xi > 0\}$. Note that the chain must start from some configuration that is feasible, i.e., from such ξ_0 that $\pi_{\xi_0} > 0$. Denote its transition matrix by $\mathbf{P}$.

Proposition 6.3.5 (Gibbs Sampler) *Let (X_n) be a Markov chain with one-step transitions given by Algorithm 44 (Gibbs sampler). Denote by $\mathbf{P}$ the corresponding transition matrix. Assume that the chain is ergodic (i.e., $\mathbf{P}$ is regular). Then (X_n) has the stationary distribution π.*

Proof We will show that the DBC (6.2.5) holds. Let Z be a random element with distribution π. Let $v \in V$ and let $\xi, \xi' \in C^V$ such that $\xi(w) = \xi'(w)$ for $w \neq v$. If $\xi(v) = \xi'(v)$, then (6.2.5) holds trivially. Otherwise, we have:

$$\pi_\xi \, p_{\xi\xi'} = \pi_\xi \frac{1}{|V|} \mathbb{P}(Z(v) = \xi'(v) \mid Z(w) = \xi(w), \, w \neq v)$$

$$= \pi_\xi \frac{1}{|V|} \frac{\mathbb{P}(Z(v) = \xi'(v), \, Z(w) = \xi(w), \, w \neq v)}{\sum_{c \in C} \mathbb{P}(Z(v) = c, \, Z(w) = \xi(w), \, w \neq v)}$$

$$= \frac{1}{|V|} \frac{\pi_\xi \pi_{\xi'}}{\sum_{c \in C} \mathbb{P}(Z(v) = c, \, Z(w) = \xi(w), \, w \neq v)} = \pi_{\xi'} \, p_{\xi'\xi}.$$

$\square$

It can be shown that the Gibbs sampler is a special case of the Metropolis-Hastings algorithm (we leave this as an exercise) with an acceptance rate of 1.

Update Function for Gibbs Sampler In Proposition 6.2.4, we stated that for any transition matrix $\mathbf{P}$, there always exists an update function $\varphi : \mathcal{S} \times [0, 1] \to \mathcal{S}$ that allows for simulating $(X_n) \sim \mathbf{P}$ using $U_1, U_2, \ldots$, i.i.d. $\mathcal{U}[0, 1)$ random variables. Specifically, given X_0, we may sample the chain recursively as:

$$X_{n+1} = \varphi(X_n, U_{n+1}).$$

The Gibbs sampler may change a value at most in one vertex (in a single step). Given a configuration ξ, it is convenient to denote a configuration with the same values at all vertices $w \neq v$ but with a value c at vertex v. We use the following notation:

$$\xi(v \to c) = \left\{ \xi' : \xi'(v) = c, \; \xi_{-v} = \xi'_{-v} \right\}. \tag{6.3.8}$$

Note that, in particular, we may have $\xi(v \to c) \equiv \xi$ (this happens when the configuration ξ already assigns the value c to vertex v, i.e., when $\xi(v) = c$).

The transition matrix of the Gibbs sampler is represented by Algorithm 44. In each step, we always choose $v \in V$ uniformly at random. Thus, we can define (slightly abusing notation, we still call it φ) an update function that takes the current configuration $\xi \in \mathcal{S}$ and a vertex v, and changes ξ only at v based on $u \in (0, 1)$ according to the conditional distribution π. In other words, we may define

$$\varphi : \mathcal{S} \times V \times [0, 1] \to \mathcal{S},$$

to simulate the Gibbs sampler recursively as:

$$X_{n+1} = \varphi(X_n, W_{n+1}, U_{n+1}), \tag{6.3.9}$$

where W_{n+1} are i.i.d. random variables uniformly distributed on V, and $U_1, U_2, \ldots$, are i.i.d. $\mathcal{U}[0, 1)$ random variables. A straightforward example (cf. (6.2.4)) of the update function is as follows (recall, the update function is not unique):

$$\varphi(\xi, v, u) = \begin{cases} \xi(v \to 1), & \text{if } u \in \left[0, p_{\xi,v}(1)\right), \\[6pt] \xi(v \to 2), & \text{if } u \in \left[p_{\xi,v}(1), p_{\xi,v}(1) + p_{\xi,v}(2)\right), \\[6pt] \vdots \\[6pt] \xi(v \to c), & \text{if } u \in \left[\sum_{i=1}^{c-1} p_{\xi,v}(i), \sum_{i=1}^{c} p_{\xi,v}(i)\right), \\[6pt] \vdots \\[6pt] \xi(v \to r), & \text{if } u \in \left[\sum_{i=1}^{r-1} p_{\xi,v}(i), 1\right), \end{cases} \tag{6.3.10}$$

where $p_{\xi,v}(c)$ is defined in (6.3.7).

In the subsequent subsections, we will present specific applications of the Gibbs sampler to particular problems.

6.3.2.1   Graph Coloring

Let $G = (V, E)$ be a graph, and let $r \geq 2$ be an integer. A proper coloring of the graph G is an assignment of "colors," $C = \{1, \ldots, r\}$, to the vertices such that no two adjacent vertices share the same color. In this way, we define the set of feasible configurations Λ, which consists of all proper colorings of G. On the set Λ, we define the distribution

$$\pi_\xi = \begin{cases} \frac{1}{z_G}, & \text{if } \xi \in \Lambda, \\ 0, & \text{otherwise,} \end{cases} \tag{6.3.11}$$

which represents the uniform distribution over Λ.

In this example, we will demonstrate both the Gibbs sampling and Metropolis algorithms.

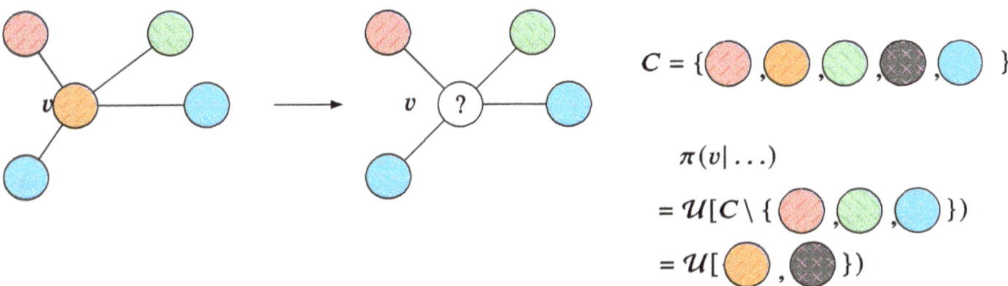

Fig. 6.3 Conditional distribution π at v given colors of its neighbors

Gibbs Sampler for Graph Coloring   Let us compute the conditional distribution at vertex v given that all other vertices are colored according to ξ_{-v}. The conditional probability is expressed as:

$$\mathbb{P}(Z(v) = c \mid Z(w) = \xi(w), w \neq v)$$

$$= \frac{\mathbb{P}(Z(v) = c, Z(w) = \xi(w), w \neq v)}{\sum_{c' \in C} \mathbb{P}(Z(v) = c', Z(w) = \xi(w), w \neq v)}$$

$$= \frac{1/z_G}{\sum_{c' \in C} \frac{1}{z_G} \mathbb{1}(\{\xi'(v) = c', \xi'(w) = \xi(w), w \neq v\} \text{ is a proper coloring })},$$

which is the uniform distribution over all colors available at v.

The Gibbs sampler for this model is presented in Algorithm 45. A sample configuration and the conditional distribution π at v given its neighbors are depicted in Fig. 6.3.

We now consider the evolution of a Markov chain (X_k) on Λ defined by this algorithm.

Algorithm 45 The Gibbs sampler for sampling graph coloring (1 step)

1: Assume the current configuration is $X_k = \xi$
2: Pick a vertex $v \in V$ uniformly at random
3: Set $X_{k+1}(w) = X_k(w)$ for all $w \neq v$
4: Let $C_{\xi,v} = \bigcup_{w \in \mathcal{N}(v)} X_k(w)$ (the set of colors used by the neighbors of v)
5: Pick $X_{k+1}(v)$ uniformly from $\{1, \dots, r\} \setminus C_{\xi,v}$

Note that the update function for graph coloring directly follows the form given in (6.3.10), with $p_{\xi,v}(c)$ set to 0 for colors used by neighbors of v and to a uniform distribution otherwise:

$$
p_{\xi,v}(c) = \begin{cases} 0, & \text{if } c \in C_{\xi,v}, \\[2mm] \dfrac{1}{r - |C_{\xi,v}|}, & \text{if } c \in \{1, \dots, r\} \setminus C_{\xi,v}. \end{cases}
$$

Remark 6.3.6

- We implicitly assume that at least one r-coloring of G exists (to initialize X_0 with a proper coloring) and that the resulting Markov chain $\mathbf{P}$ is ergodic. This assumption is not always valid. For instance:

  - If $r \geq \Delta + 1$, a proper coloring always exists.
  - If $r = \Delta$, a proper coloring exists if $\Delta \geq 3$ and G does not contain a clique of size $\Delta + 1$ as a connected component (Brooks' Theorem).
  - If $r < \Delta$, determining whether a proper coloring exists is NP-complete.

  We focus on cases where $r \geq \Delta + 1$.
- Finding the total number of proper r-colorings (i.e., computing z_G) is a #P-complete problem. Randomized polynomial-time approximation schemes exist for this problem. These schemes, with high probability, return a value within $(1 - \varepsilon)z_G$ and $(1 + \varepsilon)z_G$, with runtime polynomial in ε and the size of the graph.

  The Gibbs sampler can be employed for such schemes, but it requires knowledge of the convergence rate (distance between the distribution of the chain at time n and the stationary distribution) unless unbiased samples from π are obtainable. More details are provided in Sect. 6.5.

∎

Remark 6.3.7 Note that the assumption of regularity for the transition matrix $\mathbf{P}$ of a Gibbs sampler is essential, as the resulting Markov chain may otherwise be periodic or reducible. Consider, for example, coloring a complete graph $G = (V, E)$ with vertices $V = \{v_1, v_2, v_3\}$ and $r = 3$ colors. The graph is regular with a vertex degree $\Delta = 2$, so the condition $r \geq \Delta + 1 = 3$ is satisfied. However, in this case, the Gibbs sampler is reducible: once the process starts with a particular coloring, it will never change. This scenario is illustrated in Fig. 6.4. ∎

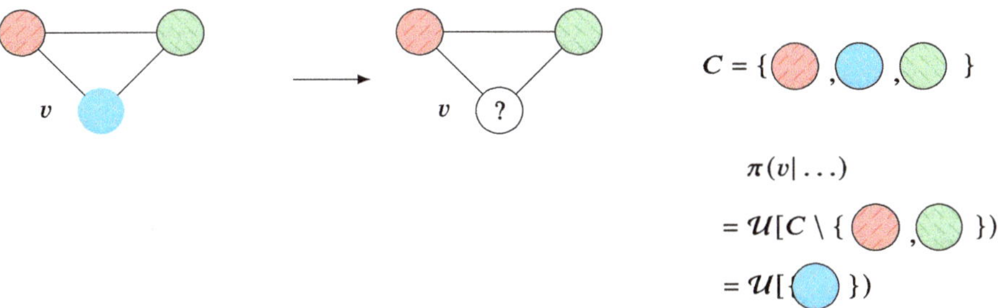

Fig. 6.4 Graph coloring: reducible Gibbs sampler

Metropolis Algorithm for Graph Coloring Let us revisit the Metropolis algorithm, which can also be applied to the graph coloring problem. While the Gibbs sampler may seem like a "more natural" choice for coloring, we aim to demonstrate that the Metropolis algorithm is equally applicable and leads to a slightly different Markov chain. The state space $S = C^V = \{1, \ldots, r\}^V$ represents all possible colorings of the graph. In the Metropolis framework, a state corresponds to a graph coloring. First, we need to define the candidate-generating matrix. Given a current coloring ξ, we choose a vertex $v \in V$ uniformly at random (with probability $1/|V|$) and assign it a new color $c \in C$, also chosen uniformly at random (with each color having probability $1/r$). Let the resulting configuration be denoted by ξ'. Note that ξ' is not necessarily a proper coloring. The candidate-generating probabilities are given by:

$$q_{\xi\xi'} = \begin{cases} \frac{1}{r|V|}, & \text{if } \xi'(v) \neq \xi(v), \xi(w) = \xi'(w), w \neq v, \\ \frac{1}{r}, & \text{if } \xi' = \xi, \\ 0, & \text{otherwise.} \end{cases} \tag{6.3.12}$$

For configurations ξ and ξ' such that $q_{\xi\xi'} > 0$, the acceptance ratio is:

$$\frac{\pi_{\xi'}}{\pi_{\xi}} = \begin{cases} 1, & \text{if } \xi' \text{ is a proper coloring,} \\ 0, & \text{otherwise.} \end{cases}$$

The one-step transition probabilities of the Metropolis algorithm, as defined in (6.3.1), are then:

$$
p_{\xi\xi'} = \begin{cases}
\dfrac{1}{r|V|}, & \text{if } \xi'(v) \neq \xi(v),\, \xi'(w) = \xi(w) \text{ for } w \neq v,\, \text{ and } \xi' \in \Lambda, \\[2ex]
1 - \displaystyle\sum_{v \in V}\sum_{c \in C} \dfrac{1}{r|V|}\, \mathbb{1}(\{\xi''(v) = c,\, \xi''(w) = \xi(w),\, w \neq v\} \in \Lambda), & \text{if } \xi' = \xi.
\end{cases}
$$

In simpler terms, the Metropolis procedure can be described as follows. Given the current coloring $X_k = \xi$:

- Choose a vertex $v \in V$ and a color $c \in C$ uniformly at random.
- Colour the vertex v with c. If the resulting coloring ξ' is a proper coloring of the graph, set $X_{k+1} = \xi'$; otherwise, set $X_{k+1} = X_k$.

The procedure is formalized in Algorithm 46.

Algorithm 46 Metropolis algorithm for sampling graph coloring (1 step)

1: Assume the current configuration is $X_k = \xi$
2: Pick a vertex $v \in V$ and a color $c \in C$ uniformly at random
3: Set $\xi'(v) = c$, and $\xi'(w) = \xi(w)$ for all $w \neq v$
4: **if** ξ' is a proper coloring **then**
5:     $X_{k+1} = \xi'$
6: **else**
7:     $X_{k+1} = \xi$
8: **end if**
9: **return** X_{k+1}

It is important to note the differences between the Gibbs sampler (Algorithm 45) and the Metropolis algorithm (Algorithm 46) with candidate-generating transitions defined in (6.3.12). The main difference lies in the probability of staying in the same configuration:

- In the Gibbs sampler, we uniformly sample from all possible colors that can be assigned to vertex v, including the current color $\xi(v)$.
- In the Metropolis algorithm, the probability of not changing the color is proportional to the number of colors that cannot be assigned to v.

6.3.2.2 Hard-core Models

Consider a connected graph $G = (V, E)$. Each vertex in V is assigned a token, either '0' or '1'. A configuration is thus an element of the set $\{0, 1\}^V$. An intuitive interpretation of the model is the following: the value '1' at a vertex represents the presence of a particle, while '0' indicates its absence. We are interested in the *hard-core* model, which imposes a restriction that no two neighboring vertices can simultaneously have a value of '1'. We call such a configuration *feasible*, and the set of all feasible configurations is denoted by Λ. Figure 6.5 illustrates feasible configurations for a simple graph G. Additionally, the neighborhood structure in this model is distinct from the original graph and is denoted by $\mathcal{N}_\Lambda$.

For example, in Fig. 6.5, we have:

- Neighborhoods as defined by the original graph structure $G = (V, E)$:

$$\mathcal{N}_G(v_1) = \{v_2, v_3\}, \qquad \mathcal{N}_G(v_2) = \{v_1, v_3\},$$
$$\mathcal{N}_G(v_3) = \{v_1, v_2, v_4\}, \qquad \mathcal{N}_G(v_4) = \{v_3\}.$$

- Neighborhoods as defined by feasible configurations that differ in the value of a single vertex:

$$\mathcal{N}_\Lambda(\xi_1) = \{\xi_2, \xi_3, \xi_4, \xi_5\},$$
$$\mathcal{N}_\Lambda(\xi_2) = \{\xi_1, \xi_6\}, \quad \mathcal{N}_\Lambda(\xi_3) = \{\xi_1, \xi_6, \xi_7\}, \quad \mathcal{N}_\Lambda(\xi_4) = \{\xi_1, \xi_4\},$$
$$\mathcal{N}_\Lambda(\xi_5) = \{\xi_1\}, \quad \mathcal{N}_\Lambda(\xi_6) = \{\xi_2, \xi_3\}, \quad \mathcal{N}_\Lambda(\xi_7) = \{\xi_3, \xi_4\}.$$

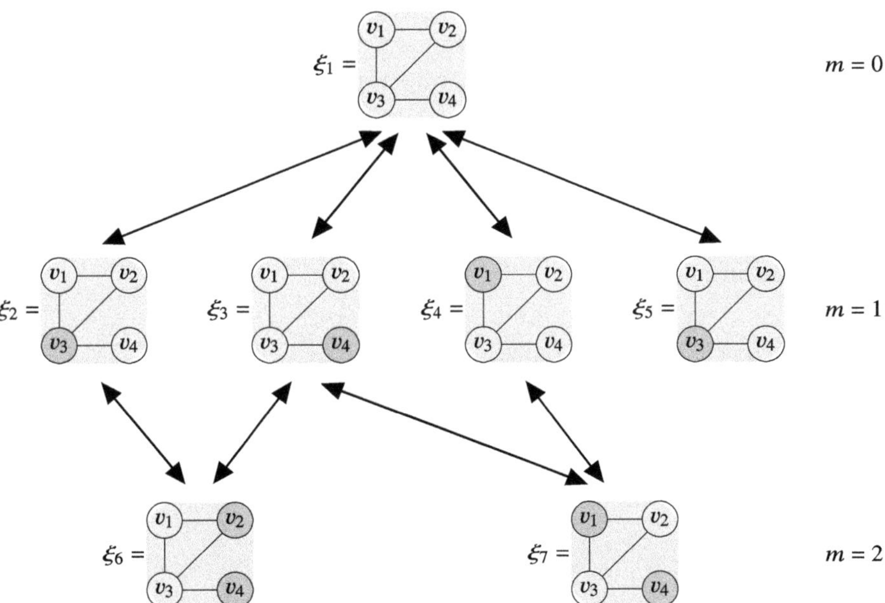

Fig. 6.5 A sample graph and all $z_G = 7$ of its feasible configurations. Neighborhoods defined by a graph structure $\mathcal{N}_G$ as well as neighborhoods defined by feasible configurations $\mathcal{N}_\Lambda$ are depicted. The number of particles is denoted by m

Let z_G be the number of all feasible configurations, i.e., $z_G = |\Lambda|$. On Λ, we define the following distribution:

$$
\pi_\xi = \begin{cases} \frac{1}{z_G}, & \text{if } \xi \in \Lambda, \\ 0, & \text{otherwise.} \end{cases}
\tag{6.3.13}
$$

A random element following this distribution takes values in the space Λ of feasible configurations. For complex graphs, z_G is often very large and computationally difficult to determine. To address this, we use the Gibbs sampler presented in Algorithm 47. In this approach, we randomly select a vertex $v \in V$, and then calculate:

$$
\mathbb{P}(Z(v) = c \mid Z_{-v} = \xi_{-v}), \qquad c \in \{0, 1\}.
$$

Consider the following two scenarios:

- For every $w \in \mathcal{N}(v)$, we have $\xi(w) = 0$ (i.e., all neighbors of vertex v have the value 0). Then:

$$
\mathbb{P}(Z(v) = 0 \mid Z_{-v} = \xi_{-v}) = \mathbb{P}(Z(v) = 1 \mid Z_{-v} = \xi_{-v}) = \frac{1}{2}.
$$

- There exists a $w \in \mathcal{N}(v)$ such that $\xi(w) = 1$ (i.e., one of the neighbors of vertex v has the value 1). Then:

$$
\mathbb{P}(Z(v) = 0 \mid Z_{-v} = \xi_{-v}) = 1, \qquad \mathbb{P}(Z(v) = 1 \mid Z_{-v} = \xi_{-v}) = 0.
$$

The Gibbs algorithm for this example is summarized in Algorithm 47. Note that there is no difficulty in choosing X_0, as configurations with no particles (or just one particle) are always feasible.

Algorithm 47 Gibbs sampler for the hard-core model on graph G (one step)

Require: Graph G with neighborhood structure $\mathcal{N}_G(v)$
1: Choose a vertex $v \in V$ uniformly at random and flip a fair coin
2: **If** the coin lands on "heads" and all neighbors of v have value 0, **then** set the value at v to 1; otherwise, set the value at v to 0
3: Leave the values of other vertices $v' \neq v$ unchanged

Remark 6.3.8 The Metropolis algorithm can also be applied to this example. However, note that two different graphs are involved here:

- A graph $G = (V, E)$ is used to define the distribution in (6.3.13) on configurations (specifically, the neighborhood structure $\mathcal{N}_G(v)$).

- The Markov chain (X_n) operates on all configurations $\{0, 1\}^V$. Here, the neighborhood is defined on configurations: $\xi' \in \mathcal{N}_\Lambda(\xi)$ if ξ and ξ' differ only at one vertex and both configurations are feasible. See Fig. 6.5 for an illustration of the distinction between $\mathcal{N}_G$ and $\mathcal{N}_\Lambda$.

To apply the Metropolis algorithm, we first define the candidate-generation matrix $\mathbf{Q} = (q_{\xi\xi'})$ as follows:

$$q_{\xi\xi'} = \begin{cases} \frac{1}{2}, & \text{if } \xi' = \xi, \\[2mm] \frac{1}{2|V|}, & \text{if } \xi'(v) = 1 - \xi(v) \text{ and } \xi'(w) = \xi(w) \text{ for all } w \neq v, \\[2mm] 0, & \text{otherwise.} \end{cases} \tag{6.3.14}$$

The matrix $\mathbf{Q}$ is symmetric, so we can use Proposition 6.3.1 and Algorithm 42. For infeasible ξ', we have $\pi_{\xi'} = 0$, and for feasible ξ, $\pi_{\xi'}/\pi_\xi = 1$. Therefore, the acceptance probability α in Algorithm 42 is:

$$\alpha = \begin{cases} 1, & \text{if } \xi' \text{ is feasible,} \\[2mm] 0, & \text{otherwise.} \end{cases}$$

The procedure works as follows. If the chain is at $X_k = \xi$, flip a coin:

- If the coin lands on "heads," do nothing ($q_{\xi,\xi} = \frac{1}{2}$).
- Otherwise, select a vertex at random and change its value to the opposite value. If the resulting configuration ξ' is feasible, set $X_{k+1} = \xi'$; otherwise, set $X_{k+1} = X_k$.

Interestingly, the resulting Metropolis algorithm is equivalent to the Gibbs algorithm described in Algorithm 47. This equivalence is not surprising, as the Gibbs algorithm is a special case of the Metropolis algorithm. See Exercise 6.T.20 on page 493.

6.3.2.3   Generalized Hard-core Model

We now consider the hard-core model with a modified formula for the distribution π, referred to as the *generalized hard-core model*. In the earlier formulation, we constructed a Markov chain with the uniform distribution over all feasible configurations. For a given configuration ξ (assigning a value $\xi(v) \in \{0, 1\}$ to vertex v), let $m(\xi) = \sum_{v \in V} \xi(v)$ denote the number of vertices with the value 1, i.e., the number of particles (cf. Fig. 6.5). In this section, we present the Gibbs sampler for the distribution $\pi = \{\pi_\xi : \xi \in \Lambda\}$ defined as:

$$\pi_\xi = \begin{cases} \frac{1}{z_{G,\lambda}} \lambda^{m(\xi)}, & \text{if } \xi \in \Lambda, \\ 0, & \text{otherwise,} \end{cases} \tag{6.3.15}$$

where

$$z_{G,\lambda} = \sum_{\xi \in \Lambda} \lambda^{m(\xi)},$$

and $\lambda > 0$ is a constant. For $\lambda = 1$, this reduces to the hard-core model from Sect. 6.3.2.2.

According to the Gibbs sampler, assuming the current configuration is $X_k = \xi$, we uniformly at random select a vertex $v \in V$ and compute the conditional π distribution of the value at v, given that all other vertices take values according to X_k. Let Z be a random variable with distribution π. We distinguish two cases:

- If there exists $w \in \mathcal{N}_G(v)$ such that $\xi(w) = 1$, then:

$$\mathbb{P}(Z(v) = 0 \mid Z_{-v} = \xi_{-v}) = 1.$$

- If $\xi(w) = 0$ for all $w \in \mathcal{N}_G(v)$, then:

$$\mathbb{P}(Z(v) = 1 | Z_{-v} = \xi_{-v})$$

$$= \frac{\mathbb{P}(Z(v) = 1, Z_{-v} = \xi_{-v})}{P(Z(v) = 0, Z_{-v} = \xi_{-v}) + \mathbb{P}(Z(v) = 1, Z_{-v} = \xi_{-v})}$$

$$= \frac{\lambda^{m(\xi)+1}}{\lambda^{m(\xi)} + \lambda^{m(\xi)+1}} = \frac{\lambda}{\lambda + 1}, \tag{6.3.16}$$

and

$$\mathbb{P}(Z(v) = 0 | Z_{-v} = \xi_{-v}) = \frac{\lambda^{m(\xi)}}{\lambda^{m(\xi)} + \lambda^{m(\xi)+1}} = \frac{1}{\lambda + 1}. \tag{6.3.17}$$

To explicitly write the transition matrix of the Gibbs sampler, we use the following notation:

$$m_0^0(\xi) = |\{v : \xi(v) = 0 \text{ and } \forall w \in \mathcal{N}_G(v), \ \xi(w) = 0\}|,$$

$$m_0^1(\xi) = |\{v : \xi(v) = 0 \text{ and } \exists w \in \mathcal{N}_G(v), \ \xi(w) = 1\}|,$$

$$m_1(\xi) = |\{v : \xi(v) = 1\}| = m(\xi).$$

The probability of staying at configuration ξ is given by:

$$p_{\xi\xi} = \sum_{v \in V} \mathbb{P}(X_{k+1}(v) = X_k(v) \mid X_k = \xi)$$

$$= \frac{1}{|V|} \left(\frac{\lambda}{\lambda + 1} m(\xi) + \frac{1}{\lambda + 1} m_0^0(\xi) + m_0^1(\xi) \right).$$

Since $|V| = m(\xi) + m_0^0(\xi) + m_0^1(\xi)$ for any ξ, this can be rewritten as:

$$p_{\xi\xi} = \sum_{v \in V} \mathbb{P}(X_{k+1}(v) = X_k(v)|X_k = \xi)$$

$$= \frac{1}{|V|} \left(|V| - \frac{\lambda}{\lambda + 1} m_0^0(\xi) - \frac{1}{\lambda + 1} m(\xi) \right)$$

$$= 1 - \frac{\lambda}{\lambda + 1} \frac{m_0^0(\xi)}{|V|} - \frac{1}{\lambda + 1} \frac{m(\xi)}{|V|}.$$

The transition matrix for the Gibbs sampler is then:

$$p_{\xi\xi'} = \begin{cases} \frac{\lambda}{\lambda+1} \frac{1}{|V|}, & \text{if } \xi' \in \mathcal{N}_\Lambda(\xi), \ \xi(v) = 0, \ \xi'(v) = 1, \\[2mm] \frac{1}{\lambda+1} \frac{1}{|V|}, & \text{if } \xi' \in \mathcal{N}_\Lambda(\xi), \ \xi(v) = 1, \ \xi'(v) = 0, \\[2mm] 1 - \frac{\lambda}{\lambda+1} \frac{m_0^0(\xi)}{|V|} - \frac{1}{\lambda+1} \frac{m(\xi)}{|V|}, & \text{if } \xi' = \xi, \\[2mm] 0, & \text{if } |m(\xi) - m(\xi')| \geq 2. \end{cases}$$

The update function φ in (6.3.10) is:

- If $\forall w : (w, v) \in E, \ \xi(w) = 0$, then:

$$\varphi(\xi, v, u) = \begin{cases} \xi(v \to 0), \text{ if } u \in \left[0, \frac{1}{\lambda+1} \right), \\[2mm] \xi(v \to 1), \text{ if } u \in \left[\frac{1}{\lambda+1}, 1 \right]. \end{cases} \tag{6.3.18}$$

- If $\exists w : (w, v) \in E, \ \xi(w) = 1$, then:

$$\varphi(\xi, v, u) = \xi, \quad \text{for all } u \in [0, 1]. \tag{6.3.19}$$

In other words, if there exists a neighbor w of v with value 1, the configuration remains unchanged.

Note that for $\lambda = 1$, the distribution (6.3.15) coincides with (6.3.13). When $\lambda > 1$, the probabilities π_ξ favor configurations with more particles, while for $\lambda < 1$, the probabilities favor configurations with fewer particles.

Remark 6.3.9 Note that the hard-core model is **not** a special case of graph coloring. If we interpret the value 0 as, say, a white color, and the value 1 as, say, a black color, then the feasible configurations in the hard-core model allow two vertices connected by an edge to both have white colors, but not both black colors. In contrast, graph coloring forbids both cases, as no two adjacent vertices can share the same color. ∎

6.3.2.4 Ising Model

As before, we consider a graph $G = (V, E)$, with the standard neighborhood system $\mathcal{N}_G(v)$, $v \in V$, defined by edges. A configuration is represented as $\xi \in \mathcal{S} = \{-1, 1\}^V$, and the energy function is defined as:

$$H(\xi) = - \sum_{v,v':v'\in\mathcal{N}_G(v)} \xi(v)\xi(v').$$

The values ± 1 assigned to vertices are called *spins*, representing positive and negative states, respectively. The Gibbs distribution (6.3.6) on the set of configurations is defined as (where $\beta \in \mathbb{R}$ is a parameter):

$$\pi_\xi = \frac{e^{-\beta H(\xi)}}{z_\beta},$$

where

$$z_\beta = \sum_{\xi \in \mathcal{S}} e^{-\beta H(\xi)}$$

is the partition function. If $\beta > 0$, the model is called the **ferromagnetic Ising model** (in this case, β is interpreted as the inverse of temperature). If $\beta < 0$, the model is called the **anti-ferromagnetic Ising model**.

Let (X_n) be a Markov chain on $\mathcal{S}$, with transitions determined by the Gibbs sampler, i.e., one-step transitions according to Algorithm 44. Suppose $X_n = \xi$ and a vertex $v^\bullet$ is selected uniformly at random. We then compute:

$$\mathbb{P}(Z(v^\bullet) = c \mid Z_{-v^\bullet} = \xi_{-v^\bullet}), \quad c \in \{-1, 1\},$$

where $Z \sim \pi$. We have:

$$\mathbb{P}(Z(v^\bullet) = 1 \mid Z_{-v^\bullet} = \xi_{-v^\bullet})$$

$$= \frac{\exp\left(-\beta \sum_{v' \in \mathcal{N}(v^\bullet)} \xi(v') - \beta \sum_{v \neq v^\bullet, v' \in \mathcal{N}(v)} \xi(v)\xi(v')\right)}{\exp\left(-\beta \sum_{v' \in \mathcal{N}(v^\bullet)} \xi(v') - \beta \sum_{v \neq v^\bullet, v' \in \mathcal{N}(v)} \xi(v)\xi(v')\right) + \exp\left(\beta \sum_{v' \in \mathcal{N}(v^\bullet)} \xi(v') + \beta \sum_{v \neq v^\bullet, v' \in \mathcal{N}(v)} \xi(v)\xi(v')\right)}.$$

This simplifies to:

$$\mathbb{P}(Z(v^\bullet) = 1 \mid Z_{-v^\bullet} = \xi_{-v^\bullet}) = \frac{1}{1 + e^{-2\beta \sum_{v' \in \mathcal{N}(v^\bullet)} \xi(v')}} = \alpha(\xi, v^\bullet). \tag{6.3.20}$$

Let $m_+(v, \xi)$ be the number of neighbors of v with spin $+1$ in configuration ξ, and let $m_-(v, \xi)$ be the number of neighbors of v with spin -1. Using this, the function α in the RHS of (6.3.20) can be written as:

$$\alpha(\xi, v) = \frac{1}{1 + \exp\left(-2\beta \left(m_+(v, \xi) - m_-(v, \xi)\right)\right)}. \tag{6.3.21}$$

Hence, we can formulate the following algorithm for the dynamics of the Gibbs sampler in the Ising model:

Algorithm 48 Gibbs sampler (one step)

1: $X_k = \xi$
2: Pick $v \in V$ uniformly at random and $U_{k+1} \sim \mathcal{U}[0, 1)$
3: Set:

$$X_{k+1}(v) = \begin{cases} +1, & \text{if } U_{k+1} < \alpha(\xi, v), \\ -1, & \text{otherwise.} \end{cases}$$

4: Set $X_{k+1}(w) = \xi(w)$ for all $w \neq v$
5: **return** X_{k+1}

The update function (6.3.10) for the Ising model is given by:

$$\varphi(\xi, v, u) = \begin{cases} \xi(v \to -1), & \text{if } u \in [0, \alpha(\xi, v)), \\ \xi(v \to +1), & \text{if } u \in [\alpha(\xi, v), 1]. \end{cases} \tag{6.3.22}$$

6.4   Perfect Simulation: Sampling Exactly from π

MCMC methods allow *approximate* sampling from a given distribution, that is, sampling from a distribution close to the target distribution π. In practice, to make the method useful, knowledge of the rate of convergence of the resulting Markov chain must often be known. Typically, knowledge of (or a bound on) the mixing time is desired. Recall the mixing time (6.2.2):

$$\tau(\varepsilon) = \min_{n \in \mathbb{N}} \{ d_{\mathrm{TV}}(\mu \mathbf{P}^n, \pi) \le \varepsilon \}.$$

In many cases, determining this is a hard task, often infeasible (the methods for studying the rate of convergence are beyond the scope of this book). **Perfect simulation** provides a way to address this issue. The term refers to the art of converting a Markov chain into an algorithm that returns an unbiased sample from its stationary distribution. No knowledge of the rate of convergence is required.

6.4.1 *Coupling from the Past (CFTP) Algorithm*

Coupling from the past is probably the most famous algorithm for perfect simulation. It was introduced in groundbreaking work by Propp and Wilson [171]. The algorithm ingeniously realizes the chain as a stochastic flow evolving *from the past*, rather than *into the future*.

In this section, we consider an ergodic (that is, irreducible and aperiodic) Markov chain (X_n) with a transition matrix $\mathbf{P}$ and the stationary distribution π. An update function $\varphi(x, u)$ (recall Eq. (6.2.3)) for this chain is also given (or one needs to find one with specific properties), such that $X_{n+1} = \varphi(X_n, U_{n+1})$, where (U_n) is an i.i.d. sequence of $\mathcal{U}[0, 1)$-distributed random variables. It is important to note that indices can be negative.

The very rough idea of the CFTP algorithm is as follows: assume that at some time in the past, say at $-N_n$, a chain is started for each $s \in \mathcal{S}$. Later, using the same update function and the same uniform random variables driving the chains, all the chains coalesce before (or at) time 0. Observe that at the coalescence time, all chains converge to the same state, regardless of their initial states. If the chains had been started even earlier, at "minus infinity," but from $-N_n$ onward the same uniform random variables were used, they would still converge to the same state at time 0. From "minus infinity" to 0, the chain has surely reached stationarity. Thus, the output of the algorithm is a random variable with the stationary distribution of the chain. For more technical details, see [171]. A bit of theory is also enclosed in Sect. 6.4.2.

We now present the Propp-Wilson algorithm. Usually, it is recommended to set $N_n = 2^{n-1}$ for $n = 1, 2, \ldots$. From Exercise 6.T.23, we see that $N_n = n$ is not an optimal choice. Given an increasing sequence $N_1, N_2, \ldots$ of positive integers, the procedure is given in Algorithm 49.

In the next section, we will prove Theorem 6.4.2, which justifies Algorithm 49.

Remark 6.4.1 (From Häggström [73], p. 83, Exercise 10.2.) Unfortunately, the condition "Assume that Algorithm 49 stops with probability 1" cannot be removed,

---

Algorithm 49 Coupling from the past (CFTP)

Require: State space $\mathcal{S}$, ergodic chain (X_n) with an update rule φ
Ensure: Exact sample from the stationary distribution of (X_n)
1: Set $n = 1$.
2: For each $s \in \mathcal{S}$, simulate the Markov chain starting from time $-N_n$ in state s and run it until time 0 using the same update rule φ and i.i.d. random variables $U_{-N_n+1}, U_{-N_n+2}, \ldots, U_{-1}, U_0$ uniformly distributed on $[0, 1]$ (the same for each chain).
3: If all chains in the previous step end up in the same state s_0 at time 0, then output s_0 and **stop**.
4: Set $n = n + 1$ and go to Step 2 (retain previously used (U_i), $i = -N_n + 1, \ldots, 0$, for the new n).

---

as the following example shows: Let

$$
\mathbf{P} = \begin{pmatrix} 0.5 & 0.5 \\ 0.5 & 0.5 \end{pmatrix}.
$$

Suppose we run Algorithm 49 for $N_j = 2^{j-1}$. There are different possibilities for choosing update functions:

$$
\varphi_1(i, u) = \begin{cases} 1 & \text{for } u \in (0, 0.5), \\ 2 & \text{for } u \in (0.5, 1), \end{cases}
$$

or

$$
\varphi_2(1, u) = \begin{cases} 1 & \text{for } u \in (0, 0.5), \\ 2 & \text{for } u \in (0.5, 1), \end{cases} \qquad \varphi_2(2, u) = \begin{cases} 2 & \text{for } u \in (0, 0.5), \\ 1 & \text{for } u \in (0.5, 1). \end{cases}
$$

If the update function is φ_1, the algorithm stops at $N = 1$, but if the update function is φ_2, the algorithm never stops. The reader is encouraged to verify this statement. $\blacksquare$

6.4.2   A Bit of Theory Related to the CFTP Algorithm

Here we address some questions that may arise when reading Algorithm 49 for the first time.

In the literature, it is well known that independent coupling can be used to prove the ergodic theorem for Markov chains (see, e.g., the proof of Theorem 8.6 in [14]). Specifically, for two independent Markov chains (X_n) and (X'_n) with the same regular transition matrix $\mathbf{P}$ and initial distributions $X_0 \sim \pi$ and $X'_0 \sim \delta_i$, the (forward) coupling time

$$
\tau_{\mathrm{f}} = \min\{n : X_n = X'_n\}
$$

is finite with probability one. Moreover, the following inequality holds:

$$d_{\mathrm{TV}}(\pi \mathbf{P}^k, \delta_i \mathbf{P}^k) \leq \mathbb{P}(\tau_{\mathrm{f}} > k). \tag{6.4.1}$$

The above inequality is a special case of a more general result. To state it, we first introduce the following definition. A **coupling** of two Markov chains (X_n) and (X'_n) with the same transition matrix $\mathbf{P}$ is a joint process (X_n, X'_n) such that each marginal process follows the Markov dynamics governed by $\mathbf{P}$, but their dependence structure may be arbitrary. In other words, X_n and X'_n are marginally Markov chains with the same transition matrix $\mathbf{P}$.

The inequality (6.4.1) is a special case of a more general result, known as the **coupling inequality**, which states that for any initial distributions μ and v, if (X_n, X'_n) is a coupling of two Markov chains with $X_0 \sim \mu$ and $X'_0 \sim v$, then

$$d_{\mathrm{TV}}(\mu \mathbf{P}^k, v \mathbf{P}^k) \leq \mathbb{P}(\tau_{\mathrm{f}} > k). \tag{6.4.2}$$

In contrast, in Algorithm 49, we start chains from different states $s \in \mathcal{S}$, but they are not independent. Moreover, the chains are defined recursively by an update function φ, such that

$$X_{n+1} = \varphi(X_n, U_{n+1}).$$

To run the chain from step j (which may be negative), we need an initial state $s \in \mathcal{S}$. We denote the chain by $X_i^j(s)$, $i = j, j+1, \ldots$. Formally:

$$X_j^j(s) = s, \quad X_{i+1}^j(s) = \varphi(X_i^j(s), U_{i+1}), \ i = j, j+1, \ldots,$$

$$X_0^j(s) = \varphi(X_{-1}^j(s), U_0).$$

It is convenient to introduce:

$$\varphi_i(s) = \varphi(s, U_i).$$

With this notation, we can express multiple compositions. For example, suppose $X_{-10}^{-10}(s) = s$, then:

$$X_{-9}^{-10}(s) = \varphi(X_{-10}^{-10}(s), U_{-9}) = \varphi_{-9}(X_{-10}^{-10}(s)) = \varphi_{-9}(s),$$

$$X_{-8}^{-10}(s) = \varphi(X_{-9}^{-10}(s), U_{-8}) = \varphi_{-8}(\varphi_{-9}(X_{-10}^{-10}(s))) = \varphi_{-8}(\varphi_{-9}(s)).$$

To formalize the problem, we introduce the following:

- Let $(\ldots, U_{-1}, U_0, U_1, \ldots)$ be a double-sided sequence of i.i.d. random variables uniformly distributed on $\mathcal{U}[0, 1)$.

- For integers $j < i$:

$$X_i^j(s) = \varphi_i(\ldots \varphi_{j+2}(\varphi_{j+1}(s))) = (\varphi_i \circ \cdots \circ \varphi_{j+1})(s).$$

Note that the random variable $X_i^j(s)$ depends on $U_i, \ldots, U_{j+1}$ and s, and the function depends only on $i - j$.
- The forward coupling time τ_f is generally not sufficient to ensure stationarity. For example, in Exercise 6.T.24, we consider a two-state chain that stops in one of the states with probability 1. In this case, the distribution of $X_{\tau_f}^0(s)$ is not the stationary distribution π.
- Consider the backward procedure in Algorithm 49. Using the notation of this section, the algorithm proceeds as follows: for $n = 1, 2, \ldots$, we run the chain from the past starting at $-N_n$. We say the chains are coupled at $i = 0$ if:

$$X_0^{-N_n}(s_1) = \cdots = X_0^{-N_n}(s_N) = Z. \tag{6.4.3}$$

- Define the first K for which the chains started at $-N_K$ meet:

$$K = \min\left\{ n \geq 1 : X_0^{-N_n}(s_1) = \cdots = X_0^{-N_n}(s_N) \right\}.$$

Let:

$$\tau_{\mathrm{cftp}} = -N_K.$$

Note that τ_{cftp} is a random variable.
- An important property is that:

$$X_0^j(s_1) = \cdots = X_0^j(s_N), \qquad \text{for } j \leq \tau_{\mathrm{cftp}}.$$

We now state a theorem justifying the procedure in Algorithm 49. The Markov chain is $X_l(s) = X_l^{-N_j}(s)$ with transition matrix $(p_{ss'})$, where $p_{ss'} = \mathbb{P}(X_{i+1}(s) = s' \mid X_i(s) = s)$.

Theorem 6.4.2 *Suppose that τ_{cftp} is finite with probability 1. Then $X_0^{\tau_{\mathrm{cftp}}}(s)$ has the distribution π.*

Proof Suppose s is an initial state. For any $i > 0$:

$$\mathbb{P}(X_i^0(s) = s') = \mathbb{P}(X_0^{-i}(s) = s').$$

From the ergodic theorem (Corollary 6.2.3), we have:

$$\lim_{i \to \infty} \mathbb{P}(X_i^0(s) = s') = \pi_{s'},$$

which implies:

$$\lim_{i \to \infty} \mathbb{P}(X_0^{-i}(s) = s') = \pi_{s'}. \tag{6.4.4}$$

Observe now that for $0 < i < N_j$:

$$X_0^{-N_j}(s) = X_0^{-i}(X_{-i}^{-N_j}(s)),$$

and setting $i = -\tau_{\text{cftp}}$:

$$X_0^{-N_j}(s) = X_0^{\tau_{\text{cftp}}}(s),$$

whenever $-N_j \le \tau_{\text{cftp}}$, which is independent of s. In view of (6.4.4):

$$\mathbb{P}(X_0^{\tau_{\text{cftp}}}(s) = s') = \pi_{s'}.$$

$\square$

Another question arises: why do we need to reuse random variables $(U_i)_{0 \le i \le -N_n+1}$ from the previous step in step $n + 1$? In Exercise 6.T.30, we propose a two-state example demonstrating that if new random variables (U_i) are used at each step, the result will not follow the stationary distribution.

Note that the algorithm presented so far is highly inefficient for large state spaces, as it requires running as many chains as there are states. However, in some cases, it is sufficient to run only two chains, significantly reducing computational effort.

6.4.3   Realizable Monotone Markov Chains

From now on, we assume that the state space $\mathcal{S} = \{s_1, \ldots, s_N\}$ is partially ordered by $\preceq$ and that for all $s \in \mathcal{S}$, we have $s_1 \preceq s \preceq s_N$, i.e., s_1 is the minimum and s_N is the maximum. We will use the notation $s_{\min} = s_1$ and $s_{\max} = s_N$. Recall that a partial order is a binary relation that is reflexive (each element is comparable to itself), antisymmetric (no two different elements precede each other), and transitive. For example, consider a hypercube as in Example 6.2.8, where the coordinate-wise ordering is given (recall $\mathcal{S} = \{0, 1\}^d$, with $s = (e_1, \ldots, e_d)$, $e_i \in \{0, 1\}$ for $i = 1, \ldots, d$):

$$s = (e_1, \ldots, e_d) \preceq s' = (e'_1, \ldots, e'_d) \iff e_1 \le e'_1, \ldots, e_d \le e'_d.$$

Definition 6.4.3 *A Markov chain X with transition matrix* **P** *is **realizable monotone** with respect to the partial ordering $\preceq$ if there exists a monotone update function φ, such that:*

$$\forall(u \in [0, 1]) \; \forall(s \preceq s') \; \varphi(s, u) \preceq \varphi(s', u).$$

Remark 6.4.4 If $\varphi(s, u)$ satisfies the above definition and U is a $\mathcal{U}[0, 1)$ random variable, then for random variables $Z_s = \varphi(s, U)$, we have with probability 1:

$$\forall(s \preceq s') \; Z_s \preceq Z_{s'}. \tag{6.4.5}$$

In the fundamental work by Fill and Machida [54], it was stated that a family of distributions $F_v, v \in V$, is monotone realizable if there exists a collection of random elements $(Z_v, v \in V)$ on a probability space $(\Omega, \mathcal{F}, \mathbb{P})$, such that $Z_v \sim F_v$ and (6.4.5) holds. $\blacksquare$

For a realizable monotone Markov chain (X_n), we can achieve a radical improvement when running the perfect simulation Algorithm 49. Assume that at some time step $-k$, we have two chains $X_{-k} = s \preceq X'_{-k} = s'$. Then:

$$X_{-k+1} = \varphi(X_{-k}, U_{-k+1}) \preceq \varphi(X'_{-k}, U_{-k+1}) = X'_{-k+1}.$$

This implies that we only need to start two chains: one starting at $s_{\min}$ and the other at $s_{\max}$. If the chains coalesce before (or at) time 0, all other chains would also coalesce. Thus, if the chain is realizable monotone (i.e., it has a monotone update rule) and has a minimum and maximum, we can use an efficient version of the CFTP procedure, summarized in Algorithm 50.

Algorithm 50 Efficient coupling from the past

Require: State space $\mathcal{S}$, ergodic chain X, a monotone update rule φ
Ensure: Exact sample from the stationary distribution of (X_n)
1: Set $n = 1$.
2: Start two chains at time $-N_n$: one at the minimum $s_{\min}$ and the other at the maximum $s_{\max}$. Run the chains until time 0 using the same update rule φ and i.i.d. random variables $U_{-N_n+1}, U_{-N_n+2}, \ldots, U_{-1}, U_0$ uniformly distributed on $[0, 1]$ (the same for both chains).
3: If both chains in the previous step end up in the same state s_0 at time 0, then output s_0 and **stop**.
4: Set $n = n + 1$ and go to Step 2 (retain previously used $(U_i), i = -N_n + 1, \ldots, 0$, for the new n).

Proposition 6.4.5 *Suppose that a Markov chain is realizable monotone with an update function φ. Then, with probability 1, Algorithm 50 stops.*

Proof Using the notation from Sect. 6.4.2, let:

$$X_i^j(s) = \varphi_i(\ldots \varphi_{j+2}(\varphi_{j+1}(s))) = (\varphi_i \circ \cdots \circ \varphi_{j+1})(s).$$

Note that $X_i^j(s)$ depends on $U_i, \ldots, U_{j+1}$ and s, and only on the difference $i - j$. For $l > j$, consider the events:

$$A_l^j = \{X_l^j(s_{\max}) = X_l^j(s_{\min})\}.$$

Due to the monotone update rule, we always have $X_l^j(s_{\min}) \preceq X_l^j(s_{\max})$ with probability 1. Since the Markov chain is ergodic, it cannot be the case that $\mathbb{P}(A_l^j) = 0$ for all $j \geq 0$ and $l > j$. Thus, there exists $L < 0$ such that:

$$\mathbb{P}(A_0^L) > 0.$$

Furthermore, the events $A_0^L, A_L^{2L}, \ldots$ have the same positive probability (since the randomness, i.e., the sequence U_i, is independent for these events). By the second Borel-Cantelli lemma, with probability 1, infinitely many events $A_{kL}^{(k+1)L}$ will occur. When these events occur, we have:

$$X_{kL}^{(k+1)L}(s_{\min}) = X_{kL}^{(k+1)L}(s_{\max}),$$

and consequently, for all $s \in \mathcal{S}$:

$$X_{kL}^{(k+1)L}(s_{\min}) = X_{kL}^{(k+1)L}(s) = X_{kL}^{(k+1)L}(s_{\max}).$$

Thus, τ_{cftp} is finite with probability 1. $\qquad\qquad\qquad\qquad\qquad\qquad\qquad\qquad\square$

Another way to understand the efficiency of CFTP is as follows: assume that two chains, one starting at $s_{\min}$ (denoted as chain (X_k')) and the other at $s_{\max}$ (denoted as chain (X_k)), end up in the same state s_0. Both chains use the same update function φ and the same sequence of random variables $U_{\tau_{\text{cftp}}+1}, \ldots, U_0$. The situation is depicted in Fig. 6.6. In this figure, we observe the following:

- When the chains were started at $-N_n$, they did not meet. Specifically, the first chain (X_k) ended up in state s', while the second chain (X_k') ended up in state $s \neq s'$.
- However, when the chains were restarted at τ_{cftp} (using the same randomness from $-N_n$ as in the previous step), both chains ended up in the same state s'.

Now imagine starting a third chain (Y_k) such that $Y_{\tau_{\text{cftp}}} \sim \pi$. Clearly, $s_{\min} \preceq Y_{\tau_{\text{cftp}}} \preceq s_{\max}$. If we apply the same update function φ and use the same sequence of random variables (U_i), this chain will also end up in the same state s_0. Since $Y_k \sim \pi$ for all $k \geq \tau_{\text{cftp}}$, it follows that $s_0 \sim \pi$.

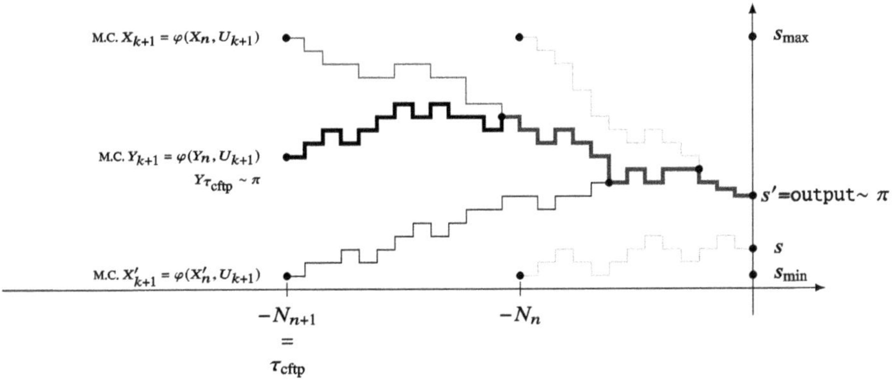

Fig. 6.6 Coupling from the past—schematic realization. See description in text

Note that identifying a monotone update function is often a challenging task, and proving that no such function exists can be even more difficult.

Example 6.4.6 Consider a symmetric random walk on $\mathcal{S} = \{0, 1\}^2 = \{0 \equiv (0, 0), 1 \equiv (1, 0), 2 \equiv (0, 1), 3 \equiv (1, 1)\}$. Let $0 < p \le \frac{1}{2}$ and $r_1 = (1, 0)$, $r_2 = (0, 1)$. Define a Markov chain (X_n) with a transition matrix:

$$
p_{ss'} = \begin{cases} p & \text{if } s' = s \pm r_i, \ i = 1, 2, \\ 1 - 2p & \text{if } s' = s, \\ 0 & \text{otherwise.} \end{cases} \tag{6.4.6}
$$

Roughly speaking, being in state $s = (e_1, e_2)$, where $e_i \in \{0, 1\}$ for $i = 1, 2$, the chain changes one of the coordinates to the opposite value ($0 \to 1$ or $1 \to 0$) with probability p, or it does nothing with the remaining probability.

Let us introduce coordinate-wise ordering. For $i = (e_1^i, e_2^i)$ and $j = (e_1^j, e_2^j)$, define:

$$
i \preceq j \ \Leftrightarrow \ e_1^i \le e_1^j \ \text{and} \ e_2^i \le e_2^j.
$$

Note that $0 \equiv (0, 0)$ is the minimum and $3 \equiv (1, 1)$ is the maximum. Additionally, we say that $j \in \mathcal{N}(i)$ if:

$$
|e_1^i - e_1^j| + |e_2^i - e_2^j| = 1.
$$

The Hasse diagram[1] for this ordering is shown in Fig. 6.7.

---

[1] https://en.wikipedia.org/wiki/Hasse_diagram.

Every update function can be represented in the following form (where all E_{ij} are subsets of $[0, 1]$):

$$\varphi(i, u) = \sum_{j \in \mathcal{N}(i)} j \mathbb{1}(u \in E_{ij}) + i \mathbb{1}\left(u \in \left(\bigcup_{j \in \mathcal{N}(i)} E_{ij}\right)^c\right), \qquad (6.4.7)$$

where for each i and $j \in \mathcal{N}(i)$, we have $|E_{ij}| = p$, and for each i, the sets E_{ij} (with $j \neq i$) are disjoint.

We will analyze the cases $p = \frac{1}{4}$, $p < \frac{1}{4}$, and $p \in (\frac{1}{4}, \frac{1}{2}]$ separately.

- **Case $p = \frac{1}{4}$.** Take:

$$E_{01} = \left(\frac{1}{2}, \frac{3}{4}\right], \quad E_{02} = \left(\frac{3}{4}, 1\right],$$

$$E_{10} = \left(0, \frac{1}{4}\right], \quad E_{13} = \left(\frac{3}{4}, 1\right],$$

$$E_{20} = \left(\frac{1}{4}, \frac{1}{2}\right], \quad E_{23} = \left(\frac{1}{2}, \frac{3}{4}\right], \qquad (6.4.8)$$

$$E_{31} = \left(\frac{1}{4}, \frac{1}{2}\right], \quad E_{32} = \left(0, \frac{1}{4}\right].$$

We leave it to the reader to verify the required monotonicity of φ.

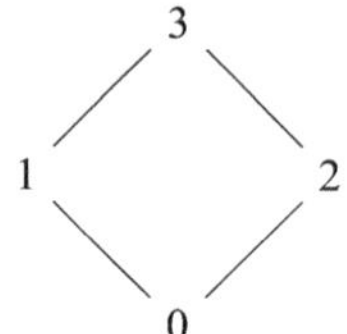

Fig. 6.7 Hasse diagram for coordinate-wise ordering on $\mathcal{S} = \{0, 1\}^2$

- **Case $p < \frac{1}{4}$.** For $0 < \alpha \leq 1$, define the α-shrinking of an interval $E = (a, b]$ as:

$$\alpha E = \left(\frac{a + b - \alpha(b - a)}{2}, \frac{a + b + \alpha(b - a)}{2}\right].$$

Let E_{ij} be as defined in (6.4.8). Note that $|E_{ij}| = \frac{1}{4}$. Setting $\alpha = 4p$, we define:

$$\varphi(i, u) = \sum_{j \in \mathcal{N}(i)} j \mathbb{1}(u \in \alpha E_{ij}) + i \mathbb{1}\left(u \in \left(\bigcup_{j \in \mathcal{N}(i)} \alpha E_{ij}\right)^c\right).$$

This defines a monotone update function that satisfies Definition 6.4.3.

- **Case** $p \in (\frac{1}{4}, \frac{1}{2}]$.

  For $p > \frac{1}{4}$, consider the update function $\varphi(0, u)$ and suppose $\varphi(0, u) = 1$ for $u \in E_{01}$. Since $0 \preceq 2$, we must have for $u \in E_{01}$ that:

  $$\varphi(0, u) = 1 \preceq \varphi(2, u),$$

  which is only possible if $\varphi(2, u) = 3$ (for $u \in E_{01}$). Since $|E_{01}| = |E_{23}| = p$, we conclude that $E_{01} = E_{23}$.

  Similarly, since $0 \preceq 1$, for $u \in E_{02}$ we must have:

  $$\varphi(0, u) = 2 \preceq \varphi(1, u),$$

  which is only possible if $\varphi(1, u) = 3$ (for $u \in E_{02}$). Since $|E_{02}| = |E_{13}| = p$, we conclude that $E_{02} = E_{13}$.

  Note that E_{23} and E_{13} are disjoint (as E_{01} and E_{02} are disjoint). Since both $1 \preceq 3$ and $2 \preceq 3$, we must have:

  $$E_{33} \subseteq E_{13} \cup E_{23}. \tag{6.4.9}$$

  However, $|E_{33}| = 1 - 2p$ and $|E_{13} \cup E_{23}| = 2p$, so (6.4.9) holds only if:

  $$1 - 2p \geq 2p,$$

  i.e., $p \leq \frac{1}{4}$. Thus, a monotone update function does not exist for $p > \frac{1}{4}$.

  Summarizing, the symmetric random walk on the cube is realizable monotone for $p \leq \frac{1}{4}$, but not for $p \in (\frac{1}{4}, \frac{1}{2}]$.

Note that different notions of monotonicity for Markov chains exist. It has been questioned whether the so-called *stochastic monotonicity*, defined in Exercise 6.T.26, is equivalent to realizable monotonicity. However, realizable monotonicity is stronger, and these two notions are not equivalent; see Exercise 6.T.26. $\Diamond$

Example 6.4.7 (Ferromagnetic Ising Model)   We continue the considerations from Sect. 6.3.2.4. Our aim is to show that a Markov chain (X_n) defined by Algorithm 48 with $\beta > 0$ is realizable monotone. Specifically, we will demonstrate that the update function φ given in (6.3.22) is monotone. Let us consider $\xi_0 \preceq \xi_0'$ and a fixed vertex v. We need to show that:

$$\forall (u \in (0, 1)) \quad \xi_1 = \varphi(\xi_0, v, u) \preceq \varphi(\xi_0', v, u) = \xi_1'.$$

For $w \neq v$, it is clear that $\xi_1(w) \leq \xi_1'(w)$, since only the value at v could possibly change.

Now consider the value at v. Note that $\xi_1(v) = +1$ if $u \geq \alpha(\xi_0, v)$, and $\xi_1'(v) = +1$ if $u \geq \alpha(\xi_0', v)$. Thus, the update function φ is monotone if:

$$\alpha(\xi_0, v) \geq \alpha(\xi_0', v). \tag{6.4.10}$$

Recall from (6.3.21) that:

$$\alpha(\xi, v) = (1 + \exp(-2\beta(m_+(v, \xi) - m_-(v, \xi))))^{-1}.$$

Thus, inequality (6.4.10) is equivalent to:

$$\exp\left(-2\beta\left(m_+(v, \xi_0') - m_-(v, \xi_0')\right)\right) \geq \exp\left(-2\beta\left(m_+(v, \xi_0) - m_-(v, \xi_0)\right)\right),$$

which simplifies to:

$$m_+(v, \xi_0') - m_-(v, \xi_0') \leq m_+(v, \xi_0) - m_-(v, \xi_0).$$

This inequality always holds because $\xi_0 \preceq \xi_0'$. Hence, the update function φ is monotone, proving that the Markov chain (X_n) is realizable monotone. $\Diamond$

Example 6.4.8 (Generalized Hard-core Model Continued)   In Sect. 6.3.2.3, we analyzed the so-called generalized hard-core model with stationary distribution π_ξ uniform on the set Λ of feasible configurations defined in (6.3.15) as

$$\pi_\xi = \begin{cases} \frac{1}{z_{G,\lambda}} \lambda^{m(\xi)} & \text{if } \xi \in \Lambda, \\ 0 & \text{otherwise.} \end{cases} \tag{6.4.11}$$

As we will see below, it is not possible to construct a monotone update function for this model. However, with some modifications, Algorithm 47 can still be adapted to provide useful results.

Recall the update function: Assume that a vertex v was chosen. If there exists a neighbor w of v with value 1, then:

$$\varphi(\xi, v, u) = \xi, \quad \forall u \in [0, 1].$$

Otherwise, if all neighbors of v have value 0 (i.e., $\forall w : (w, v) \in E, \xi(w) = 0$), the update function is given by (6.3.18):

$$\varphi(\xi, v, u) = \begin{cases} \xi(v \to 0), & \text{if } u \in \left[0, \frac{1}{\lambda+1}\right), \\ \xi(v \to 1), & \text{if } u \in \left[\frac{1}{\lambda+1}, 1\right]. \end{cases}$$

Now, consider configurations ξ and ξ' where v was chosen, as shown in Fig. 6.8.

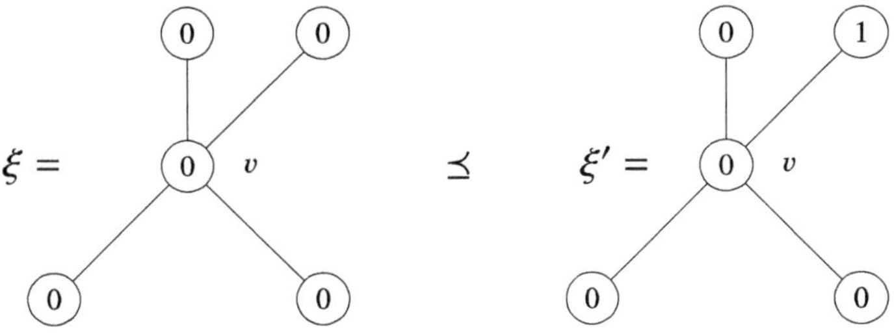

Fig. 6.8 Sample configurations $\xi \preceq \xi'$ in a generalized hard-core model

If $u \geq \frac{1}{\lambda+1}$, then $\varphi(\xi, v, u) \not\preceq \varphi(\xi', v, u)$. This shows that the update function is not monotone. In fact, the chain is **not realizable monotone** (i.e., a monotone update function does not exist). We leave the proof of this statement as an exercise.

$\Diamond$

6.4.4 A Bit of Theory Related to Monotone and Anti-monotone Systems

To utilize the efficient CFTP Algorithm 50, we require a monotone update function. In Example 6.4.7, we demonstrated that the update function for the ferromagnetic Ising model, arising from the Gibbs sampler, is monotone. However, finding such a function in general is challenging. In this section, we present perfect sampling algorithms where, instead of directly finding a monotone function, we only need to verify whether the distribution π (from which we aim to sample) possesses a specific property. Namely, we check if π is monotone or anti-monotone (definitions provided below). This property is relatively straightforward to verify. If the distribution is monotone, we can derive a direct formula for a monotone update function, and if it is anti-monotone, a modified CFTP algorithm can be used. It is worth noting that π may be neither monotone nor anti-monotone.

In this section (which is based on Häggström and Nelander [74]) we consider distributions on $\mathcal{S} = C^V$, where both C and V are finite sets. That is, the setup is the same as for the Gibbs sampler. Usually $C = \{1, 2, \ldots, r\}$ (in some cases, the numbering starts from 0, and in others—such as the Ising model, we have $C = \{-1, 1\}$). The set $V = \{v_1, \ldots, v_M\}$ represents the vertices. There is an underlying graph $G = (V, E)$ that is used for specifying the distribution π. The state (configuration) is denoted by $\xi \in \mathcal{S}$.

We use a natural coordinate-wise partial ordering for this state space:

$$\xi \preceq \xi' \iff \xi(v) \leq \xi'(v) \text{ for all } v \in V.$$

Definition 6.4.9 *Let $Z \sim \pi$. A probability measure π on $S = C^V$ is **monotone** if for $\xi \preceq \xi'$, and for all $v \in V$ such that $\pi_{\xi_{-v}} > 0$ and $\pi_{\xi'_{-v}} > 0$, we have*

$$P(Z(v) \leq c \mid Z_{-v} = \xi_{-v}) \geq P(Z(v) \leq c \mid Z_{-v} = \xi'_{-v}) \text{ for all } c \in C. \qquad (6.4.12)$$

We have the following lemma.

Lemma 6.4.10 *If π is monotone, then there exists a realizable monotone Markov chain with stationary distribution π. The monotone update function is defined as follows:*

$$\varphi(\xi, v, u) =$$

$$\begin{cases} \xi(v \to 1) \text{ if } u \in [0, P(Z(v) = 1 \mid Z_{-v} = \xi_{-v})), \\ \vdots \\ \xi(v \to c) \text{ if } u \in [P(Z(v) \leq c - 1 \mid Z_{-v} = \xi_{-v}), \\ \qquad\qquad P(Z(v) \leq c \mid Z_{-v} = \xi_{-v})), \\ \vdots \\ \xi(v \to r) \text{ if } u \in [P(Z(v) \leq r - 1 \mid Z_{-v} = \xi_{-v}), 1], \end{cases} \qquad (6.4.13)$$

where $Z \sim \pi$.

Proof Let $Z \sim \pi$. Assume $\xi \preceq \xi'$ and $u \in [P(Z(v) \leq c - 1 \mid Z_{-v} = \xi_{-v}), P(Z(v) \leq c \mid Z_{-v} = \xi_{-v}))$. If $\varphi(\xi', v, u)(v) = c'$, then $\xi_{-v} \preceq \xi'_{-v}$. Equation (6.4.12) implies that $c \leq c'$, and thus

$$\varphi(\xi, v, u) \preceq \varphi(\xi', v, u),$$

i.e., this update rule is monotone. $\qquad\qquad\qquad\qquad\qquad\qquad\qquad\qquad\qquad\square$

   In other words, the update function given in (6.3.10), (which is a "first one that comes to mind" for a Gibbs sampler) is monotone, provided that π is monotone.
   Similarly, we define:

Definition 6.4.11 *Let $Z \sim \pi$. A probability measure π on $S = C^V$ is **anti-monotone** if for $\xi \preceq \xi'$, and for all $v \in V$ such that $\pi_{\xi_{-v}} > 0$ and $\pi_{\xi'_{-v}} > 0$, we have*

$$P(Z(v) \leq c \mid Z_{-v} = \xi_{-v}) \leq P(Z(v) \leq c \mid Z_{-v} = \xi'_{-v}) \text{ for all } c \in C. \qquad (6.4.14)$$

Example 6.4.12 (Generalized Hard-core Model) Recall that the generalized hard-core model is a distribution given in (6.4.11), i.e.,

$$\pi_\xi = \begin{cases} \frac{1}{z_{G,\lambda}} \lambda^{m(\xi)} \text{ if } \xi \text{ is a feasible configuration,} \\ 0 \qquad\qquad \text{otherwise.} \end{cases}$$

Take $\xi \preceq \xi'$ and some fixed v. In this case, $C = \{0, 1\}$, and it suffices to consider $c = 0$ only. Consider the cases:

- All neighbors of v in both ξ and ξ' have value 0. Then

$$P(Z(v) = 0 \mid Z_{-v} = \xi_{-v}) = \frac{1}{1 + \lambda} = P(Z(v) = 0 \mid Z_{-v} = \xi'_{-v}).$$

- There exist neighbors of v in both ξ and ξ' which have value 1. Then

$$P(Z(v) = 0 \mid Z_{-v} = \xi_{-v}) = 1 = P(Z(v) = 0 \mid Z_{-v} = \xi'_{-v}).$$

- For all neighbors $w \in \mathcal{N}_G(v)$, we have $\xi(w) = 0$, but there exists $w' \in \mathcal{N}_G(v)$ such that $\xi'(w') = 1$. Then

$$P(Z(v) = 0 \mid Z_{-v} = \xi_{-v}) = \frac{1}{1 + \lambda} \leq 1 = P(Z(v) = 0 \mid Z_{-v} = \xi'_{-v}).$$

Thus, we have that

$$P(Z(v) = 0 \mid Z_{-v} = \xi_{-v}) \leq P(Z(v) = 0 \mid Z_{-v} = \xi'_{-v}),$$

i.e., the distribution π is anti-monotone. $\Diamond$

Example 6.4.13 (The Ising Model) Recall that the Ising model with parameter $\beta \in \mathbb{R}$ on a graph $G = (V, E)$ is a distribution

$$\pi_\xi = \frac{1}{z_\beta} \exp\left(\beta \sum_{v, v' \in \mathcal{N}_G(v)} \xi(v)\xi(v') \right),$$

for a configuration $\xi \in \{-1, +1\}^V$. In (6.3.20) and (6.3.21), we computed that

$$\mathbb{P}(Z(v) = 1 \mid Z_{-v} = \xi_{-v}) = \frac{1}{1 + e^{-2\beta(m_+(v,\xi) - m_-(v,\xi))}} = \alpha(\xi, v),$$

where $m_+(v, \xi)$ $(m_-(v, \xi))$ is the number of neighbors of v having spin $+1$ (-1). If $\xi \preceq \xi'$, then

$$m_+(v, \xi) - m_+(v, \xi') \leq m_-(v, \xi) - m_-(v, \xi').$$

Consequently,

$$\alpha(\xi, v) \leq \alpha(\xi', v) \text{ if } \beta > 0,$$
$$\alpha(\xi, v) \geq \alpha(\xi', v) \text{ if } \beta < 0.$$

Thus, π is monotone for $\beta > 0$ (i.e., the ferromagnetic Ising model), and it is anti-monotone for $\beta < 0$ (i.e., the anti-ferromagnetic Ising model). The uniform distribution ($\beta = 0$) is both monotone and anti-monotone.

Consequently, the update function for the ferromagnetic case given in (6.3.22) is monotone (by Lemma 6.4.10), which we also verified directly in Example 6.4.7. $\Diamond$

Remark 6.4.14 The ferromagnetic Ising model is an example of a monotone system. The anti-ferromagnetic Ising model, the hard-core (thus, independent sets) model, and the (somewhat trivial) 2-coloring model are examples of anti-monotone systems. Note that the terms monotone and anti-monotone align with the concepts of monotone increasing and monotone decreasing. There exist distributions π that are neither monotone nor anti-monotone. For example, graph q-coloring for $q \geq 4$ is neither monotone nor anti-monotone (see Exercise 6.T.33). $\blacksquare$

Modified Version of CFTP Algorithm for Anti-monotone Systems We have seen examples (e.g., the anti-ferromagnetic Ising model and the hard-core model) where π is anti-monotone. In this section, we present a modification of the CFTP algorithm for such systems.

Consider a bivariate Markov chain (X_k, X'_k) on the state space $\tilde{S} = C^V \times C^V$. Its evolution is determined by an update function of the form

$$\tilde{\varphi} : (C^V \times C^V) \times V \times [0, 1] \to C^V \times C^V,$$

where

$$\tilde{\varphi}((\xi, \xi'), v, u) = (\hat{\varphi}((\xi, \xi'), v, u), \ \hat{\varphi}((\xi', \xi), v, u)), \tag{6.4.15}$$

and $\hat{\varphi}((\xi, \xi'), v, u) =$

$$\begin{cases} \xi(v \to 0) \text{ if } u \in \left[0, P(Z(v) = 0 \mid Z_{-v} = \xi'_{-v})\right), \\[4pt] \vdots \\[4pt] \xi(v \to c) \text{ if } u \in \left[P(Z(v) \leq c - 1 \mid Z_{-v} = \xi'_{-v}), P(Z(v) \leq c \mid Z_{-v} = \xi'_{-v})\right), \\[4pt] \vdots \\[4pt] \xi(v \to r) \text{ if } u \in \left[P(Z(v) \leq r - 1 \mid Z_{-v} = \xi'_{-v}), 1\right]. \end{cases}$$
$$\tag{6.4.16}$$

The update function has the following property.

Lemma 6.4.15 *Assume that $\xi_0 \preceq \xi'_0$ and define*

$$(\xi_1, \xi'_1) = \tilde{\varphi}((\xi_0, \xi'_0), v, u), \quad v \in V, u \in (0, 1),$$

where $Z \sim \pi$. If π is anti-monotone, then $\xi_1 \preceq \xi'_1$ for all $v \in V$ and all $u \in (0, 1)$.

Proof Fix $v \in V$ and $u \in (0, 1)$. Note that $\tilde{\varphi}$ leaves vertices $w \neq v$ unchanged in both ξ_1 and ξ_1', which means that $\xi_{1,-v} \preceq \xi_{1,-v}'$, and we only need to show that

$$\xi_1(v) \leq \xi_1'(v).$$

Let $c_1 = \xi_1(v)$. This implies that

$$u \in \left[P(Z(v) \leq c_1 - 1 \mid Z_{-v} = \xi_{0,-v}'), P(Z(v) \leq c_1 \mid Z_{-v} = \xi_{0,-v}') \right).$$

Similarly, let $c_1' = \xi_1'(v)$, which implies that

$$u \in \left[P(Z(v) \leq c_1' - 1 \mid Z_{-v} = \xi_{0,-v}), P(Z(v) \leq c_1' \mid Z_{-v} = \xi_{0,-v}) \right).$$

The anti-monotonicity condition (6.4.14) implies that $c_1 \leq c_1'$, thus $\xi_1 \preceq \xi_1'$. $\quad\square$

Based on Lemma 6.4.15, we now construct—in the spirit of the CFTP algorithm—a bivariate chain starting at time $-N_n$ in the following way. At time $-N_n$, it starts at $(\xi_{\min}, \xi_{\max})$, i.e.,

$$Y_{-N_n}^{-N_n} = (\xi_{\min}, \xi_{\max}) =: (X_{-N_n}^{\xi_{\min}, -N_n}, X_{-N_n}^{\xi_{\max}, -N_n}).$$

In subsequent steps, it evolves as follows:

$$Y_{k+1}^{-N_n} = \tilde{\varphi}(Y_k^{-N_n}, W_{k+1}, U_{k+1}),$$
$$=: (X_{k+1}^{\xi_{\min}, -N_n}, X_{k+1}^{\xi_{\max}, -N_n}), \quad k = -N_n, \ldots, -1,$$

where U_k is an i.i.d. sequence of $\mathcal{U}[0, 1)$ random variables, and W_k is a sequence of i.i.d. random variables uniformly distributed on V. Note that Lemma 6.4.15 directly implies that

$$X_k^{\xi_{\min}, -N_n} \preceq X_k^{\xi_{\max}, -N_n}.$$

Remark 6.4.16 Although $Y_k^{-N_n} = (X_k^{\xi_{\min}, -N_n}, X_k^{\xi_{\max}, -N_n})$ is a Markov chain, marginally neither $X_k^{\xi_{\min}, -N_n}$ nor $X_k^{\xi_{\max}, -N_n}$ are Markov chains, in general. Given $X_k^{\xi_{\min}, -N_n}$, the new state $X_{k+1}^{\xi_{\min}, -N_n}$ may differ only at v, but the probabilities of setting a value at v depend on the configuration $X_k^{\xi_{\max}, -N_n}$. $\quad\blacksquare$

Note, however, that once both coordinates meet, i.e., $X_k^{\xi_{\max}, -N_n} = X_k^{\xi_{\min}, -N_n}$, they stay together until $k = 0$. Moreover, from this point onward, each coordinate of $(X_k^{\xi_{\min}, -N_n}, X_k^{\xi_{\max}, -N_n})$ evolves according to φ given in (6.4.13). This is because

$$\tilde{\varphi}((\xi, \xi), v, u) = (\hat{\varphi}((\xi, \xi), v, u), \hat{\varphi}((\xi, \xi), v, u)),$$

and the function $\hat{\varphi}((\xi, \xi), v, u)$ is equivalent to the usual update function for the Gibbs sampler given in (6.3.10). Thus, each coordinate evolves according to a chain with stationary distribution π. The procedure is summarized in Algorithm 51.

Algorithm 51 Efficient coupling from the past for anti-monotone π

Require: State space $S = C^V$, an anti-monotone π
1: Set $n = 1$
2: Start a bivariate chain $Y_k^{-N_n} = (X_k^{\xi_{\min}, -N_n}, X_k^{\xi_{\max}, -N_n})$ at time $-N_n$ with $Y_{-N_n}^{-N_n} = (\xi_{\min}, \xi_{\max})$. Run the chain until time 0 using the update rule $\tilde{\varphi}$ given in (6.4.15) and i.i.d. random variables $U_{-N_n+1}, U_{-N_n+2}, \ldots, U_{-1}, U_0$ uniformly distributed on $[0, 1]$, and i.i.d. $V_{-N_n+1}, V_{-N_n+2}, \ldots, V_{-1}, V_0$ uniformly distributed on V.
3: **if** at time 0 we have $Y_0^{-N_n} = (\xi, \xi)$ **then return** ξ and **stop**.
4: **else** set $n = n + 1$ and go to Step 2 (for the new n, reuse previously generated U_i and V_i for $-N_n + 1 \le i \le 0$).
5: **end if**

The procedure given in Algorithm 51 is very similar to the "*Efficient coupling from the past*" presented in Algorithm 50. The processes $X_k^{\xi_{\min}, -N_n}$ and $X_k^{\xi_{\max}, -N_n}$ play the role of the two chains simulated in Algorithm 50. The main difference is that $X_k^{\xi_{\min}, -N_n}$ and $X_k^{\xi_{\max}, -N_n}$ are *not* Markov chains (see Remark 6.4.16), but due to Lemma 6.4.15, we still have $X_k^{\xi_{\min}, -N_n} \preceq X_k^{\xi_{\max}, -N_n}$ for all $k = -N_n, \ldots, 0$. We now prove that the output ξ resulting from Algorithm 51 has distribution π.

We may also apply CFTP for anti-monotone systems where not all configurations ξ from $S = C^V$ have positive probability π_ξ. For example, in the hard-core model, the configuration $\xi_{\max}$ assigning values 1 to all vertices is not feasible, i.e., $\pi_{\xi_{\max}} = 0$. Recall the conditional π distribution $p_{\xi,v}(c)$ given in (6.3.7):

$$p_{\xi,v}(c) = \mathbb{P}(Z(v) = c \mid Z_{-v} = \xi_{-v}), \quad c \in C,$$

where $Z \sim \pi$. For some ξ with $\pi_\xi = 0$, we still need to define the conditional distribution such that

$$\sum_{c \le c'} p_{\xi,v}(c) = \sum_{c \le c'} \mathbb{P}(Z(v) = c \mid Z_{-v} = \xi_{-v})$$

is increasing in ξ. Häggström and Nelander [74] suggest setting

$$f_\xi(c) = \max_{\substack{\xi' \preceq \xi \\ \pi_{\xi'} > 0}} \left\{ \sum_{c' \le c} \mathbb{P}(Z(v) = c' \mid Z_{-v} = \xi'_{-v}) \right\}$$

$$= \max_{\substack{\xi' \preceq \xi \\ \pi_{\xi'} > 0}} \left\{ \mathbb{P}(Z(v) \le c \mid Z_{-v} = \xi'_{-v}) \right\},$$

where the maximum is taken over all feasible configurations ξ' such that $\xi' \preceq \xi$.

Finally, we set

$$p_{\xi,v}(1) = f_\xi(1), \quad p_{\xi,v}(c) = f_\xi(c) - f_\xi(c-1), \quad c = 2, \ldots, r.$$

For the hard-core model, this extends the definitions in (6.3.16) and (6.3.17) to non-feasible configurations.

Proposition 6.4.17 *Assume that π is an anti-monotone distribution on $\mathcal{S} = C^V$, and either $\pi_{\xi_{\min}} > 0$ or $\pi_{\xi_{\max}} > 0$ (i.e., at least one of the configurations $\xi_{\min}$ or $\xi_{\max}$ is feasible). Then:*

(i) Algorithm 51 terminates with probability one.
(ii) The output of Algorithm 51 has distribution π.

Proof We separately show *(i)* and *(ii)*. The proof of assertion *(i)* is similar to the proof of Proposition 6.4.5, while the proof of assertion *(ii)* differs slightly from the proof of Theorem 6.4.2.

(i)   Denote by K the first time such that the chain Y started at $-N_K$ satisfies

$$Y_0^{-N_K} = (\xi, \xi),$$

i.e., Algorithm 51 terminates for the first time. Define $\tau_{\text{cftp}} = -N_K$. For simplicity, let

$$\tilde{\varphi}_i((\xi, \xi')) := \tilde{\varphi}((\xi, \xi'), W_i, U_i).$$

We can write, for $j < i \leq 0$:

$$Y_i^j(\xi, \xi') = \tilde{\varphi}(Y_{i-1}^j, W_i, U_i) = \tilde{\varphi}_i(Y_{i-1}^j)$$
$$= (\tilde{\varphi}_i \circ \tilde{\varphi}_{i-1} \circ \cdots \circ \tilde{\varphi}_{j+1})(\xi, \xi'),$$

here $Y_i^j(\xi, \xi')$ denotes the state of the bivariate chain at step i assuming that at step i it was (ξ, ξ'). Note that the random variable $Y_i^j(\xi, \xi')$ depends only on the difference $i - j$. For $l \geq j$, consider events:

$$A_l^j = \{\xi_0 = \xi_0', \text{ where } (\xi_0, \xi_0') = Y_l^j(\xi_{\min}, \xi_{\max})\}.$$

By Lemma 6.4.15, we always have with probability one:

$$\xi_0 \preceq \xi_0', \quad \text{where } (\xi_0, \xi_0') = Y_l^j(\xi_{\min}, \xi_{\max}).$$

Since the events A_0^L, A_L^{2L}, ... are independent and have positive probability, the second Borel-Cantelli lemma implies that infinitely many events $A_{kL}^{(k+1)L}$ occur with probability 1. Thus, τ_{cftp} is finite with probability 1, and the algorithm stops with probability 1.

(ii) Let π' denote the distribution of the output of Algorithm 51. From *(i)*, we know that for any $\varepsilon > 0$, we can pick K sufficiently large so that:

$$\mathbb{P}(\tau_{\text{cftp}} < -N_K) \leq \varepsilon.$$

Let $Z \sim \pi$ be chosen independently of everything else. Consider another bivariate chain $(Y_k'^{-N_K})$ starting at $-N_K$ in state (Z, Z):

$$Y_{-N_K}'^{-N_K} = (Z, Z),$$

evolving according to the update function (6.4.15). When $\xi' = \xi$, the update function for any $v \in V$ and $u \in [0, 1]$ satisfies:

$$\tilde{\varphi}((\xi, \xi), v, u) = (\hat{\varphi}((\xi, \xi), v, u), \hat{\varphi}((\xi, \xi), v, u)),$$

and $\hat{\varphi}((\xi, \xi), v, u)$ (given in (6.4.16)) is equivalent to $\varphi(\xi, v, u)$ from (6.4.13). Thus, both components of $Y_k'^{-N_K}$ are identical and marginally evolve as a Markov chain with stationary distribution π. Since we started with $Z \sim \pi$, we have $Y_0'^{-N_K} \sim \pi$, and the coupling inequality (6.4.2) gives:

$$d_{\text{TV}}(\pi, \pi') \leq \mathbb{P}(Y_0'^{-N_K} \neq Y_0^{\tau_{\text{cftp}}}) \leq \mathbb{P}(\tau_{\text{cftp}} < -N_K) \leq \varepsilon.$$

Since $\varepsilon > 0$ was arbitrary, it follows that $d_{\text{TV}}(\pi, \pi') = 0$, and thus $\pi' = \pi$.

$\square$

The situation described in the proof of Proposition 6.4.17 is illustrated in Fig. 6.9. This figure showcases the evolution of the bivariate chain under the anti-monotone setup. For comparison, refer to Fig. 6.6, which depicts the corresponding situation for monotone systems. The differences between the two cases highlight the adjustments necessary for handling anti-monotone distributions effectively in the CFTP algorithm.

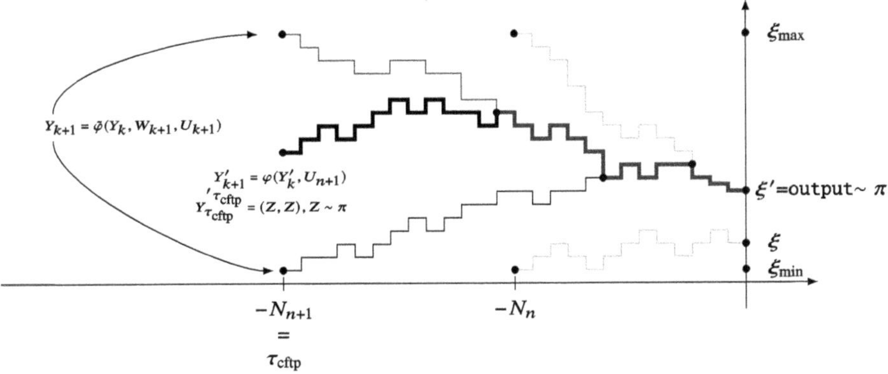

Fig. 6.9 Coupling from the past—schematic realization. See description in text

6.4.4.1   Anti-ferromagnet Ising Model

Recall the update function for the Ising model (6.3.22):

$$\varphi(\xi, v, u) = \begin{cases} \xi(v \to -1) & \text{if } u \in [0, \alpha(\xi, v)), \\ \xi(v \to +1) & \text{if } u \in [\alpha(\xi, v), 1]. \end{cases}$$

For any ξ and any v, we have $0 < \alpha(\xi, v) < 1$ (formula given in (6.3.21)). The update function for the bivariate chain $Y_k^{-N_n}$ is

$$\tilde{\varphi}((\xi, \xi'), v, u) = (\hat{\varphi}((\xi, \xi'), v, u), \hat{\varphi}((\xi', \xi), v, u)).$$

Explicitly:

$$\hat{\varphi}((\xi, \xi'), v, u) = \begin{cases} \xi(v \to -1) & \text{if } u \in \left[0, \alpha(\xi', v)\right), \\ \xi(v \to +1) & \text{if } u \in \left[\alpha(\xi', v), 1\right], \end{cases}$$

$$\hat{\varphi}((\xi', \xi), v, u) = \begin{cases} \xi'(v \to -1) & \text{if } u \in [0, \alpha(\xi, v)), \\ \xi'(v \to +1) & \text{if } u \in [\alpha(\xi, v), 1]. \end{cases}$$

Algorithm 52 presents the CFTP for the anti-monotone anti-ferromagnet (i.e., with parameter $\beta < 0$) Ising model. Recall that $\xi_{\min}$ assigns -1 to all vertices, and $\xi_{\max}$ assigns 1 to all vertices.

Algorithm 52 Efficient coupling from the past for anti-ferromagnet Ising model with parameter $\beta < 0$

Require: Graph $G = (V, E)$, parameter $\beta < 0$
1: Set $n = 1$
2: Set $Y_{-N_n}^{-N_n} = (\xi_{\min}, \xi_{\max})$
3: **for** $k = -N_n + 1$ to 0 **do**
4:     $(\xi_a, \xi_a') = Y_{k-1}^{-N_n}$
5:     Sample $U_k \sim \mathcal{U}[0, 1)$ and $W_k \sim \mathcal{U}(V)$ (if they were sampled earlier, reuse those previously used)
6:     Denote $v = W_k$
7:     **if** $U_k < \alpha(\xi_a', v)$ **then** $\xi_b = \xi_a(v \to -1)$ **else** $\xi_b = \xi_a(v \to +1)$
8:     **if** $U_k < \alpha(\xi_a, v)$ **then** $\xi_b' = \xi_a'(v \to -1)$ **else** $\xi_b' = \xi_a'(v \to +1)$
9:     Set $Y_k^{-N_n} = (\xi_b, \xi_b')$
10: **end for**
11: **if** $Y_0^{-N_n} = (\xi, \xi)$ **then return** ξ **else** set $n := n + 1$ and **goto** line 2

6.4.4.2 Generalized Hard-core Model Continued Again

The update function for the generalized hard-core model, $\varphi(\xi, v, u)$, takes a different form depending on whether, in configuration ξ, all neighbors of v have values 0, or if there exists a neighbor with value 1, as described in (6.3.18) and (6.3.19).

The function $\varphi(\xi, v, u)$ can be written compactly as:

$$
\varphi(\xi, v, u) = \begin{cases} \xi(v \to 0) & \text{if } u \in [0, \theta(\xi, v)), \\ \xi(v \to 1) & \text{if } u \in [\theta(\xi, v), 1], \end{cases} \tag{6.4.17}
$$

where

$$
\theta(\xi, v) = \begin{cases} \frac{1}{\lambda + 1} & \text{if } \forall (w \sim v)\, \xi(w) = 0, \\ 1 & \text{if } \exists (w \sim v)\, \xi(w) = 1. \end{cases}
$$

Roughly speaking, if $\theta(\xi, v) = 1$, then $\varphi(\xi, v, u) = \xi$ for all $u \in [0, 1]$. Recall that the update function of the bivariate chain $Y_k^{-N_n}$ is $\tilde{\varphi}((\xi, \xi'), v, u) = (\hat{\varphi}((\xi, \xi'), v, u), \hat{\varphi}((\xi', \xi), v, u))$. Explicitly, we have:

$$
\hat{\varphi}((\xi, \xi'), v, u) = \begin{cases} \xi(v \to 0) & \text{if } u \in \big[0, \theta(\xi', v)\big), \\ \xi(v \to 1) & \text{if } u \in \big[\theta(\xi', v), 1\big], \end{cases}
$$

$$
\hat{\varphi}((\xi', \xi), v, u) = \begin{cases} \xi'(v \to 0) & \text{if } u \in [0, \theta(\xi, v)), \\ \xi'(v \to 1) & \text{if } u \in [\theta(\xi, v), 1]. \end{cases}
$$

Algorithm 53 Efficient coupling from the past for anti-monotone generalized hard-core model

---

Require: Graph $G = (V, E)$, parameter $\lambda > 0$
1: Set $n = 1$
2: Initialize $Y_{-N_n}^{-N_n} = (\xi_{\min}, \xi_{\max})$
3: **for** $k = -N_n + 1$ to 0 **do**
4:     Set $(\xi_a, \xi_a') = Y_{k-1}^{-N_n}$
5:     Sample $U_k \sim \mathcal{U}[0, 1)$ and $W_k \sim \mathcal{U}(V)$ (reuse previously sampled values if available)
6:     Set $v = W_k$
7:     **if** $\forall (w \sim v)\ \xi_a'(w) = 0$
8:         **if** $U_k < \frac{\lambda}{\lambda+1}$ **then** $\xi_b = \xi_a(v \to 0)$ **else** $\xi_b = \xi_a(v \to 1)$
9:     **else** set $\xi_b = \xi_a$
10:
11:     **if** $\forall (w \sim v)\ \xi_a(w) = 0$
12:         **if** $U_k < \frac{\lambda}{\lambda+1}$ **then** $\xi_b' = \xi_a'(v \to 0)$ **else** $\xi_b' = \xi_a'(v \to 1)$
13:     **else** set $\xi_b' = \xi_a'$
14:
15:     Set $Y_k^{-N_n} = (\xi_b, \xi_b')$
16: **end for**
17: **if** $Y_0^{-N_n} = (\xi, \xi)$ **then return** ξ **else** set $n := n + 1$ and **goto** line 2

Algorithm 53 presents the CFTP for the anti-monotone generalized hard-core model. Recall that $\xi_{\min}$ assigns 0 to all vertices, and $\xi_{\max}$ assigns 1 to all vertices.

6.5 Approximate Counting

How many proper colorings does a given graph have? In how many ways can we pack items into a knapsack so that the total maximal (prescribed) weight is not exceeded? How many possible hard-core configurations does it have? These are examples of questions that are of great importance (in computer science, physics, biology, and more). It turns out that these types of problems are *difficult*—we cannot provide an exact answer in a reasonable time and must resort to some kind of *approximation*.

The setup in this chapter (which can apply to more general cases) is similar to the one we had with the Gibbs sampler. We are given a graph $G = (V, E)$ and a set $C = \{1, \ldots, r\}$. We assign values from C to each vertex, and the set of all configurations is thus C^V. We are also given a set $\Lambda \subset C^V$ of *feasible configurations*. This set is usually not provided as an enumerated list. Instead, Λ is typically defined through a neighboring system determined by edges (e.g., the set of proper colorings of a graph, the set of possible hard-core configurations, etc.) or some equation (e.g., in the knapsack problem). In previous sections, we provided methods for *approximate* (and, in the case of coupling from the past, exact) sampling from the uniform distribution on Λ. It turns out that if such sampling is *rapid* (mixing time $\tau(\kappa)$ is polynomial in the problem size and $1/\kappa$), then it may be possible to provide an estimator $\hat{z}_G$ of $z_G = |\Lambda|$ such that the relative error is bounded:

$$\mathbb{P}(|\hat{z}_G - z_G| \leq \varepsilon z_G) \geq 1 - \delta,$$

for given $\varepsilon \in (0, 1)$ and $\delta < 1/2$. Moreover, the procedure can yield $\hat{z}_G$ in a number of steps that is polynomial in $|G|$, $1/\varepsilon$, and $\log(1/\delta)$.

Before proceeding, let us see some specific examples in more detail:

- **Number of proper graph colorings.** Given a graph $G = (V, E)$, the set of proper graph colorings (feasible configurations) Λ is the set of assignments of colors $C = \{1, \ldots, r\}$ to vertices such that two neighboring vertices have different colors. The task is to determine the total number of proper graph colorings, i.e., $z_G = |\Lambda|$.
- **Number of feasible knapsack solutions.** We are given the weights of d items $\mathbf{w} = (w_1, \ldots, w_d)$ and a total maximal weight W allowed in the knapsack. Let $\mathbf{x} = (x_1, \ldots, x_d)$, where $x_i \in \{0, 1\}$, represent a configuration—if $x_i = 1$, it means item i is included in the knapsack. A configuration $\mathbf{x}$ is *feasible* if the total weight does not exceed W, i.e., $\sum_{j=1}^{d} w_j x_j \leq W$. We can interpret this as a graph $G = (V, E)$, where $V = \{1, \ldots, d\}$ and $C = \{0, 1\}$ (edges are irrelevant in this example). Specifically, the set of all feasible configurations is:

$$\Lambda = \left\{ \mathbf{x} \in \{0, 1\}^d : \sum_{j=1}^{d} w_j x_j \leq W \right\}.$$

The task is to determine the total number of feasible configurations, i.e., $z_G = |\Lambda|$.

The classical knapsack problem, discussed later in Sect. 7.1, aims at maximizing the total *value* of selected items. Here, in contrast, the present task is to count the number of feasible configurations.

- **Number of possible hard-core configurations (independent sets).** We have two equivalent formulations of this problem:

  - Given a graph $G = (V, E)$, what is the total number z_G of feasible hard-core configurations, i.e., what is the constant z_G in (6.3.13)?
  - A subset $F \subset V$ is an *independent set* if no two vertices in F share an edge. This is another formulation of the hard-core model, where the set of vertices with token 1 is independent. We denote by z_G the total number of independent sets.

In this section, we will focus on the last problem (using the terminology of *independent sets*). We will then identify the crucial aspects of this problem and outline how to adapt the approach to graph coloring and the knapsack problem.

6.5.1   *Independent Sets*

Throughout this section, let us denote $|V|$ by M and assume that the ratios $|V|/|E|$ and $|E|/|V|$ are both bounded, i.e., as the graph size grows, the number of vertices and edges scale proportionally. Sometimes this size is denoted by $|G|$, referring to either the number of vertices or edges. Since we assume $\deg(v) \leq \Delta$ for all $v \in V$, these measures are equivalent asymptotically (as $|G|$ grows while Δ remains fixed). Of course, we have $0 \leq z_G \leq 2^M$, because 2^M is the total number of subsets of V, and if E is empty, each subset is independent.

Finding the exact value of z_G in polynomial time is a *hard* problem—more precisely, it is **#P**-complete.[2] However, we can seek approximations using randomized algorithms.

6.5.1.1   Naïve Approximation

A **naïve approximation** algorithm could proceed as follows: Simulate R random subsets of V uniformly. The probability of successfully drawing an independent set is $p = z_G/2^M$. To generate a random subset of V, enumerate the vertices as $V = \{v_1, v_2, \ldots, v_M\}$ and include each vertex with probability $1/2$. This ensures that each subset arises with probability $1/2^M$.

Define $\hat{Y}_R = \frac{1}{R}\sum_{i=1}^{R} Y_i$, where $Y_i = 1$ if the i-th subset is an independent set, and $Y_i = 0$ otherwise. Checking whether a subset is independent is straightforward. Then:

$$\mathbb{E}(\hat{Y}_R) = \frac{z_G}{2^M}.$$

Thus, an unbiased estimator of z_G is:

$$\hat{z}_G = 2^M \hat{Y}_R. \tag{6.5.1}$$

We could say that it is a **naïve** (or **crude**) estimator of z_G. However, this estimator has some limitations, which we will now point out. Let us consider what we would ideally like from such an estimator. We want the probability that $\hat{z}_G$ differs from the true value z_G by some small amount, say ε, to be large, say $1 - \delta$, where $\varepsilon \in (0, 1)$ and $\delta \in (0, 1/2)$. Rephrasing, we want the probability of a large difference to be small. Since we are dealing with relative error, this requirement can be expressed as follows:

$$\mathbb{P}(|\hat{z}_G - z_G| > \varepsilon z_G) \leq \delta, \quad \text{i.e.,} \quad \mathbb{P}(|\hat{z}_G - z_G| \leq \varepsilon z_G) \geq 1 - \delta. \tag{6.5.2}$$

---

[2] See https://en.wikipedia.org/wiki/#P-complete.

Equivalently, this can be written as:

$$\mathbb{P}(z_G(1 - \varepsilon) \leq \hat{z}_G \leq z_G(1 + \varepsilon)) \geq 1 - \delta.$$

We call an estimator $\hat{z}_G$ that satisfies (6.5.2) an (ε, δ)-**approximation scheme** for z_G.

We indeed have the following result:

Lemma 6.5.1 *A naïve estimator $\hat{z}_G$ of z_G, as defined in (6.5.1), is an (ε, δ)-approximation scheme for any $\varepsilon \in (0, 1)$ and $\delta \in (1/2, 1)$.*

Proof Recall that $\hat{z}_G = 2^M \hat{Y}_R$, where $Y_i = 1$ if the i-th simulation (randomly generated subset of V) is an independent set, and $Y_i = 0$ otherwise. Using Corollary 4.4.2 (for i.i.d. indicator random variables $Y_1, \ldots, Y_R$ with mean p, we have $\mathbb{P}\left(\left|\hat{Y}_R - p\right| \geq \varepsilon p\right) \leq \delta$ for $R \geq \frac{3\ln(2/\delta)}{\varepsilon^2 p}$), we observe the following (with $p = \mathbb{E}(Y_i) = z_G/2^M$):

$$\mathbb{P}\left(|\hat{z}_G - z_G| \geq \varepsilon z_G\right) = \mathbb{P}\left(\left|2^M \hat{Y}_R - z_G\right| \geq \varepsilon z_G\right) = \mathbb{P}\left(\left|\hat{Y}_R - \frac{z_G}{2^M}\right| \geq \varepsilon \frac{z_G}{2^M}\right) \leq \delta,$$

i.e., (6.5.2) holds. $\qquad\qquad\square$

A **problem** with the naïve estimator $\hat{z}_G$ is the following: in many cases, the probability p (of obtaining a feasible configuration by assigning values randomly to vertices) is very small. If we suppose that p is exponentially small (in terms of the graph size $M = |G|$), then the required number of samples R becomes exponentially large.

For example, in the case of independent sets, let us assume that the graph is Δ-regular. As shown in [218], there are at most $(2^{\Delta+1} - 1)^{M/(2\Delta)}$ independent sets. By bounding further, we have:

$$z_G \leq (2^{\Delta+1} - 1)^{M/(2\Delta)} \leq 2^{\frac{M(\Delta+1)}{2\Delta}} \leq 2^{\frac{3M}{4}}.$$

The probability that a randomly chosen subset of V is an independent set is therefore:

$$p = \frac{z_G}{2^M} \leq 2^{-\frac{M}{4}}.$$

To make the naïve estimator $\hat{z}_G$ an (ε, δ)-approximation scheme, we would need:

$$R \geq \frac{3\ln\frac{2}{\delta}}{\varepsilon^2 p} \geq \frac{3\ln\frac{2}{\delta}}{\varepsilon^2 2^{-\frac{M}{4}}} = \frac{3\ln\frac{2}{\delta}}{\varepsilon^2} 2^{\frac{M}{4}},$$

which means exponentially many steps are required.

Summing up, while being an (ε, δ)-approximation scheme is a desirable property, it is not sufficient in practice. We also want to compute $\hat{z}_G$ *rapidly*. This motivates the Definition 6.5.2 below. ·

6.5.1.2   Fully Polynomial Randomized Approximation Scheme (FPRASC)

To address these limitations, we introduce a stronger notion of efficiency:

Definition 6.5.2 *We say that $\hat{z}_G$ is an (ε, δ)-**FPRASC** (**fully polynomial randomized approximation scheme**) for z_G if it is an (ε, δ)-approximation scheme and its running time is polynomial in $|G|$, $1/\varepsilon$, and $\log(1/\delta)$.*

Whether a polynomial running time such as m^{100} can be considered *rapid* is debatable, but we adopt this definition for formal purposes. In practice, however, when we encounter an FPRASC, the degree of the polynomial is typically relatively small (e.g., a single-digit number).

The aim of this section is to present an (ε, δ)-FPRASC for independent sets. The core tool for such an approximate counting algorithm is the ability to sample from an approximate uniform distribution $\pi_I = 1/z_G$ for any independent set I. This aligns with the MCMC framework. However, since we run the chain for only finitely many steps, it is essential to quantify *how far* the resulting distribution is from the uniform distribution. We measure this in terms of total variation distance. More precisely, recall the definition of mixing time (6.2.2):

$$\tau(\kappa) = \min_{n \in \mathbb{N}} \left\{ d_{\mathrm{TV}}(\boldsymbol{\mu}\mathbf{P}^n, \boldsymbol{\pi}) \leq \kappa \right\},$$

i.e., the number of steps required to make the total variation distance smaller than κ.

Definition 6.5.3 *Let (X_n) be a Markov chain with stationary distribution π. We say that it is **rapidly mixing** if the mixing time $\tau(\kappa)$ is polynomial in the problem size (e.g., $|G|$ in our examples) and in $\log(1/\kappa)$.*

Rapidly mixing Markov chains are closely related to approximate counting. Beyond the case of counting independent sets, the presented scheme applies to calculating z_G for other models. Later, we will demonstrate its application to approximating the number of proper graph colorings and the number of feasible knapsack solutions.

Suppose $\xi \in \Lambda$ is a feasible configuration. Below, we show that once we have a rapidly mixing Markov chain with uniform stationary distribution $\pi_\xi = 1/z_G$, we also have an FPRASC for z_G. Interestingly, the reverse is also true: given an FPRASC for z_G, we can approximately sample from π in polynomial time. The full relationship between exact and approximate counting and sampling was first established by Jerrum, Valiant, and Vazirani [89] for a class of problems that are self-reducible in the sense of Schnorr. This relationship can be summarized as follows (assuming all operations occur in polynomial time):

$$\begin{array}{ccc}
\text{Computing } z_G \text{ exactly} & \Rightarrow & \text{Sampling from } \pi_\xi = 1/z_G \text{ exactly} \\
\Downarrow & & \Downarrow \\
\text{Computing } z_G \text{ approximately} & \Leftrightarrow & \text{Sampling from } \pi_\xi = 1/z_G \text{ approximately}
\end{array}$$

The proofs of these relationships are conceptually similar to those in [153]. However, we present them differently, with some constants and bounds modified. Additionally, in Sect. 6.5.2, we outline a general procedure for approximate counting in other problems.

Proposition 6.5.4 *Let (X_n) be a rapidly mixing Markov chain with uniform stationary distribution $\pi_\xi = 1/z_G$ on the independent sets Λ of G, with mixing time $\tau(\kappa)$. Then there exists an (ε, δ)-FPRASC for z_G with a running time of:*

$$\frac{200m^3}{\varepsilon^2} \ln\left(\frac{2m}{\delta}\right) \tau\left(\frac{\varepsilon}{16m}\right).$$

The idea is to sample from different domains, ensuring that p (from Corollary 4.4.2) is large. Let $m = |E|$, and let $e_1, \ldots, e_m$ be the edges in some fixed enumeration. Recall $M = |V|$. Consider graphs $G_i = (V, E_i)$, where $E_i = \{e_1, \ldots, e_i\}$. Let $G = G_m$ and $G_0 = (V, \emptyset)$. Denote by Λ_{G_i} the family of all possible independent sets (feasible configurations) of G_i, and let $z_{G_i} \equiv z_i = |\Lambda_{G_i}|$ denote their count. We can decompose:

$$z_G = \frac{z_m}{z_{m-1}} \times \frac{z_{m-1}}{z_{m-2}} \times \cdots \times \frac{z_1}{z_0} \times z_0. \tag{6.5.3}$$

Defining $\varrho_i = \frac{z_i}{z_{i-1}}$ and noting that $z_0 = 2^M$, we have:

$$z_G = 2^M \prod_{i=1}^{m} \varrho_i.$$

Our procedure proceeds in two steps:

1. We first construct an (ε, δ)-FPRASC for z_G, assuming that we have an $\left(\frac{\varepsilon}{2m}, \frac{\delta}{m}\right)$-FPRASC for each ϱ_i.
2. We then construct an $\left(\frac{\varepsilon}{2m}, \frac{\delta}{m}\right)$-FPRASC for each ϱ_i.

The procedure for estimating each ϱ_i is detailed in Algorithm 54.

Using Algorithm 54, an (ε, δ)-FPRASC for z_G is detailed in Algorithm 55.

Algorithm 54 $\left(\frac{\varepsilon}{2m}, \frac{\delta}{m}\right)$-FPRASC for ϱ_i

---

Require: Graphs $G_{i-1} = (V, E_{i-1})$ and $G_i = (V, E_i)$; parameters δ, ε, $m = |E|$; a rapidly mixing Markov chain $(X_n^{(i)})$ with uniform stationary distribution $\pi^{(i)}$ on Λ_{G_i} and mixing time $\tau(\cdot)$.

1: **for** $j = 1$ to $R = \frac{200m^2}{\varepsilon^2} \ln\left(\frac{2m}{\delta}\right)$ **do**
2:     Generate approximate sample from the uniform distribution on $\Lambda_{G_{i-1}}$:
3:     Run $(X_n^{(i-1)})$ for $\tau\left(\frac{\varepsilon}{16m}\right)$ steps; denote the resulting state as ξ.
4:     **if** $\xi \in \Lambda_{G_i}$ **then** set $\tilde{Y}_j^{(i)} = 1$ **else** set $\tilde{Y}_j^{(i)} = 0$.
5: **end for**
6: **return** $\tilde{\varrho}_i = \frac{1}{R} \sum_{j=1}^{R} \tilde{Y}_j^{(i)}$.

Algorithm 55 (ε, δ)-FPRASC for z_G

Require: Graphs $G_i = (V, E_i)$, $E_i = \{e_1, \ldots, e_i\}$; parameters δ, ε, $m = |E|$.
1:
2: **for** $i = 1$ to m **do**
3:     Estimate ϱ_i via $\tilde{\varrho}_i$ using Algorithm 54.
4: **end for**
5: **return** $\hat{z}_G = 2^M \prod_{i=1}^{m} \tilde{\varrho}_i$.

To complete the procedure, we prove the following two lemmas:

Lemma 6.5.5 *Algorithm 54 is an* $\left(\frac{\varepsilon}{2m}, \frac{\delta}{m}\right)$*-FPRASC for* ϱ_i.

Lemma 6.5.6 *Algorithm 55 is an* (ε, δ)*-FPRASC for* z_G.

We start with the proof of Lemma 6.5.6.

Proof of Lemma 6.5.6 Assume that Algorithm 54 is an $\left(\frac{\varepsilon}{2m}, \frac{\delta}{m}\right)$-FPRASC for ϱ_i for any $i = 1, \ldots, m$, i.e.,

$$\mathbb{P}\left(|\tilde{\varrho}_i - \varrho_i| > \frac{\varepsilon}{2m}\varrho_i\right) \le \frac{\delta}{m}.$$

Denote the event $\left\{|\tilde{\varrho}_i - \varrho_i| > \frac{\varepsilon}{2m}\varrho_i\right\}$ by A_i. By the union bound:

$$\mathbb{P}\left(\bigcup_{i=1}^{m} A_i\right) \le \sum_{i=1}^{m} \mathbb{P}(A_i) \le \sum_{i=1}^{m} \frac{\delta}{m} = \delta.$$

Equivalently:

$$1 - \mathbb{P}\left(\bigcup_{i=1}^{m} A_i\right) = \mathbb{P}\left(\bigcap_{i=1}^{m} A_i^c\right) \ge 1 - \delta.$$

This implies that $|\tilde{\varrho}_i - \varrho_i| \leq \frac{\varepsilon}{2m}\varrho_i$ for all $i = 1, \ldots, m$, i.e.,

$$\mathbb{P}\left(\forall i \in \{1, \ldots, m\} : \ 1 - \frac{\varepsilon}{2m} \leq \frac{\tilde{\varrho}_i}{\varrho_i} \leq 1 + \frac{\varepsilon}{2m}\right) \geq 1 - \delta.$$

By Exercise 6.T.22, this yields:

$$\mathbb{P}\left(1 - \varepsilon \leq \prod_{i=1}^{m} \frac{\tilde{\varrho}_i}{\varrho_i} \leq 1 + \varepsilon\right) \geq 1 - \delta.$$

Concluding:

$$\mathbb{P}\left((1 - \varepsilon)2^M \prod_{i=1}^{m} \varrho_i \leq 2^M \prod_{i=1}^{m} \tilde{\varrho}_i \leq (1 + \varepsilon)2^M \prod_{i=1}^{m} \varrho_i\right) \geq 1 - \delta,$$

which implies that $\hat{z}_G = 2^M \prod_{i=1}^{m} \tilde{\varrho}_i$, as given in Algorithm 55, is an (ε, δ)-approximation for $z_G = 2^M \prod_{i=1}^{m} \varrho_i$.

Finally, since Algorithm 54 is an $\left(\frac{\varepsilon}{2m}, \frac{\delta}{m}\right)$-FPRASC for ϱ_i, Algorithm 55 is an (ε, δ)-FPRASC for z_G (its running time is m times the running time of Algorithm 54). $\qquad\square$

The harder part is to prove Lemma 6.5.5. Before proceeding to its proof, let us first show that $\varrho_i \geq 1/2$.

Lemma 6.5.7 *We have $\varrho_l = \frac{z_i}{z_{i-1}} \geq \frac{1}{2}$.*

Proof Assume that $(w, v) \in G_i$ but $(w, v) \notin G_{i-1}$, i.e., the graphs differ only in the edge (w, v). Any independent set in Λ_{G_i} is also an independent set in $\Lambda_{G_{i-1}}$, meaning $\Lambda_{G_i} \subseteq \Lambda_{G_{i-1}}$.

Now, consider any $\xi \in \Lambda_{G_{i-1}} \setminus \Lambda_{G_i}$, which contains both w and v. For each such ξ, we can associate an independent set $\xi \setminus \{v\} \in \Lambda_{G_i}$. In this mapping, each independent set $\xi' \in \Lambda_{G_i}$ is associated with no more than one independent set $\xi \cup \{v\} \in \Lambda_{G_{i-1}} \setminus \Lambda_{G_i}$. Thus, we have:

$$|\Lambda_{G_{i-1}} \setminus \Lambda_{G_i}| \leq |\Lambda_{G_i}|.$$

This implies:

$$\varrho_i = \frac{z_i}{z_{i-1}} = \frac{|\Lambda_{G_i}|}{|\Lambda_{G_i}| + |\Lambda_{G_{i-1}} \setminus \Lambda_{G_i}|} \geq \frac{1}{2}.$$

See Fig. 6.10 for an example. $\qquad\square$

Fig. 6.10 Example: $E = \{v_1, v_2, v_3, v_4, v_5, v_6\}$, $\Lambda_{G_5} \subseteq \Lambda_{G_4}$, $w = v_4$, $v = v_5$

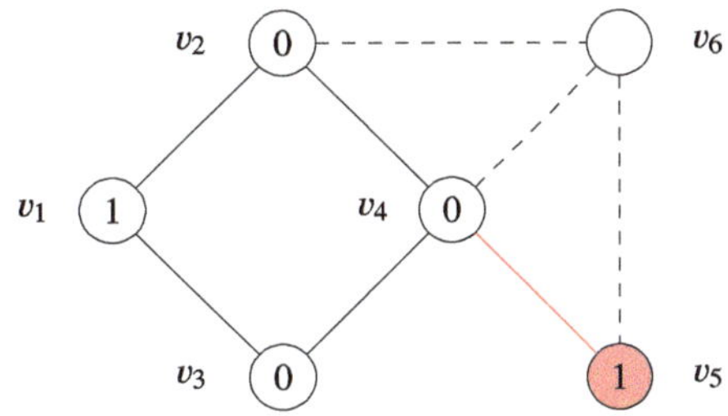

We also need the following lemma:

Lemma 6.5.8 *Let $x, y \in (0, 1)$ and assume that $\mathbb{P}(1 - x \le X \le 1 + x) \ge 1 - y$. Then, for $a, b \in (0, 1)$ such that $1 - a \le b \le 1 + a$, we have:*

$$\mathbb{P}((1 - x)(1 - a) \le bX \le (1 + x)(1 + a)) \ge 1 - y.$$

Proof We note:

$$\mathbb{P}((1 - x)(1 - a) \le bX \le (1 + x)(1 + a))$$
$$= \mathbb{P}\left(\frac{1 - a}{b} \cdot (1 - x) \le X \le \frac{1 + a}{b} \cdot (1 + x)\right).$$

Since $\frac{1+a}{b} \ge 1$ and $\frac{1-a}{b} \le 1$, the assertion follows. $\square$

We are now ready to prove Lemma 6.5.5.

Proof of Lemma 6.5.5 From the assumption, a Markov chain $(X_n^{(i-1)})$ is rapidly mixing with mixing time $\tau(\kappa)$. Let

$$\kappa = \frac{\varepsilon}{16m}.$$

Let $\tilde{\pi}^{(i-1)}$ be the distribution of $(X_n^{(i-1)})$ at step $\tau(\kappa)$. Recall, $\pi^{(i-1)}$ is the uniform distribution on $\Lambda_{G_{i-1}}$, i.e., $\pi_\xi^{(i-1)} = 1/z_{i-1}$. We thus have

$$d_{\mathrm{TV}}(\tilde{\pi}^{(i-1)}, \pi^{(i-1)}) = \frac{1}{2} \sum_{\xi \in \Lambda_{G_{i-1}}} \left|\tilde{\pi}_\xi^{(i-1)} - \frac{1}{z_{i-1}}\right| \le \kappa. \tag{6.5.4}$$

Note that in particular it implies that for any $\xi \in \Lambda_{G_{i-1}}$ we have

$$\left|\tilde{\pi}_\xi^{(i-1)} - \frac{1}{z_{i-1}}\right| \le 2\kappa. \tag{6.5.5}$$

Consider $j = 1, \ldots, R$. Let $Y_j^{(i)}$ be the indicator variable: $Y_j^{(i)} = 1$ if ξ chosen from $\pi^{(i-1)}$ (uniformly from $\Lambda_{G_{i-1}}$) is an independent set in G_i (i.e., $\xi \in \Lambda_{G_i}$), $Y_j^{(i)} = 0$ otherwise. Similarly, let $\tilde{Y}_j^{(i)} = 1$ if ξ chosen from $\tilde{\pi}^{(i-1)}$ (approximately uniformly from $\Lambda_{G_{i-1}}$) is an independent set in G_i, $\tilde{Y}_j^{(i)} = 0$ otherwise—as defined in line 4 of Algorithm 54. We have $\mathbb{P}(Y_j^{(i)} = 1) = \varrho_i$ and $\mathbb{P}(\tilde{Y}_j^{(i)} = 1) = \tilde{\varrho}_i$. Note that

$$|\mathbb{E}\tilde{Y}_j^{(i)} - \mathbb{E}Y_j^{(i)}| = \sum_{\xi \in \Lambda_{G_{i-1}}} |\tilde{\pi}_\xi^{(i-1)} - \pi_\xi^{(i-1)}| \cdot 1(\xi \in \Lambda_{G_i}) \leq 2\kappa,$$

where the last inequality follows from (6.5.4).

Since $\mathbb{P}(Y_j^{(i)} = 1) = \varrho_i$ and $\mathbb{P}(\tilde{Y}_j^{(i)} = 1) = \mathbb{E}\tilde{Y}_j^{(i)} = \mathbb{E}\frac{1}{R}\sum_{j=1}^R \tilde{Y}_j^{(i)} = \mathbb{E}\tilde{\varrho}_i$, we also have

$$|\mathbb{E}\tilde{\varrho}_i - \varrho_i| \leq 2\kappa. \tag{6.5.6}$$

Rewriting it, we have

$$\varrho_i - 2\kappa \leq \mathbb{E}\tilde{\varrho}_i \leq \varrho_i + 2\kappa,$$

$$1 - \frac{2\kappa}{\varrho_i} \leq \frac{\mathbb{E}\tilde{\varrho}_i}{\varrho_i} \leq 1 + \frac{2\kappa}{\varrho_i}.$$

Exploiting fact that $\varrho_i \geq 1/2$ (Lemma 6.5.7) we have (recall, $\kappa = \varepsilon/(16m)$)

$$1 - \frac{\varepsilon}{4m} = 1 - 4\kappa \leq \frac{\mathbb{E}\tilde{\varrho}_i}{\varrho_i} \leq 1 + 4\kappa = 1 + \frac{\varepsilon}{4m}. \tag{6.5.7}$$

Consequently, the r.h.s. of (6.5.7)—using again $\varrho_i \geq 1/2$—implies

$$\mathbb{E}\tilde{\varrho}_i \geq \varrho_i(1 - 4\kappa) \geq \frac{1}{2} - 2\kappa = \frac{1}{2} - \frac{\varepsilon}{8m} \geq \frac{1}{2} - \frac{1}{8} = \frac{3}{8}.$$

Applying Corollary 4.4.2 to $\tilde{\varrho}_i = \frac{1}{R}\sum_{i=1}^R \tilde{Y}_j^{(i)}$, we have that for

$$R \geq \frac{3\ln\frac{2}{(\delta/m)}}{(\varepsilon/(5m))^2(3/8)} = \frac{200m^2}{\varepsilon^2}\ln\frac{2m}{\delta}$$

we have

$$\mathbb{P}\left(|\tilde{\varrho}_i - \mathbb{E}\tilde{\varrho}_i| \geq \frac{\varepsilon}{5m}\mathbb{E}\tilde{\varrho}_i\right) = \mathbb{P}\left(\left|\frac{\tilde{\varrho}_i}{\mathbb{E}\tilde{\varrho}_i} - 1\right| \geq \frac{\varepsilon}{5m}\right) \leq \frac{\delta}{m},$$

i.e.,

$$\mathbb{P}\left(1 - \frac{\varepsilon}{5m} \le \frac{\tilde{\varrho}_i}{\mathbb{E}\tilde{\varrho}_i} \le 1 + \frac{\varepsilon}{5m}\right) \ge 1 - \frac{\delta}{m}. \tag{6.5.8}$$

Now, applying Lemma 6.5.8 to (6.5.8) and (6.5.7), we have

$$\mathbb{P}\left(\left(1 - \frac{\varepsilon}{5m}\right)\left(1 - \frac{\varepsilon}{4m}\right) \le \frac{\tilde{\varrho}_i}{\mathbb{E}\tilde{\varrho}_i}\,\frac{\mathbb{E}\tilde{\varrho}_i}{\varrho_i} \le \left(1 + \frac{\varepsilon}{5m}\right)\left(1 + \frac{\varepsilon}{4m}\right)\right) \ge 1 - \frac{\delta}{m}.$$

Now note that

$$\left(1 + \frac{\varepsilon}{5m}\right)\left(1 + \frac{\varepsilon}{4m}\right) = 1 + \frac{\varepsilon}{5m} + \frac{\varepsilon}{4m} + \frac{\varepsilon^2}{20m^2} \le 1 + \frac{4\varepsilon + 5\varepsilon + \varepsilon}{20m} = 1 + \frac{\varepsilon}{2m},$$

$$\left(1 - \frac{\varepsilon}{5m}\right)\left(1 - \frac{\varepsilon}{4m}\right) = 1 - \frac{\varepsilon}{5m} - \frac{\varepsilon}{4m} + \frac{\varepsilon^2}{20m^2} \ge 1 - \frac{4\varepsilon + 5\varepsilon + \varepsilon}{20m} = 1 - \frac{\varepsilon}{2m},$$

and consequently

$$\mathbb{P}\left(1 - \frac{\varepsilon}{2m} \le \frac{\tilde{\varrho}_i}{\varrho_i} \le 1 + \frac{\varepsilon}{2m}\right) \ge 1 - \frac{\delta}{m}.$$

Equivalently,

$$\mathbb{P}\left(\left(1 - \frac{\varepsilon}{2m}\right)\varrho_i \le \tilde{\varrho}_i \le \left(1 + \frac{\varepsilon}{2m}\right)\varrho_i\right) \ge 1 - \frac{\delta}{m},$$

i.e., $\tilde{\varrho}_i$ is a $\left(\frac{\varepsilon}{2m}, \frac{\delta}{m}\right)$-approximation scheme for ϱ_i. The number of replications R is polynomial in m and $1/\varepsilon$ as well as in $\ln(1/\delta)$, the mixing time $\tau(\varepsilon/(16m))$ is also polynomial (since it is rapidly mixing), thus, the total running time

$$R\tau\left(\frac{\varepsilon}{16m}\right)$$

is also polynomial, which means that the whole procedure is $\left(\frac{\varepsilon}{2m}, \frac{\delta}{m}\right)$-FPRASC.
$\square$

To summarize, we present the proof of Proposition 6.5.4.

Proof of Proposition 6.5.4 In Lemma 6.5.6, we demonstrated the construction of an (ε, δ)-FPRASC for z_G. What remains is to analyze its total running time. Algorithm 55 invokes Algorithm 54 m times. Therefore, the total running time is given by:

$$mR\tau\left(\frac{\varepsilon}{16m}\right) = \frac{200m^3}{\varepsilon^2}\ln\left(\frac{2m}{\delta}\right)\tau\left(\frac{\varepsilon}{16m}\right).$$

$\square$

Studying the mixing time of Markov chains is beyond the scope of this book. However, it has been shown in [44] that the Gibbs sampler for the uniform distribution on independent sets, as given in Algorithm 47, is rapidly mixing for $\Delta \leq 4$. Since the formula for its mixing time is quite complex (requiring the combination of results from Theorems 5.5 and 3.1 in [44]), we consider a slight modification of the algorithm, for which a more explicit formula for the mixing time is available. This modified algorithm works similarly to the classical Gibbs sampler but includes an additional operation often referred to as a *drag* move.

At step k, a vertex $v \in V$ is chosen uniformly at random. If $X_k(v) = 0$ and exactly one neighbor of v, say w, has a value of 1 (i.e., $X_k(w) = 1$), then with probability $1/8$, the values are updated as $X_k(v) = 1$ and $X_k(w) = 0$. The procedure is described in Algorithm 56. In Exercise 6.T.19, readers are asked to prove that the corresponding Markov chain is ergodic and has a uniform stationary distribution over all independent sets.

Algorithm 56 Sampler for the hard-core model on graph G (one step)

Require: Graph G with a given family of neighborhoods $\mathcal{N}_G(v)$
1: Choose a vertex $v \in V$ uniformly at random
2: **if** $X_k(v) = 1$ **then** set $X_{k+1}(v) = 0$ with probability $1/2$
3: **if** $(X_k(v) = 0$ and $\forall w \in \mathcal{N}(v), X_k(w) = 0)$ **then** set $X_{k+1}(v) = 1$ with probability $1/2$
4: **if** $(X_k(v) = 0$ and $\exists! w \in \mathcal{N}(v), X_k(w) = 1)$ **then** set $X_{k+1}(v) = 1$, $X_{k+1}(w) = 0$ with probability $1/8$
5: Leave the values of all other vertices $v' \neq v$ unchanged

---

From Theorem 3.1 in [44], we can state the following:

Lemma 6.5.9 *Let G be a graph with maximum degree $\Delta = 4$ and $M = |V|$ vertices. The Markov chain described in Algorithm 56 is rapidly mixing, and the mixing time $\tau(\kappa)$ satisfies:*

$$\tau(\kappa) \leq 12M^2(\ln M + 1)\ln\left(\frac{1}{\kappa}\right).$$

Using Algorithm 56 within Algorithm 54, along with Proposition 6.5.4 and Lemma 6.5.9, we conclude the following:

Lemma 6.5.10 *Let $G = (V, E)$ be a graph with maximum degree $\Delta = 4$, $m = |E|$ edges, and $M = |V|$ vertices. There exists an (ε, δ)-FPRASC for the number of independent sets z_G, with running time:*

$$\frac{200m^3}{\varepsilon^2}\ln\left(\frac{2m}{\delta}\right)\tau\left(\frac{\varepsilon}{16m}\right)$$

$$\leq \frac{2400 m^3 M^2}{\varepsilon^2} \ln\left(\frac{2m}{\delta}\right) (\ln M + 1) \ln\left(\frac{16m}{\varepsilon}\right).$$

$\square$

Although the restriction $\Delta = 4$ may seem limiting, this case still includes many interesting models, such as graphs representing a torus or a lattice (see, e.g., Fig. 6.17).

Boosting (ε, δ)-FPRASC Once we have an (ε, δ)-FPRASC for some $\delta \in (0, 1/2)$, we can relatively easily construct an (ε, δ')-FPRASC with $\delta' < \delta$. This property is why the definition is often stated as an (ε)-FPRASC, with the implicit assumption that $\delta = 1/4$. We now formalize this result:

Lemma 6.5.11 *Assume that the procedure for computing $\hat{z}_G$ is an $(\varepsilon, 1/4)$-FPRASC for z_G. Fix $\delta' < 1/4$. Run the procedure*

$$R = 24 \log(1/\delta')$$

times, obtaining $\hat{z}_G^1, \ldots, \hat{z}_G^R$, and define

$$\hat{z}_G' = \text{median}(\hat{z}_G^1, \ldots, \hat{z}_G^R).$$

Then $\hat{z}_G'$ is an (ε, δ')-FPRASC for z_G.

Proof The assumption that the procedure for computing $\hat{z}_G$ is an $(\varepsilon, 1/4)$-FPRASC for z_G implies that $\hat{z}_G$ is "good" (satisfying $|\hat{z}_G - z_G| \leq \varepsilon z_G$) with probability at least $p = 3/4$. Let Y_i be the indicator of whether $\hat{z}_G^i$ is "good," i.e.,

$$Y_i = \begin{cases} 1, & \text{if } |\hat{z}_G^i - z_G| \leq \varepsilon z_G, \\ 0, & \text{otherwise.} \end{cases}$$

Then $Y_1, \ldots, Y_R$ are i.i.d. Bernoulli random variables with $\mathbb{P}(Y_i = 1) \geq 3/4$. The median $\hat{z}_G' = \text{median}(\hat{z}_G^1, \ldots, \hat{z}_G^R)$ is "good" if at least half of the $\hat{z}_G^i$ values are "good." The median is "good" if and only if $S_R = \sum_{i=1}^R Y_i > R/2$. From Exercise 6.T.35, we know that for $p > 1/2$, the following holds:

$$\mathbb{P}(S_R > R/2) \geq 1 - \delta', \quad \text{where } R = \frac{\log(1/\delta')2p}{(p - 1/2)^2}.$$

Substituting $p = 3/4$, we compute:

$$R = \frac{\log(1/\delta')2(3/4)}{((3/4) - 1/2)^2} = 24 \log(1/\delta').$$

Thus, for $R \geq 24\log(1/\delta')$, the median $\hat{z}'_G = \text{median}(\hat{z}^1_G, \ldots, \hat{z}^R_G)$ satisfies:

$$\mathbb{P}(|\hat{z}'_G - z_G| \leq \varepsilon z_G) \geq 1 - \delta'.$$

The computational cost of the procedure for computing $\hat{z}'_G$ remains polynomial in $|G|$, $1/\varepsilon$, and $\log(1/\delta')$. Therefore, $\hat{z}'_G$ is an (ε, δ')-FPRASC for z_G, as required.

$\square$

Remark 6.5.12 One of the main ingredients for approximate counting is approximate sampling from the uniform distribution. Note that we could use perfect sampling for this purpose: in Algorithm 54, in line 3, instead of running $(X_n^{(i-1)})$, we could use perfect sampling (e.g., the introduced CFTP algorithm or its version for anti-monotone distributions). In such a case, in line 4, we would obtain unbiased $Y_j^{(i)}$, i.e., satisfying $\mathbb{P}(Y_j^{(i)} = 1) = \varrho_i$.

While the analysis in this case would be simpler, there are several issues with this approach:

- First, for efficient CFTP, it is necessary to find a monotone update function, which is often difficult or even impossible.
- Second, the running time of the entire procedure would become random (since the running time of CFTP is itself random), which does not align with the definition of an (ε, δ)-FRPASC.

To address this, we could define a new class of algorithms, called (ε, δ)-EFRPASC, where **E** stands for **Expected**. These algorithms would still satisfy the (ε, δ)-approximation scheme requirements, but with the additional condition that the *expected* running time is polynomial in the size of the problem, $1/\varepsilon$, and $\ln(1/\delta)$.

The main challenge in this context is proving that the expected running time of the perfect sampler is polynomial. In the case of CFTP, a necessary condition for polynomial expected running time is that the individual Gibbs sampler mixes rapidly. However, this condition is by no means sufficient.

For bounds on the expected running time of a CFTP procedure applied to generalized hard-core models (subject to certain restrictions on the maximum degree and λ), see [75].   ∎

6.5.2 Outline of a Procedure for Approximate Counting in Other Problems

We aim to estimate the number of feasible configurations $z_G = |\Lambda|$, defined as a subset of $\{1, \ldots, r\}^V$ for a given graph $G = (V, E)$. In the previous section, we focused on the problem of approximate counting for the number of independent sets. However, note that actually the following components were *crucial* for this specific example:

- The ability to decompose z_G as (recall (6.5.3)):

$$z_G = \frac{z_m}{z_{m-1}} \times \frac{z_{m-1}}{z_{m-2}} \times \cdots \times \frac{z_1}{z_0} \times z_0,$$

with z_0 computable directly.

- The existence of a rapidly mixing Markov chain $(X_n^{(i)})$ with uniform stationary distribution $\pi_\xi^{(i)} = \frac{1}{z_i}\mathbb{1}(\xi \in \Lambda_{G_i})$.
- Ensuring $\varrho_i \geq 1/2$, where $\varrho_i = z_i/z_{i-1}$. Recall that $z_i = |\Lambda_{G_i}|$ and $G_i = (V, E_i)$ is the graph with the first i edges from E (in some fixed ordering).

We can generalize these ideas as follows. Assume we have a sequence $\Lambda_0, \Lambda_1, \ldots, \Lambda_t \equiv \Lambda$ of sets of configurations, where t is an integer polynomial in the size of the graph $|G|$. For the independent sets example, we had $t = m$ and Λ_i was the set of feasible configurations for $G_i = (V, E_i)$, with E_i being the first i edges of E. Denoting $|\Lambda_i|$ by z_i, we can decompose:

$$z_G = \frac{z_t}{z_{t-1}} \times \frac{z_{t-1}}{z_{t-2}} \times \cdots \times \frac{z_1}{z_0} \times z_0. \tag{6.5.9}$$

If $\varrho_i = z_i/z_{i-1} \geq 1/2$ and we have rapidly mixing chains with uniform distributions on Λ_i, the procedure remains applicable. We summarize this in the following proposition.

Proposition 6.5.13 (Generalized Version of Proposition 6.5.4) *Let $G = (V, E)$ be a graph, and let C be a finite set of values assigned to each vertex. The set of all possible configurations is C^V. Assume the following:*

- *There exists a sequence $\Lambda_0, \Lambda_1, \ldots, \Lambda_t \equiv \Lambda$ of sets of configurations, where t is polynomial in $|G|$, and the exact value of z_0 is known.*
- *The decomposition in (6.5.9) holds, and for each $i = 1, \ldots, t$:*

  1. There exists a rapidly mixing Markov chain $(X_n^{(i)})$ with uniform stationary distribution $\pi_\xi^{(i)} = \frac{1}{z_i}\mathbb{1}(\xi \in \Lambda_i)$ and mixing time $\tau(\kappa)$.
  2. $\varrho_i := z_i/z_{i-1} \geq 1/2.$

Then there exists an (ε, δ)-FPRASC for $z_G = |\Lambda|$ with a running time of:

$$\frac{200t^3}{\varepsilon^2} \ln\left(\frac{2t}{\delta}\right) \tau\left(\frac{\varepsilon}{16t}\right),$$

which is asymptotically $O\left(\frac{t^3}{\varepsilon^2} \ln\left(\frac{2t}{\delta}\right) \tau\left(\frac{\varepsilon}{16t}\right)\right)$.

Now we present two additional examples illustrating the application of this proposition, where the results on the mixing time of the underlying Markov chains are cited from existing works.

6.5.2.1　Number of Proper Graph Colorings

Recall the problem defined in Sect. 6.3.2.1. Let $G = (V, E)$ be a graph, and let $r \geq 2$ be an integer. A proper coloring of G assigns "colors" $\{1, \ldots, r\}$ to vertices such that no two adjacent vertices share the same color. Assume $r > \max_v |\mathcal{N}(v)| = \Delta$ (ensuring the existence of at least one proper coloring). The set of feasible configurations Λ consists of all proper colorings. On Λ, we define the uniform distribution as in (6.3.11):

$$
\pi_\xi = \begin{cases} \frac{1}{z_G}, & \text{if } \xi \in \Lambda, \\[2mm] 0, & \text{otherwise.} \end{cases}
$$

Define $G_i = (V, E_i)$ as in Proposition 6.5.4, with E_i containing the first i edges of E. Recall, $m = |E|$. Let Λ_i be the set of proper colorings of G_i, and let $z_i = |\Lambda_i|$. The sequence $\Lambda_0, \ldots, \Lambda_m = \Lambda$ is thus defined. Define $\varrho_i = z_i / z_{i-1}$.

Lemma 6.5.14 *Assume the graph has maximum degree $\Delta \geq 2$, and the number of colors satisfies $r > 2\Delta^2$. Then $\varrho_i \geq 1/2$.*

Proof Let $(w, v) \in G_i$ but $(w, v) \notin G_{i-1}$ (see the pink edge in Fig. 6.10, with $v \equiv v_4$ and $w \equiv v_5$), i.e., $e_i = (w, v)$. By definition, z_i is the number of proper colorings of G_i, and proper colorings $\xi \in \Lambda_{G_i}$ are precisely those colorings of G_{i-1} for which $\xi(v) \neq \xi(w)$. Thus, ϱ_i is the probability that a random coloring X of G_{i-1} satisfies $X(v) \neq X(w)$.

To express this formally: let $X^{(i-1)}$ be a random variable chosen according to the uniform distribution $\pi^{(i-1)}$ on $\Lambda_{G_{i-1}}$, i.e.,

$$
\mathbb{P}(X^{(i-1)} = \xi) = \pi_\xi^{(i-1)} = \frac{1}{z_{i-1}} \mathbb{1}(\xi \in \Lambda_{G_{i-1}}).
$$

Then:

$$
\varrho_i = \mathbb{P}(X^{(i-1)}(v) \neq X^{(i-1)}(w)) = 1 - \mathbb{P}(X^{(i-1)}(v) = X^{(i-1)}(w)).
$$

Now, consider the coloring $X^{(i-1)}$ of all vertices and the restriction $X^{(i-1)}(V \setminus \{v\})$, which is the coloring of all vertices except v. The conditional distribution of the color at v, given $X^{(i-1)}(V \setminus \{v\})$, is uniform over all colors not assigned to any of v's neighbors. Since there are at most Δ neighbors of v, at least $r - \Delta$ colors remain available for v. Therefore, the probability that v receives the same color as w is at most:

$$
\mathbb{P}(X^{(i-1)}(v) = X^{(i-1)}(w)) \leq \frac{1}{r - \Delta}.
$$

Using the assumption $r > 2\Delta^2$, we have:

$$\varrho_i = 1 - \mathbb{P}(X^{(i-1)}(v) = X^{(i-1)}(w)) \geq 1 - \frac{1}{r - \Delta}.$$

Substituting $r > 2\Delta^2$:

$$r - \Delta > 2\Delta^2 - \Delta \geq 2\Delta - \Delta = \Delta.$$

Thus:

$$\frac{1}{r - \Delta} \leq \frac{1}{\Delta}.$$

Finally, we compute:

$$\varrho_i \geq 1 - \frac{1}{r - \Delta} \geq 1 - \frac{1}{\Delta} \geq \frac{1}{2}.$$

This completes the proof. $\qquad\square$

As mentioned earlier, studying the rate of convergence to stationarity is beyond the scope of this book. However, the Metropolis sampler for graph colorings presented in Algorithm 46 is known to be rapidly mixing for $r \geq 2\Delta + 1$ (see Lemma 1 in [85]).

Lemma 6.5.15 *Consider a graph $G = (V, E)$ with $M = |V|$ vertices, maximum degree Δ, and r colors. If $r \geq 2\Delta + 1$, the Markov chain with one-step transitions given in Algorithm 46 is rapidly mixing, with the mixing time bounded by:*

$$\tau(\kappa) \leq \frac{r}{r - 2\Delta} M \ln\left(\frac{M}{\kappa}\right).$$

The condition $r > 2\Delta^2$ in Lemma 6.5.14 implies $r > 2\Delta + 1$ in Lemma 6.5.15 above, thus Proposition 6.5.13 applied to graph coloring yields:

Lemma 6.5.16 *Assume $r > 2\Delta^2$. Then there exists an (ε, δ)-FPRASC for the number of proper colorings with a running time bounded by:*

$$\frac{200m^3}{\varepsilon^2} \ln\left(\frac{2m}{\delta}\right) \tau\left(\frac{\varepsilon}{16m}\right) \leq \frac{200m^3}{\varepsilon^2} \ln\left(\frac{2m}{\delta}\right) \frac{r}{r - 2\Delta} M \ln\left(\frac{16mM}{\varepsilon}\right).$$

Remark 6.5.17 In [85], it is shown that $\varrho_i = z_i/z_{i-1} \geq 1/2$ for $r > 2\Delta + 1$. Therefore, Lemma 6.5.16 also holds under this weaker condition. $\qquad\blacksquare$

6.5.2.2 Number of Feasible Knapsack Solutions/Configurations

Recall that we are given the weights of d items, $\mathbf{w} = (w_1, \ldots, w_d)$, and a total maximal weight W. Assume that $w_i > 0$ for $i = 1, \ldots, d$, and that W is an integer. The set of feasible configurations (so-called knapsack solutions) is:

$$\Lambda = \left\{ \mathbf{x} \in \{0, 1\}^d : \sum_{j=1}^{d} w_j x_j \leq W \right\}.$$

Define:

$$W_0 = 0, \quad W_i = \min \left\{ W, \sum_{j=0}^{i-1} w_j \right\}, \quad i = 1, \ldots, d,$$

and:

$$\Lambda_i = \left\{ \mathbf{x} \in \{0, 1\}^d : \sum_{j=1}^{d} w_j x_j \leq W_i \right\}, \quad i = 0, \ldots, d.$$

Thus, $z_i = |\Lambda_i|$ is the number of feasible knapsack solutions for the problem where W is replaced by W_i. We may decompose:

$$z_G = \frac{z_d}{z_{d-1}} \times \frac{z_{d-1}}{z_{d-2}} \times \cdots \times \frac{z_1}{z_0} \times z_0,$$

where $z_0 = 1$ (since $\mathbf{x} = (0, \ldots, 0)$ is the only solution with weight $W_0 = 0$).

Note that if the total weight of items is not greater than W_{i-1}, then it is also not greater than W_i, i.e., $\Lambda_{i-1} \subseteq \Lambda_i$, and consequently:

$$z_{i-1} \leq z_i \quad \Rightarrow \quad \varrho_i = \frac{z_i}{z_{i-1}} \geq 1 \geq \frac{1}{2}.$$

We presented a Gibbs sampler for sampling knapsack solutions with the distribution $\pi_\xi = \exp(\mathbf{v} \cdot \xi)/z_T$ in Algorithm 63, where $\mathbf{v} = (v_1, \ldots, v_d)$ was a vector of item *values* (not used here). For the uniform distribution $\pi_\xi^{(i)} = \frac{1}{z_i} \mathbf{1}(\xi \in \Lambda_i)$, it simplifies to Algorithm 57.

In [156], the authors showed that Algorithm 57 yields a rapidly mixing Markov chain.

Lemma 6.5.18 ([156]) *The Markov chain on feasible solutions of a 0-1 knapsack problem given in Algorithm 57 is rapidly mixing, with mixing time:*

$$\tau(\kappa) = O(d^{9/2+\kappa}).$$

Algorithm 57 Gibbs sampler for 0-1 knapsack problem (1 step), uniform distribution, max weight W_i

---

1: Suppose $X_k = \mathbf{x}$.
2: Sample $i \in \{1, \ldots, d\}$ uniformly at random.
3: Set $\mathbf{x}' = (x'_1, \ldots, x'_d)$, where $x'_j = x_j$ for $j \neq i$ and $x'_i = 1$.
4: **if** $\mathbf{x}' \cdot \mathbf{w} > W_i$ **then**
5:      Set $x'_i = 0$.
6: **else**
7:      Sample $U \sim \mathcal{U}(0, 1)$ and set:

$$x'_i = \begin{cases} 1, & \text{if } U \leq \frac{1}{2}, \\[2mm] 0, & \text{if } U > \frac{1}{2}. \end{cases}$$

8: **end if**
9: Set $X_{k+1} = \mathbf{x}'$.

Summarizing, Proposition 6.5.13 (with $t = d$) applied to this problem yields:

Lemma 6.5.19 *There exists an (ε, δ)-FPRASC for the number of solutions of the 0-1 knapsack problem with running time:*

$$\frac{200 d^3}{\varepsilon^2} \ln\left(\frac{2d}{\delta}\right) \tau\left(\frac{\varepsilon}{16d}\right) = O\left(\frac{d^3}{\varepsilon^2} \ln\left(\frac{2d}{\delta}\right) d^{9/2 + \varepsilon/(16d)}\right),$$

$$= O\left(\frac{d^{15/2 + \varepsilon/(16d)}}{\varepsilon^2} \ln\left(\frac{2d}{\delta}\right)\right).$$

6.6   Computational Biology: Motif Finding Problem Example

This section is devoted to motif finding, one of the important problems in computational biology. It is primarily written for mathematical readers, and we aim to abstract away from biological terminology as much as possible. In Sect. 6.6.2, we present a simplified version of the problem.

6.6.1   Simple Coin Example

Consider the following scenario to build some intuition about the problem at hand. Assume we have $w + 1$ coins, where the probability of obtaining heads with the i-th coin is $\theta_i > 0$ for $i = 0, \ldots, w$. Fix $n > 0$ and repeat the following procedure k times. The i-th step $(i = 1, \ldots, k)$ proceeds as follows:

- Flip a coin with head probability θ_0 n times, obtaining $r_{i,1}, r_{i,2}, \ldots, r_{i,n}$.
- Choose $\xi_i^\bullet$ uniformly at random from the set $\{1, \ldots, n - w + 1\}$.
- Replace $(r_{i,\xi_i^\bullet}, \ldots, r_{i,\xi_i^\bullet+w-1})$ with $(s_{i,1}, s_{i,2}, \ldots, s_{i,w})$, where $s_{i,j}$ is the result of flipping a coin with head probability θ_j.

THTHHT HHTTHHT HHTHHTTHTH
THTHTHHH HTTTHTT HTHTHHHH
TH HHTTTHH HHTTTHTHHTHHHT

Fig. 6.11 Example with $k = 3$ sequences of length $n = 16$, where motifs of length $w = 7$ are injected at positions $\xi^\bullet = (7, 9, 3)$

Roughly speaking, we start with k rows of independent random variables generated by flipping coins with head probability θ_0. Then, in each row, a random segment of fixed width w is replaced by new independent coin tosses. These tosses follow a fixed order: the j-th coin in this segment is flipped with head probability θ_j. See the example in Fig. 6.11. Denote the chosen positions $\xi_i^\bullet$, $i = 1, \ldots, k$ collectively as $\xi^\bullet = (\xi_1^\bullet, \ldots, \xi_k^\bullet)$.

Problem 1 Having observed k sequences of length n, knowing w, and knowing that exactly one sequence of length w was injected into each row, estimate $\theta_0, \theta_1, \ldots, \theta_w$ and $\xi^\bullet$.

Note that by "estimate," we mean "find some *best* estimate," which is not precisely defined at this stage (it will be clarified later).

A simpler version of the problem is the following:

Problem 2 Similarly to Problem 1, the goal is to find or approximate $\xi^\bullet$ alone.

6.6.2 Motif Finding: Problem Formulation

The motif finding problem is a natural generalization of the procedure described in Sect. 6.6.1. We still have k sequences, each of length n. However, this time each $r_{ij} \in \Sigma$, where Σ is an alphabet (in the coin example, we had $\Sigma = \{H, T\}$). Denote $|\Sigma|$ by d. For clarity, we will use the example[3] where $\Sigma = \{A, C, G, T\} \equiv \{1, 2, 3, 4\}$, making $d = 4$. Storing elements of Σ as numbers is convenient. The procedure to produce the i-th row (repeated k times) is as follows:

- Simulate n random variables $r_{i1}, \ldots, r_{in}$ independently with distribution:

$$P(r_{ij} = a) = \theta_{a,0}, \quad a = 1, \ldots, d.$$

---

[3] Typically $d = 4$ (nucleotides in DNA) or $d = 20$ (types of amino acids in protein).

- Choose $\xi_i^\bullet$ uniformly at random from the set $\{1, \ldots, n - w + 1\}$.
- Replace $(r_{i,\xi_i}, \ldots, r_{i,\xi_i+w-1})$ with $(s_{i,1}, s_{i,2}, \ldots, s_{i,w})$, where $s_{i,1}, \ldots, s_{i,w}$ are independent and:

$$P(s_{i,j} = a) = \theta_{a,j}, \quad a = 1, \ldots, d.$$

After this injection, the resulting sequence is again denoted $(r_{i1}, \ldots, r_{in})$. The injected terms are assumed to be independent observations from a *product-multinomial* model (known as a *motif*), and the remaining parts of the sequence are referred to as the *background* (see Fig. 6.11, where greyed and non-greyed parts represent motifs and background, respectively). We denote the *background parameter* as:

$$\boldsymbol{\theta}_0 = (\theta_{1,0}, \ldots, \theta_{d,0})^T,$$

and residue frequencies for each position j within an injected segment as:

$$\boldsymbol{\theta}_j = (\theta_{1,j}, \ldots, \theta_{d,j})^T, \quad j = 1, \ldots, w.$$

In the biological sciences, the matrix:

$$\boldsymbol{\Theta} = (\boldsymbol{\theta}_1, \ldots, \boldsymbol{\theta}_w)$$

is referred to as the *position weight matrix.*

We also denote all observed data as a matrix:

$$\mathbf{R} = (r_{ij}), \quad i = 1, \ldots, k, \quad j = 1, \ldots, n.$$

The most general problem can now be formulated as follows:

Problem 1' Given k observed sequences of length n, knowing w and that exactly one sequence of length w was injected into each row, estimate $\boldsymbol{\theta}_j$ ($j = 0, \ldots, w$) and $\xi^\bullet$.

A less general formulation is:

Problem 2' Similarly to Problem 1', the goal is only to find or approximate $\xi^\bullet$.

In this section, we focus on Problem 2', which is simpler than Problem 1'. However, there are $(n - w + 1)^k$ possible configurations of ξ, making the problem computationally challenging. To address this, we will apply stochastic optimization using a Gibbs sampler. Problem 1' is typically addressed with (variations of) the Expectation-Maximization (EM) algorithm (see, e.g., [155], [37]), which is beyond the scope of this book.

From now on, we assume that the parameters $\boldsymbol{\theta}_j$ ($j = 0, \ldots, w$) and w are known, but the positions $\xi^\bullet$, where motifs were injected, are unknown.

Remark 6.6.1 Note that if for each j, $\boldsymbol{\theta}_j = (0, \ldots, 1, \ldots, 0)$ with 1 at a fixed position and 0 otherwise, then the motif corresponds to a fixed sequence. For example, let $\Sigma = \{A, C, G, T\}$, $w = 5$, and:

$$\boldsymbol{\theta}_1 = \boldsymbol{\theta}_2 = (0, 0, 1, 0), \quad \boldsymbol{\theta}_3 = (0, 1, 0, 0), \quad \boldsymbol{\theta}_4 = (0, 0, 0, 1), \quad \boldsymbol{\theta}_5 = (1, 0, 0, 0).$$

Then, in each row, the same sequence $GGCTA$ is injected. The problem reduces to finding the starting positions ξ_i of this sequence in each row i ($i = 1, \ldots, k$). ■

Given the model and candidates for the starting positions of the motif in each row, i.e., given ξ, what is the probability that the model would output $\mathbf{R}$?

Let $\xi = (\xi_1, \ldots, \xi_k)$, where $\xi_i \in \{1, \ldots, n - w + 1\}$. We call such ξ a **configuration**. Before proceeding, given the observed data $\mathbf{R}$ and an initial configuration ξ (a candidate for the motif's starting positions), let us define two matrices, $\mathbf{A}_\xi$ and $\mathbf{C}_\xi$.

Matrix $\mathbf{A}_\xi$ This is a $k \times w$ matrix. The i-th row is a sequence of length w extracted from the i-th row of $\mathbf{R}$ at positions (columns) $\xi_i, \xi_i + 1, \ldots, \xi_i + w - 1$, i.e.,

$$\mathbf{A}_\xi(i, j) = \mathbf{R}(i, \xi_i + j - 1), \quad j = 1, \ldots, w.$$

Matrix $\mathbf{C}_\xi$ This is a $d \times w$ matrix. Entry $\mathbf{C}_\xi(i, j)$ represents the number of occurrences of i (recall that letters are identified by their numerical encoding) at position j in $\mathbf{A}_\xi$, i.e.,

$$\mathbf{C}_\xi(i, j) = \#\{l : \mathbf{A}_\xi(l, j) = i\}.$$

See Fig. 6.12 for an example of how these matrices are computed.

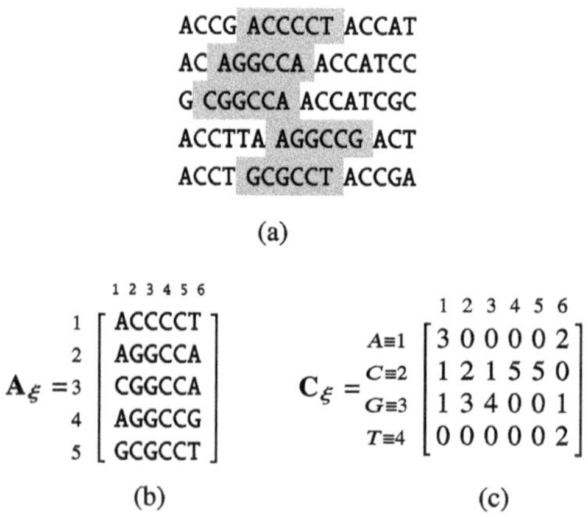

Fig. 6.12 Creating matrices $\mathbf{A}_\xi$ and $\mathbf{C}_\xi$ from given sequences and ξ. (**a**) $k = 5$ sequences of length $n = 15$ with candidates for motif's (of length $w = 6$) positions $\xi = (5, 3, 2, 7, 5)$. (**b**) Sequences of length $w = 5$ extracted from all sequences based on ξ. $\mathbf{A}_\xi$ is of size $k \times w$. (**c**) $\mathbf{C}_\xi(i, j)$ is the number of i's at position j. $\mathbf{C}_\xi$ is of size $|\Sigma| \times w = d \times w$

(a)

$$\mathbf{A}_\xi = \begin{array}{c} 1 \\ 2 \\ 3 \\ 4 \\ 5 \end{array}
\begin{bmatrix} \text{ACCCCT} \\ \text{AGGCCA} \\ \text{CGGCCA} \\ \text{AGGCCG} \\ \text{GCGCCT} \end{bmatrix}
\qquad
\mathbf{C}_\xi = \begin{array}{c} A\equiv1 \\ C\equiv2 \\ G\equiv3 \\ T\equiv4 \end{array}
\begin{bmatrix} 3 & 0 & 0 & 0 & 0 & 2 \\ 1 & 2 & 1 & 5 & 5 & 0 \\ 1 & 3 & 4 & 0 & 0 & 1 \\ 0 & 0 & 0 & 0 & 0 & 2 \end{bmatrix}$$

(b)   (c)

Let us disregard "background" for a moment and ask the following question: What is the probability that, assuming the motif's starting positions are indeed given by ξ, the model with prescribed parameters $\boldsymbol{\theta}_1, \ldots, \boldsymbol{\theta}_w$ would yield $\mathbf{A}_\xi$?

Consider the matrix $\mathbf{A}_\xi$ from Fig. 6.12. The probability that the first letter is A, C, G, or T is given respectively by $\theta_{1,1}, \theta_{2,1}, \theta_{3,1}$, and $\theta_{4,1}$. In $\mathbf{A}_\xi$, we see that A occurs 3 times, C and G each occur once, and T does not occur. The first column of $\mathbf{C}_\xi$ records these occurrences. Thus, the probability of observing the first column of $\mathbf{A}_\xi$, or equivalently obtaining the first column of $\mathbf{C}_\xi$, is given by the multinomial distribution and is proportional to:

$$\prod_{i=1}^{d} \theta_{i,1}^{\mathbf{C}_\xi(i,1)}.$$

Taking all positions into account, we derive the likelihood function:

$$g(\xi) = \prod_{j=1}^{w} \prod_{i=1}^{d} \theta_{i,j}^{\mathbf{C}_\xi(i,j)}.$$

We now consider the *background*. Assuming the motif's positions are given by ξ, the remaining data in $\mathbf{R}$ is treated as independent realizations of a random variable taking values in Σ with probabilities $\theta_{a,0}$ for $a = 1, \ldots, d$.

Let $\mathbf{R}_{\{\xi\}^c}$ be a $k \times (n - w)$ matrix obtained from $\mathbf{R}$ by removing, from each row, a segment of length w starting at positions given by ξ (i.e., the non-grey elements in Fig. 6.12a). Similarly to how $\mathbf{C}_\xi$ is created from $\mathbf{A}_\xi$, we define a vector $\mathbf{D}_\xi$ from $\mathbf{R}_{\{\xi\}^c}$ that counts occurrences of each letter:

$$\mathbf{D}_\xi(a) = \#\{(i, j) : \mathbf{R}_{\{\xi\}^c}(i, j) = a\}.$$

The likelihood that $\mathbf{R}_{\{\xi\}^c}$ was generated using $\boldsymbol{\theta}_0$ is:

$$h(\xi) = \prod_{i=1}^{d} \theta_{i,0}^{\mathbf{D}_\xi(i)}.$$

Combining g and h, and denoting the set of parameters by $\boldsymbol{\Theta} = (\boldsymbol{\theta}_0, \ldots, \boldsymbol{\theta}_w)$, we can write the complete-data likelihood as:

$$f(\xi, \mathbf{R}, \boldsymbol{\Theta}) \equiv f(\xi) = h(\xi) \times g(\xi)$$

$$= \prod_{i=1}^{d} \theta_{i,0}^{\mathbf{D}_\xi(i)} \times \prod_{j=1}^{w} \prod_{i=1}^{d} \theta_{i,j}^{\mathbf{C}_\xi(i,j)}.$$

For numerical stability (as $f(\xi)$ is usually a very small positive number), it is more convenient to consider $\log f(\xi)$. Note that:

$$\operatorname*{argmax}_{\xi} f(\xi) = \operatorname*{argmax}_{\xi} \log f(\xi).$$

We define **the energy of configuration** ξ as:

$$H(\xi) = -\log f(\xi).$$

Finally, on the set of configurations ξ, we define the Gibbs distribution (recall (6.3.6)) with parameter $T > 0$ (where $z(T)$ is the normalizing constant):

$$
\begin{aligned}
\pi_\xi &= \frac{1}{z(T)} e^{-H(\xi)/T} \\
&= \frac{1}{z(T)} \exp\left(\frac{1}{T} \log f(\xi)\right) \\
&= \frac{1}{z(T)} \exp\left(\frac{1}{T} \log\left[\prod_{i=1}^{d} \theta_{i,0}^{\mathbf{D}_\xi(i)} \times \prod_{j=1}^{w}\prod_{i=1}^{d} \theta_{i,j}^{\mathbf{C}_\xi(i,j)}\right]\right) \\
&= \frac{1}{z(T)} \exp\left(\frac{1}{T}\left[\sum_{i=1}^{d}\mathbf{D}_\xi(i) \log \theta_{i,0} + \sum_{j=1}^{w}\sum_{i=1}^{d}\mathbf{C}_\xi(i,j) \log \theta_{i,j}\right]\right).
\end{aligned}
\tag{6.6.1}
$$

Note that $H(\xi) = -\log f(\xi) = -\log g(\xi) - \log h(\xi)$. Intuitively, "*good*" configurations ξ (i.e., starting positions) have large π_ξ. The task now reduces to finding or approximating:

$$\xi^* \approx \operatorname*{argmax}_{\xi} \pi_\xi,$$

where we expect $\xi^* \approx \xi^\bullet$. Recall that the number of possible configurations ξ is $(n - w + 1)^k$, which grows exponentially with k. In typical applications, both $n - w$ and k are large. Thus, one can use an MCMC approach to simulate ξ from the approximate distribution π_ξ.

Gibbs Sampler for Motif Finding   Recall that the Gibbs sampler was introduced for sampling distributions on $\mathcal{S} = C^V$, where $C = \{1, \ldots, r\}$. We identify V with the sequence indices, *i.e.*, $V = \{1, 2, \ldots, k\}$. Each vertex can be "assigned" one of $C = \{1, \ldots, n - w + 1\}$ values, representing the starting position of a motif. The value assigned to a vertex i is denoted by ξ_i. Thus, $\xi = (\xi_1, \ldots, \xi_k)$ is a configuration, i.e., an assignment of values from C to each vertex in V.

The Gibbs sampler operates as follows: it randomly picks one of the k sequences (vertices). Suppose sequence v is chosen. The primary computation is:

$$\mathbb{P}(Z(v) = c \mid Z_{-v} = \xi_{-v}), \quad c = 1, \ldots, n - w + 1, \tag{6.6.2}$$

where ξ_{-v} is the vector ξ with the v-th component excluded, Z is a random variable with distribution π_ξ, and Z_{-v} refers to Z restricted to all vertices (sequences) except v.

In essence, Eq. (6.6.2) represents the conditional probability of π_ξ that the motif in row/sequence v starts at position c.

For a given $\xi = (\xi_1, \ldots, \xi_k)$, let $\xi(v, r)$ denote the vector ξ with value r assigned to position v, i.e.,

$$\xi(v, r) = (\xi_1, \ldots, \xi_{v-1}, r, \xi_{v+1}, \ldots, \xi_k).$$

We compute:

$$\mathbb{P}(Z(v) = c \mid Z_{-v} = \xi_{-v}) = \frac{\mathbb{P}(Z(v) = c, Z_{-v} = \xi_{-v})}{\displaystyle\sum_{r=1}^{n-w+1} \mathbb{P}(Z(v) = r, Z_{-v} = \xi_{-v})}.$$

Using π_ξ, this simplifies to:

$$\mathbb{P}(Z(v) = c \mid Z_{-v} = \xi_{-v}) = \frac{\pi_{\xi(v,c)}}{\sum_{r=1}^{n-w+1} \pi_{\xi(v,r)}}. \tag{6.6.3}$$

In practical implementation, the above fraction must be reduced to avoid numerical issues. See Sect. 6.6.3 for implementation details.

Finally, the Gibbs algorithm for the motif finding problem is presented in Algorithm 58.

Algorithm 58 Gibbs sampler for motif finding problem: transition from step t to $t + 1$

1: Matrix $\mathbf{R}$ is given.
2: Assume $X_t = \xi$.
3: Compute $\mathbf{A}_\xi$, $\mathbf{C}_\xi$, $\mathbf{R}_{\{\xi\}^c}$, and $\mathbf{D}_\xi$ (needed for computing $\pi(\cdot)$ below).
4: Pick $v \in \{1, \ldots, k\}$ uniformly at random.
5: Compute $\pi_{\xi(v,c)}$ for $c = 1, \ldots, n - w + 1$ using (6.6.2), and calculate:

$$p(c) = \frac{\pi_{\xi(v,c)}}{\sum_{r=1}^{n-w+1} \pi_{\xi(v,r)}}.$$

6: Pick c with the above distribution $p(c)$.
7: Set $X_{t+1} = \xi(v, c)$.

6.6.3   Implemented Example

Consider the alphabet $\Sigma = \{A, C, G, T\}$ and the background parameter vector:

$$\boldsymbol{\theta}_0 = (0.25, 0.25, 0.25, 0.25)^T,$$

and a position weight matrix (representing frequencies of letters at the j-th position of a motif of length $w = 13$):

$$(\boldsymbol{\theta}_1, \ldots, \boldsymbol{\theta}_{13}) =$$

$$\begin{pmatrix} 0.7 & 0.7 & 0.18 & 0.17 & 0.22 & 0.12 & 0.15 & 0.15 & 0.18 & 0.14 & 0.15 & 0.65 & 0.75 \\ 0.15 & 0.12 & 0.01 & 0.02 & 0.02 & 0.7 & 0.65 & 0.7 & 0.01 & 0.04 & 0.03 & 0.01 & 0.01 \\ 0.13 & 0.13 & 0.7 & 0.7 & 0.65 & 0.05 & 0.05 & 0.04 & 0.11 & 0.12 & 0.12 & 0.11 & 0.11 \\ 0.02 & 0.05 & 0.11 & 0.11 & 0.11 & 0.13 & 0.15 & 0.11 & 0.7 & 0.7 & 0.7 & 0.23 & 0.13 \end{pmatrix}.$$

Since $\boldsymbol{\theta}_0^T = (0.25, 0.25, 0.25, 0.25)$, the background letters are chosen randomly. The first and second letters of the motif are most likely A (with probability 0.7), the third is most likely G (also with probability 0.7), and so on.

We simulate $k = 30$ sequences of length $n = 20$, collected in a $k \times n$ matrix $\mathbf{R}$, shown in Fig. 6.14. Random starting positions $\xi^{\bullet} = (\xi_1^{\bullet}, \ldots, \xi_{30}^{\bullet})$ are simulated, and motifs are injected into each row starting at these positions.

The simulated $\xi^{\bullet}$ is:

$$(30, 9, 23, 26, 26, 28, 26, 12, 12, 26, 2, 17, 28, 30, 24, 22, 34, 0,$$

$$10, 5, 6, 31, 15, 0, 20, 33, 11, 21, 11, 17) \ .$$

The energy calculated for the above "*optimal*" $\xi^{\bullet}$ is $H(\xi^{\bullet}) = 12.62$.

Simulating some uniformly random sample configurations $\xi^{(1)}, \xi^{(2)}, \xi^{(3)}, \xi^{(4)}$, we obtained:

$$H(\xi^{(1)}) = 28.16, \quad H(\xi^{(2)}) = 26.65, \quad H(\xi^{(3)}) = 27.95, \quad H(\xi^{(4)}) = 27.08.$$

Recall, our goal is to find $\xi^* \approx \xi^{\bullet}$, where " $\approx$" means similar energy.

Reduction   Recall, the main computation in the Gibbs sampler for this example is of the conditional probabilities given by (6.6.3):

$$\frac{\pi_{\xi(v,s)}}{\sum_{r=1}^{n-w+1} \pi_{\xi(v,r)}}.$$

For a given $\xi = (\xi_1, \ldots, \xi_k)$, we denoted by $\xi(v, r)$ the vector ξ with value r at position v, i.e.,

$$\xi(v, r) = (\xi_1, \ldots, \xi_{v-1}, r, \xi_{v+1}, \ldots, \xi_k).$$

Note that in (6.6.3), the constant $z(T)$ cancels out. Recall also the final formula (6.6.1) for π:

$$\pi_\xi = \frac{1}{z(T)} \exp\left(\frac{1}{T} \left[\sum_{i=1}^{d} \mathbf{D}_\xi(i) \log \theta_{i,0} + \sum_{j=1}^{w} \sum_{i=1}^{d} \mathbf{C}_\xi(i, j) \log \theta_{i,j} \right] \right).$$

There are many terms in common in the numerator and the denominator of the fraction (6.6.3). For large n, d, and w, it is necessary to reduce these common terms to mitigate numerical issues.

$$\left(\frac{\pi_{\xi(v,s)}}{\sum_{r=1}^{n-w+1} \pi_{\xi(v,r)}} \right)^{-1}$$

$$= \left(\frac{\exp\left(\frac{1}{T} \left[\sum_{i=1}^{d} \mathbf{D}_{\xi(v,s)}(i) \log \theta_{i,0} + \sum_{j=1}^{w} \sum_{i=1}^{d} \mathbf{C}_{\xi(v,s)}(i, j) \log \theta_{i,j} \right] \right)}{\sum_{r=1}^{n-w+1} \exp\left(\frac{1}{T} \left[\sum_{i=1}^{d} \mathbf{D}_{\xi(v,r)}(i) \log \theta_{i,0} + \sum_{j=1}^{w} \sum_{i=1}^{d} \mathbf{C}_{\xi(v,r)}(i, j) \log \theta_{i,j} \right] \right)} \right)^{-1}$$

$$= \sum_{r=1}^{n-w+1} \exp\left(\frac{1}{T} \left[\sum_{i=1}^{d} \left(\underbrace{\mathbf{D}_{\xi(v,r)}(i) - \mathbf{D}_{\xi(v,s)}(i)}_{=\kappa(\mathbf{D},s,r,i)} \right) \log \theta_{i,0} + \right.\right.$$

$$\left.\left. \sum_{j=1}^{w} \sum_{i=1}^{d} \left(\underbrace{\mathbf{C}_{\xi(v,r)}(i, j) - \mathbf{C}_{\xi(v,s)}(i, j)}_{=\eta(\mathbf{C},s,r,i,j)} \right) \log \theta_{i,j} \right] \right).$$

We have

$$\mathbf{C}_{\xi(v,s)}(i, j) = \#\{l : \mathbf{A}_{\xi(v,s)}(l, j) = i\}$$
$$= \#\{l \neq v : \mathbf{A}_{\xi(v,s)}(l, j) = i\} + \mathbb{1}(\mathbf{A}_{\xi(v,s)}(v, j) = i).$$

Note that

$$\#\{l \neq v : \mathbf{A}_{\xi(v,s)}(l, j) = i\} = \#\{l \neq v : \mathbf{A}_{\xi(v,r)}(l, j) = i\}$$

and $\mathbf{A}_{\xi(v,s)}(v, j) = \mathbf{R}(v, s + j - 1)$, thus

$$\mathbf{C}_{\xi(v,s)}(i, j) = \#\{l \neq v : \mathbf{A}_{\xi(v,r)}(l, j) = i\} + \mathbb{1}(\mathbf{R}(v, s + j - 1) = i),$$

$$\mathbf{C}_{\xi(v,r)}(i, j) = \#\{l \neq v : \mathbf{A}_{\xi(v,r)}(l, j) = i\} + \mathbb{1}(\mathbf{R}(v, r + j - 1) = i)$$

and thus

$$\eta(\mathbf{C}, s, r, i, j) = \mathbb{1}(\mathbf{R}(v, r + j - 1) = i) - \mathbb{1}(\mathbf{R}(v, s + j - 1) = i) .$$

For the background, we have

$$\mathbf{D}_{\xi(v,s)}(a) = \#\{(i, j) : \mathbf{R}_{\{\xi(v,s)\}^c}(i, j) = a\}$$

$$= \#\{(i, j) : i \neq v, \mathbf{R}_{\{\xi(v,s)\}^c}(i, j) = a\}$$

$$+\#\{(v, j) : \mathbf{R}_{\{\xi(v,s)\}^c}(v, j) = a\}.$$

Of course for $i \neq v$ we have

$$\#\{(i, j) : i \neq v, \mathbf{R}_{\{\xi(v,s)\}^c}(i, j) = a\} = \#\{(i, j) : i \neq v, \mathbf{R}_{\{\xi(v,r)\}^c}(i, j) = a\}$$

and simply

$$\#\{(v, j) : \mathbf{R}_{\{\xi(v,s)\}^c}(v, j) = a\} = \#\{j : \mathbf{R}_{\{\xi(v,s)\}^c}(v, j) = a\},$$

thus

$$\mathbf{D}_{\xi(v,s)}(a) = \#\{(i, j) : i \neq v, \mathbf{R}_{\{\xi(v,r)\}^c}(i, j) = a\} + \#\{j : \mathbf{R}_{\{\xi(v,s)\}^c}(v, j) = a\}.$$

Similarly,

$$\mathbf{D}_{\xi(v,r)}(a) = \#\{(i, j) : i \neq v, \mathbf{R}_{\{\xi(v,r)\}^c}(i, j) = a\} + \#\{j : \mathbf{R}_{\{\xi(v,r)\}^c}(v, j) = a\},$$

thus

$$\kappa(\mathbf{D}, s, r, i) = \#\{j : \mathbf{R}_{\{\xi(v,r)\}^c}(v, j) = i\} - \#\{j : \mathbf{R}_{\{\xi(v,s)\}^c}(v, j) = i\}$$

In Fig. 6.13, we can see how the energy changed during 200 consecutive steps of our implementation of the Gibbs sampler.

Fig. 6.13 x-axis: Gibbs step number; y-axis: energy $H(\xi^{\text{step}})$

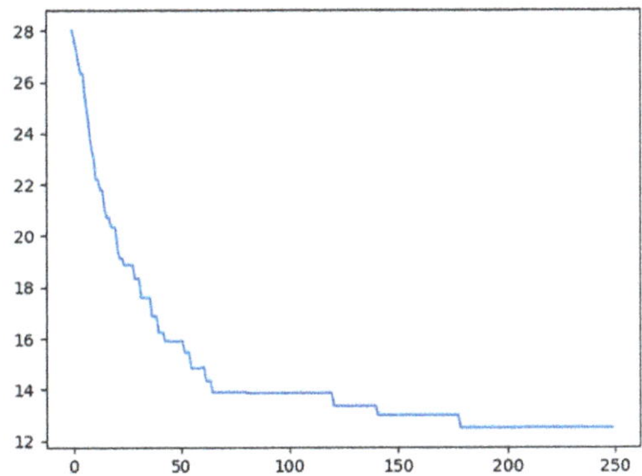

The configuration in step 200 was the following:

$$\xi = (30, \underline{8}, 23, 26, 26, 28, 26, \underline{13}, 12, 26, 2, 17, 28, 30, 24, 22, 34, 0,$$
$$10, 5, 6, 31, 15, 0, 20, 33, 11, 21, 11, 17)\,,$$

compare with the original one:

$$\xi^{\bullet} = (30, 9, 23, 26, 26, 28, 26, 12, 12, 26, 2, 17, 28, 30, 24, 22, 34, 0,$$
$$10, 5, 6, 31, 15, 0, 20, 33, 11, 21, 11, 17)\,.$$

Thus, the original configuration $\xi^{\bullet}$ was "*almost*" recovered (only two rows differ by one position). The energy of the final configuration $H(\xi^{200}) = 12.64$ (recall, $H(\xi^{\bullet}) = 12.62$).

The observed data matrix **R**, along with the initial configuration $\xi^{\bullet}$ and final configuration ξ^{200}, is shown in Fig. 6.14.

6.7 Estimating $I = Ef(X),\ X \sim \pi$

Our main goal is to compute or estimate $I = \mathbb{E}f(X)$, where f is a real-valued function and X follows a distribution π. Explicitly,

$$I = \sum_{s \in S} f(s)\pi_s,$$

where π is the stationary distribution of some Markov chain with a transition matrix **P**. We assume that $\pi_s > 0$ for all $s \in S$. This setup is typical in MCMC: while direct sampling from π is generally infeasible, we can construct a Markov chain whose stationary distribution is π. Here, we assume the state space is finite, simplifying statements and proofs.

TGTGGATACCTATTAGACATCTTTAAGAGCAGTGGCAAATAAAACGGCGG
TCGACATCTATGGCCCATTAAAGGAGAGCTAAGGGTCTGTGACTCACCGC
CCCTTGAGGTCAGATTCTATCTTAAGGGCCCTGTAAAAGCCTCTGGACTT
AAGTATTGCCCTCCTTTTAAGTTGTATAGGGCCCTTGGAGTCTACGCTTA
AATCCCACAGAGTCTAGGTTGTTGTTAAGAGCCCGTGTATCGGAAATACT
GATCACGAGGACAAGCTACGACGACGCTAAGGGCCCTTTGACGCGGACCA
GCCTTCCGATCAGGAGTAGCCAGTCATGGGGCCCATTTACACCGTGCGAA
GCCACGTAGCGTACGAGACTCTTAATGGCTTCCGCACGGGGGTTTGTTTC
CAATCAAGGTGCGGAGGCCCTTTAAGAATACATTATGACGGCACTGACTC
TAGTTCTGCAGTGCCCTCCCACTTCCAATGGCACTTTGATCATTCGACGT
TGGTGGGCCCGTTAAGAACACGCATCACTTAATTTTTTAAAATTGAATTG
GGGTCAAGAATTAATTGAAGAGCCATTGTAGTTAAATGCGACCGATAAAT
CGGCCGAACCAAGTGACGAAACGTAACCCATGGCCCTATAAACGGAGGAC
CAGAGGTCTCCGGCAGTTGTACCCCGTGGGAAAGGCCCTGTAAGATAAAA
AGTCACAGGGTCGCGCTCGATGAGCTTGGCGATATGACGCACATTGACTG
CCAAGGAGCAGTGCACGGCATACGGGGCCCATGAACAAGATCTGCCGTCG
CAAAACTTCTTCGTGGACCTATGGTTGCCACGCTCAAGTCACATTAATGG
TCGTGCCCGTTGACAGTAGCTCATCCGCGGGGCGACAGTTGGTACCCTTG
GACGTGGTGTGCGGACCATTTAGTTGGACAGCAGTATCTGGCTGTTACGA
ATTTCAAGAGCCATATAATGTTGACATGACCGCTCTGACCTCGCTCGTAC
GGGTCCAAAGGCCCTTTAAGATACTGATCACGTGCTTCAATTGGTTGGGC
TTGACCAGACAGAATCGAAGCCAAGGGTCTTAAAGACCTTTGAACCACGT
GTTATCGCCCTCTTTGAAGACCCTTCAAACGGATTTCTCCGAAACTATTT
AAGGGTCCTTTGAAGAGTAATGTGCCGTCACAATAATATCACACGTGCCC
AAATTGCGAGCTTTAAATGACGGAGCCTTGTAAATACGCAAGATTAACAT
AGGGCGCAATCGTAAGCCCCTGACTTCTCAGCAAAGTACTCAGCTACGGC
GCCGAACTACTAGGTACCCTTTAATCAAGCGTAATTTCCCTAGCCACACG
CCATATGAACACCAGAATTTGAAGGGTTCTTTAACTCCTAATGCGACACT
CTGGATAGGAGAAGAACCCATTTACTAATGGTACCTAAGAAGCGTTGCCC
GGCATGACACCCGAATGAAGAAACTTTTGACGACGCGTTTAAGACGTTGG

Fig. 6.14 Observed data: $k = 30$ sequences of length $n = 20$ collected in matrix **R**. A motif of length $w = 13$ was injected into each row, represented in *red* (configuration $\xi^{\bullet}$). The "*recovered*" motif is represented in *green* (configuration ξ^{200}). Overlapping configurations are shown in *brown*. The motif location was incorrectly identified in only two rows, with an offset of one position

The most straightforward method to estimate I is as follows. Run a Markov chain (X_n) for R steps and define

$$\hat{Y}_R = \frac{1}{R} \sum_{j=1}^{R} Y_j,$$

where $Y_j = f(X_j)$. Unless X_0 is initialized with the stationary distribution π, $\hat{Y}_R$ is not an unbiased estimator of I. However, under irreducibility assumptions, the estimator is strongly consistent. This result follows from the ergodic theorem, as stated in Proposition 6.7.1.

A typical MCMC application is as follows. Consider a simple hard-core model on a grid $V = \{1, \ldots, n_1\} \times \{1, \ldots, n_2\} \subset \mathbb{Z}^2$ with a natural neighboring system (neighbors are within a distance of one). A proper configuration on V represents the presence (denoted by '1') or absence (denoted by '0') of particles at each point, subject to the restriction that no two adjacent points can simultaneously have a particle. Let $\xi_0, \xi_1, \ldots$ be a sequence of proper configurations sampled by Algorithm 47. Our goal is to calculate the expected number of particles I in a proper configuration, i.e., the expected number of '1's in a configuration chosen uniformly at random from all feasible configurations. Let $f(\xi)$ denote the number of particles in a proper configuration ξ. By the ergodic theorem for Markov chains, we can estimate the mean number of particles I using the estimator $\hat{Y}_R$. Simulation results for this example are provided later in Example 6.7.13. This and analogous estimators for I are studied in more detail below.

6.7.1 *Ergodic Theorem and CLT for Markov Chains*

We now state the ergodic theorems for regular Markov chains. While some results (e.g., Proposition 6.7.1) hold for a broader class of irreducible chains, these are sometimes referred to as the strong law of large numbers. The proofs leverage the regenerative structure of Markov chains, which we briefly outline below. Throughout, we assume that the chain $(X_n, n \geq 0)$ operates on a finite state space and that X_0 has some initial distribution μ.

Proposition 6.7.1 *With probability 1,*

$$\lim_{R \to \infty} \hat{Y}_R = \sum_{s \in S} f(s)\pi_s = I.$$

Since $\hat{Y}_R$ is not unbiased, we can examine its asymptotic bias. The convergence rate of $\mathbb{E}\hat{Y}_R$ to I is given in the following proposition.

Proposition 6.7.2 *For some finite constant β, we have*

$$\mathbb{E}\hat{Y}_R = I + \frac{\beta}{R} + o\left(\frac{1}{R}\right).$$

Proof Set $b_j = \mathbb{E}Y_j - I$. We know $\sum_j |b_j| < \infty$ since S is finite (proof omitted). Then,

$$\mathbb{E}\hat{Y}_R = \frac{1}{R}\sum_{j=0}^{R}(\mathbb{E}Y_j - I + I) = I + \frac{1}{R}\sum_{j=0}^{R-1} b_j = I + \frac{1}{R}\sum_{j=0}^{\infty} b_j - \frac{1}{R}\sum_{j=R}^{\infty} b_j$$

Since $\frac{1}{R}\sum_{j=R}^{\infty} b_j = o(1/R)$, the proof is complete. $\qquad\square$

Remark 6.7.3 The constant $\beta \equiv \beta_\mu$ depends on the initial distribution μ. $\qquad\blacksquare$

Remark 6.7.4 Proposition 6.7.2 is valid for finite-state Markov chains. For general cases, it requires the assumption that $\sum_j |b_j| < \infty$. $\qquad\blacksquare$

We now present the CLT for Markov chains.

Proposition 6.7.5 *If (X_n) is an ergodic finite Markov chain and f is a real-valued function, then with probability 1*

$$\hat{Y}_R \to I = \sum_{s \in S} \pi_s f(s), \qquad R \to \infty, \tag{6.7.1}$$

and

$$\sqrt{R}(\hat{Y}_R - I) \xrightarrow{\mathcal{D}} \mathcal{N}(0, \varsigma^2),$$

where

$$0 < \varsigma^2 = \lim_{R \to \infty} R \mathrm{Var} \hat{Y}_R < \infty. \tag{6.7.2}$$

The limits in (6.7.1) and (6.7.2) are independent of the initial distribution μ.

A natural proof approach uses the regenerative structure of $(\hat{Y}_R)$.

Regenerative Structure of Markov Chains   Instead of providing a formal proof of Proposition 6.7.1, we will outline its key idea. Let $s_0 \in S$ and set $Y_j = f(X_j)$. Since (X_k) is aperiodic and irreducible, it visits the state s_0 infinitely often, a result that follows from the strong Markov property. Assume $X_0 = s_0$. Let $\gamma_0 = 0$, $\gamma_l, l = 1, 2, \ldots$ represent the moments of consecutive visits of (X_k) to s_0. Define the moment of the last *regeneration* during the first R as $v(R)$, given by:

$$v(R) = \max\{i : \gamma_i \le R\}.$$

This means regenerations occur at $\gamma_0, \gamma_1, \ldots, \gamma_{v(R)}$. The lengths of intervals between consecutive returns are denoted by η_l, and the length of the interval between R and the last regeneration is denoted by B_R, where:

$$\eta_l = \gamma_l - \gamma_{l-1}, \quad l = 1, 2, \ldots, \quad B_R = R - \gamma_{v(R)}.$$

This setup defines the *regeneration structure*, as the following **cycles**

$$(\eta_l; Y_{\gamma_l}, Y_{\gamma_l+1}, \ldots, Y_{\gamma_{l+1}-1})$$

are independent. Figure 6.15 illustrates the concept of the regenerative structure.
   Define:

$$W_l = \sum_{j=\gamma_{l-1}}^{\gamma_l - 1} f(X_j), \quad l = 1, \ldots, v(R), \tag{6.7.3}$$

which are i.i.d. random variables. From Markov chain theory, it is known that $\mathbb{E}\eta_1^2 < \infty$, and consequently, $\mathbb{E}W_1^2 < \infty$. To see this, we may assume without loss of generality that $\mathbf{P} > \mathbf{0}$ (i.e., each entry is positive). Set $q_i = \sum_{i \neq s_0} p_{s_0 i}$ and $\bar{q} = \max_i q_i$. Note that $\bar{q} < 1$. We have

$$\mathbb{P}(\gamma_1 > k) = \sum_{i_1,\ldots,i_k \neq s_0} p_{s_0 i_1} p_{i_1 i_2} \cdots p_{i_{k-1} i_k} \leq \bar{q}^k.$$

Hence, all moments of γ_1 are finite. Note that for a finite state space $\mathcal{S}$, we have $||f||_\infty = \max_{s \in \mathcal{S}} |f(s)| < \infty$, thus

$$\mathbb{E}|W_1| \leq \mathbb{E}\gamma_1 ||f||_\infty < \infty.$$

Finiteness of $\mathbb{E}\gamma_1$ implies that $\mathbb{E}\eta_1 < \infty$ and $\mathbb{V}\mathrm{ar}\,\eta_1 < \infty$, thus

$$\mathbb{E}W_1^2 \leq ||f||_\infty^2 \mathbb{E}\eta_1^2 < \infty.$$

For a denumerable state space, we need to assume that $\mathbb{E}W_1 < \infty$.

Denoting the remaining sum of $f(X_j)$ by:

$$r_R = \sum_{j=\gamma_{\nu(R)}}^{R} f(X_j)$$

(which is also independent of the W_l's), we obtain the decomposition:

$$Y_0 + Y_1 + \cdots + Y_R = W_1 + \cdots + W_{\nu(R)} + r_R. \tag{6.7.4}$$

This representation is key to further analysis.

To proceed, we state without proof a regenerative formula for the stationary distribution π. Define the vector $\mathbf{x} = (x_s, s \in \mathcal{S})$, where:

$$x_s = \mathbb{E}\left[\sum_{j=0}^{\gamma_1 - 1} \mathbb{1}(X_j = s) \right].$$

We have the following proposition:

Proposition 6.7.6 *[21, Chapter 3, Theorem 2.1] Suppose* $\mathbf{P}$ *is regular. The vector* $\mathbf{x}$ *is invariant with respect to the t.m.* $\mathbf{P}$*. Hence, the stationary distribution is given by* $\pi_s = x_s / \sum_{v \in \mathcal{S}} x_v$, $s \in \mathcal{S}$*, with* $\pi_s > 0$ *for all* $s \in \mathcal{S}$*.*

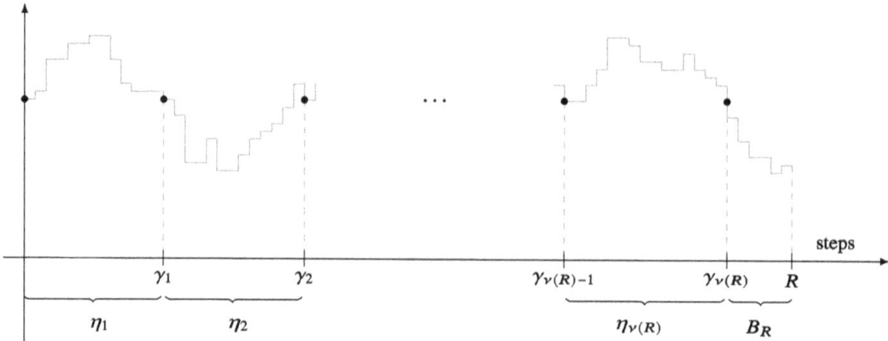

Fig. 6.15 Regenerative structure of Markov chain

In the proof of Proposition 6.7.1, we will use the following lemma.

Lemma 6.7.7 *[21, Chapter 3, Theorem 2.1 and formula (4.4)]. Let $X_0 = s_0 \in S$. Then,*

$$\lim_{R \to \infty} \frac{1}{\nu(R)} \sum_{j=0}^{R} f(X_j) = \sum_{s \in S} f(s) x_s,$$

with probability 1. In particular, setting $f \equiv 1$, we obtain $R/\nu(R) \to \sum_{x \in S} x_s = \mathbb{E}\gamma_1 = \mathbb{E}\eta_1$ with probability 1.

We now prove the ergodic theorem for Markov chains.

Proof of Proposition 6.7.1 Consider the representation (6.7.4). We outlined the proof that $\mathbb{E}W_1 < \infty$, which is equivalent to $\mathbb{E}\gamma_1 < \infty$. Let $(W_1, W_2, \ldots)$ be an i.i.d. sequence with $W_i \sim W_l$ as defined in (6.7.3). With probability 1,

$$\lim_{n \to \infty} \frac{1}{n} \sum_{l=1}^{n} W_l = \mathbb{E}W_1.$$

Now,

$$\mathbb{E}W_1 = \mathbb{E}\left[\sum_{j=0}^{\gamma_1-1} f(X_j) \right] = \mathbb{E}\left[\sum_{j=0}^{\gamma_1-1} \sum_{s \in S} f(s)\mathbb{1}(X_j = s) \right] = \sum_{s \in S} f(s) x_s.$$

$$(6.7.5)$$

Notice that

$$\sum_{l=1}^{\nu(R)} W_l \leq \sum_{j=1}^{R} Y_j \leq \sum_{l=1}^{\nu(R)+1} W_l,$$

and hence

$$\frac{\nu(R)}{R}\frac{1}{\nu(R)}\sum_{l=1}^{\nu(R)} W_l \le \frac{1}{R}\sum_{j=1}^{R} Y_j \le \frac{\nu(R)+1}{R+1}\frac{R+1}{R}\frac{1}{\nu(R)+1}\sum_{l=1}^{\nu(R)+1} W_l.$$

$$(6.7.6)$$

From Lemma 6.7.7, substituting $f \equiv 1$, we have with probability 1 (note that $\gamma_1 \equiv \eta_1$),

$$\lim_{R\to\infty}\frac{R}{\nu(R)} = \sum_{s\in S} x_s = \mathbb{E}\gamma_1 = \mathbb{E}\eta_1.$$

Passing to the limit in (6.7.6) as $R \to \infty$, we obtain with probability 1,

$$I = \lim_{R\to\infty}\hat{Y}_R = \frac{\mathbb{E}\left[\sum_{j=0}^{\eta_1-1} f(X_j)\right]}{\mathbb{E}\eta_1} = \frac{\mathbb{E}W_1}{\mathbb{E}\eta_1} = \sum_{s\in S} f(s)\pi_s. \qquad (6.7.7)$$

$\square$

To prove a version of the CLT for Markov chains, we must recall the Anscombe CLT for random sums.

Proposition 6.7.8 (Random Index CLT; see Durrett [43], p.114) *Let $V_1, V_2, \ldots$ be i.i.d. with $\mathbb{E}V_1 = 0$ and $\mathbb{E}V_1^2 = \sigma^2 \in (0, \infty)$. Let N_R be a sequence of non-negative integer-valued random variables and a_R a sequence of integers with $a_R \to \infty$ and $N_R/a_R \to 1$ in probability. Then, for $R \to \infty$,*

$$\frac{\sum_{l=1}^{N_R} V_l}{\sqrt{a_R}} \xrightarrow{\mathcal{D}} \mathcal{N}(0, \sigma^2).$$

Proof of Proposition 6.7.5 Without loss of generality, we consider $\sum_{j=0}^{R} Y_j/R$ instead of $\sum_{j=0}^{R} Y_j/(R+1)$. We also assume that $X_0 = s_0$, although the initial distribution can be arbitrary. Note that R can be rewritten as $R = \sum_{l=1}^{\nu(R)} \eta_l + B_R$. Using the decomposition in (6.7.4), we have

$$\sqrt{R}\left(\frac{\sum_{j=0}^{R} Y_j - RI}{R}\right) = \frac{\sum_{l=1}^{\nu(R)} W_l - \sum_{l=1}^{\nu(R)} I\eta_l + r_R - IB_R}{\sqrt{R}}$$

$$\underbrace{\frac{\sum_{l=1}^{\nu(R)}(W_l - I\eta_l)}{\sqrt{R}}}_{(I)} + \underbrace{\frac{r_R - IB_R}{\sqrt{R}}}_{(II)}.$$

- For part (I), we apply Anscombe's Theorem 6.7.8 with

$$N_R = v(R), \quad a_R = \frac{R}{\mathbb{E}\eta_1}, \quad V_l = W_l - I\eta_l.$$

From Lemma 6.7.7, we have $R/v(R) \to \mathbb{E}\eta_1$ as $R \to \infty$, so

$$\frac{N_R}{a_R} = \frac{v(R)}{R}\mathbb{E}\eta_1 \to \frac{1}{\mathbb{E}\eta_1}\mathbb{E}\eta_1 = 1 \quad \text{as } R \to \infty.$$

From (6.7.7), we know that $\mathbb{E}W_1 = I\mathbb{E}\eta_1$, and hence

$$\mathbb{E}V_l = \mathbb{E}W_l - I\mathbb{E}\eta_l = I\mathbb{E}\eta_l - I\mathbb{E}\eta_l = 0.$$

Thus, the assumptions of Anscombe's Theorem 6.7.8 are satisfied, leading to

$$\frac{\sum_{l=1}^{v(R)}(W_l - I\eta_l)}{\sqrt{R}} = \frac{1}{\sqrt{\mathbb{E}\eta_1}}\frac{\sum_{l=1}^{N_R} V_l}{\sqrt{a_R}} \xrightarrow{\mathcal{D}} \mathcal{N}\left(0, \frac{\mathbb{E}(W_1 - I\eta_1)^2}{\mathbb{E}\eta_1}\right).$$

- For part (II), observe that r_R can be bounded as $r_R \leq \sup_{s \in S}|f(s)|B_R$. It is known that if $\mathbb{E}\eta_1 < \infty$, then B_R converges to a proper distribution. Specifically (see [21, Section 4.4.2]),

$$\lim_{R \to \infty} \mathbb{P}(B_R = i) = \frac{\mathbb{P}(\eta_1 > i)}{\mathbb{E}\eta_1}.$$

This implies that $B_R/\sqrt{R}$ converges in probability to zero. This completes the proof.

$\square$

We have also derived the asymptotic variance ς^2 from Proposition 6.7.5, namely:

$$\varsigma^2 = \frac{\mathbb{V}\mathrm{ar}(W_1 - I\eta_1)}{\mathbb{E}\eta_1}. \tag{6.7.8}$$

Now consider another estimator for I:

$$\hat{I}_n = \frac{\sum_{l=1}^{n} W_l}{\sum_{l=1}^{n} \eta_l}.$$

The main difference between $\hat{Y}_R$ and $\hat{I}_n$ is the following: for $\hat{Y}_R$, we run the chain for a fixed number of steps R, whereas for $\hat{I}_n$, we run it for a random number of steps—specifically, until n cycles, i.e., until the chain returns to X_0 for the n-th time. Since

$$\hat{I}_n = \frac{\sum_{l=1}^{n} W_l/n}{\sum_{l=1}^{n} \eta_l/n},$$

from the strong law of large numbers, we have $\hat{I}_n \to \mathbb{E}W_1/\mathbb{E}\eta_1 = I$, showing that $\hat{I}_n$ is a consistent estimator. Moreover,

$$\sqrt{n}(\hat{I}_n - I) = \frac{\sum_{l=1}^n (W_l - I\eta_l)/\sqrt{n}}{\sum_{l=1}^n \eta_l/n} \xrightarrow{\mathcal{D}} N\left(0, \frac{\mathbb{E}(W_1 - I\eta_1)^2}{(\mathbb{E}\eta_1)^2}\right).$$

Hence, $\hat{I}_n$ has an asymptotic variance $\bar{\varsigma}^2 = \mathbb{E}(W_1 - I\eta_1)^2/(\mathbb{E}\eta_1)^2$, which is smaller than ς^2 from (6.7.8).

The Concept of Stationarity We now discuss the concept of stationarity for the random sequence (Y_n). Recall that a sequence of random variables $(Y_n, n \in \mathbb{Z}_+)$ is *stationary* if the following condition is fulfilled:

(k) For any $k = 1, 2, \ldots$ and $0 < h_1 < \cdots < h_{k-1}$, the joint distributions of $(Y_j, Y_{j+h_1}, \ldots, Y_{j+h_{k-1}})$ are independent of j.

The implications of this definition are as follows:

($k = 1$) The distributions of Y_j are identical for all $j \in \mathbb{Z}_+$. In particular, all $\mathbb{E}Y_j$ are equal, provided $I = \mathbb{E}Y_0$ exists.

($k = 2$) For any $h = 1, \ldots$, the joint distributions of (Y_j, Y_{j+h}) are independent of j. In particular, the covariance $\rho_h = \text{Cov}\,(Y_j, Y_{j+h})$ depends only on h, provided that the second moment exists.

Since in our case $Y_j = f(X_j)$, where X_j takes only a finite number of values, there is no problem with the existence of moments. Let π be a stationary distribution and $(X_n, n = 0, 1, \ldots)$ be a Markov chain with the initial distribution $\mu = \pi$ and the transition matrix $\mathbf{P}$. Then $(Y_n)_{n=0,1,\ldots}$, where $Y_n = f(X_n)$, is a stationary sequence with the correlation function

$$\rho_k = \text{Corr}\,(Y_n, Y_{n+k}).$$

Since the state space is finite and $\mathbf{P}$ is regular, we have $\sum_j |\rho_j| < \infty$. This requires a proof, for example, using the Perron-Frobenius theorem. Recall the quantity ς^2 defined in (6.7.2):

$$\varsigma^2 = \lim_{R \to \infty} R\text{Var}\hat{Y}_R.$$

Thus, it is called the **TAVC** (*time average variance constant*) or the **asymptotic variance.** We now show a formula for the TAVC in terms of the covariance sequence.

Lemma 6.7.9 *If* $\mathbf{P}$ *is regular, then*

$$\text{Var}\hat{Y}_R = \frac{\text{Var}Y_0}{R}\left(1 + 2\sum_{j=1}^{R-1}\left(1 - \frac{j}{R}\right)\rho_j\right),$$

and hence

$$\varsigma^2 = \mathbb{V}\mathrm{ar}Y\left(1 + 2\sum_{k=1}^{\infty}\rho_k\right).$$

Proof Since ς^2 is independent of the initial distribution μ, we can consider a stationary sequence (Y_R). Without loss of generality, we may assume that $\mathbb{E}Y_j = 0$. Then

$$R\mathbb{V}\mathrm{ar}\hat{Y}_R = \frac{1}{R}\mathbb{V}\mathrm{ar}(Y_0 + \cdots + Y_{R-1})$$

$$= \frac{1}{R}\left(R\mathbb{V}\mathrm{ar}Y_0 + 2\sum_{0\le i<j\le R-1}\mathrm{Cov}\,(Y_i, Y_j)\right)$$

$$= \rho_0 + \frac{2}{R}\sum_{0\le i<j\le R-1}\rho_{j-i}$$

$$= \rho_0 + \frac{2}{R}\sum_{i=0}^{R-1}\sum_{h=1}^{i}\rho_h$$

$$\sim \rho_0 + 2\sum_{h=1}^{\infty}\rho_h.$$

$\square$

Fundamental Matrix and Asymptotic Variance ς^2 Recall that $Y_j = f(X_j)$. If $X_0 \sim \pi$, then the Markov chain (X_j) is stationary with values in $\mathcal{S}$. From Lemma 6.7.9, we know that

$$\varsigma^2 = \rho_0 + 2\sum_{k=1}^{\infty}\rho_k.$$

We now calculate $\rho_k = \mathrm{Cov}\,(Y_0, Y_k)$. Namely, $\mathrm{Cov}\,(Y_0, Y_k) = \mathbb{E}Y_0Y_k - (\mathbb{E}Y_0)^2$, and

$$\mathbb{E}Y_0Y_k = \sum_{i=1}^{m}\pi_i\mathbb{E}[f(X_0)f(X_k) \mid X_0 = s_i] = \sum_{i,j}\pi_i f(s_i)f(s_j)\mathbf{P}^k(s_i, s_j).$$

Denoting by $\mathbf{f} = (f(s_1), \ldots, f(s_m))$ and $\mathrm{diag}(\mathbf{f})$ a diagonal matrix with vector $\mathbf{f}$ on the diagonal, we can rewrite $\mathbb{E}Y_0Y_k$ in matrix form:

$$\mathbb{E}Y_0Y_k = \pi\,\mathrm{diag}(\mathbf{f})\mathbf{P}^k\mathbf{f}^T.$$

In particular, $(\mathbb{E}Y_0)^2 = (\sum_{j=1}^{m} \pi_j f(s_j))^2 = \sum_{i,j} \pi_i \pi_j f(s_i) f(s_j)$, and hence

$$\text{Cov}\,(Y_0, Y_k) = \pi \, \text{diag}(\mathbf{f})(\mathbf{P}^k - \pi)\mathbf{f}^T, \qquad (6.7.9)$$

where π is an $m \times m$ matrix in which each row is $(\pi_1, \ldots, \pi_m)$. We now introduce the so-called **fundamental matrix**,

$$\mathbf{Z} = \mathbf{I} + \sum_{k=1}^{\infty}(\mathbf{P}^k - \mathbf{\Pi}).$$

Lemma 6.7.10 *For a regular transition matrix* $\mathbf{P}$, *the fundamental matrix is of the form:*

$$\mathbf{Z} = \mathbf{I} + \sum_{k=1}^{\infty}(\mathbf{P}^k - \mathbf{\Pi}) = (\mathbf{I} - (\mathbf{P} - \mathbf{\Pi}))^{-1}. \qquad (6.7.10)$$

Proof We first prove that

$$\mathbf{P}^k - \pi = (\mathbf{P} - \pi)^k.$$

This is true for $k = 1$. Suppose it holds for $k - 1$. Noting that $\pi\mathbf{P} = \mathbf{P}\pi = \pi^2 = \pi$, we have:

$$(\mathbf{P} - \pi)^k = (\mathbf{P} - \pi)^{k-1}(\mathbf{P} - \pi) = (\mathbf{P}^{k-1} - \pi)(\mathbf{P} - \pi)$$

$$= \mathbf{P}^k - \pi\mathbf{P} - \mathbf{P}^{k-1}\pi + \pi^2 = \mathbf{P}^k - \pi.$$

For a matrix $\mathbf{A}$, we have:

$$(\mathbf{I} - \mathbf{A})(\mathbf{I} + \mathbf{A} + \cdots + \mathbf{A}^{k-1}) = \mathbf{I} - \mathbf{A}^k.$$

If $\mathbf{A}^k \to \mathbf{0}$, then the right-hand side of the above equality converges to the identity matrix $\mathbf{I}$, which implies that $\mathbf{I} - \mathbf{A}$ is invertible.

Take $\mathbf{A} = \mathbf{P} - \mathbf{\Pi}$. We have shown that $\mathbf{A}^k = (\mathbf{P} - \mathbf{\Pi})^k = \mathbf{P}^k - \mathbf{\Pi}$. Since we assume that $\mathbf{P}$ is regular, then from the Perron-Frobenius theorem [4] we have $\mathbf{P}^k \to \mathbf{\Pi}$, which implies $\mathbf{A}^k = \mathbf{P}^k - \mathbf{\Pi} \to \mathbf{0}$. Thus, $\mathbf{I} - (\mathbf{P} - \mathbf{\Pi})$ is invertible, and its inverse can be written as:

$$(\mathbf{I} - (\mathbf{P} - \mathbf{\Pi}))^{-1} = \mathbf{I} + (\mathbf{P} - \mathbf{\Pi}) + (\mathbf{P} - \mathbf{\Pi})^2 + \cdots = \mathbf{Z}.$$

This completes the proof. $\qquad\square$

---

[4] See Lemma 7.2.2 in Rolski et al. [180].

Using the fundamental matrix, the sum of covariances in (6.7.9) can be computed as:

$$\sum_{k=1}^{\infty} \text{Cov}\,(Y_0, Y_k) = \pi\,\text{diag}(\mathbf{f})\sum_{k=1}^{\infty}(\mathbf{P}^k - \pi)\mathbf{f}^T,$$

and from (6.7.10), we have:

$$\sum_{k=1}^{\infty}(\mathbf{P}^k - \pi) = \mathbf{Z} - \mathbf{I},$$

thus:

$$\sum_{k=1}^{\infty} \text{Cov}\,(Y_0, Y_k) = \pi\,\text{diag}(\mathbf{f})(\mathbf{Z} - \mathbf{I})\mathbf{f}^T.$$

Hence, we obtain:

Lemma 6.7.11 *The asymptotic variance ς^2 can be expressed in terms of the fundamental matrix $\mathbf{Z}$ as:*

$$\varsigma^2 = \rho_0 + 2\pi\,\text{diag}(\mathbf{f})(\mathbf{Z} - \mathbf{I})\mathbf{f}^T = \rho_0 + 2\pi\,\text{diag}(\mathbf{f})((\mathbf{I} - (\mathbf{P} - \mathbf{\Pi}))^{-1} - \mathbf{I})\mathbf{f}^T.$$

Statistics for Estimating I Note that the sequence (Y_j) does not need to be stationary. The estimator $\hat{Y}_R$ is consistent (that is, it converges with probability 1 to $I = \sum_{s\in\mathcal{S}}^{m} \pi_s f(s)$), but if the initial distribution μ of X_0 is not the stationary one, the estimator is not unbiased.

How can we estimate ρ_k and ς^2? The covariances can be estimated using the following estimators (for a fixed N):

$$\hat{\rho}_k = \frac{1}{N-k}\sum_{j=1}^{N-k}(Y_j - \hat{Y}_N)(Y_{j+k} - \hat{Y}_N), \qquad (6.7.11)$$

and for large N, we can approximate ς^2 using:

$$\hat{\varsigma}_N^2 = \sum_{k=-N}^{N}\hat{\rho}_k = \hat{\rho}_0 + 2\sum_{k=1}^{N}\hat{\rho}_k \approx \varsigma^2. \qquad (6.7.12)$$

An alternative non-parametric method for estimating ς^2 is based on batching. Let $R = nm$, and divide the sequence into m batches of size n: $(Y_0, \ldots, Y_{n-1})$, $(Y_n, \ldots, Y_{2n-1})$, $\ldots$. The corresponding estimator is given by:

$$\hat{\varsigma}^2 = \frac{n}{m-1} \sum_{k=1}^{m} \left(\left[\frac{1}{n} \sum_{i=(k-1)n}^{kn-1} Y_i \right] - \hat{Y}_R \right)^2 .$$

As noted in [72, p.148], this estimator is approximately unbiased for ς^2 as $n, m \to \infty$. For a more detailed explanation of batching methods and their use in estimating I, see the paragraph *Method of batch means* below.

Burn-in Period We may disregard the first R_0 samples (R_0 is called the *burn-in period*) and estimate I from the average of samples at times R_0+1, R_0+2, $\ldots$, R_0+R:

$$\hat{Y}_{R_0,R} = \frac{1}{R} \sum_{j=1}^{R} Y_{j,R_0} := \frac{1}{R} \sum_{j=1}^{R} f(X_{j+R_0}).$$

We define $\hat{Y}_R \equiv \hat{Y}_{0,R}$ and $Y_{j,0} \equiv Y_j$. While we discussed $\hat{Y}_R$, in practice, one usually uses some burn-in period $R_0 > 0$. Estimating R_0 is important, as it represents the time needed to *reach* stationarity. If the mixing time τ is known, it makes sense to take $R_0 = \tau$. However, τ is often difficult to compute or even bound. Thus, using R_0 is more of a *rule of thumb* than a formally justified principle. For methods to estimate R_0, see, for example, Sahlin [188] (Section 2) and references therein.

Subsampling Another option is to use subsampling. Run R independent chains, each for the same predefined number of steps t_0, obtaining $\mathbb{X}^i = (X_0^i, \ldots, X_{t_0}^i)$, $i = 1, \ldots, R$. Then, treat $(X_{t_0}^1, \ldots, X_{t_0}^R)$ as an approximate sample from π and estimate I via

$$\hat{I}_R^{(0)} = \frac{1}{R} \sum_{i=1}^{R} Y_{t_0}^i := \frac{1}{R} \sum_{i=1}^{R} f(X_{t_0}^i).$$

There are two sources of error: (1) $X_{t_0}^i$ approximates the π-distribution, introducing a bias. To quantify this, we need bounds on the rate of convergence, such as the mixing time (6.2.2), which measures the L^1 distance between the distribution of $X_{t_0}^i$ and π. (2) Confidence intervals for $\hat{I}_R^{(0)}$ require knowledge or bounds on $\mathbb{V}\mathrm{ar}(X_{t_0}^i)$. Running $R \cdot t_0$ simulations can be computationally intensive, as both R and t_0 are typically large. If *perfect simulation* is possible, the analysis simplifies. In that case, we obtain $(X_{t_1}^1, \ldots, X_{t_R}^R)$, an unbiased sample of size R from π, where t_i, the runtime for the i-th simulation, is random. This method is computationally demanding, so $\hat{Y}_R$ or $\hat{Y}_{R_0,R}$ is often used instead.

Method of Batch Means Suppose $Y_0, Y_1, \ldots$ is a stationary sequence, not necessarily a Markov chain. Let $I = \mathbb{E}Y_0$, the value we aim to estimate. Let $\gamma_j = \mathrm{Cov}\,(Y_n, Y_{n+j})$, $j = 0, \ldots$, be the covariance function and $\rho_j = \gamma_j/\gamma_0$ its correlation function. To estimate I, we can use the *method of batch means*. Consider $Y_0, \ldots, Y_{R-1}$, which we split into m batches of size n:

$$(Y_0, \ldots, Y_{n-1}), (Y_n, \ldots, Y_{2n-1}), \ldots, (Y_{(m-1)n}, \ldots, Y_{mn-1}).$$

Define a new sequence of batch means:

$$\hat{Y}_k(n) = \frac{1}{n} \sum_{i=(k-1)n}^{kn-1} Y_i, \qquad k = 1, \ldots, m. \tag{6.7.13}$$

The method of batch means treats the batch means as independent and approximately normally distributed. The confidence interval is estimated using t-statistics. Let $\bar{Y}_R = (Y_0 + \cdots + Y_{R-1})/R$, equal to the sample mean of the batch means:

$$\overline{Y}_m(n) = \frac{\hat{Y}_1(n) + \cdots + \hat{Y}_m(n)}{m}.$$

Now set:

$$V_m^2(n) = \frac{1}{m-1} \sum_{j=1}^{m} (\hat{Y}_j(n) - \bar{Y}_R)^2.$$

To estimate I with confidence bounds at level α, we use:

$$\hat{Y}_R \pm t_{n-1,1-\alpha/2} \sqrt{\frac{V_m^2(n)}{n}},$$

where $t_{d,\alpha}$ is the quantile of size α from the t-Student distribution with d degrees of freedom. The theory behind this recommendation is as follows: Clearly,

$$I = \mathbb{E}\hat{Y}_1(n) = \mathbb{E}\hat{Y}_R = \mathbb{E}\overline{Y}_m(n).$$

For the variance of $\hat{Y}_1(n)$, Lemma 6.7.9 gives:

$$\mathrm{Var}\hat{Y}_1(n) = \frac{\sigma_Y^2}{n}\left(1 + 2\sum_{j=1}^{n-1}\left(1 - \frac{j}{R}\right)\rho_j\right).$$

The estimate with confidence bounds in (6.7.13) follows from the limiting theorem by Glynn and Iglehart [64]:

$$\frac{\hat{Y}_R - I}{\sqrt{V_m^2(n)/n}} \xrightarrow{\mathcal{D}} t_{n-1}.$$

Note that fixed n presents a challenge: $V_m^2(n)$ is not consistent with the asymptotic variance ς. This issue is discussed in Damerdji [30], and Alexopoulos and Seila [4].

Example 6.7.12 (Two-state Markov Chain) Consider a chain (X_n) on $\mathcal{S} = \{0, 1\}$ with transition matrix

$$\mathbf{P} = \begin{pmatrix} 1 - \alpha & \alpha \\ \beta & 1 - \beta \end{pmatrix} \tag{6.7.14}$$

for some $\alpha, \beta \in (0, 1)$. The chain is ergodic, with a stationary distribution

$$\pi = (\pi_0, \pi_1) = \left(\frac{\beta}{\alpha + \beta}, \frac{\alpha}{\alpha + \beta} \right).$$

Let us simply take $Y_k = f(X_k) = X_k$. Let $Z \sim \pi$, we have

$$I = \mathbb{E}f(Z) = \sum_{i=0}^{1} \pi_i f(i) = \pi_1 = \frac{\alpha}{\alpha + \beta}.$$

Assume we want to estimate I using a stationary chain (X_n), i.e., with the initial distribution $\mu = \pi$. We want to compute the asymptotic variance

$$\varsigma^2 = \rho_0 + 2 \sum_{k=1}^{\infty} \rho_k = \mathbb{V}\mathrm{ar}(Y_0) + 2 \sum_{k=1}^{\infty} \mathrm{Cov}\,(Y_0, Y_k).$$

Let us compute ρ_0 first. We have

$$\rho_0 = \mathbb{V}\mathrm{ar}(Y_0) = \mathbb{E}Y_0^2 - (\mathbb{E}Y_0)^2 = I - I^2 = \frac{\alpha}{\alpha + \beta} \left(1 - \frac{\alpha}{\alpha + \beta} \right) = \frac{\alpha\beta}{(\alpha + \beta)^2}.$$

We will compute ς^2 using two different methods. First, we will compute it directly (computing $\mathbb{E}(Y_0 Y_k)$ exploiting diagonalization of $\mathbf{P}^k$); then we will compute it using Lemma 6.7.11 by computing the fundamental matrix $\mathbf{Z}$.

- **Method 1: Diagonalization of $\mathbf{P}^k$.**

  We need to compute

$$\mathrm{Cov}\,(Y_0, Y_k) = \mathbb{E}Y_0 Y_k - \left(\frac{\alpha}{\alpha + \beta} \right)^2,$$

which means we need to compute (recall $Y_j = f(X_j)$ and $Y_j \in \{0, 1\}$):

$$\mathbb{E}(Y_0 Y_k) = \mathbb{P}(Y_0 = Y_k = 1) = \mathbb{P}(Y_0 = Y_k = 1 | Y_0 = 1)\mathbb{P}(Y_0 = 1).$$

Thus,

$$\mathbb{E}(Y_0 Y_k) = \mathbb{P}(Y_k = 1 | Y_0 = 1)\frac{\alpha}{\alpha + \beta} = \mathbf{P}^k(1, 1)\frac{\alpha}{\alpha + \beta}.$$

The eigenvalues of $\mathbf{P}$ are 1 and $1 - \alpha - \beta$. Computing the eigenvectors, we have the following diagonalization:

$$\mathbf{P} = \mathbf{A}\mathbf{D}\mathbf{A}^{-1} := \begin{pmatrix} 1 & -\frac{\alpha}{\beta} \\ 1 & 1 \end{pmatrix} \begin{pmatrix} 1 & 0 \\ 0 & 1 - \alpha - \beta \end{pmatrix} \begin{pmatrix} \frac{\beta}{\alpha+\beta} & \frac{\alpha}{\alpha+\beta} \\ -\frac{\beta}{\alpha+\beta} & \frac{\beta}{\alpha+\beta} \end{pmatrix},$$

from which we obtain:

$$\mathbf{P}^k = \frac{1}{\alpha + \beta} \begin{pmatrix} \beta & \alpha \\ \beta & \alpha \end{pmatrix} + \frac{(1 - \alpha - \beta)^k}{\alpha + \beta} \begin{pmatrix} \alpha & -\alpha \\ -\beta & \beta \end{pmatrix}.$$

We thus have:

$$\mathbf{P}^k(1, 1) = \frac{\alpha + \beta(1 - \alpha - \beta)^k}{\alpha + \beta},$$

consequently:

$$\mathbb{E}(Y_0 Y_k) = \frac{\alpha^2 + \alpha\beta(1 - \alpha - \beta)^k}{(\alpha + \beta)^2},$$

and

$$\mathrm{Cov}\,(Y_0, Y_k) = \frac{\alpha^2 + \alpha\beta(1 - \alpha - \beta)^k}{(\alpha + \beta)^2} - \left(\frac{\alpha}{\alpha + \beta}\right)^2 = \frac{\alpha\beta(1 - \alpha - \beta)^k}{(\alpha + \beta)^2}.$$

Finally, we compute:

$$\varsigma^2 = \frac{\alpha\beta}{(\alpha + \beta)^2} + \frac{2\alpha\beta}{(\alpha + \beta)^2} \sum_{k=1}^{\infty} (1 - \alpha - \beta)^k$$

$$= \frac{\alpha\beta}{(\alpha + \beta)^2} + \frac{2\alpha\beta}{(\alpha + \beta)^2} \frac{1 - \alpha - \beta}{\alpha + \beta}$$

$$= \frac{\alpha\beta}{(\alpha+\beta)^2}\left(1 + \frac{2(1-\alpha-\beta)}{\alpha+\beta}\right)$$

$$= \frac{\alpha\beta}{(\alpha+\beta)^2}\frac{2-\alpha-\beta}{\alpha+\beta} = \frac{\alpha\beta(2-\alpha-\beta)}{(\alpha+\beta)^3}.$$

- **Method 2: Using the fundamental matrix.**

  Notice that we have

  $$\pi = \frac{1}{\alpha+\beta}\begin{pmatrix} \beta & \alpha \\ \beta & \alpha \end{pmatrix},$$

  thus, we may compute the fundamental matrix using Lemma 6.7.10, namely

  $$\mathbf{Z} = (\mathbf{I} - (\mathbf{P} - \boldsymbol{\pi}))^{-1} = \left(\mathbf{I} - \frac{1-\alpha-\beta}{\alpha+\beta}\begin{pmatrix} \alpha & -\alpha \\ -\beta & \beta \end{pmatrix}\right)^{-1}$$

  $$= \frac{1}{(\alpha+\beta)^2}\begin{pmatrix} \beta(\alpha+\beta)+\alpha & -\alpha(1-\alpha-\beta) \\ -\beta(1-\alpha-\beta) & \alpha(\alpha+\beta)+\beta \end{pmatrix}.$$

Therefore,

$$\mathbf{Z} - \mathbf{I} = \frac{1}{(\alpha+\beta)^2}\begin{pmatrix} \alpha(1-\alpha-\beta) & \alpha(1-\alpha-\beta) \\ -\beta(1-\alpha-\beta) & \beta(1-\alpha-\beta) \end{pmatrix}.$$

  We also have:

  $$\mathbf{f} = (0, 1), \qquad \operatorname{diag}(\mathbf{f}) = \begin{pmatrix} 0 & 0 \\ 0 & 1 \end{pmatrix}.$$

Compute:

$$\boldsymbol{\pi}\operatorname{diag}(\mathbf{f}) = \left(\frac{\beta}{\alpha+\beta}, \frac{\alpha}{\alpha+\beta}\right)\begin{pmatrix} 0 & 0 \\ 0 & 1 \end{pmatrix} = \left(0, \frac{\alpha}{\alpha+\beta}\right).$$

Finally, using Lemma 6.7.11, we have:

$$\varsigma^2 = \rho_0 + 2\pi\operatorname{diag}(\mathbf{f})(\mathbf{Z} - \mathbf{I})\mathbf{f}^T$$

$$= \frac{\alpha\beta}{(\alpha+\beta)^2} + \frac{2}{(\alpha+\beta)^2}\pi\operatorname{diag}(\mathbf{f})\begin{pmatrix} \alpha(1-\alpha-\beta) & \alpha(1-\alpha-\beta) \\ -\beta(1-\alpha-\beta) & \beta(1-\alpha-\beta) \end{pmatrix}\begin{pmatrix} 0 \\ 1 \end{pmatrix}$$

$$= \frac{\alpha\beta}{(\alpha+\beta)^2} + \frac{2}{(\alpha+\beta)^2}\left(0, \frac{\alpha}{\alpha+\beta}\right)\begin{pmatrix} \alpha(1-\alpha-\beta) \\ \beta(1-\alpha-\beta) \end{pmatrix}$$

$$= \frac{\alpha\beta}{(\alpha+\beta)^2} + \frac{2\alpha\beta(1-\alpha-\beta)}{(\alpha+\beta)^3} = \frac{\alpha\beta(\alpha+\beta) + 2\alpha\beta(1-\alpha-\beta)}{(\alpha+\beta)^3}$$

$$= \frac{\alpha\beta(2-\alpha-\beta)}{(\alpha+\beta)^3}.$$

Summing up, the estimator $\hat{Y}_R = \frac{1}{R}\sum_{i=1}^{R} Y_i$ has an approximately normal distribution with mean $\frac{\alpha}{\alpha+\beta}$ and variance

$$\frac{1}{R}\frac{\alpha\beta(2-\alpha-\beta)}{(\alpha+\beta)^3}.$$

The confidence interval at level $\gamma \in (0, 1)$ is given by:

$$\mathbb{P}\left(I \in \left[\hat{Y}_R - z_{1-\gamma/2}\sqrt{\frac{\alpha\beta(2-\alpha-\beta)}{R(\alpha+\beta)^3}}, \ \hat{Y}_R + z_{1-\gamma/2}\sqrt{\frac{\alpha\beta(2-\alpha-\beta)}{R(\alpha+\beta)^3}}\right]\right) \approx 1 - \gamma.$$

$$\Diamond$$

Comparing with i.i.d. Random Variables Now let $X_0^{\mathrm{ind}}, X_1^{\mathrm{ind}}, \ldots,$ be i.i.d. random variables with distribution

$$\mathbb{P}(X_i^{\mathrm{ind}} = s) = \begin{cases} \frac{\beta}{\alpha+\beta}, & \text{if } s = s_1, \\[2mm] \frac{\alpha}{\alpha+\beta}, & \text{if } s = s_2. \end{cases}$$

By the way, we can think of this sequence as a Markov chain with transition matrix

$$\mathbf{P}^{\mathrm{ind}} = \begin{pmatrix} \frac{\beta}{\alpha+\beta} & \frac{\alpha}{\alpha+\beta} \\[2mm] \frac{\beta}{\alpha+\beta} & \frac{\alpha}{\alpha+\beta} \end{pmatrix}.$$

Again, taking

$$Y^{\mathrm{ind}} = f(X^{\mathrm{ind}}) = X^{\mathrm{ind}},$$

we have

$$\mathbb{E}Y^{\mathrm{ind}} = \mathbb{E}X^{\mathrm{ind}} = \frac{\alpha}{\alpha+\beta}, \quad \mathbf{Var}Y^{\mathrm{ind}} = \mathbf{Var}X^{\mathrm{ind}} = \frac{\alpha\beta}{(\alpha+\beta)^2}.$$

Thus, we estimate I via

$$\hat{Y}_R^{\text{ind}} = \frac{1}{R} \sum_{i=1}^{R} Y_i^{\text{ind}}.$$

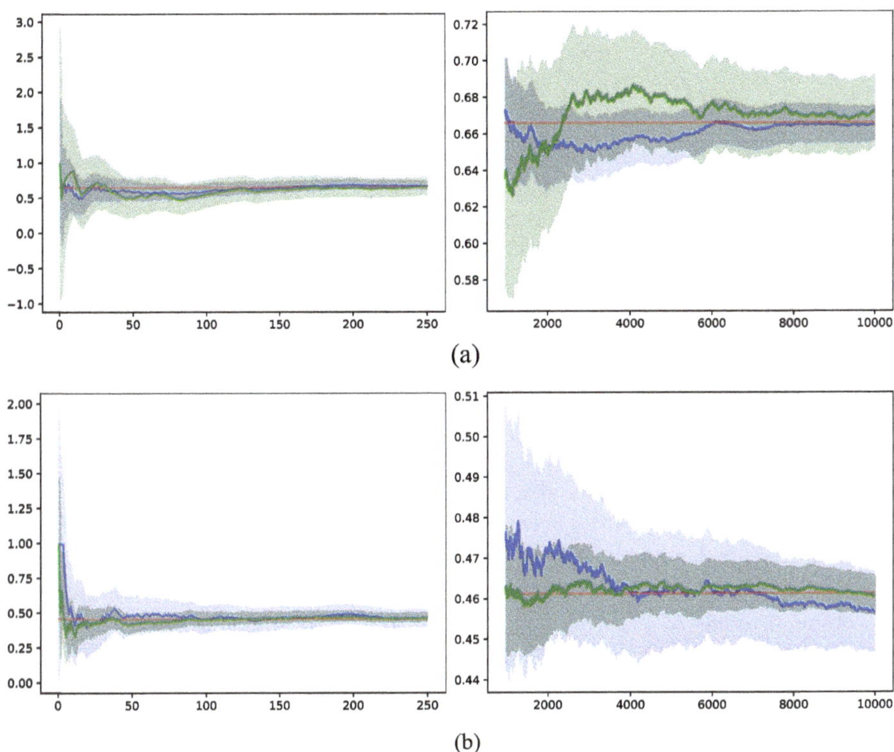

Fig. 6.16 Values of estimators of $\alpha/(\alpha + \beta)$ (*red* line) for two sets of parameters: $\hat{Y}_R^{\text{ind}}$ (i.i.d. replications)—*blue* line; $\hat{Y}_R$ (estimator based on Markov chain)—*green* line. Corresponding confidence intervals are shown in the respective colors. R ranges from 1 to 10^3; left plots show the first 250 steps, right plots show the last 9000 steps. The same randomness was used for simulating $\hat{Y}_R^{\text{ind}}$ and $\hat{Y}_R$. (**a**) Parameters: $\alpha = 1/4$, $\beta = 1/8$ (thus $\alpha + \beta = 3/8 < 1$). (**b**) Parameters: $\alpha = 3/4$, $\beta = 7/8$ (thus $\alpha + \beta = 13/8 > 1$)

From the *fundamental relation* (1.10), we have

$$b = z_{1-\alpha/2} \frac{\sqrt{\mathbf{V}\mathrm{ar}Y^{\text{ind}}}}{\sqrt{R}} = z_{1-\alpha/2} \frac{1}{\sqrt{R}} \frac{\sqrt{\alpha\beta}}{\alpha + \beta},$$

and the corresponding confidence interval at level $\gamma \in (0, 1)$ is

$$\mathbb{P}\left(I \in \left[\hat{Y}_R^{\mathrm{ind}} - z_{1-\gamma/2}\frac{1}{\sqrt{R}}\frac{\sqrt{\alpha\beta}}{\alpha + \beta}, \ \hat{Y}_R^{\mathrm{ind}} + z_{1-\gamma/2}\frac{1}{\sqrt{R}}\frac{\sqrt{\alpha\beta}}{\alpha + \beta}\right]\right) \approx 1 - \gamma.$$

Which of the estimators $\hat{Y}_R^{\mathrm{ind}}$ or $\hat{Y}_R$ has smaller variance depends on $\alpha + \beta$, since we have

$$\frac{\varsigma^2}{\mathbb{V}\mathrm{ar}\hat{Y}_R^{\mathrm{ind}}} = \frac{\alpha\beta(2 - \alpha - \beta)}{(\alpha + \beta)^3} \cdot \frac{(\alpha + \beta)^2}{\alpha\beta} = \frac{2 - \alpha - \beta}{\alpha + \beta},$$

which is smaller than 1 if $\alpha + \beta > 1$ and larger than 1 if $\alpha + \beta < 1$. For $\alpha + \beta = 1$, the variances are the same—which is no surprise since then $\mathbf{P}^{\mathrm{ind}} = \mathbf{P}$. Sample simulations for these two cases are depicted in Fig. 6.16. The example is implemented in `ch6_2state_Markov_chain_CLT.py`.

In many cases, we will not be able to compute the asymptotic variance and will need to resort to methods for estimating ς^2.

Example 6.7.13 (CLT for Number of Particles in the Hard-core Model) Consider a simple hard-core model on a grid $V = \{1, \ldots, n\} \times \{1, \ldots, n\} \subset \mathbb{Z}^2$ with a natural neighboring system (constituting a graph G). In the context of particle systems, let $\xi(v) \in \{0, 1\}$, where $\xi(v) = 1$ indicates the presence of a particle at vertex v, and $\xi(v) = 0$ its absence. A configuration $\xi = \{\xi(v) : v \in V\}$ is proper (feasible) if no two adjacent vertices contain particles. The set of all such configurations is denoted as Λ, and its cardinality as $z_G = |\Lambda|$.

Define $f(\xi) = \sum_{v \in V} \mathbf{1}(\xi(v) = 1)$ as the number of particles in a configuration. We aim to estimate

$$I = \mathbb{E}Z,$$

where $Z \sim \pi$ with $\pi_\xi = 1/z_G$ for $\xi \in \Lambda$ (see (6.3.13)). Let $\xi_0, \xi_1, \ldots$ be a sequence of proper configurations sampled using Algorithm 47. Define $Y_k = f(\xi_k)$. We performed $R = 3000$ steps, starting from the configuration $\xi_0(v) = 0$ for all $v \in V$, and computed

$$\hat{Y}_R = \frac{1}{R}\sum_{j=1}^{R} Y_j,$$

resulting in $\hat{Y}_R = 24.36767$, which serves as our estimate for I. The value at the last step was $Y_R = 25$. The states at steps 2, 20, and 3000 are depicted in Fig. 6.17.

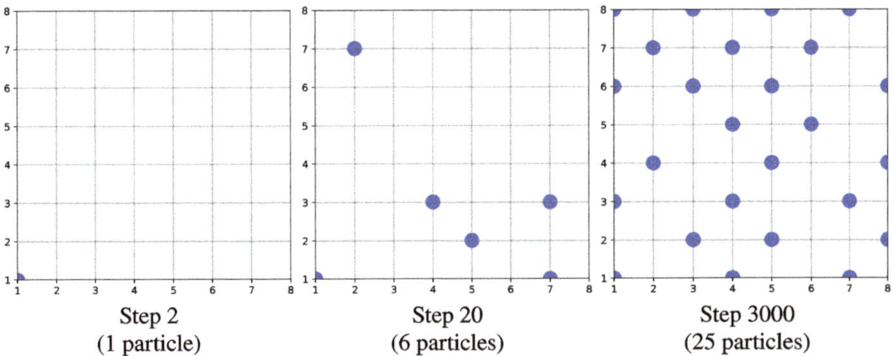

Fig. 6.17 Configurations (independent sets/feasible hard-core configurations) at steps 2, 20, and 3000

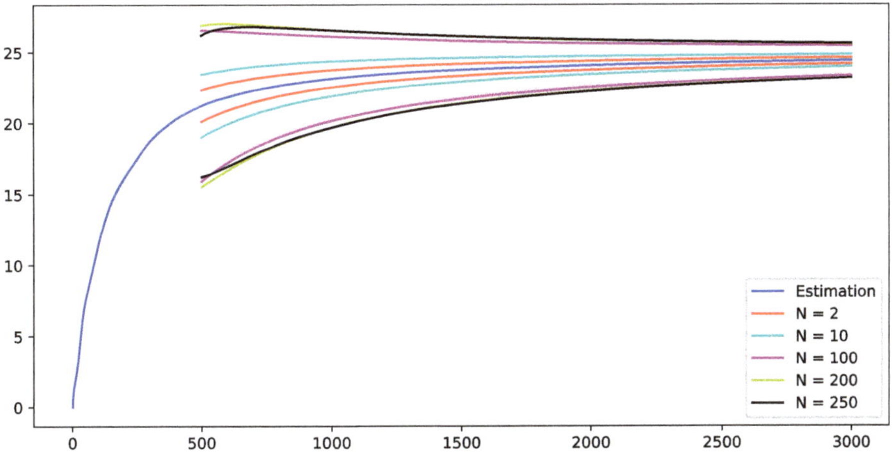

Fig. 6.18 Confidence intervals for $N \in \{2, 10, 100, 200, 250\}$

Values of $\hat{Y}_k$ for $k = 1, \ldots, R$ are depicted in Fig. 6.18 (*blue* line). Recall that the asymptotic variance can be estimated as follows:

$$\hat{\varsigma}^2 = \sum_{k=-N}^{N} \hat{\rho}_k, \quad \hat{\rho}_k = \frac{1}{N-k} \sum_{j=1}^{N-k} (Y_j - \hat{Y}_N)(Y_{j+k} - \hat{Y}_N).$$

The challenge lies in selecting an appropriate value for N. In Fig. 6.18, we present 95% confidence intervals for I, computed starting at step $k_0 = 500$ for $N \in \{2, 10, 100, 200, 250\}$. We observe that the intervals for $N = 2$ and $N = 10$ differ significantly, while those for $N = 200$ and $N = 250$ are quite similar, suggesting that $N = 250$ is sufficiently large. The final confidence intervals for $N = 250$ are depicted in Fig. 6.19. The final confidence interval for $\hat{Y}_R$, computed with $N = 250$, is

$$(23.14907,\ 25.58627)\,.$$

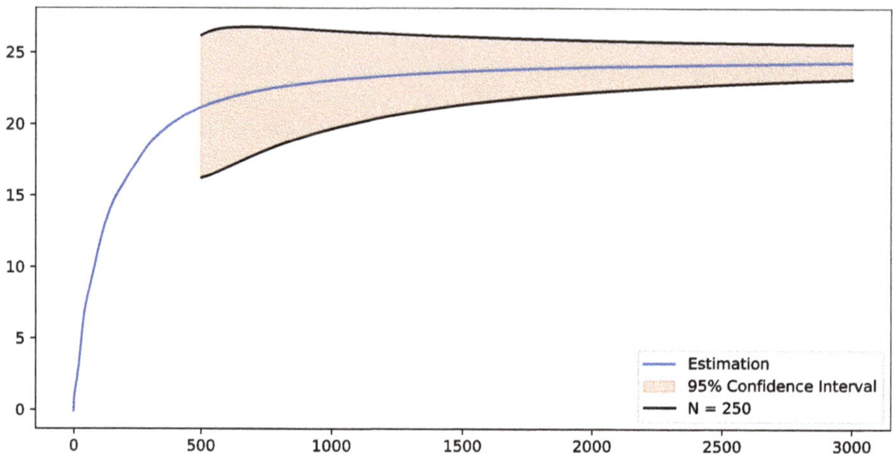

Fig. 6.19 Confidence interval for $N = 250$

The Figs. 6.17, 6.18, and 6.19 are generated using the script `ch6_hardcore_model_clt.py`. Use `--help` to view available options for adjusting model parameters (e.g., grid dimension).

6.8 Stable Simulation Scheme

In this section, we consider a very specific problem that, in certain situations, improves the analysis of stochastic systems. In Sect. 6.2, we studied Markov chains and their stationary distribution π. Suppose we aim to determine the value of $\pi(A) = \sum_{s \in A} \pi_s$ for some $A \subseteq S$. In the proposed method, the structure of S and the form of A are essential.

Assume that $S = \{s_1, \ldots, s_m\}$ is partially ordered by $\preceq$, with s_1 as the minimum and s_m as the maximum. Consider an ergodic chain (X_n) on S with transition matrix $\mathbf{P}_X$ and stationary distribution π. For $A \subseteq S$, define $\mathbf{P}_X(s, A) := \sum_{s' \in A} \mathbf{P}_X(s, s')$. We will need the following definitions: an up-set $\{s\}^{\uparrow} := \{s' \in S : s \preceq s'\}$, a down-set $\{s\}^{\downarrow} := \{s' \in S : s' \preceq s\}$, and $\delta(s, s') = \mathbf{1}(s = s')$.

To make this more specific, later in Example 6.8.4, we present a finite nonstandard tandem of two servers. The state space is $S = \{0, 1, \ldots, N\}^2$; the stationary distribution of the tandem is unknown. The method described in this section allows us to answer, for example, the following question (approximately):

What is the stationary probability that the queue lengths at both servers do not exceed k_1 and k_2, respectively?

In other words, we will show how to estimate $\pi(A)$ for $A = \{(x, y) : x \leq k_1, y \leq k_2\}$.

An example of such problems is the Gordon-Newell closed queueing networks [69], for which the form of the stationary distribution is known up to a constant. Therefore, simulations of a Markov chain with stationary distribution π can be performed using the Metropolis method. This means we have $X_0, X_1, \ldots$, and a natural estimator of $\pi(A)$ is the proportion of visits to A:

$$\hat{\pi}(A) = \frac{1}{n} \sum_{i=1}^{n} \mathbf{1}(X_i \in A).$$

By the way—such an estimator was considered in [11], estimating the stationary distribution at one state was considered in [123].

However, the quality of the estimation depends strongly on the rate of convergence of the Markov chain (X_n) to its stationary distribution. Notice also that the estimator is not unbiased.

We now present a *stable simulation scheme* for approximating $\pi(A)$, specifically when A is a down-set. Note that this method does not require any knowledge about, e.g., the rate of convergence. The method utilizes Siegmund duality and the notion of Möbius monotonicity of a chain.

6.8.1   Siegmund Duality

We say that a Markov chain (Z_n) with transition matrix $\mathbf{P}_Z$ is the **Siegmund dual** of (X_n) if

$$\forall(s_i, s_j \in S), \ \forall(n \geq 0), \quad \mathbf{P}_X^n(s_i, \{s_j\}^{\downarrow}) = \mathbf{P}_Z^n(s_j, \{s_i\}^{\uparrow}). \tag{6.8.1}$$

Given the transition matrix $\mathbf{P}_X$, we can construct a subprobability kernel $\mathbf{P}_Z$ that satisfies (6.8.1) for all $s_i, s_j \in S$. Note that (6.8.1) implies s_m is an absorbing state. Taking $s_i = s_j = s_m$, we have $\mathbf{P}_X(s_m, S) = 1 = \mathbf{P}_Z(s_m, s_m)$. Moreover, $\mathbf{P}_Z$ is a strictly substochastic matrix: there exists at least one $s \in S$ such that $\sum_{s' \in S} \mathbf{P}_Z(s, s') < 1$.

We add an extra state, s_0, and for every $s \in S$, define

$$\mathbf{P}_Z(s, s_0) = 1 - \sum_{s' \in S} \mathbf{P}_Z(s, s').$$

Thus, a Markov chain (Z_n) with transition matrix $\mathbf{P}_Z$ on $S^* = \{s_0\} \cup S$ has two absorbing states. The key idea is the following observation: Taking the limits as $n \to \infty$ on both sides of (6.8.1), we obtain:

$$\pi(\{s_j\}^{\downarrow}) = \lim_{n \to \infty} \mathbf{P}_Z^n(s_j, \{s_i\}^{\uparrow})$$

$$= \mathbb{P}(Z \text{ is absorbed in } s_m \mid Z_0 = s_j). \tag{6.8.2}$$

This relates the stationary distribution of (X_n) to the absorption probabilities of its Siegmund dual (Z_n).

6.8.2   Finding the Siegmund Dual

To construct a Siegmund dual for a given chain (X_n) and partial ordering $\preceq$, we use the notion of Möbius monotonicity. Let $\mathbf{C}(s_i, s_j) = \mathbf{1}(s_i \preceq s_j)$, often called the *ordering matrix*. By enumerating states such that $s_i \preceq s_j$ implies $i \leq j$, the $m \times m$ matrix $\mathbf{C}$ becomes a 0-1 upper triangular matrix, which is invertible. Its inverse, $\mathbf{C}^{-1}$, is known as the *Möbius function*.

Definition 6.8.1 *The Markov chain (X_n) with transition matrix $\mathbf{P}_X$ is Möbius monotone with respect to the partial order $\preceq$ if $\mathbf{C}^{-1}\mathbf{P}_X\mathbf{C} \geq 0$ (i.e., each entry is non-negative). In terms of the transition probabilities, this means that*

$$\forall (s_i, s_j \in S) \quad \sum_{s \succeq s_i} \mu(s_i, s)\mathbf{P}_X(s, \{s_j\}^{\downarrow}) \geq 0.$$

The condition (6.8.1) can be rephrased using $\mathbf{C}$:

$$\mathbf{P}_X^n \mathbf{C} = \mathbf{C}(\mathbf{P}_Z^n)^T,$$

which leads to the following lemma.

Lemma 6.8.2 *Let (X_n) be an ergodic Markov chain on $S = \{s_1, \ldots, s_m\}$ with a partial ordering $\preceq$, having the minimum s_1 and the maximum s_m. Then there exists a Siegmund dual (Z_n) of (X_n) on $S^* = \{s_0\} \cup S$ if and only if (X_n) is Möbius* *monotone. The transition probabilities of the dual Markov chain (Z_n) are as follows:*

$$\mathbf{P}_Z(s_j, s_i) = (\mathbf{C}^{-1}\mathbf{P}_X\mathbf{C})(s_i, s_j), \quad s_i, s_j \in \mathcal{S},$$

$$\mathbf{P}_Z(s_0, s_j) = \delta(s_0, s_j), \qquad\qquad s_j \in \mathcal{S}^*, \qquad\qquad (6.8.3)$$

$$\mathbf{P}_Z(s_k, s_0) = 1 - \sum_{s \in \mathcal{S}} \mathbf{P}_Z(s_k, s), \qquad s_k \in \mathcal{S}.$$

Thus, we have the following **stable simulation scheme** for approximating $\pi(A)$ when $A = \{s\}^{\downarrow}$ (note that there is some freedom in choosing the partial ordering $\preceq$):

Algorithm 59 Stable simulation scheme for approximating $\pi(\{s^{\downarrow}\})$

Require: Möbius monotone (w.r.t. partial ordering $\preceq$) chain (X_n) with stationary distribution π; set $\{s^{\downarrow}\}$; number of replications R

1: Calculate Siegmund dual Z of X on $\mathcal{S}^* = \{s_0\} \cup \mathcal{S}$
2: Assume we can simulate Z. For a given $s \in \mathcal{S}$ simulate R chains $Z^{(1)}, \ldots, Z^{(R)}$ independently, each starting at s (each simulated until absorption)
3: Estimate $\pi(\{s\}^{\downarrow})$ via (6.8.2) with the crude Monte Carlo estimator

$$\hat{\pi}(\{s\}^{\downarrow}) = \frac{1}{R} \sum_{i=1}^{R} Y_j,$$

where $Y_j = \mathbf{1}(Z^{(j)} \text{ is absorbed in } s_m | Z_0^{(j)} = s)$

In other words, we estimate the mean of the Bernoulli random variable Y with $\pi(\{s\}^{\downarrow})$ as its success probability (and thus its mean). From the central limit theorem, we know that $\pi(\{s\}^{\downarrow})$ lies in the interval

$$\left(\hat{Y}_R - z_{1-\alpha/2} \frac{\sigma_Y}{\sqrt{R}}, \ \hat{Y}_R + z_{1-\alpha/2} \frac{\sigma_Y}{\sqrt{R}} \right)$$

with probability (approximately) $1 - \alpha$, where $\sigma_Y^2 = \pi(\{s\}^{\downarrow})(1 - \pi(\{s\}^{\downarrow}))$ is the variance of Y, and $z_{1-\alpha/2}$ is the α-th quantile of the standard normal distribution. For a typical $\alpha = 0.05$ (then $z_{1-0.025} = 1.96$), we want to control the error $b = \frac{1.96\sigma_Y}{\sqrt{R}}$ (cf. (1.10)). Since $\sigma_Y^2 \leq \frac{1}{4}$, it is sufficient to run at least

$$R = \left(\frac{1.96}{2b} \right)^2 \approx b^{-2} \qquad\qquad (6.8.4)$$

replications for any s. The estimator $\hat{Y}_R$ is unbiased for $\pi(\{s\}^{\downarrow})$.

Remark 6.8.3 Sometimes we need to know π_s instead of $\pi(\{s\}^{\downarrow})$. We have the following formula:

$$\pi_s = \sum_{s' \preceq s} \mathbf{C}^{-1}(s', s)\pi(\{s'\}^{\downarrow}), \tag{6.8.5}$$

which means that we must calculate $\pi(\{s'\}^{\downarrow})$ for all $s' \preceq s$ such that $\mathbf{C}^{-1}(s', s) \neq 0$. The number of such s' depends on the specific example (and the partial ordering).

Formula (6.8.5) follows from the algebra below. We can express $\pi(\{s\}^{\downarrow})$ as:

$$\pi(\{s\}^{\downarrow}) = \sum_{s' \preceq s} \pi_{s'} = \sum_{s'} \pi_{s'} \mathbf{C}(s', s) = (\boldsymbol{\pi}\mathbf{C})(s).$$

Thus, in matrix form, denoting $(\pi(\{s_1\}^{\downarrow}), \ldots, \pi(\{s_m\}^{\downarrow}))$ by $\boldsymbol{\pi}^{\downarrow}$, we have:

$$\boldsymbol{\pi}^{\downarrow} = \boldsymbol{\pi}\mathbf{C},$$

and hence $\boldsymbol{\pi} = \boldsymbol{\pi}^{\downarrow}\mathbf{C}^{-1}$, i.e.,

$$\pi_s = (\boldsymbol{\pi}^{\downarrow}\mathbf{C}^{-1})(s),$$

as given in (6.8.5). ∎

Example 6.8.4 (Nonstandard Tandem Network of Two Servers) We present a stable simulation scheme for estimating the stationary distribution of a two-node closed tandem network with unknown stationary distribution, considering specific parameter values.

Consider a tandem network of two stations, each with a finite buffer of size N. Denote the chain by (X_n) with state space $\mathcal{S} = \{0, \ldots, N\}^2$, where $s = (x, y)$ represents a state in which there are x customers at server 1 and y customers at server 2.

When $x > 0$ and $y < N$, the system operates as a classical tandem network:

- With probability λ_1, there is an arrival at station 1 (if $x < N$).
- With probability λ_2, there is an arrival at station 2 (if $y < N$).
- With probability μ_1, a customer moves from server 1 to server 2.
- With probability μ_2, a customer leaves the network from server 2 (if $y > 0$).
- With the remaining probability, no change occurs.

Without loss of generality, we assume $\lambda_1 + \mu_1 + \lambda_2 + \mu_2 = 1$. However, at the boundaries ($x = 0$ or $y = N$), the system operates differently:

- When $x = 0$, the probability of arrival at station 2 is $\lambda_2 + \mu_1$ (if $y < N$).
- When $y = N$, there is a departure from server 1 with probability μ_1 (if $x > 0$).

This system can be interpreted as a modified Gordon-Newell closed queueing network; see [69]. A similar modification, applied to Jackson networks (with a countable state space), was studied in Foley and McDonald [60], where rough asymptotics and large deviations were derived. Importantly, even a minor modification at the boundaries significantly alters the stationary distribution.

The transitions of the chain are as follows:
$$\mathbf{P}_X((x, y), (x', y')) =$$

$$
\begin{cases}
\lambda_1 & \text{if } (x' = x + 1, y' = y) \text{ or } (x = x' = N, 0 < y' = y < N), \\[2mm]
\lambda_2 & \text{if } (y' = y + 1, x' = x > 0) \text{ or } (0 < x' = x < N, y' = y = N), \\[2mm]
\mu_1 & \text{if } (x' = x - 1, y' = y + 1) \text{ or } (x' = x - 1, y' = y = N), \\[2mm]
\mu_2 & \text{if } (y' = y - 1, x' = x) \text{ or } (x' = x < N, y' = y = 0), \\[2mm]
\mu_1 + \lambda_2 & \text{if } (y' = y + 1, x' = x = 0) \text{ or } (x' = x = 0, y' = y = N), \\[2mm]
\mu_2 + \lambda_1 & \text{if } x' = x = N, y' = y = 0, \\[2mm]
\lambda_1 + \lambda_2 & \text{if } x' = x = y' = y = N.
\end{cases}
$$

$$(6.8.6)$$

The coordinate-wise ordering is a natural one for this state space. Specifically,

$$\mathbf{C}((x, y), (x', y')) = \mathbf{1}((x, y) \preceq (x', y')) = \mathbf{1}(x \leq x' \text{ and } y \leq y').$$

The inverse of $\mathbf{C}$ is given by:

$$
\mathbf{C}^{-1}((x, y), (x', y')) =
\begin{cases}
1 \;\;\text{if} & (x' = x \text{ and } y' = y) \text{ or} \\
& (x' = x + 1 \text{ and } y' = y + 1), \\[2mm]
-1 \;\text{if} & (x' = x + 1 \text{ and } y' = y) \text{ or} \\
& (x' = x \text{ and } y' = y + 1), \\[2mm]
0 \;\;\; \text{otherwise.}
\end{cases}
$$

$$(6.8.7)$$

The proof of (6.8.7) is left as Exercise 6.T.34. Next, we compute the transition matrix $\mathbf{P}_Z$ for the Siegmund dual Z:

$$\mathbf{P}_Z((x_1, y_1), (x_2, y_2)) \tag{6.8.8}$$

$$= \sum_{(x,y) \succeq (x_2,y_2)} \mathbf{C}^{-1}((x_2, y_2), (x, y)) \mathbf{P}_Y((x, y), (x_1, y_1)^{\downarrow})$$

$$= \mathbf{P}_X((x_2, y_2), (x_1, y_1)^{\downarrow}) - \mathbf{P}_X((x_2 + 1, y_2), (x_1, y_1)^{\downarrow})$$

$$- \mathbf{P}_X((x_2, y_2 + 1), (x_1, y_1)^{\downarrow}) + \mathbf{P}_X((x_2 + 1, y_2 + 1), (x_1, y_1)^{\downarrow}),$$

where

$$\mathbf{P}_X((x, y), (x_1, y_1)^{\downarrow}) := \sum_{(x',y') \preceq (x_1,y_1)} \mathbf{P}_X((x, y), (x', y')).$$

For terms such as $\mathbf{P}_X((x_2, y_2 + 1), (x_1, y_1)^{\downarrow})$ or $\mathbf{P}_X((x_2 + 1, y_2 + 1), (x_1, y_1)^{\downarrow})$, they should be understood as 0 when $y_2 = N$.

Consider the case $x_1, x_2, y_1, y_2 > 1$ and $x_1, x_2, y_1, y_2 < N$. Then, using (6.8.8), we have, for instance:

$$\mathbf{P}_Z((x_1, y_1), (x_1 - 1, y_1))$$

$$= \mathbf{P}_X((x_1 - 1, y_1), (x_1, y_1)^{\downarrow}) - \mathbf{P}_X((x_1, y_1), (x_1, y_1)^{\downarrow})$$

$$- \mathbf{P}_X((x_1 - 1, y_1 + 1), (x_1, y_1)^{\downarrow}) + \mathbf{P}_X((x_1, y_1 + 1), (x_1, y_1)^{\downarrow})$$

$$= \lambda_1 + \mu_2 - \mu_2 - \mu_2 + \mu_2 = \lambda_1.$$

We also have $\mathbf{P}_Z((x_1, y_1), (x_1 - k, y_1)) = 0$ for $k = 1, \ldots, x_1 - 1$.

Considering possible transitions to the border of the state space:

$$\mathbf{P}_Z((x_1, y_1), (0, y_1 - 1))$$

$$= \mathbf{P}_X((0, y_1 - 1), (x_1, y_1)^{\downarrow}) - \mathbf{P}_X((1, y_1 - 1), (x_1, y_1)^{\downarrow})$$

$$- \mathbf{P}_X((0, y_1), (x_1, y_1)^{\downarrow}) + \mathbf{P}_X((1, y_1), (x_1, y_1)^{\downarrow})$$

$$= 1 - 1 - (\lambda_1 + \mu_2) + (\mu_2 + \lambda_1) = 0.$$

We will skip the remaining (lengthy) calculations. However, by considering it case by case, we find that X is Möbius monotone.

We add one extra state, denoted $(+\infty, +\infty)$. The transitions of the Siegmund dual are as follows:

$$\mathbf{P}_Z((x, y), (x', y')) =$$

$$
\begin{cases}
\lambda_1 & \text{if } (N - 1 > x' = x - 1,\, y' = y) \\
& \text{or } (x' = x = N,\, y' = y < N) \\
& \text{or } (x = 0,\, y > 0,\, x' = +\infty,\, y' = +\infty), \\[4pt]
\lambda_2 & \text{if } N - 1 > y' = y - 1,\, x' = x < N, \\[4pt]
\mu_1 & \text{if } x' = x + 1,\, N - 1 > y' = y - 1 \\
& \text{or } x' = x + 1,\, y' = y = N, \\[4pt]
\mu_2 & \text{if } x' = x,\, y' = y + 1, \\[4pt]
\mu_1 + \lambda_2 & \text{if } (N - 1 > y' = y - 1,\, x' = x = N) \\
& \text{or } (x > 0,\, y = 0,\, x' = +\infty,\, y' = +\infty), \\[4pt]
\mu_2 + \lambda_2 & \text{if } x' = x < N,\, y' = y = N, \\[4pt]
\lambda_1 + \lambda_2 + \mu_1 & \text{if } x = 0,\, y = 0,\, x' = +\infty,\, y' = +\infty, \\[4pt]
1 & \text{if } (x' = x = y' = y = N) \\
& \text{or } (x' = x = y' = y = +\infty).
\end{cases}
\tag{6.8.9}
$$

Note that outside the borders, the transitions resemble a *reversed* network (although this is not the usual time reversal). The behavior at the borders is distinct.

First, the chain transitions to $(+\infty, +\infty)$ only from states of the form $(0, y)$ or $(x, 0)$. Second, once the process is on the 'upper' (x, N) or the 'right' (N, y) border, it behaves like a standard absorbing birth-and-death chain with possible absorption only at $(0, N)$ and $(N, 0)$. On the upper border, the birth rate is μ_1, and the death rate is λ_1, whereas on the right border, the birth rate is μ_2, and the death rate is $\mu_1 + \lambda_2$.

Stable Simulation Scheme We consider the tandem for $N = 5$ (thus the state space is $S = \{0, \ldots, 5\}^2$ with $|S| = 36$) with parameters $\lambda_1 = \frac{3}{16}$, $\lambda_2 = \frac{1}{16}$, $\mu_1 = \mu_2 = \frac{6}{16}$. We estimate $\pi(\{x, y\}^{\downarrow})$ for all $(x, y) \in S$ with accuracy $b = 0.005$. Using (6.8.4), we calculate the required number of simulations:

$$\left(\frac{1.96}{0.01}\right)^2 = 3.8416 \cdot 10^4 \le 4 \cdot 10^4 =: R.$$

Table 6.1 Results of $R = 4 \cdot 10^4$ simulations for the nonstandard tandem queue system with $N = 5, \lambda_1 = \frac{3}{16}, \lambda_2 = \frac{1}{16}, \mu_1 = \mu_2 = \frac{6}{16}$. The estimators $\hat{\pi} \equiv \hat{\pi}(x, y), x, y \in \{0, \ldots, N\}^{\downarrow}$ of $\pi(\{x, y\}^{\downarrow})$ within accuracy $b = 0.005$ are presented. Additionally, the estimators of the expected time $(\hat{\tau})$ and the variance $(\hat{\sigma}^2)$ of the absorption time of a Siegmund dual are included

$y \backslash x$		0	1	2	3	4	5
0	$\hat{\pi}$	**0.030349**	**0.081442**	**0.145442**	**0.235395**	**0.350883**	**0.501939**
	$\hat{\tau}$	3.432581	6.516953	8.838628	10.490814	11.159698	10.898512
	$\hat{\sigma}^2$	37.126458	71.878733	92.804187	107.71656	110.579045	107.214705
1	$\hat{\pi}$	**0.057977**	**0.138508**	**0.248754**	**0.380847**	**0.549661**	**0.762624**
	$\hat{\tau}$	5.174186	9.379605	12.625	14.243023	14.688047	13.80893
	$\hat{\sigma}^2$	59.11922	89.951922	104.240892	102.786725	99.409276	96.261684
2	$\hat{\pi}$	**0.078907**	**0.181252**	**0.310265**	**0.467963**	**0.651847**	**0.883826**
	$\hat{\tau}$	5.92793	10.416093	13.449814	14.923465	14.533163	12.371256
	$\hat{\sigma}^2$	68.774475	94.621951	97.120298	91.983189	90.545936	84.547763
3	$\hat{\pi}$	**0.093628**	**0.207089**	**0.353242**	**0.511000**	**0.709644**	**0.945623**
	$\hat{\tau}$	6.109372	10.44014	13.141	13.863535	12.545814	9.115233
	$\hat{\sigma}^2$	70.331045	92.117536	88.790347	84.163846	83.647753	72.533966
4	$\hat{\pi}$	**0.102358**	**0.224470**	**0.371553**	**0.539710**	**0.743490**	**0.980702**
	$\hat{\tau}$	5.829372	9.947884	12.078233	12.073163	9.741093	4.85114
	$\hat{\sigma}^2$	62.136261	80.03354	81.073393	77.194024	74.588772	44.553598
5	$\hat{\pi}$	**0.111491**	**0.235068**	**0.383094**	**0.558042**	**0.762473**	**1.000000**
	$\hat{\tau}$	5.579488	9.188023	10.855837	10.235651	6.744186	0.0
	$\hat{\sigma}^2$	55.045706	70.475751	71.311736	73.753556	66.707552	0.0

The values of $\hat{\pi}(x, y)$ are highlighted in bold

Estimating each $\pi(\{x, y\}^{\downarrow})$ by simulating R times the Siegmund dual starting at (x, y), we also compute the mean absorption time $\hat{\tau}(x, y)$ and the variance of the absorption time $\hat{\sigma}^2(x, y)$. The simulation results are presented in Table 6.1, they are

Once we estimate $\pi(\{x, y\}^{\downarrow})$ by $\hat{\pi}(\{x, y\}^{\downarrow})$, we estimate $\pi_{(x,y)}$ by (see Remark 6.8.5)

$$\hat{\pi}_{(x,y)} = \sum_{(x',y') \preceq (x,y)} \mathbf{C}^{-1}((x', y'), (x, y))\hat{\pi}(\{(x', y'\}^{\downarrow})$$
$$= \hat{\pi}(\{(x, y)\}^{\downarrow}) - \hat{\pi}(\{(x - 1, y)\}^{\downarrow}) \tag{6.8.10}$$
$$- \hat{\pi}(\{(x, y - 1)\}^{\downarrow}) + \hat{\pi}(\{(x - 1, y - 1)\}^{\downarrow}).$$

These values are presented in Table 6.2. Note that we do not claim what the errors of estimations of $\pi_{(x,y)}$ are.

All computations in these examples are based on [130, Table 1]. $\qquad\qquad\diamond$

Table 6.2 Estimations $\hat{\pi}_{(x,y)}$ of $\pi_{(x,y)}$ computed from estimations $\hat{\pi}(\{x, y\}^{\downarrow})$ of $\pi(\{x, y\}^{\downarrow})$, given in Table 6.1

$y\backslash x$	0	1	2	3	4	5
0	0.030349	0.051093	0.064000	0.089953	0.115488	0.151056
1	0.027628	0.029438	0.046246	0.042140	0.053326	0.061907
2	0.020930	0.021814	0.018767	0.025605	0.015070	0.019016
3	0.014721	0.011116	0.017140	0.000060	0.014760	0.004000
4	0.008730	0.008651	0.000930	0.010399	0.005136	0.001233
5	0.009133	0.001483	0.000925	0.006791	0.000651	0.000315

6.9 Bibliographical Notes

The theory of Markov chains is well documented in numerous texts, including the classical books of Feller [52], Kemeny and Snell [98] (for finite state space), as well as the modern works by Asmussen [7] and Brémaud [21]. Spectral properties and rates of convergence are discussed in Seneta [190] and Roberts and Rosenthal [178]. Some relevant references for more advanced theory of Markov processes are Meyn and Tweedie [152] and Glynn and Meyn [65]. Asymptotic variance formulas in finite Markov chains are detailed in Rolski et al. [180, Chapter 7]. Ergodic theorems and convergence results are presented in Doob [41], Asmussen [7], and Levin et al. [124]. For the central limit theorem for Markov chains, see Billingsley [14], Glynn and Meyn [65], and Jones [94].

MCMC methods for sampling from complex distributions are treated in Brémaud [21], Asmussen and Glynn [8], Rubinstein and Kroese [186], Madras [138], and Häggström [73].

The Metropolis algorithm was introduced by Metropolis et al. [151], while its simple yet highly useful Metropolis–Hastings variation was developed by Hastings [80]. These algorithms, along with the Gibbs sampler, are discussed in Robert and Casella [176] and Brooks et al. [22], with additional theoretical details provided by Roberts and Rosenthal [178] and Meyn and Tweedie [152].

The first perfect simulation algorithm was proposed by Asmussen et al. [10], but it was prohibitively inefficient in terms of computational time. Subsequent developments led to more practical methods, including Coupling from the Past (CFTP), introduced by Propp and Wilson [171] and further explored by Fill [53]. Wilson [215] proposed the *forward algorithm*, a modification of CFTP requiring less storage. Other interesting papers are Foss and Tweedie [61] and Møller [154]. Coupling techniques and monotonicity, which play a crucial role in perfect simulation, are discussed in Lindvall [127] and Thorisson [201, 203, 204]. Extensions by Fill [53] and Murdoch and Green [161] are also notable, as is the method of Raftery and Lewis [172] for convergence monitoring. Duality-based perfect simulation is developed in Diaconis and Fill [36], Lorek [130], and Lorek and Markowski [131]. The application of these methods in statistical physics and Gibbs measures is discussed in Häggström and Nelander [75] and Kendall [99].

Approximate counting methods in combinatorics and statistical physics are extensively studied in Jerrum and Sinclair [87], Jerrum [86], and H"aggstr"om [73]. Mixing times and conductance arguments are analyzed in Sinclair [194] and Lov'asz and Vempala [136]. Fully Polynomial Randomized Approximation Schemes (FPRAS) for counting problems are presented in Jerrum et al. [89] and Jerrum and Sinclair [88], with specific applications to the 0-1 knapsack problem discussed in Morris and Sinclair [156]. Jerrum [85] provides a procedure for approximating the number of graph colorings in low-degree graphs. Approximate estimation of the permanent of a matrix is explored in Jerrum et al. [90], while the volume of a convex set is addressed in Dyer et al. [45].

In computational biology, MCMC methods are applied as described in Durbin et al. [42] and Ewens and Grant [51]. Bayesian phylogenetic inference is covered in Yang [216], and Gibbs sampling for DNA motif discovery is presented in Lawrence et al. [114] and MacKay [137]. Graphical models are introduced in Koller and Friedman [110].

Siegmund duality is introduced in Siegmund [193], with further developments in Dette et al. [32] and Huillet [84]. Generalizations for partially ordered state spaces are developed in Lorek [130], with applications to gambler's ruin problems in Lorek [129] and Lorek and Markowski [132].

6.10   Exercises

Theoretical Exercises

6.T.1 Let $\mathbf{P} = (p_{ij})_{i,j \in S}$ and $\mathbf{P}' = (p'_{ij})_{i,j \in S}$ be stochastic matrices, where $p_{ij} > 0$ if and only if $p'_{ij} > 0$ for all $i, j \in S$. Show that if $\mathbf{P}$ is regular, then $\mathbf{P}'$ is also regular.

6.T.2 Consider a deck of n cards. In one step of a Top-To-Random transposition, the top card is placed in any of the n possible positions with equal probability.

   (a) Formalize this process by providing the transition matrix for the corresponding Markov chain.

   (b) Show that the uniform distribution is the stationary distribution of this Markov chain.

6.T.3 (Simple random walk on a line, equivalently birth and death chain) Let (X_k) be a Markov chain on the state space $S = \{0, 1, \ldots, N\}$ with the following transition matrix:

$$
\mathbf{P} =
\begin{bmatrix}
r_0 & p_0 & 0 & 0 & 0 & 0 & \cdots & & q_0 \\
q_1 & r_1 & p_1 & 0 & 0 & 0 & \cdots & & 0 \\
0 & q_2 & r_2 & p_2 & 0 & 0 & \cdots & & 0 \\
& & \ddots & \ddots & \ddots & & & & \\
0 & 0 & 0 & q_i & r_i & p_i & 0\cdots & & 0 \\
& & & & \ddots & \ddots & \ddots & & \\
0 & \cdots & & & & q_{N-1} & r_{N-1} & p_{N-1} \\
p_N & \cdots & & & & 0 & q_N & r_N
\end{bmatrix},
\qquad (6.10.1)
$$

where $q_0 = p_N = 0$ and for all $i \in \{1, \dots, N-1\}$, $p_i > 0$, $q_i > 0$, and $p_i + q_i + r_i = 1$.

Find the formula for the stationary distribution of this chain.

Hint: Find $w_i > 0$ fulfilling the detailed balance condition (6.2.5) and use Proposition 6.2.5.

6.T.4 (Simple random walk on a circle) Let (X_k) be a Markov chain on the state space $\mathcal{S} = \{0, 1, \dots, N\}$ with the transition matrix $\mathbf{P}$ given in (6.10.1), where $p_i > 0$, $q_i > 0$, and $p_i + q_i + r_i = 1$ for all $i \in \{0, \dots, N\}$. (Note that in the simple random walk on a line, we had $q_0 = p_N = 0$.)

- What conditions must the parameters $p_i, q_i, i = 0, \dots, N$, satisfy for the chain (X_k) to be reversible?
- In cases where (X_k) is reversible, find the stationary distribution.

6.T.5 Let (X_k) be a Markov chain with transition matrix $\mathbf{P}^X$ on $\mathcal{S} = \{s_1, \dots, s_M\}$.

We construct a sequence (Y_n) based on (X_k) in such a way that we record only *new* values of (X_k). That is, if the chain X_k stays in the same state for some steps, the value of Y_n does not change; it only updates when (X_k) transitions to a different state. For example:

$$
k = \ 0 \ 1 \ 2 \ 3 \ 4 \ 5 \ 6 \ 7 \ 8 \ 9 \ 10
$$

$$
X_k = s_1 \ s_1 \ s_1 \ s_2 \ s_2 \ s_1 \ s_3 \ s_3 \ s_3 \ s_2 \ s_1
$$

$$
n = \ 0 \qquad\quad 1 \quad\ \ 2 \ 3 \qquad\ 4 \ 5
$$

$$
Y_n = s_1 \qquad\quad s_2 \quad\ s_1 \ s_3 \qquad s_2 \ s_1
$$

Show that (Y_n) is a Markov chain and provide its transition probability matrix $\mathbf{P}^Y$ in terms of the transition probabilities $\mathbf{P}^X$ of (X_k).

6.T.6 Consider the following Markov chain (X_k). Let $a, b \geq 2$ be integers. The state space is:

$$
\mathcal{S} = \{(x, y) : x \in \{0, \dots, a-1\}, y \in \{0, \dots, b-1\}\}.
$$

The dynamics of the chain are as follows: Being in (x, y) at step k, the chain transitions as follows in the next step:

- $((x + 1) \mod a, y)$ with probability $\frac{1}{2}$, or
- $(x, (y + 1) \mod b)$ with probability $\frac{1}{2}$.

(a) Show that (X_k) is irreducible.

(b) Show that (X_k) is aperiodic if and only if $\gcd(a, b) = 1$.

6.T.7 A lone chess king stands on a standard 8×8 board and moves randomly to one of the adjacent squares, with each move being equally likely.

(a) Is the Markov chain corresponding to this random walk aperiodic and/or irreducible?

(b) Find the stationary distribution of this chain.

6.T.8 Recall that for two probability vectors $\boldsymbol{\mu} = (\mu_s)_{s \in S}$ and $\boldsymbol{v} = (v_s)_{s \in S}$ on S, the *total variation distance* is defined as:

$$d_{\mathrm{TV}}(\boldsymbol{\mu}, \boldsymbol{v}) = \frac{1}{2} \sum_{s \in S} |\mu_s - v_s|.$$

Show that:

(a) $d_{\mathrm{TV}}(\boldsymbol{\mu}, \boldsymbol{v}) = \sum_{s : \mu_s > v_s} (\mu_s - v_s) = \sum_{s : v_s > \mu_s} (v_s - \mu_s).$

(b) $d_{\mathrm{TV}}(\boldsymbol{\mu}, \boldsymbol{v}) = \max_{A \subseteq S} |\mu(A) - v(A)|$, where $\mu(A) := \sum_{s \in A} \mu_s$.

6.T.9 Let (X_k) be a Markov chain corresponding to **Riffle Shuffle** (i.e., on $S = S_n$) with a t.m. $\mathbf{P}_X$ and let $(\hat{X}_k)$ Markov chain corresponding to time reversed **Riffle Shuffle**, denote its t.m. by $\mathbf{P}_{\hat{X}}$. Show that

$$\forall(\boldsymbol{\mu} : \text{distribution on } S_n) \ \forall(k \geq 0) \qquad d_{\mathrm{TV}}(\boldsymbol{\mu}\mathbf{P}_X^k, \boldsymbol{\pi}) = d_{\mathrm{TV}}(\boldsymbol{\mu}\mathbf{P}_{\hat{X}}^k, \boldsymbol{\pi}),$$

where π is the uniform distribution on S_n.

6.T.10 Prove that if the transition matrix $\mathbf{P}$ of a Markov chain (X_k) is *doubly stochastic* (i.e., the sum of the elements in each column is also equal to 1), then the stationary distribution of the chain is uniform.

6.T.11 Consider a graph $G = (V, E)$ with $2n$ vertices $v_1, \ldots, v_{2n}$ and with n edges (v_{2i}, v_{2i-1}) for $i = 1, \ldots, n$. Given $q \geq 2$ colors, show that there are $q^n(q - 1)^n$ proper colorings of the graph.

6.T.12 (Los Angeles model of weather). Consider a Markov chain on the state space $S = \{s_1, s_2\} = \{\text{rainy, sunny}\}$ with the transition matrix:

$$\mathbf{P} = \begin{bmatrix} 1/2 & 1/2 \\ 1/10 & 9/10 \end{bmatrix}.$$

The initial distribution is given by $\mu = (p, 1 - p)$ for $p \in [0, 1]$. Using the spectral decomposition of $\mathbf{P}$, derive a formula for $\mu^k = (\mathbb{P}(X_k = \text{rainy}), \mathbb{P}(X_k = \text{sunny}))$. Once you obtain μ^k, justify that $\lim_{k \to \infty} \mu^k = \pi$, where π is the stationary distribution of the chain. Additionally, determine the mixing time:

$$\tau(\varepsilon) = \min\{k : d_{\mathrm{TV}}(\mu \mathbf{P}^k, \pi) \leq \varepsilon\}.$$

6.T.13 Consider n cards. Let μ be the uniform distribution on all n permutations of the cards. Let ν be the uniform distribution on all permutations where the first card is fixed. Compute the total variation distance $d_{\mathrm{TV}}(\mu, \nu)$.

6.T.14 For a Markov chain (X_k) with transition matrix $\mathbf{P}$, define:

$$\bar{d}(k) = \max_{s', s''} d_{\mathrm{TV}}(\delta_{s'} \mathbf{P}^k, \delta_{s''} \mathbf{P}^k), \qquad d(k) = \max_s d_{\mathrm{TV}}(\delta_s \mathbf{P}^k, \pi).$$

Here, $\delta_s \mathbf{P}^k$ denotes the distribution of X_k given $X_0 = s$. Show that:

$$d(k) \leq \bar{d}(k) \leq 2d(k).$$

6.T.15 Let (X_k) and (Y_k) be two Markov chains on the same state space $\mathcal{S} = \{1, 2\}$ with transition matrix:

$$\mathbf{P} = \begin{bmatrix} 1 - \alpha & \alpha \\ \beta & 1 - \beta \end{bmatrix},$$

where $\alpha, \beta \in (0, 1)$. Assume that $X_0 = 1$ and $Y_0 = 2$.

Define the *coupling time T*, the first time the chains meet:

$$T = \min\{k \geq 0 : X_k = Y_k\}.$$

Determine the distribution of T in the following cases:

(a) (X_k) and (Y_k) evolve independently according to $\mathbf{P}$. Denote the corresponding coupling time by T_{ind}.
(b) (X_k) and (Y_k) are coupled via the update function:

$$\varphi(1, u) = \begin{cases} 1 \text{ for } u \in [0, 1 - \alpha], \\ 2 \text{ for } u \in (1 - \alpha, 1], \end{cases} \qquad \varphi(2, u) = \begin{cases} 1 \text{ for } u \in [0, \beta], \\ 2 \text{ for } u \in (\beta, 1]. \end{cases}$$

Denote the corresponding coupling time by T_c.

Analyze the cases $\alpha + \beta < 1$ and $\alpha + \beta \geq 1$ separately. Using the *coupling inequality* (6.4.2), provide bounds on the mixing time using T_{ind}

and T_c. For what values of α, β does coupling yield better results? Compute the expected coupling times $\mathbb{E}T_{\mathrm{ind}}$ and $\mathbb{E}T_c$.

6.T.16 Algorithm 60 describes a single step of the Markov chain (Y_k) on $\{1, \dots, n\}$. Determine $\lim_{k\to\infty} \mathbb{P}(Y_k = i)$.

Algorithm 60

1: Assume $Y_k = i \in \{1, \dots, n\}$
2: Generate $U_1 \sim \mathcal{U}[0, 1)$ and set $X = \min(\max((i + 2 \cdot \lfloor 2 \cdot U \rfloor - 1), 1), n)$
3: Set $\alpha = \min(1, (\frac{i}{X})^{2.5})$
4: Generate $U_2 \sim \mathcal{U}[0, 1)$
5: **if** $U_2 \le \alpha$ **then**
6:     $Y_{k+1} = X$
7: **else**
8:     $Y_{k+1} = Y_k$
9: **end if**
10: **return** Y_{k+1}

6.T.17 Let $B = \sum_{i=1}^{\infty} i^{-2}$. This quantity is known ($B = \pi^2/6$), but its exact value is not relevant for this task. Construct the Metropolis algorithm for simulating a random variable with the distribution $\pi_i = (Bi^2)^{-1}$, for $i = 1, 2, \dots$.

Although the state space is infinite, we may proceed as in the finite-state case. Use a symmetric random walk on the positive integers as the candidate-generating process with transition probabilities:

$$
q_{ij} = \begin{cases} \frac{1}{2}, & \text{if } j = i + 1 \text{ or } j = i - 1, \text{ or } i = j = 0, \\ 0, & \text{otherwise.} \end{cases}
$$

6.T.18 Show that the transition matrix $p_{\xi\xi'}$ of the Metropolis algorithm for a hard-core model, with the candidate-generating matrix given in (6.3.14), is given by:

$$
p_{\xi\xi'} = \begin{cases} \frac{1}{2|V|} & \text{if } \xi' \in \mathcal{N}_\Lambda(\xi), \\ \frac{1}{2} + \frac{m^1(\xi)}{2|V|} & \text{if } \xi' = \xi, \\ 0 & \text{if } |m(\xi) - m(\xi')| \ge 2, \end{cases}
$$

where recall that $m(\xi)$ is the number of vertices with value 1, i.e.,

$$
m(\xi) = |\{v : \xi(w) = 1\}|
$$

and

$$
m^1(\xi) = |\{v : \exists w \in \mathcal{N}_G(v), \, \xi(w) = 1\}|.
$$

6.T.19 For a given graph $G = (V, E)$, let (X_n) be a Markov chain on $\mathcal{S} = \{0, 1\}^V$ with one-step transitions defined by Algorithm 56. Show that the chain is ergodic and has the stationary distribution π given in (6.3.13), i.e., the uniform distribution over all feasible hard-core configurations.

6.T.20 Show that the Gibbs sampler is a special case of the Metropolis-Hastings algorithm.

6.T.21 Show that the Boltzmann distribution defined in (7.1.1) is a stationary distribution of a Markov chain with a transition matrix given by:

$$p_{vv'}^T = q_{vv'}\alpha_{vv'}^T. \tag{6.10.2}$$

Further, show that if $(Q_{vv'})$ is regular, then $(p_{vv'}^T)$ is also regular.

6.T.22 Prove the following lemma:

Lemma 6.10.1 *Let $\varepsilon \in [0, 1]$, let k be a positive integer, and let $a_1, \ldots, a_k$ and $b_1, \ldots, b_k$ be non-negative numbers satisfying*

$$\left(1 - \frac{\varepsilon}{2k}\right) \le \frac{a_j}{b_j} \le \left(1 + \frac{\varepsilon}{2k}\right), \qquad j = 1, \ldots, k.$$

Then, the following bound holds:

$$1 - \varepsilon \le \frac{a}{b} \le 1 + \varepsilon,$$

where $a = \prod_{j=1}^{k} a_j$ and $b = \prod_{j=1}^{k} b_j$.

6.T.23 Consider the CFTP Algorithm 49. Recall from Sect. 6.4.2 that the coupling time is given by $\tau_{\text{cftp}} = -N_K$, where

$$K = \min\left\{n \ge 1 : X_0^{-N_n}(s_1) = \ldots = X_0^{-N_n}(s_N)\right\}.$$

Show that:

- If $N_j = j$, then the total number of steps satisfies:

$$1 + 2 + 3 + \cdots + |\tau_{\text{cftp}}| \ge \frac{1}{2}(\tau_{\text{cftp}})^2.$$

- If $N_j = 2^{j-1}$, then the total number of steps is bounded above by $2|\tau_{\text{cftp}}|$.

6.T.24 (Häggström [73], p. 81, "Coupling to the Future") Consider a two-state Markov chain on $\mathcal{S} = \{1, 2\}$ with transition matrix:

$$\mathbf{P} = \begin{pmatrix} 0.5 & 0.5 \\ 1 & 0 \end{pmatrix}$$

Suppose we have two Markov chains (X_k) and (Y_k) with this transition matrix, initialized as $X_0 = 1$ and $Y_0 = 2$.

- Show that the chain is reversible and determine its stationary distribution π.
- Let $\tau_f = \inf\{k \geq 1 : X_k = Y_k\}$ be the first coalescence time. Find the distribution of τ_f. Note that X_k and Y_k may be dependent (i.e., we consider a coupling (X_k, Y_k)). Does this dependence affect the distribution of τ_f in this case?
- Show that X_{τ_f} does not follow the stationary distribution π.

6.T.25 Consider two Markov chains (X_k) and (Y_k) (potentially dependent) with a common transition matrix $\mathbf{P}$. Suppose the forward coupling time $\tau_f = \min\{k : X_k = Y_k\}$ satisfies the following property:

> There exist constants $k_0 > 0$ and $\alpha \in (0, 1)$ such that for all (pairs of) initial states $X_0 = s$ and $Y_0 = s'$, we have
>
> $$\mathbb{P}(\tau_f \leq k_0) \geq \alpha.$$

Show that this implies

$$\mathbb{E}[\tau_f] \leq \frac{k_0}{\alpha}.$$

6.T.26 Let $S = \{s_1, \ldots, s_m\}$ be the state space of a Markov chain with transition matrix $\mathbf{P} = (p_{ij})$, and suppose that S is partially ordered by $\preceq$. A subset $A \subset S$ is called an **upset** if

$$(s_i \preceq s_j, s_i \in A) \Rightarrow s_j \in A.$$

A Markov chain with transition matrix $\mathbf{P}$ is **stochastically monotone** if

$$\forall (s_i \preceq s_j), \forall (\text{upset } A), \quad \mathbf{P}(s_i, A) := \sum_{s \in A} \mathbf{P}(s_i, s) \leq \mathbf{P}(s_j, A) := \sum_{s \in A} \mathbf{P}(s_j, s).$$

Recall (Definition 6.4.3) that a Markov chain (X_n) with transition matrix $\mathbf{P}$ is **realizably monotone** with respect to the partial order $\preceq$ if there exists a monotone update function φ, i.e., such that

$$\forall (u \in [0, 1]), \forall (s \preceq s'), \quad \varphi(s, u) \preceq \varphi(s', u).$$

(a) Show that if (X_k) is realizably monotone, then it is stochastically monotone.
(b) Show that if $\preceq$ is a total (linear) order, i.e., $\preceq := \leq$, then stochastic monotonicity and realizable monotonicity are equivalent.

6.T.27 (Non-symmetric random walk on a square) In Example 6.4.6, we considered a symmetric random walk on the state space $\mathcal{S} = \{0, 1\}^2 = \{0 \equiv (0, 0), 1 \equiv (1, 0), 2 \equiv (0, 1), 3 \equiv (1, 1)\}$. Now, let $\alpha \in (0, 1/2)$ and $\beta \in (0, 1/2)$. Define a non-symmetric random walk (X_k) on $\mathcal{S}$ with the following transition matrix:

$$
\mathbf{P} = \begin{pmatrix}
1 - 2\alpha & \alpha & \alpha & 0 \\
\beta & 1 - \alpha - \beta & 0 & \alpha \\
\beta & 0 & 1 - \alpha - \beta & \alpha \\
0 & \beta & \beta & 1 - 2\beta
\end{pmatrix}.
$$

Roughly speaking, when the chain is in state $s = (e_1, e_2)$, where $e_i \in \{0, 1\}$, it can transition by flipping a coordinate from 0 to 1 with probability α or from 1 to 0 with probability β. The chain remains in the same state with the remaining probability.

Define a coordinate-wise partial ordering on $\mathcal{S}$. For $i = (e_1^i, e_2^i)$ and $j = (e_1^j, e_2^j)$, set:

$$
i \preceq j \quad \Leftrightarrow \quad e_1^i \leq e_1^j \text{ and } e_2^i \leq e_2^j.
$$

In this ordering, the state $0 \equiv (0, 0)$ is the minimum, and the state $3 \equiv (1, 1)$ is the maximum.

Show that (X_k) is realizably monotone if $\alpha + \beta \leq 1/2$, and explicitly construct a monotone update function φ.

6.T.28 Reconsider the random walk (X_k) from Exercise 6.T.27. Determine the values of α and β for which the chain is stochastically monotone.

6.T.29 Consider the birth-and-death chain from Exercise 6.T.3. Determine the conditions on p_i and q_i under which the chain is stochastically monotone with respect to the total ordering $\preceq := \leq$.

6.T.30 [Häggstrom [73], p. 82] Consider again a Markov chain (X_k) on $\mathcal{S} = \{1, 2\}$ with the same transition matrix as in Example 6.T.24. Now, consider a modification of Algorithm 49 (or Algorithm 50, which is equivalent for two-state chains), where the update function is defined as follows:

$$
\varphi(1, u) = \begin{cases} 1 \text{ for } u \in [0, 0.5] \\ 2 \text{ for } u \in (0.5, 1] \end{cases}, \qquad \varphi(2, u) = 1 \text{ for } u \in [0, 1].
$$

Additionally, modify step 4 of the algorithm as follows:

4: Set $n = n + 1$ and return to Step 2, using a <u>new</u> sequence of uniform random variables $\{U_i\}_{-N_{n+1} \leq i \leq 0}$.

This modification implies that at each step, a fresh sequence of U_i values is used. Let Z' be the output of the modified algorithm. Show that $P(Z' = i) \neq \pi_i$ for $i = 1, 2$, meaning the modified algorithm does not sample from the stationary distribution.

Hint: Let M be the largest m such that the algorithm stops the chains starting at time $-N_m$. Then:

$$P(Z' = 1) = \sum_{m=1}^{\infty} P(M = m, Y = 1).$$

Since we use the typical choice $N_r = 2^{r-1}$, we obtain the bound:

$$P(Z' = 1) \leq P(M = 1, Y = 1) + P(M = 2, Y = 1). \tag{6.10.3}$$

Compute the right-hand side of (6.10.3) and show that $P(Z' = 1) > \pi_1$, confirming that the algorithm does not correctly sample from the stationary distribution.

6.T.31 (The random-cluster model) For a graph $G = (V, E)$, a configuration is defined as $\xi \in \{0, 1\}^E$. We start with a fully connected graph G and retain each edge $e \in E$ independently with probability p, deleting it with probability $1 - p$. The probability of a specific configuration is given by:

$$\mu_\xi = \prod_{e \in E} p^{\xi(e)} (1 - p)^{1 - \xi(e)}.$$

Let $C(\xi)$ denote the number of connected components in the resulting random graph (including isolated vertices). For $q > 0$, define:

$$\pi_\xi = \frac{1}{z_{p,q}} q^{C(\xi)} \prod_{e \in E} p^{\xi(e)} (1 - p)^{1 - \xi(e)},$$

where $z_{p,q}$ is a normalizing constant. Note that for $q = 1$, the distribution simplifies to $\pi_\xi = \mu_\xi$.

  Show that the distribution π is monotone (as per Definition 6.4.9) if $q > 1$ and anti-monotone (as per Definition 6.4.11) if $q < 1$.

6.T.32 Consider the generalized hard-core model from Sect. 6.3.2.3. Show that the distribution π given in (6.3.15) is anti-monotone.

6.T.33 Consider the star-shaped graph G shown in Fig. 6.20. Let π be the uniform distribution over all feasible 4-colorings of G. Show that this distribution is neither monotone (Definition 6.4.9) nor anti-monotone (Definition 6.4.11).

Fig. 6.20 A star-shaped graph $G = (V, E)$

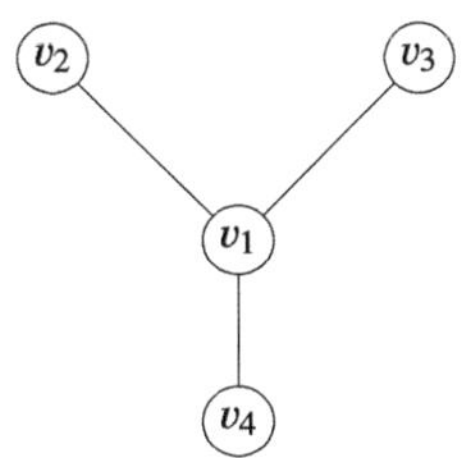

6.T.34 Suppose that the state space is $\mathcal{S} = \{0, \ldots, N\}^2$, equipped with the coordinate-wise ordering:

$$\mathbf{C}((x, y), (x', y')) = \mathbf{1}((x, y) \leq (x', y')) = \mathbf{1}(x \leq x' \text{ and } y \leq y').$$

Show that the inverse function $\mathbf{C}^{-1}$ is given by:

$$\mathbf{C}^{-1}((x, y), (x', y')) = \begin{cases} 1 & \text{if} & (x' = x \text{ and } y' = y) \text{ or} \\ & & (x' = x + 1 \text{ and } y' = y + 1), \\ -1 & \text{if} & (x' = x + 1 \text{ and } y' = y) \text{ or} \\ & & (x' = x \text{ and } y' = y + 1), \\ 0 & \text{otherwise.} \end{cases}$$

6.T.35 Consider $S_R = Y_1 + \ldots + Y_R$, where Y_i are i.i.d. indicator random variables with $\mathbb{P}(Y_i = 1) = p,\quad \mathbb{P}(Y_i = 0) = 1 - p$. Assume that $p > 1/2$.

(a) Show that:

$$\mathbb{P}\left(S_R > \frac{R}{2}\right) \geq 1 - \exp\left(-\frac{R(p - 1/2)^2}{2p}\right).$$

(b) Show that for

$$R = \frac{2p \log(1/\delta)}{(p - 1/2)^2},$$

we have:

$$\mathbb{P}\left(S_R > \frac{R}{2}\right) \geq 1 - \delta.$$

Hint: Use the Chernoff lower tail bound (4.2) and choose an appropriate value of ε.

6.T.36 In Lemma 6.5.11, we showed that if an $(\varepsilon, 1/4)$-FPRASC exists for z_G, we can construct an (ε, δ')-FPRASC for z_G with $\delta' < 1/4$ (the proof relied on

Exercise 6.T.35). Show that, in general, if an (ε, δ)-FPRASC is given for some $\delta \in (0, 1/2)$, then we can construct an (ε, δ')-FPRASC with $\delta' < \delta$ in a similar manner—by taking the median of R independent estimations obtained from the (ε, δ)-FPRASC. Determine the appropriate value of R and explicitly outline the full procedure.

Lab Exercises

6.L.1 The following Python code generates a binary image, `mouse_image`, shown in Fig. 6.21a.

Python Listing 6.1 Create image of Mickey Mouse

```python
import numpy as np
import matplotlib.pyplot as plt

x = np.linspace(-1,1, 400)
y = np.linspace(-1,1, 400)
X, Y = np.meshgrid(x, y)

head = (X**2 + Y**2) < 0.5**2
ear_right = ((X-0.4)**2 + (Y+0.5)**2) < 0.3**2
ear_left  = ((X+0.4)**2 + (Y+0.5)**2) < 0.3**2

mouse_image = head | ear_left | ear_right
```

Now, construct two "corrupted" versions of `mouse_image`, where each pixel is independently flipped with probability $\alpha \in \{0.025, 0.15\}$, producing images illustrated in Fig. 6.21b and c. Given one of these corrupted images, the objective is to reconstruct the original image in Fig. 6.21a.

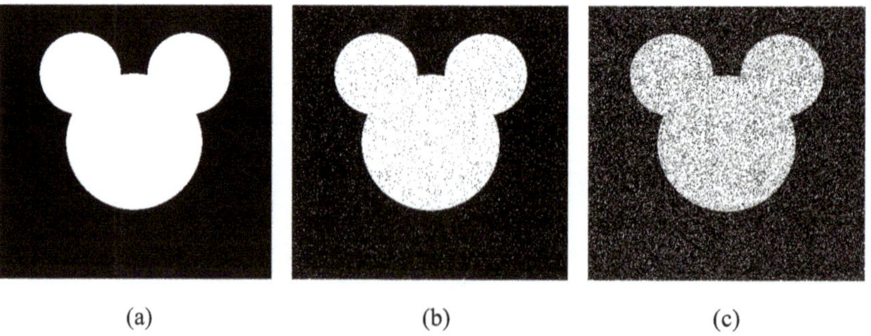

(a)　　　　　　(b)　　　　　　(c)

Fig. 6.21 Original and broken images. (**a**) Original image. (**b**) Broken: $\alpha = 0.025$. (**c**) Broken: $\alpha = 0.15$

One possible approach is the *Bayesian approach*: Denote the $n \times n$ binary image as a $d = n^2$-dimensional vector $\mathbf{x} = (x_1, \ldots, x_d)$, where pixels take values -1 (black) or $+1$ (white), i.e., $\mathbf{x} \in \{-1, +1\}^d$.

We observe $\mathbf{z}$, a corrupted version of $\mathbf{x}$, where each pixel is flipped independently with probability α. The likelihood of observing $\mathbf{z}$ given the true image $\mathbf{x}$ is:

$$
P(\mathbf{Z} = \mathbf{z} \mid \mathbf{X} = \mathbf{x}) = \frac{1}{C_2}(1 - \alpha)^{\lambda_2|\{i:x_i = z_i\}|}\alpha^{\lambda_2|\{i:x_i \neq z_i\}|},
$$

where $\lambda_2 = 1$ (though this may change later).

To incorporate prior knowledge that *"neighboring pixels tend to have similar values,"* we model $\mathbf{X}$ using the Ising model with parameter $\lambda_1 > 0$:

$$
P(\mathbf{X} = \mathbf{x}) = \frac{1}{C_1} \exp\left(\lambda_1 \sum_{(i,j) \in K} x_i x_j \right),
$$

where $G = (V, K)$ is the underlying graph, with V representing pixels and K encoding neighborhood relationships (see Fig. 6.22).

Given an observation $\mathbf{z}$, the posterior distribution is:

$$
\pi_\mathbf{x} = P(\mathbf{X} = \mathbf{x}|\mathbf{Z} = \mathbf{z})
$$

$$
= \frac{1}{C} \exp\left(\lambda_1 \sum_{(i,j) \in K} x_i x_j + \lambda_2[|\{i : x_i = z_i\}| \ln(1 - \alpha) \right.
$$

$$
\left. + |\{i : x_i \neq z_i\}| \ln \alpha] \right).
$$

Both parameters $\lambda_1, \lambda_2 > 0$. Note that if $\lambda_1 \gg \lambda_2$, the model favors configurations where neighboring pixels have similar values, giving less weight to similarity with $\mathbf{z}$, while if $\lambda_1 \ll \lambda_2$, the model prioritizes similarity to $\mathbf{z}$, paying less attention to local smoothness.

Task: Implement a Gibbs sampler for the posterior distribution $\pi_\mathbf{x}$. Tune the parameters λ_1 and λ_2 to achieve satisfactory reconstruction of the original image from a corrupted one. Estimate the number of iterations required for the method to converge.

Fig. 6.22 Lattice graph
structure

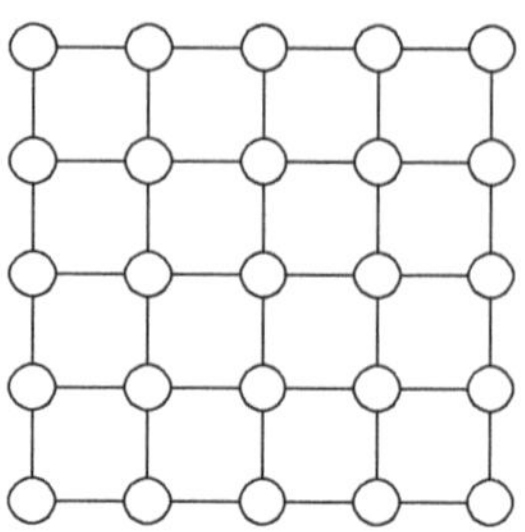

6.L.2 Consider the birth and death chain from Exercise 6.T.3 with $N = 50$ and transition probabilities:

$$p_i = \frac{1}{2(i+1)}, \quad i = 0, \ldots, 49, \qquad q_i = \frac{1}{2(i-1)}, \quad i = 1, \ldots, 50.$$

The chain is stochastically monotone (see Exercise 6.T.29) and therefore also realizable monotone (by Exercise 6.T.26). Implement the Coupling from the Past (CFTP) algorithm (Algorithm 49) for this chain and run it 1000 times.

- Plot a histogram of the results.
- Compare the obtained distribution with the known stationary distribution using, for example, the chi-square test.
- Estimate the expected value of τ_{cftp} from these 1000 simulations.

6.L.3 Consider a birth and death chain (X_k) on the state space $\mathcal{S} = \{1, \ldots, N\}$ with constant birth and death rates:

$$p_i = p, \quad i = 1, \ldots, N, \qquad q_i = q, \quad i = 1, \ldots, N.$$

Find $\mathbf{P}_Z$, the transition matrix of (Z_k), the Siegmund dual of (X_k) with respect to total ordering. That is, find a Markov chain (Z_k) such that for any $i, j \in \mathcal{S}$ and any $n \geq 0$, we have (corresponding to Eq. (6.8.1) for total ordering):

$$\mathbb{P}(X_n \leq j \mid X_0 = i) = \mathbb{P}(Z_n \geq i \mid Z_0 = j).$$

Hint: The one-step transition probabilities can be derived directly from the above formula:

$$\mathbf{P}_Z(i, j) = \mathbb{P}(Z_1 \geq j \mid Z_0 = i) - \mathbb{P}(Z_1 \geq j - 1 \mid Z_0 = i)$$

$$= \mathbb{P}(X_1 \leq j \mid X_0 = i) - \mathbb{P}(X_1 \leq j - 1 \mid X_0 = i).$$

The resulting chain (Z_k) is a birth and death chain on $\mathcal{S}^* = \{0, 1, \ldots, N\}$ with states 0 and N being absorbing.

Fix $N = 50$, $p = 0.4$, and $q = 0.5$. Estimate π_{25} by computing the difference:

$$\pi(\{1, \ldots, 25\}) - \pi(\{1, \ldots, 24\}).$$

Use the stable simulation scheme (Algorithm 59) to estimate $\pi(\{1, \ldots, 25\})$ and $\pi(\{1, \ldots, 24\})$.

- Estimate the winning probabilities of (Z_k) starting at $Z_0 = 25$ and $Z_0 = 24$ using equation (6.8.2).
- Choose an appropriate number of replications R.
- Provide the final estimation along with confidence bounds for $\pi(\{1, \ldots, 25\})$ and $\pi(\{1, \ldots, 24\})$.

6.L.4 Let $\mathcal{S} = \mathcal{S}_n$ be the set of all permutations of n elements (cards). Consider the following proposal distribution:

$$\mathbf{Q}(\sigma, \sigma') = \begin{cases} \frac{1}{2}, & \text{if } \sigma' = \sigma, \\[2ex] \frac{1}{2\binom{n}{2}}, & \text{if } \sigma' \in \mathcal{N}(\sigma), \\[2ex] 0, & \text{otherwise,} \end{cases}$$

where $\mathcal{N}(\sigma)$ consists of all permutations (neighbors) obtained from σ by swapping two elements. Construct a Metropolis algorithm using the above proposal distribution for sampling from the target distribution:

$$\pi_\sigma = \frac{\theta^{f(\sigma)}}{z},$$

where $z = \sum_\sigma \theta^{f(\sigma)}$ is the normalization constant, and the function $f(\sigma)$ is defined as:

$$f(\sigma) = \sum_{k=1}^{n} k \cdot \sigma_k.$$

For example, if $\sigma = (3, 5, 4, 1, 2)$, then

$$f((3, 5, 4, 1, 2)) = 1 \cdot 3 + 2 \cdot 5 + 3 \cdot 4 + 4 \cdot 1 + 5 \cdot 2 = 39.$$

Implement the algorithm for $n = 50$ and estimate the expected number of fixed points in a permutation sampled from the distribution π_σ.

Chapter 7
Stochastic Optimization

In this chapter, we aim to survey some Monte Carlo methods for the following problem:

- Given a function $h : S \to \mathbb{R}$, find

$$\gamma^{\text{opt}} = \max_{x \in S} h(x), \qquad x^{\text{opt}} = \arg \max_{x \in S} h(x).$$

Some methods are based on the MCMC studied in the previous Chap. 6, but we also include others based on cross-entropy (CE) or locally informed proposals. The roots of these methods are the importance sampling method from Chap. 5.1.7 and Metropolis methods.

7.1 Metropolis and Gibbs Based Methods

To be consistent with graph-based notation the state space will be denoted by $S = V$.

Consider a graph $G = (V, E)$ with a cost function $H : V \to \mathbb{R}_+$. Assume we are to find an argument (i.e., the vertex) for which H attains its minimum. The idea of stochastic optimization using MCMC methods is to sample from some distribution π (approximately, via MCMC) which assigns, in some natural way, larger mass to solutions with a small value of H. The state space is $S = V$. Sampling from π favours better solutions. Often the Boltzmann distribution is used (for any $T > 0$):

$$\pi_v^T = \frac{e^{-H(v)/T}}{\sum_{v' \in V} e^{-H(v')/T}}, \qquad v \in V. \tag{7.1.1}$$

P. Lorek, T. Rolski, *Lectures on Monte Carlo Theory*, Probability Theory and Stochastic Modelling 108, https://doi.org/10.1007/978-3-032-01190-9_7

We then have

$$\operatorname*{argmin}_{v \in V} H(v) = \operatorname*{argmax}_{v \in V} \pi_v^T.$$

If we are to find a maximum of some function $f : V \to \mathbb{R}_+$, then we may use above the Boltzmann distribution with cost function

$$H(v) = -f(v).$$

We will exchangeably use $\mathbf{x}$ or v for arguments of the functions we aim to maximize. In context of optimization they are called *states* or *solutions*. From the Metropolis-Hastings algorithm (see Proposition 6.3.2), to construct a Markov chain with stationary distribution π^T (we may sample from this distribution then) we must move from v to v' with probability $p_{vv'}^T = q_{vv'}\alpha_{vv'}^T$, where $\mathbf{Q} = (q_{vv'})$ is some candidate-generating matrix. Let us take $\mathbf{Q}$ corresponding to the simple random walk on the graph G as it was considered in Example 6.2.7. Assume that G is such that $\mathbf{Q}$ is regular. The acceptance probability is then given by

$$\alpha_{vv'}^T = \min\left(1, \frac{\deg(v)}{\deg(v')} e^{(H(v)-H(v'))/T}\right), \qquad v, v' \in V$$

and the transition probability is

$$p_{vv'}^T = \frac{1}{\deg(v)} \min\left(1, \frac{\deg(v)}{\deg(v')} e^{(H(v)-H(v'))/T}\right), \qquad v, v' \in V, q_{vv'} > 0.$$

Summarizing, the Metropolis-Hastings algorithm for Boltzmann distribution (with $\mathbf{Q}$ as in Example 6.2.7) is provided in Algorithm 61.

Algorithm 61 Metropolis-Hastings algorithm for Boltzmann distribution on V with $\mathbf{Q}$ corresponding to simple random walk on $G = (V, E)$ (see Example 6.2.7 (one step)

1: Assume $Y_k = v$
2: Choose $v' \in \mathcal{N}(v)$ uniformly at random
3: Compute $\min\left(1, \frac{\deg(v)}{\deg(v')} e^{(H(v)-H(v'))/T}\right)$
4: Generate $U \sim \mathcal{U}[0, 1)$
5: **if** $U \leq \alpha$ **then**
6:      $Y_{k+1} = v'$
7: **else**
8:      $Y_{k+1} = Y_k$
9: **end if**

Note that it is up to the user to choose the parameters, i.e., $T > 0$ and the underlying graph defining the neighborhood and thus $\mathbf{Q}$. For any choice of $\mathbf{Q}$ and T

the argmax π_v^T is attained for the same v^* (or set, if it is not unique); however, the actual "quality" of the simulated sample may (and usually will) differ—we usually will sample $\tilde{v}$ such that $\pi_{\tilde{v}}^T$ is "*close*" to $\pi_{v^*}^T$. However, this "closeness" will differ for different parameters. Say for some $\tilde{v}$ we have $H(\tilde{v}) \approx H(v^*)$ then depending on parameters $\pi_{\tilde{v}}^T$ may or may not be closer than $\pi_{v^*}^T$. Another issue is the rate of convergence. In practice, we simulate only finitely many steps and treat X_n as an approximate sample from π^T. For a given n, the distribution of X_n may be far from or close to the stationary distribution, depending on those parameters.

Substitution Cipher One day, a psychologist from a local prison came to the consulting service of the Statistics Department at Stanford University. He asked to decrypt prisoners' messages; an example is shown in Figure 1 of [35].

A student guessed that the code was a simple substitution cipher, i.e., this is a 1-1 correspondence between symbols used in messages and symbols used in written English (letters, numbers, spaces, punctuation marks). Denote the sets of the symbols used in encrypted messages by Σ^{pr} and symbols used in the original message in English by Σ^{eng}. Let $m_1, \ldots, m_L$, where $m_i \in \Sigma^{\mathrm{eng}}$ is the original message and $c_1, \ldots, c_L$ is this message encrypted in alphabet Σ^{pr}. Thus $c_i = \sigma^*(m_i)$.

The goal is to determine a 1-1 mapping

$$\sigma^* : \Sigma^{\mathrm{pr}} \to \Sigma^{\mathrm{eng}}.$$

Assume that $|\Sigma^{\mathrm{pr}}| = |\Sigma^{\mathrm{eng}}| = n$, without loss of generality, we can identify then the symbols as $1, 2, \ldots, n$. We need to store symbols-to-numbers mappings for both Σ^{pr} and Σ^{eng}. In such a case we may identify $\Sigma^{\mathrm{pr}} = \Sigma^{\mathrm{eng}} = \{1, \ldots, n\}$, σ^* is a permutation of these elements. The example is taken from [35], where in the described case, the size of Σ^{pr} could be smaller than the size of Σ^{eng} (i.e., several symbols from Σ^{eng} are not used).

The stochastic optimization idea is then as follows. From some known text, we can record first-order transitions between symbols in Σ^{eng}. In this example, Tolstoy's text *"War and Peace"* was used. It gives a transition matrix $\mathbf{M}(x, y)$, the proportion of text symbols y appearing after symbol x. For the given encrypted message $(c_1, \ldots, c_L)$ (notice that $c_j \in \{1, \ldots, n\}$) we define the likelihood function[1]

$$l(\sigma) = \prod_{i=1}^{L-1} \mathbf{M}(\sigma(c_i), \sigma(c_{i+1})). \tag{7.1.2}$$

It is clear that for the "correct decryption" function σ^*, the likelihood $l(\sigma^*)$ should be the largest, under the assumption that the encrypted message is "similar" in a way to the text used for learning the matrix $\mathbf{M}$. Hence it is reasonable to think that

---

[1] We skip the details related to the symbols-to-numbers mapping.

the larger the $l(\sigma)$, the better the σ. Define a probability measure on all $\sigma \in \mathcal{S}_n$

$$\pi_\sigma = \frac{1}{z} l(\sigma) = \frac{1}{z} \prod_{i=1}^{L-1} \mathbf{M}(\sigma(c_i), \sigma(c_{i+1})),$$

where $z = \sum_{\sigma \in \mathcal{S}_n} \prod_{i=1}^{L-1} \mathbf{M}(\sigma(c_i), \sigma(c_{i+1}))$ is an unknown normalizing constant. Sampling from π will yield "good" decryption function σ with high probability. Unknown normalizing constant z and large size of the state space $n!$ are no obstacles for the Metropolis algorithm. To proceed, we must define a proposal distribution $q_{\sigma\sigma'}, \sigma, \sigma' \in \mathcal{E} = \mathcal{S}_n$. For example, we can define the t.m. $\mathbf{Q}$ as follows. Being in σ we can only randomly (uniformly) swap two symbols (that is, to make a transposition):

$$q_{\sigma\sigma'} = \begin{cases} \frac{2}{n(n-1)} & \text{if } \sigma'(k) = \sigma(k), k \notin \{a, b\}, \sigma(a) = \sigma'(b), \sigma(b) = \sigma'(a), \\ 0 & \text{otherwise.} \end{cases}$$

For σ, σ' such that $q_{\sigma\sigma'} > 0$ we need to calculate

$$\frac{\pi_{\sigma'}}{\pi_\sigma} = \prod_{i:c_i=a,c_{i+1}=a} \frac{\mathbf{M}(\sigma(b), \sigma(b))}{\mathbf{M}(\sigma(a), \sigma(a))} \prod_{i:c_i=a,c_{i+1}=b} \frac{\mathbf{M}(\sigma(b), \sigma(a))}{\mathbf{M}(\sigma(a), \sigma(b))}$$

$$\times \prod_{i:c_i=b,c_{i+1}=b} \frac{\mathbf{M}(\sigma(a), \sigma(a))}{\mathbf{M}(\sigma(b), \sigma(b))} \prod_{i:c_i=b,c_{i+1}=a} \frac{\mathbf{M}(\sigma(a), \sigma(b))}{\mathbf{M}(\sigma(b), \sigma(a))}$$

$$= \left(\frac{\mathbf{M}(\sigma(b), \sigma(b))}{\mathbf{M}(\sigma(a), \sigma(a))} \right)^{\gamma(a,a)-\gamma(b,b)} \left(\frac{\mathbf{M}(\sigma(b), \sigma(a))}{\mathbf{M}(\sigma(a), \sigma(b))} \right)^{\gamma(a,b)-\gamma(b,a)},$$

$$(7.1.3)$$

where, for $\alpha, \beta \in \{a, b\}$ we define $\gamma(\alpha, \beta) = \#\{i : c_i = \alpha, c_{i+1} = \beta\})$.

Finally, Algorithm 42 for this example is following:

Simulation Results We computed the matrix $\mathbf{M}$ based on an English translation of Tolstoy's *"War and Peace"* and encrypted the following paragraph from another Tolstoy's novel *"Anna Karenina"* (removing non-letters and lower-casing everything first)[2]

---

[2] https://www.gutenberg.org/files/2600/2600-h/2600-h.htm.

Algorithm 62 Decrypting substitution cipher (one step)

1: Assume $Y_k = \sigma$
2: Generate uniformly at random two distinct numbers $a, b \in \{1, \ldots, n\}$.
3: Set $\sigma'(k) = \sigma(k)$, $k \notin \{a, b\}$, $\sigma'(a) = \sigma(b)$, $\sigma'(b) = \sigma(a)$.
4: Compute $\alpha = \min(1, \pi_{\sigma'}/\pi_\sigma)$ from (7.1.3).
5: Generate $U \sim \mathcal{U}[0, 1)$;
6: **if** $U \le \alpha$ **then**
7: $\quad Y_{k+1} = \sigma'$
8: **else**
9: $\quad Y_{k+1} = Y_k$
10: **end if**

```
the whole of that day anna spent at home thats to say at the oblonskys
and received no one though some of her acquaintances had already
heard of her arrival and came to call the same day anna spent the
whole morning with dolly and the children she merely sent a brief
note to her brother to tell him that he must not fail to dine at
home come god is merciful she wrote
```

We took (chosen randomly) the following permutation:

'a' → 'r'	'b' → 'a'	'c' → 'w'	'd' → 'v'	'e' → 'x'	'f' → 'l'
'g' → 'b'	'h' → 'g'	'i' → 'd'	'j' → 'm'	'k' → 'p'	'l' → 'n'
'm' → 't'	'n' → 'y'	'o' → 'u'	'p' → 'i'	'q' → 'f'	'r' → 'e'
's' → 'k'	't' → 's'	'u' → 'z'	'v' → 'h'	'w' → 'j'	'x' → ' '
'y' → 'o'	'z' → 'q'	' ' → 'c'			

This results in the following encrypted message:

```
yckqzchskqhaqyc yqw pq tt qrektyq yqchxkqyc yrqyhqr pq yqyckqhvs
htrmprq twqjkuklikwqthqhtkqychgfcqrhxkqhaqckjq udg lty tukrqc wq
sjk wpqck jwqhaqckjq jjli sq twqu xkqyhqu ssqyckqr xkqw pq tt qr
ektyqyckqzchskqxhjtltfqzlycqwhsspq twqyckquclswjktqrckqxkjkspqrk
tyq qvjlkaqthykqyhqckjqvjhyckjqyhqykssqclxqyc yqckqxgryqthyqa ls
qyhqwltkq yqchxkquhxk
```

We ran the decryption algorithm for 3000 steps (starting with a random permutation). The beginning of the decrypted message, along with the corresponding log-likelihoods for a few initial steps as well as for steps 1000, 2000, and 3000, are presented in Table 7.1. The Metropolis-based decryption algorithm is implemented in `ch7_cipher_decrypt_metropolis.py`. There are several files `ch7_cipher_*`—encryption, generating initial permutation, Tolstoy's *War and Peace* and message. The decrypted message (with incorrectly decrypted letters underlined) is following:

```
the whole of that day anna skent at home thats to say at the oplonsbys
and received no one though some of her acquaintances had already
heard of her arrival and came to call the same day anna skent the
whole morning with dolly and the children she merely sent a prief
```

Table 7.1 Metropolis algorithm applied to substitution cipher

Step	Decrypted message	Log likelihood
1	yckqzchskqhaqyc yqw pq tt qrektyq yqchxkqyc yrqyhqr pq yqyckqhvshtrmpr	-4274.98
2	yckqzchskqhaqyc yqv pq tt qrektyq yqchxkqyc yrqyhqr pq yqyckqhwshtrmpr	-4304.07
3	yckqzchskqhaqyc yqv pq tt qrektyq yqchxkqyc yrqyhqr pq yqyckqhwshtrmpr	-4431.40
4	yckqschzkqhaqyc yqv pq tt qrektyq yqchxkqyc yrqyhqr pq yqyckqhwzhtrmpr	-4643.92
5	yckqschzkqhaqyc yqv pq tt qmektyq yqchxkqyc ymqyhqm pq yqyckqhwzhtmrpm	-4726.83
6	yckvschzkvhavyc yvq pv tt vmektyv yvchxkvyc ymvyhvm pv yvyckvhwzhtmrpm	-3519.67
7	yckvschzkvhavyc yvq pv tt vmgktyv yvchxkvyc ymvyhvm pv yvyckvhwzhtmrpm	-3512.30
8	kcyvschzyvhavkc kvq pv tt vmgytkv kvchxyvkc kmvkhvm pv kvkcyvhwzhtmrpm	-3334.56
9	kcavschzavhyvkc kvq pv tt vmgatkv kvchxavkc kmvkhvm pv kvkcavhwzhtmrpm	-3059.72
1000	the whore of that das anna clent at hove thatc to cas at the oproncksc	-896.09
2000	the whole of that day anna skent at hobe thats to say at the oplonsmys	-826.07
3000	the whole of that day anna skent at home thats to say at the oplonsbys	-815.71

note to her prother to tell him that he must not fail to dine at home come god is merciful she wrote

The evolution of log-likelihood values is presented in Fig. 7.1.

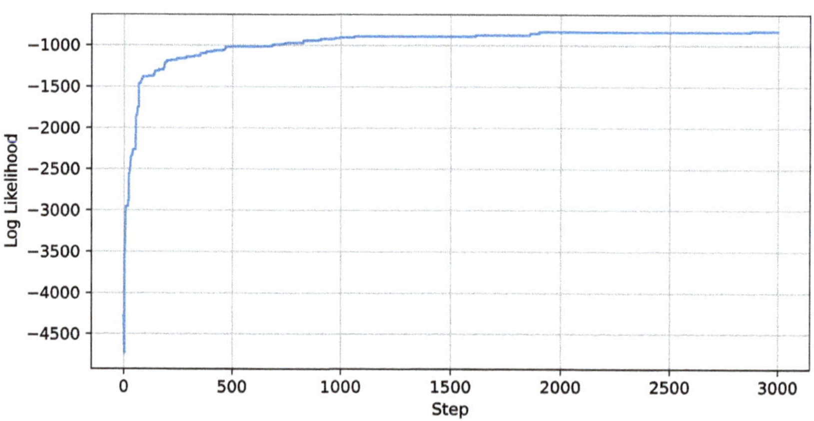

Fig. 7.1 Evolution of log-likelihood values $\log l(Y_k)$ for the sequence of permutations $Y_1, \dots, Y_{3000}$ generated by the Metropolis algorithm

Remark 7.1.1 Instead of the likelihood function l defined in (7.1.2), in [24] it was proposed to take another score function. Consider pairs (i, j), where $i, j \in \Sigma^{\text{eng}}$ and let $m(i, j)$ be the number of times the pair (i, j) appears in the reference text, let $f(\sigma(i), \sigma(j))$ be the number of times the pair appears when the cypher text is decrypted using the decryption key σ. Then, the score function is defined by

$$\pi_\sigma = \prod_{(i,j)} m(i,j)^{f(\sigma(i),\sigma(j))}.$$

During computation, it is recommended to use logarithms.   ∎

The Knapsack Problem   We are given d items with weights $\mathbf{w} = (w_1, \ldots, w_d)$ and values $\mathbf{v} = (v_1, \ldots, v_d)$, we assume all weights and values are non-negative. We have a knapsack with a maximum total weight (capacity) of W. The problem is deciding which of d items to take, so that we will maximize their total value. Formally, let $\mathbf{x} = (x_1, \ldots, x_d)$, where $x_i \in \{0, 1\}$. The aim is to maximize $f(\mathbf{x}) = \mathbf{v} \cdot \mathbf{x} = \sum_{i=1}^{d} v_i x_i$ with a constraint:

$$\text{find } \operatorname*{argmax}_{\mathbf{x}} f(\mathbf{x}) = \operatorname*{argmax}_{\mathbf{x}} \sum_{i=1}^{d} v_i x_i,$$

$$\text{subject to } \mathbf{w} \cdot \mathbf{x} = \sum_{i=1}^{d} w_i x_i \leq W.$$

Thus, our state space is $\mathcal{S} = \{0, 1\}^d$ and we want to construct a Markov chain with stationary distribution (for some $T > 0$)

$$\pi_{\mathbf{x}} = \begin{cases} \dfrac{\exp(\mathbf{v} \cdot \mathbf{x}/T)}{z_T} & \text{if } \mathbf{w} \cdot \mathbf{x} \leq W, \\ 0 & \text{otherwise} \end{cases}$$

where z_T is a normalizing constant. We will provide both the Gibbs sampler and the Metropolis algorithm for this problem.

Gibbs Sampler for the 0-1 Knapsack Problem   We can think of assigning 0 or 1 to vertices $V = \{1, \ldots, d\}$. The main part of constructing this sampler is to calculate the following conditional probability; assume that Z has distribution $\pi_{\mathbf{x}}$ and that $X_k = \mathbf{x}$:

$$\mathbb{P}(Z(i) = 1 \mid Z_{-i} = \mathbf{x}_{-i})$$

$$= \frac{\mathbb{P}(Z(i) = 1, Z_{-i} = \mathbf{x}_{-i})}{\mathbb{P}(Z(i) = 1, Z_{-i} = \mathbf{x}_{-i}) + \mathbb{P}(Z(i) = 0, Z_{-i} = \mathbf{x}_{-j})}$$

$$= \frac{\exp\left(\frac{1}{T}\left(\sum_{j \neq i} v_j x_j + v_i\right)\right)}{\exp\left(\frac{1}{T}\left(\sum_{j \neq i} v_j x_j + v_i\right)\right) + \exp\left(\frac{1}{T}\left(\sum_{j \neq i} v_j x_j\right)\right)} = \frac{1}{1 + \exp\left(-\frac{1}{T} v_i x_i\right)}.$$

The Gibbs sampler for the 0-1 knapsack problem is the following, being in state X_k is presented in Algorithm 63.

Algorithm 63 Gibbs sampler for the 0-1 knapsack problem (one step)

1: Suppose $Y_k = \mathbf{x}$
2: Pick $i \in \{1, \ldots, d\}$ uniformly at random
3: Set $\mathbf{x}' = (x_1', \ldots, x_d')$, $x_j' = x_j$, $j \neq i$, $x_i' = 1$
4: **if** $\mathbf{x}' \cdot \mathbf{w} > W$ **then**
5:     Set $x_i' = 0$.
6: **else**
7:     Simulate $U \sim U(0, 1)$ and set

$$
x_i' = \begin{cases} 1 \text{ if } U \leq \dfrac{1}{1+\exp\left(-\frac{1}{T} v_i x_i\right)}, \\[2mm] 0 \text{ otherwise.} \end{cases}
$$

8: **end if**
9: Set $Y_{k+1} = \mathbf{x}'$
10: **return** Y_{k+1}

Algorithm 64 Metropolis algorithm for the 0-1 knapsack problem (one step)

1: Suppose $X_k = \mathbf{x}$
2: Pick $i \in \{1, \ldots, d\}$ uniformly at random
3: Set $\mathbf{x}' = (x_1', \ldots, x_d')$, $x_j' = x_j$, $j \neq i$, $x_i' = 1 - x_i$
4: **if** $\mathbf{x}' \cdot \mathbf{w} \leq W$ **then**
5:     Set $\alpha = \min\left(1, \exp\left(\frac{1}{T}(1 - 2x_i)v_i\right)\right)$
6: **else**
7:     Set $\alpha = 0$
8: **end if**
9: Generate $U \sim \mathcal{U}[0, 1)$
10: **if** $U \leq \alpha$ **then**
11:     $Y_{k+1} = \mathbf{x}'$
12: **else**
13:     $Y_{k+1} = \mathbf{x}$
14: **end if**
15: **return** Y_{k+1}

Metropolis Algorithm for the 0-1 Knapsack Problem Let us take a proposal distribution which flips one (randomly chosen) coordinate of current state $\mathbf{x}$ to the opposite value. I.e., given $X_k = \mathbf{x}$, a candidate $\mathbf{x}' = (x_1', \ldots, x_d')$, $x_j' = x_j$ for $j \neq i$ and $x_i' = 1 - x_i$, with i chosen uniformly from $\{1, \ldots, d\}$. We have

$$
\frac{\pi_{\mathbf{x}'}}{\pi_{\mathbf{x}}} = \exp\left(\frac{1}{T} \sum_{j=1}^{d} (x_i' - x_i)v_i\right) \mathbf{1}(\mathbf{x}' \cdot \mathbf{w} \leq W)
$$

$$= \exp\left(\frac{1}{T}(1 - 2x_i)v_i\right) \mathbf{1}(\mathbf{x}' \cdot \mathbf{w} \le W).$$

Note that if $x_i = 0$ we have $\alpha = \min(1, \pi_{\mathbf{x}'}/\pi_{\mathbf{x}}) = 1$ and if $x_i = 1$ then $\alpha = \exp\left(\frac{1}{T}(1 - 2x_i)v_i\right)$ if $\mathbf{x}' \cdot \mathbf{w} \le W$ and 0 otherwise. The procedure is presented in Algorithm 64.

Note that in both the Gibbs sampler and the Metropolis algorithm we can always take $(0, \ldots, 0)$ (i.e., an empty knapsack) as X_0 (however, there can be better choices).

Simulation Results Consider $d = 100$ items, maximal knapsack capacity $W = 3000$ and

$$w_i = i, \quad v_i = i^{1.2}, \quad i = 1, \ldots, d.$$

In Fig. 7.2, we present the results for simulating the Metropolis algorithm (the results for the Gibbs sampler are similar) for this problem for 150 steps: horizontal axis = step number, vertical axis = $X_k \cdot \mathbf{w}$ (blue) and $X_k \cdot \mathbf{v}$ (red). For final $k = 150$ we obtained: $X_{150} \cdot \mathbf{w} = 2977$ and $X_{150} \cdot \mathbf{v} = 6745.53$ (the results may be slightly improved by tweaking the temperature T). The example is implemented in `ch7_knapsack_metropolis.py`.

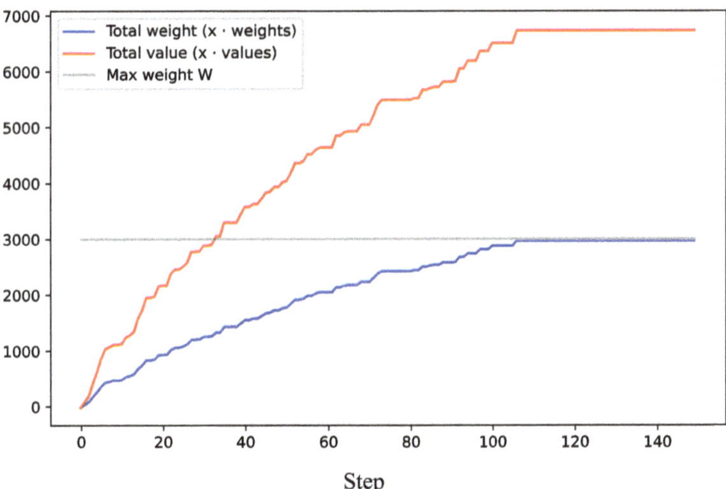

Fig. 7.2 Simulation results (Metropolis algorithm) for the 0-1 knapsack problem

7.2   Simulated Annealing

Recall our objective is to solve the following optimization problem: given a graph $G = (V, E)$ and a cost function $H : V \to \mathbb{R}_+$ we would like to find argument(s) for which H attains a minimum. Denote the set of *optimal solutions*

$$D^* = \left\{ v' \in V : H(v') = \min_{v \in V} H(v) \right\}.$$

The main idea for the stochastic optimization was to consider a distribution which "favours" better solutions (and then construct an MC with this stationary distribution and sample from it). Since we are going to present a result from Hajek [76] we need the following setting: the distribution of the form

$$\pi_v^{T_k} = \frac{\deg(v) \exp(-\frac{H(v)}{T_k})}{z_{T_k}}, \qquad v \in V, \tag{7.2.1}$$

where z_{T_k} is the normalizing constant, is the stationary distribution for the t.m. $\mathbf{P}^{T_k} = (p_{vv'}(k))$, where

$$p_{vv'}(k) = \frac{1}{\deg(v)} \min \left(1, \exp \left(-\frac{1}{T_k}(H(v') - H(v)) \right) \right). \tag{7.2.2}$$

As we see below, this fact follows directly from Proposition 6.3.2, when choosing an acceptance matrix $\mathbf{Q}$ of the form

$$q_{vv'} = \begin{cases} \beta & \text{if } v = v', \\ \frac{1-\beta}{\deg(v)} & \text{if } v' \text{ is a neighbor of } v, \\ 0 & \text{otherwise.} \end{cases}$$

for some $0 < \beta < 1$. We need to suppose that the graph structure is such that the t.m. $\mathbf{Q}$ and as a result $\mathbf{P}$ is irreducible, and $\mathbf{Q}$ is aperiodic, which indeed with this choice of $\mathbf{Q}$ holds.

Remark 7.2.1 Notice that if $\deg(v) = \text{const}$ for all $v \in V$ and $T_k \equiv T$, then the distribution π^T in (7.2.1) reduces to the Boltzmann distribution defined in (7.1.1). We will meet such a situation in *Simulated annealing for the travelling salesman problem (TSP)* below on page 7.2. ∎

Then from Proposition 6.3.2 we have that the distribution π^{T_k} is the stationary one for the t.m. $p_{vv'}$ as in (7.2.2).

The following fact suggests an idea for a simulated annealing algorithm. Let (X_n) be an MC with t.m. $\mathbf{P}^{T_k}$.

Proposition 7.2.2 *For a non-homogeneous Markov chain (X_k) with transition matrix (7.2.2), we have*

$$\lim_{T \searrow 0} \lim_{\substack{k \to \infty \\ T_k = T}} \mathbb{P}(X_k \in D^*) = 1.$$

Proof Since the t.m. $\mathbf{P}^{T_k}$ is ergodic, we clearly have

$$\lim_{\substack{k \to \infty \\ T_k = T}} \mathbb{P}(X_k \in D^*) = \sum_{v \in D^*} \pi_v^T = \sum_{v \in D^*} \frac{\deg(v) \exp(-\frac{H(v)}{T})}{\sum_v \deg(v) \exp(-\frac{H(v)}{T})}.$$

Fix any $v^* \in D^*$. Then

$$\sum_{v \in D^*} \frac{\deg(v) \exp(-\frac{H(v)}{T})}{\sum_{v'} \deg(v') \exp(-\frac{H(v')}{T})}$$

$$= \sum_{v \in D^*} \frac{\deg(v^*) \exp(-\frac{H(v^*)}{T})}{\sum_{v' \in V \setminus D^*} \deg(v') \exp(-\frac{H(v')}{T}) + \sum_{v' \in D^*} \deg(v') \exp(-\frac{H(v^*)}{T})}$$

$$= \sum_{v \in D^*} \frac{\deg(v^*)}{\sum_{v' \in V \setminus D^*} \deg(v') \exp(-\frac{(H(v') - H(v'))}{T}) + \sum_{v' \in D^*} \deg(v')}$$

and next passing to the limit as $T \searrow 0$, the proof is completed. $\qquad\square$

From Proposition 7.2.2 the following idea follows. Let us change the temperature T_k in (7.2.1) in each step according to a cooling schedule $T_0 \geq T_1 \geq \cdots$ and $T_k \searrow 0$. In this way we obtain an MC (Y_n) such that the transition in step k is governed by the t.m. $\mathbf{P}^{T_k}$. It turns out that (Y_j) is a non-homogeneous Markov chain, a concept which will be introduced formally in *Some theory of non-homogeneous Markov chains* below.

Now the question is: does the non-homogeneous Markov chain (Y_n) with the family of transition matrices given in (7.2.2), converge to a distribution concentrated on optimal solutions? I.e., do we have $\lim_{k \to \infty} \mathbb{P}(Y_k \in D^*) = 1$? The answer is: it does, provided a proper choice of a *cooling scheme* (T_k). It must converge to zero; the question now becomes: *how slowly* must (T_k) converge to zero?

Let us make the following consideration. Suppose v is a local minimum, that is $H(v) < H(v')$ for $v' \in \mathcal{N}(v)$. Set $c = \min_{v' \in \mathcal{N}(v)}(H(v') - H(v))$. Then for $v' \in \mathcal{N}(v)$

$$p_{vv'}(k) = \frac{1}{\deg(v)} \exp\left(-\frac{1}{T_k}(H(v') - H(v))\right) \leq \frac{1}{\deg(v)} e^{-c/T_k},$$

and hence

$$p_{vv}(k) = 1 - \sum_{v' \in \mathcal{N}(v)} p_{vv'}(k) \geq 1 - e^{-c/T_k}.$$

Suppose the chain visits v, which is a local minimum but not a global one, at k. The probability that the chain stays in this state forever is

$$\prod_{j=0}^{\infty} p_{vv}(k + j).$$

We have

$$\prod_{j=0}^{l} p_{vv}(k + j) \geq \prod_{j=0}^{l} (1 - e^{-c/T_j}),$$

and provided

$$\sum_{j} e^{-c/T_j} < \infty$$

we have $\lim_{l \to \infty} \prod_{j=0}^{l}(1 - e^{c/T_j}) > 0$, which yields

$$\lim_{l \to \infty} p_{vv}(k + j) > 0.$$

Above we used the fact that the infinite product $\prod_{j=1}^{\infty}(1 - a_j)$ converges if and only if $\sum_{j=1}^{\infty} a_j$ converges too. This means that in this cooling scheme, the chain can stay in a local minimum. The above considerations demonstrated that a cooling cannot be too fast.

An algorithm based on this idea is called simulated annealing (SA). One step from time k to $k + 1$ is given in Algorithm 65.

Algorithm 65 Simulated Annealing (SA); transition for step k to step $k + 1$

1: $Y_k = v$
2: Generate $v' \in \mathcal{N}(v)$ with probability $q_{vv'}$
3: Set $\alpha_k = \min(1, \exp(-(H(v') - H(v))/T_k))$
4: Generate $U \sim \mathcal{U}[0, 1)$
5: **if** $U \leq \alpha_k$ **then**
6: $\quad Y_{k+1} = v'$
7: **else**
8: $\quad Y_{k+1} = X_k$
9: **end if**
10: **return** Y_{k+1}

These convergence results are often mainly of theoretical interest. However, they also suggest the *type/rate* of the cooling scheme (e.g., $d \log(k)$ for some d), which we apply for finitely many steps only, as will be done below.

Some Theory of Non-homogeneous Markov Chains The Metropolis algorithm was originally introduced for studying the equilibrium properties of statistical-mechanic systems at a given temperature. The simulated annealing (SA) algorithm is a modification of the Metropolis algorithm in which the temperature is changing with time according to a cooling scheme. Recently SA has been used as an optimization technique for a number of combinatorial problems such as, for example, the travelling salesman problem. SA is a special case of the theory of non-stationary Markov chains. We give here some basic definitions.

In Sect. 6.2 a (time-homogeneous) Markov (X_n) chain was determined by an initial distribution μ and a transition matrix $\mathbf{P}$. In general, a sequence (Y_n) is a **time non-homogeneous Markov chain** on a state space $S = \{s_1, \dots, s_m\}$ with an initial distribution μ and a collection of transition matrices $\mathbf{P}(n)$, $n = 1, 2, \dots$ if for any $s_{i_0}, \dots, s_{i_k} \in S$

$$\mathbb{P}(Y_0 = s_{i_0}, Y_1 = s_{i_1}, \dots, Y_k = s_{i_k}) = \mu_{s_{i_0}} p_{s_{i_0} s_{i_1}}(1) \cdots p_{s_{i_{k-1}} s_{i_k}}(k).$$

Thus

$$p_{s_i s_j}(k) = \mathbb{P}(Y_{k+1} = s_j | Y_k = s_i)$$

is the probability that the chain at step $k + 1$ will be in state s_j provided that at step Y_k it was in state s_i.

Let us define $\mathbf{P}(k, l) = \mathbf{P}(k)\mathbf{P}(k + 1) \cdots \mathbf{P}(l)$ for $k < l$. By $\mathbf{P}(0)$ we mean the identity matrix. Of course, for $k < l$

$$\mathbb{P}(Y_l = s_j | Y_k = s_i) = (\mathbf{P}(k, l))_{s_i s_j}.$$

Similarly as (6.2.1), for non-homogeneous chains we have

$$(\mathbb{P}(X_l = s_1), \mathbb{P}(X_l = s_2), \dots, \mathbb{P}(X_l = s_m)) = \boldsymbol{\mu} \mathbf{P}(0, l), \qquad (7.2.3)$$

where $\boldsymbol{\mu} = (\mathbb{P}(X_0 = s_1), \mathbb{P}(X_0 = s_2), \dots, \mathbb{P}(X_0 = s_m))$ is the initial distribution.

Remark 7.2.3 Note that thus a simulated annealing, i.e., Algorithm 65, is a non-homogeneous Markov chain, with transition probabilities

$$p_{vv'}(k) = q_{vv'}\alpha_k.$$

$\blacksquare$

Simulated Annealing for the Travelling Salesman Problem (TSP) In one of the introductory Sect. 1.2.2, we mentioned the so-called travelling salesman problem.

A matrix $\mathbf{M}$ of size $n \times n$ is given, entry $M(i, j) = M(j, i)$ is a distance between i-th and j-th city. We want to start at some city, visit all the cities and return to the initial one. Let $\sigma = (\sigma(1), \ldots, \sigma(n)) \in \mathcal{S}_n$ be a permutation of $\{1, \ldots, n\}$. The permutation represents here our travelling ordering: we start in city $\sigma(1)$, then go to $\sigma(2)$ and so on. Ultimately, we go from $\sigma(n)$ to $\sigma(1)$. For fixed σ the total distance travelled (cf. the *likelihood* function from Sect. 7.1 (*Substitution cipher*)) is our score function

$$l(\sigma) = \sum_{k=1}^{n-1} \mathbf{M}(\sigma(k), \sigma(k+1)) + \mathbf{M}(\sigma(n), \sigma(1)).$$

We seek

$$\sigma^* = \underset{\sigma}{\operatorname{argmin}}\, l(\sigma)$$

by sampling from the Boltzmann distribution with $f(\sigma) = -l(\sigma) = -H(\sigma)$, i.e.,

$$\pi_\sigma^T = \frac{1}{z} e^{f(\sigma)/T}.$$

Let us take the same proposal distribution $q_{\sigma\sigma'}$, $\sigma, \sigma' \in \mathcal{S}_n$ as in Sect. 7.1, i.e., being in σ we can only randomly (uniformly) make a transposition:

$$q_{\sigma\sigma'} = \begin{cases} \frac{2}{n(n-1)} & \text{if } \sigma'(k) = \sigma(k), k \notin \{i, j\}, \sigma(i) = \sigma'(j), \sigma(j) = \sigma'(i), \\ 0 & \text{otherwise.} \end{cases}$$

For σ, σ' such that $q_{\sigma\sigma'} > 0$ we need to calculate

$$\frac{\pi_{\sigma'}}{\pi_\sigma} = e^{(l(\sigma) - l(\sigma'))/T}.$$

Note that the difference $l(\sigma) - l(\sigma')$ is easy to calculate. Let us observe that for our transition matrix $(q_{\sigma\sigma'})$, the modified Boltzmann and Boltzmann distribution are the same. Finally, the Metropolis algorithm for this example is provided in Algorithm 66.

Remark 7.2.4 Instead of choosing a neighbor of σ via transposition, we can swap longer paths in the following way.

Choose uniformly at random $i < j$, assume that

$$\sigma = (\sigma(1), \ldots, \sigma(i-1), \underbrace{\sigma(i), \ldots, \sigma(j-1)\sigma(j)}_{\text{to be reversed}}, \sigma(j+1) \ldots, \sigma(n)).$$

Algorithm 66 Metropolis algorithm for TSP (one step)

1: Assume $Y_k = \sigma$
2: Generate uniformly at random two distinct numbers $i, j \in \{1, \ldots, n\}$
3: Set $\sigma'(k) = \sigma(k), k \notin \{i, j\}, \sigma'(i) = \sigma(j), , \sigma'(j) = \sigma(i)$
4: Compute $\alpha = \min\left(1, e^{(l(\sigma)-l(\sigma'))/T_k}\right)$
5: Generate $U \sim \mathcal{U}[0, 1)$
6: **if** $U \leq \alpha$ **then**
7: $\quad Y_{k+1} = X$
8: **else**
9: $\quad Y_{k+1} = Y_k$
10: **end if**

Then, define a neighbor as

$$\sigma' = (\sigma(1), \ldots, \sigma(i-1), \underbrace{\sigma(j), \sigma(j-1), \ldots, \sigma(i)}_{\text{reversed}}, \sigma(j+1) \ldots, \sigma(n)).$$

∎

For simulated annealing with cooling scheme $T_0 \geq T_1 \geq \cdots$, we roughly take distributions

$$\pi_v^{T_k} = \frac{1}{z} e^{f(\sigma)/T_k}.$$

Consequently, the transition between σ and σ' which differ by transposition, at step k, happens with probability

$$p_{\sigma\sigma'}(k) = \frac{2}{n(n-1)} \min\left(1, e^{(f(\sigma)-f(\sigma'))/T_k}\right).$$

The corresponding SA algorithm for TSP is given in Algorithm 67.

Algorithm 67 Simulated Annealing algorithm for TSP, cooling scheme $\{T_k\}_{k \geq 0}$ (one step)

1: Assume $Y_k = \sigma$
2: Generate uniformly at random two distinct numbers $i, j \in \{1, \ldots, n\}$.
3: Set $\sigma'(k) = \sigma(k), k \notin \{i, j\}, \sigma'(i) = \sigma(j), \sigma'(j) = \sigma(i)$
4: Compute $\alpha = \min\left(1, e^{(l(\sigma)-l(\sigma'))/T_k}\right)$
5: Generate $U \sim \mathcal{U}[0, 1)$;
6: **if** $U \leq \alpha$ **then**
7: $\quad Y_{k+1} = \sigma'$
8: **else**
9: $\quad Y_{k+1} = Y_k$
10: **end if**
11: **return** Y_{k+1}

Simulation Results In Fig. 1.4, locations of 13 USA cities are presented. The distances between these cities are given in the following matrix:

Fig. 7.3 Distances between 13 USA cities. Taken from https://developers.google. com/optimization/routing/tsp

$$
\mathbf{M} = \begin{bmatrix}
0 & 2451 & 713 & 1018 & 1631 & 1374 & 2408 & 213 & 2571 & 875 & 1420 & 2145 & 1972 \\
2451 & 0 & 1745 & 1524 & 831 & 1240 & 959 & 2596 & 403 & 1589 & 1374 & 357 & 579 \\
713 & 1745 & 0 & 355 & 920 & 803 & 1737 & 851 & 1858 & 262 & 940 & 1453 & 1260 \\
1018 & 1524 & 355 & 0 & 700 & 862 & 1395 & 1123 & 1584 & 466 & 1056 & 1280 & 987 \\
1631 & 831 & 920 & 700 & 0 & 663 & 1021 & 1769 & 949 & 796 & 879 & 586 & 371 \\
1374 & 1240 & 803 & 862 & 663 & 0 & 1681 & 1551 & 1765 & 547 & 225 & 887 & 999 \\
2408 & 959 & 1737 & 1395 & 1021 & 1681 & 0 & 2493 & 678 & 1724 & 1891 & 1114 & 701 \\
213 & 2596 & 851 & 1123 & 1769 & 1551 & 2493 & 0 & 2699 & 1038 & 1605 & 2300 & 2099 \\
2571 & 403 & 1858 & 1584 & 949 & 1765 & 678 & 2699 & 0 & 1744 & 1645 & 653 & 600 \\
875 & 1589 & 262 & 466 & 796 & 547 & 1724 & 1038 & 1744 & 0 & 679 & 1272 & 1162 \\
1420 & 1374 & 940 & 1056 & 879 & 225 & 1891 & 1605 & 1645 & 679 & 0 & 1017 & 1200 \\
2145 & 357 & 1453 & 1280 & 586 & 887 & 1114 & 2300 & 653 & 1272 & 1017 & 0 & 504 \\
1972 & 579 & 1260 & 987 & 371 & 999 & 701 & 2099 & 600 & 1162 & 1200 & 504 & 0
\end{bmatrix}
$$

First, choosing σ uniformly at random, we obtained an average distance 15937.625 (with a standard deviation 1789.17). We ran Algorithm 66 and 67 with cooling scheme $T = 1/\log(k)$ for above matrix $\mathbf{M}$ for 100 steps. The results of a single simulation of each algorithm is presented in Fig. 7.4 (there are also results of other algorithms, which will be described later). As we can see, simulated annealing found the best route. However, running these algorithms 100 times, each time for 100 steps, we found that the performance of both of them was similar. However, the SA algorithm finds good solution "faster"—running these algorithms 100 times, each time for 10 steps only, SA found better solutions in 60% of cases.

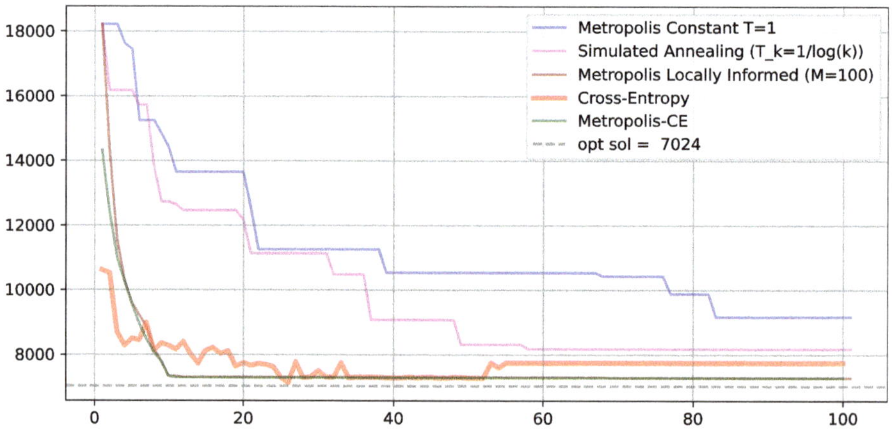

Fig. 7.4 100 steps of various Algorithms for TSP: Metropolis Algorithm 66 with $T_k = 1$ (*blue*), SA Algorithm 67 with $T_k = 1/\log(k)$ (*magenta*), locally informed proposals (LIP) (*brown*) algorithm (Example 7.3.2), cross-entropy Algorithm 71 (*red*) and Metropolis-CE algorithm (Example 7.5.2) (*green*). Horizontal axis: time step k, vertical axis: total distance $l(\sigma)$

7.3 Locally Informed Proposals

In this section we will present some of the main ideas behind *locally informed proposals*. Although providing details of these methods is out of the scope of this book, the information provided will allow us to implement them in a variety of problems.

We are in a setup where we want to find

$$s^{\mathrm{opt}} = \operatorname*{argmax}_{s \in \mathcal{S}} \pi_s, \tag{7.3.1}$$

where π is a distribution on $\mathcal{S}$ (most often we want to minimize the energy $H(s)$ and then consider $\pi_s = \frac{1}{z_T} e^{-H(s)/T}$ for some $T > 0$). Recall the Metropolis-Hastings Algorithm 43, which works in the following way. Given the current state $Y_k = s$, it samples a *candidate* s' from some fixed proposal distribution using a candidate-generating matrix $\mathbf{Q}$, i.e., sampling from $\mathbf{q}_s = (q_{ss_1}, \ldots, q_{ss_N})$. Then, the candidate is accepted with probability $\alpha = \min\left(1, \frac{\pi_{s'} q_{s's}}{\pi_s q_{ss'}}\right)$. In principle, the Metropolis-Hastings algorithm can be applied to any target distribution, however its efficiency strongly depends on the proposal distribution, i.e., the candidate-generating matrix $\mathbf{Q}$. Note that in all previous examples $\mathbf{Q}$ was independent from the target distribution π: in TSP (see page 7.2) a candidate of the current permutation was a permutation with two entries swapped; in the 0-1 knapsack problem a candidate of the current state $\mathbf{x} \in \{0, 1\}^d$ was a state with one coordinate changed to the opposite. Intuitively it is clear that, in many cases, for a given state there is only a relatively small fraction of candidates that are "good" (i.e., with larger mass π), and it may take many steps of the Metropolis-Hastings algorithm to consider such a candidate. That is why there are many methods for making $\mathbf{Q}$ dependent on π in such a way that "better" candidates have a greater chance of being chosen. The methods were first developed for continuous frameworks (which we do not consider in this book), the corresponding candidate-generating matrices were called *informed proposals*. The *locally informed proposals* (LIP) are presented in Zanella [217]. Namely, for a given (uninformed) $\mathbf{Q}$ we consider

$$\mathbf{Q}^{\mathrm{LIP}} = (q_{ss'}^{\mathrm{LIP}})_{s,s' \in \mathcal{S}}, \qquad q_{ss'}^{\mathrm{LIP}} = \frac{1}{Z_g} g\left(\frac{\pi_{s'}}{\pi_s}\right) q_{ss'},$$

where $g : \mathbb{R}_+ \to \mathbb{R}_+$, such that $g(t) = tg(1/t)$ (see [217, Theorem 1]). Note that $g(t) = 1$ corresponds to the usual uninformed $\mathbf{Q}^{\mathrm{LIP}} = \mathbf{Q}$, whereas $g(t) = t$ corresponds to "naively informed" $q_{ss'}^{\mathrm{LIP}} \propto q_{ss'} \pi_s$. Some typical functions are $g(t) = \sqrt{t}$ and $g(t) = t/(1 + t)$. Note that the unknown constant Z_g is not a problem, it cancels out (similarly to the constant in π) in computing the acceptance α.

Efficiency Note that sampling from $\mathbf{Q}^{\mathrm{LIP}}$ will usually be more computationally demanding: we need to compute $q_{ss'}^{\mathrm{LIP}}$ for all pairs s, s' such that $q_{ss'} > 0$, then normalize them. For example in the travelling salesman problem we were sampling simply two indices and swapped entries in the current state (permutation), thus we needed only to sample 2 numbers uniformly from the set $\{1, \ldots, n\}$. However, in LIP we have to evaluate $g\left(\frac{\pi'_s}{\pi_s}\right) q_{ss'}$ for all $\binom{n}{2}$ pairs s, s'. Additionally, in some cases even evaluating π_s is computationally demanding, however this may be mitigated in many cases using gradient-based approximations to locally-informed proposals, see Grathwohl et al. [71] (titled *"Oops I took a gradient: Scalable sampling for discrete distributions"*).

Algorithm 68 Locally informed Metropolis-Hastings algorithm (one step)

Require: Distribution π, proposal distribution $(q_{ss'})$, size of neighborhood n, fraction β, function $g(\cdot)$.

1: Compute $M = \lceil \beta n \rceil$.
2: Assume that $Y_k = s$
3: Generate M candidates $s^{(1)}, \ldots, s^{(M)}$ with distr. $\mathbf{q}_s = (q_{ss_1}, \ldots, q_{ss_N})$
4: Compute locally-informed proposal distribution (i.e., normalize the following)

$$\mathbf{q}_s^{\text{LIP}} \propto \left(g\left(\frac{\pi_{s^{(1)}}}{\pi_s}\right) q_{ss^{(1)}}, \ldots, g\left(\frac{\pi_{s^{(M)}}}{\pi_s}\right) q_{ss^{(M)}} \right)$$

5: Generate X according to distribution $\mathbf{q}_s^{\text{LIP}}$
6: Set $\alpha = \min\left(1, \dfrac{\pi_X q_{Xs}^{\text{LIP}}}{\pi_s q_{sX}^{\text{LIP}}}\right)$
7: Generate $U \sim \mathcal{U}[0, 1)$
8: **if** $U \leq \alpha$ **then**
9:     $Y_{k+1} = X$
10: **else**
11:     $Y_{k+1} = Y_k$
12: **end if**
13: **return** Y_{k+1}

Some Heuristic Efficiency-oriented Improvements Slightly inspired by the cross-entropy method (Sect. 7.4), we propose the following modification of the locally-informed proposal. Let $n_s = \#\{s' : q_{ss'} > 0\}$ be the number of "neighbors" of s. Assume that $n_s = n$ does not depend on s (which is the case in all our examples, e.g., $n = \binom{n}{2}$ in TSP and $n = d$ in the 0-1 knapsack problem). Let $\beta \in (0, 1]$ and set $M = \lceil \beta n \rceil$ (or in the implementation we may directly provide M). Afterwards, sample M different candidates $s^{(1)}, \ldots, s^{(M)}$ using $\mathbf{Q}$, compute

$$\mathbf{Q}^{\text{LIP}} = (q_{ss^{(j)}}^{\text{LIP}})_{s, \in \mathcal{S}, j \in \{1, \ldots, M\}}, \qquad q_{ss^{(j)}}^{\text{LIP}} = \frac{1}{Z_g} g\left(\frac{\pi_{sj}}{\pi_s}\right) q_{ss^{(j)}}$$

and proceed as with the usual LIP method. Thus, the difference is that in each step we need to evaluate $q_{ss'}$ only for a fraction of "neighbors". Note that for $\beta = 1$ we have a full LIP method. The method is summarized in Algorithm 68. In practice, the algorithm is slightly worse than full LIP in terms of finding a good/best solution, but of course it is better in terms of execution time.

Example 7.3.1 (The 0-1 Knapsack Problem Continued) We continue the 0-1 knapsack problem (see page 7.1) with the same parameters. We performed simulations comparing three algorithms: usual (uninformed) Metropolis, LIP full and LIP with $\beta = 0.3$ (i.e., in each iteration $0.3 \cdot 100 = 30$ candidates are sampled). We considered $\mathbf{Q}$ as in usual Metropolis (i.e., candidates have only one coordinate swapped), for LIP algorithms we used $g(t) = t$. We considered a distribution π with constant temperature $T = 100$. A typical result is presented in Fig. 7.5, note that the figure presents results of more algorithms, which will be described later on. Running all three algorithms 100 times, in 60% of cases LIP full was the best (i.e., it found a solution with largest knapsack value among all steps), and in 40% of cases LIP with $\beta = 0.3$ was the best (the usual Metropolis algorithm was always inferior to LIP algorithms). The example is implemented in `ch7_knapsack_metropolis_SA_LIP_CE.py`. By default, it runs one iteration of all 5 algorithms. To see the results of the 3 algorithms considered so far, and win counts of each algorithm from 100 iterations, run it with parameters `-- algs Metropolis,LIP,LIP_full --n_iter 100`.

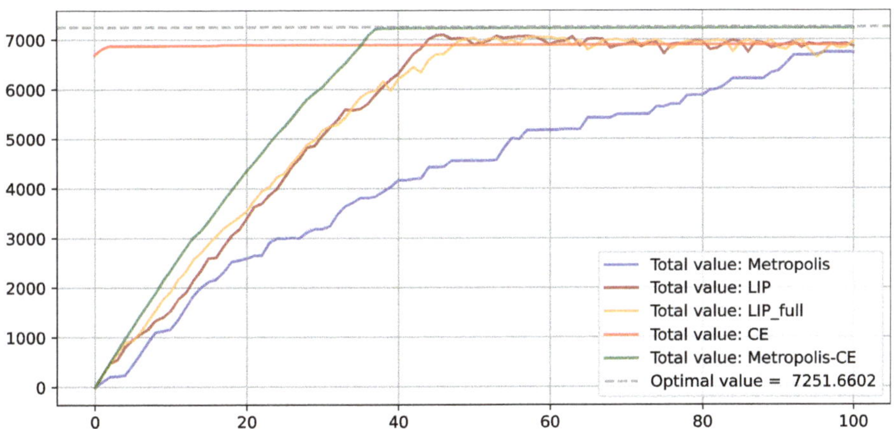

Fig. 7.5 100 steps of various Algorithms for the knapsack problem: Metropolis Algorithm 64 (*blue*), current (Example 7.3.1) LIP algorithm with $\beta = 1$ (full version, *orange*) and $\beta = 0.1$ (*brown*), cross-entropy Algorithm 7.4.1 (*red*) and Metropolis-CE (Example 7.5.1) (*green*). Horizontal axis: step number k, vertical axis: total value of a knapsack

Note that in this example proposal distribution $q_{\mathbf{xx'}} = \frac{1}{d}$ for $\mathbf{x'}$ which differ in one coordinate only, thus the proposal distribution for LIP can be written as (after sampling $\mathbf{x}^{(1)}, \ldots, \mathbf{x}^{(M)}$) $q_{\mathbf{x}}^{\text{LIP}} \propto$

$$\left(\exp\left(\frac{1}{T} \sum_{j=1}^{d} x_j^{(1)} v_j \right) \mathbf{1}(\mathbf{x}^{(1)} \cdot \mathbf{w} \leq W), \ldots, \exp\left(\frac{1}{T} \sum_{j=1}^{d} x_j^{(M)} v_j \right) \right)$$

$$\times \; \mathbf{1}(\mathbf{x}^{(M)} \cdot \mathbf{w} \leq W) \Bigg) .$$

$\Diamond$

Example 7.3.2 (TSP Continued) We continue the travelling salesman problem (distances provided in Fig. 7.3). We consider the same setup as earlier, we take a distribution π with constant temperature

$$\pi_\sigma = \frac{1}{z} e^{-l(\sigma)/T} .$$

The proposal distribution $\mathbf{Q}$ corresponds to simply swapping two randomly chosen elements in a given permutation. We compare usual (uninformed) Metropolis with LIP with $g(t) = t$, where instead of setting β we simply set $M = 100$ (i.e., in each iteration we sample 100 candidates $\sigma^{(1)}, \ldots, \sigma^{(100)}$ for current state σ). Since $q_{\sigma\sigma'}$ is constant (it is $\binom{n}{2}^{-1}$ for σ' which differs by 1 swap from σ), we simply have

$$\mathbf{q}_\sigma^{\text{LIP}} \propto \left(e^{-l(\sigma^{(1)})}, \ldots, e^{-l(\sigma^{(M)})} \right) .$$

The results are (single run for 100 iterations) depicted in Fig. 7.4. The figure presents the results of several algorithms, but comparing only those described so far, i.e., regular (uninformed) Metropolis, simulated annealing and LIP version (with fixed $T = 20$), the respective number of times each algorithm won was 2, 2 and 96 respectively (out of 100 simulations), showing that LIP outperforms the previous solutions (at not much larger cost, at each step we sampled $M = 100$ pairs). $\Diamond$

7.4 The Cross-entropy Method

In Sect. 5.1.8, we outlined the cross-entropy method for *estimating* $I = \mathbb{E}k(\mathbf{X})$ with $\mathbf{X} \sim f$. Roughly speaking, the method aims to estimate the parameter θ^* of the "best" proposal distribution for importance sampling, assuming the proposals are drawn from a parametric family $\{f_\theta\}$. Later, in Sect. 5.1.8.2, we considered rare-event estimation, i.e., estimation of I with $k(\mathbf{X}) = \mathbb{1}(h(\mathbf{X}) > \gamma)$, and introduced an adaptive approach for estimating θ^*.

The cross-entropy method for *optimization* is closely related to its estimation counterpart. We therefore recommend revisiting Sect. 5.1.8 as well as the preceding Sect. 7.3 on locally informed proposals.

Since here we are dealing with a discrete space $\mathcal{S}$, we will use a probability function p or p_θ instead of a density f or f_θ. Thus, e.g., $\mathbb{E}_\theta$ or $\mathbb{P}_\theta$ denote computing the quantities for X having probability function p_θ.

Assume we have some family of proposal distributions $\{p_\theta\}$, $\theta \in \Theta$, with some distinguished θ_0 corresponding to the initial distribution. For example, in TSP, given the current state σ, sampling from p_θ means sampling from candidates which differ by swapping two elements. The distribution of samples is not necessarily uniform, it is parametrized by θ (and θ_0 could be the initial uniform distribution over all pairs). In the 0-1 knapsack problem, a family $\{p_\theta\}$ can be some distribution on $\{1, \ldots, d\}$ w.r.t. which we sample a coordinate (to be changed to the opposite); similarly the initial θ_0 could be a uniform distribution. Keep in mind that in previous sections on Metropolis-based algorithms we used $\mathbf{Q} = (q_{ss''})_{ss' \in S}$ for a proposal distribution, which was a distribution on *neighbors* (of the current s), here in the CE algorithm for optimization we use p_θ, a distribution on the whole of S, which is parametrized via θ (thus, the notation is also consistent with Sect. 5.1.8 on the CE method for estimation).

Setup: We still aim to find $s^{\mathrm{opt}} = \mathrm{argmax}_{s \in S}\, \pi_s$. Let $h : S \to \mathbb{R}$ be such that

$$s^{\mathrm{opt}} = \underset{s \in S}{\mathrm{argmax}}\, \pi_s = \underset{s \in S}{\mathrm{argmax}}\, h(s) \tag{7.4.1}$$

and we are able to evaluate h. Denote the corresponding value by

$$\gamma^{\mathrm{opt}} = \max_{s \in S} h(s) \in \mathbb{R}. \tag{7.4.2}$$

In our examples $h(s)$ may be minus the energy $h(s) = -H(s)$ in the case of $\pi_s = \frac{1}{z_T}\exp(-H(s)/T)$, the value of the knapsack in the 0-1 knapsack problem, or minus the total distance in TSP. The function $h(\cdot)$ is called a *performance function* (in Sect. 5.1.8 the function s played the role of a performance function). We will use it to compare solutions.

We can relate optimization problem (7.4.1) to the estimation problem: consider a collection of indicators $I_t = \mathbf{1}(h(s) \geq \gamma_t)$ for varying *levels* $\gamma_t \in \mathbb{R}$. Let $\{p_\theta\}$ be a family of discrete p.d.f.s on S, parametrized via $\theta \in \Theta$. For some fixed $\theta_0 \in \Theta$ we relate (7.4.1) to the problem of estimating

$$I(\gamma) = \mathbb{P}_{\theta_0}(h(\mathbf{X}) \geq \gamma) = \sum_{s \in S} \mathbf{1}(h(s) \geq \gamma) p_{\theta_0}(s) = \mathbb{E}_{\theta_0}\mathbf{1}(h(\mathbf{X}) \geq \gamma). \tag{7.4.3}$$

Note that if $\gamma = \gamma^{\mathrm{opt}}$ and p_{θ_0} is a uniform distribution on S, then

$$I(\gamma^{\mathrm{opt}}) = \mathbb{P}_{\theta_0}(h(\mathbf{X}) \geq \gamma^{\mathrm{opt}})$$

$$= \sum_{s \in S} \mathbf{1}(h(s) \geq \gamma^{\mathrm{opt}})\frac{1}{|S|}$$

$$= \frac{\#\{s : h(s) = \gamma^{\mathrm{opt}}\}}{|S|},$$

which typically is very small (most often the numerator is 1 and the denominator is huge). This suggests estimating it with estimator (5.1.41) with likelihood ratio

$L(\mathbf{x}, \boldsymbol{\theta}_0, \boldsymbol{\theta}^*)$ with a parameter

$$\boldsymbol{\theta}^* = \underset{\boldsymbol{\theta}}{\operatorname{argmax}} \ \mathbb{E}_{\boldsymbol{\theta}_0} \mathbf{1}(h(\mathbf{X}) \geq \gamma^{\mathrm{opt}}) \ln p_{\boldsymbol{\theta}}(\mathbf{X}),$$

which can be estimated (cf. Algorithm 34, deterministic version of CE for estimation with Algorithm 35, stochastic version) by sampling $\mathbf{X}_1, \ldots, \mathbf{X}_M$ from $p_{\boldsymbol{\theta}_0}$ and computing

$$\hat{\boldsymbol{\theta}}^* = \underset{\boldsymbol{\theta}}{\operatorname{argmax}} \ \frac{1}{M} \sum_{i=1}^{M} \mathbf{1}(\pi_{\mathbf{X}_i} \geq \gamma^{\mathrm{opt}}) \log p_{\boldsymbol{\theta}}(\mathbf{X}_i).$$

The above procedure could be carried out for $\gamma \approx \gamma^{\mathrm{opt}}$, however, for other γ's (and most probably for the initial, usually random, one) too many indicators (if not all) would be 0. Thus, we proceed as with the CE method for estimation (its **multi-level** approach): we can construct a sequence of levels $\gamma_1, \ldots, \gamma_T$ and parameters of proposal distributions $\boldsymbol{\theta}'_1, \ldots \boldsymbol{\theta}'_T$ such that γ_t approaches γ^{opt} and at the same time $\boldsymbol{\theta}'_t$ approaches the *optimal proposal parameter* parameter $\boldsymbol{\theta}^*$. Moreover, while updating $\boldsymbol{\theta}'_t$ we may take into account only samples with best performances. The procedure is provided in Algorithm 69.

From Theorem 3 in [27] we may conclude that the distribution $p_{\boldsymbol{\theta}'_t}$ converges with probability 1 to a unit mass located at some random $s' \in S$. Moreover, the probability that $h(s') = \gamma^*$ can be made arbitrarily close to 1 by selecting sufficiently small value of smoothing parameter α.

Before providing details, let us point out that the method is, in a sense, similar to the introduced variation of locally informed proposals (LIP), where in each step we sample a fraction of possible candidates. The main differences are:

- In the CE algorithm, the proposal distribution, denoted here by $p_{\boldsymbol{\theta}}$, is parametrized by a parameter $\boldsymbol{\theta} \in \Theta$, which evolves with each step of the algorithm. Unlike Metropolis-based algorithms, this parameter is updated iteratively (as $\boldsymbol{\theta}'_t$ in the cross-entropy method for estimation).
- The CE algorithm uses an elite set of the best-performing samples to update the parameters of the distribution, focusing the search on promising regions. Metropolis-based algorithms work without focusing exclusively on the best examples. The acceptance criterion is based on probabilities derived from the target distribution rather than selecting best-performing states.
- In the CE algorithm, at each step, we identify the best solution from a fixed-size sample, but this serves mainly for presentation purposes (to plot a solution at each iteration). The actual *state* in the CE method is the entire *distribution* over S, in contrast to Metropolis-based algorithms, where a single state is updated at each step. The output of the CE algorithm is the final distribution, from which we can generate a sample to find the best solution.

Note that in contrast to Metropolis-based algorithms, the final candidate in each step is not sampled—it is chosen from a group (called in this context *elite*). The algorithm searches for the best solution, trying to improve "itself" in each step.

Algorithm 69 is based on Algorithm 2.3 from [31]. Note there are variations of the algorithm. For example Algorithm 4.1 from [31] is a refined version of this algorithm—sampling size M is adaptively changing (between fixed $M_{\min}$ and $M^{\max}$) to ensure the reliability of the whole procedure. The algorithm is called FACE (*fully automated cross-entropy*), its flowchart is presented in Fig. 8 in [31].

Algorithm 69 The cross-entropy method for optimization

Require: Family of probability functions $\{p_\theta\}$, $\theta \in \Theta$, fixed θ_0, performance function $h(\cdot)$, parameter $\rho \in (0, 1)$, sample size M, smoothing parameter $\alpha \in (0, 1)$, integer parameter d (by default $d = 5$)

1: Define $\theta'_0 = \theta_0$, set $t = 0$

2: **repeat**

3:     $t = t + 1$

4:     **Update** γ_t: generate samples $\mathbf{X}_1, \ldots, \mathbf{X}_M$ from $p_{\theta'_{t-1}}(\cdot)$,
        compute their performances $\eta_1 = h(\mathbf{X}_i), \ldots, \eta_M = h(\mathbf{X}_M)$,
        sort them $\eta_{(1)} \leq \cdots \leq \eta_{(M)}$ and compute $(1 - \rho)$ sample quantile

$$\gamma_t = \eta_{(\lceil (1-\rho)M \rceil)} \tag{7.4.4}$$

5:     **Update** θ'_t: from **the same** samples $\mathbf{X}_1, \ldots, \mathbf{X}_M$ create the **elite set** $\mathcal{E}_t = \{\mathbf{X}_i : h(\mathbf{X}_i) \geq \gamma_t\}$, solve

$$\theta' = \operatorname*{argmax}_{\theta} \frac{1}{M} \sum_{i:\mathbf{X}_i \in \mathcal{E}_t} \log p_\theta(\mathbf{X}_i) \tag{7.4.5}$$

$$= \operatorname*{argmax}_{\theta} \frac{1}{M} \sum_{i=1}^{M} \mathbf{1}(h(\mathbf{X}_i) \geq \gamma_t) \log p_\theta(\mathbf{X}_i)$$

and apply smoothed update

$$\theta'_t = \alpha\theta' + (1 - \alpha)\theta'_{t-1}, \tag{7.4.6}$$

6: **until** $\gamma_t = \gamma_{t-1} = \ldots = \gamma_{t-d}$

7: Set $T = t$, sample $\mathbf{X}_1, \ldots, \mathbf{X}_M$ from $p_{\theta'_T}$. Compute their performances $\eta_1 = \pi^*_{\mathbf{X}_1}, \ldots, \eta_M = \pi^*_{\mathbf{X}_M}$, sort them $\eta_{(1)} \leq \cdots \leq \eta_{(M)}$, denote the set of best solutions $\mathcal{E} = \{\mathbf{X}_i : \eta_i = \eta_{(M)}\}$

8: **return** solutions $\mathcal{E}$

It is worth pointing out the differences between Algorithm 35 (CE for estimation) and Algorithm 69 (CE for optimization):

- In the estimation problem, a parameter θ_0 was given—we were to estimate the probability that performance (in this section denoted by $h(\mathbf{X})$) is larger than the prescribed γ—however in optimization θ_0 is a parameter of the initial distribution (our choice).

- In the estimation problem, the optimal parameter was associated with $\mathbb{P}_{\boldsymbol{\theta}_0}(h(\mathbf{X}) \geq \gamma_t)$, whereas in optimization we associate it with $\mathbb{P}_{\boldsymbol{\theta}'_{t-1}}(h(\mathbf{X}) \geq \gamma_t)$. As a result, in optimization we do not use the likelihood ratio W, Eq. (7.4.5) is the Eq. (5.1.43) with $W \equiv 1$.

Example 7.4.1 (The 0-1 Knapsack Problem Continued) We continue the 0-1 knapsack problem (see page 7.1). As a proposal distribution p_θ we simply consider a distribution on $\mathcal{S} = \{0, 1\}^d$ corresponding to independent sampling of each item. Consider $\boldsymbol{\Theta} = \{(\theta_1, \ldots, \theta_d) : \theta_i \in [0, 1], i = 1, \ldots, d\}$. If $\boldsymbol{\theta} = (\theta_1, \ldots, \theta_d)$ then we sample each coordinate of $\mathbf{x} = (x_1, \ldots, x_d)$, the probability that $x_i = 1$ is θ_i and the probability that $x_i = 0$ is $1 - \theta_i$, in other words

$$p_{\boldsymbol{\theta}}(\mathbf{x}) = \prod_{j=1}^{d} \theta_j^{x_j} (1 - \theta_j)^{1-x_j}.$$

Let us simply take $\boldsymbol{\theta}_0 = \left(\frac{1}{2}, \ldots, \frac{1}{2}\right)$.

To compute $\boldsymbol{\theta}'$ we will compute the derivative of (7.4.5) w.r.t. each θ) and set it to 0. We have

$$\frac{\partial}{\partial \theta_k} \log p_{\boldsymbol{\theta}}(\mathbf{x}) = \sum_{j=1}^{d} \frac{\partial}{\partial \theta_k} \log \left(\theta_j^{x_j}(1 - \theta_j)^{1-x_j}\right) = \frac{x_k}{\theta_k} - \frac{1 - x_k}{1 - \theta_k} = \frac{x_k - \theta_k}{\theta_k(1 - \theta_k)}.$$

Denoting $(X_i(1), \ldots, X_i(d))$ by $\mathbf{X}_i$, we have

$$\frac{\partial}{\partial \theta_k} \frac{1}{M} \sum_{i=1}^{M} \mathbf{1}(h(\mathbf{X}_i) \geq \gamma_t) \log p_{\boldsymbol{\theta}}(\mathbf{X}_i)$$

$$= \frac{1}{M\theta_k(1 - \theta_k)} \sum_{i=1}^{M} \mathbf{1}(h(\mathbf{X}_i) \geq \gamma_t)(X_i(k) - \theta_k) = 0$$

yielding

$$\theta'_k = \frac{\sum_{i=1}^{M} \mathbf{1}(h(\mathbf{X}_i) \geq \gamma_t)X_i(k)}{\sum_{i=1}^{M} \mathbf{1}(h(\mathbf{X}_i) \geq \gamma_t)}.$$

In vector notation

$$\boldsymbol{\theta}' = \frac{\sum_{i=1}^{M} \mathbf{1}(h(\mathbf{X}_i) \geq \gamma_t)\mathbf{X}_i}{\sum_{i=1}^{M} \mathbf{1}(h(\mathbf{X}_i) \geq \gamma_t)}. \tag{7.4.7}$$

Consider a small example:

- $d = 4, M = 3, \rho = 1/2, \alpha = 3/4$ and $\boldsymbol{\theta}'_0 = (1/2, 1/2, 1/2, 1/2)$. Say we sampled $\mathbf{X}_1 = (1, 0, 0, 1)$, $\mathbf{X}_2 = (0, 1, 0, 1)$ and $\mathbf{X}_3 = (1, 1, 0, 1)$, assume that the corresponding performances are $\eta_1 = 3, \eta_2 = 0$ (for example, the total weight could be greater than the total capacity W) and $\eta_3 = 2$. We compute the sample quantile ($\rho = 1/2$ means median) $\gamma_1 = 2$. I.e., in (7.4.5) we consider only $\mathbf{X}_1$ and $\mathbf{X}_3$.
- Thus, from (7.4.7) we compute

$$\boldsymbol{\theta}' = \frac{\sum_{i=1}^{M} \mathbf{1}(h(\mathbf{X}_i) \geq \gamma_1)\mathbf{X}_i}{\sum_{i=1}^{M} \mathbf{1}(h(\mathbf{X}_i) \geq \gamma_1)} = \frac{\mathbf{X}_1 + \mathbf{X}_3}{2} = \frac{(2, 1, 0, 2)}{2} = (1, 1/2, 0, 1).$$

- Thus, according to $\boldsymbol{\theta}'$ we should always take the first and last item (probabilities 1), with probability $1/2$ we should take the second item, and we should never take the third item. However, we apply the smoothed update (7.4.6) with smoothing parameter $\alpha = 3/4$ resulting in

$$\boldsymbol{\theta}'_1 = \frac{3}{4} \cdot \left(\frac{1}{2}, \frac{1}{2}, \frac{1}{2}, \frac{1}{2} \right) + \frac{1}{4} \cdot (1, 1/2, 0, 1) = \left(\frac{5}{8}, \frac{1}{2}, \frac{3}{8}, \frac{5}{8} \right).$$

Simulation Results Recall the knapsack parameters $d = 100, w_i = i, v_i = i^{1.2}$. For the CE algorithm we used $M = 100, \rho = 0.35, \alpha = 0.7$ and simply ran the algorithm for 100 steps (to be consistent with previous algorithms). The result of a single simulation is depicted in Fig. 7.5. Note that it is not directly comparable to previous Metropolis-based algorithms—here in each step sampling any $\mathbf{x}$ is possible, whereas in the Metropolis-based algorithms at each step we changed at most one coordinate (except Metropolis-CE, to be described later, which is based both on Metropolis and cross entropy). $\diamond$

Example 7.4.2 (TSP Continued) Recall the setup: we are given a symmetric $n \times n$ matrix $\mathbf{M}$, where $\mathbf{M}(i, j)$ is the distance between the i-th and j-th city. Given a permutation $\sigma = (\sigma_1, \ldots, \sigma_n)$, the total distance travelled is $l(\sigma) = \sum_{k=1}^{n-1} \mathbf{M}(\sigma_k, \sigma_{k+1}) + \mathbf{M}(\sigma_n, \sigma_1)$. We aim to find

$$\sigma^{\text{opt}} = \underset{\sigma \in S_n}{\text{argmax}}\, h(\sigma), \quad \text{where } h(\sigma) = -l(\sigma).$$

We will use the following proposal distribution. To generate a random permutation $\mathbf{X} = \sigma = (\sigma_1, \ldots, \sigma_n)$ we proceed as follows. First of all, note that we may always start from the first city, i.e., $\sigma_1 = 1$, then we choose σ_2 with some distribution on $[n] \setminus \{\sigma_1\}$ (i.e., all cities except the first one), then σ_3 with some distribution on $[n] \setminus \{\sigma_1, \sigma_2\}$ (i.e., all cities except the first two) and so on. More formally: the distribution is parametrized by an $n \times n$ stochastic matrix $\boldsymbol{\theta} = (\theta_{ij})_{i,j \in \{1,\ldots,n\}}$ with zeros on the diagonal $\theta_{ii} = 0$ for all $i \in [n]$. To generate a permutation $\sigma =$

$(\sigma_1, \ldots, \sigma_n)$ we proceed as follows. As mentioned we set $\sigma_1 = 1$. Assume we already sampled $\sigma_1, \sigma_2, \ldots, \sigma_k$. Then, we consider a row $\boldsymbol{\theta}(\sigma_k, \cdot)$, we set positions corresponding to already visited cities to 0, then normalize this row (to sum up to 1) and sample σ_{k+1} from this row. The procedure is described in Algorithm 70.

Algorithm 70 Sampling permutation from $\boldsymbol{\theta}$

Require: Stochastic $n \times n$ matrix $\boldsymbol{\theta} = (\theta_{ij})$ with $\theta_{ii} = 0, i = 1, \ldots, n$
1: Set $\sigma_1 = 1$
2: **for** $k = 1, \ldots, n - 1$ **do**
3:     Let $\mathbf{r}^{k+1} = (r_1^{k+1}, \ldots, r_n^{k+1}) = \boldsymbol{\theta}(\sigma_k, \cdot)$ be k-th row of $\boldsymbol{\theta}$
4:     Set $\mathbf{r}^{k+1}$ at positions corresponding to already visited cities to 0, i.e.,

$$r_j^{k+1} = 0 \quad \text{for } j \in \{\sigma_1, \ldots, \sigma_k\}.$$

5:     Normalize vector $\mathbf{r}^{k+1}$ to get a probability vector $\mathbf{q}^{k+1} = (q_1^{k+1}, \ldots, q_n^{k+1})$:

$$q_j^{k+1} = \frac{r_j^{k+1}}{\sum_{i=1}^{n} r_i^{k+1}}.$$

6: Sample σ_{k+1} from distribution $\mathbf{q}^{k+1}$
7: **end for**
8: **return** $\boldsymbol{\sigma} = (\sigma_1, \ldots, \sigma_n)$

In Algorithm 69 the proposal distribution depends on the time step. Assume that $\mathbf{X}_1, \ldots, \mathbf{X}_M$ are permutations sampled from $\boldsymbol{\theta}'_{t-1}$ and we need to compute $\boldsymbol{\theta}'$ from (7.4.5). Denoting $(\sigma_1^{(i)}, \ldots, \sigma_n^{(i)})$ by $\mathbf{X}_i$ (recall we always start at city nr 1, i.e., $\sigma_1^{(i)} = 1$). We need to find

$$\boldsymbol{\theta}' = \underset{\boldsymbol{\theta}}{\operatorname{argmax}} \frac{1}{M} \sum_{i=1}^{M} \mathbf{1}(h(\mathbf{X}_i) \geq \gamma_t) \log p_{\boldsymbol{\theta}}(\mathbf{X}_i).$$

Let us first write down $p_{\boldsymbol{\theta}}(\mathbf{X}_i)$, normalizing each row at each step. Denote the normalizing constants by

$$c(\sigma^{(i)}, k) = \sum_{r \notin \{\sigma_1^{(i)}, \sigma_2^{(i)}, \ldots, \sigma_k^{(i)}\}} \boldsymbol{\theta}(\sigma_1^{(i)}, \sigma_r^{(i)}).$$

We have (note that $c(\sigma^{(i)}, 1) = 1$)

$$p_{\boldsymbol{\theta}}(\mathbf{X}_i) = \frac{\boldsymbol{\theta}(\sigma_1^{(i)}, \sigma_2^{(i)})}{c(\sigma^{(i)}, 1)} \cdot \frac{\boldsymbol{\theta}(\sigma_2^{(i)}, \sigma_3^{(i)})}{c(\sigma^{(i)}, 2)} \cdots \frac{\boldsymbol{\theta}(\sigma_{n-1}^{(i)}, \sigma_n^{(i)})}{c(\sigma^{(i)}, n-1)} = \prod_{k=1}^{n-1} \frac{\boldsymbol{\theta}(\sigma_k^{(i)}, \sigma_{k+1}^{(i)})}{c(\sigma^{(i)}, k)}.$$

Thus,

$$\log p_{\boldsymbol{\theta}}(\mathbf{X}_i) = \sum_{k=1}^{n-1} \log \left(\frac{\boldsymbol{\theta}(\sigma_k^{(i)}, \sigma_{k+1}^{(i)})}{c(\sigma^{(i)}, k)} \right) = \sum_{k=1}^{n-1} \log \boldsymbol{\theta}(\sigma_k^{(i)}, \sigma_{k+1}^{(i)}) - \sum_{k=1}^{n-1} \log c(\sigma^{(i)}, k).$$

$$(7.4.8)$$

To compute $\boldsymbol{\theta}'$ we will use Lagrange multipliers—we want to maximize $\sum_{i=1} \mathbf{1}(h(\mathbf{X}_i) \geq \gamma_t) \log p_{\boldsymbol{\theta}}(\mathbf{X}_i)$ under the constraint that each row of $\boldsymbol{\theta}$ sums up to 1, i.e., $\sum_{j=1}^{n} \theta_{ij} = 1$. Denoting $(\lambda_1, \ldots, \lambda_n)$ by $\boldsymbol{\lambda}$, the Lagrangian function $\mathcal{L}(\boldsymbol{\theta}, \boldsymbol{\lambda})$ is thus the following

$$\mathcal{L}(\boldsymbol{\theta}, \boldsymbol{\lambda}) = \frac{1}{M} \sum_{i=1}^{M} \mathbf{1}(h(\mathbf{X}_i) \geq \gamma_t) \sum_{k=1}^{n-1} \log \left(\frac{\boldsymbol{\theta}(\sigma_k^{(i)}, \sigma_{k+1}^{(i)})}{c(\sigma^{(i)}, k)} \right) + \sum_{a=1}^{n} \lambda_a \left(1 - \sum_{b=1}^{n} \theta_{ab} \right).$$

Compute the derivative of $\mathcal{L}(\boldsymbol{\theta}, \boldsymbol{\lambda})$ w.r.t. θ_{ab}. Note that then, the normalization term $\log c(\sigma^{(i)}, k)$ is constant w.r.t. specific transition probabilities θ_{ab}, as it is a sum of all remaining probabilities excluding the visited cities. Thus

$$\frac{\partial \log f_{\boldsymbol{\theta}}(\mathbf{X}_i)}{\partial \theta_{ab}} = \sum_{k=1}^{n-1} \frac{\mathbf{1}(\sigma_k^{(i)} = a, \sigma_{k+1}^{(i)} = b)}{\theta_{ab}}$$

and in consequence

$$\frac{\partial \mathcal{L}(\boldsymbol{\theta}, \boldsymbol{\lambda})}{\partial \theta_{ab}} = \frac{1}{M} \sum_{i=1}^{M} \mathbf{1}(h(\mathbf{X}_i) \geq \gamma_t) \sum_{k=1}^{n-1} \frac{\mathbf{1}(\sigma_k^{(i)} = a, \sigma_{k+1}^{(i)} = b)}{\theta_{ab}} - \lambda_a.$$

Setting it to zero, we have

$$\theta_{ab}' = \frac{\sum_{i=1}^{M} \mathbf{1}(h(\mathbf{X}_i) \geq \gamma_t) \sum_{k=1}^{n-1} \mathbf{1}(\sigma_k^{(i)} = a, \sigma_{k+1}^{(i)} = b)}{\lambda_a}.$$

$$(7.4.9)$$

Compute the derivative of $\mathcal{L}(\boldsymbol{\theta}, \boldsymbol{\lambda})$ w.r.t. λ_i and set it to 0:

$$\frac{\partial \mathcal{L}(\boldsymbol{\theta}, \boldsymbol{\lambda})}{\partial \lambda_i} = 1 - \sum_{j=1}^{n} \theta_{ij} = 0,$$

i.e., $\sum_{j=1}^{n} \theta_{ij} = 1$, which means that λ_a in (7.4.9) is a normalization constant. Finally, we have $\boldsymbol{\theta}' = (\theta_{ab}')$, where

$$\theta_{ab}' = \frac{\sum_{i=1}^{M} \mathbf{1}(h(\mathbf{X}_i) \geq \gamma_t) \sum_{k=1}^{n-1} \mathbf{1}(\sigma_k^{(i)} = a, \sigma_{k+1}^{(i)} = b)}{\sum_{b=1}^{n} \sum_{i=1}^{M} \mathbf{1}(h(\mathbf{X}_i) \geq \gamma_t) \sum_{k=1}^{n-1} \mathbf{1}(\sigma_k^{(i)} = a, \sigma_{k+1}^{(i)} = b)}.$$

Roughly speaking, if in many $\mathbf{X}_i$ (which are "good", i.e., $h(\mathbf{X}_i) \geq \gamma_t$) there was a transition from a to b, then θ'_{ab} will be high compared to other transitions from a. Finally we apply a smoothed update. As $\boldsymbol{\theta}_0$ we may take a matrix with uniform distribution, i.e., $\boldsymbol{\theta}_0(i, j) = \frac{\mathbf{1}(i \neq j)}{n-1}$ (note that this means that for $t = 1$ samples $\mathbf{X}_1, \ldots, \mathbf{X}_M$ are uniformly random permutations). The whole procedure is presented in Algorithm 71.

$\diamond$

Algorithm 71 The cross-entropy method for TSP

Require: distance matrix $\mathbf{M}$, parameter, $\rho \in (0, 1)$, sample size M, smoothing parameter $\alpha \in (0, 1)$, integer parameter d (by default $d = 5$)

1: Define $\boldsymbol{\theta}'_0 = \boldsymbol{\theta}_0$, set $t = 0$
2: **repeat**
3:     $t = t + 1$
4:     **Update** γ_t: generate samples $\mathbf{X}_1, \ldots, \mathbf{X}_M$ from $f_{\boldsymbol{\theta}_{t-1}}(\cdot)$ using Algorithm 70, compute their performances (minus total distances) $\eta_1 = -l(\mathbf{X}_1), \ldots, \eta_M = -l(\mathbf{X}_M)$, sort them $\eta_{(1)} \leq \cdots \leq \eta_{(M)}$ and compute $(1 - \rho)$ sample quantile

$$\gamma_t = \eta_{(\lceil (1-\rho)M \rceil)}$$

5:     **Update** $\boldsymbol{\theta}'_t$: from the same samples $\mathbf{X}_1, \ldots, \mathbf{X}_M$ compute $\boldsymbol{\theta}' = (\theta_{ab})$, where

$$\theta'_{ab} = \frac{\sum_{i=1}^{M} \mathbf{1}(h(\mathbf{X}_i) \geq \gamma_t) \sum_{k=1}^{n-1} \mathbf{1}(\sigma_k^{(i)} = a, \sigma_{k+1}^{(i)} = b)}{\sum_{b=1}^{n} \sum_{i=1}^{M} \mathbf{1}(h(\mathbf{X}_i) \geq \gamma_t) \sum_{k=1}^{n-1} \mathbf{1}(\sigma_k^{(i)} = a, \sigma_{k+1}^{(i)} = b)}.$$

and apply the smoothed update

$$\boldsymbol{\theta}'_t = \alpha \boldsymbol{\theta}' + (1 - \alpha)\boldsymbol{\theta}'_{t-1},$$

6: **until** $\gamma_t = \gamma_{t-1} = \ldots = \gamma_{t-d}$
7: Set $T = t$, sample $\mathbf{X}_1, \ldots, \mathbf{X}_M$ from $f_{\boldsymbol{\theta}_T}$. Compute their performances $\eta_1 = -l(\mathbf{X}_1), \ldots, \eta_M = -l(\mathbf{X}_M)$, sort them $\eta_{(1)} \leq \cdots \leq \eta_{(M)}$, denote the set of best solutions by $\mathcal{E} = \{\mathbf{X}_i : \eta_i = \eta_{(M)}\}$
8: **return** solutions $\mathcal{E}$

Simulation Results A single run (of each algorithm) with 100 steps is depicted in Fig. 7.4. We ran 100 times each of the algorithms considered so far, i.e., regular (uninformed) Metropolis, simulated annealing, LIP algorithm and current cross-entropy Algorithm 71 with parameters $M = 100, \rho = 0.35$ and $\alpha = 0.7$ (for comparison with previous algorithms we simply ran it for 100 steps, although the stopping criterion in line 6 was executed earlier with default $d = 5$). The win count respectively was: 1, 0, 21 and 78, showing that cross-entropy outperforms previous solutions. It is debatable whether such direct comparison is fair for the same number of steps: previous algorithms at each step only swapped two elements of the permutation, whereas cross-entropy can in principle sample any permutation at each step. Nevertheless, we see that applying the cross-entropy method is fruitful. On real-worlds scenarios, one needs however to take into account running time, which in this case was $\sim 230\,$s for CE and $\sim 10\,$s for LIP (and < 1 second for Metropolis and SA). To run this experiment run `ch7_tsp_metropolis_SA_LIP_CE.py` with parameters `--algs M_const,SA,M_LIP,CE --n_iter 100`.

7.5   The Metropolis Cross-entropy: A Hybrid Approach

In the Metropolis algorithm (and its variations, including locally informed proposals) we, roughly speaking, performed some walk: we start at $\mathbf{X}^{(0)}$, then sample somehow $\mathbf{X}^{(1)}$—which is some neighbor of $\mathbf{X}^{(0)}$ (we will use here superscripts for steps, in order not to confuse them with samples $\mathbf{X}_1, \ldots, \mathbf{X}_M$), and so on. Roughly speaking, we constructed some $\mathbf{X}^{(k+1)}$ based on $\mathbf{X}^{(k)}$ sampling some neighbor of the latter.

In cross-entropy however, things were different: we sampled M solutions according to some distribution, and the distribution was changing in time (the method implies that the distribution converges to the "best" one, in some sense). Thus, a "state" at each step was a distribution (we were taking the best solution from the current sample to create the plots, but indeed the final distribution is the output of the CE method).

Algorithm 72 Metropolis-CE Method for optimization

Require: Family of distributions $\{\mathbf{Q}_\theta(\cdot, \cdot)\}$ for sampling neighbors, $\theta \in \Theta$, initial θ_0, initial state $\mathbf{x}_0$, performance function $h(\cdot)$, parameter $\rho \in (0, 1)$, sample size M, smoothing parameter $\alpha \in (0, 1)$, number of iterations T.

1: Define $\theta'_0 = \theta_0$
2: Set $\mathbf{X}^{(0)} = \mathbf{x}_0$
3: **for** $t = 1, \ldots, T$ **do**
4:    **Update** γ_t: generate $\mathbf{X}_1, \ldots, \mathbf{X}_M$ by sampling neighbors of $\mathbf{X}^{(t-1)}$ from $\mathbf{Q}_{\theta'_{t-1}}(\mathbf{X}^{(t-1)}, \cdot)$, compute their performances $\eta_1 = h(\mathbf{X}_1), \ldots, \eta_M = h(\mathbf{X}_M)$, sort them $\eta_{(1)} \leq \cdots \leq \eta_{(M)}$ and compute $(1 - \rho)$ sample quantile

$$\gamma_t = \eta_{(\lceil (1-\rho)M \rceil)}$$

5:    **Update** θ'_t: from **the same** samples $\mathbf{X}_1, \ldots, \mathbf{X}_M$ solve

$$\theta' = \underset{\theta}{\arg\max} \; \frac{1}{M} \sum_{i=1}^{M} \mathbf{1}(h(\mathbf{X}_i) \geq \gamma_t) \log \mathbf{Q}_\theta(\mathbf{X}^{(t-1)}, \mathbf{X}_i) \tag{7.5.1}$$

and apply the smoothed update

$$\theta'_t = \alpha\theta' + (1 - \alpha)\theta'_{t-1},$$

6:    Set $\mathbf{X}^{(t)}$ as the sample with best performance, i.e., $\mathbf{X}^{(t)} = \mathbf{X}_{\eta(M)}$.
7: **end for**
8: **return** $\mathbf{X}^{(T)}$ $\triangleright$ In practice we would return best overall state

---

Consider the following hybrid approach. We will have "states" $\mathbf{X}^{(k)}$ at each step like in Metropolis-based algorithms, we will also have a distribution *over neighbors* (also like in Metropolis-based algorithms), however which changes in time (like in the CE method). We therefore consider a family of distributions $\mathbf{Q}_\theta(\mathbf{x}, \mathbf{x}')$, $\mathbf{x}, \mathbf{x}' \in \mathcal{S}$ for sampling neighbors of a given $\mathbf{x}$, which is parametrized by $\theta \in \Theta$. The parameter will change in time.

Thus now, given the current state $\mathbf{X}^{(k)}$ at step k, the parameter θ is a reparametrization of a distribution *over neighbors* of the current state. For example, in TSP for a given permutation $\sigma = (\sigma_1, \ldots, \sigma_n)$ we choose a *neighbor* which is always a permutation with two indices swapped, the indices are chosen by θ (which is thus a distribution on $\binom{n}{2}$ pairs). In the 0-1 knapsack problem, for a given $\mathbf{x} = (x_1, \ldots, x_d)$ we consider a neighbor which is a state which differs from $\mathbf{x}$ in one coordinate. Sampling a candidate thus means sampling that coordinate, which may be simply parametrized by $\theta = (\theta_1, \ldots, \theta_d)$, a distribution on $\{1, \ldots, d\}$. The distribution is updated in the same way as in Algorithm 69. The details are provided in Algorithm 72, which we call Metropolis-CE.

Example 7.5.1 (The 0-1 Knapsack Problem Continued) We consider (as in previous Metropolis-based algorithms) that $\mathbf{x}'$ is a neighbor of $\mathbf{x}$ if it differs in one coordinate, i.e., $\mathbf{x} = (x_1, \ldots, x_d)$ and $\mathbf{x}' = (x'_1, \ldots, x'_d)$ are neighbors if $x'_k = 1 - x_k$ for some $k \in [d]$ and $x'_j = x_j$ for $j \neq k$. State $\mathbf{x}'$ is chosen as a neighbor of $\mathbf{x}$ with probability $\mathbf{Q}_\theta(\mathbf{x}, \mathbf{x}') = \theta_k$, where the parametrization $\theta = (\theta_1, \ldots, \theta_d)$ is a distribution on $\{1, \ldots, d\}$.

Denote $(X_i(1), \ldots, X_i(d))$ by $\mathbf{X}_i$. To compute θ' in (7.5.1) let C_k be the number of those samples which differed from $\mathbf{X}^{(t-1)}$ which were in the elite set (i.e., $h(\mathbf{X}_i) \leq \gamma_t$) and had k-th coordinate swapped, i.e.,

$$C_k = \sum_{i=1}^{M} \mathbf{1}(h(\mathbf{X}_i) \geq \gamma_t)\mathbf{1}\big(\mathbf{X}_i(k) = 1 - \mathbf{X}^{(t-1)}(k)\big).$$

Thus we may rewrite

$$\sum_{i=1}^{M} \mathbf{1}(h(\mathbf{X}_i) \geq \gamma_t) \log \mathbf{Q}_\theta(\mathbf{X}^{(t-1)}, \mathbf{X}_i) = \sum_{i=1}^{d} C_k \log \theta_k.$$

Thus, we need to maximize the above sum subject to the constraint $\sum_{k=1}^{d} \theta_k = 1$. We will use again Lagrangian multipliers:

$$\mathcal{L}(\theta, \lambda) = \sum_{i=1}^{d} C_k \log \theta_k + \lambda \left(1 - \sum_{k=1}^{d} \theta_k\right).$$

Computing the derivative w.r.t. λ and setting it to 0 yields desired $\sum_{k=1}^{d} \theta_k = 1$. Computing the derivative w.r.t. θ_k we have

$$\frac{\partial \mathcal{L}(\boldsymbol{\theta})}{\partial \theta_k} = \frac{C_k}{\theta_k} - \lambda,$$

yielding the solution

$$\theta_k' = \frac{C_k}{\sum_{j=1}^{d} C_j}.$$

We ran the Algorithm 72 with $\rho = 0.3$, sample size $M = 30$ and small learning rate $\alpha = 0.001$ for 100 steps. The result of a single run of all algorithms is presented in Fig. 7.5, showing that Metropolis-CE yielded best solution. We ran all the algorithms 100 times and this algorithm always yielded the best solution. For more on results for this example, see Sect. 7.6.2. $\Diamond$

Example 7.5.2 (TSP Continued) Recall, in Metropolis-based algorithms for a given permutation σ we considered neighbors by swapping two elements (e.g., in uninformed Metropolis we chose the indices uniformly at random). Now, given σ we choose indices $i < j$ with probability $\theta(i, j)$ and swap them. Thus, if σ' is a permutation σ with indices i, j swapped, then $\mathbf{Q}_{\boldsymbol{\theta}}(\sigma, \sigma') = \theta(i, j)$, where $\boldsymbol{\theta} = (\theta(i, j))_{i<j}$. Permutations $\mathbf{X}_1, \ldots, \mathbf{X}_M$ are neighbors of $\mathbf{X}^{(t-1)}$, denote by $a_i < b_i$ indices swapped in the permutation $\mathbf{X}_i$. We can thus rewrite

$$\frac{1}{M} \sum_{i=1}^{M} \mathbf{1}(h(\mathbf{X}_i) \geq \gamma_t) \log \mathbf{Q}_{\boldsymbol{\theta}}(\mathbf{X}^{(t-1)}, \mathbf{X}_i) = \frac{1}{M} \sum_{i=1}^{M} \mathbf{1}(h(\mathbf{X}_i) \geq \gamma_t) \log \theta(a_i, b_i).$$

We have to maximize the above expression w.r.t. $\boldsymbol{\theta}$ under the constraint $\sum_{i<j} \theta(i, j) = 1$. Similarly as in previous examples, consider the Lagrangian function

$$\mathcal{L}(\boldsymbol{\theta}, \lambda) = \frac{1}{M} \sum_{i=1}^{M} \mathbf{1}(h(\mathbf{X}_i) \geq \gamma_t) \log \theta(a_i, b_i) + \lambda \left(1 - \sum_{i<j} \theta(i, j) \right).$$

Since the derivative of $\log \theta(a, b)$ is $1/\theta(a, b)$ and this only contributes when $(a, b) = (a_i, b_i)$, we have

$$\frac{C_{(a,b)}}{\theta(a, b)} - \lambda = 0,$$

where $C_{(a,b)}$ is the count of swaps (a, b) among elite samples, i.e.,

$$C_{(a,b)} = \sum_{i=1}^{M} \mathbf{1}(h(\mathbf{X}_i) \geq \gamma_t)\mathbf{1}((a, b) \text{ was swapped in } \mathbf{X}_i).$$

Finally,

$$\theta'(a, b) = \frac{C_{(a,b)}}{\sum_{i<j} C_{(i,j)}}.$$

Simulation Results We ran the algorithm for 100 steps with $M = 100$, $\rho = 0.35$ and learning rate $\alpha = 0.7$. The typical result is depicted in Fig. 7.4. We also ran 100 times all the algorithms considered so far (all listed in the caption of the aforementioned figure): in 61 cases the cross-entropy Algorithm 71 was the winner, in 29 cases the current Metropolis-CE was the winner, 9.66 times the Metropolis-LIP algorithm (Example 7.3.2) was the winner and 0.33 times the usual Metropolis algorithm was the winner. Fractions result from the way we compute "Wins", see Sect. 7.6.1. Note there are some hyperparameters to tune (α, ρ, M), but nevertheless it shows that cross-entropy-based methods prevail in this example. $\Diamond$

7.6 Summary of Simulation Results for the Knapsack and the Travelling Salesman Problems

In this section, we summarize the results obtained by running our algorithms 100 times. For each replication, we report the best solution found, the running time, and the "winner" scores.

7.6.1 The Travelling Salesman Problem

All experiments were performed on a 13-city instance, with the distance matrix $\mathbf{M}$ given in Fig. 7.3. The travelling salesman problem is NP-hard, meaning that no polynomial-time algorithm is known to find the optimal solution. A brute-force approach would require $O(n!)$ evaluations, which is prohibitive in Python (although a highly optimized implementation in C/C++ could run in a reasonable time). The best-known exact methods use dynamic programming. For example, the Held-Karp algorithm [82] has a time complexity of $O(n^2 2^n)$, which, while still exponential, represents a significant improvement for $n = 13$. In our case, the Held-Karp algorithm found the optimal solution with a tour cost of 7024 in 0.068 s. This algorithm is implemented in `ch7_tsp_Held-Karp_optim_solution.py`.

Table 7.2 Summary of 100 replications for various TSP algorithms

Algorithm	Best distance	Running time (sec)	Wins
Metropolis ($T = 1$)	7024	0.127	0.33
Simulated annealing	7312	0.125	0.0
Metropolis-LIP	7024	9.676	9.66
Metropolis-CE	7024	48.2059	29
CE	7024	239.69	61

Table 7.2 reports the results over 100 replications (each consisting of 100 steps per algorithm). The table displays the overall best solution found (with the exception of the Simulated Annealing method, all other methods found an optimal solution). Note that although the Simulated Annealing method never emerged as the sole winner in any replication, this does not imply that SA is inferior to the standard Metropolis algorithm. In fact, as noted earlier, SA often produced better solutions than the standard Metropolis approach; however, both were consistently outperformed by the cross-entropy (CE) and Metropolis-CE methods.

Win counts are computed as follows: if, in a given replication, k out of 5 algorithms obtain the best (i.e., lowest) tour distance, then each of these algorithms receives $1/k$ point. The "Wins" column reports the total number of points accumulated over all replications, thereby accounting for ties.

As can be seen, the cross-entropy-based methods prevail in terms of wins. However, they require significantly more running time, which must be taken into account in practical applications.

The example is implemented in `ch7_tsp_metropolis_SA_LIP_CE.py`. By default it runs a single iteration of all algorithms for 100 steps. To run it 100 times with, say, all algorithms except CE, run it with parameters `--algs M_const,SA,M_LIP,M_CE --n_iter 100`. See `--help` for more details.

7.6.2   The Knapsack Problem

We considered a 0-1 knapsack problem with $d = 100$ items, each item i having weight i and value $i^{1.2}$, and a maximum capacity of $W_{\max} = 3000$. In general, the 0-1 knapsack problem is NP-hard, meaning that no polynomial-time algorithm is known to guarantee an optimal solution for arbitrary input sizes. However, it is known as a *weakly* NP-hard problem since there is a pseudo-polynomial-time dynamic programming (DP) solution with $O(d\, W_{\max})$ complexity; for our specific instance ($d = 100$, $W_{\max} = 3000$), this amounts to 300,000 state updates and is perfectly feasible in Python. An exact DP run yields an optimal value of **7251.66** for this instance. The DP implementation for this problem is available in the script `ch7_knapsack_dynamic_prog_optim_solution.py`.

Table 7.3 Summary of 100 replications for various knapsack algorithms (including Metropolis-CE). The optimal value for this instance is 7251.66

Algorithm	Best total value	Running time (sec)	Wins
Metropolis	6974.54	0.05	0.0
Metropolis-LIP	7122.88	1.36	0.0
Metropolis-LIP-full	7110.36	4.06	0.0
CE	7046.59	1.93	0.0
Metropolis-CE	**7251.66**	5.68	100

Table 7.4 Summary of 100 replications for knapsack algorithms, *excluding* Metropolis-CE

Algorithm	Best total value	Running time (sec)	Wins
Metropolis	6942.09	0.05	0.0
Metropolis-LIP	7111.80	4.54	21
Metropolis-LIP-full	7137.16	4.10	35
CE	7136.34	4.08	44

Table 7.3 reports experimental results over 100 replications, each consisting of 100 iterations per algorithm. The "Best total value" column is the highest knapsack value found among the solutions produced in that replication, and "Wins" are assigned fractionally in the case of ties (if k algorithms tie for the best solution in a replication, each receives $1/k$ point).

As we can see in Table 7.3, Metropolis-CE consistently achieves the best solution in every replication, thus accumulating all the win points. To explore the performance of other algorithms when Metropolis-CE is excluded, Table 7.4 reports analogous results.

Without Metropolis-CE, the cross-entropy method (CE) and the full LIP variant perform similarly and typically dominate Metropolis, though neither attains the true optimum in these 100-step experiments. As is evident, although Metropolis-CE consistently is the best, it also requires more running time, which is an important consideration in real-world usage.

The example is implemented in `ch7_knapsack_metropolis_SA_LIP_CE.py`. By default, it runs one iteration of all algorithms. To see the results of, say, only Metropolis and LIP, and run it for 100 iterations (to compute win counts), run it with parameters `-- algs Metropolis,LIP --n_iter 100`. See `--help` for more details.

7.7 Bibliographical Notes

The idea of applying Metropolis and Gibbs sampling to stochastic optimization has its roots in early MCMC works. Metropolis et al. [151] laid the groundwork for sampling from complex distributions, while the Gibbs sampler introduced by Geman and Geman [62] was later extended to combinatorial optimization.

Early applications include decryption of substitution ciphers and the 0-1 knapsack problem. Diaconis [35] first proposed using the Metropolis-Hastings algorithm for decrypting substitution ciphers, a notion further studied in Chen and Rosenthal [24].

Simulated annealing emerged as a natural evolution of these ideas. Kirkpatrick et al. [104] popularized the technique by introducing a gradually decreasing temperature parameter. Cooling schedule results were later established by Hajek [76] and Geman and Geman [62], and a theory of non-homogeneous Markov chains is discussed in Brémaud [21] and Vassiliou [208].

More recent work has focused on locally informed proposals that incorporate problem-specific structure into the proposal mechanism. Zanella [217] advanced these ideas by adapting proposals to the local geometry of discrete state spaces, while Grathwohl et al. [71] proposed gradient-based approximations that reduce running time.

The cross-entropy method offers an alternative approach to stochastic optimization. Introduced by Rubinstein [184] and extended in [31], it recasts optimization as a rare-event estimation problem and has been applied to various combinatorial problems. Convergence results are provided in Costa et al. [27], and comprehensive treatments can be found in Rubinstein and Kroese [185], Kroese, Taimre and Botev [112], Asmussen and Glynn [8], and Kroese et al. [113].

7.8 Exercises

Lab Exercises

7.L.1 (Generating a random matrix with fixed row and column sums) Below is an example of a 4×4 matrix where each row and each column sum to a fixed value (bolded):

	110	**274**	**84**	**118**
210	67	112	26	5
93	19	52	18	4
217	21	88	23	85
66	3	22	17	24

Let S be the set of all 4×4 matrices with natural number entries such that each row and each column sum to the values given in the table above.

Implement a Markov chain that samples matrices from the uniform distribution over S. Generate a few sample matrices and verify that they satisfy the given constraints.

Hint: You can implement the Metropolis algorithm by defining a suitable proposal distribution. Consider what can be treated as a "neighbor" of a given matrix in the Markov chain.

7.L.2 Consider the following probability distribution on the set $\mathcal{S}_n$ of permutations of $\{1, \ldots, n\}$:

$$\pi_\sigma = \frac{1}{z} \lambda^{d(\sigma, \sigma_{\mathrm{id}})}, \quad \text{where } \lambda > 0,$$

where z is a normalization constant, $\sigma_{\mathrm{id}} = (1, \ldots, n)$ is the identity permutation, and

$$d(\sigma, \sigma') = \frac{1}{n} \sum_{i=1}^{n} |\sigma(i) - \sigma'(i)|.$$

In Exercise 3.L.5, you were asked to sample from π using the acceptance-rejection method for parameters $\lambda = 0.75$ and $n = 52$.

(a) Implement the Metropolis algorithm for this distribution using the proposal distribution applied to the travelling salesman problem (see Algorithm 66). For which values of λ is the algorithm applicable? Can it be used for $\lambda = 0.1$ and/or $\lambda = 10$? Compare its performance with the acceptance-rejection method implemented in Exercise 3.L.5. Discuss the advantages, disadvantages, and limitations of both methods.

(b) Repeat the procedure using the Kendall τ distance defined in Exercise 3.L.5 instead of $d(\sigma, \sigma')$.

(c) Implement the Metropolis algorithm using the alternative proposal distribution described in Remark 7.2.4 and compare its effectiveness.

7.L.3 Consider the setup from Exercise 7.L.2 and implement an algorithm for sampling from the distribution:

$$\pi_\sigma = \frac{\lambda^{\mathrm{FP}(\sigma)}}{z},$$

where z is a normalization constant, $\lambda > 0$ is a parameter, and $\mathrm{FP}(\sigma)$ represents the number of fixed points in permutation σ, i.e.,

$$\mathrm{FP}(\sigma) = \#\{i : \sigma(i) = i\}.$$

Run the algorithm for $n = 52$ with $\lambda \in \{0.1, 0.75, 2\}$. Compare the results with the acceptance-rejection method implemented in Exercise 3.L.6.

7.L.4 In Sect. 7.1, we introduced a Gibbs sampler for sampling solutions of the 0-1 knapsack problem (Algorithm 63) and presented implementation results for the parameters:

$$d = 100, \quad w_i = i, \quad v_i = i^{1.2}, \quad i = 1, \ldots, d, \quad W = 3000.$$

(a) Re-implement the algorithm and experiment with smaller values of W, such as $W = 1000$, $W = 500$, or $W = 50$. Compare the results with an alternative approach based on sampling random permutations multiple times and checking their validity. For each $W \in \{50, 500, 1000, 3000\}$, estimate the proportion of valid permutations, store the best solution, and compare it with the Gibbs sampler results.

(b) Implement the Metropolis algorithm for this problem. Define the neighbors of $\mathbf{x}$ as states $\mathbf{x}'$ that differ from $\mathbf{x}$ at most in one position. Compare the performance of the Metropolis algorithm with the Gibbs sampler.

7.L.5 (Simulated annealing for the knapsack problem) Recall that the Gibbs sampler for the 0-1 knapsack problem, presented in Algorithm 63, defines a Markov chain with the stationary distribution:

$$\pi_{\mathbf{x}} = \frac{\exp(\mathbf{v} \cdot \mathbf{x}/T)}{z_T}, \quad \text{if } \mathbf{w} \cdot \mathbf{x} \le W, \quad \text{and } \pi_{\mathbf{x}} = 0 \text{ otherwise.}$$

Construct a simulated annealing algorithm based on this Gibbs sampler, i.e., design a time-homogeneous Markov chain with a family of transition matrices $\mathbf{P}^{T_k} = (p_{\mathbf{x}\mathbf{x}'}(T_k))$, where at step k, the stationary distribution is given by:

$$\pi_{\mathbf{x}}^{T_k} = \frac{\exp(\mathbf{v} \cdot \mathbf{x}/T_k)}{z_{T_k}}, \quad \text{for } \mathbf{w} \cdot \mathbf{x} \le W. \tag{7.8.1}$$

Consider the same parameters as in Exercise 7.L.4, and compare the performance of:

- A non-simulated annealing version, i.e., with a fixed temperature $T_k = 1$.
- A simulated annealing scheme with a temperature schedule $T_k = 1/\log(k)$.

Perform this comparison for two values of W, namely $W = 3000$ and $W = 50$.

Note: According to Remark 7.2.1, the distribution

$$\pi_{\mathbf{x}}^{T_k} = \frac{\deg(\mathbf{x}) \exp(\mathbf{v} \cdot \mathbf{x}/T_k)}{z'_{T_k}}$$

is equivalent to the distribution given in (7.8.1).

7.L.6 In Example 7.4.2, we considered the cross-entropy (CE) method for the travelling salesman problem (TSP), where the parameter $\boldsymbol{\theta}$ was a stochastic matrix of size $n \times n$ with zeros on the diagonal. In this formulation, a permutation was sampled from $\boldsymbol{\theta}$ as follows: given the current city i, the next city was chosen based on row $\boldsymbol{\theta}(i, \cdot)$, setting entries corresponding to already visited cities to zero, normalizing the remaining probabilities, and then sampling the next city. The full procedure was provided in Algorithm 70.

Consider an alternative parametrization: instead of a stochastic matrix, let $\boldsymbol{\theta} = (\theta_1, \ldots, \theta_n)$ be a probability distribution over cities. A permutation $\sigma = (\sigma_1, \ldots, \sigma_n)$ is then sampled as follows:

(a) Start with an empty tour.
(b) At each step, given that cities $c_1, \ldots, c_k$ have already been visited, modify $\boldsymbol{\theta}$ by setting $\theta_r = 0$ for all previously visited cities $r \in \{c_1, \ldots, c_k\}$, normalize the remaining values, and sample the next city from the updated distribution.
(c) Repeat until all cities are visited.

Tasks:

- Provide a formal description of this alternative sampling procedure.
- Implement the full CE method using both parametrizations for $n = 13$ and distance matrix M given in Fig. 7.3.
- Compare the performance of both parametrizations. Analyze and discuss the differences in results, such as convergence speed and solution quality.

Chapter 8
Simulation of Queues and Related Models

Simulation methods are indispensable for analyzing complex stochastic systems. While mathematical models provide powerful abstractions for systems such as queueing networks or telecommunication systems, closed-form solutions are often unavailable due to the complexity of real-world scenarios. Consequently, simulation becomes a critical tool, enabling practitioners to study the behavior of such systems under various configurations and constraints.

This chapter focuses on the simulation of queueing models and related systems, offering both theoretical insights and practical implementations. Queueing systems are pivotal in understanding a wide range of real-world applications, from call center optimization to inventory management and telecommunications. By simulating these systems, we can evaluate performance characteristics such as average waiting times, system utilization, and throughput, even in cases where analytical solutions are intractable.

The primary goal of this chapter is to provide the reader with the tools and methods necessary to simulate, analyze, and interpret results for a variety of queueing and related models. Through carefully chosen examples, we illustrate how simulations effectively bridge the gap between theoretical modeling and practical problem-solving. Beyond the simplest cases, we will not derive or prove queueing theory formulas, as their proofs often rely on specialized techniques that are beyond the scope of this book.

In this chapter, we aim to:

- introduce standard mathematical models from queueing systems, manufacturing, and telecommunication systems,
- model the dynamics of tasks, arrivals, and processing within a system,
- construct and interpret estimators for system performance characteristics,
- identify the key concepts of steady-state probabilities and time-averaged characteristics,
- demonstrate simulation techniques for these models,

© The Author(s), under exclusive license to Springer Nature Switzerland AG 2025 541
P. Lorek, T. Rolski, *Lectures on Monte Carlo Theory*, Probability Theory
and Stochastic Modelling 108, https://doi.org/10.1007/978-3-032-01190-9_8

- explain how to conduct output analysis, including confidence intervals and error bounds,
- explore the role of stationarity, ergodicity, and regeneration in simulations.

By the end of this chapter, the reader will have a comprehensive understanding of how to simulate and analyze queueing systems, while appreciating the interplay between theoretical guarantees and empirical approximations. Additionally, we focus on understanding stability under varying input parameters—a critical aspect in ensuring meaningful simulation outcomes.

8.1 Preliminaries

We now describe the basic concepts and approaches used in the simulation of the studied models. Typically, we consider a "system" characterized by:

- tasks (jobs, customers, calls) arriving at the system according to a specified mechanism,
- these tasks being processed in the system in some manner and eventually leaving it.

Depending on the problem, one defines a characteristic of interest, such as the number of tasks in the system or the waiting time for service of an arriving task. These characteristics can be represented as:

- a *sequence of random variables* in discrete time (e.g., the waiting time W_i of the i-th arriving task), or
- a *stochastic process* in continuous time (e.g., the number of tasks $L(t)$ in the system at time t).

Time Averages and Ensemble Averages Two important ways to analyze the behavior of stochastic processes are through **time averages** and **ensemble averages**, which capture different perspectives:

- **Time average:** For a process $X(t)$, the *time average* over the interval $[0, t]$ is defined as:

$$\hat{X}(t) = \frac{1}{t} \int_0^t X(s)\, ds.$$

This represents the average value of $X(t)$ observed in a single realization. In simulations, the time average often corresponds to the computed average of a system's behavior over a long runtime.

- **Ensemble average:** The *ensemble average* of $X(t)$ at a fixed time t is defined as:

$$\mathbb{E}[X(t)].$$

This represents the average value of $X(t)$ across many realizations of the process at a specific moment in time. In simulations, this corresponds to averaging over many independent simulation runs, each starting with potentially different initial conditions.

Both averages provide complementary perspectives. A natural question in simulations is whether the **time average** from a single long run is a reliable estimate of the **ensemble average**. Intuitively, for sufficiently large $t_{\max}$, the system's behavior is expected to stabilize, making the time average a meaningful estimator of the ensemble average.

Whether this holds depends on the nature of the process $X(t)$. In practice, we are often interested in verifying whether the process satisfies conditions ensuring that it "converges" to a well-defined long-term behavior. For example, we may look for guarantees that $X(t)$ stabilizes in distribution as $t \to \infty$, or that its time average converges to a deterministic value. These conditions, explored in later sections, provide the theoretical foundation for interpreting simulation results.

Roughly speaking, ergodic theorems state that under suitable conditions, the *time average* is approximately equal to the *ensemble average*. In ideal scenarios, the ensemble average corresponds to $\mathbb{E}X^*(t)$, where $X^*(t)$ is an ergodic version of $(X(t))$.

In such cases, the evolution of the characteristic of interest, which changes over time (e.g., the number of pending tasks at time t or inventory at time t), is a random process. This variable of interest can be scalar or vector-valued. It may change continuously, as in the case of Brownian motion (studied in Sect. 5.2.1), or only at specific random moments.

Stochastic processes like Brownian motion are typically simulated at intervals $0, h, 2h, \ldots$, leading to what is called a *synchronous simulation*. However, in this chapter, we focus on models where evolution occurs at separate random moments, leading to what is called an *asynchronous simulation*. This approach is explored through *discrete-event simulations*, described in Sect. 8.5.

First, in Sect. 8.2, we discuss how to model arrival processes or, more generally, points randomly scattered over a state space. The main object is the Poisson process: both homogeneous and non-homogeneous.

In Sect. 8.3, we explore time-continuous processes evolving in a "discrete-events simulation" manner. This includes birth-and-death (B&D) processes. By specifying input parameters, we define a class of so-called *Markovian queueing processes* and examine the effects of their simulations based on chosen parameters.

In the second part of this chapter, we discuss definitions of process stability. We derive the asymptotic variance (or TAVC) using the covariance function of the considered process. Several examples are presented, along with error analyses for each case.

Many problems involve requesting entities or units arriving in a "random" manner. To model such streams, frameworks based on the theory of stochastic processes are essential. The next section describes the Poisson process (both homogeneous and non-homogeneous), often used to model arrival streams (e.g., calls, customers, etc.).

8.2 Modeling Arrival Streams; Simulation of Poisson and Renewal Processes

This section introduces fundamental methods for modeling and simulating streams of arrivals, which can be represented as points randomly distributed over time. Such models are widely used in queueing theory, telecommunications, and insurance for analyzing systems where events occur over time.

Key tools for describing these processes include counting processes and sequences of arrival times, which we detail below.

Counting Processes and Arrival Times To describe arrival streams, we often use a *counting process* $N(t)$:

- $N(t)$ counts the number of points (arrivals) up to time t ($t \geq 0$).
- It is a random process characterized by:

  - **Piecewise constant realizations:** $N(t)$ is non-decreasing, has unit jumps, and satisfies $N(0) = 0$.
  - **Right continuity:** As a function of t, $N(t)$ is right-continuous.

- $N(t)$ represents the total number of arrivals by time t, while $N(t + h) - N(t)$ gives the number of arrivals in the interval $(t, t + h]$. For simplicity, we assume single arrivals per event.

Another approach to describing arrivals is through a sequence (A_n), where $0 < A_1 < A_2 < \cdots$. Here,

$$\tau_1 = A_1, \quad \tau_i = A_i - A_{i-1}, \qquad (i = 2, \ldots) \tag{8.2.1}$$

denotes the *sequence of inter-arrival times*. The relationship between the counting process $N(t)$ and the arrival instants (A_n) is given by:

$$N(t) = \#\{i \geq 1 : A_i \leq t\}. \tag{8.2.2}$$

For $s < t$, we define

$$N(s, t] := N(t) - N(s) = \#\{i \geq 1 : s < A_i \leq t\},$$

i.e., the number of points in the interval $(s, t]$. Note that thus $N(t) \equiv N(0, t]$.

8.2.1   *Homogeneous Poisson Process*

We now introduce a fundamental type of counting process, known as the *homogeneous Poisson process*.

Definition 8.2.1 A counting process $(N(t))$ is called a *homogeneous Poisson process* with intensity $\lambda > 0$ if its inter-arrival times (τ_j) are independent and identically distributed random variables following an exponential distribution $\mathrm{Exp}(\lambda)$.

The following characterization provides an equivalent definition:

Proposition 8.2.2 *A counting process $N(t)$ is a Poisson process with intensity λ if and only if:*

1. *The numbers of points in disjoint intervals are independent.*
2. *The number of points in the interval $(t, t + h]$ follows a Poisson distribution with mean λh, denoted* $\mathrm{Poi}(\lambda h)$.

Note that above proposition implies that $N(t)$ has a $\mathrm{Poiss}(\lambda t)$ distribution.

A proof of Proposition 8.2.2 can be found in Billingsley [14], Section 26. Note that:

$$\mathbb{P}(N(t, t + h] = 1) = \lambda h e^{-\lambda h} = \lambda h + o(h), \tag{8.2.3}$$

which explains why λ is referred to as the *intensity* of the Poisson process. This interpretation is further supported by the fact that $\mathbb{E}N(t, t + h] = \lambda h$, indicating that λ represents the average arrival rate per unit time. The intensity λ does not depend on t, which is why this process is called *homogeneous*.

Recall the result mentioned in Sect. 3.2: if $\tau_1, \tau_2, \ldots$ are independent random variables following the $\mathrm{Exp}(1)$ distribution and

$$N = \#\{i \geq 1 : \tau_1 + \cdots + \tau_i \leq \lambda\},$$

then N follows a Poisson distribution with mean λ. In fact, the cumulative sums of independent $\mathrm{Exp}(\lambda)$-distributed random variables define the arrival times of a Poisson process:

$$A_i = \tau_1 + \cdots + \tau_i, \tag{8.2.4}$$

and the counting process (8.2.2) can be expressed as:

$$N(t) = \#\{i \geq 1 : \tau_1 + \cdots + \tau_i \leq t\}, \quad t \geq 0.$$

Algorithm 73 Algorithm PoiPro(λ); simulate Poisson process on $[0, t_{\max}]$

1: Create an empty list A
2: Generate $U \sim \mathcal{U}[0, 1)$, set $t = -\log(U)/\lambda$ $\triangleright \tau \sim \text{Exp}(\lambda)$
3: **while** $t \leq t_{\max}$ **do**
4:     A.append($[t]$)
5:     Generate $U \sim \mathcal{U}[0, 1)$, set $\tau = -\log(U)/\lambda$
6:     $t = t + \tau$
7: **end while**
8: **return** A

The procedure for generating a homogeneous Poisson process in the interval $(0, t_{\max}]$ is outlined in Algorithm 73. A visualization of the generated process is shown in Fig. 8.1.

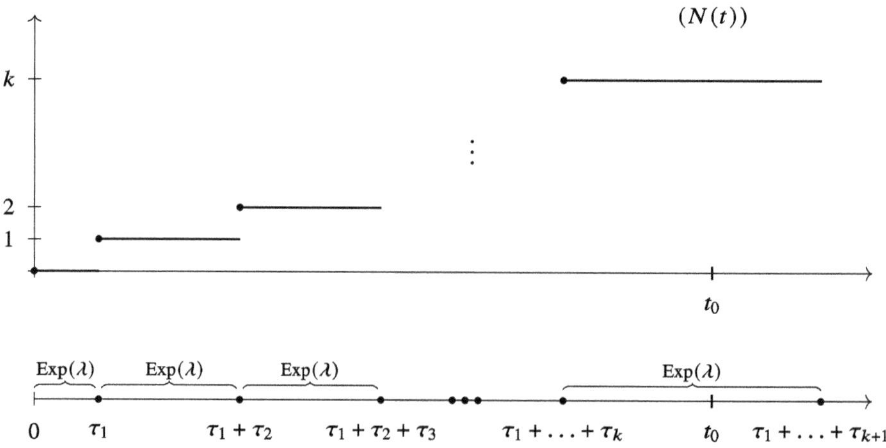

Fig. 8.1 Top: homogeneous Poisson ($N(t)$). Bottom: corresponding τ_i and thus arrival times $A_i = A_{i-1} + \tau_i$ depicted. Sample t_0 with $N(t_0) = k$ depicted

Given that $N(t) = 1$, i.e., that a single point occurred before time t, the conditional distribution of when it occurred is as follows:

$$\mathbb{P}(A_1 \leq s \mid N(t) = 1) = \frac{\mathbb{P}(A_1 \leq s, N(t) = 1)}{\mathbb{P}(N(t) = 1)} = \frac{\mathbb{P}(N(s) = 1, N(t) - N(s) = 0)}{\lambda t e^{-\lambda t}}$$

$$= \frac{\mathbb{P}(N(s) = 1)\mathbb{P}(N(t) - N(s) = 0)}{\lambda t e^{-\lambda t}}$$

$$= \frac{s\lambda e^{-\lambda s} e^{-\lambda(t-s)}}{\lambda t e^{-\lambda t}} = \frac{s}{t},$$

i.e., A_i, given $N(t) = 1$, has uniform distribution $\mathcal{U}(0, t]$. Similarly, one may show that given $N(t) = n$, the *unordered* set of n arrival events are i.i.d. uniformly $\mathcal{U}(0, t]$ distributed. This insight leads to another procedure for simulating a homogeneous Poisson process, described in Algorithm 74.

Algorithm 74 Algorithm PoiProSorted(λ); simulate Poisson process on $[0, t_{\max}]$

1: Generate $K \sim \text{Poi}(\lambda t_{\max})$
2: Generate $A_1, \ldots, A_K$ i.i.d. uniformly distributed $\mathcal{U}(0, t_{\max}]$
3: Create list A consisting of sorted values $(A_{(1)}, \ldots, A_{(K)})$
4: **return** A

Finally, let us note that the definition of a homogeneous Poisson process via Eqs. (8.2.1) or (8.2.4) is a special case of a *renewal process*. The sequence $A_1, A_2, \ldots$ defined in (8.2.1) is called a *renewal process*. In particular, when the inter-arrival times (τ_j) are exponentially distributed, the process becomes a homogeneous Poisson process. However, in general, the renewal process does not exhibit the property of stationary increments, meaning that we lose the feature where the vectors

$$(N(s + t_2) - N(s + t_1), \ldots, N(s + t_n) - N(s + t_{n-1})) \tag{8.2.5}$$

have a distribution that is independent of $s > 0$ (for $n = 2, 3, \ldots$ and $0 \le t_1 < t_2 < \cdots$). As an exercise, one can verify that the homogeneous Poisson process does have the property of stationary increments, i.e., (8.2.5) holds. Note that these increments are independent.

8.2.2   *Non-homogeneous Poisson Process*

A time-homogeneous Poisson process does not always accurately model real-world situations. In some cases, we need a process where the intensity depends on time t. For example, during a 24-hour cycle, the arrival intensity can naturally vary based on the time of day. We now define a *non-homogeneous Poisson process* with an intensity function $\lambda(t)$. If the intensity function is constant, the resulting process is a homogeneous Poisson process.

Definition 8.2.3 A counting process $N(t)$ is called a *non-homogeneous Poisson* process with intensity $\lambda(t)$ if:

1. the number of points in disjoint intervals are independent,
2. the number of points in the interval $(t, t + h]$ follows a Poisson distribution with mean $\int_t^{t+h} \lambda(s)\, ds$.

The following result provides the foundation for simulating non-homogeneous Poisson processes:

Proposition 8.2.4 *Suppose that* $(N_{\mathrm{h}}(t))$ *is a homogeneous Poisson process with intensity* $\lambda_{\max}$, *and* $\lambda(t) \leq \lambda_{\max}$ *for* $t \in (0, t_{\max})$. *Each point of* (N_{h}) *is retained with probability* $p(t) = \lambda(t)/\lambda_{\max}$, *and otherwise, it is rejected. All decisions are assumed to be independent. Then the resulting process* $(N_{\mathrm{nh}}(t))$ *is a non-homogeneous Poisson process with intensity* $\lambda(t)$.

We leave it to the reader to prove Proposition 8.2.4 after reviewing the general construction of the Poisson process in Sect. 8.2.3. This acceptance-rejection procedure in point process theory is called *thinning*. Thus, a non-homogeneous Poisson process can be obtained by independent thinning.

The procedure follows a two-step approach. First, generate a Poisson process with constant intensity $\lambda_{\max}$, and then independently thin each point, where the probabilities depend only on the location of each point.

The procedure for generating a non-homogeneous Poisson process with intensity $\lambda(t)$ and a given $\lambda_{\max}$ on $[0, t_{\max})$ is outlined in Algorithm 75. A sample realization is depicted in Fig. 8.2.

Algorithm 75 Algorithm NH-Poi-thinning($\lambda(t)$); simulate non-homogeneous Poisson process with intensity $\lambda(t)$ on $[0, t_{\max}]$

Require: $\lambda_{\max}$
 1: Create an empty list A
 2: Generate $U \sim \mathcal{U}(0, 1)$, set $t = -\log(U)/\lambda_{\max}$
 3: **while** $t \leq t_{\max}$ **do**
 4:     Compute $p = \lambda(t)/\lambda_{\max}$
 5:     Generate $U_1 \sim \mathcal{U}(0, 1)$
 6:     **if** $U_1 \leq p$ **then**
 7:         A.append($[t]$)
 8:     **end if**
 9:     Generate $U_2 \sim \mathcal{U}(0, 1)$, set $\tau = -\log(U_2)/\lambda_{\max}$ $\triangleright\ \tau \sim \mathrm{Exp}(\lambda_{\max})$
10:     $t = t + \tau$
11: **end while**
12: **return** A

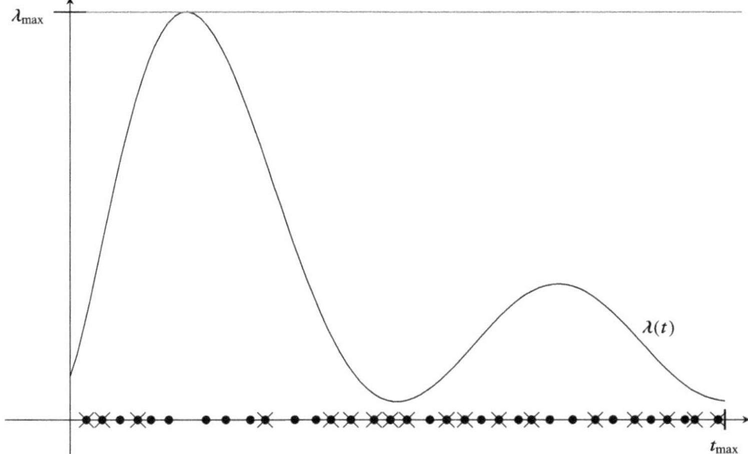

Fig. 8.2 Example: Sample function $\lambda(t)$ for $t \in (0, t_{\max})$ with $\lambda_{\max} = \max_{t \in (0, t_{\max})} \lambda(t)$ depicted. Bullets $\bullet$ represent the arrival times of a homogeneous Poisson process $(N_h(t))$ with parameter $\lambda_{\max}$. Each arrival time A_i is crossed out with probability $1 - \lambda(A_i)/\lambda_{\max}$ (and thus kept with probability $\lambda(A_i)/\lambda_{\max}$). The remaining points constitute the arrival times of a non-homogeneous Poisson process $(N_{nh}(t))$ with intensity $\lambda(t)$

Let us compute the conditional distribution of the first arrival A_1 given that $N(t) = 1$, i.e., that the event occurred before t (recall, for a homogeneous Poisson process, we computed that this distribution is uniform $\mathcal{U}(0, t])$. Let us define

$$h(t) = \int_0^t \lambda(s)\, ds.$$

We have

$$\mathbb{P}(A_1 \leq s \mid N(t) = 1) = \frac{\mathbb{P}(A_1 \leq s, N(t) = 1)}{\mathbb{P}(N(t) = 1)} = \frac{\mathbb{P}(N(s) = 1, N(t) - N(s) = 0)}{h(t)e^{-h(t)}}$$

$$= \frac{\mathbb{P}(N(s) = 1)\mathbb{P}(N(t) - N(s) = 0)}{h(t)e^{-h(t)}}$$

$$= \frac{h(s)e^{-h(s)}e^{-(h(t) - h(s))}}{h(t)e^{-h(t)}} = \frac{h(s)}{h(t)}.$$

Its p.d.f. is thus

$$f_{A_1 \mid N(t) = 1}(s) = \frac{\lambda(s)}{\int_0^t \lambda(r)\, dr}, \quad 0 \leq s \leq t.$$

In Exercise 8.T.1, the reader is asked to prove the following:

Lemma 8.2.5 *Given that $N(t) = n$, i.e., that n arrival events occurred before time t, the arrival times $0 < A_1 < \cdots < A_n < t$ are distributed as the sorted n i.i.d. random variables having distribution*

$$F(s) = \frac{\int_0^s \lambda(r)\, dr}{\int_0^t \lambda(r)\, dr}, \quad f(s) = \frac{\lambda(s)}{\int_0^t \lambda(r)\, dr}, \quad 0 \le s \le t.$$

Lemma 8.2.5 leads to an alternative (to the thinning procedure) method for simulating a non-homogeneous Poisson process, summarized in Algorithm 76.

Algorithm 76 Algorithm NH-Poi($\lambda(t)$); simulate non-homogeneous Poisson process with intensity $\lambda(t)$ on $[0, t_{\max}]$

1: Sample $K \sim \text{Poi}\left(\int_0^{t_{\max}} \lambda(s)\, ds\right)$
2: Compute $F(s) = \frac{\int_0^s \lambda(r)\, dr}{\int_0^{t_{\max}} \lambda(r)\, dr}$ and its generalized inverse $F^{\leftarrow}$
3: Sample $U_1, \ldots, U_K$ i.i.d. $\mathcal{U}(0, 1)$
4: Compute $A_i = F^{\leftarrow}(U_i), \ i = 1, \ldots, K$
5: Create list A consisting of sorted $(A_{(1)}, \ldots, A_{(K)})$
6: **return** A

Note that Algorithm 76 requires the computation of the generalized inverse of $F(s)$, whereas no such extra computation is needed in the thinning procedure of Algorithm 75. This makes the thinning procedure generally more widely used in practice.

8.2.3 What Is a Poisson Process on E and How to Simulate It?

In this section, we explain what a point process on $E \subset \mathbb{R}^d$ is and how to simulate it. Let E be a state space. As usual, $(\Omega, \mathcal{F}, \mathbb{P})$ is an abstract probability space. In the case of Poisson processes from previous subsections, we have $E = \mathbb{R}_+$ or $E = [0, t_{\max}]$.

Consider a measure space $(E, \mathcal{A}, \Lambda)$. A point process $N(\cdot)$ is called a point process on E with intensity measure $\Lambda(\cdot)$ if:

- For all $B \in \mathcal{A}$, $N(B)$ is a random variable taking values in $\{0, 1, 2, \ldots\}$.
- For each ω, $N_\omega(\cdot)$ is a measure.

Definition 8.2.6 A point process N on E is called a *Poisson point process* with intensity measure Λ if:

1. For all disjoint sets $B_1, \ldots, B_n \in \mathcal{A}$, the random variables $N(B_1), \ldots, N(B_n)$ are independent.

2. $N(B) \sim \mathrm{Poi}(\Lambda(B))$.

In previously considered cases, $E = [0, t_{\max}]$, the intensity of the non-homogeneous Poisson process is $\Lambda(dt) = \lambda(t)dt$, and in the case of a homogeneous one, $\Lambda(dt) = \lambda dt$. For general $E \subset \mathbb{R}^d$, a homogeneous Poisson process has an intensity measure of the form $\Lambda(d\mathbf{t}) = \lambda d\mathbf{t}$.

A mathematical construction of the Poisson point process provides a tool for simulating it on any state space.

Construction of a Poisson Point Process on E with Intensity Λ First, assume $\Lambda(E) < \infty$. In such a case, we simulate a Poisson random variable T with mean $\Lambda(E)$, and then simulate T points $\xi_1, \ldots, \xi_T$ independently from E with distribution $\Lambda(\cdot)/\Lambda(E)$. We can write $N(d\mathbf{x}) = \sum_{j=1}^{T} \mathbb{1}(\xi_j \in d\mathbf{x})$. The point process N is the desired Poisson process with intensity Λ. The procedure is given in Algorithm 77. Note that this generalizes Algorithm 76 (and thus Algorithm 74) from $E = [0, t_{\max}]$ to more general $E \subset \mathbb{R}^d$.

Algorithm 77 Poisson process on E with finite intensity $\Lambda(\cdot)$

1: Generate $T \sim \mathrm{Poi}(\Lambda(E))$.
2: Generate independently T points $\xi_1, \ldots, \xi_T$ from E with distribution $\Lambda(dx)/\Lambda(E)$.
3: **return** points $\xi_1, \ldots, \xi_T$.

A sample Poisson point process on $[0, 3]^2$ is depicted in Fig. 8.3.

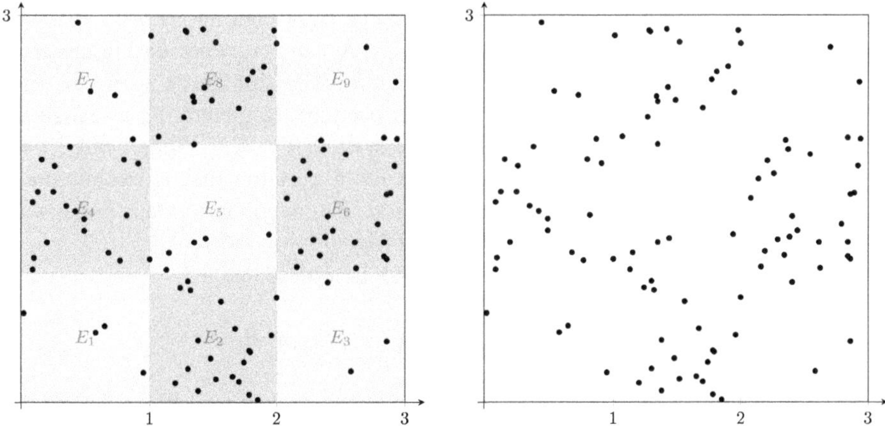

Fig. 8.3 Example: $E = E_1 \cup E_2 \cup \cdots \cup E_9$. On grey areas, the intensity is 20, and on light areas, the intensity is 5

Now, assume that $\Lambda(E) = \infty$, but there exist disjoint sets $E_1, E_2, \ldots$ such that $E = \bigcup_{j=1}^{\infty} E_j$ and $\Lambda(E_j) < \infty$ for each $j = 1, 2, \ldots$. On each subspace E_j, we independently construct a Poisson process with intensity measure $\Lambda_j(\cdot) = \Lambda(\cdot \cap E_j)$. It is important to ensure that the simulations on each subspace are independent. The resulting process N on E is the sum of Poisson processes on E_j, $j = 1, 2, \ldots$.

Example 8.2.7 Suppose $E = \mathbb{R}^2$ and $\Lambda(A) = \lambda \mathrm{Vol}(A)$ for all Borel sets A, where $\lambda > 0$ is the intensity and $\mathrm{Vol}(A)$ denotes the Lebesgue measure of A. Clearly, $\mathrm{Vol}(\mathbb{R}^2) = \infty$. To handle this, we partition $\mathbb{R}^2$ into disjoint squares:

$$E_{ij} = \{(x_1, x_2) : i \leq x_1 < i+1,\ j \leq x_2 < j+1,\ i, j \in \mathbb{Z}\}.$$

On each square E_{ij}, we simulate a Poisson process with intensity λ. Combining these independent processes yields a homogeneous Poisson process with intensity λ over the entire plane $\mathbb{R}^2$. $\Diamond$

8.3 Birth and Death Processes and Markovian Queues

In this section, we adopt a standard approach commonly found in the simulation literature to construct a stochastic process $(X(t) : 0 \leq t \leq t_{\max})$, known as a *birth and death process*. For our purposes, we restrict the state space of the process to $\mathbb{Z}_+$, representing non-negative integers.

A birth and death process is characterized by:

- *birth rates* λ_j, defined for $j = 0, 1, \ldots$,
- *death rates* μ_j, defined for $j = 1, 2, \ldots$, with $\mu_0 = 0$.

Realizations of these processes are assumed to be right-continuous with left-hand limits. State changes occur at times $0 < A_1 < A_2 < \cdots$, representing either an upward jump (*birth*) or a downward jump (*death*). The simulation spans the time horizon $[0, t_{\max}]$. Algorithm 78 outlines the procedure for generating a realization of the process $X(t)$, starting from an initial state $X(0) = i$. Specifically, once the process reaches state i, it remains there for a duration that is exponentially distributed with rate $\lambda_i + \mu_i$. After this period, it transitions to state $i + 1$ with probability $\frac{\lambda_i}{\lambda_i + \mu_i}$ or to state $i - 1$ with probability $\frac{\mu_i}{\lambda_i + \mu_i}$.
The output of the algorithm is the list $\mathtt{JS}$, which is represented as:

$$\mathtt{JS} = [(a_0, y_0), (a_1, y_1), \ldots, (a_k, y_k)],$$

where $k = \max\{i : a_i \leq t_{\max}\}$ is the random number of state transitions. Using this sequence, we construct a realization $x(t)$ of the birth and death process $(X(t))$ as:

$$x(t) = \begin{cases} y_0 & \text{if } a_0 \leq t < a_1, \\ y_i & \text{if } a_i \leq t < a_{i+1}. \end{cases}$$

Here, $a_0 = 0$ and y_0 is the initial state of the process.

Algorithm 78 Algorithm BD; generating B&D process on $[0, t_{\max}]$

Require: Parameters $(\lambda_k, \mu_k)_{k\geq 0}$; initial state $y_0 = i$
1: Create an empty list JS
2: Set $t = 0$
3: **while** $t \leq t_{\max}$ **do**
4:     JS.append($[t, i]$)
5:     Generate U_1, U_2 i.i.d. $\mathcal{U}[0, 1)$
6:     **if** $U_1 \leq \lambda_i/(\lambda_i + \mu_i)$ **then** $\xi = +1$ **else** $\xi = -1$ **end if**
7:     Set $\tau = -\log(U_2)/(\lambda_i + \mu_i)$ and $t = t + \tau$ $\triangleright\ \tau \sim \mathrm{Exp}(\lambda_i + \mu_i)$
8:     Set $i = (i + \xi)_+$ $\triangleright$ Next state
9: **end while**
10: **return** JS

Steady-state Characteristics We briefly recall key concepts about *discrete-time Markov chains* discussed earlier in Sect. 6.2. Assuming the chain (X_k) is aperiodic, irreducible, and, in the case of an infinite state space, positive recurrent, it converges to the limiting (stationary) distribution π. Formally:

$$\lim_{k \to \infty} \mathbb{P}(X_k = i) = \pi(i).$$

Let X^* denote a random variable with stationary distribution π. The *steady-state mean* and *steady-state variance* of the chain (X_k) are defined as $\mathbb{E}[X^*]$ and $\mathbb{V}\mathrm{ar}[X^*]$, respectively. Markov chain theory guarantees:

$$\mu = \mathbb{E}[X^*] = \lim_{k \to \infty} \frac{1}{k} \sum_{j=1}^{k} X_j, \qquad \sigma^2 = \mathbb{V}\mathrm{ar}[X^*] = \lim_{k \to \infty} \frac{1}{k} \sum_{j=1}^{k} X_j^2 - \mu^2,$$

allowing us to define (biased) estimators:

$$\hat{\mu}(k) = \frac{1}{k} \sum_{j=1}^{k} X_j, \qquad \hat{\sigma}^2(k) = \frac{1}{k} \sum_{j=1}^{k} (X_j^2 - \hat{\mu}(k))^2.$$

Note that, unlike in cases discussed in Chap. 5 (on variance reduction techniques), we do not have independent replications here. Instead, we estimate μ and σ^2 based on a single realization of the chain. These concepts extend to *continuous-time Markov processes* and more general steady-state functions $\mathbb{E}[f(X^*)]$, estimated as:

$$\hat{Y}_R = \frac{1}{R} \sum_{j=1}^{R} Y_j, \quad Y_j = f(X_j).$$

For continuous-time Markovian processes, irreducibility and finite state space ensure a limiting distribution. The steady-state mean and variance are defined as:

$$\mathbb{E}[X^*] = \lim_{t \to \infty} \frac{1}{t} \int_0^t X(s)\,ds, \quad \mathbb{V}\mathrm{ar}[X^*] = \lim_{t \to \infty} \frac{1}{t} \int_0^t (X(s) - \mathbb{E}[X^*])^2\,ds.$$

For infinite state spaces, irreducibility of $X(t)$ is not sufficient to ensure the existence of a limiting distribution. Examples 8.3.1 and 8.3.2 below illustrate such cases. For non-Markovian processes, steady-state characteristics exist under specific conditions, discussed in Sect. 8.7.

Now let us go back to birth and death processes. The next examples present birth and death process for which steady-state characteristics do not exist.

Example 8.3.1 Suppose that $\lambda_i = i^2$ and $\mu_i = 0$. In this case, with probability 1, the process $(X(t))$ will explode in finite time. This occurs because the birth rate grows rapidly as the state increases, causing the process to reach infinity in a finite time. $\Diamond$

Example 8.3.2 For $\lambda_i = \lambda$ and $\mu_i = 0$, we obtain the homogeneous Poisson process studied in Sect. 8.2.1, providing an alternative way to define such processes. Notice that $(X(t))$ diverges to infinity with probability 1 as $t \to \infty$, as no deaths occur and arrivals accumulate indefinitely. $\Diamond$

In this section, we focus on estimating the steady-state mean:

$$\bar{l} = \lim_{t \to \infty} \frac{1}{t} \int_0^t X(s)\,ds.$$

A natural estimator of $\bar{l}$ can be obtained from a single realization $x(t)$ of the process $(X(t))$ over the interval $[0, t_{\max}]$, which may be generated using Algorithm 78:

$$\frac{\int_0^{t_{\max}} x(s)\,ds}{t_{\max}} = \sum_{j=0}^{k} y_j (a_{j+1} - a_j), \tag{8.3.1}$$

where we set $a_{k+1} = t_{\max}$.

We tacitly assume that a variant of the Central Limit Theorem holds, in the form:

$$\sqrt{t} \left(\int_0^t X(s)\,ds - \bar{l} \right) \xrightarrow{\mathcal{D}} \mathcal{N}(0, \varsigma^2),$$

for some asymptotic variance ς^2. A formula for ς^2, along with specific examples, will be derived from the covariance function of a stationary process $(X(t))$ later in Sect. 8.7.1.

Kendall's Notation for Markovian Queues In Kendall's notation, Markovian queues, which are defined by birth and death processes, are denoted as M/M/c/n, where:

- M: defines that the inter-arrival times are exponential, which corresponds to homogeneous Poisson arrivals.
- M: specifies that the service time distribution is exponential.
- c: represents the number of servers.
- n: denotes the total capacity of the system, including servers and buffers. Each server may have a finite or infinite buffer (maximum queue length), and n is the sum of all buffers and servers.

Thus, the state space (total number of tasks in the system) is $\{0, 1, \ldots, n\}$. If only three symbols are used, M/M/c, it denotes that all buffers are infinite, making $n = \infty$. For example, M/M/∞ represents a system with infinitely many servers, and thus infinite capacity.

In Sect. 8.4 we outline key concepts of Kendall's notations for general queues. Some classical queueing systems modeling birth and death processes include:

- M/M/1/n if $\lambda_i = \lambda$, $i = 0, \ldots, n-1$ and $\mu_i = \mu$, $i = 1, \ldots, n$,
- M/M/1 if $\lambda_i = \lambda$, $i = 0, \ldots$ and $\mu_i = \mu$, $i = 1, \ldots$,
- M/M/c if $\lambda_i = \lambda$, $i = 0, \ldots$, and $\mu_i = (i \wedge c)\mu$, $i = 1, \ldots$,
- M/M/c/c if $\lambda_i = \lambda$ for $i \leq c$ and 0 otherwise, and $\mu_i = i\mu$, $i = 1, 2, \ldots, c-1$,
- M/M/∞ if $\lambda_i = \lambda$, $i = 0, 1, \ldots$ and $\mu_i = i\mu$, $i = 1, \ldots$.

From now on, we will denote the total number of tasks in the system by $L(t)$ (instead of the general process $X(t)$), following the traditional notation used in queueing theory.

Our primary focus will be on certain queueing characteristics, such as the steady-state mean queue length. It is important to note—as discussed earlier—that the limiting distribution of $L(t)$ does not necessarily exist in general. In such cases, we mean that the steady-state mean does not exist.

M/M/1 Queueing System Simulations The behavior of realizations in an M/M/1 queueing system is closely related to the ratio $\rho = \lambda/\mu$. Specifically, when $\rho < 1$, the realizations remain bounded, ensuring that steady-state conditions hold. This can be observed in the simulation results presented in Example 8.3.3. From queueing theory, it is known that under the condition $\rho < 1$, the mean queue length $\bar{l}$ satisfies, with probability 1:

$$\lim_{t \to \infty} \frac{\int_0^t L(s)\, ds}{t} = \bar{l} = \frac{\lambda}{\mu - \lambda},$$

see Exercise 8.T.10. Further analysis of M/M/1 system simulations, including error analysis, will be discussed later in Sect. 8.7.1.

Example 8.3.3 (M/M/1 Queue) Consider the M/M/1 queue and the process $L(t)$, which represents the number of tasks in the system. The process $L(t)$ in an M/M/1 queue is modeled as a birth-and-death (B&D) process with state space $\{0, 1, \ldots\}$, constant birth rate λ, and constant death rate μ. Stability of the system requires $\lambda < \mu$.

We simulate $L(t)$ over the time horizon $t_{\max} = 10^4$, starting with $L(0) = 5$, and analyze the following cases, all with $\mu = 1$:

(a) $\lambda = \rho = 1/2$. An illustrative simulation is displayed in Fig. 8.4a. The estimated steady-state mean is $\hat{l} = 0.9822$, close to the theoretical value $\bar{l} = \frac{\lambda}{\mu - \lambda} = 1$, the relative error is 1.7785%. For $t_{\max} = 10^5$ (not shown) the relative error is 1.6705%.

(b) $\lambda = \rho = 8/9$. An illustrative simulation is displayed in Fig. 8.4b. The estimated steady-state mean is $\hat{l} = 6.669786$, close to the theoretical value $\bar{l} = 8$, the relative error is 16.6276%. For $t_{\max} = 10^5$ (not shown) the relative error is 2.25305%.

(c) $\lambda = \rho = 1$. An illustrative simulation is displayed in Fig. 8.4c . The estimated mean is $\hat{l} = 40.1226$. Note that $\rho = 1$ implies the system is critically loaded, leading to instability and a steady-state mean that diverges.

(d) $\lambda = \rho = 1.2$. An illustrative simulation is displayed in Fig. 8.4d. The estimated mean is $\hat{l} = 971.4601$. Since $\rho > 1$, the system is unstable, resulting in unbounded growth of the queue length.

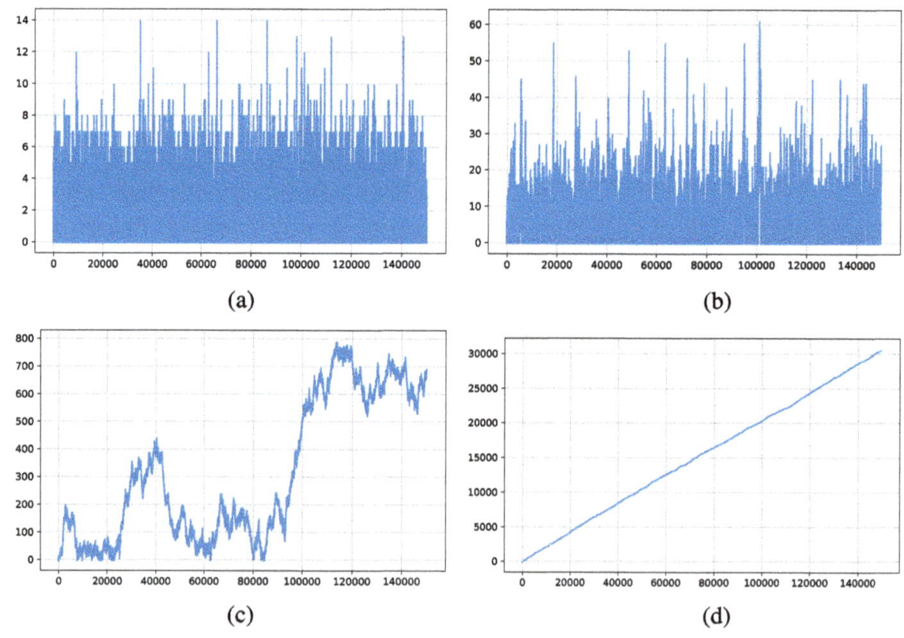

Fig. 8.4 Simulations of $L(t)$ in M/M/1 queue for $\mu = 1$ and different values of λ. (**a**) $\lambda = \rho = 1/2$. (**b**) $\lambda = \rho = 8/9$. (**c**) $\lambda = \rho = 1$. (**d**) $\lambda = \rho = 1.2$

These results highlight the different behaviors of the M/M/1 queue depending on the traffic intensity ρ:

- When $\rho < 1$, the system is stable, and the estimated steady-state mean aligns well with the theoretical value $\bar{l}$.
- At $\rho = 1$, the system becomes critically loaded, and the queue length exhibits instability, leading to a divergence in the steady-state mean.
- When $\rho > 1$, the system is unstable, and the queue length grows without bound.

The example is implemented in `ch8_mm1_queue.py`. $\Diamond$

8.4   Basic Concepts of Queueing Theory

A queueing system is characterized by an arrival process of tasks (e.g., calls) that bring tasks to be completed. It consists of a number of service channels ($1 \leq s \leq \infty$), often referred to as servers or processors, along with a specified service discipline. If a channel is free, an arriving task is immediately served. However, if all channels are occupied, tasks may either wait in a waiting room or be lost. Tasks that cannot be accommodated in the waiting room are lost.

The key components of such a system are:

- **Arrival instants:** $0 \leq A_1 \leq A_2 \leq \cdots$, where the i-th task arrives at time A_i. Multiple arrivals at the same instant, known as batch arrivals, are possible.
- **Service times:** $S_1, S_2, \ldots$, where the i-th task requires a service time of S_i.
- **Service channels:** The resources (servers) available to process the tasks.
- **Capacity of the waiting room (buffer):** $1 \leq k \leq \infty$, representing the number of tasks that can wait in line if all servers are occupied.
- **Service discipline:** The rule that determines the order in which tasks are served, such as First Come First Served (FCFS) or Last In First Out (LIFO).

In what follows, unless otherwise stated, considered systems work according to the First Come First Served (FCFS) discipline.

It is important to note that, unlike the birth and death processes considered in Sect. 8.3, the distributions of inter-arrival and service times in queueing systems do not necessarily have to be exponential.

Kendall's Notation for Queueing Systems   Queueing systems are often described using the notation A/S/c/n, which was introduced in Sect. 8.3 for the specific case where A=M and S=M (i.e., inter-arrival and service times are exponentially distributed). The remaining cases for A and S are as follows:

- A=GI—Arrival times follow a renewal process.
- A=D—Inter-arrival times are deterministic.
- S=G—Service times follow a general distribution G.
- S=D—Service times are deterministic.

We say that arrival times A_i follow a *renewal process* if the inter-arrival times

$$\tau_1 = A_1, \quad \tau_i = A_i - A_{i-1}, \quad i = 2, \dots$$

are i.i.d. This concept already appeared in the context of a homogeneous Poisson process (see (8.2.1)). For example, GI/G/1 refers to a single-server system with a renewal arrival process and general service times. Standard queueing characteristics considered here include:

- the number of tasks in the system, and
- the waiting time (or sojourn time) of the i-th task in discrete time.

Figures 8.5 and 8.6 illustrate typical queueing systems with c servers.

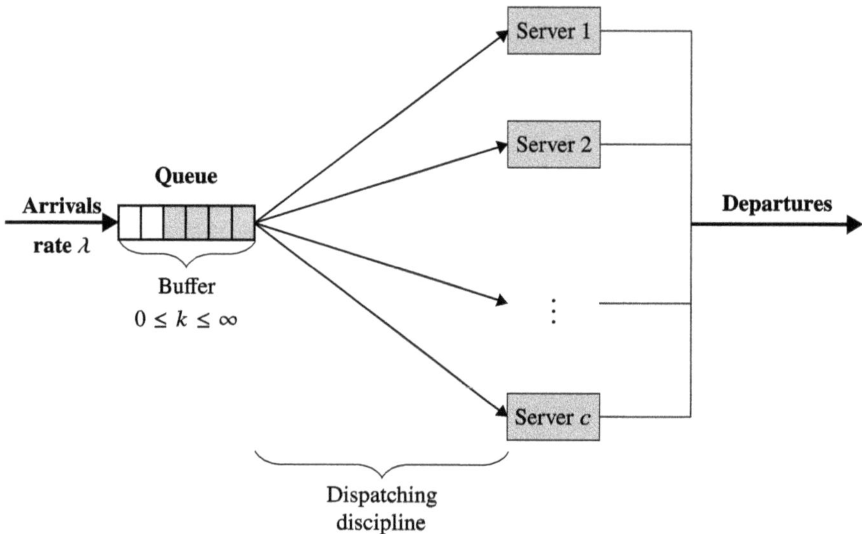

Fig. 8.5 Sample queueing system: c servers, one common queue

Arrival and service times are random, with their distributions specified accordingly. We consider two types of characteristics:

- **Discrete-time characteristics:** Often related to the sequence of arrivals and departures.
- **Continuous-time characteristics:** Typically focused on the number of tasks in the system, represented by the process $L(t)$.

We assume that all processes are *càdlàg* (right-continuous with left-hand limits). When the i-th task of size S_i arrives, it waits in the waiting room for a time W_i. Thus, it stays in the system during the inteval $[A_i, D_i)$, where $D_i = A_i + W_i + S_i$ (departure of the i-th task) The quantities W_i and $W_i + S_i$ are referred to as the *actual waiting time* and the *sojourn time*, respectively, of the i-th task.

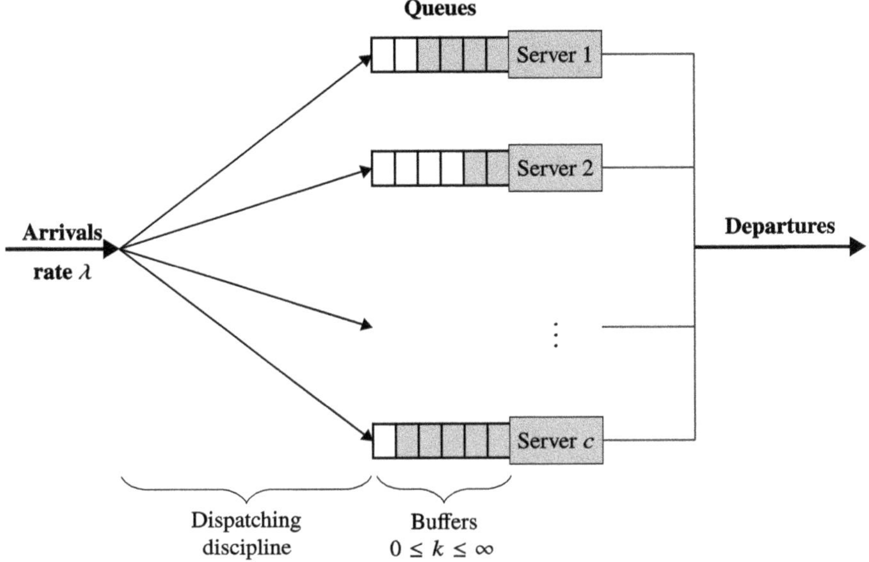

Fig. 8.6 Sample queueing system: c servers, c queues

To simulate such systems, we also require an initial condition. The simplest case assumes $L(0) = 0$, meaning the system is initially empty. In this case, the number of tasks in the system at time t is given by:

$$L(t) = \sum_{i=1}^{\infty} \mathbb{1}(A_i \leq t < D_i), \qquad t \geq 0.$$

As observed, $L(t)$ increases (with jumps) at arrival instants and decreases at departure instants. For simplicity, we assume:

- No multiple arrivals occur, so arrival times are strictly increasing, denoted by $\{0 < A_1 < A_2 < \cdots\}$.
- The service time distribution is continuous.

Single-server System GI/G/1 This is a single-server system with general inter-arrival times and service times. Specifically:

- The arrival instants $A_1 < A_2 < \cdots$ are determined by inter-arrival times (τ_k), where $A_1 = \tau_1$, $A_2 = \tau_1 + \tau_2$, and so on.
- The inter-arrival times (τ_k) are independent random variables with a common general distribution F.
- The service times (S_k) are i.i.d. random variables with a common general distribution G.

Clearly, τ_i, $S_i \geq 0$.

Efficient Estimation of Steady-state Mean Waiting Time For most steady-state characteristics of the system (e.g., steady-state queue length), simulating the entire system is necessary, especially when service disciplines differ from First Come First Served (FCFS), such as Last In First Out (LIFO). However, for specific metrics like the steady-state mean waiting time $\bar{w}$, it is possible to directly use a simple recurrence to sample i.i.d. waiting times $W_1, \ldots, W_n$ and estimate their mean. This approach is valid only under the FCFS discipline.

Consider the waiting time W_i of the i-th task under the FCFS discipline. Suppose the system is initially empty, and the first task arrives at A_1. Then $W_1 = 0$. For subsequent tasks, the waiting times are determined by the recurrence:

$$W_{i+1} = (W_i + S_i - \tau_{i+1})_+. \tag{8.4.1}$$

This recurrence is valid for any sample path and does not rely on specific distributional assumptions. Under these assumptions, the sequence W_i is Markovian. If $\mathbb{E}[S_1] < \mathbb{E}[\tau_1]$, then the distribution of W_i converges to a steady-state distribution, i.e., $\lim_{k \to \infty} \mathbb{P}(W_k \leq x)$ exists. Moreover, if $\mathbb{E}[S_1^2] < \infty$, the steady-state mean waiting time satisfies:

$$\bar{w} = \lim_{i \to \infty} \mathbb{E}[W_i] < \infty.$$

With probability 1:

$$\bar{w} = \lim_{k \to \infty} \frac{1}{k} \sum_{i=1}^{k} W_i.$$

Thus, to estimate $\bar{w}$, we can sample $W_1, \ldots, W_k$ from recurrence (8.4.1) and compute:

$$\hat{w} = \frac{1}{k} \sum_{i=1}^{k} W_i. \tag{8.4.2}$$

The Pollaczek-Khinchin formula provides the steady-state mean waiting time for M/G/1 queues when inter-arrival times τ_j are exponentially distributed with $\mathbb{E}[\tau_1] = 1/\lambda$. It states:

$$\bar{w} = \frac{\lambda \mathbb{E}[S_1^2]}{2(1 - \lambda \mathbb{E}[S_1])}. \tag{8.4.3}$$

Note that computing (W_i) from recurrence (8.4.1) also requires generating A_i, S_i, τ_i, which enables us to calculate additional system metrics. For instance, the number of tasks in the system at any time t can be computed as:

$$L(t) = \sum_{i=1}^{\infty} \mathbb{1}(A_i \le t < A_i + W_i + S_i).$$

The steady-state mean number of tasks $\bar{l}$ can be estimated via:

$$\hat{l}(t_{\max}) = \frac{1}{t_{\max}} \int_0^{t_{\max}} L(s)\, ds = \frac{1}{t_{\max}} \sum_{A_i \le t_{\max}} \min(A_i + W_i + S_i, t_{\max}) - A_i.$$

Additionally, we may be interested in the *virtual waiting time (workload)* defined as:

$$V(t) = \sum_{i=1}^{\infty} (W_i + S_i - t)_+ \mathbb{1}(A_i \le t < A_{i+1}), \quad t \in [0, t_{\max}]. \tag{8.4.4}$$

This metric can also be obtained from the simulated sequences (S_i), (A_i), and (W_i).

Systems with an Infinite Number of Servers; M/G/∞ Systems   In such systems, each arriving task is immediately assigned to a server. There is no waiting time. The sojourn time, which is the duration from the arrival instant to the departure of the task, is simply its service time.

We can inquire about the number $L(t)$ of tasks in the system at time t. Tasks arrive at times $0 \le A_1 < A_2 < \cdots$ with sizes $S_1, S_2, \ldots$. Thus, the i-th task remains in the system during the interval $[A_i, A_i + S_i)$. Suppose that $L(0) = 0$. Formally, we can define:

$$L(t) = \sum_{i=1}^{\infty} \mathbb{1}(A_i \le t < A_i + S_i), \quad t \ge 0.$$

A special case of interest is the M/G/∞ system. In such systems, arrivals form a homogeneous Poisson process with intensity λ. Each arrival brings work to the system, with task sizes $S_1, S_2, \ldots$, which are i.i.d. with distribution G. Assume $\mathbb{E}S_1 = \mu^{-1} < \infty$.

For this system, we can obtain explicit formulas for the steady-state number of tasks in the system. When considering steady-state quantities like L^* (the number of tasks in the system), only $\rho = \lambda \mathbb{E}S = \lambda/\mu$ matters, regardless of the distribution G. It is known that:

$$\mathbb{E}L^* = \lim_{t \to \infty} \frac{1}{t} \int_0^t \mathbb{1}(L(t) = k) = \frac{\rho^k}{k!} e^{-\rho}, \tag{8.4.5}$$

$$\lim_{t \to \infty} \frac{1}{t} \int_0^t L(t) = \rho, \tag{8.4.6}$$

both holding with probability 1. Similarly, it is known that:

$$\lim_{t \to \infty} \mathbb{P}(L(t) = k) = \frac{\rho^k}{k!} e^{-\rho}, \tag{8.4.7}$$

$$\lim_{t \to \infty} \mathbb{E}L(t) = \rho. \tag{8.4.8}$$

We will show simulations of M/G/∞ systems later in Sect. 8.9.

8.5 Discrete-event Simulations

Many processes in management science, operations research, and telecommunications exhibit a special structure that allows the use of specific simulation methods. These systems evolve from event to event, with no changes occurring between events. Processes such as birth and death processes or M/G/∞ systems are examples of systems that can be simulated using this approach. In such systems, we employ *discrete-event simulation* to study their performance. To illustrate this concept, we present a simple example.

Example 8.5.1 (*On-off* Process) The simplest process to simulate and analyze is the so-called *on-off* process. This is a process that takes values in $E = \{0, 1\}$, alternating between 0 and 1, representing *on* times and *off* times. It is a basic model with many known system characteristics. We assume that the *on* times and *off* times are independent, with *on* times following distribution F_{on} and *off* times following distribution F_{off}.

The system stays in state 1 during an *on* time and in state 0 during an *off* time. Suppose that at time $t = 0$, the system starts in state 1 (*on*). The length of time it stays *on* is sampled from the distribution F_{on}. When this time ends, an event $\mathcal{A}_1 = \{\text{switching from } on \text{ to } off\}$ occurs, and the system switches to state 0. The system then remains in state 0 for a duration sampled independently from F_{off} until another event $\mathcal{A}_0 = \{\text{switching from } off \text{ to } on\}$ occurs, and the system switches back to state 1.

Formally, we define the *on-off* process as follows. Let (T_i^{on}) be an i.i.d. sequence of *on*-time durations with distribution F_{on}, and (T_i^{off}) an i.i.d. sequence of *off*-time durations with distribution F_{off}, where both sequences are independent. Since we start in state 1 at time zero, define the consecutive event times as: $S_0 = 0$,

$$S_n = \begin{cases} S_{n-1} + T_k^{\mathrm{on}} & \text{if } n = 2k - 1, \\ S_{n-1} + T_k^{\mathrm{off}} & \text{if } n = 2k. \end{cases}$$

The *on-off process* $K(t)$ is then formally defined as:

$$K(t) = \begin{cases} 1 \text{ for } S_{n-1} \le t < S_n \quad \text{and } n \text{ is odd,} \\[2mm] 0 \text{ for } S_{n-1} \le t < S_n \quad \text{and } n \text{ is even.} \end{cases}$$

We assume that F_{on} and F_{off} are continuous with finite means. To simulate this process, we need to keep track of the time of the current state change, its type, and the time until the next event. The times of change define the *clock*, which keeps track of the current simulation time, and the pending time, which is the time until the next event.

One interesting characteristic is the steady-state probability

$$p = \lim_{t \to \infty} \mathbb{P}(K(t) = 1)$$

that the system is *on*. It is known that (recall, we assumed that $\mathbb{E}T^{\text{on}} < \infty$ and $\mathbb{E}T^{\text{off}} < \infty$)

$$p = \frac{\mathbb{E}T^{\text{on}}}{\mathbb{E}T^{\text{on}} + \mathbb{E}T^{\text{off}}}.$$

To estimate p by simulations, we run the system up to time $t_{\max}$ and define the estimator:

$$\hat{p}(t_{\max}) = \frac{1}{t_{\max}} \int_0^{t_{\max}} K(s)\, ds. \tag{8.5.1}$$

It is known that for continuous distributions F_{on} and F_{off}, this estimator is strongly consistent, though in general it is not unbiased. A sample *on-off* process is depicted in Fig. 8.7. $\qquad\qquad\Diamond$

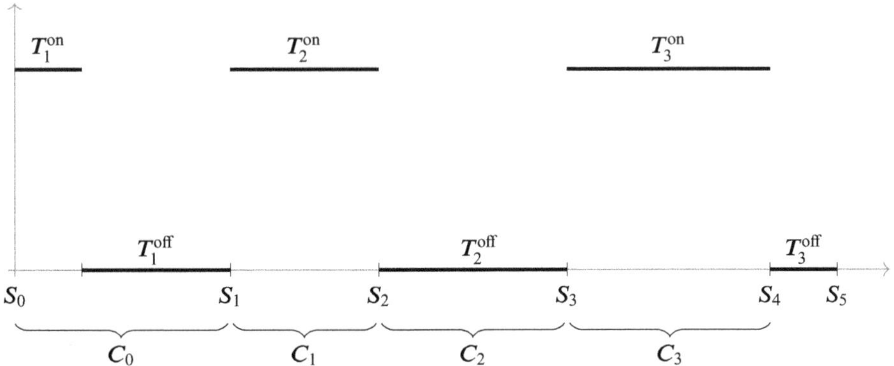

Fig. 8.7 Sample *on-off* process with $K(0) = 1$ (*on*). Regeneration cycle lengths $C_0, \dots, C_5$ depicted (see Sect. 8.8)

Example 8.5.2 (Repairman Problem) In a device, there are N elements, each with a working time distributed according to F. When an element breaks, it must be repaired. The system has M repairmen, each responsible for fixing a failed element. The repair time follows the distribution G, and after being repaired, the element is immediately ready to work again. If there are more than M failed elements, the additional ones must wait for a repairman to become available.

The system tracks the number of working elements $L(t)$. The number of failed elements is given by $K(t) = N - L(t)$. The failed elements can be divided into:

- **Elements Under Repair**: Let $s = \min(K(t), M)$ be the number of elements currently being repaired.
- **Elements Waiting for Repair**: The remaining $q = K(t) - s$ failed elements are waiting in a queue for a repairman to become available.

Define:

- l is the steady-state mean number of working elements,
- m is the steady-state mean number of elements being serviced ($m = \mathbb{E}[s]$),
- w is the steady-state mean number of elements waiting for service ($w = \mathbb{E}[q]$).

Clearly, we have:

$$l + m + w = N.$$

For example, say we have $N = 5$ elements and $M = 1$ repairman. If there are currently $L(t) = 3$ working elements, then there are $K(t) = N - L(t) = 5 - 3 = 2$ elements failed, $s = \min(K(t), M) = \min(2, 1)$ is being repaired and $q = K(t) - s = 2 - 1 = 1$ is waiting for a repair. This is the situation in Fig. 8.8 in intervals (t_2, t_3), (t_4, t_5) and (t_6, t_7).

One may also be interested in the steady-state mean number of unoccupied repairmen, denoted by $r = M - m$.

The process $L(t)$ changes due to failures and repairs of elements. To simulate $L(t)$—we will record only the times at which some event occurs and the corresponding number of elements—the pending event list in the simulation should include:

- The actual times to failure of each working element: $\tau_1^{\mathrm{f}}, \ldots, \tau_k^{\mathrm{f}}$, where $k = L(t)$.
- The actual times to repair completion for each element under repair: $\tau_1^{\mathrm{r}}, \ldots, \tau_s^{\mathrm{r}}$, where $s = \min(K(t), M)$.

By "actual", we mean the time relative to the current simulation clock. Discrete-event simulation is an excellent tool for simulating the process $L(t)$ over the time interval $[0, t_{\max}]$, we will present a pseudocode for such a simulation. We use the following data structures:

- EL—an event list (priority queue)—each element consists of a triple (event time, event type, element identifier), where event type is `Failure` or `Repair_Completion`,
- Queue—a queue of failed elements waiting for repair.

- JS—a list of pairs $[t(i), k(i)]$, recording the times of events and the number of working elements after each event.

At each event time, the simulation processes either a failure or a repair completion event. For a failure event, the number of working elements decreases, and if a repairman is available, the failed element begins repair; otherwise, it is added to the repair queue. For a repair completion event, the number of working elements increases, and if there are elements waiting for repair, the next one is assigned to a repairman. We assume that at time $t = 0$, all N elements are working, so $k = N$, $s = 0$, and $q = 0$. We initialize the failure times for all elements. The procedure is depicted in Algorithm 79.

Algorithm 79 Discrete-event simulation of $L(t)$ in the Repairman Problem

---

Require: N, M, $t_{\max}$, distributions F and G
1: $t := 0, k := N, s := 0, q := 0$
2: Set: EL = empty list, Q = empty queue, JS $:= [[0, k]]$
3: **for** $i = 1$ **to** N **do**
4:     Generate working time $\tau_i^{\mathrm{f}} \sim F$
5:     EL.append($[t + \tau_i^{\mathrm{f}}, \mathtt{Failure}, i]$)
6: **end for**
7: **while** $t < t_{\max}$ **and** EL is not empty **do**
8:     $[t_{\mathrm{event}}, \mathtt{event\_type}, i] \leftarrow$ Remove event with smallest time from EL
9:     $t := t_{\mathrm{event}}$
10:     JS.append($[t, k]$)   ▷ Record the current state
11:     **switch** $\mathtt{event\_type}$ **do**
12:         **case** $\mathtt{Failure}$
13:             $k := k - 1$ ▷ One element has failed
14:             **if** $s < M$ **then**   ▷ Repairman is available
15:                 $s := s + 1$
16:                 Generate repair time $\tau^{\mathrm{r}} \sim G$
17:                 EL.append($[t + \tau^{\mathrm{r}}, \mathtt{Repair\_Completion}, i]$)
18:             **else**   ▷ All repairmen are busy
19:                 $q \leftarrow q + 1$
20:                 Q.enqueue(i)
21:             **end if**
22:         **case** $\mathtt{Repair\_Completion}$
23:             $k := k + 1, \quad s := s - 1$ ▷ One element is repaired and working
24:             Generate working time $\tau_i^{\mathrm{f}} \sim F$
25:             EL.append($[t + \tau_i^{\mathrm{f}}, \mathtt{Failure}, i]$)
26:             **if** $q > 0$ **then**   ▷ Elements are waiting for repair
27:                 $j \leftarrow$ Q.dequeue()
28:                 $q \leftarrow q - 1, \quad s \leftarrow s + 1$
29:                 Generate repair time $\tau^{\mathrm{r}} \sim G$
30:                 EL.append($[t + \tau^{\mathrm{r}}, \mathtt{Repair\_Completion}, j]$)
31:             **end if**
32: **end while**
33: JS.append($[t_{\max}, k]$)   ▷ Record the final state

In Fig. 8.8, a sample repairman problem is depicted.

Note that using Algorithm 79 we may estimate the steady-state mean of the number of working elements l, the steady-state mean of the number of elements queueing for the service w, and the steady-state mean of the number of elements queueing for the service w respectively by

$$\hat{l} = \frac{1}{t} \int_0^t L(s)\,ds, \quad \hat{m} = \frac{1}{t} \int_0^t (N - L(s))\,ds, \quad \hat{w} = N - \hat{l} - \hat{m}.$$

Similarly, we may estimate the steady-state mean number of unoccupied repairmen using $\hat{r} = M - \hat{m}$. However, some modifications would be needed to estimate

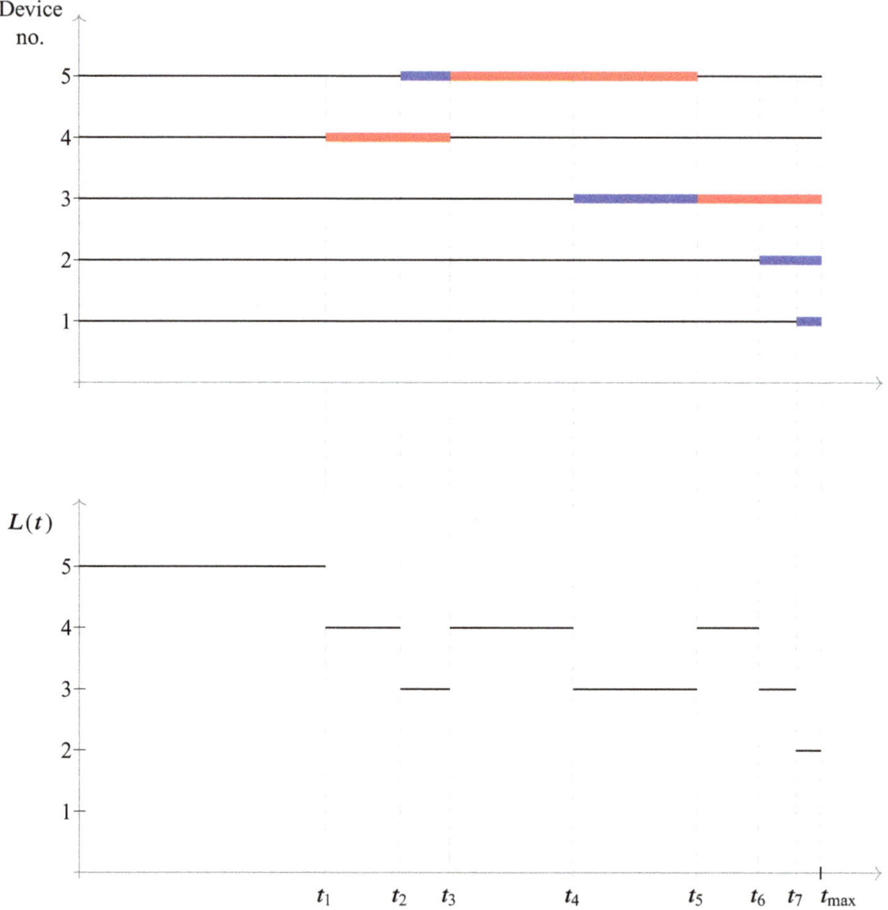

Fig. 8.8 Repairman problem: $N = 5$ elements, $M = 1$ repairman. Top: status of each device: *black line*=working, *red*=under repair, *blue*=waiting for repair(man). Bottom: Number of working devices $L(t)$

the steady-state mean sojourn time, i.e., the average time from failure to repair completion (including waiting and servicing) of an element, see Exercise 8.L.6.

Exponential Working and Repair Times   Note that if the working times of elements are exponentially distributed with parameter λ, and the repair times are exponentially distributed with parameter μ, the simulation of the process $L(t)$ is significantly easier, since it is a birth and death process. If we have k working elements, the death rate (failure rate) is $\lambda_k = k\lambda$, and the birth rate (repair completion rate) is $\mu_k = \min(N - k, M)\mu$. We thus may simply use Algorithm 78 then. $\Diamond$

Having explored specific instances of discrete-event systems, such as the *on-off* process and the repairman problem, we now turn to a general framework that underpins these simulations. This framework provides a systematic approach to modeling and analyzing a wide array of discrete-event systems.

General Discrete-event Simulation Framework   A general description of the method is as follows. Suppose we want to simulate n—the *system state*—up to time $t_{\max}$, the simulation horizon. The only stochastic changes in the system occur at specific times $T_1 < T_2 < \cdots$. Between these times, the system behaves deterministically; we assume it remains constant, although some models allow for changes, known as piecewise deterministic Markov processes.

Algorithm 80 General Scheme for Discrete-Event Simulation on $[0, t_{\max}]$

1: Set $t := 0$
2: Initialize system state and counters
3: Initialize pending event list EL with initial events
4: Initialize data recording structures (e.g., JS for state changes)
5: **while** $t < t_{\max}$ **and** EL is not empty **do**
6:     Select the next event E with the smallest scheduled time from EL
7:     Update simulation time $t \leftarrow$ time of event E
8:     Record current system state in JS
9:     **switch** event_type E **do**
10:        **case** event_type_1
11:            Update system state and counters accordingly
12:            Generate and add new events to EL
13:        **case** event_type_2
14:            Update system state and counters accordingly
15:            Generate and add new events to EL
16:        $\vdots$
17:        **case** event_type_N
18:            Update system state and counters accordingly
19:            Generate and add new events to EL
20: **end while**
21: **return** output $\triangleright$ e.g., recorded metrics from JS

Suppose we have N event types, denoted $\mathcal{A}_i$ ($i = 1, \ldots, N$). The idea is to create a *pending list*, which records the moments when the next events will occur relative to the current clock time. For an event $\mathcal{A}_i$, the next occurrence time is recorded as $t_{\mathcal{A}_i}$. If an event cannot occur in the next step, its time is not recorded in the pending list. For example, this can happen when the event time is beyond the given time horizon $t_{\max}$.

The algorithm finds the nearest event, updates the time variable, performs the necessary actions, and so on. Of course, appropriate assumptions regarding independence are necessary for the simulation to be valid. The general procedure for discrete-event simulation is summarized in Algorithm 80.

Having established the general framework for discrete-event simulation, we now delve into a specific application by simulating a single server system.

Discrete-event Simulation of GI/G/1 Note that under the FCFS discipline, many characteristic can be estimated from sequences (S_i), (A_i), (τ_i) and (W_i) obtained from recurrence (8.4.1). However, here we will illustrate how to estimate the characteristics using the discrete-event simulation method. Note that in the case of some changes, e.g., from FCFS to LIFO, the recurrence (8.4.1) may not hold, but discrete-event simulations would be very similar (thus, estimations of the characteristics would be possible using only this method).

The sojourn time of the i-th task is $W_i + S_i$. We approach the theory in an algorithmic way, using discrete-event simulation. Let $t_{\max}$ be the simulation horizon. Thus, we consider:

- the number of tasks in the system, $L(t)$, at time t for $0 \le t \le t_{\max}$;
- the sequence of sojourn times $W_i + S_i$.

If we are interested only in waiting times (e.g., in estimating the mean waiting time), one may directly use the recurrence (8.4.1) to sample $W_1, W_2, \ldots$. For estimating other characteristics, such as the mean number of customers in the system or the mean sojourn time, the discrete-event method provides a way to simulate the process.

We now present fragments of the algorithm for simulating the number of tasks $L(t)$ in a GI/G/1 system. We assume that the service discipline is *work-conserving*, meaning the system always tries to keep the processor(s) busy if there are tasks ready to be served. Moreover, the service discipline is First Come First Served (FCFS), also known as First In First Out (FIFO) in single-server queues. There are two events:

- *Arrival—A*;
- *Departure—D*.

The algorithm relies on the *pending event list—EL*, which is of the form $[t_A, t_D]$, where $t_A \in (0, \infty)$ and $t_D \in \{\emptyset\} \cup (0, \infty)$. Here, t_A is the next arrival time, and t_D is the next departure time if $t_D \ne \emptyset$. If $t_D = \emptyset$, it means that the queue is currently empty, and the next event is an arrival (its service time will be generated upon arrival).

Table 8.1 Exemplary data
for FCFS queue

Task nr i	τ_i	A_i	S_i	W_i	D_i
1	1.0	1	2.5	0	3.5
2	1.5	2.5	2.5	1	6
3	4.5	7	0.5	0	7.5
4	1.0	8	5	0	13
5	1.0	9	1	4	14

We denote by

- t—the time variable (clock);
- n—the number of tasks in the system.

Data is collected in the output variables:

- $t(i), n(i)$—the moment of the i-th consecutive event and the state immediately after this moment, collected in the list JS of pairs [jump_times, system_sizes]. The first element of JS is always [0,0].

Suppose F is the distribution of the inter-arrival time τ, and G is the distribution of the task size S. For example, in the M/M/1 system, F is exponential with parameter λ, and G is exponential with parameter μ. We assume that F and G are continuous distributions. Knowledge of JS $= [[0, 0], [t(1), n(1)], \ldots, [t_{\max}, n(K)]]$ allows us to construct a simulated realization of $L(s)$ for $0 \le s \le t_{\max}$, where K denotes the random number of events.

At the beginning, we assume that at $t = 0$, the system is empty ($n = 0$). In this case, EL $= [t_A, \emptyset]$. If $t_A < t_{\max}$, we simulate its service time (and compute its departure time) and generate the time for the next arrival. If $t_A \ge t_{\max}$, we immediately end the procedure. The complete procedure is provided in Algorithm 81.

Example 8.5.3 In Table 8.1, we present exemplary inter-arrival times τ_i and the corresponding service times S_i. The table also includes the computed values of the corresponding arrival times A_i, waiting times W_i, and departure times D_i. These values result in the following list of jump_times, system_sizes, namely JS $=$

$$[[0, 0], \ [1, 1], \ [2.5, 2], \ [3.5, 1], \ [6, 0], \ [7, 1],$$

$$[7.5, 0], \ [8, 1], \ [9, 2], \ [13, 1], \ [14, 0], \ [15, 0]].$$

A plot of the queue length, along with detailed information on the system status, is provided in Fig. 8.9. In Fig. 8.10 the corresponding virtual waiting time process is depicted. $\Diamond$

Algorithm 81 Simulation of a single server queue

1: $t_A = 0$
2: Create a list of pairs JS, set JS := $[[0, 0]]$
3: Generate $\tau \sim F$;
4: Set $t_A := \tau$
5: Set EL := $[t_A, \emptyset]$ ▷ Event list
6: **while** $t_A < t_{\max}$ **do**
7:     **switch** EL **do**
8:        **case** EL $= [t_A, \emptyset], t_A \neq \emptyset$
9:           $n := 1$
10:           JS.append($[t_A, n]$)
11:           Generate $S \sim G$, set $t_D := t_A + S$
12:           Generate $\tau \sim F$, set $t_A := t_A + \tau$
13:           Set EL := $[t_A, t_D]$
14:        **case** EL $= [t_A, t_D], t_A, t_D \neq \emptyset$
15:           **if** $t_A < t_D$ **then**
16:              $n := n + 1$
17:              JS.append($[t_A, n]$)
18:              Generate $\tau \sim F$, set $t_A := t_A + \tau$
19:              Set EL := $[t_A, t_D]$
20:           **else if** $t_A \geq t_D$ **then**
21:              $n := n - 1$
22:              JS.append($[t_D, n]$)
23:              **if** $n > 0$ **then**
24:                 Generate $S \sim G$, set $t_D := t_D + S$
25:                 Set EL := $[t_A, t_D]$
26:              **else**
27:                 Set EL := $[t_A, \emptyset]$
28:              **end if**
29:           **end if**
30: **end while**
31: **if** $t_D < t_{\max}$ **then**
32:     $n := n - 1$
33:     JS.append($[t_D, n]$)
34: **end if**
35: JS.append($[t_{\max}, n]$)

Arbitrary Initial Conditions If the system is not empty at $t = 0$, we must know the number of tasks in the system, the time to the next arrival t_A, and the time to complete the current service t_O. This requires the pending event list EL $= [t_A, t_O]$ and the value of n, which represents the number of tasks in the system.

8.5.1 *GI/G/c Queue*

This is a system with c servers, where arrivals occur according to a renewal process. Specifically, the inter-arrival times (τ_i) $(A_1 = \tau_1, A_2 = \tau_1 + \tau_2, \ldots)$ are independent random variables with a common distribution F, and task durations (S_i) form an

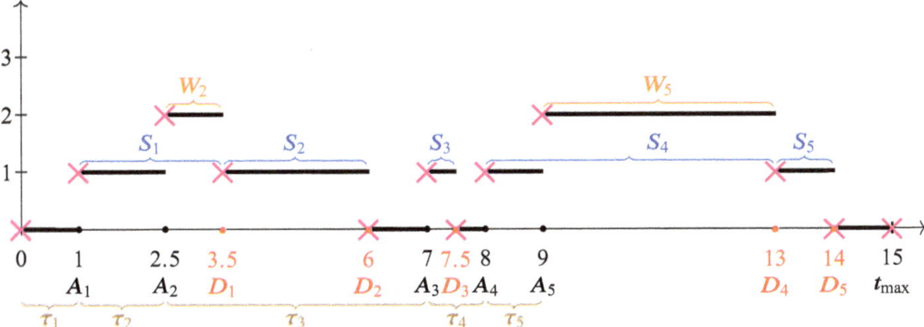

Fig. 8.9 An example path showing the number of customers in a single-server queue with $t_{max} = 15$. Arrivals are shown in black, service times in *blue*, waiting times in *orange* (waiting times equal to zero are not depicted), and departures in *red*. Inter-arrival times are shown in *brown*. Note that $A_i = \tau_1 + \cdots = \tau_i$ and $A_i + S_i + W_i = D_i$. *Magenta crosses (x)* denote the consecutive elements of the list JS, as generated by Algorithm 81. The data is taken from Table 8.1

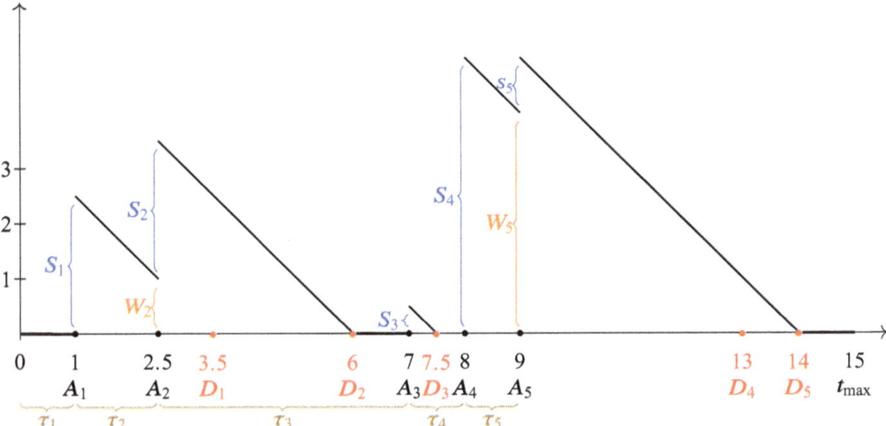

Fig. 8.10 Corresponding virtual waiting time (see definition in (8.4.4))

i.i.d. sequence, independent of the arrivals, with a common distribution G. Clearly, $\tau_i, S_i \geq 0$. Tasks wait for service in an unbounded buffer and are served in the order they arrive.

Recall that in GI/G/1, if we are interested in estimating the steady-state mean waiting time $\bar{w}$, the entire system does not need to be simulated. Instead, we can use the recursion (8.4.1) to sample $W_1, W_2, \ldots$ directly and estimate $\bar{w}$ via the sample mean (8.4.2).

A similar approach applies to GI/G/c, although it is slightly more complex. In this case, there exists a recursion—described below—that allows us to sample random variables whose sample mean provides an estimator of $\bar{w}$.

Suppose the system is initially empty, and the first task arrives at A_1. Clearly, this task does not wait, so $W_1 = 0$. Similarly, for subsequent tasks $i = 2, \ldots, c$, they also do not wait (as long as $c \geq i$), leading to $W_i = 0$ for $i \leq c$.

When the i-th task arrives ($i > c$), it must determine its waiting time. For each server, we compute the potential workload just before the arrival of the i-th task and store these values in a vector $\mathbf{V}_i = (V_{i,1}, \ldots, V_{i,c})$, where the components are ordered increasingly, i.e., $V_{i,1} \leq V_{i,2} \leq \cdots \leq V_{i,c}$. According to the FCFS discipline, the waiting time W_i of the i-th task is equal to the smallest workload among all servers, i.e.,

$$W_i = V_{i,1}.$$

The i-th task then adds its service time S_i to the workload of the selected server, increasing it to $V_{i,1} + S_i$. After the i-th task's arrival, the workloads of the remaining servers remain unchanged.

When the next task arrives at A_{i+1}, all workloads decrease by the inter-arrival time $\tau_{i+1} = A_{i+1} - A_i$, and any negative values are set to 0. Specifically:

$$\big((V_{i,1} + S_i - \tau_{i+1})_+, (V_{i,2} - \tau_{i+1})_+, \ldots, (V_{i,c} - \tau_{i+1})_+\big).$$

Since the resulting workloads are not necessarily ordered, they are reordered to obtain the main recurrence:

$$\mathbf{V}_{i+1} = \mathsf{sort}((V_{i,1} + S_i - \tau_{i+1})_+, (V_{i,2} - \tau_{i+1})_+, \ldots, (V_{i,c} - \tau_{i+1})_+). \qquad (8.5.2)$$

For $c = 1$, this recurrence reduces to the simple case (8.4.1).

The sequence $\mathbf{V}_1, \mathbf{V}_2, \ldots$ has a limiting distribution, provided that $\rho = \mathbb{E}S/\mathbb{E}\tau < c$. From this construction, the waiting time of the i-th task is $W_i = V_{i,(1)}$. Consequently, with probability 1:

$$\bar{w} = \lim_{k \to \infty} \frac{\sum_{j=1}^{k} W_j}{k},$$

which is the steady-state mean waiting time. A natural estimator for $\bar{w}$ is given by:

$$\hat{w} = \frac{\sum_{j=1}^{k} W_j}{k}.$$

While the GI/G/c system can be simulated using a general discrete-event simulation framework, such an approach is often complex and requires careful tracking of events, such as service completions across multiple servers. For this reason, we have focused on a more direct approach based on the recurrence (8.5.2). Exploring the implementation within the general framework can be left as an exercise for the reader.

8.6   Slotted ALOHA and Other Random Access Protocols

In shared communication channels, multiple transmitters attempt to send data packets over the same medium. To efficiently manage access and minimize collisions, various random access protocols have been developed. One of the foundational protocols in this category is **Slotted ALOHA**, which enhances the original ALOHA protocol by introducing time synchronization.

Slotted ALOHA operates by dividing time into discrete slots of equal duration indexed by $n = 0, 1, \ldots$. Each slot spans the interval $[n, n + 1)$. A single slot is capable of accommodating exactly one data packet. The protocol functions as follows:

- (Transmission attempt) When a packet is ready to be sent, it is transmitted at the beginning of the next available time slot.
- (Collision detection) If exactly one packet is transmitted in a slot, the transmission is successful. If two or more packets are transmitted simultaneously, a collision occurs, and none of the packets are successfully transmitted.
- (Feedback and retransmission) After a collision, all involved transmitters are notified of the collision. Each transmitter then waits for a randomly determined number of slots before attempting to retransmit the packet, thereby reducing the probability of repeated collisions.

To avoid (many) collisions and piling up packets, the idea is that after a collision, the transmission of the packet is delayed—usually in a random manner. Different protocols use different delay strategies. First, we need to develop a mathematical model capturing the dynamics of packet transmissions over time. The model includes the following components:

- (X_n)—*packets awaiting transmission.* Let X_n denote the total number of packets waiting to be transmitted at the beginning of the n-th slot. This count aggregates packets from all sources, treating each packet as an independent entity in a shared transmission queue.
- (A_n)—*new arrivals.* Let A_n represent the number of new packets arriving and ready for transmission at the beginning of the n-th slot. We assume that (A_n) is an i.i.d. sequence with: $\mathbb{P}(A_n = i) = a_i$, for $i = 0, 1, 2, \ldots$, and with $\mathbb{E}[A_n] = \lambda < \infty$.

- $\{h_i, 1 \leq i \leq K\}$—*retransmission strategy.* We assume $0 < h_i < 1$ and $K \leq \infty$. Here, h_i denotes the probability of retransmitting a packet after i unsuccessful transmission attempts.

Each packet is treated as an *independent entity* in a shared transmission queue. This means that all packets, regardless of their origin (i.e., which user they belong to), are managed collectively. A new packet always attempts to transmit in the slot it arrives. In the case of an unsuccessful transmission due to a collision, the packet undergoes a retransmission process as defined by the chosen retransmission strategy. Different retransmission strategies determine how packets are retransmitted after collisions:

- **Slotted ALOHA:** $h_i = h$ for all $i \geq 1$. After each collision, a packet is retransmitted with a fixed probability h.
- **Exponential Back-off:** $h_i = 2^{-i}$ for $i = 1, 2, \ldots$. The retransmission probability decreases exponentially with the number of unsuccessful attempts, reducing the likelihood of repeated collisions.
- **Ethernet (Finite Exponential Back-off):** $h_i = 2^{-i}$ for $i = 1, \ldots, K$, where $K < \infty$. Similar to exponential back-off but with a maximum number of retransmission attempts K. If a packet remains untransmitted after K attempts, it is discarded, enhancing protocol stability.

Let Z_n denote the number of packets transmitted in the n-th slot. If $Z_n = 0$, no transmission occurs (all previous packets were transmitted earlier, and no new packets arrived, i.e., $A_n = 0$). If $Z_n = 1$, the transmission is successful, and one packet is removed from the queue. If $Z_n \geq 2$, a collision occurs, and no packets are removed from the queue.

The queue update equation is given by:

$$X_{n+1} = X_n + A_n - \mathbb{1}(Z_n = 1), \qquad n = 0, 1, \ldots \tag{8.6.1}$$

We define the total number of successfully transmitted packets up to slot n as:

$$N_n^s = \sum_{i=1}^{n} \mathbb{1}(Z_i = 1), \quad n = 1, 2, \ldots$$

The *local throughput* at slot n and the *overall throughput* are defined as:

$$\gamma_n = \frac{N_n^s}{n}, \quad \gamma = \lim_{n \to \infty} \gamma_n.$$

In Slotted ALOHA, given $X_n = k$ are waiting to be transmitted at the beginning of slot $[n, n + 1)$, the conditional distribution of Z_n is

$$Z_n \mid X_n = k \quad \stackrel{\text{def}}{=} \quad A_n + Y_n, \quad Y_n \sim \text{Binomial}(k, h)$$

where h is the fixed retransmission probability. We will have one successful transmission (i.e., $Z_n = 1$) if none of k awaiting packages is being transmitted, i.e., $Y_n = 0$, and there is one arrival $A_n = 1$; or when exactly one of k awaiting packages are to be transmitted, i.e., $Y_n = 1$ and there is no arrivals, i.e., $A_n = 0$. We thus have

$$\mathbb{P}(Z_n = 1 \mid X_n = k) = \mathbb{P}(A_n = 0)\mathbb{P}(Y_n = 1|X_n = k)$$
$$+ \mathbb{P}(A_n = 1)\mathbb{P}(Y_n = 0|X_n = k).$$

Given that $A_n \sim \text{Poisson}(\lambda)$, we have

$$\mathbb{P}(Z_n = 1 \mid X_n = k) = e^{-\lambda} \cdot \mathbb{P}(Y_n = 1 \mid X_n = k) + \lambda e^{-\lambda} \cdot \mathbb{P}(Y_n = 0 \mid X_n = k)$$

$$= e^{-\lambda}kh(1 - h)^{k-1} + \lambda e^{-\lambda}(1 - h)^k.$$

Thus, the *conditional drift* is the following

$$\mathbb{E}[X_{n+1} - X_n \mid X_n = k] \stackrel{(8.6.1)}{=} \mathbb{E}[A_n - \mathbb{1}(Z_n = 1) \mid X_n = k]$$

$$= \mathbb{E}[A_n \mid X_n = k] - \mathbb{P}(Z_n = 1) \mid X_n = k]$$

$$= \lambda - e^{-\lambda}kh(1 - h)^{k-1} - \lambda e^{-\lambda}(1 - h)^k.$$

Since terms $kh(1 - h)^{k-1}$ and $(1 - h)^k$ converge to 0 as $k \to \infty$, we have

$$\mathbb{E}[X_{n+1} - X_n \mid X_n = k] \to \lambda \text{ as } k \to \infty,$$

indicating that for large queue lengths, the system tends to drift towards instability (i.e., the queue grows indefinitely). This suggests that Slotted ALOHA is inherently transient and can lead to unbounded queue lengths under high traffic conditions. In the theory of Markov chains, it can be formally proven that (X_n) is transient under these conditions. The typical behavior (queue length growing to infinity, or throughput converging to 0) can be seen in simulations of the model with parameters $\lambda = 0.31$ and $h = 0.1$. In Fig. 8.11a the queue length (X_n) is depicted and in Fig. 8.11b its local throughput.

Summarizing, slotted ALOHA introduces time synchronization to improve upon the original ALOHA protocol by reducing the vulnerability period for collisions. While it offers better throughput compared to pure ALOHA, its inherent transient nature under high traffic conditions necessitates the use of more sophisticated retransmission strategies to ensure system stability and efficiency.

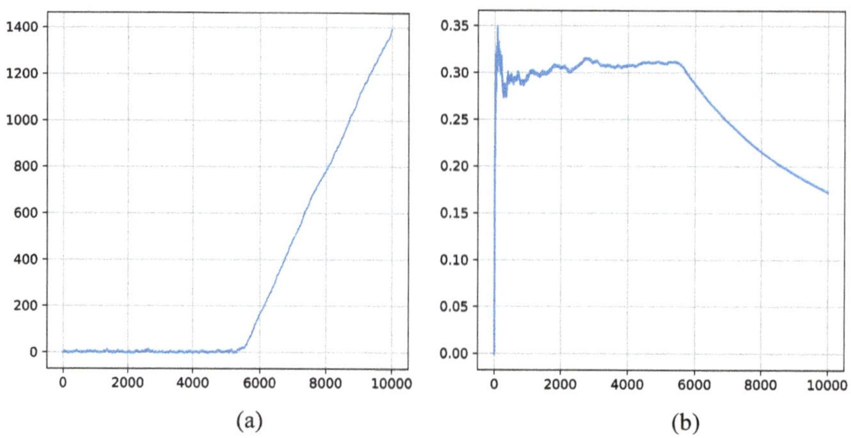

Fig. 8.11 Slotted ALOHA; $\lambda = 0.31$, $h = 0.1$. (a) Packages X_n. (b) Local throughput γ_n

Figure 8.11 is generated using the script ch8_aloha.py.

8.7 Ergodically Stable Processes and Their Output Analysis

In Sect. 8.3, we discussed **steady-state characteristics** for continuous-time Markov chains $(X(t))$. Under relatively straightforward conditions, $X(t)$ converges in distribution to some distribution, called a stationary distribution (the limiting distribution) of $X(t)$. Let X^* be a random variable with this stationary distribution. The steady-state mean of $X(t)$ is defined as $\mathbb{E}[X^*]$. Moreover, with probability 1, we have:

$$\lim_{t \to \infty} \frac{1}{t} \int_0^t X(s)\, ds = \mathbb{E}[X^*], \tag{8.7.1}$$

making $\hat{X}(t) = \frac{1}{t} \int_0^t X(s)\, ds$ a natural estimator of the steady-state mean.

Additionally, for Markov chains, a *stationary version* of the process, $(X^*(t))$, can be defined such that $X^*(t)$ has the same dynamics as $X(t)$ but starts with the limiting distribution of $X(t)$. This ensures $X^*(t)$ is ergodic, providing useful tools for analysis.

For general stochastic processes $(X(t))$, the situation is more complex. We first introduce the notions of stationarity and also ergodicity for general processes.

Stationary Processes A stochastic process $(X^*(t))$ is *stationary* if the joint distribution of $(X^*(t + h_1), \dots, X^*(t + h_d))$ does not depend on t for any $d \geq 1$ and $0 < h_1 < \cdots < h_d$. Stationary processes have the following key properties:

- The marginal distributions of $X^*(t)$ are invariant over time, implying $\mathbb{E}[X^*(t)]$ and $\mathbb{V}\mathrm{ar}[X^*(t)]$ are constant.

- The covariance function $R(h) = \text{Cov}(X^*(t), X^*(t+h))$ depends only on h and not on t.

Ergodic Processes Among stationary processes $(X^*(t))$, we distinguish a class of ergodic processes. A stationary process $(X^*(t))$ is **ergodic** if, for any function g with $\mathbb{E}[|g(X^*(h_1), \ldots, X^*(h_d))|] < \infty$, the following holds with probability 1:[1]

$$\lim_{t \to \infty} \frac{1}{t} \int_0^t g(X^*(h_1 + s), \ldots, X^*(h_d + s)) \, ds = \mathbb{E}[g(X^*(h_1), \ldots, X^*(h_d))].$$
$$(8.7.2)$$

In particular, for $d = 1, h_1 = 0$ and $g(x) = x$, we have:

$$\lim_{t \to \infty} \frac{1}{t} \int_0^t X^*(s) \, ds = \mathbb{E}[X^*(0)]. \tag{8.7.3}$$

This ensures that the time average equals the ensemble average. Verifying (8.7.3) is often sufficient to establish ergodicity for specific processes, provided their structure satisfies the general condition (8.7.2).

For simulation purposes, for an ergodic process, we additionally require that the Central Limit Theorem (CLT) holds:

$$\sqrt{t} \left(\hat{X}^*(t) - \mathbb{E}[X^*(0)] \right) \xrightarrow{\mathcal{D}} \mathcal{N}(0, \varsigma^2), \tag{8.7.4}$$

where ς^2 is the asymptotic variance. In later sections, we will illustrate how to compute ς^2 for specific examples. Note that not all ergodic processes satisfy the CLT.

Ergodically Stable Processes We now define **ergodically stable processes**, which relate non-stationary processes to stationary and ergodic counterparts. A stochastic process $(X(t))$ (not necessarily stationary or ergodic) is *ergodically stable* if there exists a stationary and ergodic process $(X^*(t))$ such that, with probability 1,

$$\lim_{t \to \infty} \frac{1}{t} \int_0^t g(X(h_1 + s), \ldots, X(h_d + s)) \, ds = \mathbb{E}[g(X^*(h_1), \ldots, X^*(h_d))],$$
$$(8.7.5)$$

for all functions $g : \mathbb{R}^d \to \mathbb{R}$ and time shifts $0 < h_1 < \cdots < h_d$, provided $\mathbb{E}|g(X^*(h_1), \ldots, X^*(h_d))| < \infty$.

---

[1] For the purposes of this book, we adopt a simplified definition of ergodicity. The standard definition involves measure-theoretic concepts such as invariant measures and metric transitivity, which are omitted here for simplicity.

Practical Implications of Ergodically Stable Processes For simulation and analysis, ergodically stable processes provide a framework to relate the long-term behavior of a possibly non-stationary process $(X(t))$ to that of a stationary and ergodic reference (or *background*) process $(X^*(t))$. This ensures consistency in estimating steady-state characteristics, even when the original process lacks stationarity or ergodicity.

For example, the steady-state k-th moment of the ergodically stable process $(X(t))$ can be expressed as (8.7.5) with $d = 1$, $g(x) = x^k$, and $h_1 = 0$, i.e.,

$$\lim_{t \to \infty} \frac{1}{t} \int_0^t (X(s))^k \, ds = \mathbb{E}[(X^*(0))^k]. \tag{8.7.6}$$

Similarly, the steady-state covariance function of the ergodically stable process $(X(t))$, which is $R(h) = \mathrm{Cov}(X^*(0), X^*(h))$, can be expressed as:

$$R(h) = \lim_{t \to \infty} \frac{1}{t} \int_0^t (X(s) - I)(X(s+h) - I) \, ds,$$

where $I = \mathbb{E}[X^*(0)]$.

The distinction between ergodically stable and ergodic processes is important: while ergodic processes guarantee that time averages converge to ensemble averages, ergodically stable processes ensure that even non-stationary dynamics are linked to a stationary and ergodic reference process for meaningful steady-state analysis.

In general, for a stochastic process $X(t)$, the following implications hold:

$$\text{Ergodically stable} \Rightarrow \text{Stationary} \Rightarrow \text{Ergodic}.$$

The following two examples demonstrate these concepts.

Example 8.7.1 (Stationarity Is Not Sufficient for Ergodicity) Consider $X(t) = \xi$, where ξ is a random variable with $\mathbb{V}\mathrm{ar}\xi > 0$. Trivially, this process is stationary. However, with probability 1,

$$\lim_{t \to \infty} \frac{1}{t} \int_0^t X(s) \, ds = \xi \neq \mathbb{E}[\xi], \tag{8.7.7}$$

which implies that it is not ergodic. $\Diamond$

Example 8.7.2 In the case of the *on-off* process $(X(t))$ from Example 8.5.1, there exists an ergodic version of the *on-off* process $(X^*(t))$. This version is discussed in Exercise 8.L.7 and further explained in Example 8.8.3. When $X(0) = i$ (where $i = 0$ or 1), the *on-off* process $(X(t))$ is neither stationary nor ergodic but is ergodically stable.

For instance, consider *on*-time and *off*-time distributions that are exponential with rates λ and μ, respectively. The ergodic background process $(X^*(t))$ is defined as starting with the distribution:

$$\mathbb{P}(X^*(0) = 1) = \frac{\mathbb{E}T^{\text{on}}}{\mathbb{E}T^{\text{on}} + \mathbb{E}T^{\text{off}}} = \frac{\mu}{\mu + \lambda},$$

$$\mathbb{P}(X^*(0) = 0) = \frac{\mathbb{E}T^{\text{off}}}{\mathbb{E}T^{\text{on}} + \mathbb{E}T^{\text{off}}} = \frac{\lambda}{\mu + \lambda}.$$

For other *on*-time and *off*-time distributions, the stationary and ergodic background process starts with the above distribution, but the first on/off times have modified distributions $\tilde{F}^{\text{on}}$ and $\tilde{F}^{\text{off}}$. Details are provided in Exercise 8.L.7. $\Diamond$

Example 8.7.3 Consider a continuous-time Markov chain (CTMC) $(X(t))$, i.e., $Y_k = X(hk)$, $k = 0, 1, \ldots$ is a discrete-time Markov chain for any $h > 0$. We assume that $(X(t))$ is irreducible, i.e., Y_k is irreducible for any $h > 0$. For a finite state space, the transition probability matrix $\mathbf{P}(t) = (p_{ij}(t))$, where $p_{ij}(t) = \mathbb{P}(X(t) = j \mid X(0) = i)$, is of the form $e^{\mathbf{Q}t}$, where $\mathbf{Q}$ is the so-called *intensity matrix* defined as:

$$\mathbf{Q} = \lim_{t \searrow 0} \frac{1}{t}(\mathbf{P}(t) - \mathbf{I}).$$

A distribution π is a stationary distribution of $(X(t))$ if $\pi \mathbf{P}(t) = \pi$ for all $t \geq 0$. This condition is equivalent to:

$$\pi (\mathbf{P}(t) - \mathbf{I}) = \mathbf{0}.$$

Dividing by t and taking the limit as $t \searrow 0$, we obtain the condition for the stationary distribution:

$$\pi \mathbf{Q} = \mathbf{0}.$$

If $X(0) \sim \pi$, then the process is ergodic. Otherwise, the process is ergodically stable (the background process is the one with the initial distribution π). Moreover, if $(X(t))$ is an irreducible CTMC—thus, as shown, it is ergodically stable—then any function $g(X(t))$ is also ergodically stable, even though it may no longer be a (continuous-time) Markov chain. $\Diamond$

Remark 8.7.4 There is a subtle terminology conflict to note: in finite Markov chains, a process is termed *ergodic* if it is aperiodic and irreducible. However, this does not imply stationarity. In contrast, in this chapter, *ergodic* refers specifically to a subclass of stationary processes.   ■

8.7.1   A Formula for Asymptotic Variance

In this section, we focus on stationary and ergodic processes $(X^*(t))$ for which the CLT (8.7.4) holds. Our aim is to express the asymptotic variance ς^2 in terms of the covariance function $R(h)$. This relationship provides insight into the long-term behavior of time-averaged processes and helps construct confidence intervals for steady-state estimates.

Lemma 8.7.5 *Suppose $\int_0^\infty |R(v)|\,dv < \infty$ and $\int_0^\infty R(s)\,ds \neq 0$. Then, as $t \to \infty$, we have:*

$$\mathbb{V}\mathrm{ar}\left(\int_0^t X^*(s)\,ds\right) \sim 2t \int_0^\infty R(v)\,dv.$$

Proof We have

$$\mathbb{E}\int_0^t X^*(s)\,ds = \int_0^t \mathbb{E}X^*(s)\,ds = t\mathbb{E}X^*(0)\,.$$

Without loss of generality, assume that $\mathbb{E}X^*(0) = 0$. Then,

$$\begin{aligned}
\mathbb{V}\mathrm{ar}\left(\int_0^t X^*(s)\,ds\right) &= \mathbb{E}\left(\int_0^t X^*(s)\,ds\right)^2 \\
&= \mathbb{E}\left[\int_0^t \int_0^t X^*(w)X^*(v)\,dw\,dv\right] \\
&= \int_0^t \int_0^t \mathbb{E}\left(X^*(w)X^*(v)\right)\,dw\,dv \\
&= \int_{(w,v)\in[0,t]^2:w\leq v} \mathbb{E}\left(X^*(w)X^*(v)\right)\,dw\,dv \\
&\quad + \int_{(w,v)\in[0,t]^2:w>v} \mathbb{E}\left(X^*(w)X^*(v)\right)\,dw\,dv.
\end{aligned}$$

For $v \leq w$, we have $\mathbb{E}X^*(w)X^*(v) = R(w - v)$. Furthermore, for $h > 0$,

$$R(-h) = \mathbb{E}X^*(s - h)X^*(s) = \mathbb{E}X^*(s)X^*(s + h) = R(h).$$

Hence,

$$\begin{aligned}
\int_{(w,v)\in[0,t]^2:w\leq v} \mathbb{E}X^*(w)X^*(v)\,dw\,dv &+ \int_{(w,v)\in[0,t]^2:w>v} \mathbb{E}X^*(w)X^*(v)\,dw\,dv \\
&= 2\int_0^t \int_0^v R(v - w)\,dw\,dv.
\end{aligned}$$

Now, $\int_0^v R(v-w)\,dw = \int_0^v R(u)\,du$, and

$$2\int_0^t\int_0^v R(v-w)\,dw\,dv = 2\int_0^t\int_0^v R(u)\,du\,dv \sim 2t\int_0^\infty R(w)\,dw \quad \text{as } t\to\infty.$$

$\square$

We have:

$$\mathbb{V}\mathrm{ar}\left(\hat{X}(t)\right) = \frac{\mathbb{V}\mathrm{ar}\left(\int_0^t X(s)\,ds\right)}{t^2} \sim \frac{2t\int_0^\infty R(v)\,dv}{t^2}, \tag{8.7.8}$$

and hence:

$$\varsigma^2 = 2\int_0^\infty R(v)\,dv.$$

Thus, we can state the following result:

Proposition 8.7.6 *Suppose an ergodic process $(X^*(t))_{t\geq 0}$ satisfies (8.7.4), with $\mathbb{E}X^*(t)^2 < \infty$, $\int_0^\infty |R(v)|\,dv < \infty$, and $\int_0^\infty R(s)\,ds \neq 0$. Then the estimator $\hat{X}^*(t)$ is unbiased, strongly consistent, and*

$$\lim_{t\to\infty} t\,\mathbb{V}\mathrm{ar}(\hat{X}^*(t)) = \varsigma^2 = 2\int_0^\infty R(s)\,ds. \tag{8.7.9}$$

The quantity $\varsigma^2 = 2\int_0^\infty R(s)\,ds$ is called the *time average variance constant* (TAVC) or *asymptotic variance*. For finite Markov chains, this appeared in Chap. 6.

Remark 8.7.7 It is known that if the CLT (8.7.4) holds, then the asymptotic variance of the limiting normal distribution must be of the form given in (8.7.9).

■

Summing up, similarly to Chap. 4 (see (1.6)), we can restate the *fundamental relationship*:

$$b = z_{1-\alpha/2}\frac{\varsigma}{\sqrt{t}}, \tag{8.7.10}$$

between the error b, a confidence level $1-\alpha$, and the simulation horizon t. The confidence interval at confidence level $1-\alpha$ is:

$$\left(\hat{X}^*(t) - \frac{z_{1-\alpha/2}\varsigma}{\sqrt{t}},\ \hat{X}^*(t) + \frac{z_{1-\alpha/2}\varsigma}{\sqrt{t}}\right),$$

which we write shortly as:

$$\left(\hat{X}^*(t) \pm \frac{z_{1-\alpha/2}\varsigma}{\sqrt{t}} \right).$$

We present a few examples where TAVC can be computed explicitly.

***On-off* Process** We continue the analysis of *on-off* processes from Example 8.5.1. Kopociński [111] derived formula (8.7.11) for TAVC using the Laplace transform of the covariance function $R(t)$ for the stationary *on-off* process:

$$\int_0^\infty e^{-st} R(t)\, dt = \frac{\rho(1-\rho)}{s} - \frac{p}{s\mathbb{E}T^{\mathrm{on}}} \frac{(1-\mathbb{E}e^{-sT^{\mathrm{on}}})(1-\mathbb{E}e^{-sT^{\mathrm{off}}})}{\mathbb{E}e^{-sT^{\mathrm{on}}}\mathbb{E}e^{-sT^{\mathrm{off}}}}.$$

Recall (8.5.1) that $\hat{p}(t) = \frac{1}{t}\int_0^t X(s)\, ds$ is the estimator of p, the steady-state probability that the system is *on*. The asymptotic variance ς^2 quantifies the variability in $\hat{p}(t)$ as $t \to \infty$, providing a measure of the long-term stability of the time-averaged fraction of time the system spends in the *on* state. For this process, the asymptotic variance is given by:

$$\varsigma^2 = \frac{(\mathbb{E}T^{\mathrm{on}})^2 \mathbb{V}\mathrm{ar}T^{\mathrm{off}} + (\mathbb{E}T^{\mathrm{off}})^2 \mathbb{V}\mathrm{ar}T^{\mathrm{on}}}{(\mathbb{E}T^{\mathrm{on}} + \mathbb{E}T^{\mathrm{off}})^3}, \tag{8.7.11}$$

see Kopociński [111], p. 294. Moreover, Theorem 3 in Section 31 from [111] states that the Central Limit Theorem holds:

$$\sqrt{t}(\hat{p}(t) - p) \xrightarrow{\mathcal{D}} \mathcal{N}(0, \varsigma^2),$$

leading to the fundamental relationship:

$$b = z_{1-\alpha/2}\frac{\varsigma}{\sqrt{t}},$$

where b is the confidence interval width at confidence level $1 - \alpha$.

In the specific case where $T^{\mathrm{on}} \sim \mathrm{Exp}(\lambda)$ and $T^{\mathrm{off}} \sim \mathrm{Exp}(\mu)$, the asymptotic variance simplifies to:

$$\varsigma^2 = 2\frac{\lambda\mu}{(\lambda + \mu)^3}, \tag{8.7.12}$$

which the reader is asked to show in Exercise 8.T.9. Using $\rho = \lambda/\mu$, this can be rewritten as:

$$\varsigma^2 = \frac{2\rho}{\mu(1 + \rho)^3}.$$

M/M/1 Queue Formula (8.9.3) can be derived using classical results on the steady-state correlation function $C(h)$ of $L(t)$, attributed to P.M. Morse [157]; see Reynolds [173]:

$$C(h) = \frac{(\mu - \lambda)^3}{\pi} \int_0^{2\pi} \frac{\sin^2\theta \exp\{-h(\lambda + \mu - 2\sqrt{\lambda\mu}\cos\theta)\}}{(\lambda + \mu - 2\sqrt{\lambda\mu}\cos\theta)^3}\, d\theta.$$

The covariance function $R(h)$ is related to $C(h)$ as:

$$C(h) = \frac{R(h)}{R(0)},$$

where the variance $R(0) = \mathbb{V}\mathrm{ar}L^*(0) = \rho/(1-\rho)^2$ (see Exercise 8.T.10). From (8.7.9), we integrate:

$$\begin{aligned}
\varsigma^2 &= 2 \int_0^\infty \frac{\rho}{(1-\rho)^2} C(h)\, dh \\
&= 2\frac{\rho}{(1-\rho)^2} \frac{(\mu-\lambda)^3}{\pi} \int_0^{2\pi} \frac{\sin^2\theta}{(\lambda + \mu - 2\sqrt{\lambda\mu}\cos\theta)^4}\, d\theta \\
&= \frac{4\rho(1-\rho)}{\mu} \frac{1}{\pi} \int_0^\pi \frac{\sin^2\theta}{(1 + \rho - \sqrt{\rho}\cos\theta)^4}.
\end{aligned}$$

Furthermore, we have:

$$\frac{1}{\pi} \int_0^\pi \frac{\sin^2\theta}{(1 + \rho - 2\sqrt{\rho}\cos\theta)^4}\, d\theta = \frac{(1+\rho)}{2(1-\rho)^5}.$$

This result can be verified through direct computation or using symbolic integration methods. Substituting this result, we obtain:

$$\varsigma^2 = \frac{2\rho(1+\rho)}{\mu(1-\rho)^4}.$$

8.7.2 Remarks on TAVC for Other Systems

Here, we list examples from the literature where the TAVC is known. Derivations are not included, as they involve advanced techniques beyond the scope of this book.

Finite Birth and Death Processes Let $(X(t))$ be a B&D process with state space $S = \{0, \ldots, n\}$, birth intensities $\lambda_i > 0$ for $i = 0, 1, \ldots, n - 1$, and death intensities $\mu_i > 0$ for $i = 1, \ldots, n$. This is a continuous-time Markov chain (see

Example 8.7.3) with the intensity matrix:

$$
\mathbf{Q} = \begin{pmatrix}
-\lambda_0 & \lambda_0 & 0 & 0 & \dots & 0 \\
\mu_1 & -(\lambda_1 + \mu_1) & \lambda_1 & 0 & \dots & 0 \\
0 & \mu_2 & -(\lambda_2 + \mu_2) & \lambda_2 & \dots & 0 \\
\vdots & \vdots & \vdots & \vdots & \ddots & \vdots \\
0 & 0 & 0 & 0 & \mu_n & -\mu_n
\end{pmatrix}.
$$

The stationary distribution π satisfies $\pi \mathbf{Q} = \mathbf{0}$, and is given by:

$$
\pi_i = \pi_0 \frac{\lambda_0 \cdots \lambda_{i-1}}{\mu_1 \cdots \mu_i}, \quad i = 1, \dots, n. \tag{8.7.13}
$$

Now, let $Y(t) = f(X(t))$ and define the time-averaged estimator:

$$
\hat{Y}(t) = \frac{\int_0^t Y(s)\, ds}{t}.
$$

The asymptotic mean is given by:

$$
I_f = \lim_{t \to \infty} \hat{Y}(t) = \sum_{i=0}^{n} f(i)\pi_i.
$$

According to Whitt [213], the TAVC is:

$$
\varsigma_f^2 = 2 \sum_{j=0}^{n-1} \frac{1}{\lambda_j \pi_j} \left[\sum_{i=0}^{j} (f(i) - I_f)\pi_i \right]^2. \tag{8.7.14}
$$

Whitt attributed this formula to D. Y. Burman.

The *on-off* process is a special case of a B&D process with $n = 1$, $\lambda_0 = \lambda$, and $\mu_1 = \mu$, where both the *on*-time and *off*-time durations are exponentially distributed. For this case, (8.7.14) simplifies to:

$$
\varsigma^2 = 2 \frac{\lambda \mu}{(\lambda + \mu)^3}.
$$

The reader is invited to verify this result in Exercise 8.T.14.

M/M/n/n Queues Consider a B&D process with $\lambda_i = \lambda$ and $\mu_i = i\mu$. The stationary distribution is:

$$
\pi_i = \frac{\rho^i}{i!} \left(\sum_{j=0}^{n} \frac{\rho^j}{j!} \right)^{-1},
$$

where $\rho = \lambda/\mu$ is the traffic intensity. Let $E_n(\rho) = \sum_{j=0}^{n} \frac{\rho^j}{j!}$. The asymptotic mean is:

$$\bar{l} = \sum_{j=1}^{n} j\pi_j = \frac{\rho E_{n-1}(\rho)}{E_n(\rho)}.$$

Using (8.7.14), the asymptotic variance simplifies to:

$$\varsigma^2 = \frac{2}{\lambda} \sum_{j=0}^{n} \frac{1}{\rho^j/j!} \frac{\left(\rho\left(E_{n-1}(\rho) - E_{j-1}(\rho)\right)\right)^2}{E_n(\rho)}.$$

8.8   Regenerative Processes

Regenerative processes form a class that covers most of the continuous-time processes considered in this chapter. We begin with a formal definition.

Definition 8.8.1 A stochastic process $(X(t), t \geq 0)$ is a *regenerative process* if there exist random instants $T_0 := 0 \leq T_1 < T_2 < \cdots$ such that the pairs of random elements (Z_i, C_i) are independent for $i \geq 0$ and identically distributed for $i \geq 1$, where

$$Z_0 = \int_0^{T_1} X(s)\,ds, \quad Z_i = \int_{T_i}^{T_{i+1}} X(s)\,ds, \quad i \geq 1,$$

and

$$C_0 = T_1, \quad C_i = T_{i+1} - T_i, \quad i \geq 1.$$

Throughout this section, we assume that the distribution of C_1 is continuous to avoid technical complications.

The key property of regenerative processes is that they can be decomposed into independent **cycles**:

$$(X(s), T_i \leq s < T_{i+1}), \qquad i = 0, 1, 2, \ldots,$$

which are independent for $i \geq 0$ and identically distributed for $i \geq 1$. This decomposition simplifies analysis significantly.

Steady-state Mean Estimation Our goal is to calculate the steady-state mean $I = \mathbb{E}X^*(0)$ of the process $(X(t))$. A natural estimator is:

$$\hat{X}(t) = \frac{1}{t} \int_0^t X(s)\,ds,$$

which satisfies:

$$I = \lim_{t \to \infty} \hat{X}(t), \quad \text{with probability 1.}$$

The validity of this approach, as well as the Central Limit Theorem (CLT), is justified under certain conditions by the following result (see Asmussen [7], Theorems VI.4.1 and VI.3.2).

Proposition 8.8.2 *If* $\mathbb{E}C_1 < \infty$ *and* $\mathbb{E}|Z_1| < \infty$, *then with probability 1:*

$$\hat{X}(t) \to I = \frac{\mathbb{E} \int_{T_1}^{T_2} X(s)\,ds}{\mathbb{E}C_1}, \quad \text{as } t \to \infty. \tag{8.8.1}$$

If additionally $\mathbb{E}C_1^2 < \infty$ *and* $\mathrm{Var}\,Z_1 < \infty$, *then:*

$$\sqrt{t}(\hat{X}(t) - I) \xrightarrow{\mathcal{D}} \mathcal{N}(0, \varsigma_A^2), \tag{8.8.2}$$

where the TAVC ς_A^2 *is given by:*

$$\varsigma_A^2 = \frac{\mathbb{E}\zeta_1^2}{\mathbb{E}C_1}, \tag{8.8.3}$$

and:

$$\zeta_k = Z_k - I C_k.$$

Non-delayed Regenerative Processes When $T_1 = 0$, the process is referred to as *non-delayed* and is denoted $(X^o(t))$. In this case, the cycles:

$$((X^o(s),\, T_i^o \le s < T_{i+1}^o),\, C_i^o), \qquad i = 1, 2, \ldots,$$

are independent and identically distributed. We use the superscript o to denote non-delayed regenerative processes.

For such processes, regenerative processes are *ergodically stable* provided $\mathbb{E}C_1 < \infty$, with the background ergodic process $(X^*(t))$ well-defined. Using (8.8.1), we have:

$$\mathbb{E}X^*(0) = \frac{\mathbb{E}\int_0^{C_1^o} X^o(s)\,ds}{\mathbb{E}C_1^o}. \tag{8.8.4}$$

Example 8.8.3 The *on-off* process $(K(t))$, defined in Example 8.5.1 (where it was denoted simply as $X(t)$), is regenerative. In this context, visits to 1 (*on* state) constitute cycles, with each cycle beginning when the process transitions to 1 and ending when it next returns to 0 (*off* state). Sample regeneration cycle lengths $C_0, \ldots, C_5$ are depicted in Fig. 8.7.

For the convergence of the estimator $\hat{K}(t)$ with probability 1 in (8.8.1), we require $\mathbb{E}C_1 = \mathbb{E}(T^{\mathrm{on}} + T^{\mathrm{off}}) < \infty$. The condition for the Central Limit Theorem in (8.8.2) is $\mathbb{E}C_1^2 = \mathbb{E}(T^{\mathrm{on}} + T^{\mathrm{off}})^2 < \infty$, which holds when $\mathbb{E}(T^{\mathrm{on}})^2 < \infty$ and $\mathbb{E}(T^{\mathrm{off}})^2 < \infty$. Note that due to an obvious bound $Z_1 \le C_1$, we do not need to check the moments of Z_1. Recall that we assumed that the process $K(t)$ starts in state 1 (*on*), i.e., $K(0) = 1$. The "dynamics" of $K^*(t)$ will be the same as $K(t)$, however, the difference is in the first cycle. Our goal is to know how to "start" an ergodic process $K^*(t)$, i.e., determine the distributions of *on*-time and *off*-time in the first cycle. For this we introduce a component $A(t)$ which is the time to the next change of the *on-off* process, i.e., if $K(t) = 1$, then $A(t)$ is the time of the next *off* event, and when $K(t) = 0$, then $A(t)$ is the time of the next *on* event.

We will derive the joint distribution of $(K^*(0), A^*(0))$. To derive it we use formula (8.8.4).

Thus we define a process $\mathbf{Y}^o(t) = (K^o(t), A^o(t))$, which in the first cycle is:

$$
\mathbf{Y}^o(t) =
\begin{cases}
(1, T^{\mathrm{on}} - t), & \text{for } 0 \le t \le T^{\mathrm{on}}, \\[2mm]
(0, T^{\mathrm{off}} - t), & \text{for } T^{\mathrm{on}} \le t < C_1^o = T^{\mathrm{on}} + T^{\mathrm{off}},
\end{cases}
$$

i.e.,

$$
K^o(t) =
\begin{cases}
1, & \text{for } 0 \le t \le T^{\mathrm{on}}, \\[2mm]
0, & \text{for } T^{\mathrm{on}} \le t < C_1^o,
\end{cases}
\quad \text{and}
$$

$$
A^o(t) =
\begin{cases}
T^{\mathrm{on}} - t, & \text{for } 0 \le t \le T^{\mathrm{on}}, \\[2mm]
T^{\mathrm{off}} - t, & \text{for } T^{\mathrm{on}} \le t < C_1^o.
\end{cases}
$$

We extend this process to all $t \ge 0$ as $\mathbf{Y}^o(t) = (K^o(t), A^o(t))$ by concatenating independent and identically distributed cycles. The first component is the *on-off* process, and the second is the time to the next jump of the process $(K^o(t))$. On the base of this process, for an arbitrary but fixed $s \ge 0$, define the process $X_s^o(t) = \mathbb{1}(K^o(t) = 1, A^o(t) > s)$ and from the stationary version $X_s^*(t) = \mathbb{1}(K^*(t) = 1, A^*(t) > s)$ it follows that

$$
\mathbb{E}X_s^*(0) = \mathbb{P}(K^*(0) = 1, A^*(0) > s). \tag{8.8.5}
$$

In a similar vein we compute $\mathbb{P}(K^*(0) = 0, A^*(0) > s)$ for an arbitrary but fixed $s \ge 0$.

From (8.8.4), we obtain:

$$\mathbb{P}(K^*(0) = 1, A^*(0) > s) =$$

$$\mathbb{E}X_s^*(0) = \frac{1}{\mathbb{E}C_1^o}\mathbb{E}\int_0^{C_1^o} X_s^o(t)\, dt$$

$$= \frac{1}{\mathbb{E}(T^{\mathrm{on}} + T^{\mathrm{off}})}\mathbb{E}\int_0^{T^{\mathrm{on}}+T^{\mathrm{off}}} \mathbb{1}(K^o(t) = 1, A^o(t) > s)\, dt$$

$$= \frac{1}{\mathbb{E}(T^{\mathrm{on}} + T^{\mathrm{off}})}\mathbb{E}\int_0^{T^{\mathrm{on}}} \mathbb{1}(T^{\mathrm{on}} - t > s)\, dt.$$

Observe that $\mathbb{1}(T^{\mathrm{on}} - t > s) = 1$ if and only if $t < (T^{\mathrm{on}} - s)_+$. Hence:

$$\mathbb{P}(K^*(0) = 1, A^*(0) > s) = \frac{1}{\mathbb{E}(T^{\mathrm{on}} + T^{\mathrm{off}})}\mathbb{E}(T^{\mathrm{on}} - s)_+$$

$$= \frac{\mathbb{E}T^{\mathrm{on}}}{\mathbb{E}(T^{\mathrm{on}} + T^{\mathrm{off}})} \cdot \frac{1}{\mathbb{E}T^{\mathrm{on}}}\mathbb{E}(T^{\mathrm{on}} - s)_+.$$

Finally, notice that:

$$\mathbb{E}(T^{\mathrm{on}} - s)_+ = \int_s^\infty \mathbb{P}(T^{\mathrm{on}} > t)\, dt.$$

Similarly, we find a formula for $\mathbb{P}(K^*(0) = 0, A^*(0) > s)$ by considering $X_s^o(t) = \mathbb{1}(K^o(t) = 0, A^o(t) > s)$.

Summarizing, we have

$$\mathbb{P}(K^*(0) = 1, A^*(0) > x) = \frac{\mathbb{E}T^{\mathrm{on}}}{\mathbb{E}(T^{\mathrm{on}} + T^{\mathrm{off}})} \cdot \frac{1}{\mathbb{E}T^{\mathrm{on}}}\int_x^\infty \mathbb{P}(T^{\mathrm{on}} > s),$$

$$\mathbb{P}(K^*(0) = 0, A^*(0) > x) = \frac{\mathbb{E}T^{\mathrm{off}}}{\mathbb{E}(T^{\mathrm{on}} + T^{\mathrm{off}})} \cdot \frac{1}{\mathbb{E}T^{\mathrm{off}}}\int_x^\infty \mathbb{P}(T^{\mathrm{off}} > s).$$

We thus see that the joint distribution of $\mathbf{Y}^*(t) = (K^*(t), A^*(t))$ is a product of independent random variables, where $\mathbb{P}(K^*(t) = 1) = p = 1 - \mathbb{P}(K^*(t) = 0) = \mathbb{E}T^{\mathrm{on}}/\mathbb{E}(T^{\mathrm{on}} + T^{\mathrm{off}})$ and the distribution of $A^*(t)$. Since $\mathbf{Y}^*(t)$ is ergodic, in particular for $t = 0$ we have that $p = \mathbb{P}(K^*(0) = 1)$. In conclusion, we have the following algorithm for simulating an ergodic (that is also stationary) *on-off* process. Define

$$\tilde{F}_{\mathrm{on}}(s) = \frac{1}{\mathbb{E}T^{\mathrm{on}}}\int_0^s \mathbb{P}(T^{\mathrm{on}} > t)\, dt, \quad \tilde{F}_{\mathrm{off}}(s) = \frac{1}{\mathbb{E}T^{\mathrm{off}}}\int_0^s \mathbb{P}(T^{\mathrm{off}} > t)\, dt.$$

1. Set $K^*(0) = 1$ with probability p and $K^*(0) = 0$ with probability $1 - p$.
2. If $K^*(0) = 1$, then draw the first *on*-time $\tilde{T}_1^{\text{on}}$ is according to distribution $\tilde{F}_{\text{on}}(x)$, otherwise the first *off*-time $\tilde{T}_1^{\text{off}}$ is according to $\tilde{F}_{\text{off}}(x)$.
3. Continue alternative independent *on*-times and *off*-times with distributions F_{on} and F_{off} respectively.

Note: The fact that after the first *on*-time or *off*-time, the next ones are independent and distributed according to the "usual" F_{on} and F_{off} needs a formal proof, which is beyond the scope of this book.

In this example, we derived the joint distribution of the ergodic *on-off* process $(K^*(t), A^*(t))$ using regenerative properties and stationary Palm theory. The resulting algorithm simulates an ergodic *on-off* process by considering the residual distributions $\tilde{F}_{\text{on}}$ and $\tilde{F}_{\text{off}}$ for the first *on*-time or *off*-time, ensuring stationarity, with subsequent cycle times independently distributed according to F_{on} and F_{off}. The ergodic nature of the process allows for computation of the steady-state probability that the system is *on* (p). $\Diamond$

The following examples highlight the regenerative structure of two key stochastic processes: a continuous-time Markov chain and the virtual waiting time process in a GI/G/1 queue. The regenerative property not only simplifies the analysis by dividing the process into independent and identically distributed cycles but also justifies the use of the estimator $\hat{X}(t) = \frac{1}{t} \int_0^t X(s)\, ds$. This ensures that the steady-state mean $I = \mathbb{E}X^*(0)$ is indeed the limit of $\hat{X}(t)$ as $t \to \infty$. Moreover, viewing the process through its regenerative cycles provides an alternative and practical way to estimate I, as explained in detail after these examples.

Example 8.8.4 Consider a continuous-time Markov chain (CTMC) $(X(t))$ with state space $\mathbb{Z}_+$ and a stable conservative intensity matrix $\mathbf{Q}$, where $q_i = -q_{ii} > 0$ and $\sum_j q_{ij} = 0$ for all i. Assume the process is irreducible and that there exists a unique stationary distribution π satisfying $\pi \mathbf{Q} = \mathbf{0}$, cf. Example 8.7.3

In the theory of CTMCs, such processes are called *ergodic*, meaning that the mean recurrence time is finite. Consequently, $(X(t))$ is a regenerative process satisfying $\mathbb{E}C_1 < \infty$ and in the case of a finite state space $\mathbb{E}Z_2 < \infty$. In the context of this chapter, this implies that the CTMC $(X(t))$ with the initial stationary distribution π is both stationary and ergodic.

From the broader perspective of stationary processes, a process is stationary if the shift operator θ_s is measure-preserving, and it is ergodic if all θ_s-invariant events have probability 0 or 1. A precise formulation of these concepts requires significant measure-theoretic details, which are beyond the scope of this chapter; for a thorough treatment, see e.g. the book by Robert [177].

For finite Markov chains, the conditions in Proposition 8.8.2 are automatically satisfied. However, for countable state spaces, additional care is required. For example, the specific structure of the intensity matrix is critical to ensure the conclusions of the proposition. $\Diamond$

Example 8.8.5 The virtual waiting time process $(V(t))$ in a GI/G/1 queue, as defined in (8.4.4), is regenerative provided $\mathbb{E}S < \mathbb{E}T$, or equivalently, $\rho < 1$. Regeneration occurs at times t where $V(t-) = 0$ and $V(t) > 0$, marking the beginning of a new busy period. In this context, the cycle duration C_1 corresponds to the *busy cycle* of the queue.

Similarly, the process representing the number of tasks in the system is also regenerative under the same conditions, with regeneration points occurring at the end of idle periods. $\Diamond$

From formula (8.8.1), another estimator for I emerges, which in practice is often better than $\hat{X}(t)$, namely:

$$\hat{I}_k = \frac{Z_1 + \cdots + Z_k}{C_1 + \cdots + C_k}. \tag{8.8.6}$$

It is clear that, with probability 1, we have:

$$\hat{I}_k \to I,$$

because:

$$\frac{Z_1 + \cdots + Z_k}{C_1 + \cdots + C_k} = \frac{(Z_1 + \cdots + Z_k)/k}{(C_1 + \cdots + C_k)/k} \to \frac{\mathbb{E}Z_1}{\mathbb{E}C_1}.$$

We then have the following theorem (see Asmussen and Glynn [8], Proposition 4.1).

Proposition 8.8.6 *Under the conditions of Proposition 8.8.2, for $k \to \infty$,*

$$\sqrt{k}(\hat{I}_k - I) \xrightarrow{\mathcal{D}} \mathcal{N}(0, \varsigma_B^2),$$

where the TAVC ς_B^2 is given by

$$\varsigma_B^2 = \frac{\mathbb{E}\zeta_1^2}{(\mathbb{E}C_1)^2}, \quad \text{where} \quad \zeta_1 = Z_1 - IC_1. \tag{8.8.7}$$

Thus, an asymptotic confidence interval at confidence level $1 - \alpha$ is given by:

$$\hat{I}_k \pm z_{1-\alpha/2}\frac{\varsigma_B}{\sqrt{k}}. \tag{8.8.8}$$

Comparing Confidence Intervals for $\hat{X}(t)$ and $\hat{I}_k$ The TAVCs ς_A^2, as given in (8.8.3), and ς_B^2, as given in Proposition 8.8.6, differ as follows: $\varsigma_B^2 < \varsigma_A^2$ if $\mathbb{E}C_1 > 1$; $\varsigma_B^2 = \varsigma_A^2$ when $\mathbb{E}C_1 = 1$; $\varsigma_B^2 > \varsigma_A^2$ if $\mathbb{E}C_1 < 1$.

Using the decomposition $t = C_1 + C_2 + \cdots + C_k$ (assuming t is the exact end of a cycle), we can approximate $t \approx k\mathbb{E}C_1$. This allows us to compare the widths of the confidence intervals for $\hat{X}(t)$ (from (8.8.2)) and $\hat{I}_k$ (from (8.8.8)):

$$z_{1-\alpha/2}\frac{\varsigma_A}{\sqrt{t}} = z_{1-\alpha/2}\frac{\sqrt{\mathbb{E}\zeta_1^2}}{\sqrt{\mathbb{E}C_1}}\cdot\frac{1}{\sqrt{t}} = z_{1-\alpha/2}\frac{\sqrt{\mathbb{E}\zeta_1^2}}{\mathbb{E}C_1}\cdot\frac{1}{\sqrt{k}},$$

$$z_{1-\alpha/2}\frac{\varsigma_B}{\sqrt{k}} = z_{1-\alpha/2}\frac{\sqrt{\mathbb{E}\zeta_1^2}}{\mathbb{E}C_1}\cdot\frac{1}{\sqrt{k}}.$$

As seen, the lengths of the confidence intervals are identical. Therefore, the choice of estimator depends on the specific context:

- The key advantage of $\hat{X}(t)$ is that it does not require explicitly identifying regeneration cycles, which can be challenging in some systems. This simplicity makes it appealing for general cases. However, $\hat{X}(t)$ may be biased for short simulations, especially if the process starts in a fixed transient state (e.g., $X(0) = 0$).
- On the other hand, $\hat{I}_k$ is unbiased, provided the regeneration cycles are correctly identified, and the first cycle is discarded. This makes $\hat{I}_k$ particularly useful for obtaining accurate estimates in shorter simulation runs.

Example 8.8.7 (Estimating p in *on-off* Process) We simulate the *on-off* process using two distinct options for the distributions of the *on* and *off* times, ensuring both have the same expected values:

- **Option 1 (Exp/Exp)**: *On* and *off* times are exponentially distributed, with $T^{\text{on}} \sim$ Exp$(1/3)$ and $T^{\text{off}} \sim$ Exp(1).
- **Option 2 (Par/Par)**: *On* times follow a Pareto distribution Par$(2.5, 1.8)$, and *off* times follow a Gamma distribution Gamma$(2, 1/2)$.

Note that in both cases, $\mathbb{E}T^{\text{on}} = 3$ and $\mathbb{E}T^{\text{off}} = 1$. The true value of the steady-state probability p, representing the fraction of time the system is in the *on* state, is given by:

$$p = \frac{\mathbb{E}T^{\text{on}}}{\mathbb{E}T^{\text{on}} + \mathbb{E}T^{\text{off}}} = \frac{3}{4}.$$

We estimate p using two estimators: $\hat{X}(t)$ and $\hat{I}_k$. The time-based estimator $\hat{X}(t)$ is computed for $t \in [0, t_{\max}]$, where $t_{\max} = 1000$. The regenerative estimator $\hat{I}_k$ is computed at the ends of regeneration cycles, excluding the first cycle to ensure unbiasedness.

Table 8.2 Simulation results for the *on-off* process

	Option 1 (Exp/Exp)	Option 2 (Par/Par)
True value of p	0.75000	0.75000
Number of cycles used (k)	257	263
Estimator $\hat{X}(t)$ at $t_{\max}$	0.72733	0.73932
Estimator $\hat{I}_k$	0.72472	0.73937
95% Confidence Interval for $\hat{X}(t)$	(0.656, 0.798)	(0.701, 0.777)
95% Confidence Interval for $\hat{I}_k$	(0.688, 0.761)	(0.720, 0.759)

The numerical results for both options are summarized in Table 8.2. Figures 8.12 and 8.13 show the corresponding plots. Both estimators are derived from the same realization.

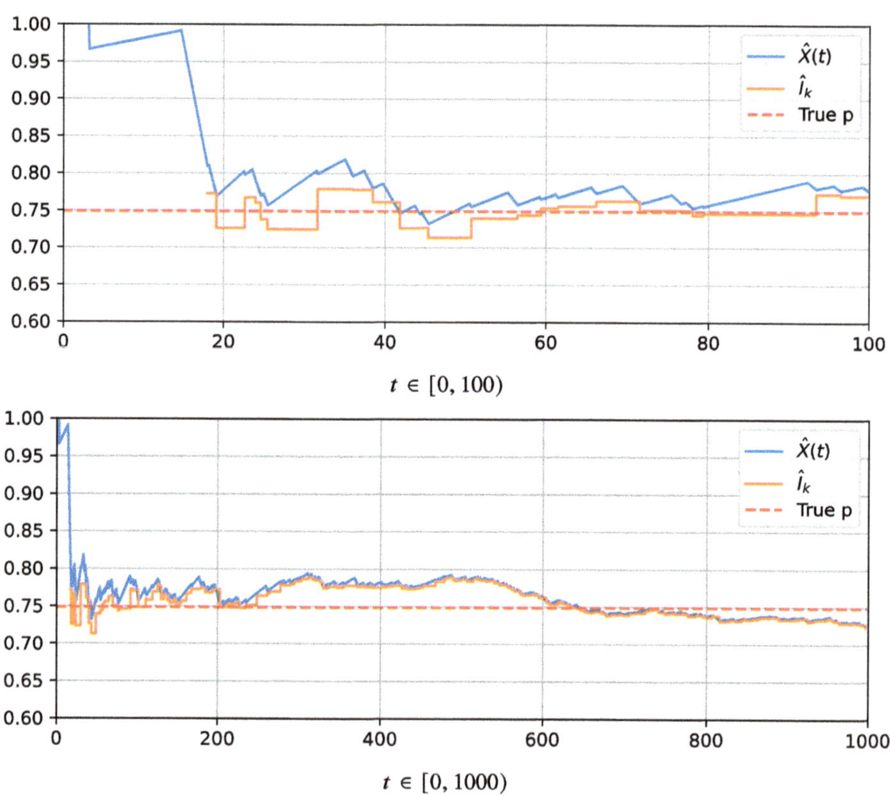

$$t \in [0, 100)$$

$$t \in [0, 1000)$$

Fig. 8.12 Estimators $\hat{X}(t)$ and $\hat{I}_k$ over time for $T^{\mathrm{on}} \sim \mathrm{Exp}(1/3)$ and $T^{\mathrm{off}} \sim \mathrm{Exp}(1)$

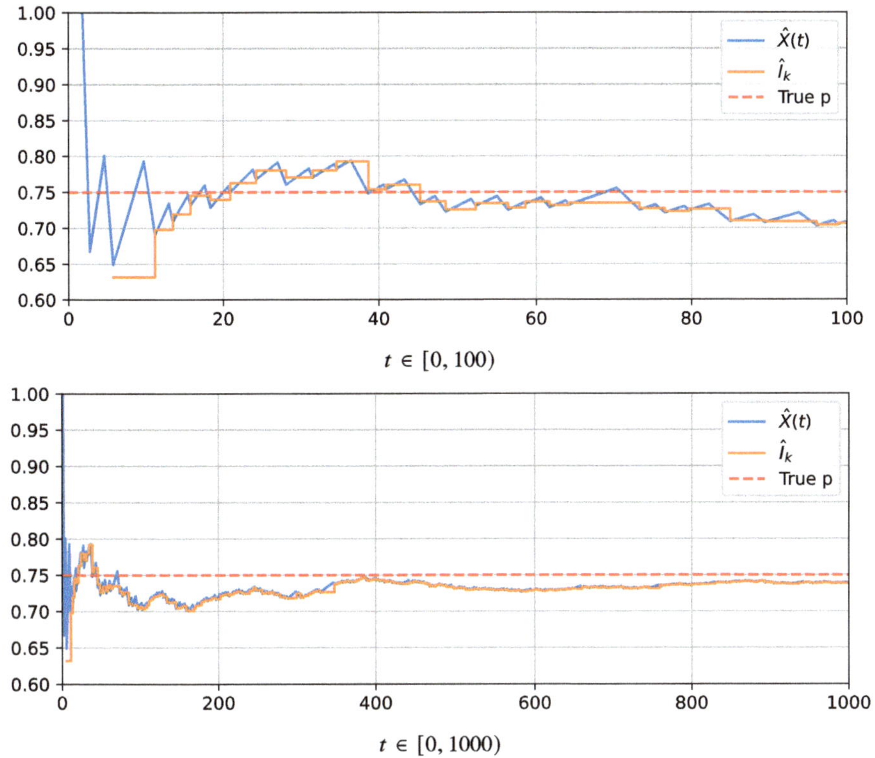

Fig. 8.13 Estimators $\hat{X}(t)$ and $\hat{I}_k$ over time for $T^{\mathrm{on}} \sim \mathrm{Par}(2.5, 1.8)$ and $T^{\mathrm{off}} \sim \mathrm{Gamma}(2, 1/2)$

Both estimators provide accurate estimates of p. However, $\hat{I}_k$ generally has narrower confidence intervals, reflecting its unbiased nature. The time-based estimator $\hat{X}(t)$, while simpler to compute, is more sensitive to the initialization of the process, particularly when regeneration cycles are not easily identifiable. This example is implemented in `ch8_on_off.py`.   ◇

Estimating Asymptotic Variances Since the formulas for the asymptotic variances ς_A^2 and ς_B^2 depend on the unknown parameter I, we propose the following estimator for ζ_j:

$$\zeta_j = Z_j - \hat{I}_k C_j, \quad j = 1, \ldots, k,$$

where $\hat{I}_k$ is defined in (8.8.6). The estimated asymptotic variance ς_A^2 is then given by:

$$\hat{\varsigma}_A^2 = \frac{1}{k-1} \sum_{j=1}^{k} (\zeta_j - \bar{\zeta})^2,$$

where $\bar{\zeta} = \frac{1}{k} \sum_{j=1}^{k} \zeta_j$ is the sample mean of ζ_j. This corresponds to the asymptotic variance ς_A^2 in Proposition 8.8.2.

For the variance ς_B^2 in Proposition 8.8.6, the scaling factor $(\mathbb{E}C_1)^2$ is replaced by its sample estimate. Specifically, we compute $\hat{C}$, the sample mean of the cycle lengths C_j:

$$\hat{C} = \frac{1}{k} \sum_{j=1}^{k} C_j.$$

The estimated asymptotic variance $\hat{\varsigma}_B^2$ is then obtained by scaling $\hat{\varsigma}_A^2$ as follows:

$$\hat{\varsigma}_B^2 = \frac{\hat{\varsigma}_A^2}{\hat{C}^2}.$$

These estimators provide practical tools for approximating asymptotic variances, which are crucial for constructing confidence intervals and ensuring the reliability of steady-state estimates.

8.9   Error Analysis of M/M/1 and M/G/∞ Systems

In this section, we analyze two examples that follow the framework of regenerative processes: the M/M/1 queue and the M/G/∞ system. These are classic models in queueing theory and stochastic processes, making them well-suited for illustrating theoretical concepts. For these models, we verify the assumptions of Proposition 8.8.2, perform simulations, and provide an error analysis. From the Central Limit Theorem, we have:

$$\mathbb{P}\left(|\hat{X}(t_{\max}) - I| \leq \frac{z_{1-\alpha/2}\varsigma}{\sqrt{t_{\max}}} \right) \approx 1 - \alpha.$$

Here, $b = |\hat{X}(t) - I|$ represents the absolute error. To maintain a given error level b, the required run length must satisfy:

$$t_{\max} = \frac{z_{1-\alpha/2}^2 \varsigma^2}{b^2}.$$

Similarly, for the relative error,

$$\mathbb{P}\left(\left| \frac{\hat{X}(t_{\max}) - I}{I} \right| \leq \frac{z_{1-\alpha/2}\varsigma}{\sqrt{I t_{\max}}} \right) \approx 1 - \alpha,$$

which leads to the requirement:

$$t_{\max} = \frac{z_{1-\alpha/2}^2}{b^2} \cdot \frac{\varsigma^2}{I^2}. \tag{8.9.1}$$

The term SACV $= \varsigma^2/I^2$ is the *squared asymptotic coefficient of variation*. In the following analysis, we assume a target error of 10%, corresponding to $b = 0.1$ for absolute error and $0.1I$ for relative error. Furthermore, we set $\alpha = 0.05$, for which $z_{1-\alpha/2}^2 \approx 4$, which yields $t_{\max} = 400 \cdot$ SACV.

M/M/1 Queues Consider a stable M/M/1 queueing system, where the traffic intensity $\rho = \lambda/\mu$ satisfies $\rho < 1$. This condition ensures that the queue remains stable.

The number of tasks in the system at time t, denoted by $L(t)$, forms a birth-and-death process with state space $\mathcal{S} = \{0, 1, 2, \ldots\}$, birth rates $\lambda_i = \lambda$ for all i, and death rates $\mu_i = i$ for $i \geq 1$. The steady-state mean queue length is given by:

$$\bar{l} = \mathbb{E}[L^*(0)] = \lim_{t \to \infty} \frac{1}{t} \int_0^t L(s)\, ds = \frac{\lambda}{\mu - \lambda} = \frac{\rho}{1 - \rho}, \tag{8.9.2}$$

where L^* denotes the steady-state number of tasks in the system. The derivation of $\bar{l}$ and the steady-state variance $\sigma_L^2 = \mathbb{V}\mathrm{ar}[L^*]$ is left as Exercise 8.T.10.

The process $L(t)$ is regenerative because it has well-defined *regeneration instants*:

$$T_1 = \inf\{t > 0 : L(t-) = 0, L(t) = 1\}$$

and for $i \geq 2$:

$$T_i = \inf\{t > T_{i-1} : L(t-) = 0, L(t) = 1\}.$$

These instants arise due to the memoryless property of the exponential distribution. When $L(t)$ reaches state zero, the time until the next arrival (the idle period) follows an exponential distribution, $\mathrm{Exp}(\lambda)$. The time from an arrival to an empty system until the departure that leaves the system empty is known as a busy period. Idle and busy periods are independent, and both have finite moments.

The regeneration cycle $C_1 = T_2 - T_1$, corresponding to the sum of an idle period and the subsequent busy period, also has finite moments. The mean busy period is $1/(\mu - \lambda)$, with variance:

$$\mathbb{V}\mathrm{ar}[\text{Busy Period}] = \frac{1 + \rho}{\mu^2(1 - \rho)^3}.$$

Similarly, $\mathbb{E}[Z_1]$ and $\mathbb{V}\mathrm{ar}[Z_1]$, where Z_1 represents the total workload during the first regeneration cycle, are also finite (see McNeil [149] for details). Consequently, the Central Limit Theorem (CLT) applies to $\hat{L}(t)$:

$$\sqrt{t}(\hat{L}(t) - \bar{l}) \xrightarrow{\mathcal{D}} \mathcal{N}(0, \varsigma_L^2),$$

where the asymptotic variance ς_L^2 is:

$$\varsigma_L^2 = \frac{2\rho(1 + \rho)}{\mu(1 - \rho)^4}. \tag{8.9.3}$$

From (8.9.3), it is evident that in heavy traffic ($\rho \to 1$), the asymptotic variance grows significantly, requiring longer simulation horizons to maintain a fixed precision. Comparing the steady-state variance $\sigma_L^2 = \mathbb{V}\mathrm{ar}[L^*(0)]$ with the asymptotic variance ς_L^2, we find:

$$\frac{\varsigma_L^2}{\sigma_L^2} = \frac{2(1 + \rho)}{(1 - \rho)^2}.$$

This shows that in heavy traffic, the two variances differ significantly. Similarly, for the squared asymptotic coefficient of variation we have:

$$\mathrm{SACV} = \left(\frac{\varsigma_L}{\bar{l}}\right)^2 = \frac{2(1 + \rho)}{\mu\rho(1 - \rho)^2}.$$

Example 8.9.1 (Error Analysis in M/M/1 Queue) In Example 8.3.3, we simulated the mean number of tasks $I = \mathbb{E}L^*(0)$ in M/M/1 systems with $\mu = 1$ and (a) $\lambda = 1/2$ and (b) $\lambda = 8/9$. Cases (c) and (d) do not have a finite steady-state mean $\bar{l}$.

In case (a), the theoretical steady-state mean queue length (from (8.9.2)) is $\bar{l} = 1$, with $\varsigma^2 = 24$, leading to a SACV of 24. This yields a required simulation length of $t_{\max} = 400 \cdot \mathrm{SACV} = 9600$. Figure 8.4 presents a sample simulation of $L(t)$ for $0 \le t \le 10^4$, which results in an estimated mean $\hat{l} = 0.9822$. The relative error is 1.7785%, which aligns with our precision requirement of 10%.

In case (b), the theoretical steady-state mean queue length is $\bar{l} = 8$, with $\varsigma^2 = 22032$, leading to a larger SACV of 344.25. This yields $t_{\max} = 400 \cdot \mathrm{SACV} = 1.377 \cdot 10^5$. From the experiment, we obtained $\hat{l} = 6.669786$, corresponding to a relative error of 16.6276%, exceeding our 10% goal, which is below the required $1.377 \cdot 10^5$, so a larger error is unsurprising. Extending the simulation to $t = 2 \cdot 10^5$ brings the relative error within 10% Thus, these findings are consistent with our error analysis.

It is important to note that this error analysis applies to steady-state simulations. In the initial phase, a significant amount of irregularity is present, which can be observed in the simulations. It is generally recommended to exclude this initial period, known as the *warm-up period*. However, we do not address this issue in this section. $\qquad\qquad\diamond$

Simulation Studies of M/G/∞ Systems   Let us analyze the process $(L(t))$, which represents the number of tasks in an infinite-server system. The arrival rate is λ, and the service times $S \sim G$ are i.i.d. random variables.

Arrivals occur at times $A_1, A_2, \ldots$, where $A_i = \tau_1 + \cdots + \tau_i$, and $\tau_1, \tau_2, \ldots$ are i.i.d. exponentially distributed random variables with rate λ, i.e., $\mathrm{Exp}(\lambda)$. For convenience, we set $A_0 = 0$. Thus, arrivals form a homogeneous Poisson process with intensity λ. Each arrival brings work to the system, with task sizes $S_1, S_2, \ldots$, drawn independently from distribution G. Assume $\mathbb{E}S_1 = \mu^{-1} < \infty$, and let $D_i = A_i + S_i$ denote the departure time of the i-th task. The number of tasks in the system at time t is given by:

$$L(t) = \sum_{i=1}^{\infty} \mathbb{1}(A_i \le t < D_i).$$

This process is regenerative. Considering its non-delayed version, we define the first cycle with duration C_1 and workload Z_1. The mean and variance of C_1 and Z_1 are known; see McNeil [149]. Hence, the conditions of Proposition 8.8.2 are satisfied, confirming that $(L(t))$ is stationary and ergodic. The corresponding stationary and ergodic process is denoted by $(L^*(t))$.

A special case occurs when G is exponential with mean $\mathbb{E}S = 1/\mu$. Then, $(L(t))$ is a birth-and-death process, and if $L(0)$ is Poisson-distributed with mean $\rho = \lambda/\mu$, then $(L^*(t))$ is stationary. For general service time distributions G, constructing the stationary version of $(L^*(t))$ involves advanced techniques from point processes; see Rolski et al. [181].

For an ergodic $(L^*(t))$, the covariance function is given by:

$$R(t) = \mathbb{E}\left[(L^*(s) - \rho)(L^*(s + t) - \rho)\right]$$
$$= \lambda \int_t^{\infty} (1 - G(v)) \, dv, \tag{8.9.4}$$

see [212]. The asymptotic variance for M/G/∞ systems is:

$$\varsigma^2 = 2 \int_0^{\infty} \lambda \int_t^{\infty} (1 - G(v)) \, dv \, dt.$$

In the special case of exponentially distributed service times $S \sim \mathrm{Exp}(\mu)$, this simplifies to:

$$\varsigma^2 = 2\lambda/\mu = 2\rho.$$

Furthermore, when the system starts empty, i.e., $L(0) = 0$, the bias is:

$$\rho - \mathbb{E}L(t) = \lambda \int_t^{\infty} (1 - G(s)) \, ds.$$

The steady-state mean is of particular interest:

$$\bar{l} = \lim_{t \to \infty} \frac{\int_0^t L(s)\, ds}{t}.$$

In simulations, $\bar{l}$ is estimated using:

$$\hat{L}(t) = \frac{1}{t} \int_0^t L(s)\, ds,$$

which, although biased if $L(0) = 0$, is strongly consistent:

$$\hat{L}(t) \to \rho$$

with probability 1 as $t \to \infty$.

For an M/D/∞ system (deterministic service times), the covariance function is:

$$R(t) = \begin{cases} \rho(1 - \mu t), & \text{for } t \le 1/\mu, \\ 0, & \text{otherwise.} \end{cases}$$

As mentioned earlier, for an M/M/∞ system (where $S \sim \text{Exp}(\mu)$), initializing $L^*(0)$ as Poisson with mean ρ ensures stationarity. The process $(L(t))$ is a birth-and-death process with birth rates $\lambda_n = \lambda$ and death rates $\mu_n = n\mu$. In this case, the asymptotic variance simplifies to:

$$\varsigma^2 = 2\rho.$$

In Sect. 8.4, we discussed basic properties of M/G/∞ systems. When simulating the process $L(t)$ over the interval $(0, t_{\max}]$, it is important to correctly handle tasks where $A_i \le t_{\max}$ and $A_{i+1} > t_{\max}$. In this case, the i-th task contributes only the interval $[A_i, \min(t_{\max}, A_i + S_i))$. To estimate the steady-state mean:

$$\bar{l} = \lim_{t \to \infty} \frac{\int_0^t L(s)\, ds}{t},$$

a natural estimator is:

$$\hat{l} = \frac{\int_0^{t_{\max}} L(s)\, ds}{t_{\max}}.$$

Expanding $\int_0^{t_{\max}} L(s)\, ds$, we obtain:

$$\int_0^{t_{\max}} L(s)\, ds = \sum_i \int_0^{t_{\max}} \mathbb{1}(A_i \le s < A_i + S_i)\, ds$$

$$= \sum_{i:A_i+S_i \leq t_{\max}} S_i + \sum_{i:A_i < t_{\max} \leq A_i+S_i} (t_{\max} - A_i)$$

$$= \sum_i [\min(t_{\max}, A_i + S_i) - A_i]. \tag{8.9.5}$$

Thus:

$$\min(t_{\max}, A_i + S_i) - A_i = \begin{cases} S_i, & \text{if } A_i + S_i \leq t_{\max}, \\ t_{\max} - A_i, & \text{if } A_i + S_i > t_{\max}. \end{cases}$$

The following Algorithm 82 describes the simulation of $\hat{l}$ for M/G/∞ systems, assuming that arrival times A_i and service times S_i are generated as described.

Algorithm 82 Simulation of $\hat{l}$ for M/G/∞ systems

1: Initialize $t_{\max}$, λ, and service time distribution G.
2: Set $L \leftarrow 0, T \leftarrow 0$.
3: **while** arrival times $A_i \leq t_{\max}$ **do**
4:     Generate $A_i \sim \text{Exp}(\lambda)$.
5:     Generate $S_i \sim G$.
6:     **if** $A_i + S_i \leq t_{\max}$ **then**
7:         $L \leftarrow L + S_i$.
8:     **else**
9:         $L \leftarrow L + (t_{\max} - A_i)$.
10:     **end if**
11: **end while**
12: Compute $\hat{l} = \frac{L}{t_{\max}}$. **return** $\hat{l}$.

We now present simulation results for different service time distributions. In all cases, the arrival rate is $\lambda = 5$, and the mean service time is $\mathbb{E}S = 1$, resulting in $\bar{l} = 5$. The distribution of S varies across different scenarios to examine its effect on system behavior.

Output Analysis Note that $\bar{l} = \rho = 5.0$. For these scenarios, we perform an output analysis. Assuming an error $b = 0.1$ and recalling that $z_{0.975} \approx 2$, formula (8.7.10) implies $t \geq 400\varsigma^2$. Here, we neglect the fact that the simulated process is not initially stationary. However, these processes exhibit a useful property: after a so-called warm-up period, they become stationary.

Figure 8.14 shows the simulated $\hat{L}(t)$ with $t_{\max} = 10^4$ and various service time distributions S. The following scenarios and their corresponding output analyses are discussed:

- $\lambda = 5$, $S \sim \text{Exp}(1)$. In this case, $R(t) = 5 \exp(-t)$, so:

$$\varsigma^2 = 2 \int_0^\infty R(s)\, ds = 10.$$

Hence, $400\varsigma^2 = 4000$. The simulation is shown in Fig. 8.14a, yielding an estimate $\hat{l} = 4.9848$, close to $\bar{l} = 5$. We can see that for $t \geq 4000$ the estimate stabilizes.

- $\lambda = 5$, $S = 2.2X$, where $X \sim \mathrm{Par}(3.2)$. Here:

$$R(t) = 5 \int_t^\infty \left(1 + \frac{s}{2.2}\right)^{-3.2} ds = 5 \left(1 + \frac{t}{2.2}\right)^{-2.2}.$$

Thus:

$$\varsigma^2 = 2 \int_0^\infty R(t)\, dt = 18.3,$$

implying $t > 7320$. The results are presented in Fig. 8.14b; indeed, the estimate stabilizes around 7000–8000 steps. The simulation yields a very good estimate $\hat{l} = 4.9925$ for $\bar{l} = 5.0$.

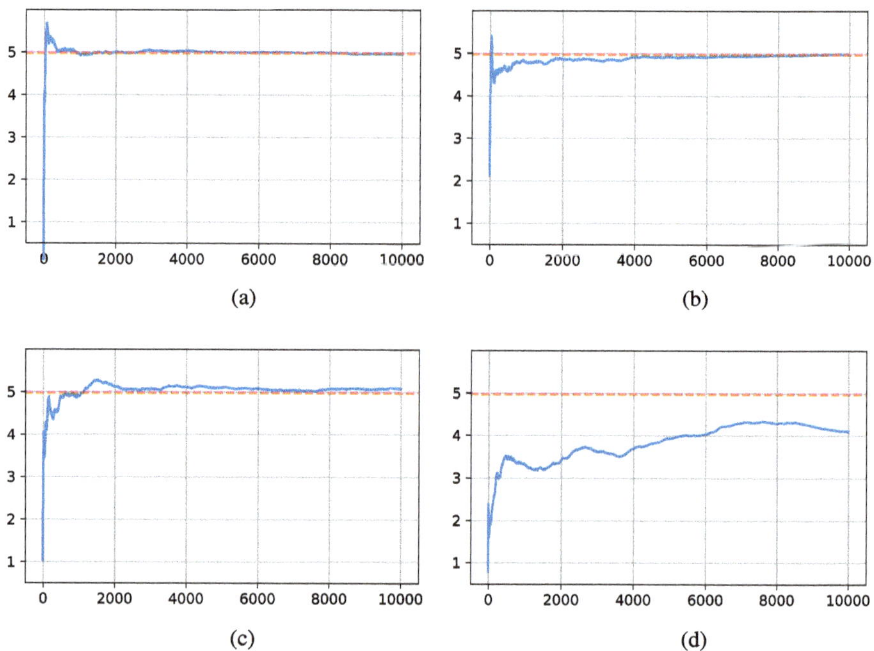

Fig. 8.14 Simulations of $\hat{L}(t)$ for $t \leq t_{\max} = 10^4$. Final estimates $\hat{l} = \hat{L}(t_{\max})$ of $\bar{l} = 5.0$ (*dashed red*) are shown. (**a**) $S \sim \mathrm{Exp}(1)$, $\hat{l} = 4.9848$. (**b**) $S = 2.2 \cdot \mathrm{Par}(3.2)$, $\hat{l} = 4.9925$. (**c**) $S = 1.2 \cdot \mathrm{Par}(2.2)$, $\hat{l} = 5.0970$. (**d**) $S = 0.2 \cdot \mathrm{Par}(1.2)$, $\hat{l} = 4.1175$

- $\lambda = 5$, $S = 1.2X$, where $X \sim \mathrm{Par}(2.2)$. Note that:

$$R(t) = 5 \int_t^\infty \left(1 + \frac{s}{1.2}\right)^{-2.2} ds = 5 \left(1 + \frac{t}{1.2}\right)^{-1.2}.$$

Hence:

$$\varsigma^2 = 2 \int_0^\infty R(t)\, dt = 60,$$

so $t > 24000$. The results are shown in Fig. 8.14c. Although the estimate is quite good ($\hat{l} = 5.0970$), by "eye inspection" it has not clearly converged to 5.0 (compare to cases in Fig. 8.14a–b, where it evidently did), which aligns with the theory, since we used $t_{\max} = 10^4 < 24000$ steps.
- $\lambda = 5$, $S = 0.2X$, where $X \sim \mathrm{Par}(1.2)$. The erratic behavior in Fig. 8.14d is explained by:

$$R(t) = 5 \int_t^\infty \left(1 + \frac{s}{0.2}\right)^{-1.2} ds = 5 \left(1 + \frac{t}{0.2}\right)^{-0.2},$$

which leads to:

$$\varsigma^2 = 2 \int_0^\infty R(t)\, dt = \infty.$$

Thus, the fundamental formula (8.7.10) does not apply, even though ergodic theory guarantees $\hat{L}(t) \to 5$ with probability 1. The simulation results show a poor estimate: $\hat{l} = 4.1175$.

In conclusion, the rate of convergence to $\bar{l} = 5$ strongly depends on the heaviness of the service time distribution tails. Although all Pareto distributions are considered heavy-tailed, the finiteness of moments depends on α. Specifically:

- If $1 \le \alpha < 2$, only the first moment is finite.
- If $2 \le \alpha < 3$, both the mean and variance are finite.
- If $\alpha > 3$, the third moment $\mathbb{E}S^3$ is finite.

All computations for this example, including the plots in Fig. 8.14, are performed using the script `ch8_mginfty.py`.

More Remarks on TAVC   In queueing theory, the concept of "traffic intensity" ϱ is widely used. For single-server systems, it is given by $\varrho = \rho = \mathbb{E}S/\mathbb{E}T$, where stability requires $\rho < 1$. In multi-server systems, such as GI/G/c, the traffic intensity is defined as $\varrho = \mathbb{E}S/(c\mathbb{E}T)$, with stability similarly requiring $\varrho < 1$.

A common observation across both single- and multi-server systems is that, in the heavy traffic regime (i.e., as $\varrho \nearrow 1$), the asymptotic variance satisfies:

$$\varsigma^2 = O((1-\rho)^{-4}). \tag{8.9.6}$$

From this fundamental relationship, we infer that to maintain a constant absolute error as $\varrho \nearrow 1$, the required simulation run length $t_{\max}$ scales as:

$$t_{\max} = z^2_{1-\alpha}\varsigma^2(\rho) = O((1-\rho)^{-4}).$$

On the other hand, to maintain a constant relative error as $\rho \nearrow 1$, the run length $t_{\max}$ scales as:

$$t_{\max} = \frac{z^2_{1-\alpha}\varsigma^2_\rho}{I^2_\rho} = O((1-\rho)^{-2}),$$

where we assume that I_ρ grows as $O((1-\rho)^{-1})$, which holds in many cases. Notably, the quantity $\frac{\varsigma_\rho}{\sqrt{t}I_\rho}$ represents the coefficient of variation of the estimator $\hat{X}_\rho(t)$.

To illustrate the drastic increase in run length required in heavy traffic, consider the following numerical example. Define:

$$k^{\mathrm{rel}}(\rho) = (1-\rho)^{-2}.$$

For a constant relative error, the required run length increases by a factor of:

$$\frac{k^{\mathrm{rel}}(0.95)}{k^{\mathrm{rel}}(0.5)} = 100.$$

For a constant absolute error, the impact is even more extreme. Setting:

$$k^{\mathrm{abs}}(\rho) = (1-\rho)^{-4},$$

we see that the run length increases by a factor of:

$$\frac{k^{\mathrm{abs}}(0.95)}{k^{\mathrm{abs}}(0.5)} = 10\,000.$$

This demonstrates the significant computational burden imposed by the heavy traffic regime.

8.10   Bibliographical Notes

The Poisson process is a fundamental concept in the theory of point processes. A recommended reference for this topic is the book by Kingman [103]. The thinning method for simulating a non-homogeneous Poisson process is introduced in [125]. Discrete-event simulation has a rich literature, with recommended readings including Fishman [56] and Bratley et al. [19].

Classic texts on queuing theory include Kleinrock [105] and Gross and Harris [38], where the basic formulas for classical queuing models are derived.

The repairman problem is discussed in the papers by Naor [163]; see also Kopociński [111].

For ALOHA and similar protocols, refer to books such as Ross [183], p. 148, and Bremaud [21], pp. 108–110, 173–178. Additionally, relevant papers include those by Kelly [96] and Kelly and MacPhee [97], as well as Aldous [3].

A series of works on simulation planning in queueing theory was authored by Whitt and colleagues [196, 211–213]. Grassmann [70] has also made significant contributions to this area. Theoretical foundations were provided in papers on the Poisson equation, e.g., [65]. These ideas were further developed for Markovian queues by Steckley and Henderson [197] and Ni and Henderson [165].

The application of the central limit theorem (CLT) in continuous-time Markov chains (CTMCs) is particularly intricate. This was studied by Liu and Zhang in [128]. The relationship between stationary processes and non-delayed regenerative processes is an example of the broader theory on Palm measures for point processes and related stochastic processes; see Rolski [179] and Baccelli and Bremaud [12]. For regenerative processes, this relation was specifically examined by Thorisson [202].

Ergodic theory provides tools for understanding the stability of simulations. For discrete time, Breiman's book [20] is an excellent introduction. The foundational literature in this area begins with Doob's classic text [41].

Theoretical Exercises

8.T.1   Let N be a nonhomogeneous Poisson process with intensity $\lambda(t)$. Given that n arrivals occurred before time t (i.e., $N(t) = n$), show that the arrival times $0 < A_1 < \cdots < A_n$ are distributed as the ordered i.i.d. random variables with c.d.f.

$$F(s) = \frac{\int_0^s \lambda(r)\,dr}{\int_0^t \lambda(r)\,dr},$$

i.e., prove Lemma 8.2.5.

8.T.2   Consider a general Poisson process N on $E \subset \mathbb{R}^d$ with intensity measure Λ, and assume that it has a density $\Lambda(d\mathbf{x}) = \lambda(\mathbf{x})d\mathbf{x}$, which is almost surely bounded by $\lambda_{\max}$. Let N_h be a homogeneous Poisson process on E with intensity $\lambda_{\max}$. Consider the following thinning procedure: if there is a point

$\mathbf{x}$ in N_h, it is retained with probability $\lambda(\mathbf{x})/\lambda_{\max}$; otherwise, it is removed. Show that the remaining points form the process N.

8.T.3 Let N_1 and N_2 be independent Poisson processes on the state space $E \subset \mathbb{R}^d$, with intensity measures Λ_1 and Λ_2, respectively. Show that $N = N_1 + N_2$ is a Poisson process on E with intensity measure $\Lambda = \Lambda_1 + \Lambda_2$. Here, $N = N_1 + N_2$ means that for all $A \subset E$,

$$N(A) = N_1(A) + N_2(A).$$

8.T.4 Consider a finite birth-and-death (B&D) process from Sect. 8.3. Show that the stationary distribution is given by:

$$\pi_j = \pi_0 \frac{\lambda_0 \lambda_1 \ldots \lambda_{j-1}}{\mu_1 \ldots \mu_j}, \quad j = 1, 2, \ldots, n.$$

8.T.5 Recall that the stationary distribution of a birth-and-death system with arbitrary birth rates $\lambda_i > 0$ and death rates $\mu_i > 0$ is given in (8.7.13).

(a) The M/M/1/n queueing system is a special case of a birth-and-death process with constant rates, i.e., $\lambda_i = \lambda > 0$ for $i = 0, \ldots, n-1$, and $\mu_i = \mu > 0$ for $i = 1, \ldots, n$. Show that the stationary distribution simplifies to:

$$\pi_j = (1 - \rho) \frac{\rho^j}{1 - \rho^{n+1}}, \quad j = 0, \ldots, n,$$

where $\rho = \lambda/\mu$.

(b) Argue that in the M/M/1 queueing system with an infinite buffer, under the condition $\rho < 1$, the stationary distribution is:

$$\pi_i = (1 - \rho)\rho^i, \quad i = 0, 1, \ldots$$

Conclude that for the stationary number of customers in the system, denoted by L^*, we have:

$$\mathbb{E}L^* = \frac{\rho}{1 - \rho}, \quad \mathbb{V}\text{ar } L^* = \frac{\rho}{(1 - \rho)^2}.$$

8.T.6 Consider the M/M/c queueing system with an infinite buffer, arrival rate λ, and service rate μ, where the stability condition $\rho = \lambda/\mu < c$ holds.

(a) Show that the stationary distribution of the number of customers L^* in the system is given by:

$$\pi_0 = S^{-1}, \quad S = \left[\sum_{i=0}^{c-1} \frac{\rho^i}{i!} + \frac{\rho^c}{c!} \frac{c}{c - \rho} \right],$$

and

$$\pi_i = \begin{cases} S^{-1}\dfrac{\rho^i}{i!}, & 0 < i < c, \\[2ex] S^{-1}\dfrac{\rho^i c^{c-i}}{c!}, & i \geq c. \end{cases}$$

(b) Show that the probability that an arriving customer must wait (i.e., all servers are occupied) is:

$$C(c, \rho) = \pi_c + \pi_{c+1} + \cdots = S^{-1}\frac{\rho^c}{c!}\frac{c}{c - \rho},$$

which can be rewritten as:

$$C(c, \rho) = \frac{1}{1 + (1 - \rho/c)\frac{c!}{\rho^c}\sum_{i=0}^{c-1}\frac{\rho^i}{i!}}.$$

This formula is known as the *Erlang C formula*.
(c) Show that:

$$\mathbb{E}L^* = S^{-1}\left[\sum_{i=1}^{c-1}\frac{i\rho^i}{i!} + \frac{\rho^c}{c!}\left(\frac{c^2}{c - \rho} + \frac{c\rho}{(c - \rho)^2}\right)\right]$$

$$= \frac{\rho}{c - \rho}C(c, \rho) + \rho.$$

Using Little's formula $\mathbb{E}L^* = \lambda(\mathbb{E}W^* + \mathbb{E}S)$, we obtain:

$$\mathbb{E}W^* = \frac{\mathbb{E}L^*}{\lambda} - \frac{1}{\mu},$$

which is known as the "$L = \lambda W$" formula, where W represents the sojourn time in the system. Since $\mathbb{E}S = 1/\mu$, the mean waiting time in the M/M/c system is given by:

$$\mathbb{E}W^* = \frac{C(c, \rho)}{c\mu - \lambda}. \tag{8.10.1}$$

8.T.7 Consider a slotted ALOHA system, where A_i follows a Poisson distribution with intensity λ. Show that for $k = 0, 1, \ldots$, we have:

$$\phi(k) = \mathbb{E}(X_{n+1} - X_n \mid X_n = k) = \lambda - e^{-\lambda}kh(1 - h)^{k-1} - e^{-\lambda}\lambda(1 - h)^k.$$

Plot $\phi(k)$ for selected values of λ and h, specifically for $\lambda \in \{0.3, 0.4\}$ and $h \in \{0.02, 0.05, 0.1\}$.

8.T.8 For the ALOHA model, given the number of awaiting packets X_n, compute:

$$g_k(\lambda) = \mathbb{P}(Z_n = 1 \mid X_n = k).$$

Determine the limit:

$$g(\lambda) = \lim_{k \to \infty} g_k(\lambda).$$

Note that for stability, it must hold that $g(\lambda) < \lambda$. Find the critical value λ_0 such that:

$$g(\lambda_0) = \lambda_0,$$

which occurs approximately at $\lambda_0 \approx 0.35$.

8.T.9 Show that for an *on-off* process with $T^{\mathrm{on}} \sim \mathrm{Exp}(\lambda)$ and $T^{\mathrm{off}} \sim \mathrm{Exp}(\mu)$, the asymptotic variance (8.7.11) simplifies to:

$$\varsigma^2 = 2\frac{\lambda\mu}{(\lambda + \mu)^3} = \frac{2\rho}{\mu(1 + \rho)^3},$$

where $\rho = \lambda/\mu$.

8.T.10 Consider an M/M/1 queueing system and the steady-state number of tasks $L^*(0)$.

(i) Derive the steady-state distribution of $L^*(0)$:

$$\mathbb{P}(L^*(0) = k) = (1 - \rho)\rho^k, \qquad k = 0, 1, \dots,$$

and verify that the stability condition requires $\rho = \lambda/\mu < 1$.

(ii) Derive the expectation and variance:

$$\mathbb{E}L^*(0) = \frac{\rho}{1 - \rho} \quad \text{and} \quad \mathbb{V}\mathrm{ar}L^* = \frac{\rho}{(1 - \rho)^2}.$$

(iii) For M/M/1 queueing systems, show that the total asymptotic variance constant (TAVC) given in (8.7.14) simplifies to:

$$\varsigma^2 = \frac{2\rho(1 + \rho)}{\mu(1 - \rho)^4}.$$

This derivation assumes that (8.7.14) remains valid for infinite-state birth-and-death processes under suitable conditions, such as the existence of a stationary distribution. Here, $\rho = \lambda/\mu < 1$, $f_i = i$, and $\pi_i = (1 - \rho)\rho^i$. This formula has already been derived from the covariance function $R(t)$ in (8.9.3).

8.T.11 For an *on-off* process with exponential *on* and *off* times, derive formulas for the steady-state probability p that the system is *on* and the total asymptotic variance constant (TAVC) ς^2.

8.T.12 For *on-off* systems from Example 8.5.1, show that the estimator $\hat{\rho}(t)$ given in (8.5.1) is consistent.

8.T.13 Consider an M/G/∞ system with service time $S = (\alpha - 1)X$, where $X \sim$ Par(α) with $\alpha > 2$. Derive the formula for the total asymptotic variance constant (TAVC) ς^2.

8.T.14 Show that the *on-off* process, where the *on*-time and *off*-time durations follow exponential distributions $T^{\mathrm{on}} \sim \mathrm{Exp}(\mu)$ and $T^{\mathrm{off}} \sim \mathrm{Exp}(\lambda)$, is a special case of a birth-and-death process with $c = 1$, $\lambda_0 = \lambda$, and $\mu_1 = \mu$. Using the TAVC formula (8.7.14):

$$\varsigma_f^2 = 2 \sum_{j=0}^{c-1} \frac{1}{\lambda_j \pi_j} \left[\sum_{i=0}^{j} (f(i) - I_f)\pi_i \right]^2 ,$$

derive the asymptotic variance:

$$\varsigma^2 = 2 \frac{\lambda\mu}{(\lambda + \mu)^3}.$$

Hint: Recall that for $c = 1$, $\pi_0 = \frac{\mu}{\lambda+\mu}$ and $\pi_1 = \frac{\lambda}{\lambda+\mu}$, and use $f(0) = 0$, $f(1) = 1$.

Lab Exercises

8.L.1 Generate $A_1, A_2, \ldots$ in the interval $[0, 1000]$, where $A_0 = 0$ and $A_1 < A_2 < \cdots$ are consecutive arrival times of a non-homogeneous Poisson process with intensity function:

$$\lambda(t) = a \left(2 - \sin\left(\frac{2\pi}{24}t\right) \right).$$

Denote the inter-arrival times as $\tau_i = A_i - A_{i-1}$, $i = 1, \ldots, 1000$. Let $A(t)$ be the number of arrivals in the interval $[0, t]$. Consider $a = 1, 2$.

- Derive the distribution of $A(t)$ along with its mean and variance. Using simulations, compute $\mathbb{E}A(1000)/1000$—the average number of arrivals per unit interval (often denoted as $\bar{\lambda}$ and referred to as the *asymptotic intensity*). Compare the results with the theoretical value:

$$\bar{\lambda} = \frac{1}{24} \int_0^{24} \lambda(t)\, dt.$$

Perform an error analysis.

- Using simulations, estimate the average distance between points, i.e.,

$$\hat{\tau} = \frac{1}{1000} \sum_{j=1}^{1000} \tau_j.$$

Note that $\{\tau_j\}$ is not a sequence of i.i.d. random variables. Compare $\bar{\lambda}$ and $\bar{\tau}^{-1}$. Perform an error analysis.

8.L.2 Let $N(t)$ be a Poisson process with intensity $\lambda(t) = 2t$. Simulate the process over the interval $[0, 10]$ using two methods: the thinning procedure given in Algorithms 75 and 76.

8.L.3 Consider the following two systems:

1. A parallel system consisting of N elements with a single repairman. Assume that the working time follows an $\text{Exp}(1/2)$ distribution, while the repair time follows an $\text{Exp}(2)$ distribution.
2. A parallel system consisting of N elements without a repairman. Again, assume that the working time follows an $\text{Exp}(1/2)$ distribution.

Using simulations, estimate (and compare for both systems) the mean and variance of the time until system failure for $N = 1, \ldots, 10$.

Compute the mean as the expectation of the maximum of N exponentially distributed random variables.

8.L.4 Consider the steady-state waiting time $W^{\text{M/G}/c}$ in the queueing system M/G/c. Kingman's law of congestion provides the approximation:

$$\mathbb{E}W^{\text{M/G}/c} \approx \frac{\text{CV}_S^2 + 1}{2} \mathbb{E}W^{\text{M/M/c}},$$

where CV_S is the coefficient of variation of the service time distribution, given by:

$$\text{CV}_S = \frac{\sqrt{\mathbb{V}\text{ar}\,S}}{\mathbb{E}S}.$$

Ward Whitt described this approximation as *"usually an excellent approximation, even given extra information about the service-time distribution.* For the M/G/1 system, this equality holds exactly.

For M/M/c systems (where $F \sim \text{Exp}(\lambda)$, $G \sim \text{Exp}(\mu)$, and $\rho = \lambda/\mu$), the mean waiting time is given by:

$$\mathbb{E}W^{\text{M/G}/c} = \frac{C(c, \rho)}{c\mu - \lambda},$$

where $p_W = C(c, \rho)$ is the probability that an arriving customer has to wait; see Exercise 8.T.6.

The variance of the waiting time W^* in the queueing system M/M/c is known, it is given by:

$$\mathbb{V}\mathrm{ar}(W^*)^2 = \frac{p_w(C(c, \rho)(2 - C(c, \rho)))}{[\mu(c - \rho)]^2}.$$

Verify Kingman's law of congestion through simulations. Try different service time distributions (both light- and heavy-tailed) and varying values of $\rho = \lambda \mathbb{E}S$. For a steady-state solution to exist, we require $\rho < c$, which can also be observed from simulations.

8.L.5 Consider random access protocols from Sect. 8.6:

- $h_i = h$—slotted ALOHA,
- $h_i = 2^{-i}, i = 1, \ldots, K$, where $K < \infty$—ETHERNET. Since ALOHA and exponential back-off are not stable, in practice, the number of retransmission attempts is limited to K. If a packet still cannot be transmitted after K attempts, it is lost.

Verify the effectiveness of these protocols through simulation. In particular, analyze the throughput for values of λ close to 0.35, both below and above this threshold. Conduct experiments for different values of $\lambda \in \{0.2, 0.3, 0.4\}$ and for $h = 0.02$ and $h = 0.005$.

Next, consider an unrealistic protocol:

- $h_i = 1/X_n$—If users have knowledge of X_n just before transmission, the throughput remains finite.

Simulate the system with aforementioned $\lambda \in \{0.2, 0.3, 0.4\}$. What observations can you make? Under what conditions does the protocol appear to be stable?

8.L.6 Consider the repairman problem described in Example 8.5.2 with $N = 10$ elements, $M = 3$ repairmen, and the following distributions:

- Working time distribution F: Uniform on $[5, 15]$.
- Repair time distribution G with density $g(x) = x/30$ for $x \in (2, 8)$.

(a) Implement Algorithm 79 with $t_{\max} = 5000$ and estimate the steady-state metrics:

  - l—number of working elements,
  - m—number of elements being serviced,
  - w—number of elements waiting for service,
  - $r = M - m$—number of unoccupied repairmen.

(b) Modify Algorithm 79 to estimate the steady-state sojourn time (i.e., the total time from failure to repair) using the same parameters.

8.L.7 The following construction provides a stationary and ergodic *on-off* process $(X^*(t))$. The process starts in state 1 with probability p and in state 0 with probability $1 - p$, where

$$p = \frac{\mathbb{E}T^{\mathrm{on}}}{\mathbb{E}T^{\mathrm{on}} + \mathbb{E}T^{\mathrm{off}}}.$$

If $X^*(0) = 1$, the process remains in state 1 for a duration of $\tilde{T}^{\mathrm{on}}$, and if $X^*(0) = 0$, it remains in state 0 for a duration of $\tilde{T}^{\mathrm{off}}$, where

$$\tilde{T}^{\mathrm{on}} \sim \tilde{F}^{\mathrm{on}}(x) = \frac{1}{\mathbb{E}T^{\mathrm{on}}} \int_0^x \bar{F}^{\mathrm{on}}(t)\,dt,$$

$$\tilde{T}^{\mathrm{off}} \sim \tilde{F}^{\mathrm{off}}(x) = \frac{1}{\mathbb{E}T^{\mathrm{off}}} \int_0^x \bar{F}^{\mathrm{off}}(t)\,dt.$$

After each state duration, the process transitions to the next state: from state 1 to state 0 or from state 0 to state 1. The time spent in the new state is generated according to the distributions F^{on} and F^{off}, respectively.

Simulate the *on-off* process assuming that F^{on} follows a uniform distribution $\mathcal{U}(0, m_0)$ and F^{off} follows a uniform distribution $\mathcal{U}(0, m_1)$. Can you observe that the process $X^*(t)$ is stationary and ergodic?

What can you say about the process $(X^*(t))$ when F^{on} and F^{off} are exponential? Find the distributions of $\tilde{T}^{\mathrm{on}}$ and $\tilde{T}^{\mathrm{off}}$.

Appendix A

A.1 Special Functions

The gamma function is defined as:

$$\Gamma(\alpha) = \int_0^\infty x^{\alpha-1} e^{-x}\, dx, \quad x > 0.$$

For a positive integer k, we have $\Gamma(k+1) = k!$. Additionally, $\Gamma(1/2) = \sqrt{\pi}$.

A.2 Families of Distributions

Random variables (r.v.s) are typically denoted by X, Y, W, Z, etc. We assume they are defined on a probability space $(\Omega, \mathcal{F}, \mathbb{P})$. Note that $\mathcal{F}$ will not be used explicitly in this book. Throughout, the cumulative distribution function (c.d.f.) of a random variable X is denoted by $F(x) = \mathbb{P}(X \leq x)$. If X has a density function (p.d.f.), we denote it by $f(x)$ (in the case of absolutely continuous distributions). Otherwise, for discrete distributions, the probability function is given by $\mathbb{P}(X = k) = p_k$.

We now introduce several classes of discrete and absolutely continuous distributions. Each of these distributions is characterized by a finite set of parameters.

Within the class of absolutely continuous distributions, we distinguish two subcategories: distributions with light tails and those with heavy tails. This distinction is fundamental in Monte Carlo theory.

A distribution with c.d.f. $F(x)$ is said to have an *exponentially bounded tail* if, for some $a, b > 0$, we have $\overline{F}(x) \leq ae^{-bx}$ for all $x > 0$. In such cases, we say the distribution has a *light tail*. Consequently, the moment-generating function satisfies $\mathbb{E}e^{sX} < \infty$ for some $0 < s < s_0$, where $s_0 > 0$ and $X \sim F$.

Otherwise, we say the distribution has a *heavy tail*, meaning that:

$$\mathbb{E}e^{sX} = \infty \quad \text{for all } s > 0.$$

We now state a useful result regarding the transformation of a random vector $\mathbf{X}$ with p.d.f. $f_{\mathbf{X}}(\mathbf{x})$.

Proposition A.2.1 *Let $\mathbf{g}(\mathbf{x}) = (g_1(\mathbf{x}), \ldots, g_d(\mathbf{x}))$, where $\mathbf{x} \in G \subset \mathbb{R}^d$. Consider a random vector $\mathbf{X} \in \mathbb{R}^d$ such that $\mathbb{P}(\mathbf{X} \in G) = 1$, with p.d.f. $f_{\mathbf{X}}(\mathbf{x})$. Assume the following conditions hold:*

1. *G is an open subset of $\mathbb{R}^d$.*
2. *The mapping $\mathbf{g} = (g_1, \ldots, g_d)^T$ is one-to-one onto the set $H = \mathbf{g}(G)$.*
3. *The first partial derivatives $\left(\frac{\partial g_i(\mathbf{x})}{\partial x_j}\right)_{i,j=1,\ldots,d}$ are continuous on G.*
4. *The Jacobian determinant $J_{\mathbf{g}}(\mathbf{x}) = \det\left(\frac{\partial g_i(\mathbf{x})}{\partial x_j}\right)_{i,j=1,\ldots,d}$ is nonzero on G.*

Let $\mathbf{h} = (h_1, \ldots, h_d)^T$ be the inverse mapping of $\mathbf{g}$, i.e., for each $\mathbf{y} \in \mathbf{g}(G)$, $\mathbf{x} = \mathbf{h}(\mathbf{y}) = (h_1(\mathbf{y}), \ldots, h_d(\mathbf{y}))^T$. The transformed random vector $\mathbf{Y} = \mathbf{g}(\mathbf{X})$ has p.d.f.:

$$f_{\mathbf{Y}}(\mathbf{y}) = f_{\mathbf{X}}(h_1(\mathbf{y}), \ldots, h_d(\mathbf{y})) \, |J_{\mathbf{h}}(\mathbf{y})| \, \mathbb{1}(\mathbf{y} \in \mathbf{g}(G)), \tag{A.1}$$

where $J_{\mathbf{h}}(\mathbf{y}) = \det\left(\frac{\partial h_i(\mathbf{y})}{\partial y_j}\right)_{i,j=1,\ldots,d}$ is the Jacobian of the transformation $\mathbf{h}$.

Note that $J_{\mathbf{h}}(\mathbf{y}) = (J_{\mathbf{g}}(\mathbf{x}))^{-1}$.

A.3 Discrete Distributions

We introduce the following distributions:

- **Degenerate distribution** δ_a, concentrated at $a \in \mathbb{R}$, with probability function $p(x) = 1$ if $x = a$ and $p(x) = 0$ otherwise.
- **Binomial distribution** $B(n, p)$, with probability function:

$$p_k = \binom{n}{k} p^k (1 - p)^{n-k}, \quad k = 0, 1, \ldots, n; \quad n \in \mathbb{Z}_+, \quad 0 < p < 1.$$

- **Poisson distribution** $Poi(\lambda)$, with:

$$p_k = \frac{e^{-\lambda}\lambda^k}{k!}, \quad k = 0, 1, \ldots; \quad 0 < \lambda < \infty.$$

Moreover, if $np \to \lambda$ as $n \to \infty$, then $B(n, p) \xrightarrow{\mathcal{D}} Poi(\lambda)$.

- **Geometric distribution** Geo(p), with:

$$p_k = (1 - p)p^k, \quad k = 0, 1, 2, \ldots; \quad 0 < p < 1.$$

- **Negative binomial distribution** or **Pascal distribution**, denoted NB(r, p), has the probability mass function:

$$p_k = \frac{\Gamma(r + k)}{\Gamma(r)\Gamma(k + 1)}(1 - p)^r p^k, \quad k = 0, 1, \ldots; \quad r > 0, \quad 0 < p < 1.$$

Using generalized binomial coefficients, it can also be written as:

$$p_k = \binom{k + r - 1}{k}(1 - p)^r p^k, \quad k = 0, 1, \ldots,$$

where the generalized binomial coefficients are given by:

$$\binom{x}{0} = 1, \quad \binom{x}{k} = \frac{x(x - 1) \cdots (x - k + 1)}{k!}, \quad \text{for } x \in \mathbb{R}, \quad k = 1, 2, \ldots.$$

Thus, we also have:

$$p_k = \binom{\alpha + k - 1}{k}(1 - p)^\alpha p^k = \binom{-\alpha}{k}(1 - p)^\alpha (-p)^k.$$

Moreover, for $r = 1, 2, \ldots$, the distribution NB(r, p) is the distribution of the sum of r i.i.d. Geo(p) random variables. The expected value is given by:

$$\mathbb{E}X = \frac{rp}{q}, \quad \text{where } q = 1 - p.$$

- **Logarithmic distribution**, denoted Log(p), has the probability mass function:

$$p_k = \frac{p^k}{-k \log(1 - p)}, \quad k = 1, 2, \ldots; \quad 0 < p < 1.$$

- **Discrete uniform distribution**, denoted UD(n), has the probability mass function:

$$p_k = \frac{1}{n}, \quad k = 1, \ldots, n; \quad n = 1, 2, \ldots.$$

A.3.1 Discrete Multivariate Distributions

- **Multinomial distribution** $M(n, \mathbf{a})$, defined by the probability mass function:

$$p_{\mathbf{k}} = \frac{n!}{k_1! \ldots k_d!} a_1^{k_1} \ldots a_d^{k_d},$$

for $\mathbf{k} = (k_1, \ldots, k_d)$ such that $k_i \in \mathbb{Z}_+$ and $k_1 + \cdots + k_d = n$, and where $\mathbf{a} = (a_1, \ldots, a_d)$ satisfies $a_i > 0$ and $a_1 + \cdots + a_d = 1$.

A.3.2 Continuous Univariate Distributions

We have:

- **Normal distribution** $\mathcal{N}(\mu, \sigma^2)$ with density function

$$f(x) = (2\pi\sigma^2)^{-1/2} e^{-(x-\mu)^2/(2\sigma^2)},$$

for all $x \in \mathbb{R}$, where $\mu \in \mathbb{R}$ and $\sigma^2 > 0$.
- **Exponential distribution** $\text{Exp}(\lambda)$ with density function

$$f(x) = \lambda e^{-\lambda x}, \quad x > 0; \ \lambda > 0.$$

- **Erlang distribution** $\text{Erl}(n, \lambda)$ with density function

$$f(x) = \frac{\lambda^n x^{n-1} e^{-\lambda x}}{(n-1)!}, \quad n = 1, 2, \ldots$$

- **Chi-square distribution** χ_n^2 with density function

$$f(x) = \frac{x^{(n/2)-1} e^{-x/2}}{2^{n/2} \Gamma(n/2)}$$

for $x > 0$ and $f(x) = 0$ for $x \leq 0$; $n = 1, 2, \ldots$. It is the p.d.f. of $\chi^2 = Z_1^2 + \cdots + Z_n^2$, where $Z_1, \ldots, Z_n$ are i.i.d. $\mathcal{N}(0, 1)$ random variables.
- **Gamma distribution** $\text{Gamma}(a, \lambda)$ with density function

$$f(x) = \frac{\lambda^a x^{a-1} e^{-\lambda x}}{\Gamma(a)}, \quad x \geq 0;$$

where $a > 0$ is the shape parameter and $\lambda > 0$ is the scale parameter. If $a = n \in \mathbb{N}$, then $\text{Gamma}(n, \lambda)$ is equivalent to $\text{Erl}(n, \lambda)$. If $\lambda = 1/2$ and $a = n/2$, then Gamma$(n/2, 1/2)$ corresponds to a chi-square distribution with n degrees of freedom.

- **Uniform distribution** $\mathcal{U}[a, b)$ with density function

$$f(x) = \frac{1}{b - a}, \quad a < x < b; \ -\infty < a < b < \infty.$$

- **Beta distribution** Beta(a, b, η) with density function

$$f(x) = \frac{x^{a-1}(\eta - x)^{b-1}}{B(a, b)\eta^{a+b-1}},$$

for $0 < x < \eta$, where $B(a, b)$ denotes the beta function, and $a, b, \eta > 0$. If $a = b = 1$, then

$$\text{Beta}(1, 1, \eta) = \mathcal{U}(0, \eta).$$

- **Extreme value distribution** $\mathrm{EV}(\gamma)$ with cumulative distribution function:

$$F(x) = \exp\left(-(1 + \gamma x)_+^{-1/\gamma}\right),$$

for all $\gamma \in \mathbb{R}$, where the expression $-(1 + \gamma x)_+^{-1/\gamma}$ is defined as e^{-x} when $\gamma = 0$. In this special case, the distribution is known as the *Gumbel distribution*. For $\gamma > 0$, we obtain the *Fréchet distribution*, which is concentrated on the half-line $(-1/\gamma, +\infty)$. Finally, for $\gamma < 0$, we obtain the *Weibull extreme value distribution*, which is concentrated on the half-line $(-\infty, -1/\gamma)$.

A.3.3 *Heavy-tailed Distributions*

We provide the following examples:

- **Log-normal distribution** $\mathrm{LN}(a, b)$ with probability density function:

$$f(x) = \frac{1}{xb\sqrt{2\pi}} \exp\left(-\frac{(\log x - a)^2}{2b^2}\right),$$

for $x > 0$, where $a \in \mathbb{R}$ and $b > 0$. If $X \sim \mathcal{N}(a, b^2)$, then $e^X \sim \mathrm{LN}(a, b)$. Moreover, if $X \sim \mathrm{LN}(a, b)$, then $\mathbb{E}X = \exp(a + b^2/2)$.

- **Weibull distribution** $W(r, c)$ with probability density function:

$$f(x) = rcx^{r-1} \exp(-cx^r),$$

for $x > 0$, where $r > 0$ is the shape parameter and $c > 0$ is the scale parameter. The tail probability is given by $\overline{F}(x) = \exp(-cx^r)$. If $r \geq 1$, then $W(r, c)$ is light-tailed; otherwise, it is heavy-tailed. The mean of $X \sim W(r, c)$ is:

$$\mathbb{E}X = \frac{\Gamma(1 + \frac{1}{r})}{c^{1/r}}.$$

- **Pareto distribution** $\mathrm{Par}(\alpha)$ with probability density function:

$$f(x) = \alpha \left(\frac{1}{1+x} \right)^{\alpha+1},$$

for $x > 0$, where $\alpha > 0$. The tail probability is given by:

$$\overline{F}(x) = \left(\frac{1}{1+x} \right)^{\alpha}.$$

- **Log-gamma distribution** $\mathrm{Log\text{-}Gamma}(a, \lambda)$ with probability density function:

$$f(x) = \frac{\lambda^a}{\Gamma(a)} (\log x)^{a-1} x^{-\lambda-1},$$

for $x > 1$, where $\lambda, a > 0$. If $X \sim \Gamma(a, \lambda)$, then e^X follows the $\mathrm{Log\text{-}Gamma}(a, \lambda)$ distribution.
- **Cauchy distribution** with probability density function:

$$f(x) = \frac{1}{\pi(1 + x^2)}, \quad x \in \mathbb{R}.$$

This distribution has no finite mean and is doubly heavy-tailed.

A.3.4 Multivariate Distributions

The most important example is the following:

- **Multivariate normal distribution** $\mathcal{N}(m, \Sigma)$. A random vector $\mathbf{X} = (X_1, \ldots, X_n)^T$ with mean $\mathbb{E}\mathbf{X} = \mu$ and covariance matrix Σ is said to have a multivariate normal distribution if every linear combination $\sum_{j=1}^{n} a_j X_j$ is normally distributed.

  Denoting $(a_1, \ldots, a_d)$ by a, the random variable $a^T \mathbf{X}$ has mean $a^T \mu$ and variance $a^T \Sigma a$.

  Let $\Sigma = \mathbf{A}\mathbf{A}^T$. If $\mathbf{Z} \sim \mathcal{N}((0, \ldots, 0), \mathbf{I})$, where $\mathbf{I}$ is the identity matrix, then $\mathbf{X} = \mathbf{A}\mathbf{Z}$ follows the $\mathcal{N}(m, \Sigma)$ distribution.

Bibliography

1. Abramowitz, M., Stegun, I.A.: Handbook of Mathematical Functions with Formulas, Graphs, and Mathematical Tables. National Bureau of Standards, Gaithersburg (1965)
2. Ahmed, S., Saleeby, E.G.: On volumes of hyper-ellipsoids. Math. Mag. **91**(1), 43–50 (2018)
3. Aldous, D.J.: Ultimate instability of exponential back-off protocol for acknowledgment-based transmission control of random access communication channels. IEEE Trans. Inf. Theory **33**, 219–223 (1987)
4. Alexopoulos, C., Seila, A.F.: Output data analysis for simulations. In: Proceeding of the 2001 Winter Simulation Conference (Cat. No. 01CH37304), vol. 1, pp. 115–122. IEEE (2001)
5. AlFardan, N., Bernstein, D.J., Paterson, K.G., Poettering, B., Schuldt, J.C.N.: On the Security of RC4 in TLS. In: On the Security of RC4 in TLS, pp. 305–320. USENIX (2013)
6. Asmussen, S.: Conditional Monte Carlo for sums, with applications to insurance and finance. Ann. Actuarial Sci. **12**, 455–478 (2018)
7. Asmussen, S.: Applied Probability and Queues, 2nd edn. Springer, New York (2003)
8. Asmussen, S., Glynn, P.W.: Stochastic Simulation: Algorithms and Analysis, vol. 57. Springer, New York (2007)
9. Asmussen, S., Rubinstein, R.Y.: Steady state rare events simulation in queueing models and its complexity properties 1. In: Advances in Queueing Theory, Methods, and Open Problems, pp. 429–462. CRC Press, Boca Raton (1995)
10. Asmussen, S., Glynn, P.W., Thorisson, H.: Stationarity detection in the initial transient problem. ACM Trans. Model. Comput. Simul. **2**(2), 130–157 (1992)
11. Athreya, K.B., Majumdar, M.: Estimating the stationary distribution of a Markov chain. Econ. Theory **21**, 729–742 (2003)
12. Baccelli, F., Brémaud, P.: Palm Probabilities and Stationary Queues, vol. 41. Springer Science & Business Media, Berlin (2012)
13. Bayer, D., Diaconis, P.: Trailing the dovetail shuffle to its lair. Ann. Appl. Probab. **2**, 294–313 (1992)
14. Billingsley, P.: Probability and Measure, 3rd edn. Wiley, Hoboken (1995)
15. Blackman, D., Vigna, S.: Scrambled linear pseudorandom number generators. ACM Trans. Math. Software (TOMS) **47**(4), 1–32 (2021)
16. Bloch, D.A., Gastwirth, J.L.: On a simple estimate of the reciprocal of the density function. Ann. Math. Stat. **39**, 1083–1085 (1968)
17. Blumenson, L.E.: Classroom notes: a derivation of n-dimensional spherical coordinates. Am. Math. Monthly **67**(1), 63–66 (1960)
18. Box, G.E.P., Muller, M.E.: A note on generating of random normal deviates. Ann. Math. Stat. **29**, 610–611 (1958)

© The Author(s), under exclusive license to Springer Nature Switzerland AG 2025

617

P. Lorek, T. Rolski, *Lectures on Monte Carlo Theory*, Probability Theory
and Stochastic Modelling 108, https://doi.org/10.1007/978-3-032-01190-9

19. Bratley, P., Fox, B.L., Schrage, L.E.: A Guide to Simulation. Springer, Berlin (1987)
20. Breiman, L.: Probability. SIAM, Philadelphia (1992)
21. Bremaud, P.: Markov Chains, Gibbs Fields, Monte Carlo Simulations, and Queues. Springer, Berlin (1999)
22. Brooks, S., Gelman, A., Jones, G., Meng, X.-L.: Handbook of Markov Chain Monte Carlo. CRC Press, Boca Raton (2011)
23. Brown, R.G., Eddelbuettel, D., Bauer, D.: Dieharder: A Random Number Test Suite. Duke University Physics Department Durham, NC 27708 (2018)
24. Chen, J., Rosenthal, J.S.: Decrypting classical cipher text using Markov chain Monte Carlo. Stat. Comput. **22**, 397–413 (2012)
25. Chu, F., Nakayama, M.K.: Confidence intervals for quantiles when applying variance-reduction techniques. ACM Trans. Model. Comput. Simul. **22** (2012)
26. Cochran, W.G.: Sampling Techniques. John Wiley & Sons Inc, Hoboken (1977)
27. Costa, A., Jones, O.D., Kroese, D.: Convergence properties of the cross-entropy method for discrete optimization. Oper. Res. Lett. **35**(5), 573–580 (2007)
28. Dagpunar, J.S.: Principle of Random Variate Generation. Clarendon Press, Oxford (1988)
29. Dagpunar, J.S.: Simulation and Monte Carlo: With Applications in Finance and MCMC. John Wiley & Sons, Hoboken (2007)
30. Damerdji, H.: Strong consistency of the variance estimator in steady-state simulation output analysis. Math. Oper. Res. **19**(2), 494–512 (1994)
31. de Boer, P.-T., Kroese, D.P., Mannor, S., Rubinstein, R.Y.: A tutorial on the cross-entropy method. Ann. Oper. Res. **134**, 19–67 (2005)
32. Dette, H., Fill, J.A., Pitman, J., Studden, W.J.: Wall and siegmund duality relations for birth and death chains with reflecting barrier. J. Theor. Probab. **10**, 349–374 (1997)
33. Devroye, L.: Nonuniform random variate generation. Handbooks in Operations Research and Management Science, vol. 13, pp. 83–121, Elsevier (2006)
34. Diaconis, P.: Mathematical developments from the analysis of riffle shuffling. Groups, Combinatorics & Geometry (Durham, 2001), pp. 73–97 (2003)
35. Diaconis, P.: The Markov chain Monte Carlo revolution. Bull. Am. Math. Soc. 1–24 (2009)
36. Diaconis, P., Fill, J.A.: Strong stationary times via a new form of duality. Ann. Probab. **18**, 1483–1522 (1990)
37. Do, C.B., Batzoglou, S.: What is the expectation maximization algorithm? Nature Biotechnol. **26**(8), 897–899 (2008)
38. Donald, G., Carl, M.H.: Fundamentals of Queueing Theory. Wiley, Hoboken (1985)
39. Dong, H., Nakayama, M.K.: Quantile estimation with Latin hypercube sampling. Oper. Res. **65**, 1678–1695 (2017)
40. Dong, H., Nakayama, M.K.: A tutorial on quantile estimation via Monte Carlo. In: Tuffin, B., L'Ecuyer, P. (eds.) Monte Carlo and Quasi-Monte Carlo Methods, pp. 3–30. Springer International Publishing, Cham (2020)
41. Doob, J.L.: Stochastic processes. Wiley, New York (1953)
42. Durbin, R.: Biological Sequence Analysis: Probabilistic Models of Proteins and Nucleic Acids. Cambridge University Press, Cambridge (1998)
43. Durrett, R.: Probability: Theory and Examples. Brook/Cole, Belmont, 3rd edition, 2005.
44. Dyer, M., Greenhill, C.: On Markov chains for independent sets. J. Algorithms **35**(1), 17–49 (2000)
45. Dyer, M., Frieze, A., Kannan, R.: A random polynomial-time algorithm for approximating the volume of convex bodies. J. ACM **38**(1), 1–17 (1991)
46. Eckhardt, R.: Stan Ulam, John von Neumann. Los Alamos Sci. 131 (1987)
47. Embrechts, P., Hofert, M.: A note on generalized inverses. Math. Methods Oper. Res. **77**, 423–432 (2013)
48. Embrechts, P., Lindskog, F., McNeil, A.: Modelling Dependence with Copulas. Rapport Technique, vol. 14, pp. 1–50. Département de Mathématiques, Institut Fédéral de Technologie de Zurich, Zurich (2001)

49. Esseen, C.-G.: On the Liapunoff limit of error in the theory of probability. Arkiv för Matematik, Astronomi och Fysik **A28**, 1–19 (1942)

50. Evans, M., Swartz, T.: Approximating Integrals via Monte Carlo and Deterministic Methods, vol. 20. Oxford University Press, Oxford (2000)

51. Ewens, W.J., Grant, G.R.: Statistical Methods in Bioinformatics: An Introduction. Springer, New York (2005)

52. Feller, W.: An Introduction to Probability Theory and Its Applications, vol. I, 3rd edn. John Wiley & Sons, Inc., New York-London-Sydney (1968)

53. Fill, J.A.: An interruptible algorithm for perfect sampling via Markov chains. Ann. Appl. Probab. **8**, 131–162 (1998)

54. Fill, J.A., Machida, M.: Stochastic monotonicity and realizable monotonicity. Ann. Probab. **29**, 938–978 (2001)

55. Fishman, G.S.: Monte Carlo: Concepts, Algorithms, and Applications. Springer, Berlin (1996)

56. Fishman, G.S.: Discrete-Event Simulation. Springer, Berlin (2001)

57. Fluhrer, S.R., McGrew, D.A.: Statistical Analysis of the Alleged RC4 Keystream Generator. In: Proceedings of the 7th International Workshop on Fast Software Encryption (FSE 2000), LNCS, pp. 19–30. Springer, Berlin (2000)

58. Fluhrer, S., Mantin, I., Shamir, A.: Weaknesses in the key scheduling algorithm of RC4. In: Proceedings of International Workshop on Selected Areas in Cryptography (SAC 2001), LNCS, pp. 1–24. Springer-Verlag, Berlin (2001)

59. Flury, B.D.: Acceptance–rejection sampling made easy. Siam Rev. **32**(3), 474–476 (1990)

60. Foley, R.D., McDonald, D.R.: Large deviations of a modified Jackson network: stability and rough asymptotics. Ann. Appl. Probab. **15**, 519–541 (2005)

61. Foss, S.G., Tweedie, R.L.: Perfect simulation and backward coupling. Stochastic Models **14**(1–2), 187–203 (1998)

62. Geman, S., Geman, D.: Stochastic relaxation, Gibbs distributions, and the Bayesian restoration of images. IEEE Trans. Pattern Anal. Mach. Intell. **6**(6), 721–741 (1984)

63. Glasserman, P.: Monte Carlo Methods in Financial Engineering. Springer, Berlin (2004)

64. Glynn, P.W., Iglehart, D.L.: Simulation output analysis using standardized time series. Math. Oper. Res. **15**(1), 1–16 (1990)

65. Glynn, P.W., Meyn, S.P.: A Liapounov bound for solutions of the Poisson equation. Ann. Probab. 916–931 (1996)

66. Glynn, P.W., Szechtman, R.: Some new perspectives on the method of control variates. In: Monte Carlo and Quasi-Monte Carlo Methods 2000: Proceedings of a Conference Held at Hong Kong Baptist University, Hong Kong SAR, China, November 27–December 1, 2000, pp. 27–49. Springer (2002)

67. Glynn, P.W., Whitt, W.: The asymptotic efficiency of simulation estimators. Oper. Res. **40**(3), 505–520 (1992)

68. Golić, J.D.: Linear statistical weakness of alleged RC4 keystream generator. In: Advances in Cryptology–EUROCRYPT'97: International Conference on the Theory and Application of Cryptographic Techniques Konstanz, Germany, May 11–15, 1997 Proceedings 16, pp. 226–238. Springer (1997)

69. Gordon, W.J., Newell, G.F.: Closed queuing systems with exponential servers. Oper. Res. **15**(2), 254–265 (1967)

70. Grassmann, W.K.: Warm-up periods in simulation can be detrimental. Probab. Eng. Inf. Sci. **22**, 415–429 (2008)

71. Grathwohl, W., Swersky, K., Hashemi, M., Duvenaud, D., Maddison, C.: Oops I took a gradient: scalable sampling for discrete distributions. In International Conference on Machine Learning, pp. 3831–3841. PMLR (2021)

72. Green, P.J., Han, X.-L.: Metropolis methods, Gaussian proposals and antithetic variables. In: Stochastic Models, Statistical Methods, and Algorithms in Image Analysis: Proceedings of the Special Year on Image Analysis, Held in Rome, Italy, 1990, pp. 142–164. Springer (1992)

73. Häggström, O.: Finite Markov Chains and Algorithmic Applications. Cambridge University Press, Cambridge (2002)
74. Häggström, O., Nelander, K.: Exact sampling from antimonotone systems. Stat. Neerlandica **52**, 360–380 (1998)
75. Häggström, O., Nelander, K.: On exact simulation of markov random fields using coupling from the past. Scand. J. Stat. **26**, 395–411 (1999)
76. Hajek, B.: Cooling schedules for optimal annealing. Math. Oper. Res. **13**(2), 311–329 (1988)
77. Hammersley, J.M., Handscomb, D.C.: Monte Carlo Methods. Chapman and Hall, Boca Raton (1964)
78. Hammersley, J.M., Handscomb, D.C.: General principles of the Monte Carlo method. In: Monte Carlo Methods, pp. 50–75. Springer, Berlin (1964)
79. Harman, R., Lacko, V.: On decompositional algorithms for uniform sampling from n-spheres and n-balls. J. Multivariate Anal. **101**(10), 2297–2304 (2010)
80. Hastings, W.K.: Monte Carlo sampling methods using Markov chains and their applications. Biometrika **57**(1), 97–109 (1970)
81. Heidelberger, P.: Fast simulation of rare events in queueing and reliability models. ACM Trans. Model. Comput. Simul. **5**(1), 43–85 (1995)
82. Held, M., Karp, R.M.: A dynamic programming approach to sequencing problems. J. Soc. Ind. Appl. Math. **10**(1), 196–210 (1962)
83. Holst, L.: Asymptotic normality and efficiency for certain goodness-of-fit tests. Biometrika **59**(1), 137–145 (1972)
84. Huillet, T.: Siegmund duality with applications to the neutral Moran model conditioned on never being absorbed. J. Phys. A Math. Theor. **43** (2010)
85. Jerrum, M.: A very simple algorithm for estimating the number of k-colorings of a low-degree graph. Random Struct. Algorithms **7**(2), 157–165 (1995)
86. Jerrum, M.: Counting, Sampling and Integrating: Algorithms and Complexity. Springer Science & Business Media, Berlin (2003)
87. Jerrum, M., Sinclair, A.: Approximating the permanent. SIAM J. Comput. **18**, 1149–1178 (1989)
88. Jerrum, M., Sinclair, A.: The Markov chain Monte Carlo method: an approach to approximate counting and integration. Approximation Algorithms for NP-hard Problems, pp. 482–520. PWS Publishing (1996)
89. Jerrum, M.R., Valiant, L.G., Vazirani, V.V.: Random generation of combinatorial structures from a uniform distribution. Theor. Comput. Sci. **43**, 169–188 (1986)
90. Jerrum, M., Sinclair, A., Vigoda, E.: A polynomial-time approximation algorithm for the permanent of a matrix with nonnegative entries. J. ACM **51**(4), 671–697 (2004)
91. Joe, H.: Multivariate models and multivariate dependence concepts. CRC Press, Boca Raton (1997)
92. Johnson, C.R., Horn, R.A.: Matrix Analysis. Cambridge University Press, Cambridge (1985)
93. Johnson, N.L., Kotz, S., Balakrishnan, N.: Continuous Univariate Distributions. Wiley Series in Probability and Mathematical Statistics: Applied Probability and Statistics, vol. 1, 2nd edn. John Wiley & Sons, Inc./A Wiley-Interscience Publication, New York (1994).
94. Jones, G.L.: On the Markov chain central limit theorem. Probab. Surveys **1**, 299–320 (2004)
95. Juneja, S., Shahabuddin, P.: Rare-event simulation techniques: An introduction and recent advances. Handbooks in Operations Research and Management Science, vol. 13, pp. 291–350. Elsevier (2006)
96. Kelly, F.P.: Stochastic models of computer communication systems. J. R. Stat. Soc. Series B (Methodological) **47**, 379–395 (1985)
97. Kelly, F.P., MacPhee, I.M.: The number of packets transmitted by collision detect random access schemes. Ann. Probab. **15** (2007)
98. Kemeny, J.G., Snell, J.L.: Finite Markov Chains, vol. 26. van Nostrand Princeton, Princeton (1969)
99. Kendall, W.S.: Perfect Simulation for the Area-Interaction Point Process, vol. 128, pp. 218–234. Springer-Verlag, Berlin (1998)

100. Khintchine, A.: Über einen Satz der Wahrscheinlichkeitsrechnung. Fundamenta Math. **6**, 9–20 (1924)
101. Kim, D.H.: A new version of first return time test of pseudorandomness. J. Korean Soc. Ind. Appl. Math. **12**, 109–118 (2008)
102. Kim, C., Choe, G.H., Kim, D.H.: Tests of randomness by the gambler's ruin algorithm. Appl. Math. Comput. **199**, 195–210 (2008)
103. Kingman, J.F.C.: Poisson Processes. Clarendon Press, Oxford (1993)
104. Kirkpatrick, S., Gelatt, Jr. C.D., Vecchi, M.P.: Optimization by simulated annealing. Science **220**(4598), 671–680 (1983)
105. Kleinrock, L.: Queueing Systems: Theory, vol. I. John Wiley & Sons, Hoboken (1975)
106. Knuth, D.E.: An algorithm for Brownian zeroes. Computing **33**(1), 89–94 (1984)
107. Knuth, D.E.: The Art of Computer Programming: Seminumerical Algorithms, vol. 2, 3rd edn. Addison-Wesley Pub. Co, Boston (1997)
108. Koçak, O., Sulak, F., Doğanaksoy, A., Uğuz, M.: Modifications of Knuth randomness tests for integer and binary sequences. Commun. Faculty Sci. Univ. Ankara Series A1 Math. Stat. **67**(2), 64–81 (2018)
109. Kolata, G.: In shuffling cards, 7 is winning number. New York Times (1990)
110. Koller, D., Friedman, N.: Probabilistic Graphical Models: Principles and Techniques. MIT Press, Cambridge (2009)
111. Kopociński, B.: Zarys teorii odnowy i niezawodności. Państwowe Wydawnictwo Naukowe, Warszawa (1973)
112. Kroese, D.P., Taimre, T., Botev, Z.I.: Handbook of Monte Carlo Methods. John Wiley & Sons, Hoboken (2013)
113. Kroese, D.P., Botev, Z., Taimre, T.: Data Science and Machine Learning: Mathematical and Statistical Methods. Chapman and Hall/CRC, Boca Raton (2019)
114. Lawrence, C.E., Altschul, S.F., Boguski, M.S., Liu, J.S., Neuwald, A.F., Wootton, J.C.: Detecting subtle sequence signals: a Gibbs sampling strategy for multiple alignment. Science **262**(5131), 208–214 (1993)
115. L'Ecuyer, P.: Random numbers for simulation. Commun. ACM **33**(10), 85–97 (1990)
116. L'Ecuyer, P.: Testing random number generators. Technical Report, Institute of Electrical and Electronics Engineers (IEEE), 1992.
117. L'Ecuyer, P.: Good parameters and implementations for combined multiple recursive random number generators. Oper. Res. **47**, 159–164 (1999)
118. L'Ecuyer, P.: Software for uniform random number generation: distinguishing the good and the bad. In: Proceeding of the 2001 Winter Simulation Conference (Cat. No. 01CH37304), vol. 1, pp. 95–105. IEEE (2001)
119. L'Ecuyer, P.: History of uniform random number generation. In: Proceedings of the 2017 Winter Simulation Conference (WSC), pp. 202–230. IEEE (2017)
120. L'Ecuyer, P., Lemieux, C.: Recent advances in randomized quasi-Monte Carlo methods. Modeling Uncertainty: An Examination of Stochastic Theory, Methods, and Applications, pp. 419–474. Springer (2002)
121. L'Ecuyer, P., Simard, R.: On the performance of birthday spacings tests with certain families of random number generators. Math. Comput. Simul. **55**(1–3), 131–137 (2001)
122. L'Ecuyer, P., Simard, R., Wegenkittl, S.: Sparse serial tests of uniformity for random number generators. SIAM J. Sci. Comput. **24**, 652–668 (2003)
123. Lee, C.E., Ozdaglar, A., Shah, D.: Approximating the Stationary Probability of a Single State in a Markov chain. Preprint (2013). arXiv:1312.1986
124. Levin, D.A., Peres, Y.: Markov Chains and Mixing Times, vol. 107. American Mathematical Society, Providence (2017)
125. Lewis, P.A.W., Shedler, G.S.: Simulation of nonhomogeneous Poisson processes by thinning. Naval Res. Logist. Q. **26**, 403–413 (1979)
126. Lewis, P.A.W., Goodman, A.S., Miller, J.M.: A pseudo-random number generator for the system/360. IBM Syst. J. **8**, 136–146 (2010)
127. Lindvall, T.: Lectures on the Coupling Method. Courier Corporation, Chelmsford (2002)

128. Liu, Y., Zhang, Y.: Central limit theorems for ergodic continuous-time Markov chains with applications to single birth processes. Front. Math. China **10**, 933–947 (2015)
129. Lorek, P.: Generalized gambler's ruin problem: explicit formulas via Siegmund duality. Methodol. Comput. Appl. Probab. **19**, 603–613 (2017)
130. Lorek, P., Siegmund duality for Markov chains on partially ordered state spaces. Probab. Eng. Inf. Sci. **32**, 495–521 (2018)
131. Lorek, P., Markowski, P.: Monotonicity requirements for efficient exact sampling with Markov chains. Markov Process. Relat. Fields **23**, 485–514 (2017)
132. Lorek, P., Markowski, P.: Absorption time and absorption probabilities for a family of multidimensional gambler models. ALEA-Latin Am. J. Probab. **19**, 125–150 (2022)
133. Lorek, P., Słowik, M., Zagórski, F.: Statistical testing of PRNG: Generalized gambler's ruin problem. In: Lecture Notes in Computer Science (Including Subseries Lecture Notes in Artificial Intelligence and Lecture Notes in Bioinformatics). LNCS, vol. 10693, pp. 425–437 (2017)
134. Lorek, P., Zagórski, F., Kulis, M.: Strong stationary times and its use in cryptography. IEEE Transactions on Dependable and Secure Computing **16**(5), 805–818 (2017)
135. Lorek, P., Łoś, G., Gotfryd, K., Zagórski, F.: On testing pseudorandom generators via statistical tests based on the arcsine law. J. Comput. Appl. Math. **380**, 112968 (2020)
136. Lovász, L., Vempala, S.: Fast algorithms for logconcave functions: sampling, rounding, integration and optimization. In: 2006 47th Annual IEEE Symposium on Foundations of Computer Science (FOCS'06), pp. 57–68. IEEE (2006)
137. MacKay, D.J.: Information theory, Inference and Learning Algorithms. Cambridge University Press, Cambridge (2003)
138. Madras, N.N.: Lectures on Monte Carlo methods. AMS, Providence (2002)
139. Mantin, I., Shamir, A.: A practical attack on broadcast RC4. Fast Software Encryption, 8th International Workshop, Yokohama, Japan, pp. 152–164 (2001)
140. Marsaglia, G.: Choosing a point from the surface of a sphere. Ann. Math. Stat. **43**(2), 645–646 (1972)
141. Marsaglia, G.: The structure of linear congruential sequences. Appl. Number Theory Numer. Anal. 249–285 (1972)
142. Marsaglia, G.: Diehard: a battery of tests of randomness (1996). http://stat.fsu.edu/geo
143. Marsaglia, T.A., Bray, G.: A convenient method for generating normal variables. SIAM Rev. **6**, 260–264 (1964)
144. Marsaglia, G., Tsang, W.W.: The ziggurat method for generating random variables. J. Stat. Software **5**, 1–7 (2000)
145. Marsaglia, G., Tsang, W.W.: Some difficult-to-pass tests of randomness. J. Stat. Software **7**, 1–9 (2002)
146. MathWorks: Matlab's rand and randn generators. Online Documentation (2006). http://www.mathworks.com/support/solutions/en/data/1-10HYAS/index.html?solution=1-10HYAS.
147. Matsumoto, M., Nishimura, T.: Mersenne twister: a 623-dimensionally equidistributed uniform pseudo-random number generator. ACM Trans. Model. Comput. Simul. **8**, 3–30 (1998)
148. Matsumoto, M., Nishimura, T.: A Nonempirical Test on the Weight of Pseudorandom Number Generators, pp. 381–395. Springer Berlin Heidelberg, Berlin/Heidelberg (2002)
149. McNeil, D.R.: Integral functionals of birth and death processes and related limiting distributions. Ann. Math. Stat. **41**(2), 480–485 (1970)
150. Metropolis, N., Ulam, S.: The Monte Carlo method. J. Am. Stat. Assoc. **44**(247), 335–341 (1949)
151. Metropolis, N., Rosenbluth, A.W., Rosenbluth, M.N., Teller, A.H., Teller, E.: Equation of state calculations by fast computing machines. J. Chem. Phys. **21**(6), 1087–1092 (1953)
152. Meyn, S.P., Tweedie, R.L.: Markov Chains and Stochastic Stability. Springer Science & Business Media, Berlin (2012)
153. Mitzenmacher, M., Upfal, E.: Probability and Computing: Randomization and Probabilistic Techniques in Algorithms and Data Analysis. Cambridge University Press, Cambridge (2017)

154. Møller, J.: Perfect simulation of conditionally specified models. J. R. Stat. Soc. Series B Stat. Methodol. **61**(1), 251–264 (1999)
155. Moon, T.K.: The expectation-maximization algorithm. IEEE Signal Process. Mag. **13**(6), 47–60 (1996)
156. Morris, B., Sinclair, A.: Random walks on truncated cubes and sampling 0-1 knapsack solutions. SIAM J. Comput. **34**(1), 195–226 (2004)
157. Morse, P.M.: Stochastic properties of waiting lines. J. Oper. Res. Soc. Am. **3**(3), 255–261 (1955)
158. Mossel, E., Peres, Y., Sinclair, A.: Shuffling by semi-random transpositions. Found. Comput. Sci. 572–581 (2004)
159. Müller, A., Stoyan, D.: Comparison Methods for Stochastic Models and Risks. Wiley Series in Probability and Statistics. John Wiley & Sons, Ltd., Chichester (2002)
160. Munkres, J.R.: Analysis on Manifolds. Addison-Wesley Publishing Company, Advanced Book Program, Redwood City (1991)
161. Murdoch, D.J., Green, P.J.: Exact sampling from a continuous state space. Scand. J. Stat. **25**(3), 483–502 (1998)
162. Nakayama, M.K.: Two-stage stopping procedures based on standardized time series. Manag. Sci. **40**(9), 1189–1206 (1994)
163. Naor, P.: On machine interference. J. R. Soc. Series B (Methodological) **18**(2), 280–287 (1956)
164. Nelsen, R.B.: An Introduction to Copulas. Springer, Berlin (2006)
165. Ni, E.C., Henderson, S.G.: How hard are steady-state queueing simulations? ACM Trans. Model. Comput. Simul. (TOMACS) **25**(4), 1–21. ACM, New York (2015)
166. Niederreiter, H.: Random Number Generation and Quasi-Monte Carlo Methods. Philadelphia, PA, SIAM (1992)
167. O'Neill, M.: PCG: A family of better random number generators. Online (2014). Accessed 19 June, 2025
168. O'Neill, M.: PCG: A family of simple fast space-efficient statistically good algorithms for random number generation. ACM Trans. Math. Software **204**, 1–46 (2014)
169. Pareschi, F., Rovatti, R., Setti, G.: Second-level NIST randomness tests for improving test reliability. In: 2007 IEEE International Symposium on Circuits and Systems, pp. 1437–1440. IEEE (2007)
170. Pareschi, F., Rovatti, R., Setti, G.: Second-level testing revisited and applications to NIST SP800-22. In: 2007 18th European Conference on Circuit Theory and Design, pp. 627–630. IEEE (2007)
171. Propp, J.G., Wilson, D.B.: Exact sampling with coupled Markov chain Monte Carlo. Random Struct. Algorithms **9**, 223–252 (1996)
172. Raftery, A.E., Lewis, S.M.: Implementing MCMC. Markov Chain Monte Carlo in Practice, pp. 115–130. CRC Press (1996)
173. Reynolds, J.F.: The covariance structure of queues and related processes–a survey of recent work. Adv. Appl. Probab. **7**(2), 383–415 (1975)
174. Ripley, B.D.: Computer generation of random variables: a tutorial. International Statistical Review/Revue Internationale de Statistique, pp. 301–319. ISI (1983)
175. Ripley, B.: Stochastic Simulation. Wiley, Hoboken (1987)
176. Robert, C.P., Casella, G.: Monte Carlo Statistical Methods, vol. 2. Springer, Berlin (1999)
177. Robert, P.: Stochastic Networks and Queues, vol. 52. Springer Science & Business Media, Berlin (2013)
178. Roberts, G.O., Rosenthal, J.S.: General state space Markov chains and MCMC algorithms. Probab. Surveys **1**, 20–71 (2004)
179. Rolski, T.: Stationary Random Processes Associated with Point Processes. Lecture Notes in Statistics, vol. 5. Springer, New York (1981)
180. Rolski, T., Schmidli, H., Schmidt, V., Teugels, J.: Stochastic Processes for Insurance and Finance. Wiley, Chichester (1999)

181. Rolski, T., Serfozo, R., Stoyan, D.: Service-time ages, residuals, and lengths in an M/GI/∞ M/GI/∞ service system. Queueing Syst. **79**, 173–181 (2015)
182. Ross, S.: A Course in Simulation. Macmillan, London (1991)
183. Ross, S.M.: Introduction to Probability Models. Academic Press, Boston (1999)
184. Rubinstein, R.Y.: Optimization of computer simulation models with rare events. Eur. J. Oper. Res. **99**(1), 89–112 (1997)
185. Rubinstein, R.Y., Kroese, D.P.: The cross-entropy method: a unified approach to combinatorial optimization. Monte-Carlo Simulation, and Machine Learning, vol. 133. Springer, Berlin (2004)
186. Rubinstein, R.Y., Kroese, D.P.: Simulation and the Monte Carlo Method, 3rd edn. Wiley, Hoboken (2016)
187. Rukhin, A., Soto, J., Nechvatal, J.: A Statistical Test Suite for Random and Pseudorandom Number Generators for Cryptographic Applications, vol. 22. National Institute of Standards and Technology, Gaithersburg (2010)
188. Sahlin, K.: Estimating convergence of Markov chain Monte Carlo simulations. Ann. Biomed. Eng. **32** (2011)
189. Savicky, P.: A strong nonrandom pattern in Matlab default random number generator. In: Proceedings of Technical Computing, Prague (2006). https://www.cs.cas.cz/~savicky/papers/. Accessed 19 June 2025
190. Seneta, E.: Non-negative Matrices and Markov Chains. Springer Science & Business Media, Berlin (2006)
191. Serfling, R.J.: Approximation Theorems of Mathematical Statistics. John Wiley & Sons, Hoboken (2009)
192. Šidák, Z.: Rectangular confidence regions for the means of multivariate normal distributions. J. Am. Stat. Assoc. **62**(318), 626–633 (1967)
193. Siegmund, D.: The equivalence of absorbing and reflecting barrier problems for stochastically monotone Markov processes. Ann. Probab. **4**, 914–924 (1976)
194. Sinclair, A.: Improved bounds for mixing rates of markov chains and multicommodity flow. Comb. Probab. Comput. **1**(4), 351–370 (1992)
195. Soboĺ, I.M.: Sensitivity estimates for nonlinear mathematical models. Math. Model. Comput. Exp. **1**, 407 (1993)
196. Srikant, R., Whitt, W.: Simulation run length planning for stochastic loss models. In: Proceedings of the 1995 Winter Simulation Conference (WSC), pp. 1384–1391 (1995)
197. Steckley, S.G., Henderson, S.G.: The error in steady-state approximations for the time-dependent waiting time distribution. Stoch. Models **23**(2), 307–332 (2007)
198. Steinhaus, H.: Liczby złote i żelazne (on golden and iron numbers). Appl. Math. **3**, 51–65 (1958)
199. Takashima, K.: Sojourn time test of m-sequences. J. Jpn. Soc. Comp. Stat. **8**, 37–46 (1995)
200. Takashima, K.: Last visit time tests for pseudorandom numbers. J. Jpn. Soc. Comput. Stat. **9**(1), 1–14 (1996)
201. Thorisson, H.: Future independent times and Markov chains. Probab. Theory Relat. Fields **78**, 143–148 (1988)
202. Thorisson, H.: Construction of a stationary regenerative process. Stoch. Process. Appl. **42**(2), 237–253 (1992)
203. Thorisson, H.: Coupling methods in probability theory. Scand. J. Stat. 159–182 (1995)
204. Thorisson, H.: Coupling, stationarity, and regeneration. Probab. Appl., Springer, New York (2000)
205. Tsang, W.W., Hui, L.C.K., Chow, K.-P., Chong, C.F., Tso, C.W.: Tuning the collision test for power. In: ACSC, vol. 4, pp. 23–30 (2004)
206. Tyurin, I.S.: Refinement of the upper bounds of the constants in Lyapunov's theorem. Russ. Math. Surv. **65**, 586–588 (2010)
207. Van der Vaart, A.W.: Asymptotic Statistics, vol. 3. Cambridge University Press, Cambridge (2000)

208. Vassiliou, P.-C.G.: Non-Homogeneous Markov Chains and Systems: Theory and Applications. CRC Press, Boca Raton (2022)
209. Vrbik, J.: Small-sample corrections to Kolmogorov–Smirnov test statistic. Pioneer J. Theor. Appl. Stat. **15**, 15–23 (2018)
210. Weiss, I.: Limiting distributions in some occupancy problems. Ann. Math. Stat. **29**(3), 878–884 (1958)
211. Whitt, W.: Planing queueing simulation. Manag. Sci. **35**, 1341–1366 (1989)
212. Whitt, W.: The efficiency of one long run versus independent replications in steady-state simulation. Manag. Sci. **37**(6), 645–666 (1991)
213. Whitt, W.: Asymptotic formulas for Markov processes with applications to simulation. Oper. Res. **40**(2), 279–291 (1992)
214. Wichura, M.J.: Algorithm as 241: the percentage points of the normal distribution. J. R. Stat. Soc. Series C (Appl. Stat.) **37**(3), 477–484 (1988)
215. Wilson, D.B.: How to couple from the past using a read-once source of randomness. Random Struct. Algorithms **16**(1), 85–113 (2000)
216. Yang, Z.: Molecular Evolution: A Statistical Approach. Oxford University Press, Oxford (2014)
217. Zanella, G.: Informed proposals for local MCMC in discrete spaces. J. Am. Stat. Assoc. **115**(530), 852–865 (2020)
218. Zhao, Y.: The number of independent sets in a regular graph. Comb. Probab. Comput. **19**, 315–320 (2010)

Code Index

© The Author(s), under exclusive license to Springer Nature Switzerland AG 2025
P. Lorek, T. Rolski, *Lectures on Monte Carlo Theory*, Probability Theory
and Stochastic Modelling 108, https://doi.org/10.1007/978-3-032-01190-9

Subject Index

A

Acceptance region, 114, 116
Acceptance-rejection method
    continuous version (AR-c), 116
    discrete version (AR-d), 114
    proposal distribution, 122
Actual simulation, 192
Algorithm
    AR-c, 117
    AR-d, 115, 116
    AR-N, 118
    BM, 127
    DB, 108
    DP, 111
    Fisher-Yates (*see* Random permutation)
    ITM-d, 106
    ITM-Exp, 104
    ITM-Par, 105
    ITR, 108
    MB, 129
Allocation
    optimal (*see* Stratified sampling)
    proportional (*see* Stratified sampling)
Anti-monotone system, 420
Antithetic variates method, 273
Approximate counting, 429
Arcsine law, 6
    test, 74, 75
Asymptotic variance, 212, 581
Asynchronous simulation, 543

B

Bessel function, 123
Bias, 223
    asymptotic, 224
    asymptotic of order, 223
Birth and death (B&D) process, 552
Birthday problem, 61, 62
Birthday spacing test, 68, 69
Bisection method, 338
Bit generator, 38, 39
Bivariate copula, 163
Black-Scholes model, 343
Bounded relative error, 315
Box-Muller method, 125
Brownian bridge, 336
Brownian motion, 334, 335
Burn-in-period, 469

C

Candidate-generating matrix, 390, 504
Card shuffling, 26
    Riffle shuffle, 27
    Top-To-Random, 27
    Top-To-Random Transposition, 27
Centered process, 337
Central limit theorem (CLT), 190
Change of variables formula, 135
Chernoff bound, 232
Chi-square
    non-central, 123
    test, 49, 50
Cholesky decomposition, 133, 201
Collision test, 63, 64
Common random numbers method, 282
Composition method, 122
Conditional inverse transform method, 161
Confidence interval, 191

GPSR Compliance
The European Union's (EU) General Product Safety Regulation (GPSR) is a set
of rules that requires consumer products to be safe and our obligations to
ensure this.

If you have any concerns about our products, you can contact us on

ProductSafety@springernature.com

In case Publisher is established outside the EU, the EU authorized
representative is:

Springer Nature Customer Service Center GmbH
Europaplatz 3
69115 Heidelberg, Germany